NON-LINEAR FINITE ELEMENT ANALYSIS OF SOLIDS AND STRUCTURES

WILEY SERIES IN COMPUTATIONAL MECHANICS

Series Advisors:

René de Borst

Perumal Nithiarasu

Tayfun E. Tezduyar

Genki Yagawa

Tarek Zohdi

Introduction to Finite Strain Theory for Continuum Elasto-Plasticity	Hashiguchi and Yamakawa	September 2012
Non-linear Finite Element Analysis of Solids and Structures: Second edition	De Borst, Crisfield, Remmers and Verhoosel	August 2012
An Introduction to Mathematical Modeling: A Course in Mechanics	Oden	November 2011
Computational Mechanics of Discontinua	Munjiza, Knight and Rougier	November 2011
Introduction to Finite Element Analysis: Formulation, Verification and Validation	Szabó and Babuška	March 2011

NON-LINEAR FINITE ELEMENT ANALYSIS OF SOLIDS AND STRUCTURES

SECOND EDITION

René de Borst
School of Engineering, University of Glasgow, UK

Mike A. Crisfield
Imperial College of Science, Technology and Medicine, UK

Joris J.C. Remmers
Eindhoven University of Technology, The Netherlands

Clemens V. Verhoosel
Eindhoven University of Technology, The Netherlands

WILEY
A John Wiley & Sons, Ltd., Publication

This edition first published 2012

Registered office
John Wiley & Sons Ltd, The Atrium, Southern Gate, Chichester, West Sussex, PO19 8SQ, United Kingdom

For details of our global editorial offices, for customer services and for information about how to apply for permission to reuse the copyright material in this book please see our website at www.wiley.com.

Library of Congress Cataloging-in-Publication Data

Non-linear finite element analysis of solids and structures. – 2nd ed. / R. de Borst ... [et al.].
p. cm.
Rev. ed. of: Non-linear finite element analysis of solids and structures / M.A. Crisfield. c1991-c1997. (2 v.)
Includes bibliographical references and index.
ISBN 978-0-470-66644-9 (hardback)
1. Structural analysis (Engineering)–Data processing. 2. Finite element method–Data processing. I. Borst, Ren? de. II. Crisfield, M. A. Non-linear finite element analysis of solids and structures.
TA647.C75 2012
624.1′71–dc23

2012011741

A catalogue record for this book is available from the British Library.

Print ISBN: 9780470666449

Typeset in 10/12pt Times-Roman by Thomson Digital, Noida, India

Contents

PART V ADVANCED DISCRETISATION CONCEPTS

Preface

When the first author was approached by John Wiley & Sons, Ltd to write a new edition of the celebrated two-volume book of Mike Crisfield, *Non-linear Finite Element Analysis of Solids and Structures*, he was initially very hesitant. The task would of course constitute a formidable amount of work. But it would also be impossible to maintain Mike's writing style, a feature which has so much contributed to the success of the books. On the other hand, it would be rewarding to provide the engineering community with a book that is as accessible as possible, that gives a broad introduction into non-linear finite element analysis, with an outlook on the newest developments, and that maintains the engineering spirit which Mike emphasised in his books. This is the philosophy behind this second edition. Indeed, although much has been changed in terms of content, it has been the intention not to change the engineering orientation with an emphasis on practical solutions.

One of the aims of the original two-volume set was to provide the user of advanced non-linear finite element packages with sufficient background knowledge, which is a prerequisite to judiciously handle modern finite element packages. A closely related aim is to make the user of such packages aware of their possibilities, but also of their limitations and pitfalls. Major developments have taken place in computational technology since Mike Crisfield wrote about the danger of the 'black-box syndrome' in the Preface to Volume 1. Therefore, his warning has gained even more strength, and provides a further justification for the publication of a second edition.

Unlike the first edition, the second edition comes as a single volume. The reduction has been achieved by omitting or reducing the discussion on developments now considered to be less central in computational mechanics, by a more compact and focused treatment, and by a removal of all Fortran code from the book. Instead, a small finite element code has been developed, written in Python, which is available through a companion website. The main purpose of the code is to illustrate the models presented in the book, and to show how abstract concepts can be translated into finite element software. To this end, the theory of the book is first transformed into algorithms, mostly listed in boxes that accompany the text. Subsequently, using ideas of literate programming, it is explained how these algorithms have been implemented in the PyFEM code, which contains the basic numerical tools needed to build a finite element code. Some of the solution techniques, element formulations, and material models treated in this book have been added. These tools are used in a series of example programs with increasing complexity.

The book comes in five parts. Part I discusses basic knowledge in mathematics and in continuum mechanics, as well as solution techniques for non-linear problems in static and dynamic analysis, and provides a first introduction into geometrical non-linearity. Some notions and concepts will be familiar, but not all, and the first chapters also serve to provide a common basis for the subsequent parts of the book. Part II contains major chapters on damage, plasticity and time-dependent non-linearities, such as creep. It contains all the material non-linearity that is treated in this book. Shell plasticity forms an exception, since it is treated in Part III, which focuses on structural elements: beams, arches and shells. Starting from a basic shallow arch formulation the discussion extends to cover modern concepts like solid-like shell theories. In Part IV first some additional continuum mechanics is provided that is needed in the remainder of this part, which focuses on large-strain elastic and elastoplastic finite element analysis. Part V, finally, gives an introduction into discretisation concepts that have become popular during the past 20 years: interface elements, discontinuous Galerkin methods, meshless methods, partition-of-unity methods, and isogeometric analysis. Particular reference is made to their potential to solve problems that arise in non-linear analysis, such as locking phenomena, damage and fracture, and non-linear shell analysis.

René de Borst
Joris Remmers
Clemens Verhoosel

Glasgow and Eindhoven

A Personal Note

Like many colleagues and friends in the community I treasure wonderful memories of my meetings and discussions with Mike. I will never forget the times that I visited him at the Transport and Road Research Laboratory, and later, at Imperial College of Science, Technology and Medicine. After a full day of intense discussions on cracking, strain softening, stability and solution techniques we normally went to his home, where Kiki, his wife, joined in and discussions broadened over a good meal.

Mike was a real scientist, and a gentleman. I hope that this Second Edition will properly preserve his legacy, and will help to keep the engineering approach alive in computational mechanics, to which he has so much contributed.

René

Series Preface

The series on *Computational Mechanics* is a conveniently identifiable set of books covering interrelated subjects that have been receiving much attention in recent years and need to have a place in senior undergraduate and graduate school curricula, and in engineering practice. The subjects will cover applications and methods categories. They will range from biomechanics to fluid-structure interactions to multiscale mechanics and from computational geometry to meshfree techniques to parallel and iterative computing methods. Application areas will be across the board in a wide range of industries, including civil, mechanical, aerospace, automotive, environmental and biomedical engineering. Practicing engineers, researchers and software developers at universities, industry and government laboratories, and graduate students will find this book series to be an indispensible source for new engineering approaches, interdisciplinary research, and a comprehensive learning experience in computational mechanics.

Non-linear Finite Element Analysis of Solids and Structures, Second Edition is based on the two original volumes by the late Mike Crisfield, who was a remarkable scholar in computational mechanics. This new edition is a greatly enriched version, written by an author team led by René de Borst, an outstanding scholar in computational mechanics, solids, and structures. The enrichments include the major developments in computational mechanics since the original version was written, such as new numerical discretization techniques, with emphasis on meshless methods and isogeometric analysis. This new edition still retains the "engineering spirit" that was emphasized by the original author, and the algorithmic explanations, which are only part of the enrichments, make it even easier to follow and more valuable in a practical context.

Non-linear Finite Element Analysis of Solids and Structures, Second Edition will serve as an excellent textbook for introductory and advanced courses in non-linear finite element analysis of solids and structures, and will also serve as a very valuable source and guide for research in this field.

Notation

Linear Algebra and Mathematical Operators

$\mathbf{a} \cdot \mathbf{b}$, $a_i b_i$	Dot-product of the vectors **a** and **b**
$\mathbf{a} \otimes \mathbf{b}$, $a_i b_j$	Tensor (or dyadic) product of the vectors **a** and **b**
$\mathbf{a} \times \mathbf{b}$, $e_{ijk} a_j b_k$	Cross-product of the vectors **a** and **b**
$\square^{\mathrm{T}}$	Transpose of matrix $\square$
$\square^{\mathrm{sym}} = (\square)^{\mathrm{sym}}$	Symmetry operator
$\mathrm{tr}(\square)$	Trace of matrix $\square$
$\lVert\square\rVert_2$	Euclidean or L_2-norm of the vector $\square$
δ_{ij}	Kronecker-delta identity
$< \square >$	MacAulay brackets/ramp function
$\nabla \cdot \mathbf{a}$, $\frac{\partial a_{ij}}{\partial x_j}$, $\mathbf{L}^{\mathrm{T}}\mathbf{a}$	Divergence of a (second-order) tensor **a**
$\mathcal{H}(\square)$	Heaviside function
$\delta\square$	Admissable variation of the quantity $\square$

Basic Continuum Mechanics

V	Arbitrary body in the current configuration
S	Boundary of an arbitrary body V in the current configuration
$\mathbf{n}$	Normal vector (to a surface S)
$\mathbf{x} = [x, y, z]^{\mathrm{T}}$	Coordinate in the physical domain
$\mathbf{u} = [u, v, w]^{\mathrm{T}}$	Displacement field
γ_{xy}, γ_{xz}, γ_{yz}	Engineering shear strains/elementary square distortions
ω_{xy}, ω_{xz}, ω_{yz}	Elementary square rotations
$\mathbf{t}$	Stress vector
$\boldsymbol{\epsilon}$ $[\boldsymbol{\mathcal{E}}]$	Infinitesimal strain tensor [matrix representation]
$\boldsymbol{\sigma}$ $[\boldsymbol{\Sigma}]$	Cauchy stress tensor [matrix representation]
$\boldsymbol{e}$ $[\boldsymbol{E}]$	Deviatoric infinitesimal strain tensor [matrix representation]
$\mathbf{s}$ $[\mathbf{S}]$	Deviatoric stress tensor [matrix representation]
$I_1^{\square}$, $I_2^{\square}$, $I_3^{\square}$	Invariants of the tensor $\square$ (Cauchy stress tensor when $\square$ is omitted)
$J_1^{\square}$, $J_2^{\square}$, $J_3^{\square}$	Invariants of the tensor $\square$ (deviatoric stress tensor when $\square$ is omitted)
p	Hydrostatic pressure
ϵ_{vol}	Volumetric infinitesimal strain
$\mathbf{T}_{\square}$	Transformation matrix for the tensor $\square$ in Voigt form
$\mathbf{D}$	Tangential stiffness tensor
$\square_{\mathrm{tan}}$	Quantity $\square$ related to the tangent stiffness

$\square^s$	Quantity $\square$ related to the secant stiffness
$\delta\mathcal{W}_{\text{int}}$	Internal virtual work
$\delta\mathcal{W}_{\text{ext}}$	External virtual work
$\mathbf{g}$	Gravity acceleration vector

Elasticity

E	Young's modulus
ν	Poisson's ratio
K	Bulk modulus
λ	Lamé's first parameter
μ, G	Lamé's second parameter/shear modulus
$\mathbf{D}^e$	Elastic stiffness matrix
$\mathbf{C}^e$	Elastic compliance matrix

Finite Element Data Structures

$\square_e, \square_{\text{elem}}$	Quantity $\square$ related to the element e
$\mathbf{Z}_e$	Element incidence (or location) matrix
$\boldsymbol{\xi} = [\xi, \eta, \zeta]^{\mathrm{T}}$	Parent element coordinates
$\mathbf{J}$	Jacobian matrix
w_i	Weight factor of parent element integration point i
$\mathbf{h}, h_i$	Finite element shape functions
$\mathbf{H}$	Displacement field interpolation matrix
$\mathbf{B}$	Strain field interpolation matrix
$\mathbf{a}$	Nodal displacement vector
$\mathbf{f}_{\text{int}}$	Internal force vector
$\mathbf{f}_{\text{ext}}$	External force vector
$\mathbf{K}$	Stiffness matrix
$\square_f$	Quantity $\square$ related to an unconstrained degree of freedom
$\square_p$	Quantity $\square$ related to a constrained/prescribed degree of freedom

Geometrically Non-linear Analysis

$\square_0$	Quantity $\square$ related to the reference configuration
$\mathbf{F}$	Deformation gradient
$\boldsymbol{l}$	Velocity gradient
$\mathbf{U}, \mathbf{V}$	Right/left pure deformation tensor/stretch tensor
$\mathbf{R}$	Rotation matrix
$\boldsymbol{\Omega}$	Rotation rate matrix
$\boldsymbol{e}$	Linear strain contribution
$\boldsymbol{\eta}$	Quadratic/non-linear strain contribution
$\square_L$	Quantity $\square$ related to linear contributions
$\square_{NL}$	Quantity $\square$ related to non-linear contributions

$\square^{\mathrm{cr}}$	Corotational contribution to quantity $\square$
$\bar{\square}$	Quantity $\square$ related to the corotational coordinate system
$\mathbf{C}, \mathbf{B}$	Right/left Cauchy–Green deformation tensor
$\boldsymbol{\gamma}$	Green–Lagrange strain tensor
$\mathbf{p}$	Nominal stress tensor
$\boldsymbol{\kappa}$	Kirchhoff stress tensor
$\boldsymbol{\tau}$	Second Piola–Kirchhoff stress tensor
$\boldsymbol{T}$	Biot stress tensor
$\square_i$	Principal values of the tensor $\square$
λ	Stretch ratio
$\overset{\diamond}{\square}$	Objective derivative of a vector $\square$/Green–Naghdi rate of $\square$
$\overset{\star}{\square}$	Truesdell rate of the tensor $\square$
$\overset{\circ}{\square}$	Jaumann rate of the tensor $\square$
$\boldsymbol{w}$	Spin tensor
$\square_{\mathrm{vol}}$	Volumetric part of quantity $\square$
$\square_{\mathrm{iso}}, \tilde{\square}$	Isochoric part of quantity $\square$
$\mathcal{E}$	Total deformation energy
e	Strain energy density
$\mathcal{W}$	Strain energy function
$\mathcal{W}^*$	Volumetric part of the strain energy function
f_p	Deviatoric part of the strain energy function
$\mathbf{T}_\sigma$	Back-transformation matrix
$\square^{\mathrm{JK}}$	Quantity $\square$ related to the Jaumann derivatives of the Kirchhoff stress
$\square^{\mathrm{JC}}$	Quantity $\square$ related to the Jaumann derivatives of the Cauchy stress
$\square^{\mathrm{TK}}$	Quantity $\square$ related to the Truesdell derivatives of the Kirchhoff stress
$\square^{\mathrm{TC}}$	Quantity $\square$ related to the Truesdell derivatives of the Cauchy stress

Incremental Iterative Analysis and Solution Techniques

$\square_0, \square^t$	Quantity $\square$ at the previous converged load step
$\Delta\square = \square^{t+\Delta t} - \square^t$	Incremental value of quantity $\square$
$\Delta\square_i$	Approximate incremental value of quantity $\square$ after i iterations
$\mathrm{d}\square_{i+1}$	Correction to the approximate incremental value $\Delta\square_i$
$\mathbf{r}$	Residual vector
$\mathbf{A}$	Constrained degrees of freedom selection matrix
λ	Scalar-valued load parameter
$\hat{\mathbf{f}}_{\mathrm{ext}}$	Unit external force vector
g	Constraint equation
Δl	Path length increment
η	Iterative procedure tolerance
N^t	Number of iterations required for convergence at time t
$\square^I, \square^{II}$	Quantity $\square$ related to a two-stage solution procedure
$\lambda_k, \mathbf{v}_k$	Eigenvalue/vector of the tangential stiffness matrix

Dynamics and Time-dependent Material Models

t	Time
$\dot{\square}$	First-order temporal derivative
$\ddot{\square}$	Second-order temporal derivative
ρ	Mass density
$\mathbf{M}$	Mass matrix
$\square^0$	Quantity $\square$ at the initial state
$\bar{\square}$	Quantity $\square$ evaluated at the time interval mid-point
θ	Generalised mid-point rule parameter
β, γ	Newmark integration parameters
α	HHT α-method integration parameter
$\mathbf{q}$	Pseudo-load vector
τ	Relaxation time
$E(t-\tilde{t})$	Response function
$J(t-\tilde{t})$	Creep function
h	Hardening/softening modulus
s	Strain-rate sensitivity
$\omega^{\max}$	Maximum natural frequency of a system
l	Internal length scale

Damage and Fracture

S_d	Discontinuity surface
$\square_d, \square_{S_d}$	Quantity $\square$ related to the discontinuity surface S_d
$\square_n, \square_s, \square_t$	Normal and shear components of quantity $\square$
$\square^+, \square^-$	Quantity $\square$ related to the positive or negative side of a discontinuity
$[\![\square]\!] = \square^+ - \square^-$	Jump operator
$\mathbf{v}$	Relative displacement across a discontinuity/crack opening
$\mathcal{D}_{S_d}$	Distance function related to the discontinuity surface S_d
$\mathcal{H}_{S_d}$	Heaviside function related to the discontinuity surface S_d
δ_{S_d}	Dirac-delta function related to the discontinuity surface S_d
$\square_{\text{eff}}, \hat{\square}$	Effective part of quantity $\square$
ω, $\boldsymbol{\omega}$, $\boldsymbol{\Omega}$	Scalar, second-order tensor, and fourth-order tensor damage parameter
$\tilde{\square}$	Scalar-valued function of the tensor $\square$
$\bar{\square}$	Spatially averaged scalar-valued function $\tilde{\square}$
l	Failure process zone length scale
$\psi(\mathbf{x}, \mathbf{y})$	Spatial averaging weight function, $\Psi(\mathbf{x}) = \int_V \psi(\mathbf{x}, \mathbf{y}) \mathrm{d}V$
c_1, c_2, c_3	(Higher-order) gradient damage parameters
f_t	Fracture strength
$\mathcal{G}_c$	Fracture energy
h	Softening modulus
f	Loading–unloading function
κ	History parameter
β	Shear retention factor

μ	Tensile stiffness damage factor
$\mathbf{A}$	Acoustic tensor
$\bar{\square} + \sum \psi_l(\mathbf{x})\tilde{\bar{\square}}_l$	Partition-of-unity decomposition of the quantity $\square$
$\boldsymbol{\lambda}$	Lagrange multiplier
$\square^{\text{con}}$	Concrete part of the quantity $\square$
$\square^{\text{cr}}$	Cracking part of the quantity $\square$
$\square^{\text{re}}$	Quantity $\square$ related to a reinforcement
$\square^{\text{rc}}$	Quantity $\square$ related to reinforced concrete
$\square^{\text{ia}}$	Quantity $\square$ related to concrete-reinforcement interaction

Plasticity

$\square^{\text{p}}$	Plastic part of quantity $\square$
$\square^{\text{e}}$	Elastic part of quantity $\square$
ψ	Dilatancy angle
ϕ	Friction angle
λ	Plastic multiplier
$\mathbf{m}$	Plastic flow direction
f	Yield function
g	Plastic potential function
$\mathbf{n}$	Yield surface normal vector
τ	Shear stress
γ	Shear deformation
c	Adhesion coefficient
$\bar{\sigma}$	Yield strength
h	Hardening modulus
$\kappa, \boldsymbol{\kappa}$	Scalar hardening parameter/vector of hardening (history) parameters
$\square_e$	Quantity $\square$ related to the trial (elastic) step
$\square_c$	Quantity $\square$ related to the corrector step
$r_\square, \mathbf{r}_\square$	Residuals for local scalar- and vector-valued quantities $\square$
$\mathbf{A}$	Stress residual tangent matrix
$\mathbf{H}$	Pseudo-elastic stiffness matrix
q	Modified J_2 stress invariant
$\mathbf{P}$	Projection matrix for the modified J_2 stress
$\mathbf{Q}$	Projection matrix for strain hardening
$\boldsymbol{\pi}$	Projection vector for the hydrostatic pressure
$\boldsymbol{\alpha}$	Back-stress tensor
$\bar{\square}$	Quantity $\square$ represented in the principal stress coordinate system
$\square_{\text{m}}$	Quantity $\square$ evaluated at the time interval mid-point
α, k, β	Drucker–Prager model parameters
θ	Lode's angle
M, p_c	Cam-clay model parameters
κ^*	Modified swelling index
λ^*	Modified compression index
ϕ^*	Void volume fraction

Structural Members

$\square_0$	Quantity $\square$ in the undeformed state
$\square_l$	Quantity $\square$ related to the centre line/mid plane
ξ, η	Centre line/mid plane parametric coordinates
ζ	Out-of-plane parametric coordinate
l	Length of the structural member
h, t	Thickness of the structural member
b	Width of the structural member
A	Cross-sectional area of the structural member
I	Moment of inertia of the structural member
$\mathbf{d}$	Director
ϕ, ψ	Rotations of the structural member
$\theta, \boldsymbol{\theta}, \boldsymbol{\Theta}$	Centre line/mid plane rotations
$\chi, \boldsymbol{\chi}$	Centre line/mid plane curvature
N	Normal force
M	Bending moment
G	Shear force
$\mathbf{a}$	Nodal variables related to the centre line/mid plane deformation
$\mathbf{w}$	Nodal variables related to the out-of-plane deformation
$\boldsymbol{\theta}$	Nodal variables related to the centre line/mid plane rotations
$\square_a$	Quantity $\square$ related to the centre line/mid plane nodal variables
$\square_w$	Quantity $\square$ related to the out-of-plane deformation nodal variables
$\square_\theta$	Quantity $\square$ related to the centre line/mid plane rotation nodal variables
$\square_c$	Quantity $\square$ related to an hierarchical mid-side node
k	Shear stiffness correction factor
w	Solid-like shell internal stretch parameter

Isogeometric Analysis

d_p	Dimension of the parameter domain
d_s	Dimension of the physical domain
$\widehat{V}$	Parameter domain
$\boldsymbol{\xi} = [\xi, \eta, \zeta]^{\mathrm{T}}$	Parametric coordinate
$\Xi_\square$	Knot vector corresponding to $\square$
$\mathbf{P} = [\mathbf{p}_1, \ldots, \mathbf{p}_N]^{\mathrm{T}}$	Control net/control points
W_i	Control point weights
$w(\boldsymbol{\xi})$	Weight function
$\mathbf{h}, h_i$	B-spline basis functions
$\mathbf{r}, r_i$	NURBS basis functions
$\mathbf{B}$	Bernstein basis functions
$\mathcal{C}_e$	Element extraction operator

About the Code

A number of models and algorithms that are discussed in this book, have been implemented in a small finite element code named PyFEM, which is available for a free download from the website that accompanies this book. The code has been written in Python, an object-oriented, interpreted, and interactive programming language. Its clear syntax allows for the development of small, yet powerful programs. A wide range of Python packages are available, which are dedicated towards numerical simulations. Many numerical libraries and software tools have been equipped with a Python interface and can be integrated within a Python program seamlessly.

In PyFEM we restrict ourselves to the use of the packages `NumPy`, `SciPy` and `Matplotlib`. The `NumPy` package contains array objects and a collection of linear algebra operations. The `SciPy` package is an extension to this package and contains additional linear algebra tools, such as solvers and sparse arrays. The `Matplotlib` package allows the user to make graphs and plots. Python and the three aforementioned packages are standard components of most Linux distributions. The most recent versions of Python for various Windows operating systems and Mac OS X can be downloaded from `www.python.org`.

The PyFEM code contains the basic numerical tools which are needed to build a finite element code. These tools are used in a series of example programs with increasing complexity. The examples that illustrate the numerical techniques presented in the first chapters of this book are basically small scripts that perform a single numerical operation and do not require an input file. These small scripts are developed further, and finally result in a general finite element program which will be presented in Chapter 4: `PyFEM.py`. This program can be considered as a stand-alone program that can carry out a variety of simulations with different element formulations and material models. In the remaining parts of this book the implementation of some solvers, elements and material models is discussed in more detail.

The directory structure of PyFEM is shown in Figure 1. The package contains the following files and directories:

- `PyFEM.py` is the main program. Executing this program requires an input file with the extension `.pro`.
- The directory `doc` contains installation notes and a short user manual of the code.
- The directory `examples` contains a number of small example programs and input files, which are stored in subdirectories `ch01`, `ch02` etc., which refer to the corresponding chapters of this book for easy reference. Some of the programs and files in these directories are discussed in detail.
- The actual finite element tools are stored in the directory `pyfem`. This directory consists of six subdirectories, including `elements`, which contains element implementations, `solvers`, which contains the solvers and `materials`, in which the material formula-

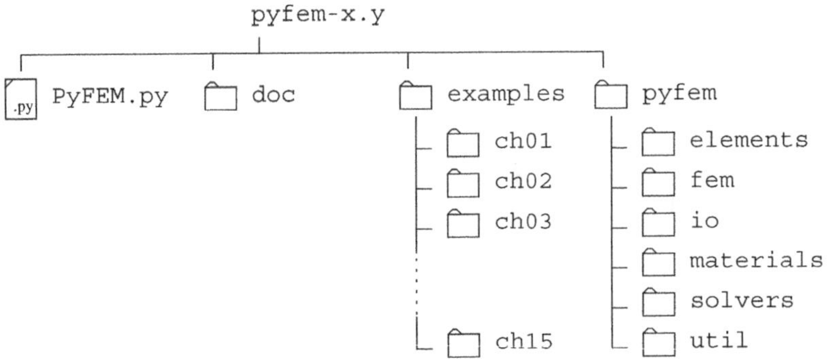

Figure 1 Directory structure of the PyFEM code. The root directory is called `pyfem-x.y`, where `x.y` indicates the version number of the code

tions are stored. A selection of files in these directories is elaborated in the book. The other three directories, `io`, `fem` and `util` contain input parsers and output writers and various finite element utility functions such as a shape function utility, which will be discussed in Chapter 2.

PyFEM is an open source code and is intended for educational and scientific use. It does not contain comprehensive libraries, e.g. of material models, but it has been designed so that it is relatively easy to implement other solvers, elements, and material models, for which the theory and the algorithmic details can be found in this book. A concise user's guide how to implement these can be found at the website.

Instead of giving full listings of classes and functions, we will use a notation that is inspired by literate programming. The main idea behind literate programming is to present a code in such a way that it can be understood by humans and by computers. An important feature of literate programming is that parts of the source code are presented as small fragments, allowing for a detailed discussion of the code. A short overview of the notation, including a system to refer to other fragments, is given in Figure 2.

The name of the fragment. The '≡' symbol indicates that this is a new fragment. When the '+≡' symbol is used, it augments a fragment that has been defined before.

In the case of a new fragment, this number refers to the page where this fragment was announced. When this fragment augments an existing fragment, this number refers to the page where the previous part of the fragment was presented.

```
⟨Solution procedure ⟩≡                                              65

  K = zeros( shape=( totDof , totDof ) )

  for elemNodes in elems:
    elemDofs    = getDofs(elemNodes)                                67
    elemCoords = coords[elemNodes,:]

    sData = getElemShapeData( elemCoords )                          49

    for iData in sData:
      ⟨Calculate integration point contribution 69⟩
```

Inclusion of another fragment in the code. The number behind the fragment name refers to the page on which the fragment has been presented. If this number is missing, the fragment is not discussed in the book.

Reference to the page on which the function or class that is mentioned in the same line in the code is described. In this case, the function `getElemShapeData` has been explained on page 49.

Figure 2 Example of a code fragment with the nomenclature and references to other code fragments

Part I

Basic Concepts and Solution Techniques

1

Preliminaries

This chapter is primarily intended to familiarise the reader with the notation we have adopted throughout this book and to refresh some of the required background in mathematics, especially linear algebra, and applied mechanics. As regards notation, we remark that most developments have been carried out using matrix-vector notation, and tensor notation is less often needed, either in indicial form or in direct form. For the benefit of readers who are less familiar with tensor notation, we have added a small section on this topic. But, first, we will give an example of non-linearity in a structural member. This example involving a simple truss element can be solved analytically, and serves well to illustrate the various procedures that are described in this book for capturing non-linear phenomena in solids and structures, and for accurately solving the ensuing initial/boundary-value problems.

1.1 A Simple Example of Non-linear Behaviour

Many features of solution techniques can be demonstrated for simple truss structures, possibly in combination with springs, where the non-linear structural behaviour can stem from geometrical as well as from material non-linearities. In this section we shall assume that the displacements and rotations can be arbitrarily large, but that the strains remain small, say less than 5%. This limitation will be dropped in Part IV of this book, where the extension will be made to large elastic and inelastic strains.

We consider the shallow truss structure of Figure 1.1. From elementary equilibrium considerations in the deformed configuration, the following expression for the force can be deduced that acts in a symmetric half of the shallow truss:

$$F_{\text{int}} = -A\sigma \sin\phi - F_s \tag{1.1}$$

where σ is the axial stress in the member, F_s is half of the force in the spring, and ϕ is the angle of the truss member with the horizontal plane in the deformed configuration. Owing to the small-strain assumption, the difference between the cross section in the current configuration, A, and that in the original configuration, A_0, is negligible. For the same reason, the difference

Non-linear Finite Element Analysis of Solids and Structures, Second Edition.
René de Borst, Mike A. Crisfield, Joris J.C. Remmers and Clemens V. Verhoosel.

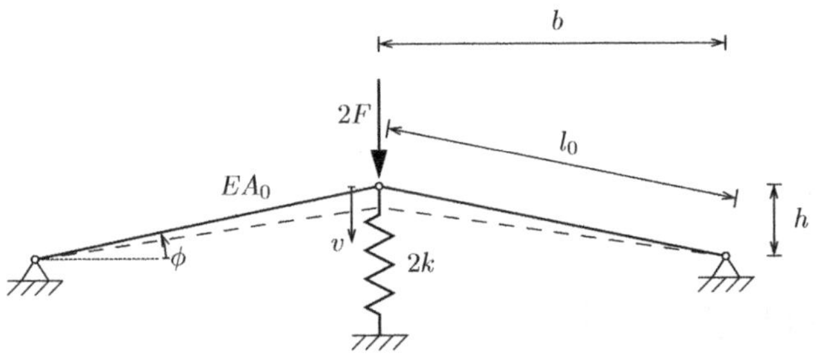

Figure 1.1 Plane shallow truss structure

between the length of the bar in the original configuration,

$$\ell_0 = \sqrt{b^2 + h^2} \tag{1.2}$$

and that in the current configuration,

$$\ell = \sqrt{b^2 + (h - v)^2} \tag{1.3}$$

can be neglected in the denominator of the expression for the strain:

$$\epsilon = \frac{\ell - \ell_0}{\ell_0} \tag{1.4}$$

or when computing the inclination angle ϕ:

$$\sin\phi = \frac{h - v}{\ell} \approx \frac{h - v}{\ell_0} \tag{1.5}$$

The dimensions b and h are defined in Figure 1.1. The vertical displacement v is taken positive in the downward sense. For half of the force in the spring we have

$$F_s = -kv \tag{1.6}$$

with k the spring stiffness, and the axial stress in the bar reads:

$$\sigma = E\epsilon \tag{1.7}$$

with E the Young's modulus. Substitution of the expressions for the stress σ, the force in the spring F_s and the angle ϕ into the equilibrium condition (1.1) yields:

$$F_{\text{int}}(v) = -EA_0 \sin\phi \frac{\ell - \ell_0}{\ell_0} + kv \tag{1.8}$$

Equation (1.8) expresses the internal force that acts in the structure as a non-linear function of the vertical displacement v. Normally, the external force at time $t + \Delta t$, $F_{\text{ext}}^{t+\Delta t}$, is given. The displacement v must then be computed such that

$$F_{\text{ext}}^{t+\Delta t} - F_{\text{int}}^{t+\Delta t} = 0 \tag{1.9}$$

The correct value of v is computed in an iterative manner, for instance using the Newton–Raphson method:

$$F_{\rm ext}^{t+\Delta t} = F_{\rm int}(v_j) + \frac{{\rm d}F_{\rm int}}{{\rm d}v}{\rm d}v + \frac{1}{2}\frac{{\rm d}^2 F_{\rm int}}{{\rm d}v^2}{\rm d}v^2 + \mathcal{O}({\rm d}v^3) \tag{1.10}$$

with j the iteration counter. In a linear approximation we have for the iterative correction to the displacement v:

$${\rm d}v = \left(\frac{{\rm d}F_{\rm int}}{{\rm d}v}\right)_j^{-1} \left(F_{\rm ext}^{t+\Delta t} - F_{\rm int}(v_j)\right) \tag{1.11}$$

The iterative process is terminated when a convergence criterion has been met, $\|F_{\rm ext}^{t+\Delta t} - F_{\rm int}(v_j)\| < \varepsilon$, with ε a small number. For the present case the derivative $\frac{{\rm d}F_{\rm int}}{{\rm d}v}$, or in computational mechanics terminology, the tangential stiffness modulus, can be evaluated from Equation (1.8) as:

$$\frac{{\rm d}F_{\rm int}}{{\rm d}v} = \frac{A_0 \sin^2\phi}{\ell_0}\left(E + \frac{{\rm d}E}{{\rm d}\ell}(\ell - \ell_0)\right) + \left(k + \frac{{\rm d}k}{{\rm d}v}v\right) + \frac{A_0\sigma}{\ell_0} \tag{1.12}$$

where, for generality, it has been assumed that the stiffness of the truss as well as that of the spring depend on how much they have been extended. If this so-called material non-linearity is not present, the terms that involve $\frac{{\rm d}E}{{\rm d}\ell}$ and $\frac{{\rm d}k}{{\rm d}v}$ cancel. The last term in Equation (1.12) is due to the inclusion of large displacement/rotation effects (geometrical non-linearity), and is linear in the stress. This term is of crucial importance when computing the stability of slender structures. Figure 1.2 shows the behaviour of the truss for different values of the spring stiffness k. The graphs directly follow from application of the closed-form expression (1.8) for the internal force, in combination with the equilibrium condition (1.9). The iterative procedure can only be applied for larger values of the spring stiffness k, i.e. when there is no local maximum in the load–displacement curve.

1.2 A Review of Concepts from Linear Algebra

In computer oriented methods in the mechanics of solids frequent use is made of the concepts of a vector and a matrix. Herein, we shall denote by a vector a one-dimensional array of scalars. A scalar is a physical quantity that has the same value, irrespective of the choice of the reference frame. When we denote scalars by italic symbols and vectors by roman, bold-faced, lower-case symbols, the vector $\mathbf{v}$ has n scalar entries $v_1, \ldots, v_n$, so that:

$$\mathbf{v} = \begin{pmatrix} v_1 \\ \cdots \\ \cdots \\ v_n \end{pmatrix} \tag{1.13}$$

In Equation (1.13) the scalar entries are written in a column format. Alternatively, it is possible to write the scalar quantities $v_1, \ldots, v_n$ as a row. This row of scalars is named the transpose

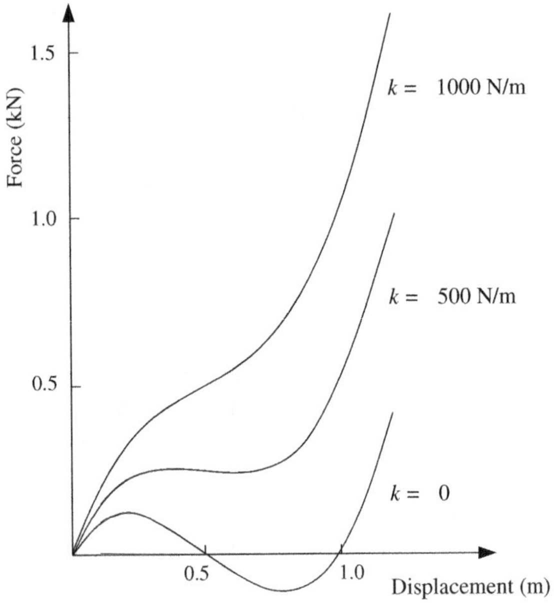

Figure 1.2 Force–displacement diagram for the shallow truss structure for different values of the spring stiffness k (b = 10 m, h = 0.5 m and EA_0 = 5 MN/m^2)

of the vector **v** and is written as:

$$\mathbf{v}^{\mathrm{T}} = (v_1, \ldots, v_n)$$

Addition of vectors is defined as the addition of their components, so that

$$\mathbf{w} = \mathbf{u} + \mathbf{v} \tag{1.14}$$

implies that $w_i = u_i + v_i$ for $i = 1, \ldots, n$. The multiplication of a vector by a scalar, say λ, is defined as:

$$\mathbf{w} = \lambda \mathbf{u} \tag{1.15}$$

with the components $w_i = \lambda u_i$.

An important operation between two vectors **u** and **v**, each with n entries, is the inner product, also named scalar product:

$$\mathbf{u}^{\mathrm{T}}\mathbf{v} = \sum_{i=1}^{n} u_i v_i \tag{1.16}$$

The scalar product of two vectors possesses the commutativity property, i.e. $\mathbf{u}^{\mathrm{T}}\mathbf{v} = \mathbf{v}^{\mathrm{T}}\mathbf{u}$ as can be verified easily from the definition (1.16). The inner product can also be useful for the definition of the norm of a vector. Several definitions of the norm of a vector are possible, but in theoretical and applied mechanics the most customary definition is the Euclidian or L_2-norm:

$$\|\mathbf{v}\|_2 = \sqrt{\mathbf{v}^{\mathrm{T}}\mathbf{v}} \tag{1.17}$$

where the subscript is often omitted. The cross product of two vectors **a** and **b**, also named the vector product, forms a vector **c**:

$$\mathbf{c} = \mathbf{a} \times \mathbf{b} \tag{1.18}$$

that is orthogonal to **a** and **b** in the three-dimensional space, and has a direction that is given by the right-hand rule, and as a consequence, is anti-symmetric: $\mathbf{b} \times \mathbf{a} = -\mathbf{a} \times \mathbf{b}$. The components of $\mathbf{c} = \mathbf{a} \times \mathbf{b}$ read:

$$\mathbf{c} = \begin{pmatrix} a_2b_3 - a_3b_2 \\ a_3b_1 - a_1b_3 \\ a_1b_2 - a_2b_1 \end{pmatrix} \tag{1.19}$$

The entries (or components) of a vector may be used to form a scalar function. Examples in mechanics are the invariants of the stress and strain tensors, or the yield function in plasticity. An operation that is often used is the calculation of the gradient of a function. Let the scalar-valued function f be a function of the components a_i of the vector **a**. Then, the gradient **b** is obtained by differentiation of f with respect to **a**

$$\mathbf{b} = \frac{\partial f}{\partial \mathbf{a}} \tag{1.20}$$

or in component form:

$$b_i = \frac{\partial f}{\partial a_i} \tag{1.21}$$

The gradient operation is such that **b** is orthogonal to the hypersurface in the n-dimensional vector space that is described by $f = c$, with c a constant that usually is taken equal to zero.

Matrices are another suitable mathematical vehicle that can be used in computational mechanics. While vectors in their most simple description are denoted as one-dimensional arrays of scalars, matrices are two-dimensional arrays of scalars. A matrix is said to have m rows and n columns. In general m does not have to be equal to n. If we think of vectors as matrices with only one column, a vector with m components can be termed a $m \times 1$ matrix. Similarly, a row vector with n entries can be named an $1 \times n$ matrix.

In this book we shall consistently denote a matrix by a bold-faced, upper-case symbol. The entries or components of the matrix **A** are, in a similar fashion as the components of a vector, denoted as a_{ij}, where, for an $m \times n$ matrix $i = 1, \ldots, m$ and $j = 1, \ldots, n$. A vector **b** of length n can be premultiplied by an $m \times n$ matrix **A**, as follows:

$$\mathbf{c} = \mathbf{A}\mathbf{b} \tag{1.22}$$

The resulting vector **c** has m components:

$$c_i = \sum_{j=1}^{n} a_{ij} b_j \tag{1.23}$$

The addition of two $m \times n$ matrices $\mathbf{A}$ and $\mathbf{B}$ is exactly analogous to the addition of vectors, as we have for each entry: $c_{ij} = a_{ij} + b_{ij}$, while the multiplication of a matrix by a scalar, say λ, is also defined similarly: $c_{ij} = \lambda a_{ij}$.

The product of two matrices is defined similar to the product of a matrix and a vector. Let $\mathbf{A}$ be an $m \times k$ matrix and $\mathbf{B}$ be a $k \times n$ matrix. The result of multiplying $\mathbf{A}$ and $\mathbf{B}$ is an $m \times n$ matrix $\mathbf{C}$, with components:

$$c_{ij} = \sum_{e=1}^{k} a_{ie} b_{ej} \tag{1.24}$$

A special matrix multiplication occurs when the number of columns of $\mathbf{A}$, and consequently also the number of rows of $\mathbf{B}$, is set equal to 1 ($k = 1$). Now, $\mathbf{A}$ and $\mathbf{B}$ reduce to vectors, say $\mathbf{a}$ and $\mathbf{b}^{\mathrm{T}}$. The resulting product is still an $m \times n$ matrix,

$$\mathbf{C} = \mathbf{a}\mathbf{b}^{\mathrm{T}} \tag{1.25}$$

with components $c_{ij} = a_i b_j$. This operation is named the dyadic or outer product of two vectors $\mathbf{a}$ and $\mathbf{b}$. The transpose operation for matrices is identical to that for vectors, i.e. $\mathbf{B} = \mathbf{A}^{\mathrm{T}}$ implies that $b_{ij} = a_{ji}$. An operation that is frequently carried out in the derivation of finite element equations is taking the transpose of a product of two matrices. For such a transpose the following relationship holds:

$$(\mathbf{A}\mathbf{B})^{\mathrm{T}} = \mathbf{B}^{\mathrm{T}}\mathbf{A}^{\mathrm{T}} \tag{1.26}$$

The most common type of matrices are square matrices, for which $m = n$. Under certain conditions, to be discussed in the following pages, an inverse $\mathbf{B} = \mathbf{A}^{-1}$ can be defined, such that

$$\mathbf{A}\mathbf{B} = \mathbf{I} \tag{1.27}$$

with $\mathbf{I}$ the unit matrix, i.e. all entries of $\mathbf{I}$ are zero with exception of the diagonal entries of $\mathbf{I}$ which are equal to 1: $\mathbf{I} = \mathrm{diag}[1, \ldots, 1]$. The inversion of matrices is required for the solution of large systems of linear equations which arise as a result of finite element discretisation. Such systems have the form

$$\begin{aligned} a_{11}x_1 + a_{12}x_2 + \ldots + a_{1n}x_n &= b_1 \\ a_{21}x_1 + a_{22}x_2 + \ldots + a_{2n}x_n &= b_2 \\ \ldots + \ldots + \ldots &= \ldots \\ a_{n1}x_1 + a_{n2}x_2 + \ldots + a_{nn}x_n &= b_n \end{aligned} \tag{1.28}$$

When the known coefficients $a_{11}, \ldots, a_{nn}$ are assembled in a matrix $\mathbf{A}$, the known components $b_1, \ldots, b_n$ in a vector $\mathbf{b}$, and the unknowns $x_1, \ldots, x_n$ in a vector $\mathbf{x}$, the system (1.28) can be written in a compact fashion

$$\mathbf{A}\mathbf{x} = \mathbf{b} \tag{1.29}$$

Formally, the vector of unknowns $\mathbf{x}$ can be obtained from

$$\mathbf{x} = \mathbf{A}^{-1}\mathbf{b} \tag{1.30}$$

provided, of course, that $\mathbf{A}^{-1}$ exists. In solid mechanics the matrix $\mathbf{A}$ is often symmetric, i.e. $a_{ij} = a_{ji}$, which facilitates the computation of $\mathbf{A}^{-1}$. However, when non-linearities are incorporated in computational models, symmetry can be lost.

An efficient manner to carry out the above operation is to decompose the matrix $\mathbf{A}$ as

$$\mathbf{A} = \mathbf{LDU} \tag{1.31}$$

with $\mathbf{L}$ a lower triangular matrix

$$\mathbf{L} = \begin{bmatrix} 1 & 0 & 0 & \dots & 0 \\ l_{21} & 1 & 0 & \dots & 0 \\ l_{31} & l_{32} & 1 & \dots & 0 \\ \dots & \dots & \dots & \dots & \dots \\ l_{n1} & l_{n2} & l_{n3} & \dots & 1 \end{bmatrix} \tag{1.32}$$

$\mathbf{U}$ an upper triangular matrix,

$$\mathbf{U} = \begin{bmatrix} 1 & u_{12} & u_{13} & \dots & u_{1n} \\ 0 & 1 & u_{23} & \dots & u_{2n} \\ 0 & 0 & 1 & \dots & u_{3n} \\ \dots & \dots & \dots & \dots & \dots \\ 0 & 0 & 0 & \dots & 1 \end{bmatrix} \tag{1.33}$$

and

$$\mathbf{D} = \text{diag}[d_{11}, \dots, d_{nn}] \tag{1.34}$$

a diagonal matrix. For symmetric matrices the identity $\mathbf{U} = \mathbf{L}^{\mathrm{T}}$ holds.

This LDU decomposition is based on Gauss elimination, and can preserve bandedness in the sense that if the matrix $\mathbf{A}$ has a band structure, as is normally the case in finite element applications, the lower and upper triangular matrices $\mathbf{L}$ and $\mathbf{U}$ also have a banded structure. Since

$$\mathbf{x} = (\mathbf{LDU})^{-1}\mathbf{b} = \mathbf{U}^{-1}(\mathbf{LD})^{-1}\mathbf{b} = \mathbf{U}^{-1}\mathbf{D}^{-1}\mathbf{L}^{-1}\mathbf{b}$$

we can now solve for $\mathbf{x}$:

$$\begin{aligned} \mathbf{c} &= \mathbf{L}^{-1}\mathbf{b} \\ \mathbf{d} &= \mathbf{D}^{-1}\mathbf{c} \\ \mathbf{x} &= \mathbf{U}^{-1}\mathbf{d} \end{aligned} \tag{1.35}$$

This equation reveals another interesting fact. While the operations $\mathbf{L}^{-1}\mathbf{b}$ and $\mathbf{U}^{-1}\mathbf{d}$ only involve multiplications, and cannot result in arithmetic problems, the operation $\mathbf{D}^{-1}\mathbf{c}$ consists of divisions, since $\mathbf{D}^{-1} = \text{diag}[d_{11}^{-1}, \dots, d_{nn}^{-1}]$. Hence, as soon as one of the diagonal entries, named pivots, of $\mathbf{D}$ is zero, $\mathbf{x}$ can no longer be computed. In such a case the matrix $\mathbf{A}$ is said to be singular and a unique decomposition no longer exists. We distinguish between three cases: all pivots of **D** are positive, one or more pivots of **D** are zero, and finally, one or more pivots of **D** are negative. When the diagonal matrix **D** has only positive pivots, the matrix **A** is called positive definite. An example is the stiffness matrix **A** which results from a displacement-method based finite element discretisation of a linear-elastic body. For positive-definite matrices the LDU decomposition is unique and round-off errors which arise are not amplified. When non-linear effects are introduced, the tangential stiffness matrix **A** can become singular (one or more zero pivots) during the loading process and eventually become indefinite (one or more negative pivots). As argued above, a singular matrix cannot be decomposed and meaningful answers cannot be obtained. However, a unique LDU decomposition can again be obtained if one or more pivots have turned negative, but are non-zero. Nevertheless, for indefinite matrices it cannot be ensured that round-off errors which arise during the decomposition are not amplified. In a non-linear analysis this observation implies that the iterative process that is necessary to solve the set of non-linear algebraic equations which then arises, can diverge.

Singularity of a matrix is also closely related to its determinant. The determinant of a matrix is defined as (Golub and van Loan 1983; Noble and Daniel 1969; Ortega 1987; Saad 1996)

$$\det\mathbf{A} = \sum_{j=1}^{n}(-1)^{i+j}a_{ij}\det\mathbf{A}_{ij} \tag{1.36}$$

where $\mathbf{A}_{ij}$ is an $(n-1)\times(n-1)$ matrix obtained by deleting the ith row and the jth column of **A**. This recursive relation is closed by $\det\mathbf{A} = a_{11}$ for $n = 1$. A useful property is that $\det(\mathbf{AB}) = \det\mathbf{A}\cdot\det\mathbf{B}$. In view of Equation (1.31) we have $\det\mathbf{A} = \det\mathbf{L}\cdot\det\mathbf{D}\cdot\det\mathbf{U}$ and from definition (1.36) we deduce that $\det\mathbf{L} = \det\mathbf{U} = 1$. We thus obtain the useful result that

$$\det\mathbf{A} = \prod_{i=1}^{n} d_i \tag{1.37}$$

which implies that the determinant of a matrix equals zero if one or more pivots are zero. In view of the discussion on pivots the matrix is then singular.

A useful result on the inversion of a special type of matrices is the Sherman–Morrison formula. Let **A** be a non-singular $n\times n$ matrix and let **u** and **v** be two vectors with n entries each. Then, the following identity holds:

$$(\mathbf{A}+\mathbf{u}\mathbf{v}^{\mathrm{T}})^{-1} = \mathbf{A}^{-1} - \frac{\mathbf{A}^{-1}\mathbf{u}\mathbf{v}^{\mathrm{T}}\mathbf{A}^{-1}}{1+\mathbf{v}^{\mathrm{T}}\mathbf{A}^{-1}\mathbf{u}} \tag{1.38}$$

A further useful result involving vectors is Gauss' divergence theorem. Using this theorem a volume integral can be transformed into a surface integral:

$$\int_V \mathrm{div}\mathbf{v}\,\mathrm{d}V = \int_S \mathbf{n}^{\mathrm{T}}\mathbf{v}\,\mathrm{d}S \tag{1.39}$$

where **n** is the outward normal to the bounding surface of the body, and div is the divergence operator:

$$\mathrm{div}\mathbf{v} = \frac{\partial v_1}{\partial x_1} + \frac{\partial v_2}{\partial x_2} + \frac{\partial v_3}{\partial x_3} \tag{1.40}$$

In the preceding, use has been made of the summation symbol Σ. A short-hand notation is to omit the Σ symbol and to suppose that a summation is implied whenever a subscript occurs twice in an expression. For instance, we can replace the summation in Equation (1.24) by the abbreviated notation (called the Einstein summation convention)

$$c_{ij} = a_{ie} b_{ej} \tag{1.41}$$

where summation with respect to the repeated index e is implied. Such an index is often called a 'dummy' index, since it is irrelevant which letter we take for this index. Indeed, the expression $c_{ij} = a_{iq} b_{qj}$ is identical. Of course, the indices i and j may not be replaced by other letters unless it is done on both sides of the equation. When rewriting Gauss' theorem in index notation, the result is:

$$\int_V \frac{\partial v_i}{\partial x_i} \mathrm{d}V = \int_S n_i v_i \mathrm{d}S$$

An important tensorial quantity is the Kronecker delta, defined as:

$$\begin{cases} \delta_{ij} = 1 & \text{if} \quad i = j \\ \delta_{ij} = 0 & \text{if} \quad i \neq j \end{cases} \tag{1.42}$$

As an example we note that $a_{ij} = a_{ik} \delta_{kj}$. Also useful is the permutation tensor e_{ijk}, which equals $+1$ for e_{123} and for even permutations thereof (e.g. e_{231}), and equals -1 for odd permutations (e.g. e_{213}). If two subscripts are identical, then $e_{ijk} = 0$.

In more recent years index notation has been gradually replaced by direct tensor notation, which, at first sight, somewhat resembles the matrix-vector notation. Now, the multiplication of Equation (1.24) is denoted as:

$$\mathbf{C} = \mathbf{A} \cdot \mathbf{B} \tag{1.43}$$

where the central dot denotes a single contraction, i.e. the summation over the dummy index. In a similar fashion, a double contraction is denoted as:

$$c = \mathbf{A} : \mathbf{B} \tag{1.44}$$

or using index notation: $c = a_{ie} b_{ei}$. Taking the gradient of a quantity is done using the ∇ symbol, as follows,

$$\mathbf{b} = \nabla f \tag{1.45}$$

which equals the gradient vector defined in Equation (1.20). This operator can also be used for vectors, and Gauss' theorem is now written as:

$$\int_V \nabla \cdot \mathbf{v} \mathrm{d}V = \int_S \mathbf{n} \cdot \mathbf{v} \mathrm{d}S$$

The dyadic product of two vectors **a** and **b** is now written as:

$$\mathbf{C} = \mathbf{a} \otimes \mathbf{b} \tag{1.46}$$

with components $c_{ij} = a_i b_j$. Finally, we define for a second-order tensor $\mathbf{A}$ the divergence operator

$$\mathbf{a} = \nabla \cdot \mathbf{A} \tag{1.47}$$

such that

$$a_j = \frac{\partial A_{ij}}{\partial x_i} \tag{1.48}$$

and its trace:

$$c = \mathrm{tr}(\mathbf{A}) \tag{1.49}$$

through $c = a_{ii}$.

1.3 Vectors and Tensors

So far, vectors have been introduced and treated as mere mathematical tools, arrays which contain a number of scalar quantities in an ordered fashion. Nonetheless, vectors can be given a physical interpretation. Take for instance the concept of force. A force not only has a magnitude, but also has a direction. It is often of interest to know how the components of a force change if the force is represented in a different coordinate system. A translation only adds the same number to all force components. A rotation of the reference frame, for instance from the x, y-coordinate system to a $\bar{x}, \bar{y}$-coordinate system, Figure 1.3, changes the components of a vector in a more complicated manner.

The components of a vector $\bar{\mathbf{n}}$ in the $\bar{x}, \bar{y}$-coordinate system can be obtained from those in the x, y-coordinate system, assembled in $\mathbf{n}$, by the transformation

$$\bar{\mathbf{n}} = \mathbf{R}\mathbf{n} \tag{1.50}$$

with $\mathbf{R}$ a transformation matrix. Since a full three-dimensional treatment is quite cumbersome, and hardly adds anything to the understanding, we will elaborate $\mathbf{R}$ only for planar conditions. Let the angle from the x, y-coordinate system to the $\bar{x}, \bar{y}$-coordinate be ϕ. For $\mathbf{n} = [1, 0]^{\mathrm{T}}$ and $\mathbf{n} = [0, 1]^{\mathrm{T}}$, respectively, the representations in the rotated coordinate system are $\bar{\mathbf{n}} = [\cos\phi, -\sin\phi]^{\mathrm{T}}$ and $\bar{\mathbf{n}} = [\sin\phi, \cos\phi]^{\mathrm{T}}$, respectively. It follows that in two dimensions the

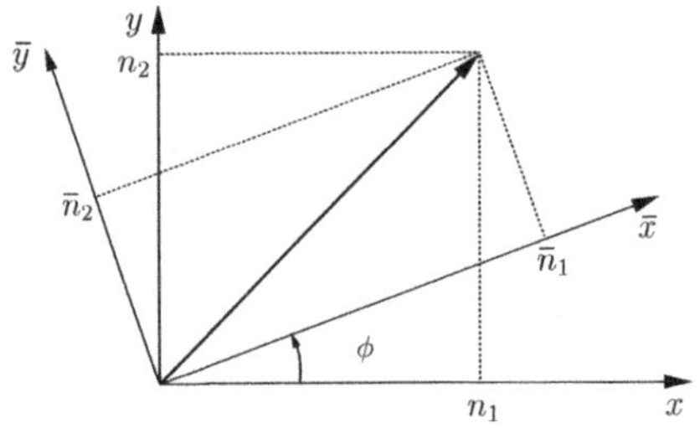

Figure 1.3 Original x, y-coordinate system and rotated x, y-coordinate system

transformation matrix **R** is given by

$$\mathbf{R} = \begin{bmatrix} \cos\phi & \sin\phi \\ -\sin\phi & \cos\phi \end{bmatrix} \tag{1.51}$$

The transformation, or rotation matrix **R** has a special structure. Inspection shows that

$$\mathbf{R}^{-1} = \mathbf{R}^{\mathrm{T}} \tag{1.52}$$

which also holds true for the general three-dimensional case. Matrices that satisfy requirement (1.52) are called orthogonal matrices, for which det(**R**) = 1.

With the aid of the transformation rules for vectors we can derive transformation rules for tensors. Tensors, or here, more precisely, second-order tensors, are physical quantities that relate two vectors. For instance, the stress tensor sets a relation between the force on a plane and the normal vector of that plane, see also the next section. A natural representation of a second-order tensor is a matrix. However, not all matrices are tensors: only matrices that obey certain transformation rules can represent tensorial quantities. Suppose that the second-order tensor **C** relates the vectors, or first-order tensors, **t** and **n**:

$$\mathbf{t} = \mathbf{C}\mathbf{n} \tag{1.53}$$

In the $\bar{x}$, $\bar{y}$ frame the second-order tensor $\bar{\mathbf{C}}$ sets a similar relation between $\bar{\mathbf{t}}$ and $\bar{\mathbf{n}}$:

$$\bar{\mathbf{t}} = \bar{\mathbf{C}}\bar{\mathbf{n}} \tag{1.54}$$

We next substitute Equation (1.50) and an identical relation for **t**, i.e. $\bar{\mathbf{t}} = \mathbf{R}\mathbf{t}$, into Equation (1.54). Comparison with Equation (1.53) shows that any second-order tensor transforms according to:

$$\bar{\mathbf{C}} = \mathbf{R}\mathbf{C}\mathbf{R}^{\mathrm{T}} \tag{1.55}$$

Using Equation (1.51) this identity can be elaborated for two dimensions as

$$\begin{aligned} \bar{c}_{11} &= c_{11}\cos^2\phi + (c_{12} + c_{21})\cos\phi\sin\phi + c_{22}\sin^2\phi \\ \bar{c}_{22} &= c_{11}\sin^2\phi - (c_{12} + c_{21})\cos\phi\sin\phi + c_{22}\cos^2\phi \\ \bar{c}_{12} &= -c_{11}\cos\phi\sin\phi + c_{12}\cos^2\phi - c_{21}\sin^2\phi + c_{22}\cos\phi\sin\phi \\ \bar{c}_{21} &= -c_{11}\cos\phi\sin\phi - c_{12}\sin^2\phi + c_{21}\cos^2\phi + c_{22}\cos\phi\sin\phi \end{aligned} \tag{1.56}$$

For symmetric second-order tensors, which will be employed here exclusively, $c_{21} = c_{12}$, and consequently also: $\bar{c}_{21} = \bar{c}_{12}$.

We observe that the components of a second-order tensor change from orientation to orientation. It is often of interest to know the extremal values of the tensor components $\bar{c}_{11}$ and $\bar{c}_{22}$, and on which plane they are attained, i.e. for which value of ϕ. For symmetric second-order tensors, there exist two mutually orthogonal planes on which $\bar{c}_{11}$ and $\bar{c}_{22}$ have a maximum and a minimum, respectively. The values in this coordinate system are commonly named the principal values. Since $\bar{c}_{11}$ and $\bar{c}_{22}$ are functions of the inclination angle ϕ these extremal values

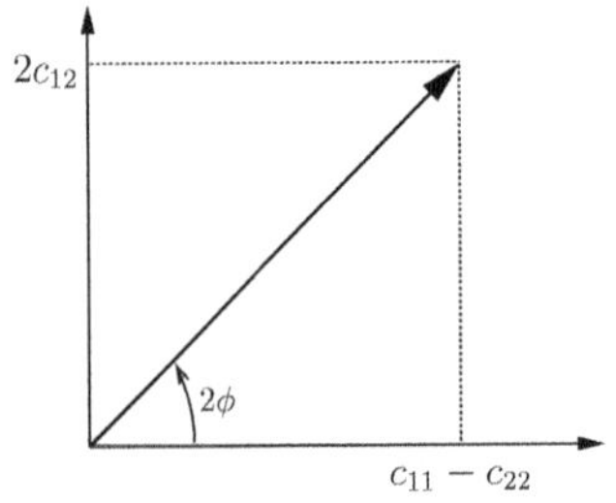

Figure 1.4 Principal directions of a second-order tensor

are obtained by requiring that

$$\frac{\partial \bar{c}_{11}}{\partial \phi} = 0 \quad \text{or} \quad \frac{\partial \bar{c}_{22}}{\partial \phi} = 0 \tag{1.57}$$

Elaborating these identities for symmetric second-order tensors we obtain that the diagonal tensor components attain extremal values for

$$\tan 2\phi = \frac{2c_{12}}{c_{11} - c_{22}} \tag{1.58}$$

To derive the principal values we first rewrite the first two equations of (1.56) as

$$\bar{c}_{11} = \frac{1}{2}(c_{11} + c_{22}) + \frac{1}{2}(c_{11} - c_{22})\cos 2\phi + c_{12}\sin 2\phi \tag{1.59}$$

From Figure 1.4, cf. Equation (1.58), we infer that

$$\begin{aligned} \sin 2\phi &= \pm\frac{2c_{12}}{\sqrt{(c_{11} - c_{22})^2 + 4c_{12}^2}} \\ \cos 2\phi &= \pm\frac{c_{11} - c_{22}}{\sqrt{(c_{11} - c_{22})^2 + 4c_{12}^2}} \end{aligned} \tag{1.60}$$

whence we obtain the following closed-form expression for the principal values:

$$\begin{cases} \bar{c}_{11} = \frac{1}{2}(c_{11} + c_{22}) - \frac{1}{2}\sqrt{(c_{11} - c_{22})^2 + 4c_{12}^2} \\ \bar{c}_{22} = \frac{1}{2}(c_{11} + c_{22}) + \frac{1}{2}\sqrt{(c_{11} - c_{22})^2 + 4c_{12}^2} \end{cases} \tag{1.61}$$

It is a property of symmetric second-order tensors (to which the treatment will be limited) that for this inclination angle also the off-diagonal tensor components are zero: $\bar{c}_{12} = 0$. This is shown most simply by rewriting the first equation of (1.56) as:

$$\bar{c}_{12} = -\frac{1}{2}(c_{11} - c_{22})\sin 2\phi + c_{12}\cos 2\phi \tag{1.62}$$

whereupon substitution of the identities (1.60) proves the assertion.

Another interpretation can be given to the coordinate system in which the principal values of the diagonal tensor components attain a maximum. Let **C** be the matrix representation of a symmetric second-order tensor. Let $\boldsymbol{e}$ be a vector. As a rule, the product $\mathbf{C}\boldsymbol{e}$ will not be parallel with $\boldsymbol{e}$. However, for every such tensor there exists a coordinate system for which the resulting vector is indeed parallel with the original vector:

$$\mathbf{C}\boldsymbol{e} = \lambda \boldsymbol{e} \tag{1.63}$$

with λ the scalar-valued eigenvalue. We can rewrite Equation (1.63) as

$$(\mathbf{C} - \lambda \mathbf{I})\boldsymbol{e} = \mathbf{0} \tag{1.64}$$

with $\mathbf{I} = \text{diag}[1, \ldots, 1]$ the unit matrix. A non-trivial solution ($\boldsymbol{e} \neq \mathbf{0}$) then exists if and only if the determinant of $\mathbf{C} - \lambda \mathbf{I}$ vanishes:

$$\det[\mathbf{C} - \lambda \mathbf{I}] = 0 \tag{1.65}$$

Elaborating Equation (1.65) then yields exactly Equation (1.58). Thus, the coordinate system in which c_{11} and c_{22} attain extremal values is the same coordinate system in which a vector $\boldsymbol{e}$ multiplied by a tensor **C** results in a vector that is a multiple of $\boldsymbol{e}$. Since the eigenvalues λ_i correspond to the principal values, the eigenvectors $\boldsymbol{e}_i$ point in the principal directions. An elaboration for a symmetric second-order tensor is given in Box 1.1. Similar to pivots, Equation (1.37), a direct relationship can be established between the product of all eigenvalues and the determinant of a matrix:

$$\det \mathbf{C} = \prod_{i=1}^{n} \lambda_i \tag{1.66}$$

which is known as Vieta's rule, and is valid for symmetric and non-symmetric matrices. From Equation (1.66) we infer that the singularity of a matrix not only implies that the determinant and one or more pivots vanish, but also that at least one eigenvalue is equal to zero.

Inverting Equation (1.55) yields

$$\mathbf{C} = \mathbf{R}^{\mathrm{T}} \bar{\mathbf{C}} \mathbf{R} \tag{1.67}$$

with, in the principal axes,

$$\mathbf{C} = \begin{bmatrix} \bar{c}_{11} & 0 \\ 0 & \bar{c}_{22} \end{bmatrix} \tag{1.68}$$

and $\bar{c}_{11} = \lambda_1$ and $\bar{c}_{22} = \lambda_2$ the principal values or eigenvalues of **C**. Elaboration of Equation (1.67) using expression (1.51) for **R** in two dimensions yields:

$$\mathbf{C} = \lambda_1 \begin{bmatrix} \cos^2\phi & \cos\phi\sin\phi \\ \cos\phi\sin\phi & \sin^2\phi \end{bmatrix} + \lambda_2 \begin{bmatrix} \sin^2\phi & -\cos\phi\sin\phi \\ -\cos\phi\sin\phi & \cos^2\phi \end{bmatrix}$$

or

$$\mathbf{C} = \lambda_1 \begin{pmatrix} \cos\phi \\ \sin\phi \end{pmatrix} (\cos\phi, \sin\phi) + \lambda_2 \begin{pmatrix} \sin\phi \\ -\cos\phi \end{pmatrix} (\sin\phi, -\cos\phi)$$

Box 1.1 Eigenvalues of a symmetric second-order tensor

For a symmetric matrix **C** the condition $\det[\mathbf{C} - \lambda \mathbf{I}] = 0$ can be elaborated as follows:

$$\begin{vmatrix} c_{11} - \lambda & c_{12} \\ c_{12} & c_{22} - \lambda \end{vmatrix} = 0 \quad \text{or} \quad (c_{11} - \lambda)(c_{22} - \lambda) - c_{12}^2 = 0$$

Solving for the eigenvalues λ yields: $\lambda_{1,2} = \frac{1}{2}(c_{11} + c_{22}) \pm \frac{1}{2}\sqrt{(c_{11} - c_{22})^2 + 4c_{12}^2}$, which are exactly the principal values of the tensor **C**, see Equation (1.61). The directions of $\boldsymbol{e}$ can be computed by inserting the principal values of the tensor **C** in either

$$(c_{11} - \lambda)e_1 + c_{12}e_2 = 0 \quad \text{or} \quad c_{12}e_1 + (c_{22} - \lambda)e_2 = 0$$

with e_1, e_2 the components of $\boldsymbol{e}$. Taking the first equation as an example, we can derive that substitution of the principal values $\lambda_{1,2}$ yields:

$$\left(\frac{1}{2}(c_{11} - c_{22}) \pm r\right) e_1 + c_{12}e_2 = 0 \quad \text{where} \quad r = \frac{1}{2}\sqrt{(c_{11} - c_{22})^2 + 4c_{12}^2}$$

Bringing the re_1 term to the right-hand side, and squaring gives:

$$\frac{e_1 e_2}{e_1^2 - e_2^2} = \frac{c_{12}}{c_{11} - c_{22}}$$

Simple goniometry shows that

$$\tan 2\phi = \frac{2\tan\phi}{1 - \tan^2\phi} = \frac{2e_1e_2}{e_1^2 - e_2^2}$$

which proves that Equation (1.58) also defines the directions of the eigenvectors $\boldsymbol{e}$. The notions of eigenvectors and principal directions, and of eigenvalues and principal values of symmetric second-order tensors coincide.

Identifying $\boldsymbol{e}_1^{\mathrm{T}} = (\cos\phi, \sin\phi)$ and $\boldsymbol{e}_2^{\mathrm{T}} = (\sin\phi, -\cos\phi)$ as the eigenvectors, we can represent **C** through the spectral decomposition

$$\mathbf{C} = \sum_{i=1}^{n} \lambda_i \boldsymbol{e}_i \otimes \boldsymbol{e}_i \tag{1.69}$$

where a generalisation to n dimensions has been made. Defining the eigenprojections

$$\boldsymbol{E}_i = \boldsymbol{e}_i \otimes \boldsymbol{e}_i \tag{1.70}$$

the spectral decomposition of a symmetric, second-order tensor can also be written as:

$$\mathbf{C} = \sum_{i=1}^{n} \lambda_i \boldsymbol{E}_i \tag{1.71}$$

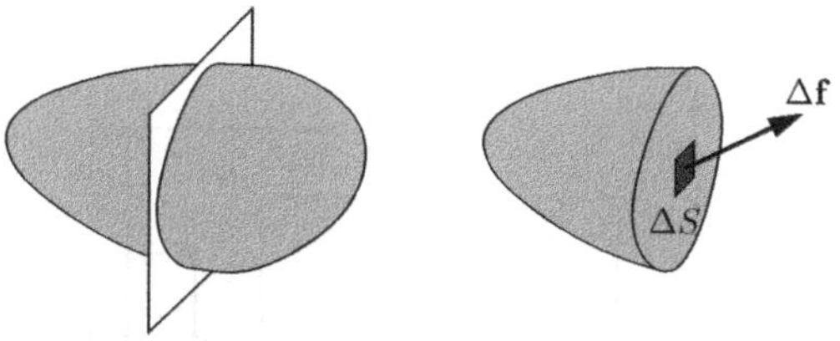

Figure 1.5 Force acting on an imaginary cut in a solid body

1.4 Stress and Strain Tensors

The basic problem of solid mechanics is to determine the response of a body to forces that are exerted onto that body. For instance, we want to know which forces act from one side of an imaginary cut in the body on the other side (Figure 1.5). It has become customary to consider a small area in that cut, say ΔS, and to investigate which force works on that area. This force is called $\Delta\mathbf{f}$. When we take the limiting case that $\Delta S \to 0$ the stress vector $\mathbf{t}$ is obtained:

$$\mathbf{t} = \lim_{\Delta S \to 0} \frac{\Delta \mathbf{f}}{\Delta S} = \frac{\mathrm{d}\mathbf{f}}{\mathrm{d}S} \tag{1.72}$$

On each plane the stress vector $\mathbf{t}$ can be decomposed in a component that acts along the normal to that plane and in two mutually orthogonal vectors which form a vectorial basis of the plane. We now choose the normal vector of this plane to coincide with the x-axis. The normal component of $\mathbf{t}$ is denoted by σ_{xx}, while the two components that lie in the plane are labelled as σ_{xy} and σ_{xz}. σ_{xy} is the stress component which acts in the direction of the y-axis and σ_{xz} is the stress component which acts in the direction of the z-axis. In accordance with the sign convention in solid mechanics the normal stress component σ_{xx} is considered positive when it points in the direction of the positive x-axis and works on a plane with a normal vector that points in the positive x-direction. In a similar fashion the shear stress σ_{xy} is taken positive when it points in the positive y-direction and acts on a plane with its normal in the positive x-direction. The definition of the other shear stress, σ_{xz}, is analogous. Along this line of reasoning the normal stress σ_{xx} is also called positive if it acts in the negative x-direction on a plane with its normal in the negative x-direction, while a positive shear stress σ_{xy} is also obtained when a shear stress acts on a plane with its normal in the negative x-direction and is directed along the negative y-axis.

In three dimensions there are nine stress components (Figure 1.6). These nine stress components fully determine the state of stress in a point of a body, and are components of the stress tensor. The stress tensor $\boldsymbol{\sigma}$ is a second-order tensor. It can be naturally expressed in matrix notation:

$$\boldsymbol{\Sigma} = \begin{bmatrix} \sigma_{xx} & \sigma_{yx} & \sigma_{zx} \\ \sigma_{xy} & \sigma_{yy} & \sigma_{zy} \\ \sigma_{xz} & \sigma_{yz} & \sigma_{zz} \end{bmatrix} \tag{1.73}$$

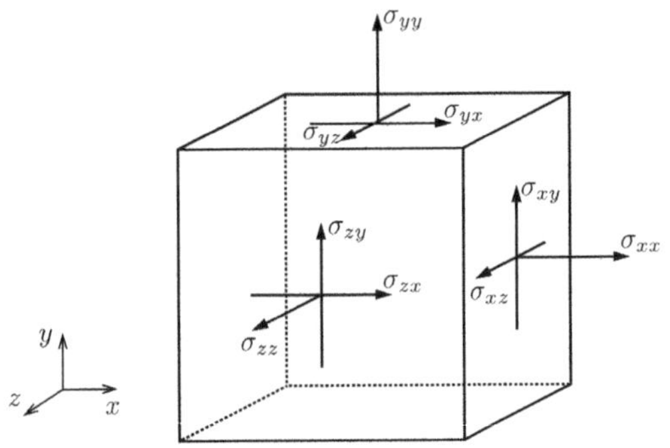

Figure 1.6 Stress components in a three-dimensional continuum

The stress tensor $\boldsymbol{\sigma}$ is related to the stress vector $\mathbf{t}$ which acts on a plane with normal $\mathbf{n}$. In matrix-vector notation, the relationship between $\boldsymbol{\Sigma}$, $\mathbf{t}$ and $\mathbf{n}$ is:

$$\boldsymbol{\Sigma}\mathbf{n} = \mathbf{t} \tag{1.74}$$

The validity of this relationship can be verified easily if the normal vector is chosen to be parallel to the x-axis ($\mathbf{n}^{\mathrm{T}} = [1, 0, 0]$), the y-axis ($\mathbf{n}^{\mathrm{T}} = [0, 1, 0]$), and the z-axis ($\mathbf{n}^{\mathrm{T}} = [0, 0, 1]$), respectively. For future use the analogue of Equation (1.74) is also given in index notation:

$$n_i \sigma_{ij} = t_j$$

and in direct tensor notation:

$$\mathbf{n} \cdot \boldsymbol{\sigma} = \mathbf{t} \tag{1.75}$$

For a non-polar or Boltzmann continuum, the balance of moment of momentum in the three directions shows that not all the stress components are independent. In particular we find for the shear stress components that

$$\begin{aligned} \sigma_{xy} &= \sigma_{yx} \\ \sigma_{yz} &= \sigma_{zy} \\ \sigma_{zx} &= \sigma_{xz} \end{aligned} \tag{1.76}$$

(see Chapter 2 for a formal proof). Accordingly, there are six independent stress components and the matrix representation of the symmetric stress tensor $\boldsymbol{\sigma}$ can be written as

$$\boldsymbol{\Sigma} = \begin{bmatrix} \sigma_{xx} & \sigma_{xy} & \sigma_{zx} \\ \sigma_{xy} & \sigma_{yy} & \sigma_{yz} \\ \sigma_{zx} & \sigma_{yz} & \sigma_{zz} \end{bmatrix} \tag{1.77}$$

The observation that there are only six independent stress components makes it also feasible to write the stress tensor in a vector form (the so-called Voigt notation):

$$\boldsymbol{\sigma}^{\mathrm{T}} = (\sigma_{xx}, \sigma_{yy}, \sigma_{zz}, \sigma_{xy}, \sigma_{yz}, \sigma_{zx}) \tag{1.78}$$

Note that for the vector representation the stress tensor is symbolically written as $\boldsymbol{\sigma}$ instead of $\boldsymbol{\Sigma}$ which is used for the matrix representation.

Often, for instance in geotechnical applications, it is convenient to decompose the normal stresses σ_{xx}, σ_{yy} and σ_{zz} into a deviatoric and a hydrostatic part. The deviatoric part then causes changes in the shape of an elementary cube, while the hydrostatic pressure causes a change in volume of the cube. The hydrostatic pressure is here defined as

$$p = \frac{1}{3}(\sigma_{xx} + \sigma_{yy} + \sigma_{zz}) \tag{1.79}$$

With the aid of the definition of p we can define the deviatoric stress tensor. In matrix representation we have

$$\mathbf{S} = \boldsymbol{\Sigma} - p\mathbf{I} \tag{1.80}$$

while in Voigt's notation the following formula is obtained:

$$\mathbf{s} = \boldsymbol{\sigma} - p\mathbf{i} \tag{1.81}$$

where

$$\begin{aligned} \mathbf{s}^{\mathrm{T}} &= (s_{xx}, s_{yy}, s_{zz}, s_{xy}, s_{yz}, s_{zx}) \\ \mathbf{i}^{\mathrm{T}} &= (1, 1, 1, 0, 0, 0) \end{aligned} \tag{1.82}$$

Stress invariants are important quantities in non-linear constitutive theories. These are functions of the stress components that are invariant with respect to the choice of the reference frame. They arise naturally if the principal stresses in a three-dimensional continuum are computed. From the previous section it is known that the principal values λ of a second-order tensor are computed from the requirement that

$$\det(\boldsymbol{\Sigma} - \lambda\mathbf{I}) = 0 \tag{1.83}$$

or, in component form:

$$\begin{vmatrix} \sigma_{xx} - \lambda & \sigma_{xy} & \sigma_{zx} \\ \sigma_{xy} & \sigma_{yy} - \lambda & \sigma_{yz} \\ \sigma_{zx} & \sigma_{yz} & \sigma_{zz} - \lambda \end{vmatrix} = 0 \tag{1.84}$$

When we introduce the identities

$$\begin{aligned} I_1 &= \sigma_{xx} + \sigma_{yy} + \sigma_{zz} \\ I_2 &= \sigma_{xx}\sigma_{yy} + \sigma_{yy}\sigma_{zz} + \sigma_{zz}\sigma_{xx} - \sigma_{xy}^2 - \sigma_{yz}^2 - \sigma_{zx}^2 \\ I_3 &= \sigma_{xx}\sigma_{yy}\sigma_{zz} + 2\sigma_{xy}\sigma_{yz}\sigma_{zx} - \sigma_{xx}\sigma_{yz}^2 - \sigma_{yy}\sigma_{zx}^2 - \sigma_{zz}\sigma_{xy}^2 \end{aligned} \tag{1.85}$$

Equation (1.84) can be reformulated as:

$$\lambda^3 - I_1\lambda^2 + I_2\lambda - I_3 = 0 \tag{1.86}$$

A crucial observation is that, since this equation has the same solution in each reference frame, I_1, I_2 and I_3 must have the same value irrespective of the choice of the reference frame. Thus, the coefficients I_1, I_2 and I_3 must be invariant under a coordinate transformation. For this reason, I_1, I_2 and I_3 are called invariants of the stress tensor. The concept of principal values and principal directions exists for any second-order tensor, and invariants can be defined for any second-order tensor, also for the strain tensor to be treated next.

Any function of invariants is an invariant itself. Such modified invariants arise naturally if the principal values of the deviatoric stress tensor are computed. These quantities are obtained by solving the cubic equation:

$$\lambda^3 - J_2\lambda - J_3 = 0 \tag{1.87}$$

where

$$J_2 = -s_{xx}s_{yy} - s_{yy}s_{zz} - s_{zz}s_{xx} + s_{xy}^2 + s_{yz}^2 + s_{zx}^2 \tag{1.88}$$

and

$$J_3 = s_{xx}s_{yy}s_{zz} + 2\sigma_{xy}\sigma_{yz}\sigma_{zx} - s_{xx}\sigma_{yz}^2 - s_{yy}\sigma_{zx}^2 - s_{zz}\sigma_{xy}^2 \tag{1.89}$$

The first invariant of the deviatoric stress tensor vanishes by definition. With the above definitions for the invariants of the stress tensor and the deviatoric stress tensor it can be shown that (Fung 1965):

$$J_2 = \frac{1}{3}I_1^2 - I_2$$
$$J_3 = I_3 - \frac{1}{3}I_1I_2 + \frac{2}{27}I_1^3 \tag{1.90}$$

We now consider an elementary cube which we deform only in the x, y-plane. The sides of the cube are denoted by Δx, Δy and Δz ($\Delta x = \Delta y = \Delta z$). Suppose that point A undergoes the displacements u and v and that points B and C displace as $[u + \Delta u^B, v + \Delta v^B]$ and $[u + \Delta u^C, v + \Delta v^C]$, respectively (Figure 1.7). In the limiting case that $\Delta x \to 0$ and $\Delta y \to 0$

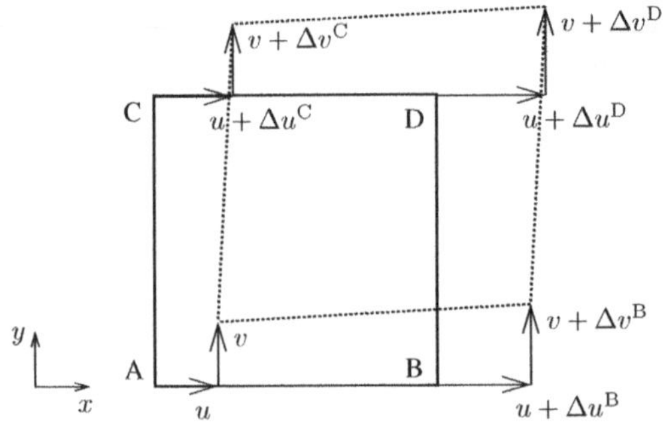

Figure 1.7 Undeformed and deformed configuration of an elementary quadrilateral

the strains in the x- and y-directions become (neglecting second-order terms):

$$\epsilon_{xx} = \lim_{\Delta x \to 0} \frac{\Delta u^B}{\Delta x} = \frac{\partial u}{\partial x}$$
$$\epsilon_{yy} = \lim_{\Delta y \to 0} \frac{\Delta v^C}{\Delta y} = \frac{\partial v}{\partial y} \quad (1.91)$$

The distortion of the elementary square in the x, y-plane is given by:

$$\gamma_{xy} = \lim_{\Delta x \to 0, \Delta y \to 0} \frac{\Delta u^C}{\Delta y} + \frac{\Delta v^B}{\Delta x} = \frac{\partial u}{\partial y} + \frac{\partial v}{\partial x} \quad (1.92)$$

while its rotation is given by:

$$\omega_{xy} = \lim_{\Delta x \to 0, \Delta y \to 0} \frac{1}{2}\left(\frac{\Delta v^B}{\Delta x} - \frac{\Delta u^C}{\Delta y}\right) = \frac{1}{2}\left(\frac{\partial v}{\partial x} - \frac{\partial u}{\partial y}\right) \quad (1.93)$$

Here γ_{xy} is the total angular distortion of the elementary cube in the x, y-plane. This measure for the shear strain is often used in engineering applications. For theoretical investigations it is more customary to adopt the tensorial shear strain component $\epsilon_{xy} = \frac{1}{2}\gamma_{xy}$. In a similar fashion to which we have introduced the normal strains ϵ_{xx}, ϵ_{yy} and the engineering shear strain γ_{xy} we can introduce the normal strain ϵ_{zz} and the shear strains γ_{yz} and γ_{zx} by considering deformations of the elementary cube in the y, z- and the z, x-planes, respectively. In accordance with the definitions (1.91) and (1.92) these strain components are defined as:

$$\epsilon_{zz} = \frac{\partial w}{\partial z}$$
$$\gamma_{yz} = \frac{\partial v}{\partial z} + \frac{\partial w}{\partial y} \quad (1.94)$$
$$\gamma_{zx} = \frac{\partial w}{\partial x} + \frac{\partial u}{\partial z}$$

where w is the displacement in the z-direction. The convention for subscripts in the strain components is exactly the same as for stress components, e.g. ϵ_{xx} defines a normal strain component in the x-direction and γ_{xy} represents a shear strain component in the x, y-plane. Also the sign convention is identical: a strain component is called positive if it is related to a positive displacement of a plane with normal in the positive direction, etc. This implies for instance that elongation is considered positive.

Similar to the stress tensor we can now introduce the strain tensor. Again, matrix and vector representations are possible. For the fully three-dimensional case we have the matrix representation

$$\mathcal{E} = \begin{bmatrix} \epsilon_{xx} & \epsilon_{yx} & \epsilon_{zx} \\ \epsilon_{xy} & \epsilon_{yy} & \epsilon_{zy} \\ \epsilon_{xz} & \epsilon_{yz} & \epsilon_{zz} \end{bmatrix} \quad (1.95)$$

or, noting that the strain tensor has been defined such that it is symmetric, we can write in Voigt notation:

$$\boldsymbol{\epsilon}^{\mathrm{T}} = (\epsilon_{xx}, \epsilon_{yy}, \epsilon_{zz}, \gamma_{xy}, \gamma_{yz}, \gamma_{zx}) \tag{1.96}$$

While the use of the total distortion γ_{xy} etc. is more common in the Voigt notation, the tensorial shear strain ϵ_{xy} is normally used in the matrix representation. For future use we note that the rate of the internal energy per unit volume can be expressed equivalently in Voigt notation, direct tensor notation and index notation as:

$$\dot{\mathcal{W}}_{\mathrm{int}} = \boldsymbol{\epsilon}^{\mathrm{T}}\boldsymbol{\sigma} = \boldsymbol{\epsilon} : \boldsymbol{\sigma} = \dot{\epsilon}_{ij}\sigma_{ji} \tag{1.97}$$

In the treatment of the stress tensor the hydrostatic pressure p was introduced. Similarly, we can introduce the volumetric strain ϵ_{vol} as the sum of the normal strains:

$$\epsilon_{\mathrm{vol}} = \epsilon_{xx} + \epsilon_{yy} + \epsilon_{zz} \tag{1.98}$$

With the aid of the volumetric strain ϵ_{vol} we can define the so-called deviatoric strain tensor in a manner similar to the introduction of the deviatoric stresses:

$$\mathbf{E} = \boldsymbol{\mathcal{E}} - \frac{1}{3}\epsilon_{\mathrm{vol}}\mathbf{I} \tag{1.99}$$

or using Voigt's notation,

$$\mathbf{e} = \boldsymbol{\epsilon} - \frac{1}{3}\epsilon_{\mathrm{vol}}\mathbf{i} \tag{1.100}$$

with

$$\mathbf{e}^{\mathrm{T}} = (e_{xx}, e_{yy}, e_{zz}, \gamma_{xy}, \gamma_{yz}, \gamma_{zx}) \tag{1.101}$$

In a preceding section the transformation rule for second-order tensors was derived, cf. Equation (1.56). Using Voigt notation, these transformation rules can, for the two-dimensional case, be written as:

$$\bar{\boldsymbol{\sigma}} = \mathbf{T}_{\sigma}\boldsymbol{\sigma} \tag{1.102}$$

with

$$\mathbf{T}_{\sigma} = \begin{bmatrix} \cos^2\phi & \sin^2\phi & 2\sin\phi\cos\phi \\ \sin^2\phi & \cos^2\phi & -2\sin\phi\cos\phi \\ -\sin\phi\cos\phi & \sin\phi\cos\phi & \cos^2\phi - \sin^2\phi \end{bmatrix} \tag{1.103}$$

with the stress tensor $\boldsymbol{\sigma}^{\mathrm{T}} = [\sigma_{xx}, \sigma_{yy}, \sigma_{xy}]$ for plane-stress conditions. By substituting $(-\phi)$ in the latter equation it is seen that $\mathbf{T}_{\sigma}^{-1} = \mathbf{T}_{\sigma}^{\mathrm{T}}$, whence

$$\boldsymbol{\sigma} = \mathbf{T}_{\sigma}^{\mathrm{T}}\bar{\boldsymbol{\sigma}} \tag{1.104}$$

Since the engineering shear strain γ_{xy} is normally used in Voigt's notation, we have for the strain transformation:

$$\bar{\boldsymbol{\epsilon}} = \mathbf{T}_{\epsilon}\boldsymbol{\epsilon} \tag{1.105}$$

with

$$\mathbf{T}_\epsilon = \begin{bmatrix} \cos^2\phi & \sin^2\phi & \sin\phi\cos\phi \\ \sin^2\phi & \cos^2\phi & -\sin\phi\cos\phi \\ -2\sin\phi\cos\phi & 2\sin\phi\cos\phi & \cos^2\phi - \sin^2\phi \end{bmatrix} \tag{1.106}$$

and $\boldsymbol{\epsilon}^{\mathrm{T}} = [\epsilon_{xx}, \epsilon_{yy}, \gamma_{xy}]$. As for the stress transformation it holds that:

$$\boldsymbol{\epsilon} = \mathbf{T}_\epsilon^{\mathrm{T}} \bar{\boldsymbol{\epsilon}} \tag{1.107}$$

1.5 Elasticity

So far, we have introduced the stress tensor and we have considered kinematic relations, i.e. relations between displacements and strains. In Chapter 2 we will introduce the equations of motion. To complete the field equations we need stress–strain relations, or constitutive equations, which set a relation between the stress tensor and the strain tensor. For the simplest constitutive model, namely isotropic, linear elasticity (Hooke's law), the fourth-order elastic compliance tensor $\mathbf{C}^{\mathrm{e}}$ sets the relation between the strain tensor $\boldsymbol{\epsilon}$ and the stress tensor $\boldsymbol{\sigma}$:

$$\boldsymbol{\epsilon} = \mathbf{C}^{\mathrm{e}} : \boldsymbol{\sigma} \tag{1.108}$$

or in its inverse form:

$$\boldsymbol{\sigma} = \mathbf{D}^{\mathrm{e}} : \boldsymbol{\epsilon} \tag{1.109}$$

with $\mathbf{D}^{\mathrm{e}}$ the elastic stiffness tensor. In Voigt notation the compliance relation can be elaborated as:

$$\begin{pmatrix} \epsilon_{xx} \\ \epsilon_{yy} \\ \epsilon_{zz} \\ \gamma_{xy} \\ \gamma_{yz} \\ \gamma_{zx} \end{pmatrix} = \frac{1}{E} \begin{bmatrix} 1 & -\nu & -\nu & 0 & 0 & 0 \\ -\nu & 1 & -\nu & 0 & 0 & 0 \\ -\nu & -\nu & 1 & 0 & 0 & 0 \\ 0 & 0 & 0 & 2(1+\nu) & 0 & 0 \\ 0 & 0 & 0 & 0 & 2(1+\nu) & 0 \\ 0 & 0 & 0 & 0 & 0 & 2(1+\nu) \end{bmatrix} \begin{pmatrix} \sigma_{xx} \\ \sigma_{yy} \\ \sigma_{zz} \\ \sigma_{xy} \\ \sigma_{yz} \\ \sigma_{zx} \end{pmatrix} \tag{1.110}$$

with E the Young's modulus and ν the Poisson's ratio. Equation (1.110) can be written compactly as:

$$\boldsymbol{\epsilon} = \mathbf{C}^{\mathrm{e}} \boldsymbol{\sigma} \tag{1.111}$$

with $\mathbf{C}^{\mathrm{e}}$ the elastic compliance matrix.

Equation (1.110) gives the strain tensor $\boldsymbol{\epsilon}$ as a function of the stress tensor $\boldsymbol{\sigma}$. To obtain the inverse relation we rewrite the first three equations of (1.110) as

$$
\begin{aligned}
\epsilon_{xx} &= \frac{1+\nu}{E}\sigma_{xx} - 3\nu E p \\
\epsilon_{yy} &= \frac{1+\nu}{E}\sigma_{yy} - 3\nu E p \\
\epsilon_{zz} &= \frac{1+\nu}{E}\sigma_{zz} - 3\nu E p
\end{aligned}
\tag{1.112}
$$

Next, we add these equations and, using Equations (1.79) and (1.98) we obtain:

$$\epsilon_{\text{vol}} = K^{-1} p \tag{1.113}$$

where the bulk modulus K, which sets the relation between the volumetric strain and the hydrostatic pressure, has been introduced:

$$K = \frac{E}{3(1-2\nu)} \tag{1.114}$$

Subsequent substitution of Equation (1.113) into Equations (1.112) and inversion yields the elastic stiffness relation:

$$
\begin{pmatrix} \sigma_{xx} \\ \sigma_{yy} \\ \sigma_{zz} \\ \sigma_{xy} \\ \sigma_{yz} \\ \sigma_{zx} \end{pmatrix}
=
\begin{bmatrix}
\lambda + 2\mu & \lambda & \lambda & 0 & 0 & 0 \\
\lambda & \lambda + 2\mu & \lambda & 0 & 0 & 0 \\
\lambda & \lambda & \lambda + 2\mu & 0 & 0 & 0 \\
0 & 0 & 0 & \mu & 0 & 0 \\
0 & 0 & 0 & 0 & \mu & 0 \\
0 & 0 & 0 & 0 & 0 & \mu
\end{bmatrix}
\begin{pmatrix} \epsilon_{xx} \\ \epsilon_{yy} \\ \epsilon_{zz} \\ \gamma_{xy} \\ \gamma_{yz} \\ \gamma_{zx} \end{pmatrix}
\tag{1.115}
$$

where the two Lamé constants have been introduced:

$$
\begin{aligned}
\lambda &= \frac{\nu E}{(1+\nu)(1-2\nu)} \\
\mu &= \frac{E}{2(1+\nu)}
\end{aligned}
\tag{1.116}
$$

The latter quantity is conventionally defined as the shear modulus and is also often denoted by the symbol G. The above stiffness relation can be written as

$$\boldsymbol{\sigma} = \mathbf{D}^{\mathrm{e}} \boldsymbol{\epsilon} \tag{1.117}$$

with $\mathbf{D}^{\mathrm{e}}$ the elastic stiffness matrix. An alternative expression for this matrix in terms of the Young's modulus and the Poisson's ratio can be obtained by inserting Equations (1.116) into

Equation (1.115)

$$\mathbf{D}^{\mathrm{e}} = \frac{E}{(1+\nu)(1-2\nu)} \begin{bmatrix} 1-\nu & \nu & \nu & 0 & 0 & 0 \\ \nu & 1-\nu & \nu & 0 & 0 & 0 \\ \nu & \nu & 1-\nu & 0 & 0 & 0 \\ 0 & 0 & 0 & \frac{1-2\nu}{2} & 0 & 0 \\ 0 & 0 & 0 & 0 & \frac{1-2\nu}{2} & 0 \\ 0 & 0 & 0 & 0 & 0 & \frac{1-2\nu}{2} \end{bmatrix} \tag{1.118}$$

1.6 The PyFEM Finite Element Library

A number of the models that are discussed in this book have been implemented in a small finite element code named PyFEM, which is written in the programming language Python. In order to demonstrate some features of this programming language in a numerical simulation, the implementation of the simple non-linear calculation in Section 1.1 is discussed in this section. The file is called `ShallowTruss.py` and can be found in the directory `examples/ch01` of PyFEM.

Instead of giving complete code listings, we will use a notation that is inspired by literate programming, see e.g. Ramsey (1994). A concise overview of this notation has been given earlier in the book on page xix. Some specific details of the notation will be highlighted in this section as well.

In literate programming the complete script can be represented as a collection of code fragments:

```
⟨Shallow truss example ⟩≡

    ⟨Initialisation of the calculation 25⟩
    ⟨Step-wise calculation of the equilibrium path 27⟩
    ⟨Print results 28⟩
```

This defines a fragment ⟨*Shallow truss example*⟩. The fragment itself refers to three other fragments, which are executed one after the other. Their function within the program can be deduced from their names. At this moment, this is the appropriate abstraction level.

The number behind the name of the fragment indicates the page number in this book where this fragment is discussed. Accordingly, the fragment ⟨*Initialisation of the calculation*⟩ is discussed on page 25. The absence of a page number indicates that the fragment is not discussed explicitly. One has to study the original source code to understand its functionality,

In the first fragment of this example, the variables that set the dimensions of the simulation are declared:

```
⟨Initialisation of the calculation ⟩≡                                    25

  b = 10.
  h = 0.5
```

`b` and `h` represent the dimensions of the system, b and h, as specified in Figure 1.1. The number in the right margin indicates the page number on which the fragment is mentioned before.

Obviously, the initialisation of the shallow truss example requires more than the system dimensions. We therefore extend the fragment by defining the stiffnesses k and EA_0, Equation (1.8), by writing:

⟨*Initialisation of the calculation* ⟩+≡ 25

```
k     = 1000.
EA0   = 5.0e6
```

The +≡ symbol after the fragment name indicates that this fragment augments a fragment defined before. We can further extend the fragment by specifying the magnitude of the incremental external force in the simulation, `DF`, the number of steps `N`, the convergence tolerance `tol`, and the maximum number of iterations to reach convergence, `iterMax`.

⟨*Initialisation of the calculation* ⟩+≡ 26

```
DF        = 50
N         = 30
tol       = 1e-6
iterMax = 5
```

The actual model is defined through use of the **`lambda`** function of Python:

⟨*Initialisation of the calculation* ⟩+≡ 26

```
from math import sqrt

l    = lambda v :  sqrt(b**2+(h-v)**2)
F    = lambda v :  -EA0*(h-v)/l(v)*(l(v)-l(0))/l(0)+k*v
dFdv= lambda v :  (EA0/l(v))*((h-v)/l(v))**2+k+/
                   (EA0/l(v))*(l(v)-l(0))/l(0)
```

Please note that in this fragment we have used a function, namely the square root operator. This function is imported from the `math` module. In PyFEM, we will often use functions from the `math`, `numpy` and `scipy` modules. In order to limit the amount code listing in this book, we will omit these import statements (**`from`** `..` **`import`** `..`) when possible. When the origin of a function is not exactly clear, the import statement will be listed.

Subsequently, the length of the beam l and the reaction force F of the system are defined as functions of the unknown v, see Equations (1.3) and (1.8). In the last line the derivative of the force with respect to the unknown is given as a function of the unknown v, Equation (1.12).

Finally, the parameters that are needed during the simulation are initialised:

```
⟨Initialisation of the calculation ⟩+≡                                   26
    v      = 0.
    Dv     = 0.
    Fext   = 0.
    output = [ [0.,0.] ]
```

The variables v, Dv and Fext represent the displacement v in the last converged solution, the incremental displacement Δv and the total external force F_{ext}, respectively. A list of lists output is created to store the variables that are plotted in the load–displacement curve, see Figure 1.2.

The simulation consists of a loop over N load steps, where i is the current step number. First a header is printed to denote the current load step. Then, the iterations are prepared in the fragment ⟨*Prepare iteration*⟩. Finally, the calculations are done in the fragment ⟨*Iteration*⟩:

```
⟨Step-wise calculation of the equilibrium path ⟩≡                        25
    for i in range(N):
      print '================================='
      print '  Load step %i' % i
      print '================================='
      print '  NR iter : |Fext-F(v)|'
      ⟨Prepare iteration 27⟩
      ⟨Iteration 28⟩
```

An important feature of Python is that code blocks are defined by indentation. Code blocks are collections of statements that are executed within an **if** statement, within a **for** statement, or within a **while** loop. A block starts by indenting the code. A block ends by a reset to the original indent. The specific Python indentation rules also apply in the literate programming notation. In the above fragment, the new fragment ⟨*Prepare iteration*⟩ is executed within the loop **for** i **in** range(N):, since the fragment name is indented with respect to the **for** statement. The same holds for the fragment ⟨*Iteration*⟩.

The preparation of the iteration consists of the update of the magnitude of the external force Fext, and resetting the error and the iteration counter iiter:

```
⟨Prepare iteration ⟩≡                                                    27
    Fext  = Fext + DF
    error = 1.
    iiter = 0
```

The iteration itself is described in the following fragment:

⟨*Iteration* ⟩≡ 27

```
while error > tol:
  iiter += 1
  dv    = ( 1. / dFdv(v+Dv) ) * ( Fext - F(v+Dv) )
  Dv   += dv
  error  = abs( Fext - F(v+Dv) )
  print '  Iter', iiter, ':', error

  if iiter == iterMax:
    raise RuntimeError('Iterations did not converge!')

print '  Converged solution'
v  += Dv
Dv  = 0.
output.append( [ v, F(v) ] )
```

In this fragment, the iteration counter `iiter` is initially increased by 1. The new displacement increment `dv` is calculated in the second line and added to the total increment of this step `Dv`. The error is calculated and printed in the following lines. In order to prevent the program from entering an infinite loop, a runtime error occurs when the number of iterations exceeds the maximum number of iterations. When the error is smaller than a certain tolerance, a converged solution has been found. The total displacement `v` is updated, and this displacement and the current internal force are added to the output list.

The last fragment of the ⟨*Shallow truss example*⟩ program prints the results:

⟨*Print results* ⟩≡ 25

```
from pylab import plot, show, xlabel, ylabel

plot( [x[0] for x in output], [x[1] for x in output], 'ro' )
```

The `output` array is plotted using `pylab` (a part of the `MatPlotLib` package). In this fragment, list comprehension has been used to create two new lists that contain the data for the horizontal and the vertical axes, respectively.

For this example problem, the exact solution `F` is known and can be added to the plot:

⟨*Print results* ⟩+≡ 28

```
from numpy import arange

vrange = arange(0,1.2,0.01)
plot( vrange, [F(vval) for vval in vrange], 'b-' )
xlabel('v [m]')
ylabel('F [N]')

show()
```

The exact solution has been calculated in the range $0.0 \leq F \leq 1.2$, with increments of 0.01. After printing the labels for the horizontal and vertical axes, the graph appears on the screen by invoking the command `show()`.

References

Fung YC 1965 *Foundations of Solid Mechanics*. Prentice-Hall.
Golub GH and van Loan CF 1983 *Matrix Computations*. The Johns Hopkins University Press.
Noble B and Daniel JW 1969 *Applied Linear Algebra*. Prentice-Hall.
Ortega JM 1987 *Matrix Theory – A Second Course*. Plenum Press.
Ramsey N 1994 Literate programming simplified. *IEEE Software* **11**, 97–105.
Saad Y 1996 *Iterative Methods for Sparse Linear Systems*. International Thomson Publishing.

2

Non-linear Finite Element Analysis

The pure displacement version of the finite element method is the most convenient spatial discretisation method for the majority of the applications of non-linear constitutive relations. Its formulation is simple, and allows for a straightforward implementation of complicated constitutive relations. However, in some cases the displacement version of the finite element method ceases to give accurate results. In such cases one usually has to resort to mixed or hybrid methods, e.g. Chapters 7 or 11. These methods tend to give answers that are more accurate than pure displacement models, but there is an increased risk of improper element behaviour. This is already the case for linear elastic material behaviour, but if material non-linearities are incorporated, the possibility that spurious kinematic modes in elements are triggered is far greater. In fact, this is another reason why pure displacement-based methods are often preferred in finite element analyses involving material non-linearities such as plasticity or damage. We shall therefore now outline displacement-based finite element models and the structure of non-linear finite element codes based upon this concept.

2.1 Equilibrium and Virtual Work

The concept of stress is vital in the derivation of the equations of motion. An elegant way to derive these equations is to consider the balance of momentum of a body V with boundary S in its current configuration. With the stress vector $\mathbf{t}$ and the gravity acceleration assembled in the vector $\mathbf{g}$ the linear momentum balance reads:

$$\int_S \mathbf{t}\mathrm{d}S + \int_V \rho\mathbf{g}\mathrm{d}V = \int_V \rho\ddot{\mathbf{u}}\mathrm{d}V \tag{2.1}$$

where ρ is the mass density, and a superimposed dot denotes differentiation with respect to time. Using Equation (1.75) we can modify this equation to give:

$$\int_S \mathbf{n}\cdot\boldsymbol{\sigma}\mathrm{d}S + \int_V \rho\mathbf{g}\mathrm{d}V = \int_V \rho\ddot{\mathbf{u}}\mathrm{d}V \tag{2.2}$$

Non-linear Finite Element Analysis of Solids and Structures, Second Edition.
René de Borst, Mike A. Crisfield, Joris J.C. Remmers and Clemens V. Verhoosel.

where **n** is the outward unit vector at the boundary of the body. The divergence theorem can now be employed to give

$$\int_V (\nabla \cdot \boldsymbol{\sigma} + \rho \mathbf{g} - \rho \ddot{\mathbf{u}}) \mathrm{d}V = \mathbf{0} \tag{2.3}$$

Since this identity must also hold for each subpart of the body, we must require that locally:

$$\nabla \cdot \boldsymbol{\sigma} + \rho \mathbf{g} = \rho \ddot{\mathbf{u}} \tag{2.4}$$

It is noted that by considering the balance of moment of momentum, the symmetry of the stress tensor can be demonstrated (Box 2.1).

Box 2.1 Symmetry of the stress tensor

We start the proof by considering rotational equilibrium of a body $\mathcal{B}$. If t_k are the components of the surface traction and if g_k are the components of the gravity acceleration, we have:

$$\int_S e_{ijk} x_j t_k \mathrm{d}S + \int_V \rho e_{ijk} x_j g_k \mathrm{d}V = 0$$

We insert the relation between the stress vector and the stress tensor to obtain:

$$\int_S e_{ijk} x_j \sigma_{lk} n_l \mathrm{d}S + \int_V \rho e_{ijk} x_j g_k \mathrm{d}V = 0$$

with n_l the components of the unit normal vector on the given plane. We next apply the divergence theorem to give

$$\int_V \left(e_{ijk} \frac{\partial x_j \sigma_{lk}}{\partial x_l} + \rho e_{ijk} x_j g_k \right) \mathrm{d}V = 0$$

whence

$$\int_V \left(e_{ijk} x_j \left[\frac{\partial \sigma_{lk}}{\partial x_l} + \rho g_k \right] + e_{ijk} \delta_{jl} \sigma_{lk} \right) \mathrm{d}V = 0$$

Because of translational equilibrium the first term in the integral vanishes. Using an argument similar to that used for the derivation of the equations of translational equilibrium we require that the second integral holds pointwise: $e_{ijk} \delta_{jl} \sigma_{lk} = 0$, which directly results in the sought symmetry: $\sigma_{kl} = \sigma_{lk}$.

In the remainder of this chapter we shall adopt matrix-vector notation, and we introduce the operator matrix **L**

$$\mathbf{L}^{\mathrm{T}} = \begin{bmatrix} \frac{\partial}{\partial x} & 0 & 0 & \frac{\partial}{\partial y} & 0 & \frac{\partial}{\partial z} \\ 0 & \frac{\partial}{\partial y} & 0 & \frac{\partial}{\partial x} & \frac{\partial}{\partial z} & 0 \\ 0 & 0 & \frac{\partial}{\partial z} & 0 & \frac{\partial}{\partial y} & \frac{\partial}{\partial x} \end{bmatrix} \tag{2.5}$$

With Equation (2.5), Equation (2.4) can be recast in a compact matrix-vector format:

$$\mathbf{L}^{\mathrm{T}}\boldsymbol{\sigma} + \rho\mathbf{g} = \rho\ddot{\mathbf{u}} \tag{2.6}$$

The crucial step is to transform Equation (2.6) into a weak formulation. To this end we multiply this equation by a virtual displacement field $\delta\mathbf{u}$ and integrate over the domain V currently occupied by the body:

$$\int_V \delta\mathbf{u}^{\mathrm{T}}(\mathbf{L}^{\mathrm{T}}\boldsymbol{\sigma} + \rho\mathbf{g} - \rho\ddot{\mathbf{u}})\mathrm{d}V = 0 \tag{2.7}$$

We next apply the divergence theorem to obtain:

$$\int_V \left(\rho\delta\mathbf{u}^{\mathrm{T}}\ddot{\mathbf{u}} + (\mathbf{L}\delta\mathbf{u})^{\mathrm{T}}\boldsymbol{\sigma}\right)\mathrm{d}V = \int_V \rho\delta\mathbf{u}^{\mathrm{T}}\mathbf{g}\mathrm{d}V + \int_S \delta\mathbf{u}^{\mathrm{T}}\mathbf{t}\mathrm{d}S \tag{2.8}$$

with the boundary conditions $\boldsymbol{\Sigma}\mathbf{n} = \mathbf{t}$ or $\mathbf{u} = \mathbf{u}_{\mathrm{p}}$ prescribed on complementary parts of the body surface S, and the initial conditions: $\mathbf{u}(t_0) = \mathbf{u}_0$, $\mathbf{u}(t_0) = \mathbf{u}_0$.

Identity (2.8) is the weak form of the equation of motion and represents the principle of virtual work expressed in the current configuration. It is emphasised that in the above derivation no assumptions have been made with regard to the material behaviour, nor with respect to the magnitude of the spatial gradients of the displacements. Consequently, it is valid for linear as well as for non-linear material behaviour and for arbitrarily large displacement gradients.

2.2 Spatial Discretisation by Finite Elements

We shall now use Equation (2.8) as the starting point of the finite element approximation. We adopt a pure displacement-based formulation, in which the displacements at nodes of elements are considered as the fundamental unknowns. Upon introduction of a vector $\mathbf{a}_k$ in which the components (a_x, a_y, a_z) of the displacement vector at node k are gathered, we can approximate the continuous displacement field $\mathbf{u}$ elementwise as:

$$\mathbf{u} = \sum_{k=1}^{n} h_k(\xi, \eta, \zeta)\mathbf{a}_k \tag{2.9}$$

where h_k are the shape functions, or interpolation functions, of an element that is supported by n nodes. Normally, the interpolation functions are polynomials expressed in terms of the isoparametric coordinates (ξ, η, ζ), see Figure 2.1 and Box 2.2 for an example of an eight-noded three-dimensional element. Upon introduction of the vector $\mathbf{a}_e$ in which all the displacement degrees of freedom of the nodes connected to this element are assembled

$$\mathbf{a}_e = \begin{pmatrix} \mathbf{a}_1 \\ \mathbf{a}_2 \\ \dots \\ \dots \\ \mathbf{a}_n \end{pmatrix} \tag{2.10}$$

and the $3 \times 3n$ matrix $\mathbf{H}$,

$$\mathbf{H} = \begin{bmatrix} h_1 & 0 & 0 & h_2 & 0 & 0 & \ldots & \ldots & h_n & 0 & 0 \\ 0 & h_1 & 0 & 0 & h_2 & 0 & \ldots & \ldots & 0 & h_n & 0 \\ 0 & 0 & h_1 & 0 & 0 & h_2 & \ldots & \ldots & 0 & 0 & h_n \end{bmatrix} \tag{2.11}$$

the interpolation of the continuous displacement field for all points within an element can be written in a more compact manner:

$$\mathbf{u} = \mathbf{H}\mathbf{a}_e \tag{2.12}$$

The displacements that are contained in the element-related vector $\mathbf{a}_e$ can be related to the global displacements contained in a global displacement vector $\mathbf{a}$ via an incidence or location matrix $\mathbf{Z}_e$, which reflects the topology of the discretisation:

$$\mathbf{a}_e = \mathbf{Z}_e\mathbf{a} \tag{2.13}$$

When the system consists of N global degrees of freedom $\mathbf{Z}_e$ is a $3n \times N$ matrix. If the global coordinate system and the element coordinate system have the same axes, the matrix $\mathbf{Z}_e$ simply consists of zeros and ones, else the cosines and the sines of the transformation between both coordinate systems enter this matrix. With the aid of Equations (2.12) and (2.13) the weak form of the balance of momentum (2.8) can be reformulated as

$$\sum_{e=1}^{n_e} \int_{V_e} \rho(\mathbf{H}\mathbf{Z}_e\delta\mathbf{a})^{\mathrm{T}}\mathbf{H}\mathbf{Z}_e\ddot{\mathbf{a}}\mathrm{d}V + \sum_{e=1}^{n_e} \int_{V_e} (\mathbf{L}\mathbf{H}\mathbf{Z}_e\delta\mathbf{a})^{\mathrm{T}}\boldsymbol{\sigma}\mathrm{d}V = \sum_{e=1}^{n_e} \int_{V_e} \rho(\mathbf{H}\mathbf{Z}_e\delta\mathbf{a})^{\mathrm{T}}\mathbf{g}\mathrm{d}V + \sum_{e=1}^{n_e} \int_{S_e} (\mathbf{H}\mathbf{Z}_e\delta\mathbf{a})^{\mathrm{T}}\mathbf{t}\mathrm{d}S \tag{2.14}$$

where all integrals extend over the element domain V_e of each of the n_e elements of the finite element mesh. The global nodal virtual displacements as collected in $\delta\mathbf{a}$ are independent of the spatial coordinates, and can therefore be brought outside the integral as well as outside of the summation sign. The incidence or location matrices $\mathbf{Z}_e$ also do not depend on the spatial coordinates, but are different for each element and can consequently be brought only outside of the integral, but not outside of the summation operation. Considering that the latter equation must hold for any virtual displacement we arrive at the semi-discrete balance of momentum:

$$\mathbf{M}\ddot{\mathbf{a}} = \mathbf{f}_{\mathrm{ext}} - \mathbf{f}_{\mathrm{int}} \tag{2.15}$$

with the mass matrix,

$$\mathbf{M} = \sum_{e=1}^{n_e} \mathbf{Z}_e^{\mathrm{T}} \int_{V_e} \rho\mathbf{H}^{\mathrm{T}}\mathbf{H}\mathrm{d}V\mathbf{Z}_e \tag{2.16}$$

the external force vector,

$$\mathbf{f}_{\mathrm{ext}} = \sum_{e=1}^{n_e} \mathbf{Z}_e^{\mathrm{T}} \int_{V_e} \rho\mathbf{H}^{\mathrm{T}}\mathbf{g}\mathrm{d}V + \sum_{e=1}^{n_e} \mathbf{Z}_e^{\mathrm{T}} \int_{S_e} \mathbf{H}^{\mathrm{T}}\mathbf{t}\mathrm{d}S \tag{2.17}$$

and the internal force vector:

$$\mathbf{f}_{\text{int}} = \sum_{e=1}^{n_e} \mathbf{Z}_e^{\mathrm{T}} \int_{V_e} \mathbf{B}^{\mathrm{T}} \boldsymbol{\sigma} \mathrm{d}V \tag{2.18}$$

with

$$\mathbf{B} = \mathbf{LH} \tag{2.19}$$

the matrix that sets the relation between the strains within an element and the nodal displacements. The connotation 'semi-discrete' is used since the discretisation only applies to the spatial domain, but not (yet) to the time domain.

The evaluation of the mass matrix, and the external and internal force vectors [and also that of the tangential stiffness matrix $\mathbf{K}$ that will be introduced in Equation (2.43)] require computation of integrals over the element domain. The functions to be integrated can be quite complex if higher-order interpolation functions are used, for axisymmetric configurations where $1/r$ terms enter the integrand, if element geometries are used that are not rectangular, or if geometric non-linearities are so important that they change the shape of the element from its original rectangular form. These complexities make it virtually impossible to carry out analytical integration of the integrals in Equations (2.16)–(2.18). If material non-linearities are included, a closed-form evaluation of the integrals is impossible because of the a priori unknown manner in which the non-linearity varies over the domain. For the above reasons numerical integration is used almost exclusively when considering non-linearities, arising either from material or from geometrical non-linearity.

To demonstrate the application of numerical integration, we shall take the element internal force vector as an example. To facilitate the evaluation we will define a mapping from a cuboidal parent element in a ξ, η, ζ-coordinate system (Figure 2.1), onto the actual, arbitrary geometry of the element in the x, y, z-coordinate system. If $\boldsymbol{\xi}^{\mathrm{T}} = [\xi, \eta, \zeta]$ and if $\mathbf{x}^{\mathrm{T}} = [x, y, z]$ we have

$$\mathbf{x} = \mathbf{x}(\boldsymbol{\xi}) \tag{2.20}$$

The shape functions h_k have been defined directly on the parent element in ξ, η, ζ coordinates, cf. Equation (2.9). The derivatives that appear in Equation (2.18) through the presence of

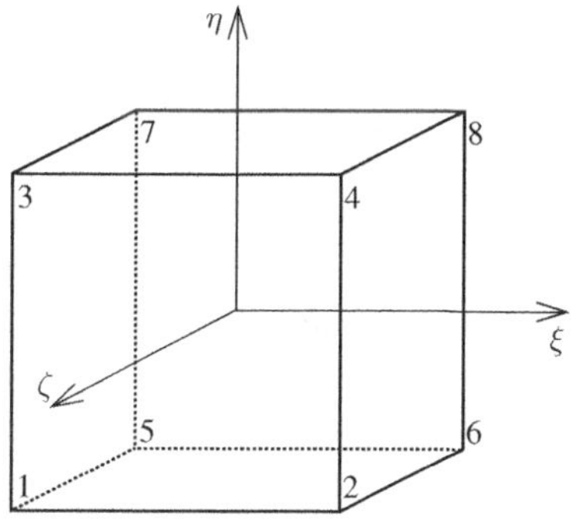

Figure 2.1 Eight-noded three-dimensional element in isoparametric coordinates

Box 2.2 Example of interpolation functions and derivatives

For an eight-noded three-dimensional element (Figure 2.1), the trilinear interpolation functions are as follows:

$$
\begin{aligned}
h_1 &= \tfrac{1}{8}(1+\xi)(1+\eta)(1+\zeta) & h_5 &= \frac{1}{8}(1-\xi)(1-\eta)(1+\zeta) \\
h_2 &= \tfrac{1}{8}(1-\xi)(1+\eta)(1+\zeta) & h_6 &= \frac{1}{8}(1-\xi)(1+\eta)(1-\zeta) \\
h_3 &= \tfrac{1}{8}(1+\xi)(1-\eta)(1+\zeta) & h_7 &= \frac{1}{8}(1+\xi)(1-\eta)(1-\zeta) \\
h_4 &= \tfrac{1}{8}(1+\xi)(1+\eta)(1-\zeta) & h_8 &= \frac{1}{8}(1-\xi)(1-\eta)(1-\zeta)
\end{aligned}
$$

The matrix $\mathbf{B} = \mathbf{LH}$ involves differentiation with respect to the global coordinates $\mathbf{x}$. However, the shape functions h_k are functions of the isoparametric coordinates $\boldsymbol{\xi}$. For this reason the chain rule is used to obtain

$$\frac{\partial h_k}{\partial \boldsymbol{\xi}} = \frac{\partial \mathbf{x}}{\partial \boldsymbol{\xi}} \frac{\partial h_k}{\partial \mathbf{x}} = \mathbf{J} \frac{\partial h_k}{\partial \mathbf{x}}$$

cf. Equation (2.22), or upon inversion,

$$\frac{\partial h_k}{\partial \mathbf{x}} = \mathbf{J}^{-1} \frac{\partial h_k}{\partial \boldsymbol{\xi}}$$

The $\mathbf{B}$ matrix is then retrieved as:

$$
\mathbf{B} = \mathbf{J}^{-1}
\begin{bmatrix}
\frac{\partial h_1}{\partial \xi} & 0 & 0 & \frac{\partial h_2}{\partial \xi} & 0 & 0 & \dots & \dots & \frac{\partial h_8}{\partial \xi} & 0 & 0 \\
0 & \frac{\partial h_1}{\partial \eta} & 0 & 0 & \frac{\partial h_2}{\partial \eta} & 0 & \dots & \dots & 0 & \frac{\partial h_8}{\partial \eta} & 0 \\
0 & 0 & \frac{\partial h_1}{\partial \zeta} & 0 & 0 & \frac{\partial h_2}{\partial \zeta} & \dots & \dots & 0 & 0 & \frac{\partial h_8}{\partial \zeta} \\
\frac{\partial h_1}{\partial \eta} & \frac{\partial h_1}{\partial \xi} & 0 & \frac{\partial h_2}{\partial \eta} & \frac{\partial h_2}{\partial \xi} & 0 & \dots & \dots & \frac{\partial h_8}{\partial \eta} & \frac{\partial h_8}{\partial \xi} & 0 \\
0 & \frac{\partial h_1}{\partial \zeta} & \frac{\partial h_1}{\partial \eta} & 0 & \frac{\partial h_2}{\partial \zeta} & \frac{\partial h_2}{\partial \eta} & \dots & \dots & 0 & \frac{\partial h_8}{\partial \zeta} & \frac{\partial h_8}{\partial \eta} \\
\frac{\partial h_1}{\partial \zeta} & 0 & \frac{\partial h_1}{\partial \xi} & \frac{\partial h_2}{\partial \zeta} & 0 & \frac{\partial h_2}{\partial \xi} & \dots & \dots & \frac{\partial h_8}{\partial \zeta} & 0 & \frac{\partial h_8}{\partial \xi}
\end{bmatrix}
$$

where the inverse of the Jacobian is now written as a 6×6 matrix, filled at appropriate entries and expressed in terms of isoparametric coordinates, see Equation (2.25).

the $\mathbf{B}$ matrix are computed via straightforward differentiation, cf. Box 2.2. Using a standard transformation the integration domain in x, y, z-coordinates can subsequently be converted into the simple cuboidal domain of the parent element:

$$\mathbf{f}_{\text{int}} = \sum_{e=1}^{n_e} \mathbf{Z}_e^{\text{T}} \int_{-1}^{+1} \int_{-1}^{+1} \int_{-1}^{+1} (\det \mathbf{J}) \mathbf{B}^{\text{T}} \boldsymbol{\sigma} \, \text{d}\xi \text{d}\eta \text{d}\zeta \tag{2.21}$$

where the advantage of defining the shape functions h_k in terms of isoparametric coordinates becomes apparent. In Equation (2.21) $\mathbf{J}$ is the Jacobian matrix of the mapping $\mathbf{x} = \mathbf{x}(\boldsymbol{\xi})$ as defined through:

$$\mathbf{J} = \frac{\partial \mathbf{x}}{\partial \boldsymbol{\xi}} \tag{2.22}$$

or, written in component form:

$$\mathbf{J} = \begin{bmatrix} \frac{\partial x}{\partial \xi} & \frac{\partial x}{\partial \eta} & \frac{\partial x}{\partial \zeta} \\ \frac{\partial y}{\partial \xi} & \frac{\partial y}{\partial \eta} & \frac{\partial y}{\partial \zeta} \\ \frac{\partial z}{\partial \xi} & \frac{\partial z}{\partial \eta} & \frac{\partial z}{\partial \zeta} \end{bmatrix} \tag{2.23}$$

Similar to the interpolation of the displacement field in Equation (2.9) we interpolate the geometry of the element via

$$\mathbf{x} = \sum_{k=1}^{n} h_k(\xi, \eta, \zeta)\mathbf{x}_k \tag{2.24}$$

with $\mathbf{x}_k^{\mathrm{T}} = (x_k, y_k, z_k)$, the set (x_k, y_k, z_k) being the spatial coordinates of node k. When the functions h_k used for the interpolation of the displacement field are the same as those used for the interpolation of the geometry we refer to the element formulation as isoparametric. When lower-order interpolation functions are used for the geometry than for the displacement field, the formulation is called subparametric, while the terminology superparametric is employed for the case that the interpolation of the geometry is done with higher-order polynomials than that of the displacement field. For simplicity we shall adhere to the isoparametric concept. Then, the Jacobian matrix takes the following simple form:

$$\mathbf{J} = \sum_{k=1}^{n} \frac{\partial h_k}{\partial \boldsymbol{\xi}} \mathbf{x}_k^{\mathrm{T}} \tag{2.25}$$

Finally, numerical integration is applied for the computation of the integral of Equation (2.21):

$$\mathbf{f}_{\mathrm{int}} = \sum_{e=1}^{n_e} \mathbf{Z}_e^{\mathrm{T}} \sum_{i=1}^{n_i} w_i(\det \mathbf{J}_i)\mathbf{B}_i^{\mathrm{T}} \boldsymbol{\sigma}_i \tag{2.26}$$

with w_i the weight factor of integration point i, and n_i the number of integration points in element e.

All matrices in Equations (2.16)–(2.18) must be evaluated separately for each individual integration point. The behaviour at an integration point is thought to be representative for the tributary area that 'belongs' to this integration point. For continuum elements it has become customary to employ Gauss integration, as it provides the highest accuracy for a given number of integration points. For the through-the-thickness integration of structural elements like beams, plates and shells, Gauss integration is less appropriate and Simpson, Lobatto, or Newton–Cotes integration rules are generally preferred, see also Chapters 9 and 10. Also for interface elements, the latter type of integration rules are superior, since in this case the application of Gauss integration rules tends to result in oscillatory stress fields, see Chapter 13.

2.3 PyFEM: Shape Function Utilities

In the previous section, the weak form of the balance of momentum equation has been discretised by means of finite element shape functions. In this section, we will take a closer look at the implementation of these shape functions in the finite element code PYFEM. The numerical integration of a domain, which is closely related to the spatial discretisation, is also dealt with in these routines.

The complete implementation of the shape functions and the corresponding numerical integration can be found in the file `shapeFunctions.py` in the directory `pyfem/util` of the PYFEM code. The file contains the data structure definitions as well as the implementation of various functions:

```
⟨Shape functions ⟩≡
  ⟨Shape function data structures 38⟩
  ⟨Shape function algorithms 40⟩
  ⟨Shape function main routine 39⟩
```

The fragment ⟨*Shape function data structures*⟩ contains the class definitions of the data structures that contain the element shape functions and the numerical integration scheme of an element. Irrespective of the kind of numerical integration technique that is used, the values of the shape functions and their derivatives are calculated in the integration points. All the data in an integration point are stored in an object of the type `shapeData`:

```
⟨Shape function data structures ⟩≡                                    38
  class shapeData:
    pass
```

The class `shapeData` is initially empty and is gradually filled with the values of the shape functions at this point, by their derivatives and by the integration weight.

It can be useful to compute the shape function and the integration data of all the integration points in an element at once. This information is stored in an object of the type `elemShapeData`:

```
⟨Shape function data structures ⟩+≡                                   38
  class elemShapeData:

    def __init__( self ):
      self.sData = []

    def __iter__( self ):
      return iter(self.sData)

    def __len__( self ):
      return len(self.sData)
```

This class contains a single member: an empty list named `sData` which will be appended with objects of the type `shapeData`. The function `__iter__` has been defined in order to be able to iterate over the items in the list `sData`. The function `__len__` returns the number of items in this list.

The main routine of the code is `getElemShapeData`, which can be used to obtain the shape function and the integration point data in an element.

⟨*Shape function main routine* ⟩≡ 38

```
def getElemShapeData( elemCoords , order = 0 , \
                      method = 'Gauss' , elemType = 'default' ):

  elemData = elemShapeData()                                          38

  if elemType == 'default':
    elemType = getElemType( elemCoords )

  (intCrds,intWghts) = getIntPoints( elemType , order , method )

  for xi,weight in zip( intCrds , intWghts ):

    try:
      sData = eval( 'getShape'+elemType+'(xi)' )                     40
    except:
      raise NotImplementedError('Unknown type :'+elemType)

    jac = dot ( elemCoords.transpose() , sData.dhdxi )

    if jac.shape[0] is jac.shape[1]:
      sData.dhdx = dot ( sData.dhdxi , inv( jac ) )

    sData.weight = calcWeight( jac ) * weight

    elemData.sData.append(sData)

  return elemData
```

This function has four arguments. The array `elemCoords` passes the coordinates of the nodes of the element. These coordinates are defined in the global reference frame. The second argument, `order`, is used to specify the order of the numerical integration scheme. The default value `0` indicates that the standard integration order is used. The argument `method` selects the integration scheme. `'Gauss'` integration is the default scheme, but `'Newton-Cotes'`, `'Lobatto'` and `'Simpson'` schemes are also available. Finally, `elemType` sets the parent element type that is used to construct the shape functions, e.g. `Line2`, `Quad4` or `Hexa8`. When this argument is set to `'default'`, the element type is derived from the number of nodes of the element and the spatial dimensions.

In this routine, an empty object `elemData` of the type `elemShapeData` is created first. If `elemType` is not specified, the type of the element is determined by the function `getElemType`. This is done by checking the dimensions of the two-dimensional array `elemCoords`. The number of rows in this array is equal to the number of nodes that support the element, while the number of columns is equal to the spatial dimension of the element. The function returns the element type as a string. For example, when the dimensions of the array `elemCoords` are equal to `(4,2)`, the function `getElemType` returns the string `Quad4`, which indicates a four-noded quadrilateral element.

The position of the integration points in the parent element and the corresponding integration weights are determined by the function `getIntPoints`. The positions and the weights depend on the element type, on the order of integration and on the integration scheme, which are the arguments of the function. The output is the two-dimensional array `intCrds`, which contains the coordinates of the integration point in the parent element coordinate system and the array `intWghts` in which the integration weights are stored. The length of these arrays specify the number of integration points in the element.

Next, a loop over the integration points is carried out. For each point `xi` the corresponding shape functions and their derivatives are calculated. For a four-noded quadrilateral element, this is done in the function `getShapeQuad4`. Note that in the code, the function call is somewhat hidden in the command `eval( 'getShape'+elemType+'(xi)')` where the string `elemType` has the value `'Quad4'`. This function returns the object `sData` of the type `shapeData` which contains the values of the shape functions and its spatial derivatives in the parent element coordinate system.

When the spatial derivatives of the physical element and the parent element are identical, i.e. when the Jacobian in Equation (2.22) is a square matrix, the derivatives of the shape functions in the physical space can be calculated and stored in `sData` as the member `dhdx`. Finally, the integration weight is calculated, which is composed of the weight factor of the integration point in the parent element `weight`, and an additional factor for the mapping to the physical element. This parameter is calculated in the function `calcWeight`. When the spatial dimensions of the parent and the physical element are identical, this factor is equal to the absolute value of the determinant of the Jacobian: `jac`. Finally, the integration point data `sData` is appended to the list of integration point data of the element.

The implementation of the function `getShapeQuad4` is given by:

⟨*Shape function algorithms* ⟩≡ 38

```
def getShapeQuad4( xi ):

  sData        = shapeData()
  sData.h      = empty(4)
  sData.xi     = xi

  sData.h[0] = 0.25*(1.0-xi[0])*(1.0-xi[1])
  sData.h[1] = 0.25*(1.0+xi[0])*(1.0-xi[1])
  sData.h[2] = 0.25*(1.0+xi[0])*(1.0+xi[1])
  sData.h[3] = 0.25*(1.0-xi[0])*(1.0+xi[1])
```

38

The argument `xi` of this function is an array of length two which represents the parent element coordinate $\boldsymbol{\xi} = [\xi, \eta]$. First, an object `sData` of the type `shapeData` is created in which the coordinate `xi` is stored. The values of the interpolation functions h_i are stored in the member `h`, which is a one-dimensional array of length four. The values of the shape function h_i in point $\boldsymbol{\xi}$ are calculated next. It is is noted that in Python the index of arrays start counting at zero. Hence, `h[0]` represents the value of the interpolation function for the first node in point $h_1(\boldsymbol{\xi})$, etc. The derivatives of the interpolation functions with respect to the coordinate $\boldsymbol{\xi}$ are calculated next:

⟨*Shape function algorithms* ⟩+≡ 40

```
sData.dhdxi = empty( shape=(4,2) )

sData.dhdxi[0,0] = -0.25*(1.0-xi[1])
sData.dhdxi[1,0] =  0.25*(1.0-xi[1])
sData.dhdxi[2,0] =  0.25*(1.0+xi[1])
sData.dhdxi[3,0] = -0.25*(1.0+xi[1])
```

⟨*Calculate derivatives of shape functions h_i with respect to η*⟩

```
return sData
```

Similar routines are available for the calculation of the shape functions and derivatives for a variety of one- two- and three-dimensional elements. The procedure in the routine `getElemShape` remains the same for all cases.

2.4 Incremental-iterative Analysis

For quasi-static processes Equation (2.15) reduces to:

$$\mathbf{f}_{\text{ext}} - \mathbf{f}_{\text{int}} = \mathbf{0} \tag{2.27}$$

For purely static processes, time plays no role anymore. Yet, also then we need a parameter to order the sequence of events. For this reason we shall continue to use the concept of 'time' also in static mechanical processes to order the loading sequence. In particular, the concept of time can be employed to apply the external load in a number of loading steps (or increments). It would be possible to impose the entire external load $\mathbf{f}_{\text{ext}}$ in a single step, but this is not a sensible approach because of the following reasons:

- The set of algebraic equations that arises from the discretisation of a non-linear continuum model is non-linear, thus necessitating the use of an iterative procedure for its solution. For very large loading steps, the case of imposing the entire load in one step being the extreme, it is usually difficult to obtain a properly converged solution, if a solution can be obtained at all. Indeed, the convergence radius is limited for most commonly used iterative procedures, including the Newton–Raphson method.

- Experiments show that most materials exhibit path-dependent behaviour. This means that different values for the stress are obtained depending on the strain path that is followed. For instance, the resulting stress can be different when we first apply tension on a panel followed by a shear strain increment or when the same strain increments are imposed in the reverse order. Evidently, the structural behaviour can only be predicted correctly if the strain increments are relatively small, so that the strain path is followed as closely as possible.

Along this line of reasoning we decompose the vector of unknown stress components at time $t + \Delta t$, denoted by $\boldsymbol{\sigma}^{t+\Delta t}$ into a stress vector $\boldsymbol{\sigma}^t$ at time t when the stress components are known, and $\Delta\boldsymbol{\sigma}$ which contains the hitherto unknown components of the stress increment:

$$\boldsymbol{\sigma}^{t+\Delta t} = \boldsymbol{\sigma}^t + \Delta\boldsymbol{\sigma} \tag{2.28}$$

Substituting this additive decomposition into Equation (2.27) and using Equation (2.18) results in:

$$\mathbf{f}_{\text{ext}}^{t+\Delta t} - \sum_{e=1}^{n_e} \mathbf{Z}_e^{\mathrm{T}} \int_{V_e} \mathbf{B}^{\mathrm{T}} \boldsymbol{\sigma}^t \mathrm{d}V - \sum_{e=1}^{n_e} \mathbf{Z}_e^{\mathrm{T}} \int_{V_e} \mathbf{B}^{\mathrm{T}} \Delta\boldsymbol{\sigma} \mathrm{d}V = \mathbf{0} \tag{2.29}$$

where the superscript t $+ \Delta t$ is attached to $\mathbf{f}_{\text{ext}}$ to emphasise that the external force vector must be evaluated at time $t + \Delta t$. The set of Equations (2.29) can be non-linear for two reasons. First, the stress increment $\Delta\boldsymbol{\sigma}$ generally depends on the displacement increment $\Delta\mathbf{u}$ in a non-linear manner. Secondly, the volume of the element V_e over which the integration extends is unknown at time $t + \Delta t$. In other words, the integrals of Equation (2.29) depend on the incremental nodal displacements $\Delta\mathbf{a}$, which are not yet known. We will show in the next chapter that this problem can be solved by an adequate mapping to a reference configuration, and presently it suffices to just note this source of geometric non-linearity, and linearise, so that the integrals of Equation (2.29) become independent of the nodal displacement increments.

We now use Equation (2.18) to rewrite Equation (2.29) as:

$$\sum_{e=1}^{n_e} \mathbf{Z}_e^{\mathrm{T}} \int_{V} \mathbf{B}^{\mathrm{T}} \Delta\boldsymbol{\sigma} \mathrm{d}V = \mathbf{f}_{\text{ext}}^{t+\Delta t} - \mathbf{f}_{\text{int}}^{t} \tag{2.30}$$

The superscript t has been attached explicitly to $\mathbf{f}_{\text{int}}$ to underscore that the internal force vector has been evaluated at time t, i.e. for $\boldsymbol{\sigma} = \boldsymbol{\sigma}^t$.

The solution of the set of non-linear Equations (2.30) requires the use of an iterative solution technique. Typically, such techniques, for instance the Newton–Raphson method which is frequently used in structural analysis, involve repeated linearisation of the governing equations. Hence, we must linearise the dependence of the stress increment $\Delta\boldsymbol{\sigma}$ on the displacement increment $\Delta\mathbf{u}$. The stress increment $\Delta\boldsymbol{\sigma}$ depends on the increment of the strain tensor, say $\Delta\boldsymbol{\epsilon}$ (we by pass which definition of the strain tensor should be used), while the increment of the strain tensor can be a non-linear function of the increment of the continuous displacement field $\Delta\mathbf{u}$:

$$\Delta\boldsymbol{\sigma} = \Delta\boldsymbol{\sigma}(\Delta\boldsymbol{\epsilon}(\Delta\mathbf{u})) \tag{2.31}$$

The stress increment, $\Delta\sigma$ can be linearised as:

$$\delta\sigma = \left(\frac{\partial\sigma}{\partial\epsilon}\right)^t \delta\epsilon \tag{2.32}$$

Defining

$$\mathbf{D} = \left(\frac{\partial\sigma}{\partial\epsilon}\right)^t \tag{2.33}$$

as the material tangential stiffness matrix, we can also write:

$$\delta\sigma = \mathbf{D}\delta\epsilon \tag{2.34}$$

We furthermore note that the second term on the left-hand side of Equation (2.8) represents the internal virtual work δW_{int}. Since by definition we also have

$$\delta W_{\text{int}} = \int_V \delta\epsilon^{\mathrm{T}}\sigma \mathrm{d}V \tag{2.35}$$

with ϵ the work-conjugate strain measure, the kinematic relation between the variation of the strain tensor and that of the continuous displacement field directly follows:

$$\delta\epsilon = \mathbf{L}\delta\mathbf{u} \tag{2.36}$$

and, for quasi-static loading conditions, Equation (2.8) is rewritten as:

$$\int_V \delta\epsilon^{\mathrm{T}}\sigma \mathrm{d}V = \int_V \rho\delta\mathbf{u}^{\mathrm{T}}\mathbf{g}\mathrm{d}V + \int_S \delta\mathbf{u}^{\mathrm{T}}\mathbf{t}\mathrm{d}S \tag{2.37}$$

Using Equation (2.36) the variation of the stress becomes:

$$\delta\sigma = \mathbf{DL}\delta\mathbf{u} \tag{2.38}$$

Inserting the interpolation (2.12) for the continuous displacements one obtains:

$$\delta\sigma = \mathbf{DLH}\delta\mathbf{a}_e \tag{2.39}$$

or using Equation (2.13):

$$\delta\sigma = \mathbf{DLHZ}_e\delta\mathbf{a} \tag{2.40}$$

Inserting this result into Equation (2.30) and using Equation (2.19) to introduce **B** yields the linearised equation for a finite load increment:

$$\sum_{e=1}^{n_e} \mathbf{Z}_e^{\mathrm{T}} \int_{V_e} \mathbf{B}^{\mathrm{T}}\mathbf{DBZ}_e\Delta\mathbf{a}\mathrm{d}V = \mathbf{f}_{\text{ext}}^{t+\Delta t} - \mathbf{f}_{\text{int}}^{t} \tag{2.41}$$

Since the incremental nodal displacements do not depend upon the spatial coordinates, they can be brought outside the integral and one obtains the following linearised set of N equations:

$$\mathbf{K}\Delta\mathbf{a} = \mathbf{f}_{\text{ext}}^{t+\Delta t} - \mathbf{f}_{\text{int}}^{t} \tag{2.42}$$

where

$$\mathbf{K} = \sum_{e=1}^{n_e} \mathbf{Z}_e^{\mathrm{T}} \left(\int_{V_e} \mathbf{B}^{\mathrm{T}} \mathbf{D} \mathbf{B} \mathrm{d}V \right) \mathbf{Z}_e \tag{2.43}$$

has been introduced, which is the tangential stiffness matrix of the structure upon a small increment of loading. Equation (2.42) can readily be solved since it is a set of N linear equations.

Often, the set (2.42) is written in a slightly different format:

$$\mathbf{K}\Delta\mathbf{a} = \Delta\mathbf{f}_{\mathrm{ext}} + \mathbf{f}_{\mathrm{ext}}^{t} - \mathbf{f}_{\mathrm{int}}^{t} \tag{2.44}$$

where the external load vector has been split into the load increment during the present step and the contribution that had already been applied to the structure at the beginning of the step. Furthermore, we often consider only a single load type during a non-linear analysis. This definitely holds during an increment, and we can rewrite the load increment as $\Delta\lambda\hat{\mathbf{f}}_{\mathrm{ext}}$ with $\Delta\lambda$ a scalar-valued incremental load parameter, and $\hat{\mathbf{f}}_{\mathrm{ext}}$ the normalised external load vector, so that:

$$\mathbf{K}\Delta\mathbf{a} = \Delta\lambda\hat{\mathbf{f}}_{\mathrm{ext}} + \mathbf{f}_{\mathrm{ext}}^{t} - \mathbf{f}_{\mathrm{int}}^{t} \tag{2.45}$$

In the above derivation, both the non-linear stress–strain relation and the non-linear relation between the (incremental) strains and the (incremental) displacements were linearised at the beginning of the loading step, i.e. at time t, while the loading step ranges from t to $t + \Delta t$. This linearisation leads to a 'drifting away' from the true equilibrium solution, especially if relatively large loadings steps are employed. A graphical illustration of this 'drifting tendency' is provided in Figures 2.2 and 2.3. The gradual departure of the numerical solution from the true solution can be prevented, or at least be made smaller, by adding equilibrium iterations within each loading step. Now, we obtain an incremental-iterative procedure instead of a pure incremental procedure. In an incremental-iterative solution method a first estimate for the displacement increment $\Delta\mathbf{a}$ is made through

$$\Delta\mathbf{a}_1 = \mathbf{K}_0^{-1}\mathbf{r}_0 \tag{2.46}$$

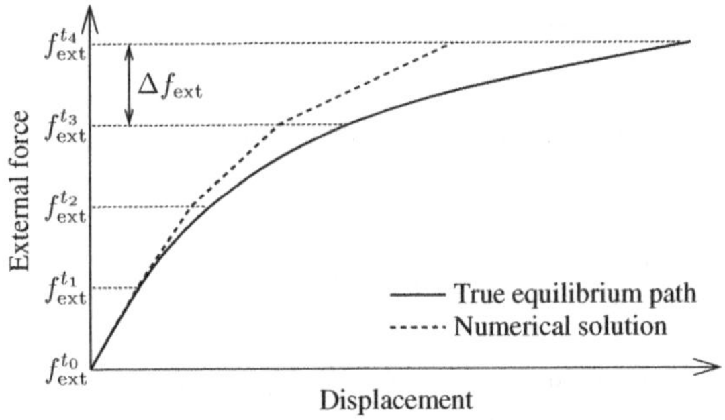

Figure 2.2 Purely incremental solution procedure

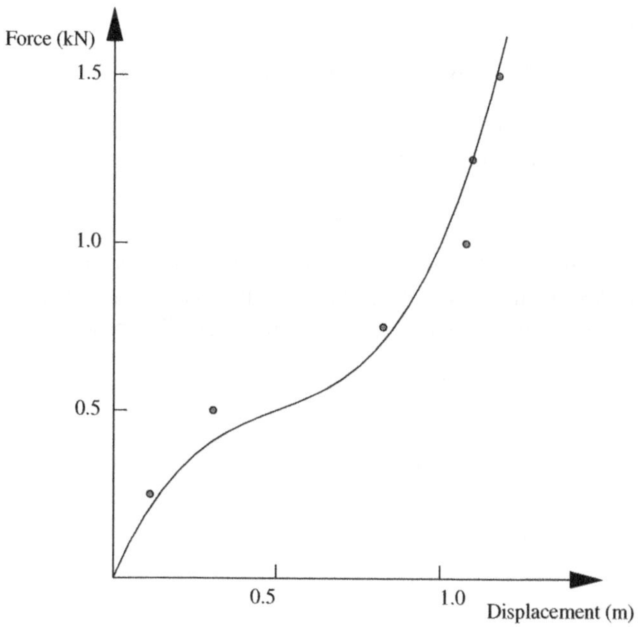

Figure 2.3 Application of incremental solution procedure to shallow truss for a spring stiffness k = 1000 N/m

with

$$\mathbf{r}_0 = \mathbf{f}_{\text{ext}}^{t+\Delta t} - \mathbf{f}_{\text{int},0} \tag{2.47}$$

the residual vector, or out-of-balance vector at the beginning of the load increment. The subscript 1 of $\Delta\mathbf{a}$ signifies that we deal with the estimate in the first iteration for the incremental displacement vector. Likewise, the subscript 0 of the internal force vector relates to the fact that this vector is calculated using the stresses at the beginning of the loading step, i.e. that are left behind at the end of the previous iteration ($\boldsymbol{\sigma}_0 = \boldsymbol{\sigma}^{\text{t}}$):

$$\mathbf{f}_{\text{int},0} = \sum_{e=1}^{n_e} \mathbf{Z}_e^{\text{T}} \sum_{i=1}^{n_i} w_i (\det \mathbf{J}_i) \mathbf{B}_i^{\text{T}} \boldsymbol{\sigma}_{i,0} \tag{2.48}$$

From the incremental displacement vector $\Delta\mathbf{a}_1$ a first estimate for the strain increment $\Delta\boldsymbol{\epsilon}_1$ can be calculated, whereupon, using the stress–strain law, the stress increment $\Delta\boldsymbol{\sigma}_1$ can be computed. The stresses after the first iteration are then given by:

$$\boldsymbol{\sigma}_1 = \boldsymbol{\sigma}_0 + \Delta\boldsymbol{\sigma}_1 \tag{2.49}$$

Generally, the internal force vector $\mathbf{f}_{\text{int},1}$ that is computed on the basis of the stresses $\boldsymbol{\sigma}_1$ is not in equilibrium with the external loads $\mathbf{f}_{\text{ext}}^{t+\Delta t}$ that have been added up to and including this loading step. For this reason a correction to the displacement increment is necessary. Denoting this correction by $\mathbf{da}_2$,

$$\mathbf{da}_2 = \mathbf{K}_1^{-1}\mathbf{r}_1 \quad \text{with} \quad \mathbf{r}_1 = \mathbf{f}_{\text{ext}}^{t+\Delta t} - \mathbf{f}_{\text{int},1} \tag{2.50}$$

and $\mathbf{K}_1$ the updated tangential stiffness matrix, the displacement increment after the second iteration in the loading step follows from

$$\Delta\mathbf{a}_2 = \Delta\mathbf{a}_1 + \mathbf{da}_2 \tag{2.51}$$

In a similar fashion to the calculation of the strain and stress increment in the first iteration the quantities $\Delta\boldsymbol{\epsilon}_2$ and $\Delta\boldsymbol{\sigma}_2$ are now computed. From the latter quantity an improved approximation for the stress at the end of the loading step, $\boldsymbol{\sigma}_2$ can be made. This process can be summarised as:

$$\begin{aligned}
\mathbf{r}_j &= \mathbf{f}_{\text{ext}}^{t+\Delta t} - \mathbf{f}_{\text{int},j} \\
\mathbf{da}_{j+1} &= \mathbf{K}_j^{-1}\mathbf{r}_j \\
\Delta\mathbf{a}_{j+1} &= \Delta\mathbf{a}_j + \mathbf{da}_{j+1} \\
\Delta\boldsymbol{\epsilon}_{i,j+1} &= \Delta\boldsymbol{\epsilon}_i(\Delta\mathbf{a}_{j+1}) \\
\Delta\boldsymbol{\sigma}_{i,j+1} &= \Delta\boldsymbol{\sigma}_i(\Delta\boldsymbol{\epsilon}_{i,j+1}) \\
\boldsymbol{\sigma}_{i,j+1} &= \boldsymbol{\sigma}_{i,0} + \Delta\boldsymbol{\sigma}_{i,j+1} \\
\mathbf{f}_{\text{int},j+1} &= \sum_{e=1}^{n_e} \mathbf{Z}_e^{\text{T}} \sum_{i=1}^{n_i} w_i (\det\mathbf{J}_i)\mathbf{B}_{i,j+1}^{\text{T}}\boldsymbol{\sigma}_{i,j+1}
\end{aligned} \tag{2.52}$$

where the operations in lines three to six have to be done for each integration point, denoted by the index i. This iterative process ultimately results in stresses that are in equilibrium internally and with the applied external loading within some user-prescribed convergence tolerance. A graphical explanation is given in Figure 2.4, while the algorithm that underlies essentially any incremental-iterative procedure in non-linear finite element analysis is given in Box 2.3.

The procedure summarised in Equations (2.52) is called the total-incremental method. Every iteration the total displacement increment within the step is computed and on the basis of

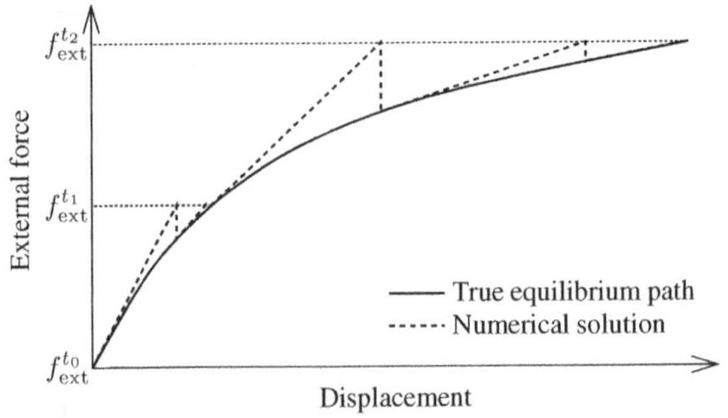

Figure 2.4 Incremental-iterative solution procedure

Box 2.3 Computational flow in a non-linear finite element code

For each loading step:

1. Initialise the data for the loading step. Set $\Delta \mathbf{a}_0 = \mathbf{0}$.
2. Compute the new external force vector $\mathbf{f}_{\text{ext}}^{t+\Delta t}$.
3. Compute the tangential stiffness matrix:

$$\mathbf{K}_j = \sum_{e=1}^{n_e} \mathbf{Z}_e^{\mathrm{T}} \sum_{i=1}^{n_i} w_i \det \mathbf{J}_i \mathbf{B}_{i,j}^{\mathrm{T}} \mathbf{D}_{i,j} \mathbf{B}_{i,j} \mathbf{Z}^e$$

4. Adjust for prescribed displacements and linear dependence relations.
5. Solve, e.g. by LDU decomposition, the linear system:

$$\mathbf{K}_j \mathrm{d}\mathbf{a}_{j+1} = \mathbf{f}_{\text{ext}}^{t+\Delta t} - \mathbf{f}_{\text{int},j}$$

6. Add the correction $\mathrm{d}\mathbf{a}_{j+1}$ to the incremental displacement vector:

$$\Delta \mathbf{a}_{j+1} = \Delta \mathbf{a}_j + \mathrm{d}\mathbf{a}_{j+1}$$

7. Compute the strain increment $\Delta \boldsymbol{\epsilon}_{i,j+1}$ for each integration point i:

$$\Delta \mathbf{a}_{j+1} \rightarrow \Delta \boldsymbol{\epsilon}_{i,j+1}$$

8. Compute the stress increment from the strain increment for each integration point i:

$$\Delta \boldsymbol{\epsilon}_{i,j+1} \rightarrow \Delta \boldsymbol{\sigma}_{i,j+1}$$

9. Add the stress increment to $\boldsymbol{\sigma}_{i,0}$ for each integration point i:

$$\boldsymbol{\sigma}_{i,j+1} = \boldsymbol{\sigma}_{i,0} + \Delta \boldsymbol{\sigma}_{i,j+1}$$

10. Compute the internal force vector:

$$\mathbf{f}_{\text{int},j+1} = \sum_{e=1}^{n_e} \mathbf{Z}_e^{\mathrm{T}} \sum_{i=1}^{n_i} w_i \det \mathbf{J}_i \mathbf{B}_{i,j+1}^{\mathrm{T}} \boldsymbol{\sigma}_{i,j+1}$$

11. Check convergence. Is $\|\mathbf{f}_{\text{ext}}^{t+\Delta t} - \mathbf{f}_{\text{int},j+1}\| < \eta$, with η a small number? If **yes**, go to next loading step, **else** go to 3.

this total displacement increment the total strain increment and the total stress increment are computed. Then, the new stresses are found as the sum of the stresses at the beginning of the step and the total stress increment. As an alternative approach, we might continue to work with corrections. Rather than first adding the correction to the total displacement increment obtained in the previous iteration, we could also proceed by calculating a correction to the strain increment $\mathrm{d}\boldsymbol{\epsilon}_{i,j+1}$, which can be used to compute a correction to the stress increment $\mathrm{d}\boldsymbol{\sigma}_{i,j+1}$.

This so-called delta-incremental method is less robust, particularly when we have materially non-linear models in which we have different behaviour in loading than in unloading, e.g. plasticity. Then, the delta-incremental update methodology can result in pseudo-unloading which impairs numerical stability.

An issue that has not been discussed yet, is the implicit assumption that the tangential stiffness matrix $\mathbf{K}_j$ is updated after each iteration, see for instance Figure 2.4. This is by no means necessary, since, as long as the stresses are determined in a proper manner and the resulting internal force vector is computed on the basis of these stresses and inserted on the right-hand side of Equation (2.15), it is in most cases less relevant which stiffness matrix is being used to iterate towards equilibrium. Indeed, it can be rather costly to compute and to decompose a stiffness matrix every iteration, as is being done within a full Newton–Raphson process, especially in computations of three-dimensional structures. This has motivated the search for methods which obviate the need to construct and decompose a tangential stiffness matrix in every iteration. Here, we will consider two classes of such methods.

In the first class, the stiffness matrix is obtained simply by setting up a new tangential stiffness only every few iterations, or only once within a loading step. It is assumed that the stiffness matrix varies so slowly that the stiffness matrix set up in an iteration serves as a reasonably accurate approximation of the tangential stiffness for a couple of subsequent iterations. It is anticipated that the slowing down of the convergence speed, actually the loss of quadratic convergence that is characteristic of Newton's method, is off set by the gain in computer time within each iteration.

An alternative to the full Newton–Raphson method is setting up and decomposing the tangential stiffness matrix only once within every loading step. Within this modified Newton–Raphson method (Figure 2.5), some alternatives exist with regard to the iteration in which the stiffness matrix is computed anew. The most classical approach is at the beginning of a loading step. An advantage is that all state variables are computed on the basis of an equilibrium state (presuming of course that a converged solution has been obtained in the preceding load increment). The variant in which the stiffness matrix is only reformulated at the beginning of the second iteration of each load increment does not have this advantage, but also does not suffer from the drawback of the first variant, namely that none of the non-linearities that arise during the loading step are incorporated in the stiffness matrix that is being used in the majority of the iterations. In particular sudden (physical) non-linearities, e.g. local unloading

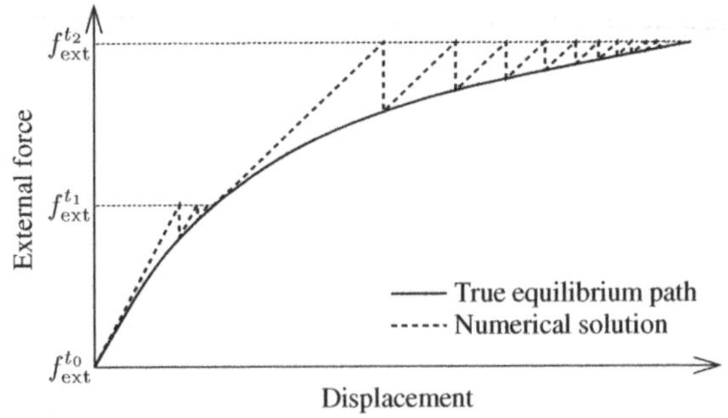

Figure 2.5 Modified Newton–Raphson iteration scheme

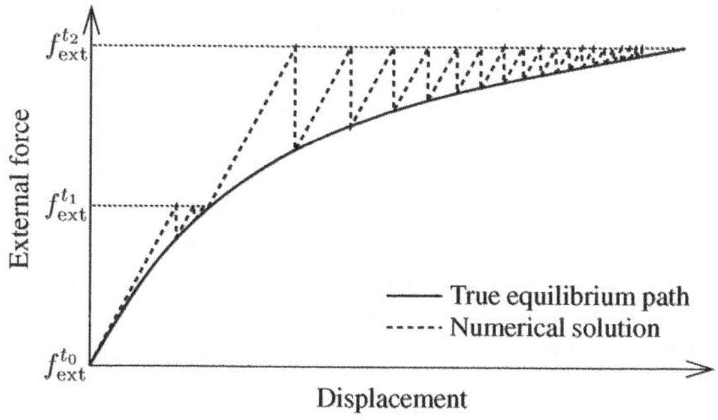

Figure 2.6 Linear elastic stiffness iteration scheme

from an inelastic to an elastic state, can cause convergence difficulties, since the conventional modified Newton–Raphson procedure cannot cope well with either local or global stiffening of structural behaviour. Using the linear-elastic stiffness matrix in the first iteration, and then setting up, decomposing and using the tangential stiffness matrix in the subsequent iterations can be even more effective. By doing so we allow for local or global unloading, so that we do not iterate with a 'too soft' stiffness matrix, which can result in divergence.

The simplest variant in this class of methods is the Initial Stiffness method, shown in Figure 2.6. In this scheme, the stiffness matrix is set up and decomposed only at the beginning of the first loading step. It is obvious that, especially when the failure load is approached and the current stiffness deviates considerably from this initial, linear-elastic stiffness, convergence becomes slow and a large number of iterations are required to obtain a reasonable accuracy. Experience shows, that often a rather slack convergence tolerance (for more information see Chapter 4) must be adopted in order for the number of iterations to remain below approximately thirty. As a result of the inability to stick to a tight convergence tolerance with this scheme the failure load can be overestimated. However, by continuing the calculation beyond the limit point of the load–deflection curve the proper failure load can sometimes be obtained, since the additional iterations that are added in these post-peak increments result in a levelling out of the load–deflection path until the true failure load has been reached. This phenomenon can be called 'numerical softening'. One has to be careful to distinguish properly between this numerical artifact and true structural softening which occurs for instance with the snapping of thin shells and for certain types of constitutive models, e.g. softening plasticity or damage.

The difference in convergence behaviour between the full Newton–Raphson method, the modified Newton–Raphson method and the Initial Stiffness method can be nicely illustrated by the truss in Figure 1.1. We take the case without a spring ($k = 0$) and as convergence criterion we take that the change in computed displacement from one iteration to the next should not be smaller than 10^{-6}. The load is applied in four equal increments of 30 N. With a failure load of 116.9 N this implies that the last loading step exceeds the limit load and the iterative process should diverge. Indeed, from Table 2.1 we observe that this is the case for all three methods. We also observe that the convergence behaviour becomes slower at each loading step for the modified Newton–Raphson method and for the elastic-stiffness method, while, until failure, with the full Newton–Raphson method convergence is attained within four to five iterations. Moreover, from Table 2.2, which gives the convergence behaviour of the Newton–Raphson

Table 2.1 Comparison of iteration methods

Load (N)	FNR	MNR	IS
30	4	8	8
60	4	10	13
90	5	13	23
120	+100	+100	+100

FNR, full Newton–Raphson method; MNR, modified Newton–Raphson method; IS, Initial Stiffness method.

Table 2.2 Full Newton–Raphson method

Step 1	Step 2	Step 3	Step 4
0.24×10^{-1}	0.28×10^{-1}	0.36×10^{-1}	0.52×10^{-1}
0.20×10^{-2}	0.32×10^{-2}	0.67×10^{-2}	0.28×10^{-1}
0.13×10^{-4}	0.42×10^{-4}	0.24×10^{-3}	0.15×10^{-1}
0.59×10^{-9}	0.68×10^{-8}	0.29×10^{-6}	0.78×10^{-2}
—	—	0.44×10^{-12}	0.42×10^{-2}
—	—	—	0.29×10^{-2}
—	—	—	0.25×10^{-1}
—	—	—	-0.13×10^{-1}

method at each step for the first eight iterations, we observe that if the error at iteration j is ε_j, we approximately have for the error at iteration $j+1$: $\varepsilon_{j+1} = C(\varepsilon_j)^2$, with C a constant. This so-called quadratic convergence is typical for this iterative process once we are within the so-called radius of convergence, i.e. when we are sufficiently close to the solution that subsequent iterations will indeed yield convergence to the solution.

The second class of methods consists of the so-called Quasi-Newton methods or Secant-Newton methods. These methods apply updates on existing tangential stiffness matrices such that the stiffness in the subsequent iteration is computed using a multi-dimensional secant approximation. A more in-depth discussion of this class of methods is given in Chapter 4.

2.5 Load *versus* Displacement Control

In the preceding section the load has been applied to the structure in a number of steps. This process is named load control. Alternatively, we can prescribe displacement increments. This so-called displacement control procedure causes a stress development within the specimen, which in turn results in nodal forces at the nodes where the displacements are prescribed. Summation of these forces gives the total reaction force, which, except for a minus sign, equals the equivalent external load that would be caused by the prescribed displacements.

Often the physics dictate which type of load application is the most obvious choice. With creep problems, for instance, the loads must be prescribed (Chapter 8). For other problems displacement control is a more natural choice, such as when a very stiff plate is pushed into a relatively soft subsoil. However, when there is no preference for either load or displacement control from a physical point of view, the latter method is often to be preferred. The reasons for the preference for displacement control are twofold:

- The tangential stiffness matrix is better conditioned for displacement control than for load control. This tends to result in a faster convergence behaviour of the iterative procedure.
- Under load control, the tangential stiffness matrix becomes singular at a limit point in the load–deflection diagram, not only when global failure occurs, but also when we have a local maximum along this curve (Figure 2.7). The tangential stiffness matrix of the displacement controlled problem, on the other hand, does not become singular.

These statements are best elucidated starting from Equation (2.42). This equation has been derived for load control, and the prescribed external load level is contained explicitly in the vector $\mathbf{f}_{\text{ext}}$. The use of displacement control does not directly cause external forces to be exerted on the structure. Rather, a number of non-zero displacements are prescribed in an incremental loading programme. We now decompose the incremental displacement vector $\Delta\mathbf{a}$ into a vector that contains only degrees of freedom that are 'free', i.e. which have to be calculated, $\Delta\mathbf{a}_{\text{f}}$, and displacement increments that have been assigned a certain non-zero value, $\Delta\mathbf{a}_{\text{p}}$:

$$\Delta\mathbf{a} = \begin{bmatrix} \Delta\mathbf{a}_{\text{f}} \\ \Delta\mathbf{a}_{\text{p}} \end{bmatrix} \tag{2.53}$$

In a similar manner the stiffness matrix can be partitioned, as follows:

$$\mathbf{K} = \begin{bmatrix} \mathbf{K}_{\text{ff}} & \mathbf{K}_{\text{fp}} \\ \mathbf{K}_{\text{pf}} & \mathbf{K}_{\text{pp}} \end{bmatrix} \tag{2.54}$$

Using Equations (2.53) and (2.54), Equation (2.42) can be replaced by the expression

$$\begin{bmatrix} \mathbf{K}_{\text{ff}} & \mathbf{K}_{\text{fp}} \\ \mathbf{K}_{\text{pf}} & \mathbf{K}_{\text{pp}} \end{bmatrix} \begin{bmatrix} \Delta\mathbf{a}_{\text{f}} \\ \Delta\mathbf{a}_{\text{p}} \end{bmatrix} = - \begin{bmatrix} (\mathbf{f}_{\text{f}})_{\text{int},0} \\ (\mathbf{f}_{\text{p}})_{\text{int},0} \end{bmatrix} \tag{2.55}$$

where it has been assumed that, apart from the prescribed displacements, no other forces act on the structure. Next, the unknown or 'free' displacement increments can be calculated by eliminating $\Delta\mathbf{a}_{\text{p}}$ from Equation (2.55). For the first iteration this elimination process yields

$$\Delta\mathbf{a}_{\text{f},1} = -\mathbf{K}_{\text{ff}}^{-1}(\mathbf{K}_{\text{fp}}\Delta\mathbf{a}_{\text{p}} + (\mathbf{f}_{\text{f}})_{\text{int},0}) \tag{2.56}$$

while in the subsequent iterations the formula for computing the unknown degrees of freedom changes into:

$$\mathbf{da}_{\text{f},j+1} = -\mathbf{K}_{\text{ff}}^{-1}(\mathbf{f}_{\text{f}})_{\text{int},j} \tag{2.57}$$

since $\mathbf{da}_{\text{p}}$ vanishes. Comparison of Equations (2.42) and (2.56) shows, that for the first iteration the external load $\mathbf{f}_{\text{ext}}^{t+\Delta t}$ must be replaced by the 'equivalent force vector' $\mathbf{K}_{\text{fp}}\Delta\mathbf{a}_{\text{p}}$ when switching from load to displacement control. In the next iterations this contribution vanishes altogether for displacement control.

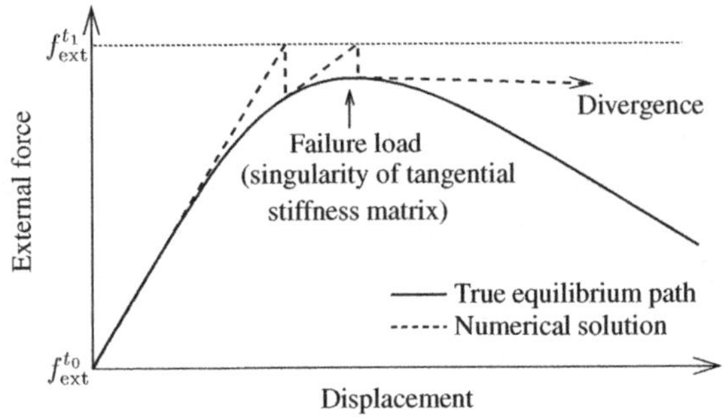

Figure 2.7 Singularity of tangential stiffness matrix at limit point and divergence of iterative procedure

The most important distinction between load and displacement control lies in the fact that load control requires the inversion (or in practice the LDU decomposition) of the matrix $\mathbf{K}$ while in the latter method only the reduced stiffness matrix $\mathbf{K}_{\text{ff}}$ needs to be inverted (or factorised). For symmetric matrices, which covers the majority of all practical computations, it is possible to show that the spectral radius of $\mathbf{K}_{\text{ff}}$, i.e. the quotient of the largest and smallest eigenvalues, is smaller than or equal to that of $\mathbf{K}$. A heuristic reasoning is that a better conditioned tangential stiffness matrix results in a faster convergence, and since displacement control involves iterating with a better conditioned tangential stiffness matrix, the result is ultimately a faster convergence for displacement control than for load control. An additional advantage of displacement control is that the tangential stiffness matrix $\mathbf{K}_{\text{ff}}$ does not become singular at a local or global peak load in the load–displacement curve, whereas the tangential stiffness matrix $\mathbf{K}$ that is used in conjunction with load control does (Figure 2.7). In fact, the problem is not so much that use of load control results in a tangential stiffness matrix that becomes singular at limit points. For instance, one might argue that use of an elastic stiffness matrix or some other Ersatz-stiffness matrix would circumvent the problem of decomposing the tangential stiffness matrix at a singular point. However, the basic problem is that in load controlled processes one tries to find an intersection between the horizontal line in the load–displacement diagram which characterises the load level that is imposed on the structure and the load–displacement path, but that there does not exist such an intersection point (Figure 2.7). The result is obviously divergence of the iterative procedure, which manifests itself in an unbounded growth of the unbalanced forces, i.e. the difference between external load and internal forces. Displacement control does not share this disadvantage, since we now strive to calculate the intersection point of the load–displacement curve with the vertical line in Figure 2.8. The latter line is the result of imposing a fixed value for one of the degrees of freedom (displacements). Nonetheless, some types of structural behaviour are still not traceable with a displacement control procedure. This is obviously the case when the physics prohibit the use of displacement control, e.g. in creep problems, or when in the course of the loading process the displacements under externally applied forces do not grow at an equal pace. Displacement control also cannot be applied for structural behaviour, as shown in Figure 2.8 (so-called snap-back behaviour which is encountered in geometrically non-linear behaviour of thin shells and for strain-softening constitutive relations). The most elegant procedure that can be used to analyse these kinds of problems properly is known as a 'path-following method', or 'arc-length control', see Chapter 4.

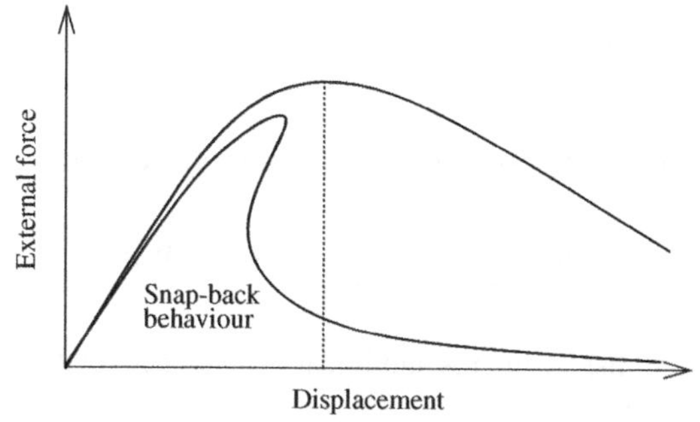

Figure 2.8 Intersection of the load–displacement curve with a line that represents a fixed value of one degree of freedom and 'snap-back' behaviour

2.6 PyFEM: A Linear Finite Element Code with Displacement Control

The implementation of a finite element simulation under displacement control is demonstrated by means of a patch test. Patch tests are simple finite element simulations which are intended to demonstrate that an element possesses some basic requirements, such as the ability to represent a uniform stress state. The following patch test is taken from MacNeal and Harder (1985).

We consider a rectangular domain, shown in Figure 2.9. The domain is discretised by five quadrilateral elements. The positions of the nodes are chosen such that all elements are skewed and are mildly distorted. The positions of the internal nodes 4, 5, 6, and 7 are given in the figure. The displacements $\mathbf{a} = (a_x, a_y)$ of the external nodes 0, 1, 2, and 3 are prescribed according to the relations:

$$a_x(x, y) = 10^{-3}(x + y/2)\,; \qquad a_y(x, y) = 10^{-3}(y + x/2) \tag{2.58}$$

Finally, a plane-stress constitutive relation is assumed with Young's modulus $E = 10^6$ and Poisson's ratio $\nu = 0.25$.

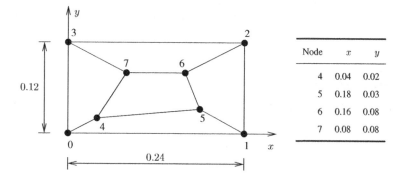

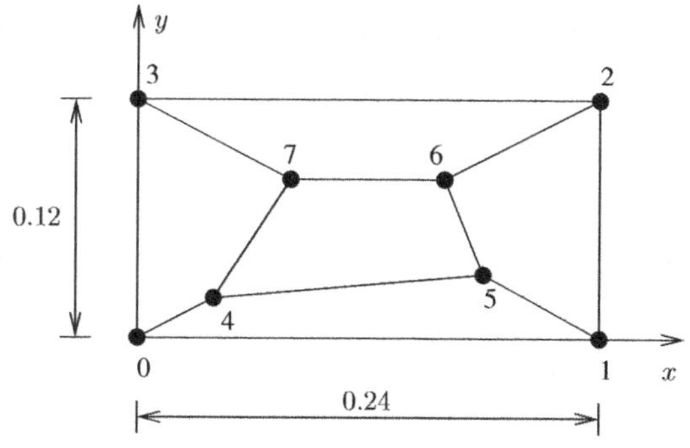

Node	x	y
4	0.04	0.02
5	0.18	0.03
6	0.16	0.08
7	0.08	0.08

Figure 2.9 Patch test for linear quadrilateral elements. The positions of the internal nodes (4,5,6,7) are given on the right

Box 2.4 Computational flow in a linear elastic finite element code

1. Compute the tangential stiffness matrix of the undeformed system:

$$\mathbf{K} = \sum_{e=1}^{n_e} \mathbf{Z}_e^{\mathrm{T}} \sum_{i=1}^{n_i} w_i \det \mathbf{J}_i \mathbf{B}_i^{\mathrm{T}} \mathbf{D} \mathbf{B}_i \mathbf{Z}^e$$

2. Adjust for prescribed displacements and linear dependence relations.
3. Solve the system: $\mathbf{K}\mathbf{a} = \mathbf{f}_{\mathrm{ext}}$
4. Compute the stresses from the total strain for each integration point i:

$$\sigma_i = \mathbf{D}\epsilon_i = \mathbf{D}\mathbf{B}_i \mathbf{a}_e.$$

5. Compute the internal force vector:

$$\mathbf{f}_{\mathrm{int}} = \sum_{e=1}^{n_e} \mathbf{Z}_e^{\mathrm{T}} \sum_{i=1}^{n_i} w_i \det \mathbf{J}_i \mathbf{B}_i^{\mathrm{T}} \sigma_i$$

In this chapter, we have focused on the solution of a non-linear problem using the Newton–Raphson iterative method. The linear solution can be considered as a further simplification of this algorithm and deals with the solution of the following system:

$$\mathbf{K}\mathbf{a} = \mathbf{f}_{\mathrm{ext}} \tag{2.59}$$

where $\mathbf{K}$ is the stiffness matrix of the undeformed structure, $\mathbf{a}$ is the total displacement vector and $\mathbf{f}_{\mathrm{ext}}$ is the external force vector. The complete procedure to solve this system and to calculate the resulting stresses is described in Box 2.4.

The code for solving the patch test can be found in the directory `examples/ch02` and is called `PatchTest.py`. The file contains all the tools for the solution, including the material and element formulation. The outline of the code is as follows:

```
⟨Patch test example ⟩ ≡

  ⟨Patch test utility functions 57⟩
  ⟨Patch test main code 55⟩
```

In the fragment ⟨*Patch test utility functions*⟩, we describe a number of common functions that will be used at different locations in the code. The program is listed in the fragment ⟨*Patch test main code*⟩:

```
⟨Patch test main code ⟩≡                                          54

  ⟨Description of geometry and boundary conditions 55⟩
  ⟨Solution procedure 58⟩
  ⟨Post processing the results⟩
```

The main code of this example can be split into three parts. In the first part, ⟨*Description of geometry and boundary conditions*⟩, the geometry, the boundary conditions and the material parameters are specified. The solution procedure for linear systems (Box 2.4), is given in ⟨*Solution procedure*⟩. Finally, the program is completed by printing the results in the fragment ⟨*Post processing the results*⟩.

A finite element mesh is described by two sets of data: (i) the position of the nodes in the mesh; and (ii) the connectivity of the elements. In this program, the position of the nodes is stored in a two-dimensional array `coords` with the dimensions `(8,2)`, which matches the number of nodes in the mesh and the spatial dimension of the problem. After the creation of this array, the coordinates of the nodes are set.

```
⟨Description of geometry and boundary conditions ⟩≡               55

  coords = zeros( shape=(8,2) )

  coords[0,:] = [0.0 ,0.0 ]
  coords[1,:] = [0.24,0.0 ]
  coords[2,:] = [0.24,0.12]
  coords[3,:] = [0.0 ,0.12]
  coords[4,:] = [0.04,0.02]
  coords[5,:] = [0.18,0.03]
  coords[6,:] = [0.16,0.08]
  coords[7,:] = [0.08,0.08]
```

The element connectivity is stored in a two-dimensional integer array called `elems`. The dimensions of this array are equal to the number of elements (five), and to the number of nodes that support an element (four).

```
⟨Description of geometry and boundary conditions ⟩+≡              55

  elems = zeros( shape=(5,4) , dtype=int )

  elems[0,:] = [ 0 , 1 , 5 , 4 ]
  elems[1,:] = [ 1 , 2 , 6 , 5 ]
  elems[2,:] = [ 2 , 3 , 7 , 6 ]
  elems[3,:] = [ 3 , 0 , 4 , 7 ]
  elems[4,:] = [ 4 , 5 , 6 , 7 ]
```

The next step is the creation of the solution space. In this example, it is assumed that each node k supports two degrees of freedom: a displacement in the x-direction, $a_{k,x}$, and a displacement in the y-direction: $a_{k,y}$. The vectorial displacement of a node is equal to $\mathbf{a}_k = (a_{k,x}, a_{k,y})$. All the nodes in this problem are connected to an element and take part in the solution space. Hence, the total number of degrees of freedom is two times the number of nodes, which in this case equals 16. The degrees of freedom in the solution space are stored node-wise.

In most finite element simulations, not all the degrees of freedom are part of the solution space. Some of the degrees of freedom have a prescribed value. In this example, all the external nodes (0, 1, 2, and 3) are prescribed in the x- and in the y-direction. In other words, the corresponding degrees of freedom, $a_{k,x}$ and $a_{k,y}$, are fixed. As a first step, the indices of these prescribed degrees of freedom must be determined. This is done by the function `getDofs()`, which is described in the fragment ⟨*Patch test utility functions*⟩. The resulting index array is stored as `presInds`. The length of this array equals `consDof`.

⟨*Description of geometry and boundary conditions* ⟩+≡ 55

```
presNodes = array([0,1,2,3])

presInds  = getDofs( presNodes )                                 57

consDof   = len(presInds)
```

When the indices and the nodes are known, the magnitude of the prescribed displacement components can be calculated according to Equation (2.58) and can be stored in the array `presVals`, which has the same length.

⟨*Description of geometry and boundary conditions* ⟩+≡ 56

```
presVals  = zeros( consDof )

upres = lambda crd : 1e-3*(crd[0]+crd[1]/2)
vpres = lambda crd : 1e-3*(crd[1]+crd[0]/2)

presVals[2*presNodes]=[upres(crd) for crd in coords[presNodes,:] ]
presVals[2*presNodes+1]=[vpres(crd) for crd in coords[presNodes,:] ]
```

The function `getDofs` is used to determine the indices of the degrees of freedom that correspond to a specific set of nodes. The array of node indices, `nodes`, is passed onto this function as an argument. Since it is assumed that all nodes have two degrees of freedom and that the node numbering is continuous, the list of indices can be determined in the following manner:

⟨*Patch test utility functions* ⟩≡ 54

```
def getDofs( nodes ):

  n = 2*len(nodes)

  dofs = zeros( n , dtype=int )

  dofs[0:n:2]=2*nodes
  dofs[1:n:2]=2*nodes+1

  return dofs
```

Note that the length of the returning array `dofs` is twice the length of the argument `nodes`.

The initialisation of the simulation is completed by the constitutive behaviour. The patch is assumed to be in a plane-stress condition. The Young's modulus is given by `E`, and the Poisson's ratio is given by `nu`.

⟨*Description of geometry and boundary conditions* ⟩+≡ 56

```
nu = 0.25
E  = 1.e6
```

Because the material behaviour is fully elastic, all stresses can be determined by means of the elastic material stiffness matrix $\mathbf{D} = \mathbf{D}^e$. This stiffness matrix is calculated as well and stored as a two-dimensional array:

⟨*Description of geometry and boundary conditions* ⟩+≡ 57

```
D  = zeros( shape = (3,3) )

D[0,0] = E / (1.0 - nu*nu )
D[0,1] = D[0,0] * nu
D[1,0] = D[0,1]
D[1,1] = D[0,0]
D[2,2] = E / (2.0 * (1+nu) )
```

In the next fragment of the example, ⟨*Solution procedure*⟩, the problem is solved according to the algorithm in Box 2.4.

```
⟨Solution procedure ⟩≡                                                    55

  K = zeros( shape = ( totDof , totDof ) )

  for elemNodes in elems:
    elemDofs    = getDofs(elemNodes)                                       57
    elemCoords = coords[elemNodes,:]

    sData = getElemShapeData( elemCoords )                                 39

    for iData in sData:
      B    = getBmatrix( iData.dhdx )                                      58
      Kint = dot ( B.transpose() , dot ( D , B ) ) * iData.weight
      K[ix_(elemDofs,elemDofs)] += Kint
```

In the first step, the system stiffness matrix **K** is computed. This stiffness matrix is stored in the array `K` with the dimensions (`totDof,totDof`). Initially, all items in this array are set equal to zero. It is filled in a step-wise manner by calculating the local stiffness matrix of each element in the model. The first **for**-loop in the fragment ⟨*solution procedure*⟩ is a loop over the elements. The variable `elemNodes` is an array that contains the indices of the nodes that support the current element. This array is used as an input of the function `getDofs` to obtain the array with the indices of the corresponding degrees of freedom. The array `elemNodes` is also used to build an array with the coordinates of the nodes of this element `elemCoords`, which serves as input for the function `getElemShapeData`. The latter provides all the shape function and the integration data for this element.

The second **for**-loop in this fragment evaluates the stiffness components at each integration point of the element. The matrix **B** is calculated by the function `getBmatrix`. In this function, the derivatives of the shape functions in the current integration point, `dhdx` are passed as an argument. The length of this array is equal to the number of nodes in the element `nNel`. The number of degrees of freedom in the element and the length of the second dimension of the array `B` are equal to two times the number of nodes.

```
⟨Patch test utility functions ⟩+≡                                          57

  def getBmatrix( dhdx ):

    B = zeros( shape = ( 3 , 2*len(dhdx) ) )

    for i,dp in enumerate(dhdx):
      B[0,i*2  ] = dp[0]
      B[1,i*2+1] = dp[1]
      B[2,i*2  ] = dp[1]
      B[2,i*2+1] = dp[0]

    return B
```

The contribution of a single integration point is stored in the array `Kint`, which in turn is added to the global stiffness matrix. Note that the assembly of the stiffness matrix is described by means of an element specific location matrix $\mathbf{Z}_e$, see Equation (2.13). In most finite element codes, this assembly is performed in a more direct way by means of an index operator. In `numpy`, the `ix_` operator can be used to add the components of an multi-dimensional array in arbitrary positions of a different array. The argument of the operator `ix_` is the array `elemDofs`, which contains the element degrees of freedom.

Similar to Equation (2.56), the prescribed degrees of freedom must be eliminated from the system of equations, and we must solve for the 'free' degrees of freedom $\mathbf{a}_\mathrm{f}$:

$$\mathbf{K}_\mathrm{ff}\mathbf{a}_\mathrm{f} = \mathbf{f}_{\mathrm{ext,f}} - \mathbf{K}_\mathrm{fp}\mathbf{a}_\mathrm{p} \tag{2.60}$$

Note that in this case, the load is applied by means of prescribed displacements only and that the external force vector $\mathbf{f}_{\mathrm{ext,f}}$ is equal to zero. In general, the prescribed degrees of freedom are not nicely lumped, but are located in an arbitrary manner in the array that contains the total degrees of freedom. As a result, we cannot simply take a block of the stiffness matrix to solve Equation (2.60). Instead, we have to construct the matrix $\mathbf{K}_\mathrm{ff}$ in a different way, using a constraint matrix $\mathbf{C}$. The constraint matrix consists of N rows – it is recalled that N is the total number of degrees of freedom – and of N_f columns, with N_f the number of 'free' degrees of freedom. The row of the matrix $\mathbf{C}$ that corresponds to a degree of freedom that is prescribed consists of 0 terms only. The rows of this matrix that correspond to degrees of freedom that are *not* prescribed contain only one term that is equal to 1, located in the first column that, so far, did not contain a term 1. In the code, the constraint matrix $\mathbf{C}$ is constructed as follows:

⟨*Solution procedure* ⟩≡ 58

```
consDof = len( presInds )

C = zeros( shape = (totDof,totDof-consDof) )

j = 0

for i in range(totDof):
  if i in presInds:
    continue
  C[i,j] = 1.
  j+=1
```

Using the constraint matrix, the reduced stiffness matrix can be written as:

$$\mathbf{K}_\mathrm{ff} = \mathbf{C}^\mathrm{T}\mathbf{K}\mathbf{C} \tag{2.61}$$

The right-hand side of the system is equal to:

$$\mathbf{f}_{\text{ext,f}} = -\mathbf{C}^{\text{T}}\mathbf{K}\mathbf{a}^{*} \tag{2.62}$$

with $\mathbf{a}^{*}$ the solution vector in which the prescribed displacements have the corresponding value. Upon solving the reduced equation, Equation (2.60), the solution vector $\mathbf{a}$ is obtained from:

$$\mathbf{a} = \mathbf{C}\mathbf{a}_{\text{f}} \tag{2.63}$$

In the example code, these operations are programmed as follows:

⟨*Solution procedure* ⟩+≡ 59

```
a = zeros(totDof)

a[presInds] = presVals

Kff = dot( dot( C.transpose(), K ), C )
rhs = dot( C.transpose(), dot( K , -a ) )

af = scipy.linalg.solve( Kff, rhs )

a = dot( C, af )
a[presInds] = presVals
```

When the new displacement field has been calculated, the stresses and the internal force vector can be calculated. In a finite element code, the stresses are only directly computed in the integration points. For post-processing, it is often necessary to have knowledge of the stresses in the nodes. This can be done by extrapolating the stresses from the integration points to the nodal coordinates. This process is the inverse of the element integration, and is, in most cases, not straightforward. In addition, a single node normally supports different elements. When using C^0 interpolation functions, the stress field is usually not continuous, and therefore not uniquely defined across element boundaries. Accordingly, the value of the stress at a node is at best the weighted average of the contributions of the stresses in the neighbouring elements.

To avoid the effort of first extrapolating the stress, and then to take an averaged value, it is common practice to simply calculate the average of the stresses in all integration points of the elements that are supported by the node of interest. To this end, a two-dimensional array `nodalStress` is created to store the stresses at integration points. Furthermore, an integer array `nodalCount` is constructed to count the number of times that a stress state is added to the array `nodalStress`. The internal forces are stored in an array `fint` which has the size `totDof`.

⟨*Solution procedure* ⟩+≡ 60

```
fint         = zeros( totDof )
nodalStress = zeros( shape = (len(coords),3) )
nodalCount  = zeros( len(coords) )

for elemNodes in elems:
  elemDofs = getDofs( elemNodes )                                    57
  sData = getElemShapeData( coords[elemNodes,:] )                    39

  for iData in sData:
    B = getBmatrix( iData.dhdx )                                     58

    strain = dot( B , a[elemDofs] )
    stress = dot( D , strain )

    fint[elemDofs] += dot(b.transpose(),stress)*iData.weight

    nodalStress[elemNodes,:] += stress
    nodalCount [elemNodes]   += ones(len(elemNodes));
```

The structure of the calculation of the internal forces is identical to the structure of the calculation of the stiffness matrix. The fragment consists of two nested **for**-loops. The first loop is over the elements in the model and the second loop is over the integration points within an element. It is noted that the stress in each integration point is stored in the row of the array `nodalStress` that corresponds to the nodes that support this element: `elemNodes`. The corresponding counters are increased by one.

In the final fragment of this example, ⟨*Post processing the results*⟩, the displacements and the stresses are printed. The results of the analysis are given, in Table 2.3. The nodal displacements, including the displacements of the internal nodes, match the prescribed field given by Equation (2.58). Furthermore, the internal forces that act on the internal nodes are equal to zero. The sum of the internal forces on the external nodes in the x- as well as in the y-direction

Table 2.3 Displacements, internal forces and stress at the nodes in the patch test

Node	$a_x[10^{-3}]$	$a_y[10^{-3}]$	$f_{\text{int},x}$	$f_{\text{int},y}$	σ_{xx}	σ_{yy}	σ_{xy}
0	0.000	0.000	−128.0	−184.0	1333.0	1333.0	400.0
1	0.240	0.120	32.0	−136.0	1333.0	1333.0	400.0
2	0.300	0.240	128.0	184.0	1333.0	1333.0	400.0
3	0.060	0.120	−32.0	136.0	1333.0	1333.0	400.0
4	0.050	0.040	0.0	0.0	1333.0	1333.0	400.0
5	0.195	0.120	0.0	0.0	1333.0	1333.0	400.0
6	0.200	0.160	0.0	0.0	1333.0	1333.0	400.0
7	0.120	0.120	0.0	0.0	1333.0	1333.0	400.0

are equal to zero since the body is in equilibrium. The same holds for the sum of moments about an arbitrary point in the domain. Finally, the stress state, which is constant over the domain, matches the analytical solution $\boldsymbol{\sigma} = [1333.0, 1333.0, 400.0]$.

Reference

MacNeal R and Harder R 1985 A proposed standard set of problems to test finite element accuracy. *Finite Elements in Analysis and Design* **1**, 3–20.

3

Geometrically Non-linear Analysis

Structural stability problems constitute an important field of application of the finite element method. Usually, structural instability is assumed to be caused by geometrically non-linear effects in spite of the fact that physically non-linear material models can also play an important role in causing destabilisation.

In the present chapter we will set up a proper description of the statics and kinematics of continuous media subjected to large deformations. To elucidate the underlying concepts, which are somewhat abstract, the appropriate stiffness matrices and load vectors are first derived for simple truss elements. Subsequently, the extension is made towards a continuum.

When we employ the terminology 'large deformations' it may not be clear what we exactly mean. Sometimes, reference is made to large strains as, for example, observed in deforming rubber materials or in extrusion processes of metals, while others only imply the large displacements and rotations that occur in, e.g. thin-walled, slender structural members. Consider for instance the cantilever beam of Figure 3.1. By increasing the stiffness EI the strains in the beam can be made arbitrarily small. But even for large values of EI, and consequently for small strains, the vertical displacement and the rotation of the tip of the beam can be made arbitrarily large if the beam is made long enough. Apparently, the notions 'large strains' and 'large displacements', or more accurately, 'large displacement gradients', do not coincide. Large strains can only occur if the displacement gradients are also large, but the reverse does not necessarily hold true: large displacement gradients can be observed in structural behaviour while the strains are still limited, smaller than, say, 2%. For many materials in engineering practice the strains are usually small. This assumption is adopted in this chapter, and will be made explicitly for the truss models in the next section. In the subsequent sections we will extend the formulations to two- and three-dimensional bodies, for which a more rigorous approach is necessary. In this approach no approximations will be made with respect to the kinematics and statics of a continuum subject to large deformations. The only restriction resides in the constitutive relations, i.e. the relation between stress and strain. For large displacement gradients, but small strains, the usual constitutive relationships, e.g. Hooke's law of linear elasticity, remain valid. However, for large strains the extraction of these relations from experimental data, for instance from uniaxial tensile tests, requires careful interpretation of test data (e.g. does the specimen

Non-linear Finite Element Analysis of Solids and Structures, Second Edition.
René de Borst, Mike A. Crisfield, Joris J.C. Remmers and Clemens V. Verhoosel.

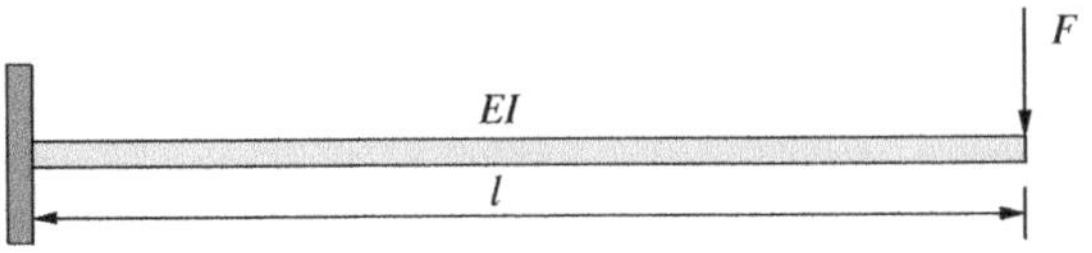

Figure 3.1 Cantilever beam

indeed deform homogeneously for such large strains) and a judicious choice of physically relevant stress and strain measures.

The treatment of continuum mechanics has been kept to a minimum. For further reading, reference is made to standard textbooks (Fung 1965; Holzapfel 2000; Malvern 1969; Ogden 1984). For a treatment of non-linear continuum mechanics in a finite element setting the reader is referred to Bonet and Wood (1997).

3.1 Truss Elements

Since we will restrict ourselves to small strains in this chapter, the following two assumptions are made in the derivation of tangential stiffness matrices and load vectors for truss elements:

$$A \approx A_0 \quad \text{and} \quad \ell \approx \ell_0 \tag{3.1}$$

These two identities state that the cross section and the length of the truss member in the undeformed configuration (A_0 and ℓ_0, respectively) are approximately equal to the cross section and the length in the deformed configuration (A and ℓ, respectively). Furthermore, we have external forces only at the nodes of a truss member. If these nodal forces are assembled in a vector $\mathbf{f}$ and if the assumptions (3.1) are invoked, the virtual work equation (2.37) at $t + \Delta t$ reduces to:

$$A_0 \int_{\ell_0} (\delta\boldsymbol{\epsilon}^{t+\Delta t})^{\mathrm{T}} \boldsymbol{\sigma}^{t+\Delta t} \mathrm{d}x = (\delta\mathbf{u}^{t+\Delta t})^{\mathrm{T}} \mathbf{f}_{\mathrm{ext}}^{t+\Delta t} \tag{3.2}$$

with $\boldsymbol{\epsilon}^{t+\Delta t}$ and $\boldsymbol{\sigma}^{t+\Delta t}$ the strain and stress at $t + \Delta t$, respectively. In truss elements the only non-vanishing stress component is the axial stress, so that:

$$A_0 \int_{\ell_0} \delta\epsilon^{t+\Delta t} \sigma^{t+\Delta t} \mathrm{d}x = (\delta\mathbf{u}^{t+\Delta t})^{\mathrm{T}} \mathbf{f}_{\mathrm{ext}}^{t+\Delta t} \tag{3.3}$$

Non-linear finite element calculations are carried out in a number of increments, i.e. the total load is applied in a stepwise fashion. The stress increment between time t and time $t + \Delta t$ is given by

$$\Delta\sigma = \sigma^{t+\Delta t} - \sigma^{t} \tag{3.4}$$

With this decomposition Equation (3.3) can be written as:

$$A_0 \int_{\ell_0} \delta\epsilon^{t+\Delta t} (\sigma^{t} + \Delta\sigma) \mathrm{d}x - (\delta\mathbf{u}^{t+\Delta t})^{\mathrm{T}} \mathbf{f}_{\mathrm{ext}}^{t+\Delta t} \tag{3.5}$$

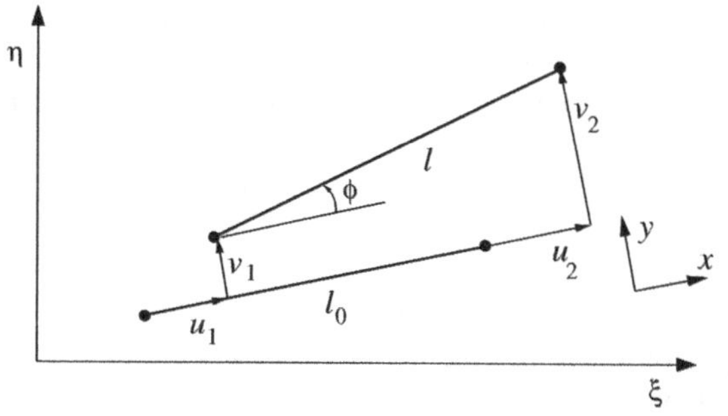

Figure 3.2 Truss member in undeformed and deformed configuration

The majority of the constitutive models are characterised by a linear relation between the stress increment $\Delta\sigma$ and the strain increment $\Delta\epsilon$:

$$\Delta\sigma = E_{\tan}\Delta\epsilon \tag{3.6}$$

with $E_{\tan}$ the instanteneous, tangential stiffness modulus. For a linear-elastic material model the tangential stiffness modulus is identical to the Young's modulus: $E_{\tan} = E$. Since we shall limit the treatment in the present chapter to geometrical non-linearity we will henceforth assume that we deal with a linear-elastic material and, consequently, the subscript 'tan' will be dropped in the constitutive relation. Combination of the last two equations yields:

$$EA_0 \int_{\ell_0} \delta\epsilon^{t+\Delta t} \Delta\epsilon \mathrm{d}x + A_0 \int_{\ell_0} \delta\epsilon^{t+\Delta t} \sigma^t \mathrm{d}x = (\delta \mathbf{u}^{t+\Delta t})^{\mathrm{T}} \mathbf{f}_{\mathrm{ext}}^{t+\Delta t} \tag{3.7}$$

We now set out to derive a proper relationship between the displacements of the truss member and the strain ϵ. The length of the deformed rod in Figure 3.2 is related to its nodal displacements by:

$$\ell^2 = (\ell_0 + u_2 - u_1)^2 + (v_2 - v_1)^2 \tag{3.8}$$

Dividing by ℓ_0^2, and expanding the right-hand side gives

$$\left(\frac{\ell}{\ell_0}\right)^2 = 1 + 2\frac{u_2 - u_1}{\ell_0} + \left(\frac{u_2 - u_1}{\ell_0}\right)^2 + \left(\frac{v_2 - v_1}{\ell_0}\right)^2 \tag{3.9}$$

While for small differences in the displacements the strain can be defined in a straightforward fashion by setting it equal to the elongation of the bar $\ell - \ell_0$ divided by the length ℓ_0,

$$\epsilon = \frac{\ell - \ell_0}{\ell_0} \tag{3.10}$$

the definition

$$\epsilon = \frac{1}{2}\left[\left(\frac{\ell}{\ell_0}\right)^2 - 1\right] \tag{3.11}$$

is more appropriate when large differences between the displacements are allowed. This strain measure is identical to the normal components of the so-called Green–Lagrange strain tensor, which will be introduced in the next section. Equation (3.11) can be rewritten as:

$$\epsilon = \frac{\ell - \ell_0}{\ell_0} + \frac{1}{2}\left(\frac{\ell - \ell_0}{\ell_0}\right)^2 \tag{3.12}$$

which, when $\ell \approx \ell_0$, approaches Equation (3.10).

Substitution of Equation (3.9) into the definition (3.11) yields a strain measure for large differences in displacements:

$$\epsilon = \frac{u_2 - u_1}{\ell_0} + \frac{1}{2}\left(\frac{u_2 - u_1}{l_0}\right)^2 + \frac{1}{2}\left(\frac{v_2 - v_1}{l_0}\right)^2 \tag{3.13}$$

The second term on the right-hand side is usually small compared with the other two contributions. It is therefore often neglected in the strain expression. Nevertheless, we will retain this contribution in the present treatment, although in Box 3.1 we present an alternative derivation

Box 3.1 Alternative derivation of the strain measure

Rather than defining the strain ϵ for large deformations as in Equation (3.11), we can also directly start from definition (3.10). Use of Equation (3.8) then results in

$$\epsilon(u, v) = \frac{1}{\ell_0}\sqrt{(\ell_0 + u)^2 + v^2} - 1$$

where the shorthand notations $u = u_2 - u_1$ and $v = v_2 - v_1$ have been utilised. Developing this expression in a Taylor series including the quadratic terms yields:

$$\epsilon = \epsilon(0, 0) + \frac{\partial \epsilon}{\partial u}u + \frac{\partial \epsilon}{\partial v}v + \frac{1}{2}\frac{\partial^2 \epsilon}{\partial u^2}u^2 + \frac{\partial^2 \epsilon}{\partial u \partial v}uv + \frac{1}{2}\frac{\partial^2 \epsilon}{\partial v^2}v^2$$

The differentials can now be evaluated as follows for small deviations from $(u, v) = (0, 0)$:

$$\frac{\partial \epsilon}{\partial u} = \frac{1}{\ell_0}\ , \quad \frac{\partial \epsilon}{\partial v} = 0\ , \quad \frac{\partial^2 \epsilon}{\partial u^2} = 0\ , \quad \frac{\partial^2 \epsilon}{\partial u \partial v} = 0\ , \quad \frac{\partial^2 \epsilon}{\partial v^2} = \frac{1}{\ell_0^2}$$

which gives the following expression for the strain measure:

$$\epsilon = \frac{u_2 - u_1}{\ell_0} + \frac{1}{2}\left(\frac{v_2 - v_1}{\ell_0}\right)^2$$

We observe that compared with the original derivation the contribution $\frac{1}{2}(u_2 - u_1)^2/\ell_0^2$ is missing. The somewhat different choice of strain measure causes this difference. However, for small strains, the quadratic term is small compared with the linear contribution $(u_2 - u_1)/\ell_0$ and can be neglected. Thus, the seemingly different expressions obtained in both derivations can be reconciled provided that we remain within the realm of small strains.

of the expression for the strain measure ϵ in which, as a consequence of the approximation $\ell \approx \ell_0$, this term cancels.

With aid of Equation (3.13) we can formulate an expression for the strain increment $\Delta\epsilon = \epsilon^{t+\Delta t} - \epsilon^t$:

$$\begin{aligned}\Delta\epsilon = \frac{\Delta u_2 - \Delta u_1}{\ell_0}\left(1 + \frac{u_2 - u_1}{\ell_0}\right) + \left(\frac{\Delta v_2 - \Delta v_1}{\ell_0}\right)\left(\frac{v_2 - v_1}{\ell_0}\right) \\ + \frac{1}{2}\left(\frac{\Delta u_2 - \Delta u_1}{\ell_0}\right)^2 + \frac{1}{2}\left(\frac{\Delta v_2 - \Delta v_1}{\ell_0}\right)^2\end{aligned} \tag{3.14}$$

Apparently, the last two terms are quadratic in the displacement increments Δu and Δv. Note that u_1, u_2, v_1 and v_2 do not give non-linear contributions to the strain increment, since they represent the values of the displacements at the beginning of the load increment, and are known. Strictly speaking, superscripts t should be attached to u_1, u_2, v_1 and v_2 to indicate that these displacements are evaluated at time t, but for notational simplicity, these superscripts are neglected.

To maintain a strict analogy between the present treatment of truss elements, and the derivation for continuum elements to be presented next, we will formally split the strain increment $\Delta\epsilon$ into a part Δe that is linear in the displacement increments,

$$\Delta e = \frac{\Delta u_2 - \Delta u_1}{\ell_0}\left(1 + \frac{u_2 - u_1}{\ell_0}\right) + \left(\frac{\Delta v_2 - \Delta v_1}{\ell_0}\right)\left(\frac{v_2 - v_1}{\ell_0}\right) \tag{3.15}$$

and a contribution that is quadratic in the displacement increments:

$$\Delta\eta = \frac{1}{2}\left(\frac{\Delta u_2 - \Delta u_1}{\ell_0}\right)^2 + \frac{1}{2}\left(\frac{\Delta v_2 - \Delta v_1}{\ell_0}\right)^2 \tag{3.16}$$

Consequently, from Equation (3.14) it follows that:

$$\Delta\epsilon = \Delta e + \Delta\eta \tag{3.17}$$

The virtual strain increments, which are also needed in the process of constructing proper tangential stiffness matrices and force vectors, are now given by:

$$\begin{aligned}\delta\Delta e &= \frac{\delta\Delta u_2 - \delta\Delta u_1}{\ell_0}\left(1 + \frac{u_2 - u_1}{\ell_0}\right) + \left(\frac{\delta\Delta v_2 - \delta\Delta v_1}{\ell_0}\right)\left(\frac{v_2 - v_1}{\ell_0}\right) \\ \delta\Delta\eta &= \left(\frac{\Delta u_2 - \Delta u_1}{\ell_0}\right)\left(\frac{\delta\Delta u_2 - \delta\Delta u_1}{\ell_0}\right) + \left(\frac{\Delta v_2 - \Delta v_1}{\ell_0}\right)\left(\frac{\delta\Delta v_2 - \delta\Delta v_1}{\ell_0}\right)\end{aligned} \tag{3.18}$$

3.1.1 Total Lagrange Formulation

Equation (3.3), which expresses the principle of virtual work for truss elements, is valid at any time. Note that, as already discussed in Chapter 2, the notion of 'time' can be used in an abstract sense, since in the present context of static deformation processes, the concept of time is merely used to order the sequence of events and has no correlation with the real time. Noting that $\delta\epsilon^t$ vanishes because the variation of a constant is zero, the identity $\delta\epsilon^{t+\Delta t} = \delta\Delta\epsilon$ holds.

Using Equation (3.17) to elaborate Equation (3.7) gives:

$$
\begin{aligned}
EA_0 \left(\int_{\ell_0} (\delta\Delta e)\Delta e \mathrm{d}x + \int_{\ell_0} (\delta\Delta\eta)\Delta e \mathrm{d}x \right) &+ \\
EA_0 \left(\int_{\ell_0} (\delta\Delta e)\Delta\eta \mathrm{d}x + \int_{\ell_0} (\delta\Delta\eta)\Delta\eta \mathrm{d}x \right) &+ \\
A_0 \left(\int_{\ell_0} (\delta\Delta\eta)\sigma^t \mathrm{d}x + \int_{\ell_0} (\delta\Delta e)\sigma^t \mathrm{d}x \right) &= (\delta\Delta\mathbf{u})^{\mathrm{T}} \mathbf{f}_{\mathrm{ext}}^{t+\Delta t}
\end{aligned}
\tag{3.19}
$$

On the left-hand side of the equation we have to arrive at a product of the tangential stiffness matrix, that represents the tangent at that point to the load–displacement curve, and a vector $\Delta\mathbf{u}$ which contains the unknown displacement increments. The tangential stiffness matrix itself is not a function of the displacement increments, which leads to the requirement that only integrals which give a linear contribution to $\Delta\mathbf{u}$ can be kept on the left-hand side. $A_0 \int (\delta\Delta e)\sigma^t \mathrm{d}x$ is not a function of $\Delta\mathbf{u}$ and, consequently, cannot contribute to the tangential stiffness matrix either. Therefore, this term is transferred to the right-hand side to form the internal force vector. Since the terms $EA_0 \int (\delta\Delta\eta)\Delta e \mathrm{d}x$ and $EA_0 \int (\delta\Delta e)\Delta\eta \mathrm{d}x$ are quadratic functions of the displacement increment vector $\Delta\mathbf{u}$, they also cannot contribute to the tangential stiffness matrix. A similar reasoning holds for $EA_0 \int (\delta\Delta\eta)\Delta\eta \mathrm{d}x$ which is cubic in the displacement increments. The latter three contributions are therefore deleted, to obtain the linearised expression:

$$
EA_0 \int_{\ell_0} (\delta\Delta e)\Delta e \mathrm{d}x + A_0 \int_{\ell_0} (\delta\Delta\eta)\sigma^t \mathrm{d}x = (\delta\Delta\mathbf{u})^{\mathrm{T}} \mathbf{f}_{\mathrm{ext}}^{t+\Delta t} - A_0 \int_{\ell_0} (\delta\Delta e)\sigma^t \mathrm{d}x \tag{3.20}
$$

It is emphasised that deleting the contributions that are quadratic and cubic in the displacement increments does not affect the accuracy of the solution, provided that an iterative solution procedure is employed to arrive at a proper equilibrium state.

When translating finite element concepts into software it is convenient to have the governing equations in matrix-vector notation. As a first step we write

$$
\Delta e = \mathbf{b}_L^{\mathrm{T}} \Delta\mathbf{u} \tag{3.21}
$$

instead of Equation (3.15). In Equation (3.21) the following definitions have been adopted:

$$
\mathbf{b}_L^{\mathrm{T}} = \frac{1}{\ell_0} \left(-\left(1 + \frac{u_2 - u_1}{\ell_0}\right), \; -\frac{v_2 - v_1}{\ell_0}, \; \left(1 + \frac{u_2 - u_1}{\ell_0}\right), \; \frac{v_2 - v_1}{\ell_0} \right) \tag{3.22}
$$

and

$$
\Delta\mathbf{u}^{\mathrm{T}} = (\Delta u_1 \,, \; \Delta v_1 \,, \; \Delta u_2 \,, \; \Delta v_2) \tag{3.23}
$$

Use of the approximation $\ell \approx \ell_0$ then gives:

$$
\cos\phi = 1 + \frac{u_2 - u_1}{\ell_0} \quad \text{and} \quad \sin\phi = \frac{v_2 - v_1}{\ell_0}
$$

and $\mathbf{b}_L^{\mathrm{T}}$ can be rewritten as:

$$
\mathbf{b}_L^{\mathrm{T}} = \frac{1}{\ell_0} (-\cos\phi \,, \; -\sin\phi \,, \; \cos\phi \,, \; \sin\phi) \tag{3.24}
$$

With the above results the calculation of the internal force vector that stems from the term $A_0 \int (\delta\Delta e)\sigma^t \mathrm{d}x$ on the right-hand side of the virtual work Equation (3.20) is straightforward. With the aid of Equation (3.21) the internal force vector at element level can be written as

$$A_0 \int_{\ell_0} (\delta\Delta e)\sigma^t \mathrm{d}x = (\delta\Delta\mathbf{u})^{\mathrm{T}} \mathbf{f}^t_{\mathrm{int}} \tag{3.25}$$

with

$$\mathbf{f}^t_{\mathrm{int}} = A_0 \int_{\ell_0} \mathbf{b}_L \sigma^t \mathrm{d}x = A_0 \ell_0 \mathbf{b}_L \sigma^t \tag{3.26}$$

since $\mathbf{b}_L$ and σ are constant along the length of the truss member. Using Equation (3.21) we can elaborate the first term on the left-hand side as:

$$EA_0 \int_{\ell_0} (\delta\Delta e)\Delta e \mathrm{d}x = EA_0 \ell_0 (\delta\Delta\mathbf{u})^{\mathrm{T}} \mathbf{b}_L \mathbf{b}_L^{\mathrm{T}} \Delta\mathbf{u} \tag{3.27}$$

Defining

$$\mathbf{K}_L = EA_0 \ell_0 \mathbf{b}_L \mathbf{b}_L^{\mathrm{T}} \tag{3.28}$$

as the first, or 'linear', contribution to the tangential stiffness matrix the first integral of Equation (3.20) results in:

$$EA_0 \int_{\ell_0} (\delta\Delta e)\Delta e \mathrm{d}x = (\delta\Delta\mathbf{u})^{\mathrm{T}} \mathbf{K}_L \Delta\mathbf{u} \tag{3.29}$$

The linear contribution to the tangential stiffness matrix is found by elaborating Equation (3.28) and results in:

$$\mathbf{K}_L = \frac{EA_0}{\ell_0} \begin{bmatrix} \cos^2\phi & \cos\phi\sin\phi & -\cos^2\phi & -\cos\phi\sin\phi \\ \cos\phi\sin\phi & \sin^2\phi & -\cos\phi\sin\phi & -\sin^2\phi \\ -\cos^2\phi & -\cos\phi\sin\phi & \cos^2\phi & \cos\phi\sin\phi \\ -\cos\phi\sin\phi & -\sin^2\phi & \cos\phi\sin\phi & \sin^2\phi \end{bmatrix} \tag{3.30}$$

The second, 'geometric' or 'non-linear', contribution to the tangential stiffness matrix, can be rewritten as:

$$A_0 \int (\delta\Delta\eta)\sigma^t \mathrm{d}x = (\delta\Delta\mathbf{u})^{\mathrm{T}} \mathbf{K}_{NL} \Delta\mathbf{u} \tag{3.31}$$

so that the second part of the tangential stiffness matrix becomes:

$$\mathbf{K}_{NL} = \frac{A_0 \sigma^t}{\ell_0} \begin{bmatrix} 1 & 0 & -1 & 0 \\ 0 & 1 & 0 & -1 \\ -1 & 0 & 1 & 0 \\ 0 & -1 & 0 & 1 \end{bmatrix} \tag{3.32}$$

The contribution (3.32) is very important in geometrically non-linear calculations, since it embodies the destabilising influence in structural members that are subjected to compressive forces.

Substitution of Equations (3.25), (3.29) and (3.31) into Equation (3.20) yields:

$$(\delta\Delta\mathbf{u})^{\mathrm{T}}(\mathbf{K}_L + \mathbf{K}_{NL})\Delta\mathbf{u} = (\delta\Delta\mathbf{u})^{\mathrm{T}}\left(\mathbf{f}_{\mathrm{ext}}^{t+\Delta t} - \mathbf{f}_{\mathrm{int}}^{t}\right) \tag{3.33}$$

Since this equation must hold for any virtual displacement vector, the following system of non-linear algebraic equations is obtained:

$$(\mathbf{K}_L + \mathbf{K}_{NL})\Delta\mathbf{u} = \mathbf{f}_{\mathrm{ext}}^{t+\Delta t} - \mathbf{f}_{\mathrm{int}}^{t} \tag{3.34}$$

In the above approach each material point of a body is monitored. It is a function of time and of the coordinates of that material point in some previous configuration, usually called the reference configuration. While this procedure, which is named a Lagrange or material description, is natural and effective in solid mechanics, this is not so for flow problems, e.g. in fluid flow or in forming processes. Then, the Euler or spatial description is often more appropriate. In this approach a constant volume of the space is considered. Matter may flow into this fixed volume as well as leave it.

3.1.2 Updated Lagrange Formulation

Two alternatives exist within the Lagrange description, namely the Total Lagrange formulation and the Updated Lagrange formulation (Bathe *et al.* 1975). While in the former method all quantities are referred to the original, undeformed configuration, the second approach utilises the configuration at the beginning of a loading step as the reference configuration during that step. For most applications in structural mechanics it is not so important which formulation is adopted. If only geometrical non-linearities are taken into account the differences between both formulations are usually in the order of only a few per cent. When physical non-linearities also play a role, the differences may become larger, but will seldom exceed 5%. Within the class of physically non-linear problems the largest differences occur when path-dependent material models are used. In such models the total deformation state depends on the sequence in which the load has been applied to the structure. Examples of such models are plasticity and damage models. For these models the Updated Lagrange formulation is more appealing, since the physical relevance of the undeformed configuration is then lost. Elasticity, including non-linear elasticity, is an example of a path-independent model, since the strain state for this class of constitutive models is unique for a given stress level, no matter in which order the loading programme has been applied.

In the case of geometrically non-linear behaviour of trusses, the assumptions that $A \approx A_0$ and $\ell \approx \ell_0$ cause both formulations to exactly coincide. To show this we will now derive the tangential stiffness matrix in the updated coordinate system (Updated Lagrange formulation). This is accomplished by introducing the quantity ϕ that sets the angle between the axes of the truss member in the deformed and undeformed configurations (Figure 3.2). For a small load increment the contribution $\bar{\mathbf{K}}_L$ to the tangential stiffness matrix in the updated coordinate system is given by:

$$\bar{\mathbf{K}}_L = \frac{EA_0}{\ell_0}\begin{bmatrix} 1 & 0 & -1 & 0 \\ 0 & 0 & 0 & 0 \\ -1 & 0 & 1 & 0 \\ 0 & 0 & 0 & 0 \end{bmatrix} \tag{3.35}$$

where the bar denotes quantities referred to the rotated or updated coordinate system. The 'linear' contribution to the tangential stiffness matrix in the undeformed configuration, $\mathbf{K}_L$, is now obtained through the transformation

$$\mathbf{K}_L = \mathbf{T}^{\mathrm{T}}\bar{\mathbf{K}}_L\mathbf{T} \tag{3.36}$$

with

$$\mathbf{T} = \begin{bmatrix} \cos\phi & \sin\phi & 0 & 0 \\ -\sin\phi & \cos\phi & 0 & 0 \\ 0 & 0 & \cos\phi & \sin\phi \\ 0 & 0 & -\sin\phi & \cos\phi \end{bmatrix} \tag{3.37}$$

which derives directly from the two-dimensional rotation matrix for vectors, cf. Equation (1.51). Carrying out these multiplications results again in Equation (3.30).

For the geometric contribution to the tangential stiffness matrix a similar result can be obtained. It is straightforward to show rotational invariance of $\mathbf{K}_{NL}$:

$$\mathbf{K}_{NL} = \mathbf{T}^{\mathrm{T}}\mathbf{K}_{NL}\mathbf{T} \tag{3.38}$$

Apparently, it is immaterial in which configuration $\mathbf{K}_{NL}$ is set up. The Total Lagrange and the Updated Lagrange formulations result in exactly the same set of equations for truss elements. For other types of elements, however, the approximations that are made in the course of the derivation can cause small differences between both formulations, which, in turn, may result in minor deviations in the numerical results.

The above statements regarding the equivalence between the Total Lagrange formulation and the Updated Lagrange formulation for truss elements can be further supported by considering again the spring-truss structure of Figure 1.1. We again consider the left half of the structure for symmetry reasons. The boundary conditions of this truss element read: $u_1 = v_1 = u_2 = 0$, and, in view of Equations (3.30) and (3.32), and after adding the spring stiffness k, we obtain

$$\frac{\mathrm{d}F}{\mathrm{d}v} = \frac{EA_0\sin^2\phi}{\ell_0} + \frac{A_0\sigma}{\ell_0} + k$$

which is exactly the stiffness expression of Equation (1.12) when materially non-linear effects are disregarded. The Lagrangian formulation of truss elements derived above thus results in an exact expression for the tangential stiffness.

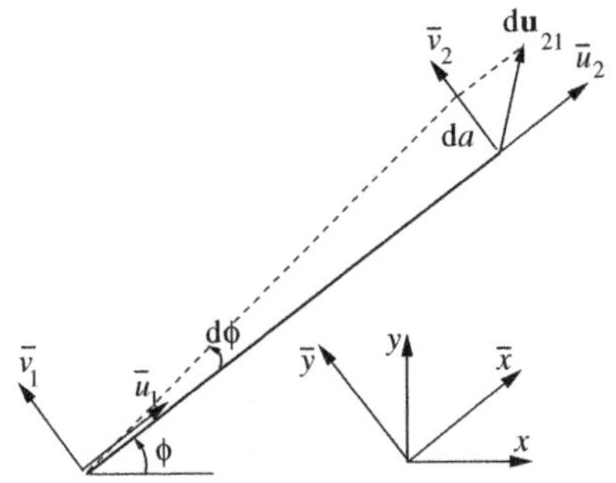

Figure 3.3 Truss element with a corotating $\bar{x}$, $\bar{y}$-reference frame

3.1.3 Corotational Formulation

A further possibility to derive the load vector and the stiffness matrix is to use a coordinate system that is attached to the truss element. This approach is called corotational, since the relevant vectors and matrices are derived in a reference frame which corotates with the truss element. When this coordinate system is named the $\bar{x}$, $\bar{y}$-system, cf. Figure 3.3, the (axial) strain is given by:

$$e = \bar{\mathbf{b}}^{\mathrm{T}} \bar{\mathbf{u}} \tag{3.39}$$

with

$$\bar{\mathbf{b}}^{\mathrm{T}} = \frac{1}{\ell}(-1\,,\ \ 0\,,\ \ 1\,,\ \ 0) \tag{3.40}$$

and with

$$\bar{\mathbf{u}}^{\mathrm{T}} = (\bar{u}_1\,,\ \ \bar{v}_1\,,\ \ \bar{u}_2\,,\ \ \bar{v}_2) \tag{3.41}$$

the vector which assembles the displacement increments in the corotated $\bar{x}$, $\bar{y}$-system. The internal virtual work, i.e. the left-hand side of Equation (3.3), can then be elaborated as:

$$A_0 \int_{\ell_0} \delta(\Delta\epsilon)\sigma^{t+\Delta t}\mathrm{d}x = A_0 \int_{\ell_0} \delta(\bar{\mathbf{b}}^{\mathrm{T}} \Delta\bar{\mathbf{u}})\sigma^{t+\Delta t}\mathrm{d}x \tag{3.42}$$

With the matrix $\mathbf{T}$ defined in Equation (3.37) we can transform this expression to the fixed x, y-coordinate system in a straightforward manner:

$$A_0 \int_{\ell_0} \delta(\Delta\epsilon)\sigma^{t+\Delta t}\mathrm{d}x = A_0 \int_{\ell_0} \delta(\bar{\mathbf{b}}^{\mathrm{T}} \mathbf{T} \Delta\mathbf{u})\sigma^{t+\Delta t}\mathrm{d}x \tag{3.43}$$

With the definitions (3.37) and (3.40) the matrix product $\bar{\mathbf{b}}^{\mathrm{T}}\mathbf{T}$ can be elaborated as:

$$\bar{\mathbf{b}}^{\mathrm{T}}\mathbf{T} = \frac{1}{\ell}(-1\,,\ 0\,,\ +1\,,\ 0)\begin{bmatrix} \cos\phi & \sin\phi & 0 & 0 \\ -\sin\phi & \cos\phi & 0 & 0 \\ 0 & 0 & \cos\phi & \sin\phi \\ 0 & 0 & -\sin\phi & \cos\phi \end{bmatrix} = \frac{1}{\ell}[-\cos\phi \quad -\sin\phi \quad \cos\phi \quad \sin\phi] = \frac{\ell_0}{\ell}\mathbf{b}_L^{\mathrm{T}} \tag{3.44}$$

Using this result, the decomposition of the stress defined in Equation (3.4), the one-dimensional constitutive relation from Equation (3.6), and restricting the use to linear elasticity for simplicity, we subsequently obtain:

$$A_0\int_{\ell_0}\delta(\Delta\epsilon)\sigma^{t+\Delta t}\mathrm{d}x = A_0\int_{\ell_0}\delta\left(\frac{\ell_0}{\ell}\mathbf{b}_L^{\mathrm{T}}\Delta\mathbf{u}\right)(\sigma^t + E\Delta\epsilon)\mathrm{d}x \tag{3.45}$$

or, upon elaboration,

$$\begin{aligned} A_0\int_{\ell_0}\delta(\Delta\epsilon)\sigma^{t+\Delta t}\mathrm{d}x &= A_0\int_{\ell_0}\delta\left(\frac{\ell_0}{\ell}\right)\mathbf{b}_L^{\mathrm{T}}\Delta\mathbf{u}(\sigma^t + E\Delta\epsilon)\mathrm{d}x + \\ A_0\int_{\ell_0}\frac{\ell_0}{\ell}\delta\mathbf{b}_L^{\mathrm{T}}\Delta\mathbf{u}(\sigma^t + E\Delta\epsilon)\mathrm{d}x &+ A_0\int_{\ell_0}\frac{\ell_0}{\ell}\mathbf{b}_L^{\mathrm{T}}\delta(\Delta\mathbf{u})(\sigma^t + E\Delta\epsilon)\mathrm{d}x \end{aligned} \tag{3.46}$$

As with the derivation for the Total and Updated Lagrange formulations, terms that are zero order in the displacement increments result in the internal force vector. This holds for the product of $\frac{\ell_0}{\ell}\mathbf{b}_L^{\mathrm{T}}\delta(\Delta\mathbf{u})$ and σ^t, which can be elaborated as:

$$A_0\int_{\ell_0}\frac{\ell_0}{\ell}\mathbf{b}_L^{\mathrm{T}}\delta(\Delta\mathbf{u})\sigma^t\mathrm{d}x = A_0\frac{\ell_0^2}{\ell}\mathbf{b}_L^{\mathrm{T}}\delta(\Delta\mathbf{u})\sigma^t = A_0\frac{\ell_0^2}{\ell}\delta(\Delta\mathbf{u})^{\mathrm{T}}\mathbf{b}_L\sigma^t \tag{3.47}$$

so that the internal force vector becomes:

$$\mathbf{f}_{\mathrm{int}}^t = A_0\frac{\ell_0^2}{\ell}\mathbf{b}_L\sigma^t \approx A_0\ell_0\mathbf{b}_L\sigma^t \tag{3.48}$$

which, under the assumption that $\ell \approx \ell_0$, is identical to the internal force vector derived in the Lagrange approaches.

The terms $\frac{\ell_0}{\ell}\mathbf{b}_L^{\mathrm{T}}\delta(\Delta\mathbf{u})E\Delta\epsilon$, $\frac{\ell_0}{\ell}\delta\mathbf{b}_L^{\mathrm{T}}\Delta\mathbf{u}\sigma^t$, and $\delta\left(\frac{\ell_0}{\ell}\right)\mathbf{b}_L^{\mathrm{T}}\Delta\mathbf{u}\sigma^t$ are all linear in the displacement increment and contribute to the tangential stiffness matrix, while the remaining terms are of higher order in terms of the displacement increments and cancel in the linearisation process. The first term contributing to the tangential stiffness can be elaborated as:

$$A_0\int\frac{\ell_0}{\ell}\mathbf{b}_L^{\mathrm{T}}\delta(\Delta\mathbf{u})E\Delta\epsilon\mathrm{d}x = \frac{EA_0\ell_0^3}{\ell^2}\delta(\Delta\mathbf{u})^{\mathrm{T}}\mathbf{b}_L\mathbf{b}_L^{\mathrm{T}}\Delta\mathbf{u} \tag{3.49}$$

The contribution to the tangential stiffness matrix that results from this term therefore equals:

$$\mathbf{K}_L = \frac{EA_0\ell_0^3}{\ell^2}\mathbf{B}_L^{\mathrm{T}}\mathbf{B}_L \approx EA_0\ell_0\mathbf{b}_L\mathbf{b}_L^{\mathrm{T}}$$

and is also equal to that derived in the Lagrange approaches. Next, the second term of Equation (3.46) is elaborated as:

$$A_0 \int_{\ell_0} \frac{\ell_0}{\ell} \delta \mathbf{b}_L^{\mathrm{T}} \Delta \mathbf{u} \sigma^t \mathrm{d}x = \frac{A_0 \sigma^t \ell_0^2}{\ell} \delta(\Delta \mathbf{u})^{\mathrm{T}} \left(\frac{\partial \mathbf{b}_L^{\mathrm{T}}}{\partial \mathbf{u}} \right) \Delta \mathbf{u} \tag{3.50}$$

Straightforward differentiation of $\mathbf{b}_L^{\mathrm{T}}$, cf. Equation (3.21), leads to:

$$\frac{\partial \mathbf{b}_L^{\mathrm{T}}}{\partial \mathbf{u}} = \frac{1}{\ell_0^2} \begin{bmatrix} 1 & 0 & -1 & 0 \\ 0 & 1 & 0 & -1 \\ -1 & 0 & 1 & 0 \\ 0 & -1 & 0 & 1 \end{bmatrix}$$

Using $\ell \approx \ell_0$ we then derive:

$$\mathbf{K}_{NL} = \frac{A_0 \sigma^t}{\ell_0} \begin{bmatrix} 1 & 0 & -1 & 0 \\ 0 & 1 & 0 & -1 \\ -1 & 0 & 1 & 0 \\ 0 & -1 & 0 & 1 \end{bmatrix}$$

as the geometric contribution to the tangential stiffness matrix, which also equals the expression obtained using the Lagrange approaches, Equation (3.32). Finally, the third term can be elaborated as follows:

$$\begin{aligned} A_0 \int \delta \left(\frac{\ell_0}{\ell} \right) \mathbf{b}_L^{\mathrm{T}} \Delta \mathbf{u} \sigma^t \mathrm{d}x = -\frac{A_0 \sigma^t \ell_0^2 \delta \ell}{\ell^2} \mathbf{b}_L^{\mathrm{T}} \Delta \mathbf{u} = \\ -\frac{A_0 \sigma^t \ell_0^2}{\ell^2} \delta(\Delta \mathbf{u})^{\mathrm{T}} \left(\frac{\partial \ell}{\partial \mathbf{u}} \right)^{\mathrm{T}} \mathbf{b}_L^{\mathrm{T}} \Delta \mathbf{u} = -\frac{A_0 \sigma^t \ell_0^4}{\ell^3} \delta(\Delta \mathbf{u})^{\mathrm{T}} \mathbf{b}_L \mathbf{b}_L^{\mathrm{T}} \Delta \mathbf{u} \end{aligned} \tag{3.51}$$

where the definition of ℓ, Equation (3.8), has been used to derive that

$$\frac{\partial \ell}{\partial \mathbf{u}} = \frac{\ell_0^2}{\ell} \mathbf{b}_L^{\mathrm{T}}$$

The third contribution to the tangential stiffness matrix thus reads:

$$\mathbf{K}_{NL}^{cr} = -\frac{A_0 \sigma^t \ell_0^4}{\ell^3} \mathbf{b}_L \mathbf{b}_L^{\mathrm{T}} \approx -A_0 \sigma^t \ell_0 \mathbf{b}_L \mathbf{b}_L^{\mathrm{T}} \tag{3.52}$$

or after elaboration

$$\mathbf{K}_{NL}^{cr} = -\frac{A_0 \sigma^t}{\ell_0} \begin{bmatrix} \cos^2 \phi & \cos \phi \sin \phi & -\cos^2 \phi & -\cos \phi \sin \phi \\ \cos \phi \sin \phi & \sin^2 \phi & -\cos \phi \sin \phi & -\sin^2 \phi \\ -\cos^2 \phi & -\cos \phi \sin \phi & \cos^2 \phi & \cos \phi \sin \phi \\ -\cos \phi \sin \phi & -\sin^2 \phi & \cos \phi \sin \phi & \sin^2 \phi \end{bmatrix} \tag{3.53}$$

This contribution is absent in the tangential stiffness matrix when adopting the Lagrange approaches. An alternative derivation of the tangential stiffness expression for truss elements using a corotational formulation is given in Box 3.2.

Box 3.2 Alternative derivation tangential stiffness matrix corotational formulation

In an engineering approach the strain in the truss element can be defined as $\epsilon = \frac{\bar{u}_2 - \bar{u}_1}{\ell_0}$, so that instead of Equation (3.40), we now have $\bar{\mathbf{b}}^{\mathrm{T}} = \frac{1}{\ell_0}(-1,\ 0,\ 1,\ 0)$, and the expression for the virtual work is elaborated as:

$$A_0 \int_{\ell_0} \delta(\Delta\epsilon)\sigma^{t+\Delta t}\mathrm{d}x = A_0 \int_{\ell_0} \bar{\mathbf{b}}^{\mathrm{T}}(\mathbf{T}\delta(\Delta\mathbf{u}) + \delta\mathbf{T}\Delta\mathbf{u})(\sigma^t + E\Delta\epsilon)\mathrm{d}x$$

The terms that are linear in the displacement increments, and thus contribute to the tangential stiffness matrix are $\bar{\mathbf{b}}^{\mathrm{T}}\mathbf{T}\delta(\Delta\mathbf{u})E\Delta\epsilon$ and $\bar{\mathbf{b}}^{\mathrm{T}}\delta\mathbf{T}\Delta\mathbf{u}\sigma^t$. The first term results in the contribution $\mathbf{K}_L$ to the tangential stiffness matrix, cf. Equation (3.28). To elaborate the second term we first observe that:

$$\delta\mathbf{T} = \frac{\partial\mathbf{T}}{\partial\phi}\delta\phi \quad \text{with} \quad \frac{\partial\mathbf{T}}{\partial\phi} = \begin{bmatrix} -\sin\phi & \cos\phi & 0 & 0 \\ -\cos\phi & -\sin\phi & 0 & 0 \\ 0 & 0 & -\sin\phi & \cos\phi \\ 0 & 0 & -\cos\phi & -\sin\phi \end{bmatrix}$$

From Figure 3.3 we have

$$\mathrm{d}\phi \approx \frac{\mathrm{d}a}{\ell} = \frac{\mathbf{n}^{\mathrm{T}}\mathrm{d}\mathbf{u}_{21}}{\ell} = \frac{1}{\ell}[-\sin\phi \quad \cos\phi]\begin{bmatrix} -1 & 0 & 1 & 0 \\ 0 & -1 & 0 & 1 \end{bmatrix}\mathrm{d}\mathbf{u}$$

whence

$$\frac{\partial\phi}{\partial\mathbf{u}} = \frac{1}{\ell}[\sin\phi \quad -\cos\phi \quad -\sin\phi \quad \cos\phi]$$

and the second term which is linear in the displacement increments can be written as:

$$A_0 \int_{\ell_0} \bar{\mathbf{b}}^{\mathrm{T}}\delta\mathbf{T}\Delta\mathbf{u}\sigma^t\mathrm{d}x = A_0\ell_0\sigma^t\delta(\Delta\mathbf{u})^{\mathrm{T}}\left(\frac{\partial\phi}{\partial\mathbf{u}}\right)^{\mathrm{T}}\bar{\mathbf{b}}^{\mathrm{T}}\frac{\partial\mathbf{T}}{\partial\phi}\Delta\mathbf{u}$$

With $\ell \approx \ell_0$ the second contribution to the tangential stiffness matrix then becomes:

$$\mathbf{K} = \frac{A_0\sigma^t}{\ell_0}\begin{bmatrix} \sin^2\phi & -\cos\phi\sin\phi & -\sin^2\phi & \cos\phi\sin\phi \\ -\cos\phi\sin\phi & \cos^2\phi & \cos\phi\sin\phi & -\cos^2\phi \\ -\sin^2\phi & \cos\phi\sin\phi & \sin^2\phi & -\cos\phi\sin\phi \\ \cos\phi\sin\phi & -\cos^2\phi & -\cos\phi\sin\phi & \cos^2\phi \end{bmatrix}$$

which exactly equals the sum of $\mathbf{K}_{NL}$ and $\mathbf{K}_{NL}^{cr}$, Equations (3.32) and (3.53).

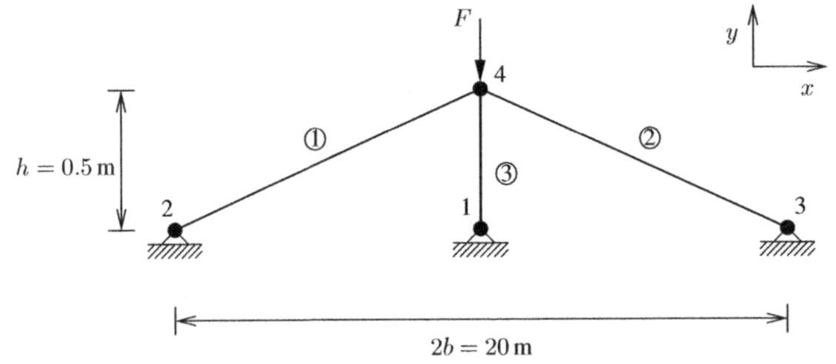

Figure 3.4 Finite element model of the non-linear shallow truss problem presented in Section 1.1. The elements 1 and 2 are non-linear trusses with a Young's modulus $E = 10^5$ N/m^2 and a cross-section $A_0 = 1.0$ m^2. Element 3 is a linear spring with a spring constant $k = 2000.0$ N/m.

3.2 PyFEM: The Shallow Truss Problem

In this section, the plane shallow truss problem presented in Section 1.1 is solved by means of a finite element simulation. The finite element discretisation of the problem is shown in Figure 3.4. The model consists of three elements which are supported by four nodes. The elements 1 and 2 are non-linear truss elements, which have been derived in the previous section. Element 3 is a linear spring. Nodes 1, 2 and 3 are constrained in the x- and in the y-direction. Hence, node 4 is the only 'free' node.

The finite element code for this problem is called `ShallowTrussFE.py` and can be found in the directory `examples/ch3`. It has the following lay-out:

⟨*Finite element shallow truss structure example* ⟩≡

⟨*Initialisation of the finite element model* 77⟩
⟨*Non-linear solution procedure* 80⟩
⟨*Post-processing the results*⟩

First, the finite element model is described in the fragment ⟨*Initialisation of the finite element model*⟩. This includes a description of the geometry, of the boundary conditions, and of the material parameters for each element. The simulation is carried out in the fragment ⟨*Non-linear solution procedure*⟩, and the results are plotted in the form of a load–displacement curve in fragment ⟨*Post-processing the results*⟩.

To be able to handle large finite element meshes, that contain many nodes, elements and degrees of freedom, we will introduce a number of new classes to store the information. The first new class is the `Properties` class, which will be used to store the material parameters of the elements in the model.

⟨*Initialisation of the finite element model* ⟩≡ 76

```
from pyfem.util.dataStructures import Properties

props = Properties()
props.TrussElem  = Properties( { 'type' : 'Truss'  , 'E' : 5.0e6 , \
                                 'Area' : 1.0 } )
props.SpringElem = Properties( { 'type' : 'Spring' , 'k' : 2000. } )
```

The `Properties` class is derived from the standard Python database. The implementation can be found in the file `Properties.py` in the directory `pyfem/util`. In this fragment, the empty instance `props` is first created. Next, two members are added, `TrussElem` and `SpringElem`, which contain the material and geometry parameters of the truss and of the spring elements, respectively. Both members contain the mandatory argument `'type'` which refers to the element type. Here, we use the element types `'Truss'` and `'Spring'`. The other arguments depend on the specific element type. The truss element requires a Young's modulus `E` and a cross-sectional area `Area` which are set equal to `5.0e6` and `1.0`, respectively. These parameters will be used in the calculation of the element stiffness matrix $\mathbf{K}_e$ and of the internal force vector $\mathbf{f}_{\text{int}}$, see page 83. The spring element has one parameter: the spring constant `k` which is set equal to `2000`.

Even though the shallow truss example consists of a small number of nodes, we will store the position of the nodes in an instance `nodes` of the class `NodeSet`. The implementation can be found in the file `NodeSet.py` in the directory `pyfem/fem`.

⟨*Initialisation of the finite element model* ⟩+≡ 77

```
from pyfem.fem.NodeSet import NodeSet

nodes = NodeSet()

nodes.add( 1 , [ 0.   , 0. ] )
nodes.add( 2 , [-10.0, 0. ] )
nodes.add( 3 , [ 10.0, 0. ] )
nodes.add( 4 , [ 0.   , 0.5] )
```

A new node can be added to this instance using the member function `add`. The first argument of this function is the node identification number, where the numbering does not have to be continuous. Internally, the node identification number is mapped to an alternative numbering that starts at zero, and is continuous. The second argument is a one-dimensional array that contains the global coordinate of the node. The length of this array can be one, two or three, depending on the spatial dimensions of the problem. In this two-dimensional example, the length of this array obviously equals two.

The element connectivity is stored in the instance `elements` of the class `ElementSet`:

⟨*Initialisation of the finite element model* ⟩+≡ 77

```
from pyfem.fem.ElementSet import ElementSet

elements = ElementSet( nodes , props )

elements.add( 1, 'TrussElem'  , [2,4]  )
elements.add( 2, 'TrussElem'  , [3,4]  )
elements.add( 3, 'SpringElem' , [1,4]  )
```

The constructor of the class `ElementSet` has two arguments: an instance of the class `NodeSet` and an instance of `Properties` that contains the parameters of the various models. Individual elements can be added by using the member function `add`, which has three arguments. The first argument is the element identification number. Similar to the node numbering, the element numbering does not have to be continuous. The second argument is a string that corresponds to the model properties. In this example, the element numbers 1 and 2 are of the type `'TrussElem'`, which has been defined previously. The final argument is an array of integers which contains the identification numbers of the nodes that support this element. The length of this array depends on the element type. In this example, the truss as well as the spring elements are supported by two nodes, hence, the length of this integer array equals two.

When the nodes and the elements have been added, the instance `dofs` of the class `DofSpace` can be created. This instance contains a database in which the connection of the global degrees of freedom and the corresponding nodes are stored.

⟨*Initialisation of the finite element model* ⟩+≡ 78

```
from pyfem.fem.DofSpace import DofSpace

dofs = DofSpace( elements )

dofs.constrain( 1, ['u','v'] )
dofs.constrain( 2, ['u','v'] )
dofs.constrain( 3, ['u','v'] )
```

The constructor of the `DofSpace` class has a single argument of the type `ElementSet`. During the initialisation of the corresponding object `dofs`, the constructor checks which elements have been added to `elements`, which nodes are connected to those elements, and which degrees of freedom must be assigned to those nodes. Boundary conditions can be added to the class by means of the function `constrain`. This function has two arguments: the node identification number and the degrees of freedom of that node, which must be set to zero. In this fragment, the displacements in the x- and in the y-direction, `'u'` and `'v'`, of the nodes 1, 2 and 3 are constrained (Figure 3.4). The function `constrain` will signal when a combination of nodes and nodal degree of freedom does not exist.

The three instances `nodes`, `elements` and `dofs` are stored in a global database `globdat` in order to provide access in other parts of the program:

⟨*Initialisation of the finite element model* ⟩+≡ 78

```
from pyfem.util.dataStructures import GlobalData

globdat = GlobalData( nodes, elements, dofs )
```

During the initialisation of the instance `globdat` the global solution vector **a** and the incremental solution vector $\Delta\mathbf{a}$ are created as the members `state` and `Dstate`, respectively. The length of these arrays is identical to the number of degrees of freedom in `dofs`.

When the total number-of-degrees of freedom is known, the external force vector can be created:

⟨*Initialisation of the finite element model* ⟩+≡ 79

```
loadDof = dofs.getForType(4,'v')
Dfext   = zeros( len(dofs) )
Dfext[loadDof] = -100.
```

In the present example, the load acts on node 4 in the negative y-direction. The member function `getForType` is used to obtain the global degree of freedom identification number from the instance `dofs`. This identification number is stored as the variable `loadDof`. The length of the external force vector evidently equals the total number of degrees of freedom. It contains zeros, except for the position that corresponds to the degree of freedom that represents the y-displacement of node 4. We adopt a value of `-100` for the external load increment.

The non-linear solver is controlled by three variables: `N`, which gives the number of load steps in the simulation, `tol`, which sets the error tolerance in the convergence criterion, and `iterMax`, which bounds the maximum number of iterations. These variables are set in the next lines of the code.

⟨*Initialisation of the finite element model* ⟩+≡ 79

```
N       = 30
tol     = 1e-6
iterMax = 10
```

Finally, the temporary vectors `a` and `Da` are created as copies of the state vectors in the global database, and an array is initialised that represents the external force:

⟨*Nonlinear solution procedure* ⟩+≡ 79

```
a    = globdat.state
Da   = globdat.Dstate

fext = zeros( len(dofs) )
```

Upon initialisation the model can be solved by means of a Newton–Raphson method, where the procedure described in Box 2.3 has been implemented.

⟨*Non-linear solution procedure* ⟩≡ 76

```
from pyfem.fem.Assembly import assembleStiffness

for iCyc in range(N):
  fext += Dfext

  error = 1.
  iiter = 0

  K,fint = assembleStiffness( props, globdat )

  ⟨Solve non-linear system of equations 80⟩
```

In this fragment, we loop over `N` steps using `iCyc` as the counter. First, the external force vector `fext` for the current step is determined. Since we use a constant step size, the external force vector is simply increased by the external force increment `Dfext`. The tangential stiffness matrix `K` is computed using the `assembleStiffness` function. The arguments of this function are the instance `props`, which contains the element parameters, `globdat`, which contains the topology of the finite element mesh, and the solution vectors `state` and `Dstate`. The function calculates the element stiffness matrices $\mathbf{K}_e$ for all elements and assembles the total stiffness matrix, as shown in Equations (2.43). As we will see later, it is highly efficient to compute the internal force vector `fint` in the same procedure. Therefore, the function returns the tangential stiffness matrix and the internal force vector `fint`.

The non-linear system of equations is solved in the fragment ⟨*Solve non-linear system of equations*⟩:

⟨*Solve non-linear system of equations* ⟩≡ 80

```
  while error > tol:
    iiter +=1
    da = dofs.solve( K, fext-fint )

    Da[:] += da[:]
    a [:] += da[:]

    K,fint = assembleStiffness( props, globdat )
```

The solution is determined iteratively. The iteration loop continues as long as the error exceeds the tolerance `tol`. The function `solve` is a member of the `DofSpace` class. This function reduces the global system of equations by removing the constrained degrees of freedom, similar to the procedure shown in the fragment ⟨*Solution procedure*⟩ on page 58. The incremental displacement vector `Da` and the total displacement vector `a` are updated, and the new internal force vector and the new tangential stiffness matrix are computed.

The norm of the residual vector is determined next:

```
⟨Solve non-linear system of equations ⟩+≡                              80
    error   = dofs.norm( fext-fint )

    if iiter == iterMax:
      raise RuntimeError('Iterations did not converge!')
```

The function `norm` determines the L_2-norm of the residual vector and is equally well a member of the `DofSpace` class, since the constrained degrees of freedom terms must be removed from the residual vector. When the norm is smaller than the tolerance, the **`while`** loop is discontinued, and the displacement increment `Da` is reset to zero:

```
⟨Solve non-linear system of equations ⟩+≡                              81
  Da[:] = zeros( len(dofs) )

  elements.commitHistory()
```

The element history is stored in the function `elements.commitHistory()`, i.e. the history parameters are replaced by their new values. Finally, the results of the simulation are plotted in the fragment ⟨*Post-processing the results*⟩. The code of this fragment is similar that of the fragment ⟨*Print results*⟩ on page 28.

The function call `assembleStiffness` collects the tangential stiffness matrices and the internal force vectors of each element, and assembles the system stiffness matrix and the internal force vector. It will be shown here for the truss element. We will confine our attention to the Total Lagrange implementation for a non-linear truss element, see Section 3.1.1.

The complete implementation of the truss element can be found in the file `Truss.py` in the directory `pyfem/elements`. The structure of the code is as follows:

```
⟨Truss element ⟩≡
  ⟨Truss class definition 81⟩
  ⟨Truss class main functions 82⟩
  ⟨Truss class utility functions 83⟩
```

The truss element is implemented as a class, which is derived from the base class `Element`:

```
⟨Truss element class definition ⟩≡                                     81
  class Truss ( Element ):

    dofTypes = ['u','v']
```

In the class definition the types of degrees of freedom that are attached to this element have to be defined. The two-dimensional truss element has two degrees of freedoms per node: a displacement u in the x-direction, and a displacement v in the y-direction. In the code, these nodal degrees of freedom types are stored as a tuple of length two: `['u','v']`. The tuple `dofTypes` is used by the `DofSpace` class to construct the global solution space in which the corresponding degrees of freedom are assigned to each node.

Additional data for the element are initialised in the `__init__` function of the class:

⟨*Truss class definition* ⟩≡ 81

```
def __init__ ( self, elnodes , props ):
  Element.__init__( self, elnodes , props )

  self.setHistory( 'sigma0', 0. )
  self.commitHistory()

  self.l0   = 0.
```

In the constructor of the base class `Element`, the element properties are read from the instance `props` and are stored as members of the class. The truss element contains two parameters, which are initialised as the members `E` and `Area`. The element has a single history parameter `sigma0`, which is initialised in this fragment as well. This history parameter represents the stress in the element in the previous step after convergence. It is also useful to initialise the variable `l0`, which represents the current length of the element as a member of the class.

For each element a number of tasks have to be executed. One of these tasks is to construct the element stiffness matrix $\mathbf{K}_e$ and the internal force vector $\mathbf{f}_{\text{int}}$. This is done in the function `getElementStiffness` which is used by the assembler function `assembleStiffness` to construct the global and the internal force vector, see also the fragment ⟨*Non-linear solution procedure*⟩ on page 80.

⟨*Truss class main fuctions* ⟩≡ 81

```
def getElementStiffness ( self, elemdat ):

  from pyfem.utils.transformations import glob2Elem

  a  = glob2Elem( elemdat.state , elemdat.coords )
  Da = glob2Elem( elemdat.Dstate, elemdat.coords )
  a0 = a - Da

  self.l0 = norm( elemdat.coords[1]-elemdat.coords[0] )

  epsilon , Depsilon = self.getStrain( a , a0 )
```

83

The function `getElementStiffness` has a single argument `elemdat`, which is an instance that contains the element data such as the nodal coordinates and the material parameters. The output of this routine, the element stiffness matrix and the internal force vector are stored as members of this class. To construct the stiffness matrix and the internal force vector, the nodal displacements must be transformed to the local frame of reference of the element. This transformation is carried out by the function `glob2Elem`. The resulting total and incremental local displacement fields are stored in the arrays `a` and `Da`, respectively. The length of the undeformed element `l0` is equal to the norm of the vector that connects the positions of the two nodes of the element. The strain and the strain increment can be determined using the local displacement arrays `a` and `a0`. This is done in the member function `getStrain`, see also Equation (3.13).

⟨*Truss class utility fuctions* ⟩≡ 81

```
def getStrain( self , a , a0 ):

  epsilon  = (a[2]-a[0])/self.l0
  epsilon += 0.5*((a[2]-a[0])/self.l0)**2
  epsilon += 0.5*((a[3]-a[1])/self.l0)**2

  epsilon0  = (a0[2]-a0[0])/self.l0 +
  epsilon0 += 0.5*((a0[2]-a0[0])/self.l0)**2
  epsilon0 += 0.5*((a0[3]-a0[1])/self.l0)**2

  Depsilon = epsilon -epsilon0

  return epsilon,Depsilon
```

The stresses are obtained by multiplying the strain increment by the Young's modulus `E` and adding the result to the stress in the previous state, see also Equations (3.4) and (3.6):

⟨*Truss class main fuctions* ⟩+≡ 82

```
Dsigma = self.E * Depsilon
sigma  = self.getHistory('sigma') + Dsigma

self.setHistory( 'sigma', sigma )
```

Note that the current value of `sigma` is set as the new history value. When the solution has converged, this value will serve as the old history parameter.

The linear and non-linear parts of the stiffness matrix, `KL` and `KNL`, respectively, are calculated according to Equations (3.28) and (3.32):

⟨*Truss class main fuctions* ⟩+≡ 83

```
    BL = self.getBL( a )
    KL = self.E * self.Area * self.l0 * outer( BL , BL )           84

    KNL = self.getKNL( sigmaA , self.Area )

    elStiff = KL + KNL

    elemdat.stiff = elem2Glob( elStiff , elemdat.coords )
```

The total stiffness matrix is transformed back to the global frame of reference using the function `elem2Glob`. The vector `BL` that is needed to construct the linear part of the stiffness matrix is calculated in the member function `getBL`, see also Equation (3.22):

⟨*Truss element utility fuctions* ⟩+≡ 83

```
  def getBL( self , a ):

    BL    = zeros( 4 )
    BL[0] = (-1./self.l0)*(1.+(a[2]-a[0])/self.l0)
    BL[1] = (-1./self.l0)*(a[3]-a[1])/self.l0
    BL[2] = -BL[0]
    BL[3] = -BL[1]

    return BL
```

Since the components of the internal force vector are already available, it is efficient to calculate this vector concurrently, see also Equation (3.26):

⟨*Truss element main fuctions* ⟩+≡ 84

```
    elFint = self.l0 * sigma * elemdat.props.Area * BL

    elemdat.fint = elem2Glob( elFint , elemdat.coords )
```

Each element in the code follows the same structure. In addition to the function `getStiffness`, two other functions are mentioned. In the function `getIntForce`, the internal force vector of an element is calculated. This function is useful in solution techniques in which the global stiffness matrix is not required, for example in explicit time integration. Another important member function is `getMass`, which constructs the element mass matrix that is used in dynamic calculations, see Chapter 5.

3.3 Stress and Deformation Measures in Continua

The formulation of tangential stiffness matrices and the associated consistent internal force vectors requires much more care in continuous media than for simple truss elements. In particular, stress and strain measures must be defined in an unambiguous, but physically meaningful manner. The most straightforward and elegant procedure is to consider the total motion of an elementary, originally square material element to be the product of a translation, a rigid rotation and a pure deformation. We suppose that the elementary particle occupies a position $\boldsymbol{\xi}$ in space in the reference configuration (usually the undeformed configuration), with the so-called material coordinates (ξ_1, ξ_2, ξ_3), and a position $\mathbf{x}$, with so-called spatial coordinates (x_1, x_2, x_3), in the deformed configuration. In line with the notions of Lagrange and Euler descriptions introduced for large deformations of truss elements the former coordinate set is also denoted by the terminology Lagrange coordinates, while the latter coordinate set is also named Euler coordinates. Clearly: $\mathbf{x} = \mathbf{x}(\boldsymbol{\xi})$.

The translation can be eliminated from the total motion through differentiation. Differentiating the spatial coordinates with respect to the material coordinates results in the deformation gradient tensor

$$\mathbf{F} = \frac{\partial \mathbf{x}}{\partial \boldsymbol{\xi}} \tag{3.54}$$

The deformation gradient, which is a second-order tensor, gives a complete description of the motion of an elementary particle up to a (rigid) translation. To proceed we consider the elementary quadrilateral element of Figure 3.5. The total deformation is decomposed into a pure deformation $\mathbf{U}$ – from configuration A to configuration B – and a rigid-body rotation $\mathbf{R}$ – from configuration B to configuration C. A line element $\mathrm{d}\boldsymbol{\xi}$ which connects two material points that are only a small, infinitesimal distance apart in the reference configuration, is transformed to a line element $\mathbf{dx}$ in the current configuration via position $\mathrm{d}\boldsymbol{\eta}$ in an intermediate configuration B. Then, $\mathrm{d}\mathbf{x}$ and $\mathrm{d}\boldsymbol{\eta}$ are related through

$$\mathrm{d}\mathbf{x} = \mathbf{R} \cdot \mathrm{d}\boldsymbol{\eta} \tag{3.55}$$

while the relationship between $\mathrm{d}\boldsymbol{\eta}$ and $\mathrm{d}\boldsymbol{\xi}$ is given by

$$\mathrm{d}\boldsymbol{\eta} = \mathbf{U} \cdot \mathrm{d}\boldsymbol{\xi} \tag{3.56}$$

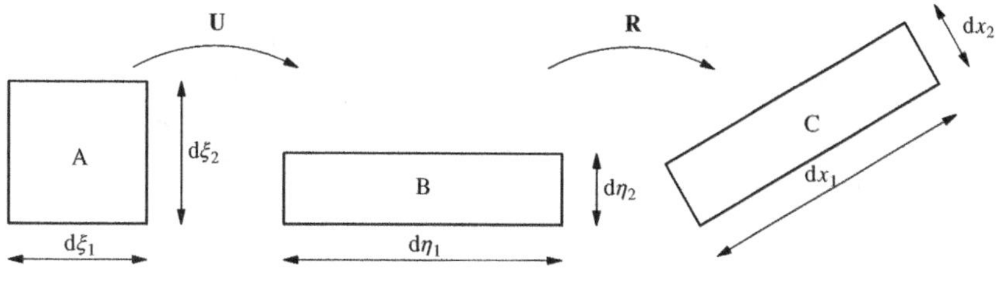

Figure 3.5 Graphical interpretation of polar decomposition: pure deformation followed by a rigid rotation

Combination of Equations (3.55) and (3.56) results in the relation

$$\mathbf{dx} = \mathbf{R} \cdot \mathrm{d}\boldsymbol{\eta} = \mathbf{R} \cdot \mathbf{U} \cdot \mathrm{d}\boldsymbol{\xi} \tag{3.57}$$

and a comparison with Equation (3.54) shows that the deformation gradient **F** can be decomposed into a pure rotation **R** and a pure deformation **U** in a multiplicative fashion:

$$\mathbf{F} = \mathbf{R} \cdot \mathbf{U} \tag{3.58}$$

This decomposition is commonly named the polar decomposition of the deformation gradient.

The right stretch tensor **U** is less convenient as a deformation measure. Its calculation involves expensive square root evaluations. A deformation measure that is easier to compute is the right Cauchy–Green deformation tensor **C**. This tensor is defined by first taking the difference between the squares of the lengths of $\mathrm{d}\boldsymbol{\xi}$ and $\mathrm{d}\mathbf{x}$:

$$\mathrm{d}\mathbf{x} \cdot \mathbf{dx} - \mathrm{d}\boldsymbol{\xi} \cdot \mathrm{d}\boldsymbol{\xi} = (\mathbf{F} \cdot \mathrm{d}\boldsymbol{\xi}) \cdot (\mathbf{F} \cdot \mathrm{d}\boldsymbol{\xi}) - \mathrm{d}\boldsymbol{\xi} \cdot \mathrm{d}\boldsymbol{\xi} = \mathrm{d}\boldsymbol{\xi} \cdot (\mathbf{F}^{\mathrm{T}} \cdot \mathbf{F} - \mathbf{I}) \cdot \mathrm{d}\boldsymbol{\xi} \tag{3.59}$$

Since the right stretch tensor **U** is a measure of a pure deformation, the right Cauchy–Green deformation tensor

$$\mathbf{C} = \mathbf{F}^{\mathrm{T}} \cdot \mathbf{F} = \mathbf{U}^2 \tag{3.60}$$

also completely defines the state of deformation, and the effects of rigid-body rotations are eliminated. The Green–Lagrange strain tensor is then defined as:

$$\boldsymbol{\gamma} = \frac{1}{2}(\mathbf{C} - \mathbf{I}) \tag{3.61}$$

Equations (3.54), (3.60) and (3.61) show that the Green–Lagrange tensor $\boldsymbol{\gamma}$ is referred to the original, undeformed configuration. It is a deformation measure that is used within the framework of a Lagrange description of the motion of a body, and reduces to the small strain tensor $\boldsymbol{\epsilon}$ for small displacement gradients.

Alternatively, the deformation process can be considered as a sequence of a rigid-body rotation **R** – from configuration A to configuration B in Figure 3.6 – and a pure deformation **V** – from configuration B to configuration C. Now, $\mathrm{d}\mathbf{x}$ and $\mathrm{d}\boldsymbol{\eta}$ are related through:

$$\mathrm{d}\mathbf{x} = \mathbf{V} \cdot \mathrm{d}\boldsymbol{\eta} \tag{3.62}$$

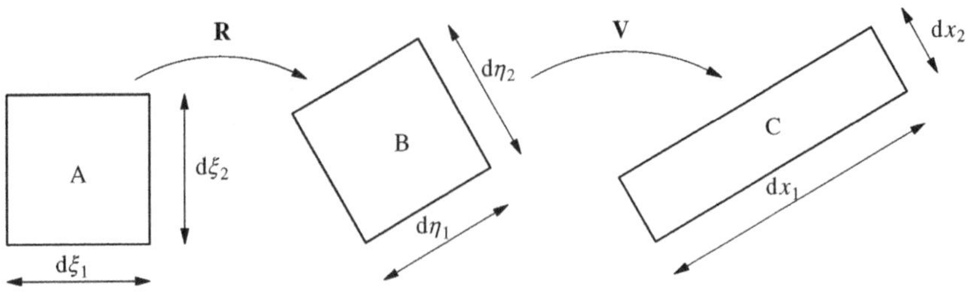

Figure 3.6 Graphical interpretation of polar decomposition: rigid rotation followed by a pure deformation

while the relationship between $\mathrm{d}\boldsymbol{\eta}$ and $\mathrm{d}\boldsymbol{\xi}$ is given by

$$\mathrm{d}\boldsymbol{\eta} = \mathbf{R} \cdot \mathrm{d}\boldsymbol{\xi} \tag{3.63}$$

Combination of Equations (3.62) and (3.63) results in the relation

$$\mathrm{d}\mathbf{x} = \mathbf{V} \cdot \mathrm{d}\boldsymbol{\eta} = \mathbf{V} \cdot \mathbf{R} \cdot \mathrm{d}\boldsymbol{\xi} \tag{3.64}$$

and a comparison with Equation (3.54) shows that the deformation gradient $\mathbf{F}$ can equivalently be decomposed multiplicatively into a pure deformation $\mathbf{V}$ and a rigid rotation $\mathbf{R}$:

$$\mathbf{F} = \mathbf{V} \cdot \mathbf{R} \tag{3.65}$$

Like the right stretch tensor $\mathbf{U}$ the left stretch tensor $\mathbf{V}$ is less convenient to use, and for this reason the left Cauchy–Green deformation tensor $\mathbf{B}$ – also sometimes called the Finger deformation tensor – is introduced in a manner similar to the right Cauchy–Green deformation tensor:

$$\mathbf{B} = \mathbf{F} \cdot \mathbf{F}^{\mathrm{T}} = \mathbf{V}^2 \tag{3.66}$$

In the remainder of this chapter we will only use the right Cauchy–Green deformation tensor and the Green–Lagrange strain tensor. We will come back to the left Cauchy–Green deformation tensor when discussing large strains, where it is useful in some formulations.

It is customary to use the displacement vector $\mathbf{u}$ rather than the vector $\mathbf{x}$ which contains the spatial coordinates of a material point in the deformed configuration. Clearly, $\mathbf{x} = \boldsymbol{\xi} + \mathbf{u}$, and we obtain

$$\mathbf{F} = \mathbf{I} + \frac{\partial \mathbf{u}}{\partial \boldsymbol{\xi}} \tag{3.67}$$

or, in component form

$$F_{ij} = \frac{\partial(\xi_i + u_i)}{\partial \xi_j} = \delta_{ij} + \frac{\partial u_i}{\partial \xi_j} \tag{3.68}$$

The components of the Green–Lagrange strain tensor thus read:

$$\gamma_{ij} = \frac{1}{2}\left(F_{ki}F_{kj} - \delta_{ij}\right) \tag{3.69}$$

or after substitution of Equation (3.68),

$$\gamma_{ij} = \frac{1}{2}\left(\frac{\partial u_i}{\partial \xi_j} + \frac{\partial u_j}{\partial \xi_i}\right) + \frac{1}{2}\frac{\partial u_k}{\partial \xi_i}\frac{\partial u_k}{\partial \xi_j} \tag{3.70}$$

When the displacement gradients remain small, the quadratic terms can be omitted from the definition for the strain, and the (classical) expression for the linear strain tensor is retrieved, cf. Equations (1.91) and (1.92). Now, the relevance of introducing the factor $1/2$ in the definition of the Green–Lagrange strain tensor becomes clear. Without introducing this factor the linear expression for the strain tensor would not have been obtained when reducing the full expression for the strain tensor. For future use we also list the increment of the Green–Lagrange strain tensor:

$$\Delta\boldsymbol{\gamma} = \frac{1}{2}\left(\mathbf{F}^{\mathrm{T}} \cdot \nabla_0(\Delta\mathbf{u}) + (\nabla_0(\Delta\mathbf{u}))^{\mathrm{T}} \cdot \mathbf{F}\right) + \frac{1}{2}(\nabla_0(\Delta\mathbf{u}))^{\mathrm{T}} \cdot \nabla_0(\Delta\mathbf{u}) \qquad (3.71)$$

which is obtained by substracting $\boldsymbol{\gamma}^t$ from $\boldsymbol{\gamma}^{t+\Delta t}$. The subscript 0 signifies that the gradient is taken with respect to the material coordinates $\boldsymbol{\xi}$. When this subscript is omitted, differentiation with respect to the spatial coordinates $\mathbf{x}$ is implied.

The concept of stress is defined as the force per unit load-carrying area. When the displacement gradients remain small compared with unity, it is not relevant in which configuration the load-carrying area is measured, the deformed or the undeformed configuration. However, when this assumption no longer applies, it must be defined unambiguously to which configuration the stresses are referred. For engineering purposes we usually wish to know the magnitude of the stresses in the current configuration, i.e. the total force divided by the current load-carrying area. The stress tensor $\boldsymbol{\sigma}$ that contains these 'true stress components' is called the Cauchy stress tensor.

In solid mechanics the Lagrange description is the most popular way of describing the statics and kinematics of a structure subject to large displacements. In this approach all kinematic and static quantities must be referred to some previous configuration. Since the Cauchy stress tensor is referred to the current, unknown, configuration, the use of an auxiliary stress measure, which refers to a reference configuration, is needed. The quantity that is most often employed in non-linear analyses is the Second Piola–Kirchhoff stress tensor $\boldsymbol{\tau}$, which is related to the Cauchy stress tensor $\boldsymbol{\sigma}$ through:

$$\boldsymbol{\sigma} = \frac{\rho}{\rho_0}\mathbf{F} \cdot \boldsymbol{\tau} \cdot \mathbf{F}^{\mathrm{T}} \qquad (3.72)$$

with ρ and ρ_0 the mass densities in the deformed and reference configurations, respectively. In index notation the relation between both stress tensors reads:

$$\sigma_{ij} = \frac{\rho}{\rho_0}F_{ik}\tau_{kl}F_{jl} \qquad (3.73)$$

The Second Piola–Kirchhoff stress tensor has no direct physical relevance. When the stresses must be determined in an analysis in addition to the displacements, Cauchy stresses have to be computed from the Second Piola–Kirchhoff stresses. For small displacement gradients $\rho \approx \rho_0$, $\mathbf{F} \approx \mathbf{I}$, and the Cauchy stress tensor and the Second Piola–Kirchhoff stress tensor coincide. We finally note that there exists also the First Piola–Kirchhoff stress tensor, see Box 3.3.

Just as the stress measure used internally in the formulation of the finite element equations must be referred to the reference configuration, so must all quantities occurring in the principle of virtual work. Normally, this principle is formulated in the current domain of the structure. Again, this is inconvenient and we will transform the virtual work equation into an expression which only involves integrals which can be evaluated in the reference configuration. The starting point for this derivation is the notion of conservation of mass of an elementary volume, which is mathematically expressed as:

$$\rho \mathrm{d}V = \rho_0 \mathrm{d}V_0 \qquad (3.74)$$

Box 3.3 First Piola–Kirchhoff stress tensor

The First Piola–Kirchhoff stress tensor $\mathbf{p}^*$ is defined through:

$$\boldsymbol{\sigma} = \frac{\rho}{\rho_0}\mathbf{p}^* \cdot \mathbf{F}^{\mathrm{T}} \quad \text{or} \quad \mathbf{p}^* = \frac{\rho_0}{\rho}\boldsymbol{\sigma} \cdot (\mathbf{F}^{-1})^{\mathrm{T}}$$

In component form we have $\sigma_{ij} = \frac{\rho}{\rho_0} p^*_{ik} F_{jk}$ or $p^*_{jk} = \frac{\rho_0}{\rho}\sigma_{ji}(F^{-1})_{ki}$. Substitution of the expression for the internal energy in the reference configuration, Equation (3.83), yields:

$$\delta W_{\mathrm{int}} = \int_{V_0} p^*_{ik}\frac{\partial \delta u_i}{\partial x_j}F_{jk}\mathrm{d}V = \int_{V_0} p^*_{ik}\frac{\partial \delta u_i}{\partial \xi_k}\mathrm{d}V = \int_{V_0} p^*_{ik}\delta F_{ik}\mathrm{d}V$$

so that the First Piola–Kirchhoff stress tensor is energetically conjugate to the variation of the deformation gradient. A disadvantage of the First Piola–Kirchhoff stress tensor is its unsymmetry, which makes it less suitable for computations. However, it can be given a physical interpretation. For this, we first derive the relation between an area $\mathrm{d}S_0$ in the undeformed configuration and that in the deformed configuration, $\mathrm{d}S$. When the normal vectors are denoted by $\mathbf{n}_0$ and $\mathbf{n}$, respectively, and an arbitrary vector $\mathrm{d}\boldsymbol{\ell}_0$, that transforms into $\mathrm{d}\boldsymbol{\ell}$ and is not orthogonal to $\mathbf{n}_0$, there exists an elementary volume $\mathrm{d}V_0 = \mathrm{d}\boldsymbol{\ell}_0 \cdot \mathbf{n}_0\mathrm{d}S_0$ which transforms into $\mathrm{d}V = \mathrm{d}\boldsymbol{\ell} \cdot \mathbf{n}\mathrm{d}S$. Since $\rho\mathrm{d}V = \rho_0\mathrm{d}V_0$ and $\mathrm{d}\boldsymbol{\ell} = \mathbf{F} \cdot \mathrm{d}\boldsymbol{\ell}_0$, one obtains: $\rho\mathbf{n} \cdot \mathbf{F} \cdot \mathrm{d}\boldsymbol{\ell}_0\mathrm{d}S = \rho_0\mathbf{n}_0 \cdot \mathrm{d}\boldsymbol{\ell}_0\mathrm{d}S_0$. This identity must hold for arbitrary $\mathrm{d}\boldsymbol{\ell}_0$, which results in Nanson's formula for the transformation of surface elements:

$$\mathbf{n}\mathrm{d}S = \frac{\rho_0}{\rho}\mathbf{n}_0(\mathbf{F}^{-1})^{\mathrm{T}}\mathrm{d}S_0 \quad \text{or;} \quad n_i\mathrm{d}S = \frac{\rho_0}{\rho}(n_0)_k(F^{-1})_{ki}\mathrm{d}S_0$$

Now, we can express the force f_j on a tributary area $\mathrm{d}S$ in the current configuration as:

$$f_j = t_j\mathrm{d}S = n_i\sigma_{ij}\mathrm{d}S = \frac{\rho_0}{\rho}(n_0)_k(F^{-1})_{ki}\sigma_{ij}\mathrm{d}S_0$$

Invoking the symmetry of the Cauchy stress tensor and the definition of the First Piola–Kirchhoff stress tensor, and defining the nominal traction as $(t_0)_j = f_j/\mathrm{d}S_0$ then results in:

$$(t_0)_j = \frac{\rho_0}{\rho}(n_0)_k\sigma_{ji}(F^{-1})_{ki} = p^*_{jk}(n_0)_k$$

The components of the First Piola–Kirchhoff stress tensor can thus be interpreted as the stresses that result from the force that acts on a surface in the undeformed configuration.

Since

$$\det\mathbf{F} = \frac{\mathrm{d}x\mathrm{d}y\mathrm{d}z}{\mathrm{d}\xi\mathrm{d}\eta\mathrm{d}\zeta} = \frac{\mathrm{d}V}{\mathrm{d}V_0} \tag{3.75}$$

in view of Equation (3.74) we also have:

$$\det\mathbf{F} = \frac{\rho_0}{\rho} \tag{3.76}$$

The variation of the internal energy, or the internal virtual work, $\delta\mathcal{W}_{\text{int}}$ is given by:

$$\delta\mathcal{W}_{\text{int}} = \int_V \delta\boldsymbol{\epsilon} : \boldsymbol{\sigma} \mathrm{d}V \tag{3.77}$$

Conservation of mass permits conversion of this integral into an integral of the body in the reference configuration:

$$\delta\mathcal{W}_{\text{int}} = \int_{V_0} \frac{\rho_0}{\rho} \delta\boldsymbol{\epsilon} : \boldsymbol{\sigma} \mathrm{d}V \tag{3.78}$$

In Equation (3.78) $\delta\boldsymbol{\epsilon}$ and $\boldsymbol{\sigma}$ are functions of the current configuration. To transform them into quantities in the reference configuration we must establish a relationship between the virtual strain field $\delta\boldsymbol{\epsilon}$ and the virtual displacement field $\delta\mathbf{u}$. From the definition of the Green–Lagrange strain tensor, Equation (3.61), the variation reads:

$$\delta\boldsymbol{\gamma} = \frac{1}{2}\delta(\mathbf{F}^{\mathrm{T}} \cdot \mathbf{F} - \mathbf{I}) = \frac{1}{2}(\delta\mathbf{F}^{\mathrm{T}} \cdot \mathbf{F} + \mathbf{F}^{\mathrm{T}} \cdot \delta\mathbf{F}) \tag{3.79}$$

Using index notation, the variation of the Green-Lagrange strain tensor can be rewritten as:

$$\begin{aligned}
\delta\gamma_{ij} &= \frac{1}{2}\left(\frac{\partial\delta u_k}{\partial\xi_i}\frac{\partial x_k}{\partial\xi_j} + \frac{\partial x_k}{\partial\xi_i}\frac{\partial\delta u_k}{\partial\xi_j}\right) \\
&= \frac{1}{2}\left(\frac{\partial\delta u_k}{\partial x_l}\frac{\partial x_l}{\partial\xi_i}\frac{\partial x_k}{\partial\xi_j} + \frac{\partial x_k}{\partial\xi_i}\frac{\partial\delta u_k}{\partial x_l}\frac{\partial x_l}{\partial\xi_j}\right) \\
&= \frac{1}{2}F_{kj}\left(\frac{\partial\delta u_k}{\partial x_l} + \frac{\partial\delta u_l}{\partial x_k}\right)F_{li}
\end{aligned}$$

so that

$$\delta\boldsymbol{\gamma} = \mathbf{F}^{\mathrm{T}}\delta\boldsymbol{\epsilon}\mathbf{F} \tag{3.80}$$

sets the relation between the variation of the Green–Lagrange strain tensor referred to the undeformed configuration, $\delta\boldsymbol{\gamma}$, and $\delta\boldsymbol{\epsilon}$ which is referred to the current configuration:

$$\delta\boldsymbol{\epsilon} = \frac{1}{2}(\delta\mathbf{F}^{\mathrm{T}} + \delta\mathbf{F}) \tag{3.81}$$

or in index notation:

$$\delta\epsilon_{ij} = \frac{1}{2}\left(\frac{\partial\delta u_j}{\partial x_i} + \frac{\partial\delta u_i}{\partial x_j}\right) \tag{3.82}$$

The latter identity can be substituted into Equation (3.78). Together with the symmetry of the Cauchy stress tensor this results in:

$$\delta\mathcal{W}_{\text{int}} = \int_{V_0} \frac{\rho_0}{\rho} \nabla(\delta\mathbf{u}) : \boldsymbol{\sigma} \mathrm{d}V \tag{3.83}$$

We next substitute the relation between the Cauchy stress tensor and the Second Piola–Kirchhoff stress tensor in Equation (3.73):

$$\int_{V_0} \frac{\rho_0}{\rho} \nabla(\delta \mathbf{u}) : \boldsymbol{\sigma} \mathrm{d}V = \int_{V_0} \mathrm{tr}(\nabla(\delta \mathbf{u}) \cdot \mathbf{F} \cdot \boldsymbol{\tau} \cdot \mathbf{F}^{\mathrm{T}}) \mathrm{d}V = \int_{V_0} \mathrm{tr}(\delta \mathbf{F} \cdot \boldsymbol{\tau} \cdot \mathbf{F}^{\mathrm{T}}) \mathrm{d}V \tag{3.84}$$

where the chain rule has been used to establish the last identity. Using the definition of the first variation of the Green–Lagrange strain tensor referred to the undeformed configuration, Equation (3.79), and the symmetry of the Second Piola–Kirchhoff stress tensor subsequently permits rewriting the latter equation as:

$$\delta \mathcal{W}_{\mathrm{int}} = \int_{V_0} \delta \boldsymbol{\gamma} : \boldsymbol{\tau} \mathrm{d}V \tag{3.85}$$

This results states that the Second Piola–Kirchhoff stress tensor is energetically conjugate to the Green–Lagrange strain tensor, and arises naturally as a convenient stress measure in numerical schemes that take some previous configuration as the reference configuration.

3.4 Geometrically Non-linear Formulation of Continuum Elements

3.4.1 *Total and Updated Lagrange Formulations*

Using Equation (3.85) we can write the principle of virtual work referred to the reference configuration in terms of matrices and vectors:

$$\int_{V_0} \delta \boldsymbol{\gamma}^{\mathrm{T}} \boldsymbol{\tau}^{t+\Delta t} \mathrm{d}V = \int_{S_0} \delta \mathbf{u}^{\mathrm{T}} \mathbf{t}_0 \mathrm{d}S + \int_{V_0} \rho_0 \delta \mathbf{u}^{\mathrm{T}} \mathbf{g} \mathrm{d}V \tag{3.86}$$

with $\mathbf{t}_0$ the nominal traction, i.e. the force divided by a surface in the undeformed configuration. Basically, the derivation of tangential stiffness matrices and consistent load vectors runs along the same lines as for the truss elements. First, the unknown stress $\boldsymbol{\tau}^{t+\Delta t}$ is decomposed into a stress $\boldsymbol{\tau}^t$ at the beginning of a load step, and a stress increment $\Delta \boldsymbol{\tau}$. Substitution into the virtual work equation, which, although referred to some previous configuration, is valid at time $t + \Delta t$, results in:

$$\int_{V_0} \delta \boldsymbol{\gamma}^{\mathrm{T}} \Delta \boldsymbol{\tau} \mathrm{d}V + \int_{V_0} \delta \boldsymbol{\gamma}^{\mathrm{T}} \boldsymbol{\tau}^t \mathrm{d}V = \int_{S_0} \delta \mathbf{u}^{\mathrm{T}} \mathbf{t}_0 \mathrm{d}S + \int_{V_0} \rho_0 \delta \mathbf{u}^{\mathrm{T}} \mathbf{g} \mathrm{d}V \tag{3.87}$$

For small strains a linear relation can be assumed between the stress increment $\Delta \boldsymbol{\tau}$ and the strain increment $\Delta \boldsymbol{\gamma}$:

$$\Delta \boldsymbol{\tau} = \mathbf{D} \Delta \boldsymbol{\gamma} \tag{3.88}$$

The matrix $\mathbf{D}$ contains the instantaneous stiffness moduli of the material model. For linear elasticity it reduces to Hooke's law. Substitution of the latter equation into Equation (3.87) gives

$$\int_{V_0} \delta\boldsymbol{\gamma}^{\mathrm{T}}\mathbf{D}\Delta\boldsymbol{\gamma}\mathrm{d}V + \int_{V_0} \delta\boldsymbol{\gamma}^{\mathrm{T}}\boldsymbol{\tau}^t\mathrm{d}V = \int_{S_0} \delta\mathbf{u}^{\mathrm{T}}\mathbf{t}_0\mathrm{d}S + \int_{V_0} \rho_0\delta\mathbf{u}^{\mathrm{T}}\mathbf{g}\mathrm{d}V \tag{3.89}$$

Again, the strain increment $\Delta\boldsymbol{\gamma}$ contains contributions that are linear and contributions that are quadratic in the displacement increment $\Delta\mathbf{u}$, cf. Equation (3.71). Assembling the terms that are linear in the displacement increment in the contribution $\Delta\mathbf{e}$ and the terms that are non-linear in the displacement increment in $\Delta\boldsymbol{\eta}$, we can formally define the following decomposition:

$$\Delta\boldsymbol{\gamma} = \Delta\mathbf{e} + \Delta\boldsymbol{\eta} \tag{3.90}$$

In the same spirit as with the derivation for truss elements the variation of the Green–Lagrange strain tensor at time t vanishes and we have $\delta\boldsymbol{\gamma}^{t+\Delta t} = \delta\Delta\boldsymbol{\gamma}$. Substitution then gives:

$$\begin{aligned} &\int_{V_0} \delta\Delta\mathbf{e}^{\mathrm{T}}\mathbf{D}\Delta\mathbf{e}\mathrm{d}V + \int_{V_0} \delta\Delta\mathbf{e}^{\mathrm{T}}\mathbf{D}\Delta\boldsymbol{\eta}\mathrm{d}V + \\ &\int_{V_0} \delta\Delta\boldsymbol{\eta}^{\mathrm{T}}\mathbf{D}\Delta\mathbf{e}\mathrm{d}V + \int_{V_0} \delta\Delta\boldsymbol{\eta}^{\mathrm{T}}\mathbf{D}\Delta\boldsymbol{\eta}\mathrm{d}V + \\ \int_{V_0} \delta\boldsymbol{\eta}^{\mathrm{T}}\boldsymbol{\tau}^t\mathrm{d}V = &\int_{S_0} \delta\mathbf{u}^{\mathrm{T}}\mathbf{t}_0\mathrm{d}S + \int_{V_0} \rho_0\delta\mathbf{u}^{\mathrm{T}}\mathbf{g}\mathrm{d}V - \int_{V_0} \delta\mathbf{e}^{\mathrm{T}}\boldsymbol{\tau}^t\mathrm{d}V \end{aligned} \tag{3.91}$$

where, just as with the derivation for the truss elements, the contribution $\int(\delta\Delta\mathbf{e})^{\mathrm{T}}\boldsymbol{\tau}^t\mathrm{d}V$, which is of degree zero in the displacement increments, has been brought to the right-hand side to form the internal force vector. For the derivation of the tangential stiffness matrix we can only use those contributions that are linear in the displacement increment. The second, third and fourth terms on the left-hand side are non-linear in the displacement increment, and linearisation yields:

$$\begin{aligned} \int_{V_0} \delta\Delta\mathbf{e}^{\mathrm{T}}\mathbf{D}\Delta\mathbf{e}\mathrm{d}V + \int_{V_0} \delta\boldsymbol{\eta}^{\mathrm{T}}\boldsymbol{\tau}^t\mathrm{d}V = \\ \int_{S_0} \delta\mathbf{u}^{\mathrm{T}}\mathbf{t}_0\mathrm{d}S + \int_{V_0} \rho_0\delta\mathbf{u}^{\mathrm{T}}\mathbf{g}\mathrm{d}V - \int_{V_0} \delta\mathbf{e}^{\mathrm{T}}\boldsymbol{\tau}^t\mathrm{d}V \end{aligned} \tag{3.92}$$

The relation between the linear part of the strain increment $\Delta\mathbf{e}$ and the displacement increment $\Delta\mathbf{u}$ formally reads

$$\Delta\mathbf{e} = \mathbf{L}\Delta\mathbf{u} \tag{3.93}$$

where the matrix **L** contains differentials which must be determined for every strain measure and every strain condition separately. For instance, for a two-dimensional configuration the Green–Lagrange strain definition gives

$$\mathbf{L} = \begin{bmatrix} F_{11}\frac{\partial}{\partial\xi_1} & F_{21}\frac{\partial}{\partial\xi_1} \\ F_{12}\frac{\partial}{\partial\xi_2} & F_{22}\frac{\partial}{\partial\xi_2} \\ F_{11}\frac{\partial}{\partial\xi_2} + F_{12}\frac{\partial}{\partial\xi_1} & F_{21}\frac{\partial}{\partial\xi_2} + F_{22}\frac{\partial}{\partial\xi_1} \end{bmatrix} \tag{3.94}$$

where the last row of this matrix reflects the fact that the engineering shear strain is used in the matrix-vector notation, which is the double of the tensorial quantity of Equation (3.71), and for a three-dimensional configuration we have:

$$\mathbf{L} = \begin{bmatrix} F_{11}\frac{\partial}{\partial\xi_1} & F_{21}\frac{\partial}{\partial\xi_1} & F_{31}\frac{\partial}{\partial\xi_1} \\ F_{12}\frac{\partial}{\partial\xi_2} & F_{22}\frac{\partial}{\partial\xi_2} & F_{32}\frac{\partial}{\partial\xi_2} \\ F_{13}\frac{\partial}{\partial\xi_3} & F_{23}\frac{\partial}{\partial\xi_3} & F_{33}\frac{\partial}{\partial\xi_3} \\ F_{11}\frac{\partial}{\partial\xi_2} + F_{12}\frac{\partial}{\partial\xi_1} & F_{21}\frac{\partial}{\partial\xi_2} + F_{22}\frac{\partial}{\partial\xi_1} & F_{31}\frac{\partial}{\partial\xi_2} + F_{32}\frac{\partial}{\partial\xi_1} \\ F_{12}\frac{\partial}{\partial\xi_3} + F_{13}\frac{\partial}{\partial\xi_2} & F_{22}\frac{\partial}{\partial\xi_3} + F_{23}\frac{\partial}{\partial\xi_2} & F_{32}\frac{\partial}{\partial\xi_3} + F_{33}\frac{\partial}{\partial\xi_2} \\ F_{13}\frac{\partial}{\partial\xi_1} + F_{11}\frac{\partial}{\partial\xi_3} & F_{23}\frac{\partial}{\partial\xi_1} + F_{21}\frac{\partial}{\partial\xi_3} & F_{33}\frac{\partial}{\partial\xi_1} + F_{31}\frac{\partial}{\partial\xi_3} \end{bmatrix} \tag{3.95}$$

While for truss elements the derivation is completed at this point, continua require the interpolation of the continuous displacement field $\mathbf{u}$. With the matrix $\mathbf{H}$ containing the interpolation functions $h_1, h_2, h_3, h_4, \ldots$ defined in Equation (2.11), the relation between $\mathbf{u}$ and the discrete nodal displacements as contained in the vector $\mathbf{a}$ can be formulated as $\mathbf{u} = \mathbf{Ha}$, and consequently $\Delta\mathbf{u} = \mathbf{H}\Delta\mathbf{a}$. With the definition $\mathbf{B}_L = \mathbf{LH}$ the relation between the linear part of the strain increment, $\Delta\mathbf{e}$, and the continuous displacement vector $\mathbf{u}$ becomes:

$$\Delta\mathbf{e} = \mathbf{B}_L\Delta\mathbf{a} \tag{3.96}$$

For example, in a two-dimensional configuration and using the Green–Lagrange strain we have:

$$\mathbf{B}_L = \begin{bmatrix} F_{11}\frac{\partial h_1}{\partial\xi_1} & F_{21}\frac{\partial h_1}{\partial\xi_1} & \ldots & \ldots \\ F_{12}\frac{\partial h_1}{\partial\xi_2} & F_{22}\frac{\partial h_1}{\partial\xi_2} & \ldots & \ldots \\ F_{11}\frac{\partial h_1}{\partial\xi_2} + F_{12}\frac{\partial h_1}{\partial\xi_1} & F_{21}\frac{\partial h_1}{\partial\xi_2} + F_{22}\frac{\partial h_1}{\partial\xi_1} & \ldots & \ldots \end{bmatrix} \tag{3.97}$$

while for the three-dimensional case we have:

$$\mathbf{B}_L = \begin{bmatrix} F_{11}\frac{\partial h_1}{\partial \xi_1} & F_{21}\frac{\partial h_1}{\partial \xi_1} & F_{31}\frac{\partial h_1}{\partial \xi_1} & \dots & \dots \\ F_{12}\frac{\partial h_1}{\partial \xi_2} & F_{22}\frac{\partial h_1}{\partial \xi_2} & F_{32}\frac{\partial h_1}{\partial \xi_2} & \dots & \dots \\ F_{13}\frac{\partial h_1}{\partial \xi_3} & F_{23}\frac{\partial h_1}{\partial \xi_3} & F_{33}\frac{\partial h_1}{\partial \xi_3} & \dots & \dots \\ F_{11}\frac{\partial h_1}{\partial \xi_2} + F_{12}\frac{\partial h_1}{\partial \xi_1} & F_{21}\frac{\partial h_1}{\partial \xi_2} + F_{22}\frac{\partial h_1}{\partial \xi_1} & F_{31}\frac{\partial h_1}{\partial \xi_2} + F_{32}\frac{\partial h_1}{\partial \xi_1} & \dots & \dots \\ F_{12}\frac{\partial h_1}{\partial \xi_3} + F_{13}\frac{\partial h_1}{\partial \xi_2} & F_{22}\frac{\partial h_1}{\partial \xi_3} + F_{23}\frac{\partial h_1}{\partial \xi_2} & F_{32}\frac{\partial h_1}{\partial \xi_3} + F_{33}\frac{\partial h_1}{\partial \xi_2} & \dots & \dots \\ F_{13}\frac{\partial h_1}{\partial \xi_1} + F_{11}\frac{\partial h_1}{\partial \xi_3} & F_{23}\frac{\partial h_1}{\partial \xi_1} + F_{21}\frac{\partial h_1}{\partial \xi_3} & F_{33}\frac{\partial h_1}{\partial \xi_1} + F_{31}\frac{\partial h_1}{\partial \xi_3} & \dots & \dots \end{bmatrix} \tag{3.98}$$

When we define

$$\mathbf{K}_L = \int_{V_0} \mathbf{B}_L^{\mathrm{T}} \mathbf{D} \mathbf{B}_L \mathrm{d}V \tag{3.99}$$

as the first contribution to the tangential stiffness matrix, it follows that:

$$\int_{V_0} (\delta\Delta\mathbf{e})^{\mathrm{T}} \mathbf{D} \Delta\mathbf{e} \mathrm{d}V = (\delta\Delta\mathbf{a})^{\mathrm{T}} \mathbf{K}_L \Delta\mathbf{a} \tag{3.100}$$

Formally, the second contribution to the left-hand side can be rewritten as

$$\int_{V_0} (\delta\Delta\boldsymbol{\eta})^{\mathrm{T}} \boldsymbol{\tau}^t \mathrm{d}V = (\delta\Delta\mathbf{a})^{\mathrm{T}} \mathbf{K}_{NL} \Delta\mathbf{a} \tag{3.101}$$

so that the geometric contribution to the tangential stiffness matrix is given by:

$$\mathbf{K}_{NL} = \int_{V_0} \mathbf{B}_{NL}^{\mathrm{T}} \mathcal{T}^t \mathbf{B}_{NL} \mathrm{d}V \tag{3.102}$$

Herein the Second Piola–Kirchhoff stress is represented in matrix form, which, for two-dimensional configuration is given by:

$$\mathcal{T} = \begin{bmatrix} \tau_{xx} & \tau_{xy} & 0 & 0 \\ \tau_{xy} & \tau_{yy} & 0 & 0 \\ 0 & 0 & \tau_{xx} & \tau_{xy} \\ 0 & 0 & \tau_{xy} & \tau_{yy} \end{bmatrix} \tag{3.103}$$

and

$$\mathbf{B}_{NL} = \begin{bmatrix} \frac{\partial h_1}{\partial \xi_1} & 0 & \frac{\partial h_2}{\partial \xi_1} & 0 & \cdots & \cdots \\ \frac{\partial h_1}{\partial \xi_2} & 0 & \frac{\partial h_2}{\partial \xi_2} & 0 & \cdots & \cdots \\ 0 & \frac{\partial h_1}{\partial \xi_1} & 0 & \frac{\partial h_2}{\partial \xi_1} & \cdots & \cdots \\ 0 & \frac{\partial h_1}{\partial \xi_2} & 0 & \frac{\partial h_2}{\partial \xi_2} & \cdots & \cdots \end{bmatrix} \tag{3.104}$$

For three-dimensional configurations the matrix form of the Second Piola–Kirchhoff stress tensor becomes:

$$\mathcal{T} = \begin{bmatrix} \tau_{xx} & \tau_{xy} & \tau_{zx} & 0 & 0 & 0 & 0 & 0 & 0 \\ \tau_{xy} & \tau_{yy} & \tau_{yz} & 0 & 0 & 0 & 0 & 0 & 0 \\ \tau_{zx} & \tau_{yz} & \tau_{zz} & 0 & 0 & 0 & 0 & 0 & 0 \\ 0 & 0 & 0 & \tau_{xx} & \tau_{xy} & \tau_{zx} & 0 & 0 & 0 \\ 0 & 0 & 0 & \tau_{xy} & \tau_{yy} & \tau_{yz} & 0 & 0 & 0 \\ 0 & 0 & 0 & \tau_{zx} & \tau_{yz} & \tau_{zz} & 0 & 0 & 0 \\ 0 & 0 & 0 & 0 & 0 & 0 & \tau_{xx} & \tau_{xy} & \tau_{zx} \\ 0 & 0 & 0 & 0 & 0 & 0 & \tau_{xy} & \tau_{yy} & \tau_{yz} \\ 0 & 0 & 0 & 0 & 0 & 0 & \tau_{zx} & \tau_{yz} & \tau_{zz} \end{bmatrix} \tag{3.105}$$

while then:

$$\mathbf{B}_{NL} = \begin{bmatrix} \frac{\partial h_1}{\partial \xi_1} & 0 & 0 & \frac{\partial h_2}{\partial \xi_1} & 0 & 0 & \cdots & \cdots \\ \frac{\partial h_1}{\partial \xi_2} & 0 & 0 & \frac{\partial h_2}{\partial \xi_2} & 0 & 0 & \cdots & \cdots \\ \frac{\partial h_1}{\partial \xi_3} & 0 & 0 & \frac{\partial h_2}{\partial \xi_3} & 0 & 0 & \cdots & \cdots \\ 0 & \frac{\partial h_1}{\partial \xi_1} & 0 & 0 & \frac{\partial h_2}{\partial \xi_1} & 0 & \cdots & \cdots \\ 0 & \frac{\partial h_1}{\partial \xi_2} & 0 & 0 & \frac{\partial h_2}{\partial \xi_2} & 0 & \cdots & \cdots \\ 0 & \frac{\partial h_1}{\partial \xi_3} & 0 & 0 & \frac{\partial h_2}{\partial \xi_3} & 0 & \cdots & \cdots \\ 0 & 0 & \frac{\partial h_1}{\partial \xi_1} & 0 & 0 & \frac{\partial h_2}{\partial \xi_1} & \cdots & \cdots \\ 0 & 0 & \frac{\partial h_1}{\partial \xi_2} & 0 & 0 & \frac{\partial h_2}{\partial \xi_2} & \cdots & \cdots \\ 0 & 0 & \frac{\partial h_1}{\partial \xi_3} & 0 & 0 & \frac{\partial h_2}{\partial \xi_3} & \cdots & \cdots \end{bmatrix} \tag{3.106}$$

In a similar spirit as for truss elements we derive that:

$$(\delta\Delta\mathbf{a})^{\mathrm{T}}(\mathbf{K}_L + \mathbf{K}_{NL})\Delta\mathbf{a} = (\delta\Delta\mathbf{a})^{\mathrm{T}}\left(\mathbf{f}_{\mathrm{ext}}^{t+\Delta t} - \mathbf{f}_{\mathrm{int}}^{t}\right) \tag{3.107}$$

with the external force vector and the internal force vector defined as:

$$\begin{cases} \mathbf{f}_{\mathrm{ext}}^{t+\Delta t} = \int_{S_0} \mathbf{H}^{\mathrm{T}}\mathbf{t}_0 \mathrm{d}S + \int_{V_0} \rho_0 \mathbf{H}^{\mathrm{T}}\mathbf{g}\mathrm{d}V \\ \mathbf{f}_{\mathrm{int}}^{t} = \int_{V_0} \mathbf{B}_L^{\mathrm{T}}\boldsymbol{\tau}^t \mathrm{d}V \end{cases} \tag{3.108}$$

Identity (3.107) must hold for any virtual displacement increment $\delta\Delta\mathbf{a}$, whence

$$(\mathbf{K}_L + \mathbf{K}_{NL})\Delta\mathbf{a} = \mathbf{f}_{\mathrm{ext}}^{t+\Delta t} - \mathbf{f}_{\mathrm{int}}^{t} \tag{3.109}$$

The above derivation applies to two- and three-dimensional continuum elements. The original, undeformed configuration has been chosen as the reference configuration, so that formally we have a Total Lagrange formulation. Nonetheless, any other reference configuration can be chosen without affecting the derivation, which thus encompasses the Updated Lagrange formulation. At variance with the formulation for truss elements, minor differences in the results may now arise between both Lagrange formulations. As mentioned already for the truss element, these differences can become more significant when inelastic effects are included, such as plasticity or damage. Then, use of the Total Lagrange formulation is less appealing, since the physical relevance of the undeformed configuration is lost.

3.4.2 *Corotational Formulation*

A crucial issue in corotational formulations is the proper definition of the local coordinate frame. For structural elements this definition is usually straightforward, as for instance with the truss element. For continuum elements several definitions are possible for the choice of the local rotation ϕ, as has been discussed by Crisfield and Moita (1996a), which has been extended to three dimensions by Crisfield and Moita (1996b). Herein, we shall restrict the discussion to a two-dimensional configuration, for which we can define the base vectors $\bar{\mathbf{n}}_1^{\mathrm{T}} = (\cos\phi, \sin\phi)$ and $\bar{\mathbf{n}}_2^{\mathrm{T}} = (-\sin\phi, \cos\phi)$ (Figure 3.7), which are attached to the element and located at its centroid. In order that the element passes the 'large-strain patch test', which means that for

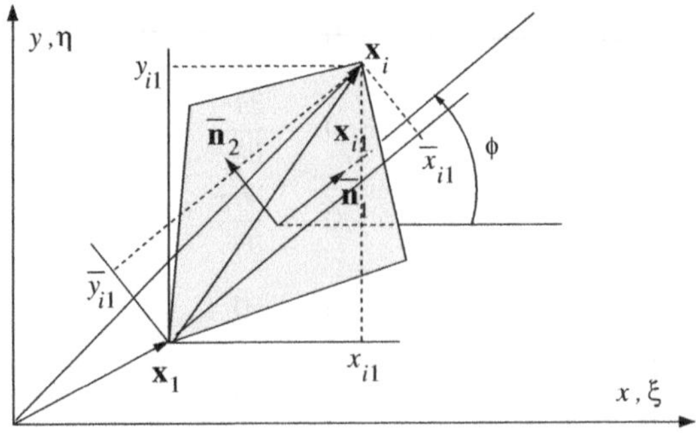

Figure 3.7 Position vectors for a deformed element in the corotational formulation

a constant deformation gradient, a patch of elements should respond with the same, constant strains at each integration point, the local spin at the centroid of the element in the current configuration should be constrained to zero (Jetteur and Cescotto 1991):

$$\bar{\Omega} \equiv \frac{\partial \bar{u}}{\partial \eta} - \frac{\partial \bar{v}}{\partial \xi} = 0 \tag{3.110}$$

where an overbar refers to the 'local', rotated coordinate frame, and ξ, η are the Lagrangian, material coordinates. The local displacements $\bar{u}$ and $\bar{v}$ are interpolated in a standard manner,

$$\begin{cases} \bar{u} = \sum_{i=1}^{n} (h_i)_0 (\bar{a}_x)_i \\ \bar{v} = \sum_{i=1}^{n} (h_i)_0 (\bar{a}_y)_i \end{cases} \tag{3.111}$$

where $(\bar{a}_x)_i$, $(\bar{a}_y)_i$ are the displacements in the local coordinate system at node i, and the subscript 0 denotes that the interpolants h_i are evaluated at the element centroid. Using these identities, Equation (3.110) transforms into:

$$\bar{\Omega} \equiv \sum_{i=1}^{n} \frac{\partial (h_i)_0}{\partial \eta} (\bar{a}_x)_i - \sum_{i=1}^{n} \frac{\partial (h_i)_0}{\partial \xi} (\bar{a}_y)_i = 0 \tag{3.112}$$

We transform the nodal displacements to the global coordinate system using Equation (1.50), with the local rotation matrix **R** given by Equation (1.51), so that:

$$\bar{\Omega} \equiv a \sin\phi + b \cos\phi = 0 \tag{3.113}$$

with

$$\begin{cases} a = \sum_{i=1}^{n} \left(\frac{\partial (h_i)_0}{\partial \xi} (a_x)_i + \frac{\partial (h_i)_0}{\partial \eta} (a_y)_i \right) \\ b = \sum_{i=1}^{n} \left(\frac{\partial (h_i)_0}{\partial \eta} (a_x)_i - \frac{\partial (h_i)_0}{\partial \xi} (a_y)_i \right) \end{cases} \tag{3.114}$$

The local rotation ϕ can then be computed from:

$$\phi = \tan^{-1}\left(-\frac{b}{a} \right) \tag{3.115}$$

For future use we also list the variation of ϕ:

$$\delta\phi = \frac{b\delta a - a\delta b}{a^2 + b^2} = \mathbf{w}^{\mathrm{T}} \delta\mathbf{a} \tag{3.116}$$

with

$$\mathbf{w} = \frac{b\mathbf{c} - a\mathbf{d}}{a^2 + b^2} \tag{3.117}$$

where $\mathbf{a} = [(a_x)_1, (a_y)_1, \ldots, (a_x)_n, (a_y)_n]$ contains the nodal displacements, and

$$\begin{cases} \mathbf{c}^{\mathrm{T}} = \left[\dfrac{\partial (h_1)_0}{\partial \xi}, \dfrac{\partial (h_1)_0}{\partial \eta}, \ldots, \dfrac{\partial (h_n)_0}{\partial \xi}, \dfrac{\partial (h_n)_0}{\partial \eta} \right] \\ \mathbf{d}^{\mathrm{T}} = \left[-\dfrac{\partial (h_1)_0}{\partial \eta}, \dfrac{\partial (h_1)_0}{\partial \xi}, \ldots, -\dfrac{\partial (h_n)_0}{\partial \eta}, \dfrac{\partial (h_n)_0}{\partial \xi} \right] \end{cases} \tag{3.118}$$

Having derived an expression for the local rotation ϕ, we proceed by expressing the relative position vector $\bar{\mathbf{x}}_{i1}$ for node i in the local, rotated coordinate system with respect to the global relative position vector $\mathbf{x}_{i1} = \mathbf{x}_i - \mathbf{x}_1$

$$\bar{\mathbf{x}}_{i1} = \mathbf{R}\mathbf{x}_{i1} \tag{3.119}$$

see Figure 3.7. Since $\delta \mathbf{x}_{i1} = \delta \mathbf{a}_{i1} = \delta \mathbf{a}_i$ (and similarly for the displacement quantities in the rotated, local coordinate system), the latter equation can be differentiated to yield:

$$\delta \bar{\mathbf{a}}_i = \mathbf{R}\delta \mathbf{a}_i + \frac{\partial \mathbf{R}}{\partial \phi}\mathbf{x}_{i1}\delta\phi \tag{3.120}$$

with (cf. Box 3.2):

$$\frac{\partial \mathbf{R}}{\partial \phi} = \begin{bmatrix} -\sin\phi & \cos\phi \\ -\cos\phi & -\sin\phi \end{bmatrix} \tag{3.121}$$

We introduce the composite rotation matrix:

$$\mathcal{R} = \begin{bmatrix} \mathbf{R} & \ldots & \mathbf{0} \\ \vdots & \ddots & \vdots \\ \mathbf{0} & \ldots & \mathbf{R} \end{bmatrix} \tag{3.122}$$

with an equivalent expression for $\frac{\partial \mathcal{R}}{\partial \phi}$, and use Equation (3.116) to give:

$$\delta \bar{\mathbf{a}} = \left(\mathcal{R} + \frac{\partial \mathcal{R}}{\partial \phi}\mathbf{x}_1\mathbf{w}^{\mathrm{T}} \right) \delta \mathbf{a} \tag{3.123}$$

with $\mathbf{x}_1^{\mathrm{T}} = [\mathbf{0}, \mathbf{x}_{21}, \ldots, \mathbf{x}_{n1}]$ the composite, relative position vector. The matrix

$$\mathbf{T} = \mathcal{R} + \frac{\partial \mathcal{R}}{\partial \phi}\mathbf{x}_1\mathbf{w}^{\mathrm{T}} \tag{3.124}$$

is thus the transformation matrix for continuum elements in the corotational formulation.

For the derivation of the tangential stiffness we exploit the invariance of the virtual work in the global and local systems to give:

$$\mathbf{f}_{\mathrm{int}} = \mathbf{T}^{\mathrm{T}}\bar{\mathbf{f}}_{\mathrm{int}} \tag{3.125}$$

so that:

$$\delta \mathbf{f}_{\mathrm{int}} = \mathbf{T}^{\mathrm{T}}\delta\bar{\mathbf{f}}_{\mathrm{int}} + \delta\mathbf{T}^{\mathrm{T}}\bar{\mathbf{f}}_{\mathrm{int}} \tag{3.126}$$

Elaboration of the first term of Equation (3.126) gives:

$$\mathbf{T}^{\mathrm{T}}\delta\bar{\mathbf{f}}_{\mathrm{int}} = \mathbf{T}^{\mathrm{T}}\int_{V_0}\mathbf{B}^{\mathrm{T}}\mathbf{D}\mathbf{B}\mathrm{d}V\,\delta\bar{\mathbf{a}} \tag{3.127}$$

with the linear **B** matrix defined in Chapter 2, which, for two-dimensional configurations, reads:

$$\mathbf{B} = \begin{bmatrix} \frac{\partial h_1}{\partial\xi} & 0 & \cdots & \cdots & \frac{\partial h_n}{\partial\xi} & 0 \\ 0 & \frac{\partial h_1}{\partial\eta} & \cdots & \cdots & 0 & \frac{\partial h_n}{\partial\eta} \\ \frac{\partial h_1}{\partial\eta} & \frac{\partial h_1}{\partial\xi} & \cdots & \cdots & \frac{\partial h_n}{\partial\eta} & \frac{\partial h_n}{\partial\xi} \end{bmatrix} \tag{3.128}$$

cf. Box 2.2 for the elaboration for a three-dimensional eight-noded element, and **D** the material tangential stiffness matrix which accounts for possible inelastic effects (Yaw *et al.* 2009), with $\mathbf{D} \equiv \mathbf{D}^{\mathrm{e}}$ for linear elasticity. Backtransformation of the nodal displacements to the global coordinate system yields:

$$\mathbf{T}^{\mathrm{T}}\delta\bar{\mathbf{f}}_{\mathrm{int}} = \mathbf{T}^{\mathrm{T}}\bar{\mathbf{K}}\mathbf{T}\delta\mathbf{a} \tag{3.129}$$

with

$$\bar{\mathbf{K}} = \int_{V_0}\mathbf{B}^{\mathrm{T}}\mathbf{D}\mathbf{B}\mathrm{d}V$$

the conventional stiffness matrix for small displacement gradients. Using Equation (3.124) we obtain for the second term:

$$\begin{aligned} \delta\mathbf{T}^{\mathrm{T}}\bar{\mathbf{f}}_{\mathrm{int}} = {} & \left(\frac{\partial\mathcal{R}}{\partial\phi}\right)^{\mathrm{T}}\bar{\mathbf{f}}_{\mathrm{int}}\delta\phi + \mathbf{w}\mathbf{x}_1^{\mathrm{T}}\left(\frac{\partial^2\mathcal{R}}{\partial\phi^2}\right)^{\mathrm{T}}\bar{\mathbf{f}}_{\mathrm{int}}\delta\phi \\ & +\mathbf{w}\delta\mathbf{a}^{\mathrm{T}}\left(\frac{\partial\mathcal{R}}{\partial\phi}\right)^{\mathrm{T}}\bar{\mathbf{f}}_{\mathrm{int}} + \delta\mathbf{w}\mathbf{x}_1^{\mathrm{T}}\left(\frac{\partial\mathcal{R}}{\partial\phi}\right)^{\mathrm{T}}\bar{\mathbf{f}}_{\mathrm{int}} \end{aligned} \tag{3.130}$$

Using Equation (3.116) the first two terms can be rewritten as:

$$\left(\frac{\partial\mathcal{R}}{\partial\phi}\right)^{\mathrm{T}}\bar{\mathbf{f}}_{\mathrm{int}}\delta\phi = \left(\frac{\partial\mathcal{R}}{\partial\phi}\right)^{\mathrm{T}}\bar{\mathbf{f}}_{\mathrm{int}}\mathbf{w}^{\mathrm{T}}\delta\mathbf{a}$$

and

$$\mathbf{w}\mathbf{x}_1^{\mathrm{T}}\left(\frac{\partial^2\mathcal{R}}{\partial\phi^2}\right)^{\mathrm{T}}\bar{\mathbf{f}}_{\mathrm{int}}\delta\phi = \mathbf{w}\mathbf{x}_1^{\mathrm{T}}\left(\frac{\partial^2\mathcal{R}}{\partial\phi^2}\right)^{\mathrm{T}}\bar{\mathbf{f}}_{\mathrm{int}}\mathbf{w}^{\mathrm{T}}\delta\mathbf{a}$$

The third and fourth terms can be rearranged to give, respectively:

$$\mathbf{w}\left(\delta\mathbf{a}^{\mathrm{T}}\left(\frac{\partial\mathcal{R}}{\partial\phi}\right)^{\mathrm{T}}\bar{\mathbf{f}}_{\mathrm{int}}\right) = \mathbf{w}\left(\left(\frac{\partial\mathcal{R}}{\partial\phi}\right)^{\mathrm{T}}\bar{\mathbf{f}}_{\mathrm{int}}\right)^{\mathrm{T}}\delta\mathbf{a} = \mathbf{w}\bar{\mathbf{f}}_{\mathrm{int}}^{\mathrm{T}}\frac{\partial\mathcal{R}}{\partial\phi}\delta\mathbf{a}$$

and

$$\delta \mathbf{w} \left(\mathbf{x}_1^{\mathrm{T}} \left(\frac{\partial \mathcal{R}}{\partial \phi} \right)^{\mathrm{T}} \bar{\mathbf{f}}_{\mathrm{int}} \right) = \left(\mathbf{x}_1^{\mathrm{T}} \left(\frac{\partial \mathcal{R}}{\partial \phi} \right)^{\mathrm{T}} \bar{\mathbf{f}}_{\mathrm{int}} \right) \mathbf{W} \delta \mathbf{a}$$

where the matrix $\mathbf{W}$ follows by straightforward differentiation of $\mathbf{w}$, Equation (3.117):

$$\mathbf{W} = \frac{2ab(\mathbf{d}\mathbf{d}^{\mathrm{T}} - \mathbf{c}\mathbf{c}^{\mathrm{T}}) + (a^2 - b^2)(\mathbf{c}\mathbf{d}^{\mathrm{T}} + \mathbf{d}\mathbf{c}^{\mathrm{T}})}{(a^2 + b^2)^2} \tag{3.131}$$

The second contribution to the tangential stiffness matrix is therefore elaborated as:

$$\delta \mathbf{T}^{\mathrm{T}} \bar{\mathbf{f}}_{\mathrm{int}} = \left[\left(\frac{\partial \mathcal{R}}{\partial \phi} \right)^{\mathrm{T}} \bar{\mathbf{f}}_{\mathrm{int}} \mathbf{w}^{\mathrm{T}} + \mathbf{w} \mathbf{x}_1^{\mathrm{T}} \left(\frac{\partial^2 \mathcal{R}}{\partial \phi^2} \right)^{\mathrm{T}} \bar{\mathbf{f}}_{\mathrm{int}} \mathbf{w}^{\mathrm{T}} + \mathbf{w} \bar{\mathbf{f}}_{\mathrm{int}}^{\mathrm{T}} \frac{\partial \mathcal{R}}{\partial \phi} + \left(\mathbf{x}_1^{\mathrm{T}} \left(\frac{\partial \mathcal{R}}{\partial \phi} \right)^{\mathrm{T}} \bar{\mathbf{f}}_{\mathrm{int}} \right) \mathbf{W} \right] \delta \mathbf{a} \tag{3.132}$$

From Equations (3.126), (3.129) and (3.132) the complete tangential stiffness matrix for continuum elements for a corotational formulation is derived as:

$$\mathbf{K} = \mathbf{T}^{\mathrm{T}} \bar{\mathbf{K}} \mathbf{T} + \left(\frac{\partial \mathcal{R}}{\partial \phi} \right)^{\mathrm{T}} \bar{\mathbf{f}}_{\mathrm{int}} \mathbf{w}^{\mathrm{T}} + \mathbf{w} \mathbf{x}_1^{\mathrm{T}} \left(\frac{\partial^2 \mathcal{R}}{\partial \phi^2} \right)^{\mathrm{T}} \bar{\mathbf{f}}_{\mathrm{int}} \mathbf{w}^{\mathrm{T}} + \mathbf{w} \bar{\mathbf{f}}_{\mathrm{int}}^{\mathrm{T}} \frac{\partial \mathcal{R}}{\partial \phi} + \left(\mathbf{x}_1^{\mathrm{T}} \left(\frac{\partial \mathcal{R}}{\partial \phi} \right)^{\mathrm{T}} \bar{\mathbf{f}}_{\mathrm{int}} \right) \mathbf{W} \tag{3.133}$$

Numerical experience shows that the last term is often less important and can be omitted.

3.5 Linear Buckling Analysis

A complete non-linear calculation can be expensive in terms of the computer time that is needed. For stability problems it is desirable to have a simple method which gives an accurate estimate of the critical load at which loss of structural stability occurs. Such a method is known as linear buckling analysis. In this method the complete non-linear analysis in which the entire load–deflection path is followed up to, and possibly beyond the critical load level, is replaced by an eigenvalue analysis. A derivation of the method, starting from the complete set of the non-linear field equations and elucidating the assumptions that are made, is given below.

The basic assumption is that prior to the point where loss of uniqueness occurs,[1] the displacement gradients remain small, e.g. $\frac{\partial v}{\partial x} \ll 1$. Under this condition, the following approximations hold:

1. The second, third and fourth members on the left-hand side of Equation (3.91) can be neglected, since they are of second order and third order in the displacement gradients.
2. The difference between the Cauchy stress tensor $\boldsymbol{\sigma}$ and the Second Piola–Kirchhoff stress tensor $\boldsymbol{\tau}$ disappears.
3. No distinction has to be made between material coordinates and spatial coordinates. Consequently, all integrals can be evaluated in the reference configuration.
4. The contribution to the increment of the Green–Lagrange strain tensor that is linear in the displacement increments, $\Delta\mathbf{e}$, can be replaced by the increment of the engineering strain. This has the effect that the $\mathbf{B}_L$ matrix, which relates the increments of the nodal point displacements $\Delta\mathbf{u}$ with the strain increment, ceases to depend upon the current displacement gradient, since the latter quantities are small compared with unity. Therefore, the $\mathbf{B}_L$ matrix as defined in Equation (3.97) can be replaced by the linear $\mathbf{B}$ matrix defined in Chapter 2.

With these assumptions we can rewrite Equation (3.91) as

$$\begin{aligned}\int_{V_0} (\delta\Delta\mathbf{u})^{\mathrm{T}}\mathbf{B}^{\mathrm{T}}\mathbf{D}\mathbf{B}\Delta\mathbf{u}\mathrm{d}V + \int_{V_0} (\delta\Delta\mathbf{u})^{\mathrm{T}}\mathbf{B}_{NL}^{\mathrm{T}}\boldsymbol{\Sigma}\mathbf{B}_{NL}\Delta\mathbf{u}\mathrm{d}V = \\ \int_{S_0} (\delta\Delta\mathbf{u})^{\mathrm{T}}\mathbf{H}^{\mathrm{T}}\mathbf{t}_0\mathrm{d}S + \int_{V_0} \rho_0(\delta\Delta\mathbf{u})^{\mathrm{T}}\mathbf{H}^{\mathrm{T}}\mathbf{g}\mathrm{d}V\end{aligned} \tag{3.134}$$

where $\boldsymbol{\Sigma}$ is a matrix representation of the Cauchy stress tensor, with a format similar to Equation (3.103) for the Second Piola–Kirchhoff stress tensor. Since the latter equation must hold for any virtual displacement increment $\delta\Delta\mathbf{u}$, the resulting set of equilibrium equations ensues:

$$\left[\int_{V_0} \mathbf{B}^{\mathrm{T}}\mathbf{D}\mathbf{B}\mathrm{d}V + \int_{V_0} \mathbf{B}_{NL}^{\mathrm{T}}\boldsymbol{\Sigma}\mathbf{B}_{NL}\mathrm{d}V\right]\Delta\mathbf{u} = \int_{S_0} \mathbf{H}^{\mathrm{T}}\mathbf{t}_0\mathrm{d}S + \int_{V_0} \rho_0\mathbf{H}^{\mathrm{T}}\mathbf{g}\mathrm{d}V \tag{3.135}$$

At the critical load level at least two solutions exist which *both* satisfy incremental equilibrium. If $\Delta\mathbf{u}_1$ denotes the incremental displacement field belonging to the first solution and if $\Delta\mathbf{u}_2$ is

[1] Note that we are not talking about loss of stability of the equilibrium, but about loss of uniqueness of the solution. With the latter terminology we mean that after a certain load level has been reached in an incremental loading programme, the differential equations that govern the next, infinitesimally small load increment, have more than one solution. All solutions that emanate from this so-called bifurcation point satisfy the differential equations. A simple example is the perfect Euler strut, which has two solutions at the buckling (bifurcation) point: one for which the strut remains perfectly straight and one for which we observe large lateral deflections (buckling). It is emphasised that, although they are intimately related for major classes of material models, the notions of loss of uniqueness and loss of stability are not synonymous. This will be discussed in Chapter 4.

the incremental displacement field of the second solution, subtraction of the solutions, which must both satisfy Equation (3.135), yields:

$$\left[\int_{V_0} \mathbf{B}^{\mathrm{T}}\mathbf{D}\mathbf{B}\mathrm{d}V + \int_{V_0} \mathbf{B}_{NL}^{\mathrm{T}}\boldsymbol{\Sigma}_c\mathbf{B}_{NL}\mathrm{d}V\right](\Delta\mathbf{u}_1 - \Delta\mathbf{u}_2) = \mathbf{0} \tag{3.136}$$

with $\boldsymbol{\Sigma}_c$ the stress matrix at the critical (buckling) load.[2]

We now label $\boldsymbol{\Sigma}_e$ as the stress matrix that is obtained in a linear-elastic calculation for a unit load, and λ the load factor that sets the relation between this elastic solution for a unit load and the stresses at the critical load $\boldsymbol{\Sigma}_c$: $\boldsymbol{\Sigma}_c = \lambda\boldsymbol{\Sigma}_e$, so that λ is the multiplication factor for the unit load to obtain the critical load at which loss of uniqueness (buckling) occurs. Substitution of this identity in Equation (3.136) yields:

$$\left[\int_{V_0} \mathbf{B}^{\mathrm{T}}\mathbf{D}\mathbf{B}\mathrm{d}V + \lambda\int_{V_0} \mathbf{B}_{NL}^{\mathrm{T}}\boldsymbol{\Sigma}_e\mathbf{B}_{NL}\mathrm{d}V\right](\Delta\mathbf{u}_1 - \Delta\mathbf{u}_2) = \mathbf{0} \tag{3.137}$$

Since by definition $\Delta\mathbf{u}_1 - \Delta\mathbf{u}_2 \neq \mathbf{0}$, this equation forms a linear eigenvalue problem for which a non-trivial solution exists if and only if the determinant of the characteristic equation vanishes:

$$\det\left(\mathbf{K}_0 + \lambda\mathbf{K}_{NL}^e\right) = 0 \tag{3.138}$$

where

$$\mathbf{K}_0 = \int_{V_0} \mathbf{B}^{\mathrm{T}}\mathbf{D}\mathbf{B}\mathrm{d}V \quad \text{and} \quad \mathbf{K}_{NL}^e = \int_{V_0} \mathbf{B}_{NL}^{\mathrm{T}}\boldsymbol{\Sigma}_e\mathbf{B}_{NL}\mathrm{d}V$$

The solution of Equation (3.138) results in n eigenvalues, which belong to the load levels at which loss of uniqueness (bifurcation) can occur. The lowest eigenvalue corresponds to the lowest load level for which a bifurcation exists. Multiplication of the elastic solution with this eigenvalue therefore gives the critical load level at which bifurcation is first possible. An example for a simple truss–spring structure is given in Box 3.4. It is noted that the load levels predicted by the higher eigenvalues can be so high that the assumptions made for the linear buckling analysis may be violated, and that these eigenvalues can therefore be unrealistic. In computations of structures that have a fine discretisation, high eigenvalues may arise that are merely artifacts of the discretisation, which is another possible source of unrealistic eigenvalues. It is finally noted that efficient algorithms exist for the computation of the lowest eigenvalues (Chatelin 1993; Golub and van Loan 1983).

[2] Note that the validity of this subtraction rests upon the assumption that at the bifurcation point, all quantities in Equation (3.135) have the same value for both solutions. For the kinematic quantities, as contained in the **B** matrices this is obviously the case, since they represent the current deformation state which is unique. The issue is not so clear for the **D** matrix. For constitutive models which have different stiffness moduli for loading and unloading as in plasticity or damage the question arises as to which stiffness modulus we have to insert in the **D** matrix. This issue will be discussed in greater depth in Chapter 4. For the case of linear elasticity, however, the stiffness moduli are unique and this problem does not arise.

Box 3.4 Buckling of a truss member supported by a spring

For the simple structure of Figure 3.8 the buckling stress is given by $\sigma_c = k\frac{\ell_0}{A_0}$, with k the stiffness of the spring, ℓ_0 the length of the truss member and A_0 the cross-sectional area of the member. The x-axis is chosen to coincide with the axis of the truss member. Further, the boundary conditions are such that $u_1 = v_1 = 0$. Then, after addition of the spring stiffness k, the linear part of the stiffness matrix, Equation (3.30), becomes:

$$\mathbf{K}_0 = \begin{bmatrix} \frac{EA_0}{\ell_0} & 0 \\ 0 & k \end{bmatrix}$$

The second, non-linear part of the stiffness matrix is given by:

$$\mathbf{K}_{NL} = \frac{A_0}{\ell_0}\begin{bmatrix} 1 & 0 \\ 0 & 1 \end{bmatrix}$$

According to Equation (3.138) loss of uniqueness (buckling) first occurs when the determinant of the matrix $(\mathbf{K}_0 + \sigma_c \mathbf{K}_{NL})$ vanishes. Since $\frac{A_0}{\ell_0}$ is always non-zero, this condition can be elaborated as: $(k - \frac{A_0\sigma_c}{\ell_0})(E - \sigma_c) = 0$. so that the two solutions for the eigenvalue problem are given by:

$$\begin{cases} \sigma_c = \frac{k\ell_0}{A_0} \\ \sigma_c = E \end{cases}$$

The first solution is related to buckling of the truss member with the spring acting as a support to prevent lateral buckling. The second solution represents the possible lateral buckling of the spring with the (non-rigid) truss member functioning as a support. Of course, this possibility can only become active if there is a horizontal load present. In any case, EA_0 is usually much larger than kl_0: $EA_0 \gg k\ell_0$. Then, the solution $\sigma_c = \frac{k\ell_0}{A_0}$ corresponds to the lowest eigenvalue and, consequently, gives the lowest (most critical) buckling load. For a rigid truss member $E \to \infty$ and the second buckling stress goes to infinity. The buckling stress that is computed on the basis of a linear buckling analysis is exact for the present case, which is due to the fact that the assumptions made in deriving the linear buckling criterion have been satisfied exactly.

3.6 PyFEM: A Geometrically Non-linear Continuum Element

In this section, we shall continue the development of the PyFEM code by adding a geometrically non-linear continuum element, described in Section 3.4. To improve the flexibility of the code, we will use an input file to specify the finite element discretisation, the boundary conditions, the material constants and the parameters of the solver. The syntax of this input file is presented in the reference manual that is included in the source code of the program.

Moreover, the program will be restructured and we will construct the code in a modular way. The code of the main routine of the program that performs a non-linear simulation is called `NewtonRaphson.py` and can be found in the directory `examples/ch03`.

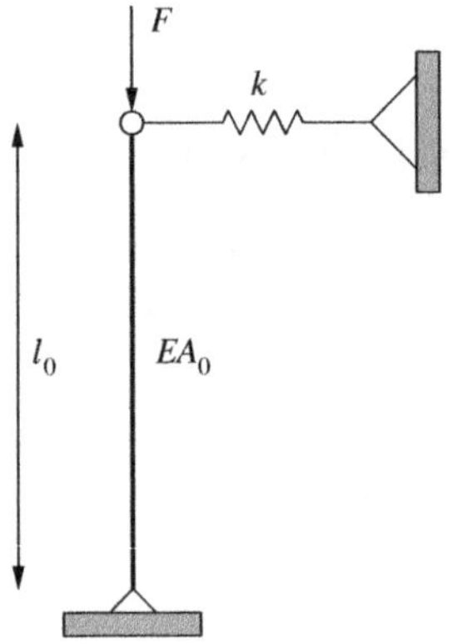

Figure 3.8 Buckling problem: truss member supported by a spring

⟨*Non-linear Newton–Raphson finite element simulation* ⟩≡

```
from pyfem.io.InputReader   import InputReader
from pyfem.io.OutputWriter import OutputWriter
from pyfem.io.MeshWriter   import MeshWriter

from pyfem.solvers.NonlinearSolver import NonlinearSolver

props,globdat    = InputReader( sys.argv )

solver         = NonlinearSolver( props,globdat )
outputWriter = OutputWriter    ( props,globdat )
meshWriter     = MeshWriter      ( props,globdat )

while globdat.active:
  solver        .run( props , globdat )
  outputWriter.run( props , globdat )
  meshWriter    .run( props , globdat )

print "The Newton-Raphson simulation terminated successfully"
```

It is noted that this routine is very small. It basically consists of a number of modular units, such as `InputReader`, `NonlinearSolver` and `MeshWriter`. The simulation is initialised in the function `InputReader`. This function reads the input file, which is passed onto this function through the argument `sys.argv`, together with some command line statements. The entire input file is stored in `props`, which is a return parameter from the function. The instance `props` has been used before in the fragment on page 77. The function furthermore constructs the instances `nodes`, `elements` and `dofs`, which contain the nodal coordinates, the element connectivity and the mapping of global degrees of freedom, respectively. These instances are stored in the data container `globdat`, similar to the fragment on page 79.

Whereas `props` is just a database representation of the input file, the data that are created afterwards, and which can be changed, are stored in the instance `globdat`. This instance not only includes `nodes`, `elements` and `dofs`, but also the current solution vectors `state` and `dstate`, and solver information, such as the current cycle number `cycle` and a boolean `active` that is initialised as `True`. This boolean indicates whether the program is running. When it is set to `False`, e.g. when the maximum number of load steps has been reached, the program is terminated.

The program consists of a single **while** loop, that is controlled by the flag `globdat.active`. Each execution of this loop represents a single step in the solution procedure. Hence, the function `solver.run` solves the non-linear system of equations for a single given external load, similar to the calculations in the fragment ⟨*Solve Non-linear Systems of Equations*⟩ on page 80. When the solution has converged, the main routine continues executing the function `outputWriter.run`, which writes the current solution to a file in a specific text format. The name of the output file, as well as additional instructions, can be given in the input file, which passes this information onto this function via the argument `props`. Finally, the geometry of the deformed mesh is written to a file in `meshWriter.run`. Here, the geometry is stored in the Visualization Toolkit data file format (`www.vtk.org`). Data that are stored in this format can be visualised by means of a variety of post-processors, including the open-source program Paraview (`www.paraview.org`).

We demonstrate the program `NewtonRaphson.py` by analysing the cantilever beam of Figure 3.1. The beam is assigned a length $l = 8.0$ mm, a height $h = 0.5$ mm, a Young's modulus $E = 100\,\text{N/mm}^2$ and a Poisson's ratio $\nu = 0.3$. The beam is loaded by a concentrated load $F = 0.2$ N, which is applied in 20 steps of $\Delta F = 0.01$ N. The finite element mesh consists of 8 eight-noded continuum elements. The input file for this problem is named `cantilever8.pro` and can be found in the same directory. The program is executed by the following command:

```
> python NewtonRaphson.py cantilever8.pro
```

The load–displacement curve of the cantilever beam in this example is shown in Figure 3.9.

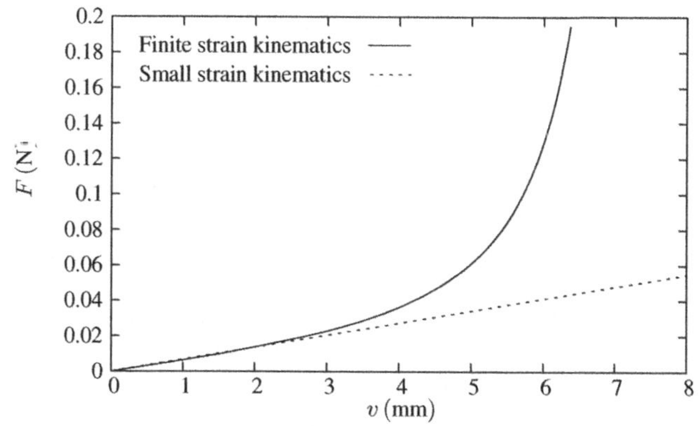

Figure 3.9 Load–displacement curve of the cantilever beam. The solution with a small strain continuum model is given as a reference

The continuum element that is used in this example, is implemented in the same manner as the truss element in Section 3.2. The source code of this element can be found in the file `FiniteStrainContinuum.py` in the directory `elements`. The main structure of this file is as follows:

```
⟨Finite strain continuum element ⟩≡

  ⟨Finite strain continuum element class definition 106⟩
  ⟨Finite strain continuum element class main functions 106⟩
  ⟨Finite strain continuum element class utility functions 107⟩
```

The constructor of the class, which is derived from the base class `Element`, is given by:

```
⟨Finite strain continuum element class definition ⟩≡                106

class FiniteStrainContinuum ( Element ):

  dofTypes = ['u','v']
```

In the constructor, the element nodal degrees of freedom are defined. Since this element is a two-dimensional element, the nodal degrees of freedom are `'u'` and `'v'`. Additional data for the element are initialised in the `__init__` function:

```
⟨Finite strain continuum element class definition ⟩≡                106

  def __init__ ( self, elnodes , props ):
    Element.__init__( self, elnodes , props )
```

The function `getElementStiffness` which is used to construct the element stiffness matrix $\mathbf{K}_e$ and the internal force vector $\mathbf{f}^e_{\rm int}$ is implemented next:

```
⟨Finite strain continuum element class main functions ⟩≡            106

  def getElementStiffness ( self, elemdat ):

    sData = getElemShapeData( elemdat.coords )                        39
```

The first action in this function is the determination of the shape functions. This is done using the function `getElemShapeData`, which has been described in Chapter 2. The single argument of this function is an array that contains the nodal coordinates of the element. This array is a member of the instance `elemdat`. The length of the array `coords` determines the shape function type. As a result, the element implementation that follows is valid for all two dimensional continuum elements. The element shape functions are stored in the instance `sData`. The length of this instance is used to loop over the integration points of the elements:

⟨*Finite strain continuum element class main functions* ⟩+≡ 106

```
for iData in sData:

  kin = self.getKinematics( iData.dhdx , elemdat.state )      107
```

The deformation gradient, Equation (3.54), and the Green–Lagrange strain tensor, Equation (3.61), are calculated in the member function `getKinematics` which uses the derivatives of the shape functions and the current element displacement as input.

⟨*Finite strain continuum element class utility functions* ⟩≡ 106

```
def getKinematics( self , dphi , elstate ):

  kin = Kinematics( 2 , 3 )                                   107

  kin.F = eye(2)

  for i in range(len(dphi)):
    for j in range(2):
      for k in range(2):
        kin.F[j,k] += dphi[i,k]*elstate[2*i+j]

  kin.E = 0.5*( dot(kin.F.transpose() , kin.F ) - eye(2) )

  kin.strain[0] = E[0,0]
  kin.strain[1] = E[1,1]
  kin.strain[2] = 2.0*E[0,1]

  return kin
```

The kinematic properties of the current integration point are stored in the instance `kin` of the class `Kinematics`. This instance contains at least the deformation gradient `F` and the Green–Lagrange strains, which are stored in tensor format, `E`, as well as in vector format, `strain`. When initialising the instance `kin`, two arguments are required, `nDim` and `nStr`, which are the spatial dimensions of the problem and the number of independent strain terms, respectively.

⟨*Kinematics class* ⟩≡

```
class Kinematics:

  def __init__( self , nDim , nStr ):

    self.F      = zeros( shape = ( nDim , nDim ) )
    self.E      = zeros( shape = ( nDim , nDim ) )
    self.strain = zeros( nStr )
```

In this example, the spatial dimension is two and the number of independent strain terms is three.

After the calculation of the strains at the integration point, the Second Piola–Kirchhoff stresses and the material tangential stiffness matrix can be calculated. This is done in the material manager `mat`, which will be discussed later. The stiffness matrix is calculated in two parts, according to Equations (3.99) and (3.102):

⟨*Finite strain continuum element class main functions* ⟩+≡ 107

```
stress,tang = self.mat.getStress( kin )                                  109

B   = self.getBmatrix    ( iData.dhdx , kin.F )                           108
Kl  = dot( B.transpose() , dot( tang , B ) ) * iData.weight

T   = self.stress2matrix( stress )                                        109
Bnl = self.getBNLmatrix( iData.dhdx )
Knl = dot( Bnl.transpose() , dot( T , Bnl ) ) * iData.weight

elemdat.stiff += Kl + Knl
```

The relation between the linear part of the Green–Lagrange strain and the displacement vector, matrix $\mathbf{B}_L$ in Equation (3.97), is calculated in an additional utility function:

⟨*Finite strain continuum element class utility functions* ⟩+≡ 107

```
def getBmatrix( self , dphi , F ):

  B = zeros( shape=( 3 , 2*len(dphi) ) )

  for i,dp in enumerate( dphi ):
    B[0,2*i  ] = dp[0]*F[0,0]
    B[0,2*i+1] = dp[0]*F[1,0]

    B[1,2*i  ] = dp[1]*F[0,1]
    B[1,2*i+1] = dp[1]*F[1,1]

    B[2,2*i  ] = dp[1]*F[0,0]+dp[0]*F[0,1]
    B[2,2*i+1] = dp[0]*F[1,1]+dp[1]*F[1,0]

  return B
```

The non-linear counterpart of this matrix, $\mathbf{B}_{NL}$ in Equation (3.104), is calculated in a similar fashion. The Second Piola–Kirchhoff stress is stored in matrix form in the member function `stress2matrix`, see also Equation (3.103):

⟨*Finite strain continuum element class utility functions* ⟩+≡ 108

```
    def stress2matrix( self , stress ):

      T = zeros( shape=( 4 , 4 ) )

      T[0,0] = stress[0]
      T[1,1] = stress[1]
      T[0,1] = stress[2]
      T[1,0] = stress[2]

      T[2:,2:] = T[:2,:2]

      return T
```

The loop over the integration points in the function `getElementStiffness` is concluded by constructing the internal force vector $\mathbf{f}_{\text{int}}$:

⟨*Finite strain continuum element class main functions* ⟩+≡ 108

```
        elemdat.fint  += dot ( B.transpose() , stress ) * iData.weight
```

In order to be able to use different material models in this element, we introduce another level of abstraction, the material model. One of the input properties of the element is a reference to a material model. In this case, the material model that is specified in the input file is `PlaneStrain`. The `elementSet` class contains a so-called material manager that connects the correct material model to a specific element. In the fragment on page 108 we have already encountered the most important function of the material model `getStress`. Now, we will focus on the implementation.

The structure of the file `PlaneStrain.py`, which contains the constitutive relation for plane-strain conditions, is as follows:

⟨*Plane strain material model* ⟩≡

⟨*Initialisation of the plain strain material class* 110⟩
⟨*Plane strain member functions* 110⟩

The material is implemented as a class, derived from the class `BaseMaterial`:

⟨*Initialisation of the plain strain class* ⟩≡ 109

```
from pyfem.materials.BaseMaterial import BaseMaterial

class PlaneStrain( BaseMaterial ):

  def __init__ ( self, props ):

    BaseMaterial.__init__( self, props )

    self.H = zeros( (3,3) )

    self.H[0,0] = self.E*(1.-self.nu)/((1+self.nu)*(1.-2.*self.nu))
    self.H[0,1] = self.H[0,0]*self.nu/(1-self.nu)
    self.H[1,0] = self.H[0,1]
    self.H[1,1] = self.H[0,0]
    self.H[2,2] = self.H[0,0]*0.5*(1.-2.*self.nu)/(1.-self.nu)
```

In the constructor of the `BaseMaterial`, the material properties are read from the instance `props` and are stored as members of the class. For example, the Young's modulus, which is labelled as `E` in the input file and in the properties data structure, is stored as `self.E` in this class. The same holds for the other parameter of this material model, the Poisson's ratio `nu`. Since the model is a linear stress–strain relation, it is advantageous to calculate the constant material tangential stiffness also in the constructor, and to store it as a member variable `self.H`

The only member function in the material model class is `getStress`:

⟨*Plane strain class member functions* ⟩≡ 109
107

```
  def getStress( self, kinematics ):

    sigma = dot( self.H, kinematics.strain )

    return sigma, self.H
```

The return values of this function are the stress, which in this case is the result of the multiplication of the material stiffness matrix and the strain vector, and the material stiffness matrix. In Chapter 6, when non-linear material models are discussed, we will return to the implementation of material models.

References

Bathe KJ, Ramm E and Wilson EL 1975 Finite element formulations for large deformation dynamic analysis. *International Journal for Numerical Methods in Engineering* **9**, 353–386.

Bonet J and Wood RD 1997 *Nonlinear Continuum Mechanics for Finite Element Analysis*. Cambridge University Press.

Chatelin F 1993 *Eigenvalues of Matrices*. John Wiley & Sons, Ltd.

Crisfield MA and Moita GF 1996a A co-rotational formulation for 2-D continua including incompatible modes. *International Journal for Numerical Methods in Engineering* **39**, 2619–2933.

Crisfield MA and Moita GF 1996b A finite element formulation for 3-D continua using the co-rotational technique. *International Journal for Numerical Methods in Engineering* **39**, 3775–3792.

Fung YC 1965 *Foundations of Solid Mechanics*. Prentice-Hall.

Golub GH and van Loan CF 1983 *Matrix Computations*. The Johns Hopkins University Press.

Holzapfel GA 2000 *Nonlinear Solid Mechanics*. John Wiley & Sons, Ltd.

Jetteur PH and Cescotto S 1991 A mixed finite element for the analysis of large inelastic strains. *International Journal for Numerical Methods in Engineering* **31**, 229–239.

Malvern L 1969 *Introduction to the Mechanics of a Continuous Medium*. Prentice-Hall.

Ogden RW 1984 *Non-Linear Elastic Deformations*. Ellis Horwood.

Yaw LL, Sukumar N and Kunnath SK 2009 Meshfree co-rotational formulation for two-dimensional continua. *International Journal for Numerical Methods in Engineering* **79**, 979–1003.

4

Solution Techniques in Quasi-static Analysis

In Chapter 2 the basic ideas were introduced regarding the solution of the set of non-linear algebraic equations which arises in a non-linear finite element analysis, namely load and displacement control, the incremental-iterative solution strategy, and the use of Newton–Raphson type methods within a load step. In this chapter advanced solution techniques will be discussed, including line searches, path-following methods, Quasi-Newton methods, and branch switching techniques at bifurcation points.

4.1 Line Searches

A drawback of any variant of the Newton–Raphson method is the limited radius of convergence. To enlarge this convergence radius, line searches have been introduced. The basic idea of the line search technique is to apply an improvement to the original incremental displacement vector $\mathrm{d}\tilde{\mathbf{a}}_{j+1}$ by scaling it by a multiplier η_{j+1} such that we arrive at the point of lowest potential energy along the search direction. The correction to the displacement increment then reads:

$$\mathrm{d}\mathbf{a}_{j+1} = \eta_{j+1}\mathrm{d}\tilde{\mathbf{a}}_{j+1} \tag{4.1}$$

where, for load control, $\mathrm{d}\tilde{\mathbf{a}}_{j+1}$ follows from

$$\mathrm{d}\tilde{\mathbf{a}}_{j+1} = \mathbf{K}_j^{-1}(\mathbf{f}_{\mathrm{ext}}^{t+\Delta t} - \mathbf{f}_{\mathrm{int},j})$$

cf. Equation (2.52). Generally, the potential energy Π of a system is a function of the displacements. Let the total displacements after iteration j of a certain load step be given by $\mathbf{a}_j$ and the correction in iteration $j+1$ be given by $\mathrm{d}\mathbf{a}_{j+1}$, such that $\mathbf{a}_{j+1} = \mathbf{a}_j + \mathrm{d}\mathbf{a}_{j+1}$. Then, a Taylor

Non-linear Finite Element Analysis of Solids and Structures, Second Edition.
René de Borst, Mike A. Crisfield, Joris J.C. Remmers and Clemens V. Verhoosel.

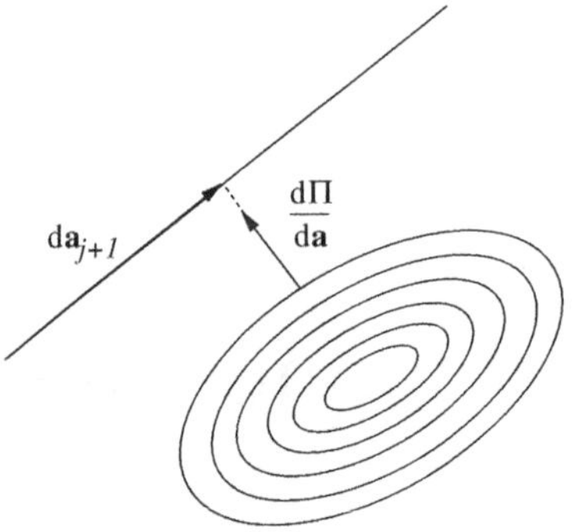

Figure 4.1 Isolines of the potential energy Π and the line-search direction

expansion of Π gives:

$$\Pi(\mathbf{a}_{j+1}) = \Pi(\mathbf{a}_j) + \left(\frac{\partial \Pi}{\partial \mathbf{a}}\right)^{\mathrm{T}} \mathrm{d}\mathbf{a}_{j+1} + \frac{1}{2}\mathrm{d}\mathbf{a}_{j+1}^{\mathrm{T}} \left(\frac{\partial^2 \Pi}{\partial \mathbf{a}\partial \mathbf{a}}\right) \mathrm{d}\mathbf{a}_{j+1} + \cdots \tag{4.2}$$

Evidently, Π is minimised by requiring

$$\left(\frac{\partial \Pi}{\partial \mathbf{a}}\right)^{\mathrm{T}} \mathrm{d}\mathbf{a}_{j+1} = 0 \tag{4.3}$$

see also Figure 4.1 for a graphical interpretation. The variation of the potential energy equals the difference of the external and the internal virtual work, $\delta\Pi = \delta W_{\mathrm{ext}} - \delta W_{\mathrm{int}}$, so that, in view of the discussion in Section 2.3, we have

$$\delta\Pi = (\mathbf{f}_{\mathrm{ext}}^{t+\Delta t})^{\mathrm{T}}\delta\mathbf{a} - (\mathbf{f}_{\mathrm{int},j+1})^{\mathrm{T}}\delta\mathbf{a}$$

Considering that

$$\delta\Pi = \left(\frac{\partial \Pi}{\partial \mathbf{a}}\right)^{\mathrm{T}} \delta\mathbf{a}$$

the derivative $\frac{\partial \Pi}{\partial \mathbf{a}}$ can be identified as

$$\frac{\partial \Pi}{\partial \mathbf{a}} = \mathbf{f}_{\mathrm{ext}}^{t+\Delta t} - \mathbf{f}_{\mathrm{int},j+1} = \mathbf{f}_{\mathrm{ext}}^{t+\Delta t} - \mathbf{f}_{\mathrm{int}}(\mathbf{a}_j + \eta_{j+1}\mathrm{d}\tilde{\mathbf{a}}_{j+1}) \tag{4.4}$$

and the line search multiplier η_{j+1} can be determined after substitution of this result into Equation (4.3):

$$\left(\mathbf{f}_{\mathrm{ext}}^{t+\Delta t} - \mathbf{f}_{\mathrm{int}}(\mathbf{a}_j + \eta_{j+1}\mathrm{d}\tilde{\mathbf{a}}_{j+1})\right)^{\mathrm{T}} \mathrm{d}\mathbf{a}_{j+1} = 0 \tag{4.5}$$

Experience shows that an exact satisfaction of this requirement, and accordingly, an accurate determination of η_{j+1}, are not necessary. An accurate determination of η_{j+1} often only has a marginal effect on the speed of the global iterative procedure, definitely when the solution has

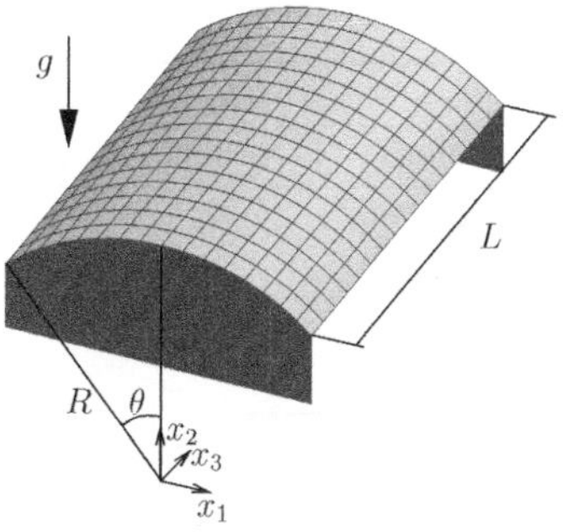

Figure 4.2 Cylindrical shell subjected to self weight loading (Scordelis-Lo roof)

come within the convergence radius of the Newton–Raphson procedure. We define

$$s = (\mathbf{f}_{\text{ext}}^{t+\Delta t} - \mathbf{f}_{\text{int}}(\mathbf{a}_j + \eta_{j+1}\mathbf{da}_{j+1}))^{\mathrm{T}}\mathbf{da}_{j+1} \tag{4.6}$$

The line-search procedure is then terminated when

$$|s| \leq \psi|s_0| \tag{4.7}$$

with the line-search tolerance ψ usually taken in the range 0.7–0.9 and

$$s_0 = (\mathbf{f}_{\text{ext}}^{t+\Delta t} - \mathbf{f}_{\text{int}}(\mathbf{a}_j))^{\mathrm{T}}\mathbf{da}_{j+1} \tag{4.8}$$

which is easily computed, since $\mathbf{f}_{\text{int}}(\mathbf{a}_j)$ is known from the previous global equilibrium iteration. With a slack tolerance ψ, the criterion (4.7) is normally fulfilled in the first line search.

The Scordelis-Lo roof is one of the classical benchmarks in shell theories (MacNeal and Harder 1985). A schematic representation of the problem is shown in Figure 4.2. The problem is considered in non-dimensional form with radius $R = 25$, length $L = 50$ and thickness $t = 0.25$. The cross section of the roof is described by a circular arc of $2\theta = 80°$. The modulus of elasticity and Poisson's ratio are taken as 4.32×10^8 and 0, respectively, with a von Mises yield contour to bound the allowable stresses, see Chapter 7 for details. The roof is loaded by a gravity load of $g = 90$ (per unit area), and is clamped at its ends by rigid diaphragms which constrain the displacements of the shell in the x_1- and in the x_2-direction. Because of symmetry considerations the calculations have been carried out for a quarter of the shell, which has been idealised using 16 eight-noded shell elements. A 2×2 Gauss integration rule in the plane and five-point Simpson integration through the thickness have been used (Chapter 9).

In Tables 4.1 and 4.2 the effect of line searches on the convergence is presented for different values of the line-search tolerance. These tables suggest that line searches are only then useful when no proper tangential stiffness matrix is adopted. In the case of a full Newton–Raphson method hardly any savings in computer time are obtained when applying line searches. It is also observed that although a line-search procedure improves the performance of the modified Newton–Raphson method, it is still not competitive with a full Newton–Raphson scheme. In fact, line searches can only enhance the performance of a full Newton–Raphson scheme in the first iterations, in order to bring the solution within the radius of convergence of the Newton–Raphson method.

Table 4.1 Effect of line searches on a full Newton–Raphson method

Tolerance	Iterations	No. of searches	Relative computing time(%)
No line search	36	0	100
0.8	34	1	102
0.6	34	1	102
0.4	34	1	102

Table 4.2 Effect of line searches on a modified Newton–Raphson method

Tolerance	Iterations	No. of searches	Relative computing time(%)
No line search	249	0	302
0.8	204	64	284
0.6	120	113	235
0.4	129	157	275

4.2 Path-following or Arc-length Methods

The crucial idea of the path-following method or arc-length method is that the load increment $\Delta\lambda$ is considered as an additional unknown, thereby augmenting the n-dimensional space of unknown displacements, collected in the array $\mathbf{a}$, to an $n+1$-dimensional space of unknowns. Let $\mathbf{a}_0$ contain the nodal displacements at the beginning of a generic load increment, let λ_0 be the value of the load parameter, and let $\Delta\mathbf{a}$ and $\Delta\lambda$ be their increments, see also Equation (2.45). Since we have $n+1$ unknowns and only n equations the system is indeterminate and an additional equation must be supplied. This is done by adding a path-following constraint (Riks 1972):

$$g(\mathbf{a}_0, \lambda_0, \Delta\mathbf{a}, \Delta\lambda, \Delta\ell) = 0 \tag{4.9}$$

where $\Delta\ell$ is the path length increment that determines the size of the load increment. The new equilibrium state of the augmented system of $n+1$ equations can now be determined by simultaneously solving

$$\begin{bmatrix} -\mathbf{r} \\ g \end{bmatrix} = \begin{bmatrix} \mathbf{0} \\ 0 \end{bmatrix} \tag{4.10}$$

Solution of this non-linear system can be achieved in a standard manner, e.g. via a Newton–Raphson method, where the set (4.10) is linearised to give:

$$\begin{bmatrix} \mathbf{f}_{\text{int},j} + \frac{\partial \mathbf{f}_{\text{int}}}{\partial \mathbf{a}} \mathrm{d}\mathbf{a}_{j+1} - \lambda_j \hat{\mathbf{f}}_{\text{ext}} - \mathrm{d}\lambda_{j+1} \hat{\mathbf{f}}_{\text{ext}} \\ g_j + \left(\frac{\partial g}{\partial \mathbf{a}}\right)^{\mathrm{T}} \mathrm{d}\mathbf{a}_{j+1} + \frac{\partial g}{\partial \lambda} \mathrm{d}\lambda_{j+1} \end{bmatrix} = \begin{bmatrix} \mathbf{0} \\ 0 \end{bmatrix} \tag{4.11}$$

The solution $(\mathbf{a}, \lambda)$ at iteration $j+1$ is then obtained by solving:

$$\begin{bmatrix} \mathbf{K} & -\hat{\mathbf{f}}_{\text{ext}} \\ \mathbf{h}^{\text{T}} & s \end{bmatrix} \begin{bmatrix} \mathbf{da}_{j+1} \\ \text{d}\lambda_{j+1} \end{bmatrix} = \begin{bmatrix} \mathbf{r}_j \\ -g_j \end{bmatrix} \tag{4.12}$$

with

$$\mathbf{K} \equiv \frac{\partial \mathbf{f}_{\text{int}}}{\partial \mathbf{a}} \tag{4.13}$$

the tangential stiffness matrix, and the array $\mathbf{h}$ and the scalar s defined as:

$$\mathbf{h} = \frac{\partial g}{\partial \mathbf{a}} \quad , \quad s = \frac{\partial g}{\partial \lambda} \tag{4.14}$$

Unless the rather exceptional case occurs that $\mathbf{h}$ is orthogonal to $\hat{\mathbf{f}}_{\text{ext}}$, the augmented stiffness matrix does not become singular, not in limit points, and neither at points of the load–deflection curve where snap-back behaviour is encountered (Figure 2.8).

Equation (4.12) destroys the symmetry and the banded nature of the tangential stiffness matrix. For this reason, solution of the coupled set of equations (4.12) is not done in a direct manner. Instead, a partitioned procedure is normally adopted, which effectively leads to a two-stage solution procedure for the system of equations (Crisfield 1981; Ramm 1981). The computational procedure for this implementation of path-following techniques is as follows. Assuming that $\mathbf{K}$ is non-singular, the following arrays are computed that contain contributions to the incremental displacements:

$$\mathbf{da}^{\text{I}}_{j+1} = \mathbf{K}^{-1}\hat{\mathbf{f}}_{\text{ext}} \tag{4.15}$$

$$\mathbf{da}^{\text{II}}_{j+1} = \mathbf{K}^{-1}\mathbf{r}_j \tag{4.16}$$

From the first equation of the set (4.11) the new estimate for the displacement increment then follows as:

$$\mathbf{da}_{j+1} = \text{d}\lambda_{j+1}\mathbf{da}^{\text{I}}_{j+1} + \mathbf{da}^{\text{II}}_{j+1} \tag{4.17}$$

while the second equation of the set (4.11) gives the new estimate for the load increment:

$$\text{d}\lambda_{j+1} = -\frac{g_j + \mathbf{h}^{\text{T}}\mathbf{da}^{\text{II}}_{j+1}}{s + \mathbf{h}^{\text{T}}\mathbf{da}^{\text{I}}_{j+1}} \tag{4.18}$$

The computational procedure is given in Box 4.1.

Geometrically, path-following techniques can be interpreted by considering a generalised $n+1$-dimensional load–displacement space, that includes the n discrete displacements and the load parameter λ. The idea is that it is always possible to construct a hypersurface g in this space that intersects with the load–displacement curve. Such a surface can be a hypersphere, as

Box 4.1 Computational flow when using a path-following technique

For each loading step:

1. Initialise the data for the loading step. Set $\Delta \mathbf{a}_0 = \mathbf{0}$.
2. Compute the tangential stiffness matrix: $\mathbf{K}_j$.
3. Adjust for prescribed displacements and linear dependence relations.
4. Solve the linear systems: $\mathrm{d}\mathbf{a}^{\mathrm{I}}_{j+1} = \mathbf{K}_j^{-1}\hat{\mathbf{f}}_{\mathrm{ext}}$ and $\mathrm{d}\mathbf{a}^{\mathrm{II}}_{j+1} = \mathbf{K}_j^{-1}\mathbf{r}_j$.
5. Compute $\mathrm{d}\lambda_{j+1}$ via a constraint equation.
6. Compute: $\Delta\mathbf{a}_{j+1} = \Delta\mathbf{a}_j + \mathrm{d}\lambda_{j+1}\mathrm{d}\mathbf{a}^{\mathrm{I}}_{j+1} + \mathrm{d}\mathbf{a}^{\mathrm{II}}_{j+1}$.
7. Compute $\Delta\boldsymbol{\epsilon}_{i,j+1}$ for each integration point i: $\Delta\mathbf{a}_{j+1} \rightarrow \Delta\boldsymbol{\epsilon}_{i,j+1}$.
8. Compute $\Delta\boldsymbol{\sigma}_{i,j+1}$ for each integration point i: $\Delta\boldsymbol{\epsilon}_{i,j+1} \rightarrow \Delta\boldsymbol{\sigma}_{i,j+1}$.
9. Add $\Delta\boldsymbol{\sigma}_{i,j+1}$ to $\boldsymbol{\sigma}_{i,0}$ for each integration point i: $\boldsymbol{\sigma}_{i,j+1} = \boldsymbol{\sigma}_{i,0} + \Delta\boldsymbol{\sigma}_{i,j+1}$.
10. Compute the internal force vector: $\mathbf{f}_{\mathrm{int},j+1}$.
11. Check convergence: is $\|\mathbf{r}_{j+1}\| < \eta$, with η a small number? If yes, go to the next loading step, else go to 2.

depicted in Figure 4.3, but hyperplanes, either fixed or updated (Figure 4.4), and hyperellipsoids have been suggested as well. The use of a path-following technique makes it possible to overcome limit points (where the tangent is horizontal) and also points at which that tangent becomes vertical (snap-back behaviour) in load–displacement curves in an elegant and robust manner. In accordance with the convergence radius of Newton–Raphson methods, the path length increment $\Delta\ell$ has to be reduced when the curvature of the load–displacement curve is strong.

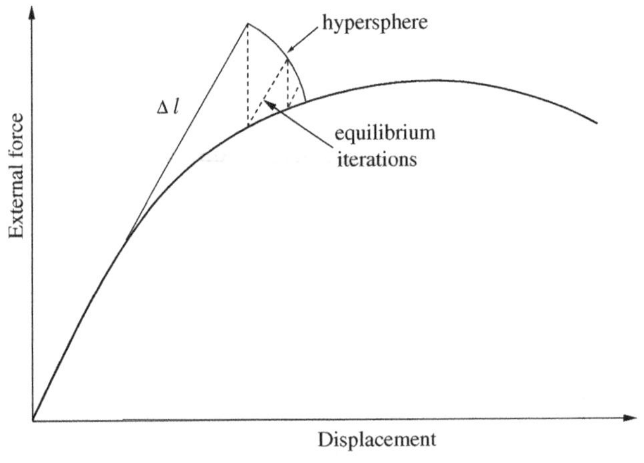

Figure 4.3 Example of a path following technique, where a hypersphere is used for the constraint function g and a modified Newton–Raphson method is used to reach equilibrium

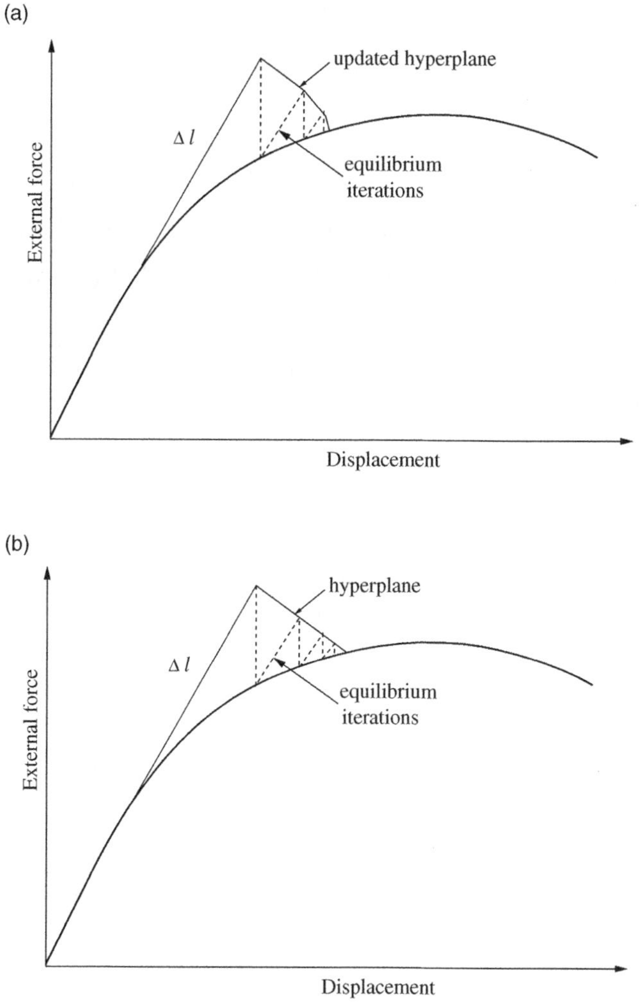

Figure 4.4 Geometrical interpretations of alternative constraint equations g. (a) Updated hyperplanes. (b) A fixed hyperplane

A constraint function that is used widely is the spherical arc-length constraint (Crisfield 1981):

$$g = \Delta\mathbf{a}_{j+1}^{\mathrm{T}}\Delta\mathbf{a}_{j+1} + \beta^2\Delta\lambda_{j+1}^2\hat{\mathbf{f}}_{\mathrm{ext}}^{\mathrm{T}}\hat{\mathbf{f}}_{\mathrm{ext}} - \Delta\ell^2 \tag{4.19}$$

with β a user-specified value that weighs the importance of the contributions that stem from the displacement degrees of freedom and the load increment. It should be chosen such that the different magnitudes of the displacement increments $\Delta\mathbf{a}_{j+1}$ and the load as collected in $\Delta\lambda\hat{\mathbf{f}}_{\mathrm{ext}}$ are properly balanced. A disadvantage of this method is that the algebraic equation in

$\Delta\lambda_{j+1}$ is quadratic. Setting $g = 0$ and employing Equation (4.17) gives:

$$a_1 \mathrm{d}\lambda_{j+1}^2 + a_2 \mathrm{d}\lambda_{j+1} + a_3 = 0 \tag{4.20}$$

with

$$\begin{aligned} a_1 &= (\mathbf{da}_{j+1}^{\mathrm{I}})^{\mathrm{T}} \mathbf{da}_{j+1}^{\mathrm{I}} + \beta^2 \hat{\mathbf{f}}_{\mathrm{ext}}^{\mathrm{T}} \hat{\mathbf{f}}_{\mathrm{ext}} \\ a_2 &= 2(\Delta\mathbf{a}_j + \mathbf{da}_{j+1}^{\mathrm{II}})^{\mathrm{T}} \mathbf{da}_{j+1}^{\mathrm{I}} + 2\beta^2 \Delta\lambda_j \hat{\mathbf{f}}_{\mathrm{ext}}^{\mathrm{T}} \hat{\mathbf{f}}_{\mathrm{ext}} \\ a_3 &= (\Delta\mathbf{a}_j + \mathbf{da}_{j+1}^{\mathrm{II}})^{\mathrm{T}} (\Delta\mathbf{a}_j + \mathbf{da}_{j+1}^{\mathrm{II}}) + \beta^2 \Delta\lambda_j^2 \hat{\mathbf{f}}_{\mathrm{ext}}^{\mathrm{T}} \hat{\mathbf{f}}_{\mathrm{ext}} - \Delta\ell^2 \end{aligned} \tag{4.21}$$

Commonly, the root of the above quadratic equation is chosen which gives an incremental displacement vector that points in the same direction as that which was ultimately obtained in the previous loading step $\Delta\mathbf{a}^t$

$$\Delta\mathbf{a}_{j+1}^{\mathrm{T}} \Delta\mathbf{a}^t > 0 \tag{4.22}$$

At strongly curved parts of the equilibrium path, especially in the presence of sharp snap-back behaviour, this method does not always work well. A particular difficulty is that two imaginary roots can be computed. The obvious remedy to this non-physical result is to decrease the increment size. As an alternative, Equation (4.19) can be linearised to give (cf. Figure 4.4):

$$\Delta\mathbf{a}_j^{\mathrm{T}} \Delta\mathbf{a}_{j+1} + \beta^2 \Delta\lambda_j \Delta\lambda_{j+1} \hat{\mathbf{f}}_{\mathrm{ext}}^{\mathrm{T}} \hat{\mathbf{f}}_{\mathrm{ext}} - \Delta\ell^2 = 0 \tag{4.23}$$

This linearisation results in the updated normal path method (Ramm 1981):

$$\mathrm{d}\lambda_{j+1} = \frac{\Delta\ell^2 - \Delta\mathbf{a}_j^{\mathrm{T}} \Delta\mathbf{a}_j - \Delta\mathbf{a}_j^{\mathrm{T}} \mathbf{da}_{j+1}^{\mathrm{II}} - \beta^2 \Delta\lambda_j^2 \hat{\mathbf{f}}_{\mathrm{ext}}^{\mathrm{T}} \hat{\mathbf{f}}_{\mathrm{ext}}}{\Delta\mathbf{a}_j^{\mathrm{T}} \mathbf{da}_{j+1}^{\mathrm{I}} + \beta^2 \Delta\lambda_j \hat{\mathbf{f}}_{\mathrm{ext}}^{\mathrm{T}} \hat{\mathbf{f}}_{\mathrm{ext}}} \tag{4.24}$$

where Equation (4.18) has been used. Noting that $\Delta\ell^2 \approx \Delta\mathbf{a}_j^{\mathrm{T}} \Delta\mathbf{a}_j$, one can, within the spirit of linearisation, approximate Equation (4.24) by:

$$\mathrm{d}\lambda_{j+1} = -\frac{\Delta\mathbf{a}_j^{\mathrm{T}} \mathbf{da}_{j+1}^{\mathrm{II}} + \beta^2 \Delta\lambda_j^2 \hat{\mathbf{f}}_{\mathrm{ext}}^{\mathrm{T}} \hat{\mathbf{f}}_{\mathrm{ext}}}{\Delta\mathbf{a}_j^{\mathrm{T}} \mathbf{da}_{j+1}^{\mathrm{I}} + \beta^2 \Delta\lambda_j \hat{\mathbf{f}}_{\mathrm{ext}}^{\mathrm{T}} \hat{\mathbf{f}}_{\mathrm{ext}}} \tag{4.25}$$

As a further simplification, we can keep constant the direction of the tangent normal to the hyperplane which is used to intersect the equilibrium path after the first iteration:

$$\Delta\mathbf{a}_1^{\mathrm{T}} \Delta\mathbf{a}_{j+1} + \beta^2 \Delta\lambda_1 \Delta\lambda_{j+1} \hat{\mathbf{f}}_{\mathrm{ext}}^{\mathrm{T}} \hat{\mathbf{f}}_{\mathrm{ext}} - \Delta\ell^2 = 0 \tag{4.26}$$

This normal path method is very simple, especially if one realises that upon subtraction of

$$\Delta\mathbf{a}_1^{\mathrm{T}} \Delta\mathbf{a}_j + \beta^2 \Delta\lambda_1 \Delta\lambda_j \hat{\mathbf{f}}_{\mathrm{ext}}^{\mathrm{T}} \hat{\mathbf{f}}_{\mathrm{ext}} - \Delta\ell^2 = 0$$

which holds in the preceding iteration, we arrive at an expression from which $\Delta\ell$ has been dropped:

$$\Delta\mathbf{a}_1^{\mathrm{T}} \mathbf{da}_{j+1} + \beta^2 \Delta\lambda_1 \mathrm{d}\lambda_{j+1} \hat{\mathbf{f}}_{\mathrm{ext}}^{\mathrm{T}} \hat{\mathbf{f}}_{\mathrm{ext}} = 0 \tag{4.27}$$

Together with Equation (4.18) the following simple formula ensues:

$$d\lambda_{j+1} = -\frac{\Delta\mathbf{a}_1^{\mathrm{T}}\mathbf{da}_{j+1}^{\mathrm{II}}}{\Delta\mathbf{a}_1^{\mathrm{T}}\mathbf{da}_{j+1}^{\mathrm{I}} + \beta^2\Delta\lambda_1\hat{\mathbf{f}}_{\mathrm{ext}}^{\mathrm{T}}\hat{\mathbf{f}}_{\mathrm{ext}}} \tag{4.28}$$

which resembles Equation (4.25), in particular when $\beta = 0$. Equation (4.28) resembles the constraint equation originally introduced by Riks (1972). It is noted that the above results for $d\lambda_{j+1}$ could also have been obtained by substituting the various expressions for the constraint equation g into Equation (4.18), while using Equation (4.14).

An argument in favour of still using Equation (4.12) is that the augmented stiffness matrix does not become singular, not in limit points, and also not at points of the load–deflection curve where snap-back behaviour is encountered. On the other hand, engineering systems usually involve so many degrees of freedom, that it is extremely seldom that one really 'hits' a point in which the tangential stiffness matrix exactly becomes singular. It is finally noted that extensions exist for cases where the external force vector is not directly available, such as in thermal loading or in out-of-plane loading of generalised plane-strain elements, see Box 4.2.

The various alternatives for determining $d\lambda_{j+1}$ normally do not give very different results, and a systematic advantage of one formula over the others does not seem to exist. Also, numerical experience indicates that the value of β does not seem to influence the performance of the method very much. The version with $\beta = 0$ seems to be robust for many cases of engineering interest.

When physical non-linearities (plasticity, damage) are involved, the situation can be different. Indeed, in such cases we observe that at incipient failure all further deformation tends to localise in narrow bands in the structure, for instance shear bands in soils and metals, dilational bands in polymers, rock faults and cracks in concrete and brittle rocks. Then, a global constraint equation to control the solution is inadequate. For instance, in the case of a single, dominant crack, it is very reasonable from a physical viewpoint – and actually, numerically also very robust and efficient – to apply the constraint condition directly on the crack opening displacement (COD). Labelling the nodes at both sides of the notch as m and n, the constraint condition then becomes

$$da_m - da_n = 0 \tag{4.29}$$

with a_m and a_n the displacements of the nodes m and n. This can be incorporated elegantly within the format of global constraint equations as outlined before by redefining the constraint equation as in Wriggers and Simo (1990):

$$\Delta\mathbf{a}_{j+1}^{\mathrm{T}}\mathbf{A}\Delta\mathbf{a}_{j+1} + \beta^2\Delta\lambda_{j+1}^2\hat{\mathbf{f}}_{\mathrm{ext}}^{\mathrm{T}}\hat{\mathbf{f}}_{\mathrm{ext}} - \Delta\ell^2 = 0 \tag{4.30}$$

where we have taken the spherical constraint equation (4.19) as an example. Similar relations can be constructed for the normal path and the updated normal path method. For COD control the matrix $\mathbf{A}$ reads:

$$\mathbf{A} = \mathrm{diag}[\ldots, a_m, \ldots, a_n, \ldots] \tag{4.31}$$

The limiting case when only a single degree of freedom is selected to control the loading process constitutes a generalisation of early attempts to control the solution by prescribing, directly or indirectly, a single degree of freedom (Batoz and Dhatt 1979). Selecting a limited number of

Box 4.2 A path-following technique for thermal loading

A path-following method can also be applied when an external load vector that arises from mechanical actions is not directly available. Examples are thermal loading, or when generalised plane-strain elements are used for which the loading direction is out-of-plane. We take the example of thermal loading, for which the constitutive relation reads:

$$d\boldsymbol{\sigma} = \mathbf{D}_{\text{tan}}(d\boldsymbol{\epsilon} - \alpha d\theta \boldsymbol{\pi})$$

with $\mathbf{D}_{\text{tan}}$ the material tangential stiffness operator, α the linear coefficient of thermal expansion, θ the temperature measured with respect to some reference temperature and $\boldsymbol{\pi}^{\text{T}} = (1, 1, 1, 0, 0, 0)$. In the absence of external actions, equilibrium at iteration $j+1$ reduces to $\mathbf{f}_{\text{int},j+1}$, or, after a first-order Taylor series expansion:

$$\int_V \mathbf{B}^{\text{T}} d\boldsymbol{\sigma} dV = -\mathbf{f}_{\text{int},j}$$

Inserting the constitutive relation and applying the kinematic relation between strains and nodal displacements, $\boldsymbol{\epsilon} = \mathbf{Ba}$, yields:

$$\int_V \mathbf{B}^{\text{T}} \mathbf{D}_{\text{tan}}(\mathbf{B} d\mathbf{a}_{j+1} - \alpha d\theta_{j+1} \boldsymbol{\pi}) dV = -\mathbf{f}_{\text{int},j}$$

Using the global tangential stiffness matrix $\mathbf{K}$ and defining the normalised 'external' force vector as:

$$\hat{\mathbf{f}}_{\text{ext}} = \int_V \alpha \mathbf{B}^{\text{T}} \mathbf{D}_{\text{tan}} \boldsymbol{\pi} dV$$

we obtain:

$$\mathbf{K} d\mathbf{a}_{j+1} = d\theta_{j+1} \hat{\mathbf{f}}_{\text{ext}} - \mathbf{f}_{\text{int},j}$$

Setting $d\mathbf{a}^{\text{I}}_{j+1} = \mathbf{K}^{-1} \hat{\mathbf{f}}_{\text{ext}}$ and $d\mathbf{a}^{\text{II}}_{j+1} = -\mathbf{K}^{-1} \mathbf{f}_{\text{int},j}$ we obtain:

$$d\mathbf{a}_{j+1} = d\theta_{j+1} d\mathbf{a}^{\text{I}}_{j+1} + d\mathbf{a}^{\text{II}}_{j+1}$$

which has the same form as Equation (4.17) and can be regarded as a constrained non-linear finite element procedure, where the constraint applies to θ.

degrees of freedom to control the loading process can be conceived as applying a constraint equation in a subspace of the $n+1$-dimensional load–displacement space. A comprehensive overview and comparison of the application of constraint equations in subspaces has been provided by Geers (1999a,b).

Recently, a constraint equation g has been proposed for dissipative failure processes, which is based on the energy release rate. This constraint appears to be very robust for inelasticity, works well for highly localised failure modes, and has the advantage that no a priori selection of degrees of freedom has to be done, since the dissipated energy is a global quantity (Verhoosel *et al.* 2009). Moreover, since such a constraint is directly related to the failure process, good convergence is observed until the last stages of the failure process. The dissipation rate of a body $\mathcal{G}$ is defined as the difference of the exerted power $\mathcal{P}$ and the rate of elastic energy $\dot{\mathcal{V}}$. Since the exerted power equals the nodal velocities times the applied external forces, the rate of dissipation becomes:

$$\mathcal{G} = \lambda \dot{\mathbf{a}}^{\mathrm{T}} \hat{\mathbf{f}}_{\mathrm{ext}} - \dot{\mathcal{V}} \tag{4.32}$$

The elastic energy stored in the solid depends on the constitutive behaviour and on the kinematic formulation. For instance, for small deformations and inelastic behaviour with unloading back to the origin along a secant branch, the stored elastic energy reads:

$$\mathcal{V} = \frac{1}{2} \int_V \boldsymbol{\epsilon}^{\mathrm{T}} \boldsymbol{\sigma} \mathrm{d}V = \frac{1}{2} \mathbf{a}^{\mathrm{T}} \mathbf{f}_{\mathrm{int}}$$

Assuming that at the end of the previous loading step an equilibrium has been obtained, so that $\mathbf{r} \equiv \mathbf{0}$, we have equivalently

$$\mathcal{V} = \frac{1}{2} \lambda \mathbf{a}^{\mathrm{T}} \hat{\mathbf{f}}_{\mathrm{ext}}$$

so that

$$\mathcal{G} = \frac{1}{2} (\lambda \dot{\mathbf{a}} - \dot{\lambda} \mathbf{a})^{\mathrm{T}} \hat{\mathbf{f}}_{\mathrm{ext}} \tag{4.33}$$

With a forward Euler integration the expression for the constraint equation g then becomes

$$g = \frac{1}{2} (\lambda_0 \Delta \mathbf{a} - \Delta \lambda \mathbf{a}_0)^{\mathrm{T}} \hat{\mathbf{f}}_{\mathrm{ext}} - \Delta \ell \tag{4.34}$$

as an example of a constraint equation based on energy dissipation. The derivatives of g with respect to $\mathbf{a}$ and λ needed in Equation (4.12) are subsequently computed in a straightforward manner.

The example of snap-through behaviour in a simple truss of Chapter 1 also lends itself well to elucidate the concept of arc-length control. For this purpose Equations (1.11) and (1.12) are rewritten as:

$$\mathrm{d}v = \mathrm{d}F \mathrm{d}v^{\mathrm{I}} + \mathrm{d}v^{\mathrm{II}}$$

with

$$\mathrm{d}v^{\mathrm{I}} = \frac{1}{(EA_0/\ell_0)\sin^2\phi + k + (A_0\sigma)/\ell_0}$$

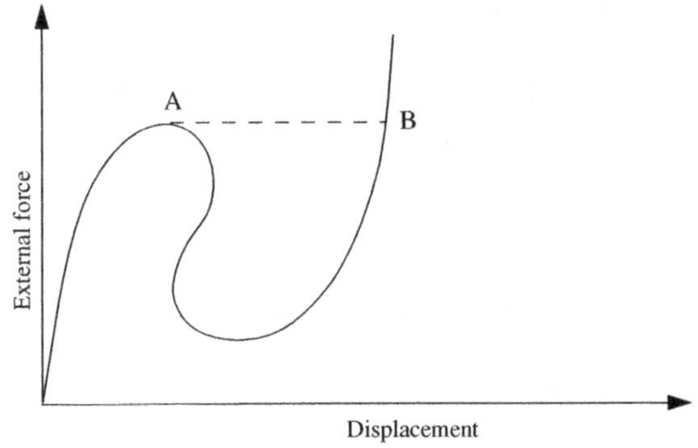

Figure 4.5 Mode jumping from point A to point B

and

$$dv^{\mathrm{II}} = \frac{F_{\mathrm{ext},j} + F(v_j)}{(EA_0/\ell_0)\sin^2\phi + k + (A_0\sigma)/\ell_0}$$

where the variable load increment dF is determined from the requirement that for $j > 1$ we have $dv = 0$, so that:

$$dF = -\frac{dv^{\mathrm{II}}}{dv^{\mathrm{I}}} = F_{\mathrm{ext},j} + F(v_j)$$

An alternative to tracing the complete equilibrium path in a static manner, is to recognise that snap-through as, for instance, observed in this truss structure, but also snap-back, are dynamic phenomena. At the onset of snap-through or snap-back behaviour, point A in Figure 4.5, a switch is made from a static to a dynamic solver, and the full dynamic equations are solved. The load remains constant during this part of the computation, the dashed line in Figure 4.5. When the equilibrium curve is hit, point B, one switches back to a static solution procedure. In general a different failure mode will have developed during snapping, and the name mode jumping has been coined for this procedure (Riks *et al.* 1996).

When line searches are applied without arc-length control, hence at a constant load level, the line search can be invoked in a straightforward manner within each equilibrium iteration and no additional operations are needed. This also holds for an (updated) normal plane method and $\beta = 0$, since then η disappears from the constraint equation. However, in combination with the quadratic constraint formulation, Equation (4.19), or when $\beta \neq 0$, the arc-length constraint will be violated because of the modification of $d\mathbf{a}$ through the scalar η and refinements are necessary (Crisfield 1983).

4.3 PyFEM: Implementation of Riks' Arc-length Solver

In the previous chapter, we have presented a general Python program to solve a system of non-linear equations using a Newton–Raphson procedure. In this chapter, we have discussed alternative procedures for the solution of the set of non-linear equations that govern the equilibrium of discretised systems. In the final version of PyFEM, which will be discussed in this section, the solution procedure can be selected in the input file. The complete main routine of PyFEM is called `PyFEM.py` and can be found in the root directory of the code. The listing is:

⟨*PyFEM Main Routine* ⟩≡

```
from pyfem.io.InputReader    import InputReader
from pyfem.io.OutputManager import OutputManager
from pyfem.solvers.Solver    import Solver

props,globdat = InputReader( sys.argv )

solver = Solver          ( props , globdat )
output = OutputManager ( props , globdat )

while globdat.active:
  solver.run( props , globdat )
  output.run( props , globdat )

print "PyFem analysis terminated successfully."
```

Obviously, the code shows many similarities with the Newton–Raphson code presented in the fragment ⟨*Non-linear Newton–Raphson finite element simulation*⟩ on page 104. Instead of the Newton–Raphson solver `NonlinearSolver` a generic solver `Solver` is now specified. This instance is, in fact, a wrapper around a collection of solvers, varying from a linear solver (`LinearSolver`) to Riks' arc-length solver (`RiksSolver`), which will be discussed here. The solver and the parameters can be selected in the input file, which is read and processed in the function `InputReader`. More information on the format of the input file is given in the user's manual that is attached to the code on the website.

A similar construction is used to specify the output modules. The instance `output` of the type `OutputManager` contains the output modules that are used in the simulation. Different from the solver, it is possible to select multiple output modules, which will be executed consecutively in the function call `output.run`. Again, the output modules and their parameters can be specified in the input file.

Riks' arc-length solver, which is implemented in the file `RiksSolver.py`, is now discussed in greater detail. The file is located in the directory `pyfem/solvers`, together with the other solution modules. The structure of the file is as follows:

⟨*Riks Arclength Solver* ⟩≡

⟨*Initialisation of the Riks solver class* 126⟩
⟨*Riks solver class main functions* 126⟩
⟨*Riks solver class utility functions*⟩

All solvers are implemented as classes, which are derived from the class `BaseModule`:

⟨*Initialisation of the Riks solver class* ⟩ ≡ 125

```
from pyfem.util.BaseModule import BaseModule

class RiksSolver( BaseModule ):

  def __init__( self , props , globdat ):

    ⟨Initialisation of default solver parameters⟩

    BaseModule.__init__( self , props )

    self.Daprev    = zeros( len(globdat.dofs) )
    self.Dlamprev  = 0.0

    globdat.lam    = 1.0
```

In the constructor of this class, the default parameters are set in the fragment ⟨*Initialisation of default solver parameters*⟩. The most important parameters are `self.tol`, which sets the tolerance for the convergence check, and `self.maxLambda`, which defines the load factor λ_{max} at which the simulation is stopped.

All parameters of the solver, including `tol` and `maxLambda`, can be specified in the input file. These values are read by the `InputReader` and are stored in the instance `props`. In the constructor of the base class `BaseModule.__init__` the parameters are turned into members of the class. For example, if in the input file the parameter `tol` is set equal to `1.0e-3`, this value is copied to the member `self.tol` in this constructor.

The array `self.Daprev` and double `self.Dlamprev` are created to store the incremental solution vector and the load factor of the previous loading step. These values are needed to predict the solution of the next load step. Finally, the global load factor λ is initialised as `lam` in the global database `globdat`.[1]

A single load step of the system is solved in the function `run`:

⟨*Riks solver class main functions* ⟩ ≡ 125

```
  def run( self , props , globdat ):

    globdat.cycle += 1

    a    = globdat.state
    Da   = globdat.Dstate
    fhat = globdat.fhat

    self.printHeader( globdat.cycle )
    error = 1.
    globdat.iiter = 0
```

[1] Since **lambda** is one of the few keywords of the programming language Python, we have to use the alternative variable name `lam` to denote the parameter λ.

First, the load cycle number is increased by one. Then, for simplicity, copies of the solution vector `state` and the incremental solution vector `Dstate` are made. The same is done for the array `fhat`, which represents the normalised, unit force vector $\hat{\mathbf{f}}_{\text{ext}}$, see Equation (2.45). After printing a short header that indicates that a new load increment has been added, the variable `error`, which represents the norm of the residual vector, is initialised to `1.` and the iteration counter is reset to `0`.

The predictor for the current load increment is calculated next:

⟨*Riks solver class main functions* ⟩+≡ 126

```
if globdat.cycle == 1:
  K,fint = assembleTangentStiffness( props, globdat )
  Da1    = globdat.dofs.solve( K , globdat.lam*fhat )
  Dlam1  = globdat.lam
else:
  Da1    = self.factor * self.Daprev
  Dlam1  = self.factor * self.Dlamprev
  globdat.lam += Dlam1

a [:] += Da1[:]
Da[:] =  Da1[:]
Dlam = Dlam1
```

In the first load increment, when `globdat.cycle==1`, the predictor $\Delta\mathbf{a}_1$ is obtained as the solution of the linearised system of equations:

$$\mathbf{K}\Delta\mathbf{a}_1 = \Delta\lambda_1\hat{\mathbf{f}}_{\text{ext}}$$

cf. Equation (4.15), with $\Delta\lambda_1$ the load factor in the first step. In the following steps, the predictor is obtained according to Equation (4.58). After calculation of the predictor, the new stiffness matrix, the internal force vector, and the new residual `res` are computed:

⟨*Riks solver class main functions* ⟩+≡ 127

```
K,fint = assembleTangentStiffness( props, globdat )

res = globdat.lam*fhat-fint
```

The actual iteration process can now be started:

⟨*Riks solver class main functions* ⟩+≡ 127

```
while error > tol:
  globdat.iiter += 1

  d1 = globdat.dofs.solve( K , fhat )
  d2 = globdat.dofs.solve( K , res )

  ddlam = -dot(Da1,d2)/dot(Da1,d1)
  dda   = ddlam*d1 + d2
```

After increasing the iteration counter `globdat.iiter`, the components `d1` and `d2` of the solution vector are calculated according to Equations (4.15) and (4.16). These solution vectors are used to determine the load increment `ddlam` and the solution increment `dda`, see Equations (4.17) and (4.28).

Subsequently, the solution vector `a` and the total load factor `globdat.lam` are updated:

⟨*Riks solver class main functions* ⟩+≡ 127

```
Dlam          += ddlam
globdat.lam += ddlam

Da[:] += dda[:]
a [:] += dda[:]
```

The new solution is used to update the stiffness matrix, the internal force vector and the residual vector, and the error at the current iteration is calculated by taking the L_2-norm of the residual vector, normalised by the absolute value of the external force vector:

⟨*Riks solver class main functions* ⟩+≡ 128

```
K,fint = assembleTangentStiffness( props, globdat )

res = globdat.lam*fhat-fint

error  = globdat.dofs.norm( res )
error  = error / globdat.dofs.norm( globdat.lam*fhat )

self.printIteration( globdat.iiter,error )

if globdat.iiter == self.iterMax:
  raise RuntimeError("Riks solver did not converge!")
```

When the error is smaller than the tolerance `self.tol`, the **`while`**-loop is stopped.

⟨*Riks solver class main functions* ⟩+≡ 128

```
self.printConverged( globdat.iiter )

globdat.elements.commitHistory()

globdat.fint = fint

if not self.fixedStep:
  self.factor = pow(0.5,0.25*(globdat.iiter-self.optiter))

self.Daprev[:]  = Da[:]
self.Dlamprev   = Dlam
```

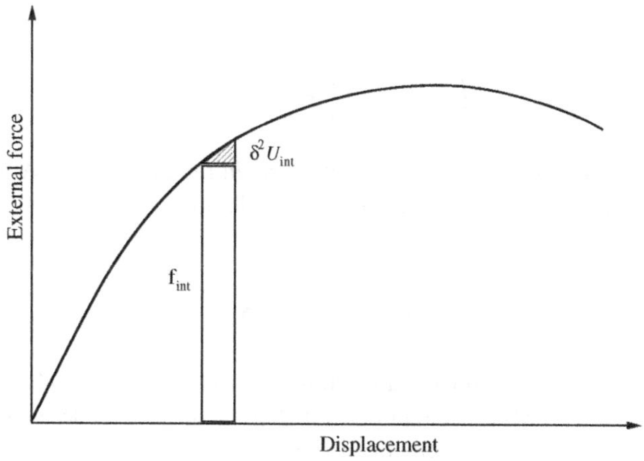

Figure 4.6 Graphical interpretation of the second-order work

After printing an iteration report, the element history is updated and the internal force vector is stored in `globdat`. The new factor that sets the magnitude of the next increment is determined next, unless the option `fixedStep` is chosen. In that case `factor` remains equal to `1.0`. In Section 4.5 we will have a closer look at load stepping. Finally, the displacement increment and load increment in this step are stored as `Daprev` and `Dlamprev` and will be used as the predictor in the next load step.

This implementation of Riks' arc-length solver has been used in the simulation of the shallow truss problem `ShallowTrussRiks.pro` in the directory `examples/ch04`. Different from the implementation of the problem in Chapter 2, this model can determine the quasi-static equilibrium path also for values of k smaller than 1000.0, resulting in the curves shown in Figure 1.2.

4.4 Stability and Uniqueness in Discretised Systems

4.4.1 Stability of a Discrete System

Several definitions of the notion of stability exist in mechanics. Most probably, the oldest is due to Diderot and d'Alembert, who, in their famous *Encyclopedie* (1778), equivalenced stability with the notions rigid and unmovable. Evidently, this definition is too narrow, and nowadays an equilibrium state is called stable if the response on a vanishingly small disturbance remains vanishingly small (Hill 1959; Koiter 1969), which is often called stability in the sense of Lyapounov.

Stability in the above sense can be difficult to prove, and another definition is normally adopted. We consider a discrete system which is in equilibrium at time $\tau = t$, and undergoes an infinitesimal displacement $\delta\mathbf{a} = \mathbf{a}\delta t$. It is further assumed that the external forces $\mathbf{f}_{\text{ext}}$ do not depend on the position (dead loading). During the infinitesimal time increment δt the increase in the internal energy minus the work of the external forces equals [up to second order (Hill

1959); Figure 4.6]:

$$\delta^2 U = \left(\mathbf{f}_{\text{int}} + \frac{1}{2} \frac{\partial \mathbf{f}_{\text{int}}}{\partial \mathbf{a}} \mathbf{a} \delta t \right)^{\mathrm{T}} \mathbf{a} \delta t - \mathbf{f}_{\text{ext}}^{\mathrm{T}} \mathbf{a} \delta t \tag{4.35}$$

along any kinematically admissible path which starts in the direction $\mathbf{a}$. At equilibrium of a discete system, Equation (2.27) holds, and in view of definition (4.13), this expression can be simplified to

$$\delta^2 U = \frac{1}{2} (\delta t)^2 \mathbf{a}^{\mathrm{T}} \mathbf{K} \mathbf{a} \tag{4.36}$$

It is now assumed that stability under dead loading is ensured if $\delta^2 U > 0$ for all kinematically admissible velocity fields, while the equilibrium state is unstable under dead loading if $\delta^2 U$ becomes negative for at least one kinematically admissible velocity field. Stability in the sense of Lyapounov and the energy criterion $\delta^2 U > 0$ are not identical, but for the restricted class of elastic materials at infinitesimal strains it can be proven that they coincide (Koiter 1969).

A discrete mechanical system is thus said to be in a state of stable equilibrium under dead loading if

$$\mathbf{a}^{\mathrm{T}} \mathbf{K} \mathbf{a} > 0 \tag{4.37}$$

for all kinematically admissible $\mathbf{a}$, while it is said to be in a critical state of neutral equilibrium if

$$\mathbf{a}^{\mathrm{T}} \mathbf{K} \mathbf{a} = 0 \tag{4.38}$$

for at least one admissible $\mathbf{a}$, which implies loss of positive definiteness of $\mathbf{K}$. If the tangential stiffness matrix $\mathbf{K}$ is symmetric, a sufficient and necessary condition for Equation (4.38) to be satisfied is that:

$$\det(\mathbf{K}) = 0 \tag{4.39}$$

Using Vieta's rule, Equation (1.66), this identity implies that at least one eigenvalue vanishes. It is noted that for the more general case of a non-symmetric tangential stiffness matrix, which arises in some plasticity and damage models, the vanishing of the lowest eigenvalue is only a sufficient condition for loss of stability, but not a necessary condition.

4.4.2 *Uniqueness and Bifurcation in a Discrete System*

If more equilibrium paths emanate from a point in the $n + 1$-dimensional load–displacement solution space, such a point is named a bifurcation point. The differential equations that govern the next, infinitesimally small load increment, then have more than one solution. Accordingly, at a bifurcation point we observe a loss of uniqueness of the incremental solution. Bifurcations and multiple equilibrium branches arise in many engineering problems. They can be caused by the inclusion of non-linear terms in the kinematic description, as for instance with buckling of slender, thin-walled members, because of the non-linearity of the material model used, or can stem from both causes.

A discussion on loss of uniqueness is best started from Equation (2.45), which represents equilibrium of the structure. We assume that the iterative process has been such that we have obtained a state of perfect equilibrium, so that the array of residuals vanishes, $\mathbf{r} \equiv \mathbf{f}_{\text{ext}}^t - \mathbf{f}_{\text{int}}^t = \mathbf{0}$. Then:

$$\mathbf{K}\Delta\mathbf{a} = \Delta\lambda\hat{\mathbf{f}}_{\text{ext}} \tag{4.40}$$

We introduce the eigenvectors $\mathbf{v}_1, \mathbf{v}_2, \ldots, \mathbf{v}_n$ of the tangential stiffness matrix $\mathbf{K}$. They correspond to the eigenvalues $\lambda_1, \lambda_2, \ldots, \lambda_n$ (in ascending order), where the eigenvalues, which have a subscript k, should not be confused with the load increment $\Delta\lambda$, which is no subscript. By standard concepts of linear algebra we have:

$$\mathbf{K}\mathbf{v}_k = \lambda_k\mathbf{v}_k \tag{4.41}$$

where no summation over repeated indices is implied. For a symmetric matrix $\mathbf{K}$ there exists the relationship $\mathbf{v}_k^{\text{T}}\mathbf{v}_l = 0,\ k \neq l$ between the eigenvectors $\mathbf{v}_k$ and $\mathbf{v}_l$. To simplify the subsequent derivations the eigenvectors $\mathbf{v}_k$ will be normalised such that $\mathbf{v}_k{}^{\text{T}}\mathbf{v}_k = 1$.

If the tangential stiffness matrix $\mathbf{K}$ is not defect, i.e. if the n eigenvectors span an n-dimensional vector space, any vector can be expressed as a linear combination of the eigenvectors:

$$\mathbf{a} = \sum_{k=1}^{n}(\mathbf{v}_k^{\text{T}}\mathbf{a})\mathbf{v}_k \tag{4.42}$$

We now apply this decomposition to the array of incremental nodal displacements $\Delta\mathbf{a}$ and to the normalised external load vector $\hat{\mathbf{f}}_{\text{ext}}$, and we substitute the result into Equation (4.40):

$$\mathbf{K}\left(\sum_{k=1}^{n}(\mathbf{v}_k^{\text{T}}\mathbf{a})\mathbf{v}_k\right) = \Delta\lambda\sum_{i=1}^{n}(\mathbf{v}_k^{\text{T}}\hat{\mathbf{f}}_{\text{ext}})\mathbf{v}_k \tag{4.43}$$

With the aid of Equation (4.41) we can modify this equation as:

$$\left(\sum_{k=1}^{n}\lambda_k(\mathbf{v}_k^{\text{T}}\mathbf{a}) - \Delta\lambda(\mathbf{v}_k^{\text{T}}\hat{\mathbf{f}}_{\text{ext}})\right)\mathbf{v}_k = \mathbf{0} \tag{4.44}$$

Since $\mathbf{K}$ is not defect, the eigenvectors $\mathbf{v}_k$ constitute a set of n linearly independent vectors, which implies that this equation can only be satisfied non-trivially if and only if

$$\lambda_k\mathbf{v}_k^{\text{T}}\mathbf{a} - \Delta\lambda\mathbf{v}_k^{\text{T}}\hat{\mathbf{f}}_{\text{ext}} = 0 \tag{4.45}$$

for each eigenvector $\mathbf{v}_k$. In particular, we have for $k = 1$ (the lowest eigenvalue):

$$\lambda_1\mathbf{v}_1^{\text{T}}\mathbf{a} - \Delta\lambda\mathbf{v}_1^{\text{T}}\hat{\mathbf{f}}_{\text{ext}} = 0 \tag{4.46}$$

Since λ_1 is the lowest eigenvalue, its vanishing implies that either the identity

$$\Delta\lambda = 0 \tag{4.47}$$

or the orthogonality condition

$$\mathbf{v}_1^{\text{T}}\hat{\mathbf{f}}_{\text{ext}} = 0 \tag{4.48}$$

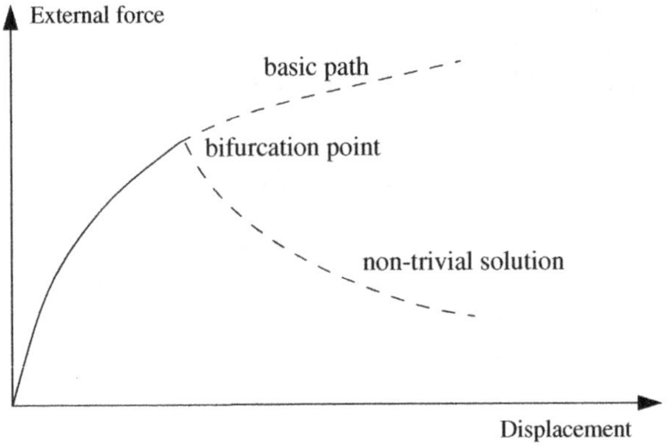

Figure 4.7 Bifurcation point and possible post-bifurcation equilibrium paths

must hold, or both. The case that λ_2, λ_3, etc. also vanish is not considered here, but the generalisation to more vanishing eigenvalues can be done. The first possibility, Equation (4.47), is referred to as limit point behaviour, since the load becomes stationary (Figure 2.7). Indeed, there is a maximum in the load–deflection curve. The second possibility, Equation (4.48), corresponds to a bifurcation point (loss of uniqueness), from which at least two equilibrium paths emanate (Figure 4.7).

At a bifurcation point, Equation (4.40) holds for a given solution $\Delta\mathbf{a}$ with, in general, a non-zero value of the incremental load parameter, $\Delta\lambda \neq 0$. Loss of uniqueness of the incremental solution implies that there exists another solution, often called the non-trivial solution, say $\Delta\mathbf{a}^*$, which also satisfies incremental equilibrium for the same tangential stiffness matrix $\mathbf{K}$:

$$\mathbf{K}\Delta\mathbf{a}^* = \Delta\lambda\hat{\mathbf{f}}_{\text{ext}} \tag{4.49}$$

Subtraction of Equation (4.40) from Equation (4.49) yields:

$$\mathbf{K}(\Delta\mathbf{a}^* - \Delta\mathbf{a}) = \mathbf{0} \tag{4.50}$$

We now premultiply this equation by $\mathbf{v}_k^{\mathrm{T}}$ and use Equation (4.41) to derive that

$$\lambda_k \mathbf{v}_k^{\mathrm{T}}(\Delta\mathbf{a}^* - \Delta\mathbf{a}) = 0 \tag{4.51}$$

for all i. Because $\lambda_1 = 0$ and because of the mutual orthogonality of the eigenvectors this equation can be satisfied non-trivially only if $\Delta\mathbf{a}^* - \Delta\mathbf{a} = \gamma\mathbf{v}_1$ with γ some scalar. Hence, all solutions

$$\Delta\mathbf{a}^* = \Delta\mathbf{a} + \gamma\mathbf{v}_1 \tag{4.52}$$

are possible and we have loss of uniqueness of the incremental solution (bifurcation). A distinction can be made between symmetric and asymmetric bifurcation points, but this classification is restricted to elastic solids under small strains and has been developed specifically for buckling phenomena in thin-walled structural members (Koiter 1945; Riks 1984).

In discrete numerical processes, limit or bifurcation points are extremely difficult to isolate. Rather, distinction is made between stable equilibrium states for which Equation (4.37) holds, and equilibrium states which are unstable under dead loading, i.e.

$$\mathbf{a}^{\mathrm{T}}\mathbf{K}\mathbf{a} < 0 \tag{4.53}$$

for at least one kinematically admissible $\mathbf{a}$. Substitution of Equation (4.42) gives

$$\sum_{k=1}^{n}\sum_{l=1}^{n}(\mathbf{v}_k^{\mathrm{T}}\mathbf{a})(\mathbf{v}_l^{\mathrm{T}}\mathbf{a})\mathbf{v}_k^{\mathrm{T}}\mathbf{K}\mathbf{v}_l < 0 \tag{4.54}$$

or using Equation (4.41) and the orthogonality relation between eigenvectors as well as their normalisation:

$$\sum_{k=1}^{n}(\mathbf{v}_k^{\mathrm{T}}\mathbf{a})^2\lambda_k < 0 \tag{4.55}$$

This inequality can be satisfied if one or more eigenvalues have become negative. We choose $\mathbf{a} = \alpha\mathbf{v}_1$, with α a scalar. Then, $\sum_{k=1}^{n}(\mathbf{v}_k^{\mathrm{T}}\mathbf{a})^2\lambda_k = \alpha^2\lambda_1 < 0$, since $\lambda_1 < 0$, and the equilibrium state is unstable.

Negative eigenvalues in the tangential stiffness matrix emerge when a limit point has been passed (and the load-carrying capacity is decreasing), when a bifurcation point has been passed, or when both a limit point and a bifurcation point have been passed, possibly at the same point in the load–displacement curve. The first case is characterised by a descending branch in the load–displacement curve and there is a single negative eigenvalue. If a bifurcation point has been passed, but not (yet) a limit point, the solution will normally, that is without perturbation, continue on the basic equilibrium path. The Euler strut is a simple example where this is encountered. Even though the load is rising, the tangential stiffness matrix will exhibit one or more negative eigenvalues, each related to a non-trivial equilibrium path. When a limit point has been passed as well, and the load is descending, one negative eigenvalue relates to the limit point and the other negative eigenvalues relate to other, non-trivial equilibrium paths.

This discussion shows the importance of monitoring the eigenvalues of the tangential stiffness matrix as they reveal whether the current solution is still the most critical one in the sense that the 'lowest' equilibrium path is followed. Incremental-iterative solution procedures normally converge towards one of the possible equilibrium states that exist in the structure. But if there are more possible solutions they will not necessarily pick the most critical branch. Monitoring the eigenvalues of the tangential stiffness matrix thus allows the analyst to assess whether the solution which has been obtained is the most critical solution.

Considering the computational effort that is necessary to compute the lowest eigenvalue at each loading step, monitoring the (lowest) eigenvalues in order to assess whether the most critical solution path is still followed, may lead to an unacceptable computational overhead. A computationally much cheaper solution is to monitor the pivots that are computed during an LDU decomposition. In Chapter 1 it has been shown that loss of positive definiteness of a matrix is equivalent to the positiveness of all pivots. Loss of positive definiteness of $\mathbf{K}$, i.e. when Equation (4.53) holds, must therefore be signalled by the emergence of at least one negative pivot. Monitoring the negative pivots that arise during the LDU decomposition is therefore equivalent to checking whether one of the eigenvalues becomes negative if the analyst wishes to assess whether the solution that is computed is unique.

While the notions of stability and uniqueness are closely related, they are not identical and can give rise to different requirements on the structural system. The stability requirement is single-valued, that is, the stress rate is associated with a unique velocity gradient. However, the uniqueness requirement is multi-valued, as both possible velocity gradient distributions can be related to stress rates by different stiffness moduli, which happens when we have different behaviour in loading and unloading as in plasticity or continuum damage relations. Strictly speaking, we have to investigate all possible combinations of loading and unloading for such a multi-valued constitutive relation in order to determine whether uniqueness ceases to hold for some combination. Indeed, when monitoring the eigenvalues, or pivots, of the tangential stiffness matrix, we will only detect bifurcations for which the material tangent moduli show loading for at least an infinitesimal instant after bifurcation. Such a solid has been named a 'linear comparison solid' (Hill 1959), and numerical experience shows that this situation is normally the most critical. For a restricted number of multi-valued constitutive relations, this can be proven rigorously.

4.4.3 Branch Switching

When a bifurcation point has been passed the trivial solution $\Delta\mathbf{a}$ can be perturbed in the sense of Equation (4.52) by adding a part of the eigenvector $\mathbf{v}_1$ which corresponds to the lowest eigenvalue λ_1 to the trivial solution. The factor γ, which is undetermined for the infinitesimal case, can be estimated for finite load increments from the following orthogonality requirement (Riks 1972; de Borst 1987):

$$\Delta\mathbf{a}^{\mathrm{T}}\Delta\mathbf{a}^{*} = 0 \tag{4.56}$$

Combination of Equations (4.52) and (4.56) gives for the perturbed displacement increment:

$$\Delta\mathbf{a}^{*} = \Delta\mathbf{a} - \frac{\Delta\mathbf{a}^{\mathrm{T}}\Delta\mathbf{a}}{\mathbf{v}_1^{\mathrm{T}}\Delta\mathbf{a}}\mathbf{v}_1 \tag{4.57}$$

Equation (4.56) states that the search direction for the non-trivial solution $\Delta\mathbf{a}^*$ is orthogonal to the trivial or basic solution path. For finite increments, the non-trivial solution will not be in the search direction. For instance, the occurrence of loading–unloading conditions in plasticity or damage models will cause deviations. However, when we add equilibrium iterations, condition (4.56) will maximise the possibility that we converge to the non-trivial solution.

4.5 Load Stepping and Convergence Criteria

In the preceding discussion attention has been focused on the determination of the value of the load increment for the second and subsequent iterations in such a fashion that the 'arc-length' $\Delta\ell$ remains more or less constant. No attention has been paid so far to the determination of its value in the first iteration of a new loading step. In other words, what should $\Delta\ell$ be to find a proper balance between a value of the load increment that is not unnecessarily small, but small enough that iterative procedure will converge? A related question is the proper choice of the sign of the load increment. At present, a well-founded method does not seem to exist for handling these two important issues in an automatic and robust manner. Some procedures that work well in numerical practice are summarised below.

In general, the predictor is obtained by a simple linear extrapolation of the previous increment of the displacement vector and of the load factor:

$$\Delta \mathbf{a}_1^{t+\Delta t} = \xi \Delta \mathbf{a}^t \,; \qquad \Delta \lambda_1^{t+\Delta t} = \xi \Delta \lambda^t \tag{4.58}$$

where

$$\xi = \frac{\Delta \ell^{t+\Delta t}}{\Delta \ell^t} \tag{4.59}$$

sets the magnitude of the new load increment, with $\Delta \ell^t$ the computed arc-length in the preceding loading step, and $\Delta \ell^{t+\Delta t}$ the desired value in the new loading step. In a heuristic procedure ξ is estimated from the requirement that the number of 'desired' iterations in the new loading step $n+1$ equals N^d. If the number of iterations needed to satisfy equilibrium in the preceding load increment equals N^t, the estimate for the new arc-length reads

$$\frac{\Delta \ell^{t+\Delta t}}{\Delta \ell^t} = \left(\frac{N^d}{N^t} \right)^{\zeta} \tag{4.60}$$

with ζ a parameter to damp/amplify the influence of the quotient N^d/N^t, and is normally chosen equal to $\frac{1}{2}$. Although heuristic, the method has been reported to work well (Crisfield 1981).

Another method is based on the fact that the second-order work

$$\Delta \mathcal{U}^2 = \frac{1}{2} \Delta \mathbf{f}_{\text{ext}}^{\text{T}} \Delta \mathbf{a} \tag{4.61}$$

gives a good indication on the stability of a structure. Obviously, $\Delta \mathcal{U}^2$ becomes zero at a limit point, i.e. when the ultimate load-carrying capacity of a structure is reached, and becomes negative afterwards. The idea is to select the load increment in the first iteration of the new loading step, $\Delta \lambda_1^{n+1}$, such that the value of the second-order work in both increments is equal (Bergan *et al.* 1978):

$$\frac{1}{2} \Delta \lambda^t \hat{\mathbf{f}}_{\text{ext}}^{\text{T}} \Delta \mathbf{a}^t = \frac{1}{2} \Delta \lambda_1^{t+\Delta t} \hat{\mathbf{f}}_{\text{ext}}^{\text{T}} \Delta \mathbf{a}_1^{t+\Delta t} \tag{4.62}$$

Assuming that we have arrived at a properly converged solution in the preceding load step, so that $\mathbf{r} \approx \mathbf{0}$, we have $\Delta \mathbf{a}_1^{t+\Delta t} \approx \Delta \lambda_1^{t+\Delta t} \Delta \mathbf{a}_1^{\text{I}}$. This gives the following approximation for the magnitude of the new load increment:

$$\Delta \lambda_1^{t+\Delta t} = \sqrt{\frac{\|\Delta \lambda^t \hat{\mathbf{f}}_{\text{ext}}^{\text{T}} \Delta \mathbf{a}^t\|}{\|\hat{\mathbf{f}}_{\text{ext}}^{\text{T}} \Delta \mathbf{a}_1^{\text{I}}\|}} \tag{4.63}$$

The application of any automatic load incrementation scheme may fail in the sense that it may produce too large or impractically small increments. To avoid this situation from happening, a lower and an upper bound on the load increment should be prescribed:

$$\Delta \lambda_{\min} \leq \|\Delta \lambda\| \leq \Delta \lambda_{\max} \tag{4.64}$$

Having determined the magnitude of the load step in the first iteration of the new increment, it remains to determine its proper sign. An obvious choice would be to monitor the lowest eigenvalue of the structural tangential stiffness matrix **K** and to reverse the sign of the load increment when the lowest eigenvalue changes sign, or alternatively, if a negative pivot appears during the decomposition of the tangential stiffness matrix. However, this procedure can give wrong indications when bifurcation points are encountered. A more heuristic procedure, which works well according to ample numerical evidence, is to adopt a sign-switching strategy similar to the one used by Crisfield (1981) for determining the correct sign in the quadratic arc-length procedure:

$$\Delta\lambda_1^{t+\Delta t} = \begin{cases} +\|\Delta\lambda_1^{t+\Delta t}\| & \text{if } (\Delta\mathbf{a}^t)^{\mathrm{T}}\Delta\mathbf{a}_1^{\mathrm{I}} > 0 \\ -\|\Delta\lambda_1^{t+\Delta t}\| & \text{if } (\Delta\mathbf{a}^t)^{\mathrm{T}}\Delta\mathbf{a}_1^{\mathrm{I}} < 0 \end{cases} \tag{4.65}$$

where the approximation $\Delta\mathbf{a}_1^{t+\Delta t} \approx \Delta\lambda^{t+\Delta t}\Delta\mathbf{a}^{\mathrm{I}}$ has again been used. It is emphasised that a proper estimation of the initial step size $\Delta\lambda$ through the new arc-length $\Delta\ell$ is of great importance for the extent to which the whole computation can be made automatic, as well as for the required computer time. Indeed, numerical evidence shows that this is far more important than the exact choice of the constraint condition that is employed in the path-following procedure.

In order to be able to assess whether an iterative procedure has converged a so-called convergence criterion is needed. Such a criterion requires that some quantity, e.g. a force or a displacement, must be approximated within some tolerance. If the error does not become smaller than this pre-set tolerance the iterative process is said to not have converged. If the quantity that is being monitored becomes unbounded the process diverges. To prevent the computer program continuing to search for a solution when either of the latter two possibities occur, a maximum number of iterations must be specified. It is difficult to provide guidelines as to the number of iterations the iterative procedure should be limited to, since this is not only problem-dependent, but also depends on the type of iterative procedure, e.g. full Newton–Raphson, modified Newton–Raphson, Initial Stiffness or Quasi-Newton method, that is employed, and on the quantity that is being monitored. In practice, as a rule-of-thumb one can set the maximum number of iterations equal to 8 for the full Newton–Raphson method, equal to 20 for the modified Newton–Raphson method, and equal to 30 for the Initial Stiffness approach.

The above guidelines are primarily valid for so-called global convergence criteria, in which a global quantity is monitored. Here, one may think of a norm of the unbalanced force vector, a norm of the displacement increments or the energy of the system. When the interest mainly lies in obtaining an impression of the global behaviour of the structure, e.g. the ultimate load-bearing capacity, such a global criterion is always sufficient. Also, in analyses in which attention is focused on local structural behaviour one can usually rely on a global convergence criterion. Yet, in critical cases it may be wise to adopt a local convergence criterion in which, for instance, the unbalanced forces of some nodes have to be zero within some tolerance. Since most finite element packages only offer options for monitoring the convergence behaviour in a global sense, we will restrict the treatment to such criteria. In particular, a force criterion, a displacement criterion and an energy criterion will be discussed, since these criteria are the most popular in existing finite element software.

The most demanding global criterion is the force criterion. With this criterion equilibrium iterations are added until the change in the norm of the unbalanced force vector is smaller than the prescribed convergence tolerance η times the value of the norm in the first iteration of that loading step. As a rule, the L_2-norm is used to measure the unbalanced force vector and iterations are terminated when:

$$\|\mathbf{f}_{\text{ext}} - \mathbf{f}_{\text{int},j}\| \leq \eta \times \|\mathbf{f}_{\text{ext}} - \mathbf{f}_{\text{int},1}\| \tag{4.66}$$

A reasonable balance between accuracy and consumption of computer time is usually achieved if the tolerance η is set equal to 10^{-3}.

Another criterion that is used frequently in non-linear finite element analysis is the energy criterion. Now, the iterations can be stopped when:

$$\mathbf{f}_{\text{int},j}^{\text{T}} \mathbf{da}_j \leq \eta \times \mathbf{f}_{\text{int},1}^{\text{T}} \Delta\mathbf{a}_1 \tag{4.67}$$

Experience shows that this convergence criterion is often somewhat easier to satisfy than the preceding criterion. To achieve the same accuracy the tolerance should be set at smaller value, e.g. 10^{-4}.

The norm of the incremental displacements (displacement criterion) is usually the easiest to comply with. Defining this criterion as

$$\|\mathbf{da}_j\| \leq \eta \times \|\Delta\mathbf{a}_1\| \tag{4.68}$$

the η parameter should normally not exceed 10^{-6} in order that a reasonably accurate solution is obtained.

The choice of a convergence criterion and the associated convergence tolerance η must be done with great care. A simple example is pure relaxation. For this problem the incremental displacements have the correct value immediately after the first iteration, whereas it may need several more iterations to allow the stresses, and consequently the internal forces, to relax to their proper values. It will be clear that in such a case any convergence criterion that involves the incremental displacement vector $\mathbf{da}$, including the energy criterion and the norm of incremental displacements, will erroneously identify the process as converged after the first iteration. In such cases only a force norm does not result in a premature termination of the iterative procedure.

The value of the convergence tolerance η must be chosen carefully. On one hand a too loose convergence may result in inaccurate and unreliable answers. On the other hand, a too strict convergence tolerance sometimes hardly improves the results while drastically increasing the required computer time. It should be realised that convergence is a relative matter. Neither of the convergence criteria discussed above warrants that the error is smaller than some prescribed value in an absolute sense.

Finally, while it is clear that a diverged solution is unreliable, it is more difficult to assess whether a non-converged solution must also be abandoned. It can be reasonable to continue a solution after a loading step in which the convergence criterion has been missed only marginally, but caution should be exercised.

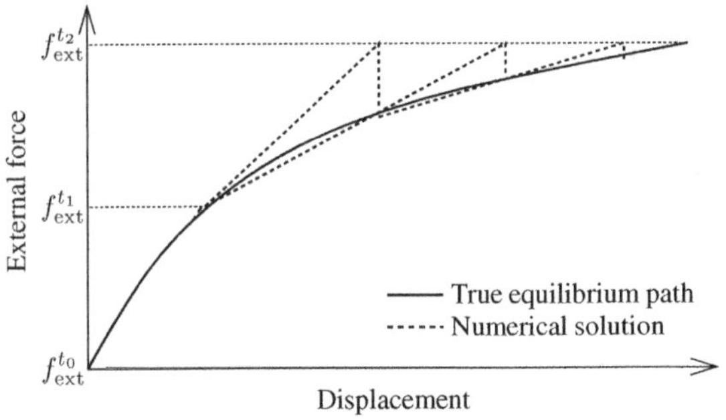

Figure 4.8 One-dimensional representation of Quasi-Newton iterative procedures

4.6 Quasi-Newton Methods

In Chapter 2 alternatives have been discussed for obtaining a solution that satisfies equilibrium, and complies with the constitutive and kinematic equations. As the Newton–Raphson method involves the rather costly formulation and the ensuing decomposition of the tangential stiffness matrix, alternatives have been proposed in which this formulation and decomposition is carried out less frequently, e.g. once every load step, or only in the elastic stage. A different strategy is to approximate the tangential stiffness matrix, for instance using a secant approach. One can elucidate this starting from the correction to the displacement vector in iteration $j + 1$:

$$\mathbf{da}_{j+1} = \mathbf{K}_j^{-1}(\mathbf{f}_{\text{ext}}^{t+\Delta t} - \mathbf{f}_{\text{int},j})$$

cf. Equation (2.50). From $\mathbf{da}_{j+1}$ a new internal force vector $\mathbf{f}_{\text{int,j}+1}$ can be computed. Subsequently a secant-like approximation $\mathbf{K}_{j+1}$ to the tangential stiffness matrix can be constructed such that it satisfies:

$$\mathbf{K}_{j+1}\mathbf{da}_{j+1} = \mathbf{df}_{\text{int},j+1} \tag{4.69}$$

where $\mathbf{df}_{\text{int},j+1} = \mathbf{f}_{\text{int},j+1} - \mathbf{f}_{\text{int},j}$, see Figure 4.8 for a graphical illustration of the one-dimensional case.

With a single degree of freedom the construction of a secant approximation is unique and straightforward. This is not so for the multi-dimensional case. By simple substitution it can be shown that all first-order approximations of the form

$$\mathbf{K}_{j+1} = \mathbf{K}_j + \frac{(\mathbf{df}_{\text{int},j+1} - \mathbf{K}_j\mathbf{da}_{j+1})\mathbf{u}_{j+1}^{\mathrm{T}}}{\mathbf{u}_{j+1}^{\mathrm{T}}\mathbf{da}_{j+1}} \tag{4.70}$$

satisfy the Quasi-Newton Equation (4.69), where $\mathbf{u}_{j+1}$ is a vector that can be chosen in order to optimise the performance of the iteration method. Equation (4.70) is a so-called rank-one update of the stiffness matrix $\mathbf{K}_j$. Rank-two updates are also used frequently in Quasi-Newton methods (Dennis and More 1977; Dennis and Schnabel 1983):

$$\mathbf{K}_{j+1} = \mathbf{K}_j + \frac{(\mathbf{df}_{\text{int},j+1} - \mathbf{K}_j\mathbf{da}_{j+1})\mathbf{u}^{\text{T}}_{j+1} + \mathbf{u}_{j+1}(\mathbf{df}_{\text{int},j+1} - \mathbf{K}_j\mathbf{da}_{j+1})^{\text{T}}}{\mathbf{u}^{\text{T}}_{j+1}\mathbf{da}_{j+1}} - \frac{(\mathbf{df}_{\text{int},j+1} - \mathbf{K}_j\mathbf{da}_{j+1})^{\text{T}}\mathbf{da}_{j+1}}{(\mathbf{u}^{\text{T}}_{j+1}\mathbf{da}_{j+1})^2}\mathbf{u}_{j+1}\mathbf{u}^{\text{T}}_{j+1} \quad (4.71)$$

Update formulas like Equations (4.70) or (4.71) avoid the need to construct the tangential stiffness matrix completely anew every iteration. However, the costly LDU decomposition still has to be carried out. A solution is to apply the update directly to the factorised tangential stiffness matrix. This can be done using the Sherman–Morrison formula (1.38). Application to the rank-one update of Equation (4.70) gives:

$$\mathbf{K}^{-1}_{j+1} = \mathbf{K}^{-1}_j + \frac{(\mathbf{da}_{j+1} - \mathbf{K}^{-1}_j\mathbf{df}_{\text{int},j+1})\mathbf{u}^{\text{T}}\mathbf{K}^{-1}_j}{\mathbf{u}^{\text{T}}\mathbf{K}^{-1}_j\mathbf{df}_{\text{int},j+1}} \quad (4.72)$$

A well-known first-order update formula is that of Broyden, which is obtained for $\mathbf{u}_{j+1} = \mathbf{da}_{j+1}$. The inverse of a second-order update formula can be derived by applying the Sherman–Morrison formula twice, leading to:

$$\mathbf{K}^{-1}_{j+1} = \mathbf{K}^{-1}_j + \frac{(\mathbf{da}_{j+1} - \mathbf{K}^{-1}_j\mathbf{df}_{\text{int},j+1})\mathbf{v}^{\text{T}}_{j+1} + \mathbf{v}_{j+1}(\mathbf{da}_{j+1} - \mathbf{K}^{-1}_j\mathbf{df}_{\text{int},j+1})^{\text{T}}}{\mathbf{v}^{\text{T}}_{j+1}\mathbf{df}_{\text{int},j+1}} - \frac{(\mathbf{da}_{j+1} - \mathbf{K}^{-1}_j\mathbf{df}_{\text{int},j+1})^{\text{T}}\mathbf{df}_{\text{int},j+1}}{(\mathbf{v}^{\text{T}}_{j+1}\mathbf{df}_{\text{int},j+1})^2}\mathbf{v}_{j+1}\mathbf{v}^{\text{T}}_{j+1} \quad (4.73)$$

with $\mathbf{v}_{j+1}$ a vector that can be chosen in order to optimise the convergence behaviour. Its relation to $\mathbf{u}_{j+1}$ is complex. The most popular rank-two formula is the Broyden–Fletcher–Goldfarb–Shanno (BFGS) update, which is obtained by setting $\mathbf{v}_{j+1} = \mathbf{da}_{j+1}$.

As the update formulas (4.72) and (4.73) destroy the bandedness and sparseness of the inverse stiffness matrix, the actual implementation is done in a different manner. For the first-order update formulas we substitute the expression for the inverse stiffness matrix (4.72) into the expression for the correction to the displacement vector in iteration $j+1$:

$$\mathbf{da}_{j+1} = \left[\mathbf{I} + \frac{(\mathbf{da}_j - \mathbf{K}^{-1}_{j-1}\mathbf{df}_{\text{int},j})\mathbf{u}^{\text{T}}_j}{\mathbf{u}^{\text{T}}_j\mathbf{K}^{-1}_{j-1}\mathbf{df}_{\text{int},j}}\right]\mathbf{K}^{-1}_{j-1}(\mathbf{f}^{t+\Delta t}_{\text{ext}} - \mathbf{f}_{\text{int},j}) \quad (4.74)$$

Defining the auxiliary vector

$$\mathbf{w}_j = \mathbf{K}^{-1}_{j-1}(\mathbf{f}^{t+\Delta t}_{\text{ext}} - \mathbf{f}_{\text{int},j}) \quad (4.75)$$

we can derive that:

$$\mathbf{w}_j = \mathbf{da}_j - \mathbf{K}^{-1}_{j-1}\mathbf{df}_{\text{int},j} \quad (4.76)$$

and the expression for the correction to the incremental displacement vector can be rewritten as:

$$\mathbf{da}_{j+1} = (1 + \alpha_j \beta_j)\mathbf{w}_j \tag{4.77}$$

with

$$\alpha_j = (\mathbf{u}_j^{\mathrm{T}} \mathbf{K}_{j-1}^{-1} \mathbf{df}_{\mathrm{int},j})^{-1} = (\mathbf{u}_j^{\mathrm{T}}(\mathbf{da}_j - \mathbf{w}_j))^{-1} \tag{4.78}$$

$$\beta_j = \mathbf{u}_j^{\mathrm{T}} \mathbf{w}_j \tag{4.79}$$

In the next iteration, $j+2$, the updated auxiliary vector $\mathbf{w}_{j+1}$ must be computed in order to obtain the correction to the displacement vector, $\mathbf{da}_{j+2}$ using Equation (4.77). This can be done in a recursive manner, as follows:

$$\begin{aligned}\mathbf{w}_{j+1} = \mathbf{K}_j^{-1}(\mathbf{f}_{\mathrm{ext}}^{t+\Delta t} - \mathbf{f}_{\mathrm{int},j+1}) &= \left[\mathbf{I} + \alpha_j \mathbf{w}_j \mathbf{u}_j^{\mathrm{T}}\right] \mathbf{K}_{j-1}^{-1}(\mathbf{f}_{\mathrm{ext}}^{t+\Delta t} - \mathbf{f}_{\mathrm{int},j+1}) \\ &= \prod_{i=1}^{j} \left[\mathbf{I} + \alpha_i \mathbf{w}_i \mathbf{u}_i^{\mathrm{T}}\right] \mathbf{K}_0^{-1}(\mathbf{f}_{\mathrm{ext}}^{t+\Delta t} - \mathbf{f}_{\mathrm{int},j+1})\end{aligned} \tag{4.80}$$

Clearly, the update reduces to simple scalar-vector operations, and no inverse has to be computed explicitly. Second-order update formulas can be handled in a similar manner. For instance, for the BFGS update, the following formulas ensue:

$$\mathbf{da}_{j+1} = (1 + \alpha_j \beta_j)\mathbf{w}_j + \alpha_j(\delta_j - \alpha_j \beta_j \gamma_j)\mathbf{v}_j \tag{4.81}$$

with:

$$\alpha_j = (\mathbf{v}_j^{\mathrm{T}} \mathbf{df}_{\mathrm{int},j})^{-1} \tag{4.82}$$

$$\beta_j = \mathbf{v}_j^{\mathrm{T}}(\mathbf{f}_{\mathrm{ext}}^{t+\Delta t} - \mathbf{f}_{\mathrm{int},j}) \tag{4.83}$$

$$\gamma_j = \mathbf{w}_j^{\mathrm{T}} \mathbf{df}_{\mathrm{int},j} \tag{4.84}$$

$$\delta_j = \mathbf{w}_j^{\mathrm{T}}(\mathbf{f}_{\mathrm{ext}}^{t+\Delta t} - \mathbf{f}_{\mathrm{int},j}) \tag{4.85}$$

and for the update of $\mathbf{w}$:

$$\mathbf{w}_{j+1} = \mathbf{K}_0^{-1}(\mathbf{f}_{\mathrm{ext}}^{t+\Delta t} - \mathbf{f}_{\mathrm{int},j+1}) + \sum_{i=1}^{j} \left[\alpha_i \beta_i \mathbf{w}_i + \alpha_i(\delta_i - \alpha_i \beta_i \gamma_i)\mathbf{v}_i\right] \tag{4.86}$$

When using a Quasi-Newton method for the solution of a set of non-linear equations, the convergence behaviour deteriorates compared with a Newton–Raphson method. The typical quadratic convergence of Table 2.2 is lost. Instead, a so-called superlinear convergence behaviour is obtained, in which $\epsilon_{k+1} = \mu \epsilon_k^{\alpha}$, with $1 \leq \alpha \leq 2$ and μ a number.

Clearly, the added computational cost is minimal, since it reduces to some additional inner products and scalar-vector multiplications. Although significant gains have been reported in terms of computer time, others report a somewhat erratic behaviour of Quasi-Newton methods with a non-monotonous convergence behaviour (Crisfield 1979; Matthies and Strang 1979). This lack of numerical stability seems to have decreased the popularity of this class of iterative methods in more recent years.

References

Batoz JL and Dhatt G 1979 Incremental displacement algorithms for nonlinear problems. *International Journal for Numerical Methods in Engineering* **14**, 1262–1267.

Bergan PG, Horrigmoe G, Krakeland B and Soreide TH 1978 Solution techniques for non-linear finite element problems. *International Journal for Numerical Methods in Engineering* **12**, 1677–1696.

Crisfield MA 1979 A faster modified Newton-Raphson iteration. *Computer Methods in Applied Mechanics and Engineering* **20**, 267–278.

Crisfield MA 1981 A fast incremental/iterative solution procedure that handles snap-through. *Computers and Structures* **13**, 55–62.

Crisfield MA 1983 An arc-length method including line searches and accelerations. *International Journal for Numerical Methods in Engineering* **19**, 1269–1289.

de Borst R 1987 Computation of post-bifurcation and post-failure behaviour of strain-softening solids. *Computers & Structures* **25**, 211–224.

Dennis JE and More JJ 1977 Quasi-Newton methods, motivation and theory. *SIAM Review* **19**, 46–89.

Dennis JE and Schnabel RB 1983 *Numerical Methods for Unconstrained Optimization and Nonlinear Equations*. Prentice-Hall.

Geers MGD 1999a Enhanced solution control for physically and geometrically non-linear problems. Part I—the subplane control method. *International Journal for Numerical Methods in Engineering* **46**, 177–204.

Geers MGD 1999b Enhanced solution control for physically and geometrically non-linear problems. Part II—comparative performance analysis. *International Journal for Numerical Methods in Engineering* **46**, 205–230.

Hill R 1959 Some basic principles in the mechanics of solids without a natural time. *Journal of the Mechanics and Physics of Solids* **7**, 209–225.

Koiter WT 1945 *Over de Stabiliteit van het Elastisch Evenwicht (On the Stability of Elastic Equilibrium)*. PhD thesis, Delft University of Technology, Delft.

Koiter WT 1969 On the thermodynamic background of elastic stability theory, in *Problems of Hydrodynamics and Continuum Mechanics* (ed. Radok JRM), pp. 423–433. SIAM.

MacNeal R and Harder R 1985 A proposed standard set of problems to test finite element accuracy. *Finite Elements in Analysis and Design* **1**, 3–20.

Matthies H and Strang G 1979 The solution of nonlinear finite element equations. *International Journal for Numerical Methods in Engineering* **14**, 1613–1625.

Ramm E 1981 Strategies for tracing ther nonlinear response near limit points, in *Nonlinear Finite Element Analysis in Structural Mechanics* (eds. Wunderlich W, Stein E and Bathe KJ) , pp. 63–89. Springer.

Riks E 1972 The application of Newton's method to the problem of elastic stability. *Journal of Applied Mechanics* **39**, 1060–1066.

Riks E 1984 Some computational aspects of the stability analysis of nonlinear structures. *Computer Methods in Applied Mechanics and Engineering* **47**, 219–259.

Riks E, Rankin CC and Brogan FA 1996 On the solution of mode jumping phenomena in thin-walled shell structures. *Computer Methods in Applied Mechanics and Engineering* **136**, 59–92.

Verhoosel CV, Remmers JJC and Gutiérrez MA 2009 A dissipation-based arc-length method for robust simulation of brittle and ductile failure. *International Journal for Numerical Methods in Engineering* **77**, 1290–1321.

Wriggers P and Simo JC 1990 A general procedure for the direct computation of turning and bifurcation points. *International Journal for Numerical Methods in Engineering* **30**, 155–176.

5

Solution Techniques for Non-linear Dynamics

In Chapters 2 and 4 the basic structure of non-linear finite element programs has been laid out and solution techniques have been described to solve quasi-static problems. In this chapter we will supplement this by explaining how the methods described in these chapters can be extended to yield solutions for dynamic problems. It is not the intention of this chapter to give an overview of the rich literature on time integration approaches, including important issues like stability, accuracy, dissipation and dispersion, that has been developed over the past decades. For that, reference is made to other books and overview articles (Bathe 1982; Belytschko 1983; Hughes 1983, 1987; Hulbert 2004). Also, we will not touch upon important issues like computational multibody dynamics, either for rigid bodies or for flexible bodies.

5.1 The Semi-discrete Equations

We pick up the discussion at Equation (2.15), the semi-discrete balance of momentum:

$$\mathbf{M}\ddot{\mathbf{a}}^{t+\Delta t} = \mathbf{f}_{\text{ext}}^{t+\Delta t} - \mathbf{f}_{\text{int}}^{t+\Delta t}$$

with $\mathbf{M}$ the mass matrix, $\mathbf{f}_{\text{ext}}$ the external force vector, and $\mathbf{f}_{\text{int}}$ the internal force vector, Equations (2.16), (2.17) and (2.18), and proceed in a manner similar to that in Chapter 2, but now including the inertia term. Equation (2.15) is used directly in explicit time integration schemes, where this equation is usually approximated by a finite difference scheme in time, and which are treated in Section 5.2. Within a time step, iterations are not carried out to rigorously satisfy the balance of momentum, and small time steps are therefore mandatory to obtain an accurate solution.

Within implicit time integration schemes, on the other hand, an iterative procedure like those treated in Chapters 2 and 4, is carried out to satisfy the balance of momentum at the end of the time step. Partitioning the stress at $t + \Delta t$, in a known quantity at t and an increment $\Delta\boldsymbol{\sigma}$, Equation (2.28), and substituting this into the expression for the internal force, Equation (2.18),

Non-linear Finite Element Analysis of Solids and Structures, Second Edition.
René de Borst, Mike A. Crisfield, Joris J.C. Remmers and Clemens V. Verhoosel.

allows us to extend expression (2.29) to:

$$\sum_{e=1}^{n_e} \mathbf{Z}_e^{\mathrm{T}} \left(\int_{V_e} \rho \mathbf{H}^{\mathrm{T}} \mathbf{H} \mathrm{d}V \mathbf{Z}_e \ddot{\mathbf{a}}^{t+\Delta t} + \int_{V_e} \mathbf{B}^{\mathrm{T}} \Delta \boldsymbol{\sigma} \mathrm{d}V \right) = \mathbf{f}_{\mathrm{ext}}^{t+\Delta t} - \mathbf{f}_{\mathrm{int},0}^{t} \tag{5.1}$$

With a linearisation similar to that following Equation (2.30), we can elaborate Equation (5.1) as:

$$\mathbf{M}\ddot{\mathbf{a}}^{t+\Delta t} + \mathbf{K}_0 \Delta \mathbf{a} = \mathbf{f}_{\mathrm{ext}}^{t+\Delta t} - \mathbf{f}_{\mathrm{int},0}^{t} \tag{5.2}$$

cf. Equation (2.42), with $\mathbf{K}_0$ the tangential stiffness matrix at the beginning of the time step, Equation (2.43). Unlike in Chapter 2 the superscripts in Equations (5.1) and (5.2) now denote the real time, and not a virtual time. Similar to Chapter 2 for quasi-static situations, the linearisation error committed in going from Equation (5.1) to Equation (5.2) leads to a drifting away from the 'dynamic' equilibrium curve, and iterations should be added to ensure that the error remains within a certain tolerance, see Section 5.3 on implicit time integration methods. For this purpose we define, similar to the quasi-static case, cf. Equation (2.47), a 'dynamic residual force vector':

$$\mathbf{r}_0^* = \mathbf{f}_{\mathrm{ext}}^{t+\Delta t} - \mathbf{f}_{\mathrm{int},0}^{t} - \mathbf{M}\ddot{\mathbf{a}}^{t+\Delta t} \tag{5.3}$$

so that a first estimate for the displacement increment, $\Delta \mathbf{a}_1$, can be obtained from solving the linearised equations:

$$\Delta \mathbf{a}_1 = \mathbf{K}_0^{-1} \mathbf{r}_0^* \tag{5.4}$$

5.2 Explicit Time Integration

One of the most popular explicit time integration schemes is the central difference scheme, which is classically written as:

$$\dot{\mathbf{a}}^{t+\Delta t} = \frac{\mathbf{a}^{t+\Delta t} - \mathbf{a}^{t-\Delta t}}{2\Delta t} \tag{5.5a}$$

$$\ddot{\mathbf{a}}^{t+\Delta t} = \frac{\mathbf{a}^{t+\Delta t} - 2\mathbf{a}^{t} + \mathbf{a}^{t-\Delta t}}{\Delta t^2} \tag{5.5b}$$

and is second-order accurate in time, i.e. the error associated with the time integration scheme decreases proportional to Δt^2. Substitution of Equation (5.5b) into the semi-discrete balance of momentum, Equation (2.15), and rearranging gives:

$$\frac{1}{\Delta t^2}\mathbf{M}\mathbf{a}^{t+\Delta t} = \mathbf{f}_{\mathrm{ext}}^{t+\Delta t} - \mathbf{f}_{\mathrm{int}}^{t+\Delta t} + \frac{1}{\Delta t^2}\mathbf{M}\left(2\mathbf{a}^{t} - \mathbf{a}^{t-\Delta t}\right) \tag{5.6}$$

Box 5.1 Central difference time integration scheme for non-linear problems

Initialise $\mathbf{a}^0$ and $\mathbf{a}^0$
Compute the mass matrix: $\mathbf{M}$
Compute: $\mathbf{a}^{-\Delta t} = \mathbf{a}^0 - \Delta t\mathbf{a}^0 + \frac{1}{2}\Delta t^2\mathbf{M}^{-1}(\mathbf{f}^0_{\text{ext}} - \mathbf{f}^0_{\text{int}})$
For each time step:

1. Solve for total displacements: $\mathbf{a}^{t+\Delta t} = \Delta t^2\mathbf{M}^{-1}\left(\mathbf{f}^{t+\Delta t}_{\text{ext}} - \mathbf{f}^{t+\Delta t}_{\text{int}}\right) + 2\mathbf{a}^t - \mathbf{a}^{t-\Delta t}$
2. Compute the displacement increment: $\Delta\mathbf{a} = \mathbf{a}^{t+\Delta t} - \mathbf{a}^t$
3. *For each integration point i:*
   - Compute the strain increment: $\Delta\mathbf{a} \to \Delta\boldsymbol{\epsilon}_i$
   - Compute the stress increment: $\Delta\boldsymbol{\epsilon}_i \to \Delta\boldsymbol{\sigma}_i$
   - Compute the total stress: $\boldsymbol{\sigma}^{t+\Delta t}_i = \boldsymbol{\sigma}^t_i + \Delta\boldsymbol{\sigma}_i$
4. Compute the internal force vector: $\mathbf{f}^{t+\Delta t}_{\text{int}} = \sum_{e=1}^{n_e}\mathbf{Z}^{\mathrm{T}}_e\sum_{i=1}^{n_i} w_i \det\mathbf{J}_i\mathbf{B}^{\mathrm{T}}_i\boldsymbol{\sigma}_i$
5. Update the velocities: $\mathbf{a}^{t+\Delta t} = \frac{\mathbf{a}^{t+\Delta t} - \mathbf{a}^{t-\Delta t}}{2\Delta t}$
6. Update the accelerations: $\ddot{\mathbf{a}}^{t+\Delta t} = \frac{\mathbf{a}^{t+\Delta t} - 2\mathbf{a}^t + \mathbf{a}^{t-\Delta t}}{\Delta t^2}$

This equation can be solved directly for the displacement at time $t + \Delta t$:

$$\mathbf{a}^{t+\Delta t} = \Delta t^2\mathbf{M}^{-1}\left(\mathbf{f}^{t+\Delta t}_{\text{ext}} - \mathbf{f}^{t+\Delta t}_{\text{int}}\right) + 2\mathbf{a}^t - \mathbf{a}^{t-\Delta t} \tag{5.7}$$

Then, the displacement increment $\Delta\mathbf{a} = \mathbf{a}^{t+\Delta t} - \mathbf{a}^t$ can be computed, which, through the kinematic relation yields the strain increment $\Delta\boldsymbol{\epsilon}$, and subsequently, using the constitutive equations, the stress increment $\Delta\boldsymbol{\sigma}$ can be computed. The stress is updated according to:

$$\boldsymbol{\sigma}^{t+\Delta t} = \boldsymbol{\sigma}^t + \Delta\boldsymbol{\sigma}$$

and the internal force vector $\mathbf{f}^{t+\Delta t}_{\text{int}}$ is computed according to Equation (2.18). The algorithm is summarised in Box 5.1. Please note that in Equation (5.7) the displacement at $t + \Delta t$ is given in terms of the displacements at time t and at time $t - \Delta t$, which implies that information on the displacement field of the two preceding time steps is required. This poses a problem for initialising the computation, since this would require knowledge of the displacement field at $-\Delta t$. To circumvent this problem Equations (5.5) are considered for $t = 0$ and $\mathbf{a}^{\Delta t}$ is eliminated to yield:

$$\mathbf{a}^{-\Delta t} = \mathbf{a}^0 - \Delta t\mathbf{a}^0 + \frac{1}{2}\Delta t^2\mathbf{M}^{-1}(\mathbf{f}^0_{\text{ext}} - \mathbf{f}^0_{\text{int}}) \tag{5.8}$$

where $\mathbf{a}^0$ and $\mathbf{a}^0$ are the initial displacement and velocity fields, respectively. The semi-discrete balance of momentum, Equation (2.15), at $t = 0$ has been used to obtain the last term. Often, this term will be zero, but not necessarily, e.g. in the presence of initial stresses.

An alternative, which is employed in most explicit computer codes, is to approximate the velocity at mid-interval (Belytschko *et al.* 1976):

$$\dot{\mathbf{a}}^{t+\frac{1}{2}\Delta t} = \frac{\mathbf{a}^{t+\Delta t} - \mathbf{a}^{t}}{\Delta t} \tag{5.9}$$

together with the following approximation for the acceleration at $t + \Delta t$:

$$\ddot{\mathbf{a}}^{t} = \frac{\dot{\mathbf{a}}^{t+\frac{1}{2}\Delta t} - \dot{\mathbf{a}}^{t-\frac{1}{2}\Delta t}}{\Delta t} \tag{5.10}$$

Upon substitution of Equation (5.9) into Equation (5.10) the central difference approximation for the acceleration, Equation (5.5b), is recovered. Using Equation (5.9) the nodal displacements at $t + \Delta t$ are obtained from the velocities at $t + \frac{1}{2}\Delta t$ as:

$$\mathbf{a}^{t+\Delta t} = \mathbf{a}^{t} + \Delta t \dot{\mathbf{a}}^{t+\frac{1}{2}\Delta t} \tag{5.11}$$

As before, the displacement increment is computed next: $\Delta\mathbf{a} = \mathbf{a}^{t+\Delta t} - \mathbf{a}^{t}$, followed by the strain increment $\Delta\boldsymbol{\epsilon}$, and subsequently, the stress increment $\Delta\boldsymbol{\sigma}$. With the updated stress, $\boldsymbol{\sigma}^{t+\Delta t} = \boldsymbol{\sigma}^{t} + \Delta\boldsymbol{\sigma}$, the internal force vector $\mathbf{f}_{\text{int}}^{t+\Delta t}$ is computed according to Equation (2.18). The accelerations are then straightforwardly computed using the semi-discrete balance of momentum, Equation (2.15):

$$\ddot{\mathbf{a}}^{t+\Delta t} = \mathbf{M}^{-1}\left(\mathbf{f}_{\text{ext}}^{t+\Delta t} - \mathbf{f}_{\text{int}}^{t+\Delta t}\right) \tag{5.12}$$

and, for the next step, the velocities at mid-time interval are computed using Equation (5.10):

$$\dot{\mathbf{a}}^{t+\frac{3}{2}\Delta t} = \dot{\mathbf{a}}^{t+\frac{1}{2}\Delta t} + \Delta t \ddot{\mathbf{a}}^{t+\Delta t} \tag{5.13}$$

As for the previous central difference scheme a special starting condition must be used. It is assumed that:

$$\dot{\mathbf{a}}^{\frac{1}{2}\Delta t} = \dot{\mathbf{a}}^{0} + \frac{1}{2}\Delta t \ddot{\mathbf{a}}^{0} \tag{5.14}$$

The algorithm is summarised in Box 5.2.

When the full, or consistent mass matrix is used, a global system of equations must be solved, cf. Equation (2.15) or Equation (5.7), which is not attractive, especially not for explicit methods, which are conditionally stable and have a rather strict requirement on the critical time step, see below. For this reason the mass matrix is often diagonalised in explicit time integration schemes. Several possibilities of this lumping process are available, such as nodal quadrature, row-sum lumping, or a 'special lumping technique' (Hinton *et al.* 1970), where only the latter method produces positive lumped masses for any element type (Hughes 1987). As an example, in row-sum lumping, the diagonal terms of the lumped mass matrix $\mathbf{M}^{\text{lumped}}$ read:

$$M_{kk}^{\text{lumped}} = \sum_{l=1}^{N} M_{kl} \tag{5.15}$$

Box 5.2 Alternative central difference time integration scheme for non-linear problems

Initialise $\mathbf{a}^0$ and $\mathbf{a}^0$

Compute the mass matrix: $\mathbf{M}$

Compute: $\mathbf{a}^{\frac{1}{2}\Delta t} = \mathbf{a}^0 + \frac{1}{2}\Delta t \ddot{\mathbf{a}}^0$

For each time step:

1. Solve for total displacements: $\mathbf{a}^{t+\Delta t} = \mathbf{a}^t + \Delta t \mathbf{a}^{t+\frac{1}{2}\Delta t}$
2. Compute the displacement increment: $\Delta \mathbf{a} = \mathbf{a}^{t+\Delta t} - \mathbf{a}^t$
3. *For each integration point i:*
   - Compute the strain increment: $\Delta \mathbf{a} \rightarrow \Delta \boldsymbol{\epsilon}_i$
   - Compute the stress increment: $\Delta \boldsymbol{\epsilon}_i \rightarrow \Delta \boldsymbol{\sigma}_i$
   - Compute the total stress: $\boldsymbol{\sigma}_i^{t+\Delta t} = \boldsymbol{\sigma}_i^t + \Delta \boldsymbol{\sigma}_i$
4. Compute the internal force vector: $\mathbf{f}_{\text{int}}^{t+\Delta t} = \sum_{e=1}^{n_e} \mathbf{Z}_e^{\mathrm{T}} \sum_{i=1}^{n_i} w_i \det \mathbf{J}_i \mathbf{B}_i^{\mathrm{T}} \boldsymbol{\sigma}_i$
5. Solve for the new accelerations: $\ddot{\mathbf{a}}^{t+\Delta t} = \mathbf{M}^{-1}\left(\mathbf{f}_{\text{ext}}^{t+\Delta t} - \mathbf{f}_{\text{int}}^{t+\Delta t}\right)$
6. Compute the velocities at new mid-time: $\mathbf{a}^{t+\frac{3}{2}\Delta t} = \mathbf{a}^{t+\frac{1}{2}\Delta t} + \Delta t \ddot{\mathbf{a}}^{t+\Delta t}$

for a system with N unknowns. After diagonalisation, the system of equations, Equation (2.15) or Equation (5.7), is no longer coupled and the acceleration for each degree of freedom k can be solved separately:

$$\ddot{a}_k^{t+\Delta t} = \frac{r_k}{M_{kk}^{\text{lumped}}} \tag{5.16}$$

The computational time of explicit time integration schemes is thus reduced significantly by using a diagonal mass matrix.

Replacing the consistent mass matrix by a lumped mass matrix can have an additional benefit. For a consistent mass matrix a higher frequency, and therefore a smaller period, is calculated compared with the exact solution. The use of a consistent mass matrix thus yields an upper bound value for the frequency. A lumped mass matrix, on the other hand, tends to result in a frequency that is below the exact solution. Analyses of time integrators for linear systems furthermore show that implicit schemes provide a lower bound to the frequency, while explicit schemes provide an upper bound. These observations suggest that optimal results with respect to the induced period errors can be obtained when pairing an implicit time integration scheme to a consistent mass matrix, and using an explicit time integrator in conjunction with a lumped mass matrix (Hughes 1987). Although this statement can be made more rigorously for linear systems, numerical experience shows that it normally also holds true for the analysis of non-linear systems.

Explicit time integration schemes can have advantages in reducing the computational effort, especially if a lumped mass matrix is used, as factorisation and storage of a system matrix are then not required. Two possible disadvantages of explicit time integration schemes are the stability of the solution and the impossibility to ascertain that the solution computes the 'dynamic equilibrium path' with sufficient accuracy. Explicit time stepping schemes are only conditionally stable, which implies that, for linear problems, the time step is bounded by the Courant–Friedrichs–Lewy (CFL) criterion (Courant *et al.* 1928):

$$\Delta t = \frac{2}{\omega^{\max}} \tag{5.17}$$

see also Park and Underwood (1980) and Underwood and Park (1980). The maximum natural frequency $\omega^{\max}$ of the system is rather expensive to compute. For this reason, it is often approximated by the maximum frequency of a single finite element $\omega_e^{\max}$, which provides an upper bound to the maximum natural frequency of the system: $\omega_e^{\max} \geq \omega^{\max}$ (Belytschko 1983; Hughes 1987), and can be calculated by assuming a displacement field of the form:

$$\mathbf{a} = \hat{\mathbf{a}} \exp(\mathrm{i}\omega_e t)$$

with $\hat{\mathbf{a}}$ the amplitude of the displacement, and ω_e the frequencies of the element. Substituting this expression into the linearised and discretised balance of momentum for the element:

$$\mathbf{M}_e \ddot{\mathbf{a}} + \mathbf{K}_e \mathbf{a} = \mathbf{0}$$

with $\mathbf{M}_e$ and $\mathbf{K}_e$ the element mass and stiffness matrices, respectively, yields the maximum frequency $\omega_e^{\max}$ by solving the eigenvalue problem:

$$\det\left(\mathbf{K}_e - \omega_e^2 \mathbf{M}_e\right) = 0 \tag{5.18}$$

Alternatively, the maximum natural frequency $\omega^{\max}$ can be approximated through the Rayleigh coefficient, where the 'current frequency' is estimated as (Bergan and Mollestad 1985):

$$\omega^2 = \frac{\Delta\mathbf{a}^{\mathrm{T}}\mathbf{K}\Delta\mathbf{a}}{\Delta\mathbf{a}^{\mathrm{T}}\mathbf{M}\Delta\mathbf{a}} \tag{5.19}$$

In explicit time integration methods usually no iterative procedure is applied to enforce that the dynamic residual force vector

$$\mathbf{r}^* = \mathbf{f}_{\mathrm{ext}}^{t+\Delta t} - \mathbf{f}_{\mathrm{int}}^{t+\Delta t} - \mathbf{M}\ddot{\mathbf{a}}^{t+\Delta t}$$

is sufficiently close to zero, measured in some norm. Indeed, a drifting tendency can occur, where the dynamic residual force vector $\mathbf{r}^*$ grows when the calculation proceeds. The use of small time steps is the only solution to mitigate this drifting tendency.

5.3 PyFEM: Implementation of an Explicit Solver

An explicit time integration solver has been implemented in `ExplicitSolver.py`. This file can be found in the directory `pyfem/solvers`. The global structure of this file is similar to the structure of Riks' arc-length solver, which has been discussed in Chapter 4. It contains the declaration of the class `ExplicitSolver` and a number of member functions:

⟨*Explicit time integration solver* ⟩≡

⟨*Initialisation of the explicit solver class* 149⟩
⟨*Explicit solver class main functions* 150⟩
⟨*Explicit solver class utility functions*⟩

The class is derived from the class `BaseModule` as can be seen in the following fragment.

⟨*Initialisation of the explicit solver class* ⟩≡ 149

```
from pyfem.fem.Assembly import assembleMass

class ExplicitSolver( BaseModule ):

  def __init__( self , props , globdat ):

    ⟨Initialisation of default solver parameters⟩

    BaseModule.__init__( self , props )

    M,self.Mlumped = assembleMass( props , globdat )

    ⟨Calculate velocity at t = ½Δt⟩
```

The parameters of the solver, such as the time step and the loading function, are read from the instance properties `props` and are stored as members of this class, `self.dtime` and `self.loadfunc`. Since it has been assumed that the mass of the system remains constant during the simulation, the mass terms need to be determined only once, at the start of the simulation, see also Box 5.2. The function `assembleMass` assembles the mass terms in a similar way as the function `assembleTangentStiffness`, which has been discussed in Chapter 2. This function returns the consistent mass matrix as well as the lumped mass matrix. The consistent mass matrix will not be used here and is stored temporarily in the local variable `M`. The lumped mass matrix is stored in vector format as a member of the class `Mlumped`.

The function `assembleMass` assembles the total mass of the system by looping over all elements and calculating their contributions through the function `getElementMass`. Most element formulations in the code contain a description of a mass matrix. For example, the implementation of this function in a continuum element that incorporates large displacement gradients is given on page 151. The constructor of the class ends with the calculation of the velocity **a** at $t = \frac{1}{2}\Delta t$, see Box 5.2.

The class `ExplicitSolver` has a member function `run` which solves the governing system of equations for a single time step.

⟨*Explicit solver class main functions* ⟩≡ 149

```
def run( self , props , globdat ):

  globdat.cycle += 1
  globdat.time  += self.dtime

  lam = self.loadfunc( globdat.time )

  disp  = globdat.state
  ddisp = globdat.dstate
  velo  = globdat.velo
  fhat  = globdat.fhat
```

After increasing the cycle number and the simulation time `globdat.time`, the current load parameter `lam` is determined as a function of time. The displacement vector is copied from the `globdat` database. Next to the total and the incremental displacement vectors, `state` and `dstate`, an additional solution vector is used, `velo`, which represents the time derivative of the solution vector. The unit external force vector is represented in `fhat`.

The updated total and incremental displacement vectors are calculated according to steps 1 and 2 in Box 5.2:

⟨*Explicit solver class main functions* ⟩+≡ 150

```
  ddisp = props.dtime * velo
  disp += ddisp

  from pyfem.fem.Assembly import assembleInternalForce

  fint  = assembleInternalForce( props, globdat )

  globdat.dofs.setConstrainFactor( lam )

  acce  = globdat.dofs.solve( self.Mlumped , lam*fhat - fint )
```

The updated state vectors are used to compute the internal force vector. In this case the function `assembleInternalForce` can be used since there is no need to compute the tangential stiffness matrix of the system. The constrained degrees of freedom are updated next. Since we are solving a system of equations where the acceleration is the unknown, we specify prescribed accelerations. The new accelerations are calculated in the function `dofs.solve`, which is an alternative implementation of the solver used in the fragments on pages 80 and 127. When the first argument is a diagonal matrix, stored in vector format, the system of equations is uncoupled and can be solved for each degree of freedom separately, according to Equation (5.16).

With the updated accelerations the new velocity field can be computed (step 6 in Box 5.2). At the end of the time step, the element history is stored.

⟨*Explicit solver class main functions* ⟩+≡ 150

```
velo += 0.5 * props.dtime * acce

elements.commitHistory()
```

The calculation of the consistent and the lumped mass matrices of the element is implemented as a member function of the `FiniteStrainContinuum` class, see Chapter 3.

⟨*Finite strain continuum element class main functions* ⟩+≡ 109

```
def getMassMatrix ( self, elemdat ):

  sData = getElemShapeData( elemdat.coords )           39

  rho = elemdat.matprops.rho

  for iData in sData:
    N  = self.getNmatrix( iData.h )
    elemdat.mass += dot ( N.transpose() , N ) * rho * iData.weight

  elemdat.lumped = sum(elemdat.stiff)
```

The function has a single argument `elemdat` which contains the element nodal coordinates. The element shape functions are obtained using these coordinates. The mass density of the element is stored in the material properties instance `matprops`, which is a member of `elemdat`. The mass matrix is constructed according to Equation (2.16). When the consistent mass matrix has been computed, the lumped mass matrix is obtained by summing over each row of the consistent mass matrix. The result is stored in the vector `elemdat.lumped`.

An example calculation using the explicit solver is presented next. We consider a block of material in a plane-strain configuration (Figure 5.1). It has the dimensions $L = 5$ mm and $W = 10$ mm. The Young's modulus $E = 3.24$ GPa, the Poisson's ratio $\nu = 0.35$, and the mass density $\rho = 1190\,\text{kg/m}^3$, which results in a dilatational wave speed $c_d = 2090$ m/s (Freund 1998). The block is not supported and is loaded by an impact velocity which acts in the positive y-direction on the top boundary of the block, at $y = +L$. The impact velocity is increased linearly to $v = 10$ m/s with a rise time $t_r = 1.0 \times 10^{-7}$ s. In the finite element model we must specify the prescribed accelerations instead. Differentiation of this velocity profile yields a constant prescribed acceleration of $10^8\,\text{m/s}^2$ during the first 0.1 μs of the simulation. Because of symmetry with respect to the y-axis only one half of the block has been modelled, using a mesh

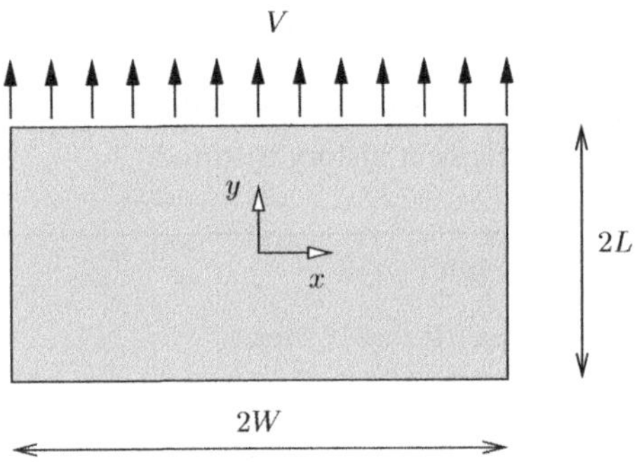

Figure 5.1 Geometry and boundary conditions of a rectangular block loaded by impact.

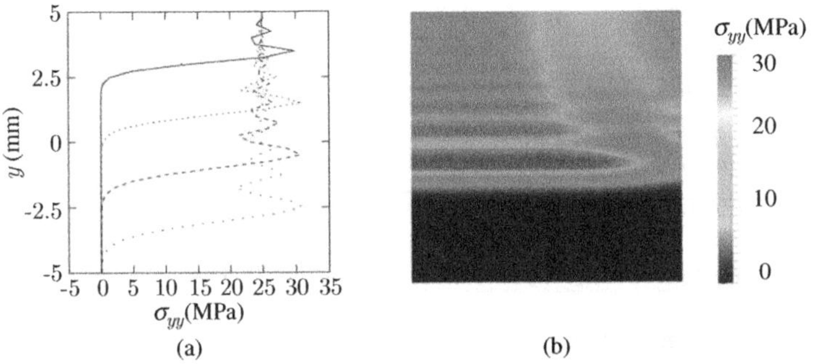

Figure 5.2 (a) σ_{yy} measured along the centre of the specimen (at $x = 0$ mm) for different times during the simulation. (b) Contours of σ_{yy} in the right-hand side of the specimen at $t = 3\ \mu$s

of 20×40 quadrilaterals. The displacements of the nodes on the symmetry axis are constrained in the x-direction, which, in the present case, implies that the prescribed acceleration of the corresponding degree of freedom is constrained to zero. The time step Δt is set to 1.0×10^{-8} which satisfies the CFL criterion, Equation (5.17).

The stress contours along the y-axis at different times are given in Figure 5.2(a). The oscillations at the wave front can be attributed to initiation effects. The amplitude of the peaks are smaller when the prescribed acceleration is smeared out over a longer period of time. The magnitude of the stress is approximately 25.0 MPa, which resembles the results presented by Xu and Needleman (1994). The speed of the stress wave equals the distance travelled by the first peak divided by the elapsed time. We can estimate this velocity as 2000 m/s, which is close to the dilatational wave speed $c_d = 2090$ m/s. Figure 5.2(b) shows the contours of the normal stress in the y-direction, σ_{yy}, at $t = 3\ \mu$s.

5.4 Implicit Time Integration

Implicit time integration schemes are more complicated, and more expensive per time step, but can allow for (significantly) larger time steps, and provide a control on the dynamic residual force vector, since they are usually used in conjunction with an iterative procedure within each time step. While explicit schemes are primarily chosen when high frequencies are present in the problem, such as in blast or impact loading, implicit time integration schemes are more often used when the accelerations are relatively low, such as in earthquake engineering.

5.4.1 The Newmark Family

One of the most widely used implicit time integration methods is due to Newmark (1959), and is based on the assumption that the acceleration varies linearly over the time step, so that:

$$\dot{\mathbf{a}}^{t+\Delta t} = \dot{\mathbf{a}}^t + \Delta t \left((1-\gamma)\ddot{\mathbf{a}}^t + \gamma \ddot{\mathbf{a}}^{t+\Delta t} \right) \tag{5.20a}$$

$$\mathbf{a}^{t+\Delta t} = \mathbf{a}^t + \Delta t \dot{\mathbf{a}}^t + \frac{1}{2}\Delta t^2 \left((1-2\beta)\ddot{\mathbf{a}}^t + 2\beta \ddot{\mathbf{a}}^{t+\Delta t} \right) \tag{5.20b}$$

The integration parameters β and γ determine the stability, accuracy, dissipative and dispersion characteristics of the system. For linear systems unconditional stability is achieved when:

$$2\beta \geq \gamma \geq \frac{1}{2} \tag{5.21}$$

and second-order accuracy is obtained for $\gamma = \frac{1}{2}$, i.e. the error decreases proportional to Δt^2. For

$$\gamma \geq \frac{1}{2} \ \wedge \ 2\beta < \gamma \tag{5.22}$$

conditional stability is obtained, where the time step is constrained by:

$$\begin{cases} \Delta t \leq \frac{\omega^{\text{crit}}}{\omega^{\text{max}}} \\ \omega^{\text{crit}} = \frac{\sqrt{2}}{\sqrt{\gamma - 2\beta}} \end{cases} \tag{5.23}$$

Several well-known time integration schemes can be conceived as special cases of the Newmark family. For $\beta = \frac{1}{4}, \gamma = \frac{1}{2}$ the average acceleration scheme, or trapezoidal rule is obtained, which is unconditionally stable and second-order accurate in the time step. Other implicit schemes are obtained for $\beta = \frac{1}{6}, \gamma = \frac{1}{2}$, the linear acceleration scheme, and for $\beta = \frac{1}{12}, \gamma = \frac{1}{2}$, the Fox–Goodwin scheme. Neither scheme, although having an implicit format, is unconditionally stable, and the time step is limited by Equation (5.23). Also some explicit integration schemes can be considered as special cases of the Newmark family, for instance the central difference scheme, which is obtained for $\beta = 0$, $\gamma = \frac{1}{2}$, so that $\omega^{\text{crit}} = 2$, thus retrieving the critical time step for explicit methods derived in the preceding section.

The above statements on unconditional stability and the time step limit for the case of conditional stability hold rigorously for linear elasticity, and the linear-elastic element stiffness matrix must be substituted in Equation (5.18). For non-linear behaviour the element tangential stiffness matrix can be substituted for $\mathbf{K}_e$, but no proof exists that this indeed provides a bound to the time step. For sufficient accuracy, significantly smaller time steps should be selected when non-linear behaviour is considered. Depending on the material model and the specific structure, the time step can be in the range 0.1–0.5 of that used in linear structural behaviour. For softening problems, i.e. when there is a negative stiffness, cf. Chapter 6, $\omega_e^{\max}$ becomes imaginary, and a value for the critical time step cannot be obtained at all.

5.4.2 *The HHT α-method*

Numerical dissipation can be desirable in a number of cases, e.g. to filter out the high-frequency modal components that are introduced by the spatial discretisation. Numerical dissipation can be introduced in the Newmark scheme, for $\gamma > \frac{1}{2}$. Unfortunately, second-order accuracy is then lost. To circumvent this problem, Hilber *et al.* (1977) developed the α-method, in which the dynamic residual force vector (5.3) is replaced by:

$$\mathbf{r}_0^* = (1+\alpha)\mathbf{f}_{\text{ext}}^{t+\Delta t} - \alpha\mathbf{f}_{\text{ext}}^{t} - \mathbf{f}_{\text{int},0}^{t} - \mathbf{M}\ddot{\mathbf{a}}^{t+\Delta t} \tag{5.24}$$

while maintaining Newmark's assumption that the acceleration varies linearly over the time step, Equations (5.20):

$$\begin{cases} \mathbf{a}^{t+\Delta t} = \dot{\mathbf{a}}^t + \Delta t\left((1-\gamma)\ddot{\mathbf{a}}^t + \gamma\ddot{\mathbf{a}}^{t+\Delta t}\right) \\ \mathbf{a}^{t+\Delta t} = \mathbf{a}^t + \Delta t\dot{\mathbf{a}}^t + \frac{1}{2}\Delta t^2\left((1-2\beta)\ddot{\mathbf{a}}^t + 2\beta\ddot{\mathbf{a}}^{t+\Delta t}\right) \end{cases}$$

Evidently, the α-method reduces to Newmark's method for $\alpha = 0$, but introduces numerical dissipation for $-\frac{1}{3} < \alpha < 0$, while second-order accuracy is preserved for $\beta = \frac{1}{4}(1-\alpha)^2$ and $\gamma = \frac{1}{2} - \alpha$. In non-linear analyses $\alpha = -0.05$ is often used, but similar to any time integration method, stability cannot be assured.

For the algorithmic implementation we proceed by solving for $\ddot{\mathbf{a}}^{t+\Delta t}$ from Equation (5.20b):

$$\ddot{\mathbf{a}}^{t+\Delta t} = \frac{1}{\beta\Delta t^2}\Delta\mathbf{a} - \frac{1}{\beta\Delta t}\mathbf{a}^t - \frac{1-2\beta}{2\beta}\ddot{\mathbf{a}}^t \tag{5.25}$$

and substitute this expression into the dynamic residual force vector, Equation (5.24), which using Equation (5.4) yields for the first estimate of the displacement increment within the time step:

$$\Delta\mathbf{a}_1 = (\mathbf{K}_0^*)^{-1}\mathbf{f}_0^* \tag{5.26}$$

with the algorithmic tangential stiffness matrix

$$\mathbf{K}_0^* = (1+\alpha)\mathbf{K}_0 + \frac{1}{\beta\Delta t^2}\mathbf{M} \tag{5.27}$$

and the right-hand side vector $\mathbf{f}_0^*$ defined as:

$$\mathbf{f}_0^* = (1+\alpha)\mathbf{f}_{\text{ext}}^{t+\Delta t} - \alpha\mathbf{f}_{\text{ext}}^{t} - \mathbf{f}_{\text{int},0}^{t} + \mathbf{M}\left(\frac{1}{\beta\Delta t}\mathbf{a}^t + \frac{1-2\beta}{2\beta}\ddot{\mathbf{a}}^t\right) \tag{5.28}$$

This first estimate for the displacement increments, $\Delta\mathbf{a}_1$, can be used – at each integration point – to compute a first estimate for the strain increment, $\Delta\boldsymbol{\epsilon}_1$, and via the constitutive relation, the stress increment $\Delta\boldsymbol{\sigma}_1$. From this, the first estimate for the stress:

$$\boldsymbol{\sigma}_1^{t+\Delta t} = \boldsymbol{\sigma}^t + \Delta\boldsymbol{\sigma}_1$$

at $t + \Delta t$ can be computed, so that first estimate for the internal force vector ensues as:

$$\mathbf{f}_{\text{int},1}^{t+\Delta t} = \sum_{e=1}^{n_e} \mathbf{Z}_e^{\mathrm{T}} \int_{V_e} \mathbf{B}^{\mathrm{T}} \boldsymbol{\sigma}_1^{t+\Delta t} \mathrm{d}V$$

With these results the accelerations can be updated according to Equation (5.25):

$$\ddot{\mathbf{a}}_1^{t+\Delta t} = \frac{1}{\beta \Delta t^2}\Delta\mathbf{a}_1 - \frac{1}{\beta\Delta t}\mathbf{a}^t - \frac{1-2\beta}{2\beta}\ddot{\mathbf{a}}^t$$

and the new estimate for the dynamic residual force vector can be computed as, cf. Equation (5.24):

$$\mathbf{r}_1^* = (1+\alpha)\mathbf{f}_{\text{ext}}^{t+\Delta t} - \alpha\mathbf{f}_{\text{ext}}^t - \mathbf{f}_{\text{int},1}^{t+\Delta t} - \mathbf{M}\ddot{\mathbf{a}}_1^{t+\Delta t}$$

With the (possibly) updated tangential stiffness matrix $\mathbf{K}_1$ the new algorithmic stiffness matrix $\mathbf{K}_1^*$ and the force vector $\mathbf{f}_1^*$ can be computed, similar to Equations (5.27) and (5.28), whereupon the correction to the displacement increment $\mathrm{d}\mathbf{a}_2$ ensues as:

$$\mathrm{d}\mathbf{a}_2 = (\mathbf{K}_1^*)^{-1}\mathbf{f}_1^*$$

and the new estimate for the displacement increment reads:

$$\Delta\mathbf{a}_2 = \Delta\mathbf{a}_1 + \mathrm{d}\mathbf{a}_2$$

from which new strain and stress increments can be computed. An algorithm for the HHT α-method is given in Box 5.3.

5.4.3 *Alternative Implicit Methods for Time Integration*

There is a host of methods available that can attain higher-order accuracy. In particular the linear multi-step (LMS) family of algorithms has gained some popularity in structural dynamics. An early LMS method that has been developed for use in structural dynamics is due to Houbolt (1950). In it, the equation of motion is augmented by the following expressions to evaluate the acceleration and the velocity:

$$\begin{cases} \ddot{\mathbf{a}}^{t+\Delta t} = \dfrac{2\mathbf{a}^{t+\Delta t} - 5\mathbf{a}^t + 4\mathbf{a}^{t-\Delta t} - \mathbf{a}^{t-2\Delta t}}{\Delta t^2} \\ \mathbf{a}^{t+\Delta t} = \dfrac{11\mathbf{a}^{t+\Delta t} - 18\mathbf{a}^t + 9\mathbf{a}^{t-\Delta t} - 2\mathbf{a}^{t-2\Delta t}}{6\Delta t} \end{cases} \tag{5.29}$$

This three-step LMS method has second-order accuracy and is unconditionally stable for linear problems. A more accurate (six-step) LMS method is Park's method (Park 1975). A disadvantage of both methods is that they require special starting procedures.

Box 5.3 HHT α time integration scheme for non-linear problems

Initialise $\mathbf{a}^0$, $\mathbf{a}^0$, and $\mathbf{f}_{\text{int}}^0$

Compute the mass matrix: $\mathbf{M}$

For each time step:

1. Initialise the displacement increment: $\Delta\mathbf{a}_0 = \mathbf{0}$, and the internal force: $\mathbf{f}_{\text{int},0}^{t+\Delta t} = \mathbf{f}_{\text{int}}^t$
2. Iterations $j = 0, \ldots$ for finding 'dynamic equilibrium' within the time step:
   (a) Compute tangential stiffness: $\mathbf{K}_j = \sum_{e=1}^{n_e} \mathbf{Z}_e^{\mathrm{T}} \sum_{i=1}^{n_i} w_i \det \mathbf{J}_i \mathbf{B}_{i,j}^{\mathrm{T}} \mathbf{D}_{i,j} \mathbf{B}_{i,j} \mathbf{Z}^e$
   (b) Compute the algorithmic stiffness matrix: $\mathbf{K}_j^* = (1+\alpha)\mathbf{K}_j + \frac{1}{\beta \Delta t^2}\mathbf{M}$
   (c) Compute $\mathbf{f}_j^* = (1+\alpha)\mathbf{f}_{\text{ext}}^{t+\Delta t} - \alpha \mathbf{f}_{\text{ext}}^t - \mathbf{f}_{\text{int},j}^{t+\Delta t} + \mathbf{M}\left(\frac{1}{\beta \Delta t}\mathbf{a}^t + \frac{1-2\beta}{2\beta}\ddot{\mathbf{a}}^t\right)$
   (d) Solve the linear system: $\mathrm{d}\mathbf{a}_{j+1} = (\mathbf{K}_j^*)^{-1}\mathbf{f}_j^*$
   (e) Update the displacement increments: $\Delta\mathbf{a}_{j+1} = \Delta\mathbf{a}_j + \mathrm{d}\mathbf{a}_{j+1}$
   (f) *For each integration point i:*
      - Compute the strain increment: $\Delta\mathbf{a}_{j+1} \rightarrow \Delta\boldsymbol{\epsilon}_{i,j+1}$
      - Compute the stress increment: $\Delta\boldsymbol{\epsilon}_{i,j+1} \rightarrow \Delta\boldsymbol{\sigma}_{i,j+1}$
      - Compute the total stress: $\boldsymbol{\sigma}_{i,j+1} = \boldsymbol{\sigma}_i^t + \Delta\boldsymbol{\sigma}_{i,j+1}$
   (g) Compute internal force: $\mathbf{f}_{\text{int},j+1}^{t+\Delta t} = \sum_{e=1}^{n_e} \mathbf{Z}_e^{\mathrm{T}} \sum_{i=1}^{n_i} w_i \det \mathbf{J}_i \mathbf{B}_{i,j+1}^{\mathrm{T}} \boldsymbol{\sigma}_{i,j+1}$
   (h) Compute accelerations: $\ddot{\mathbf{a}}_{j+1}^{t+\Delta t} = \frac{1}{\beta \Delta t^2}\Delta\mathbf{a}_{j+1} - \frac{1}{\beta \Delta t}\mathbf{a}^t - \frac{1-2\beta}{2\beta}\ddot{\mathbf{a}}^t$
   (i) Compute residual: $\mathbf{r}_{j+1}^* = (1+\alpha)\mathbf{f}_{\text{ext}}^{t+\Delta t} - \alpha \mathbf{f}_{\text{ext}}^t - \mathbf{f}_{\text{int},j+1}^{t+\Delta t} - \mathbf{M}\ddot{\mathbf{a}}_{j+1}^{t+\Delta t}$
   (j) Check convergence: if $\|\mathbf{r}_{j+1}^*\| < \eta$, with η the convergence tolerance, go to 3.
3. Compute the velocities and displacements at the end of the time step:
   - Velocities: $\mathbf{a}^{t+\Delta t} = \mathbf{a}^t + \Delta t\left((1-\gamma)\ddot{\mathbf{a}}^t + \gamma \ddot{\mathbf{a}}^{t+\Delta t}\right)$
   - Displacements: $\mathbf{a}^{t+\Delta t} = \mathbf{a}^t + \Delta\mathbf{a}$

Algorithms that involve multiple function evaluations such as the Runge–Kutta methods are usually less efficient in structural dynamics because of the increased computational costs per time step, which can become prohibitive for larger systems. We finally note that in the last two decades time-discontinuous Galerkin methods have become popular, especially for elastic wave propagation because of their favourable properties with respect to the error made with respect to wave dispersion (Cho *et al.* 2011; Hughes and Hulbert 1988), i.e. the propagation of the different harmonics of a wave with different velocities, thereby changing the wave profile (Whitham 1974).

5.5 Stability and Accuracy in the Presence of Non-linearities

As remarked before, the unconditional stability of certain time integration schemes can only be proven rigorously for linear systems. Indeed, it has been shown that integration schemes which

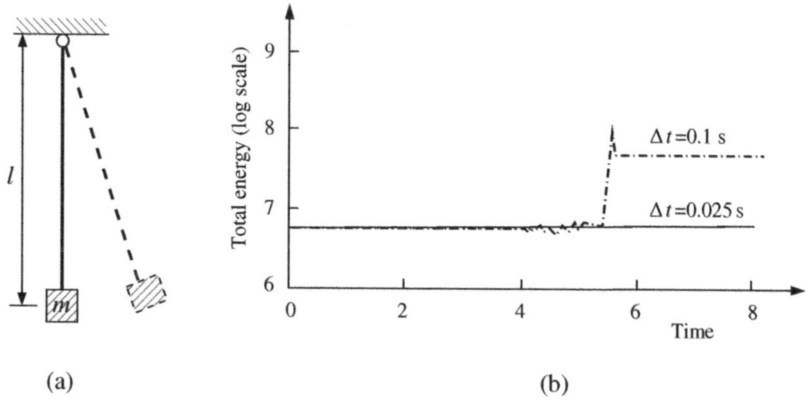

Figure 5.3 (a) Simple pendulum problem. (b) Variation of the energy with time

are unconditionally stable in the linear regime, can suffer from severe numerical instabilities when used in non-linear problems (Crisfield and Shi 1994; Galvanetto and Crisfield 1996; Simo and Tarnow 1992, 1994; Simo *et al.* 1995).

Although general analyses of stability and accuracy of time-stepping schemes for non-linear problems, and definitely for problems that involve softening, are impossible, some insight can be gained by studying simple examples (Kulkarni *et al.* 1995). We first consider the pendulum problem of Figure 5.3(a) (Bathe 1982; Crisfield and Shi 1994; Galvanetto and Crisfield 1996). In this problem the mass m is given an initial horizontal velocity v_0. The problem is solved using the trapezoidal rule (Newmark scheme with $\beta = \frac{1}{4}$ and $\gamma = \frac{1}{2}$), which is unconditionally stable for linear problems. The solution was computed for two time steps, $\Delta t = 0.1$ s and $\Delta t = 0.025$ s, which are both considerably smaller than the period (≈ 4 s). Figure 5.3(b) shows that both time steps give correct answers until approximately the end of the first period. Thereafter, the solution for the coarser time step locks on a wrong level of the total energy. It is emphasised that in this example no instability was encountered in the sense of lack of convergence. Simply the wrong solution was computed. A possible remedy in this case would be to use a time-stepping algorithm with some built-in numerical dissipation, like the HHT α-method (Hilber *et al.* 1977).

Next, we focus on the mass–spring system of Figure 5.4(a) (Sluys *et al.* 1995; Xie and Wood 1993), which has a negative spring stiffness k in order to simulate some basic characteristics of softening problems, see also Chapter 6. The advantages of this single degree of freedom system are that there is no spatial discretisation, so that no errors can stem from this source, that an analytical solution can be constructed against which the numerical solutions can be compared, and that a dispersion analysis can be carried out, which shows the dispersive properties of the various time integrators.

The mass is given an initial displacement u_0 with a zero initial velocity $\dot{u}_0 = 0$. The displacement then grows according to the analytical solution:

$$u(t) = \frac{u_0}{2}(\exp(-\omega_r t) + \exp(\omega_r t)) \tag{5.30}$$

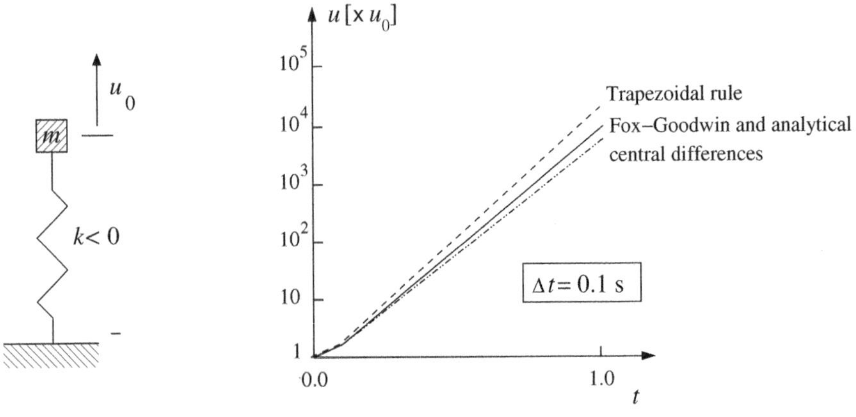

(a) (b)

Figure 5.4 (a) Mass–spring system with a negative stiffness k. (b) Exact solution and results for different time integrators

with the growth coefficient

$$\omega_r = \sqrt{-\frac{k}{m}} \tag{5.31}$$

Figure 5.4(b) shows the displacement as a function of time for the trapezoidal scheme, for the Fox–Goodwin scheme, and for the central difference scheme. For this particular case the central difference method underestimates the analytical solution, while the other time integration methods overestimate the displacement. Interestingly, the second-order accuracy of the time integrators that is assured for linear problems, is preserved (Table 5.1), while the Fox–Goodwin method even exhibits $\mathcal{O}(\Delta t^4)$ accuracy.

More insight can be gained by studying the equivalent problem of a one-dimensional bar with a bilinear stress–strain relation (Figure 5.5). With ρ the mass density and u the axial displacement, the balance of momentum for the one-dimensional bar reads:

$$\frac{\partial \sigma}{\partial x} = \rho \frac{\partial^2 u}{\partial t^2} \tag{5.32}$$

Beyond the peak strength the relation between an increment of stress and an increment of strain reads: $\Delta\sigma = H\Delta\epsilon$, with $H < 0$ the tangential stiffness. Taking into account the

Table 5.1 Absolute error in displacement for mass–spring system at $t = 1.0$ s for various time-stepping schemes

Time step(s)	Trapezoidal rule	Fox–Goodwin	Central difference
0.05	2.66×10^3	1.42×10^1	-1.06×10^3
0.025	5.95×10^2	8.95×10^{-1}	-2.81×10^2
0.0125	1.45×10^2	5.60×10^{-2}	-7.13×10^1

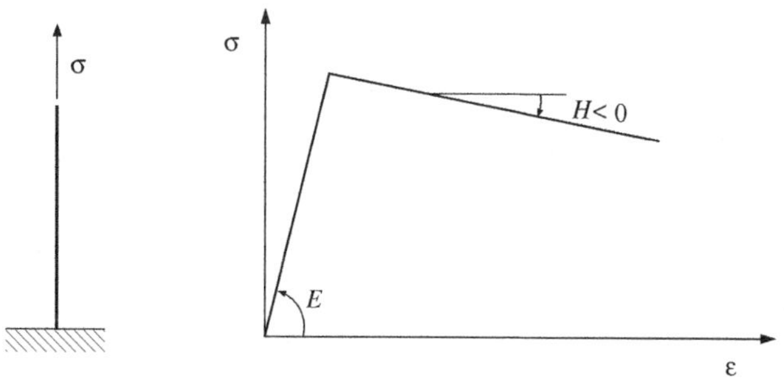

Figure 5.5 (a) One-dimensional bar. (b) Bilinear stress–strain relation

one-dimensional, linear kinematic relation, $\epsilon = \frac{\partial u}{\partial x}$, the equation of motion becomes:

$$\rho \frac{\partial^2 \Delta u}{\partial t^2} - H \frac{\partial^2 \Delta u}{\partial x^2} = 0 \tag{5.33}$$

We assume a solution of the form

$$\Delta u(x, t) = \mathrm{a}(t) \exp(ikx) \tag{5.34}$$

with k the wave number and $a(t)$ a discrete displacement that only varies in time (Whitham 1974). Substitution into Equation (5.33) eliminates the dependence on x and yields:

$$\rho \frac{\partial^2 a}{\partial t^2} + Hk^2 a = 0 \tag{5.35}$$

We consider the Newmark scheme, Equations (5.20), and the homogeneous part of Equation (5.2), i.e. $\mathbf{f}_{\mathrm{ext}}^{t+\Delta t} - \mathbf{f}_{\mathrm{int}}^{t} = \mathbf{0}$, at $t - \Delta t$, t and $t + \Delta t$, and eliminate the time derivatives to arrive at the temporally discretised equation of motion:

$$Hk^2 \Delta t^2 \left(c_1 a^{t+\Delta t} - c_2 a^t - c_3 a^{t-\Delta t} \right) = \rho \left(-a^{t+\Delta t} + 2a^t - a^{t-\Delta t} \right) \tag{5.36}$$

with the constants $c_1 = \beta$, $c_2 = 2\beta - \gamma - \frac{1}{2}$ and $c_3 = \gamma - \beta - \frac{1}{2}$. Considering that we are dealing with an instability problem we assume a non-harmonic solution for a with the growth coefficient ω_r:

$$a^t = A \exp(\omega_r t) \tag{5.37}$$

with A a constant. The values at $t + \Delta t$ follow from $a^{t+\Delta t} = A \exp(\omega_r \Delta t) \exp(\omega_r t)$ etc., and upon substitution into Equation (5.36) the dispersion relation for the temporally discretised system is found:

$$k^2 = \frac{\rho H}{\Delta t^2} \frac{-\exp(\omega_r \Delta t) + 2 - \exp(-\omega_r \Delta t)}{c_1 \exp(\omega_r \Delta t) - c_2 - c_3 \exp(-\omega_r \Delta t)} \tag{5.38}$$

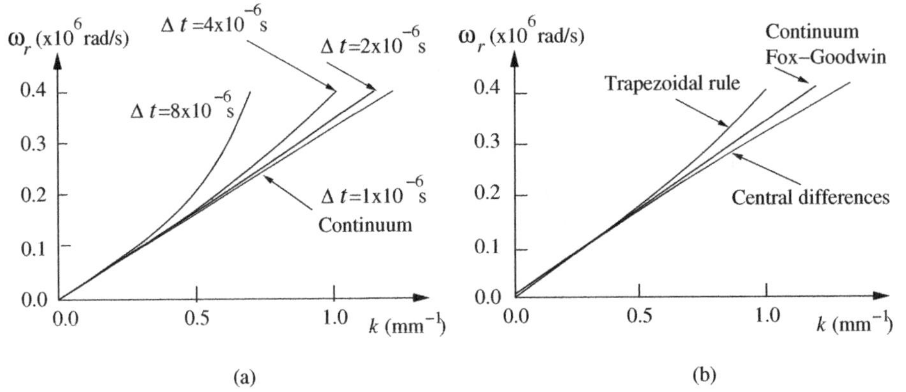

Figure 5.6 (a) Dispersion relations for trapezoidal time integration scheme for different time steps. (b) Dispersion relations for different members of the Newmark family

Figure 5.6(a) shows the dispersion curves for different values of the time step when the trapezoidal rule is used. The curves converge to the continuum dispersion curve upon refinement of the time step. For this implicit method upper bound values are obtained for ω_r, which is in agreement with the tendency observed for the mass–spring system of Figure 5.4. Figure 5.6(b) shows the dispersion curves for different members of the Newmark family. The Fox–Goodwin method gives the best approximation, while the (explicit) central difference method underestimates the growth coefficient ω_r. To determine the convergence rate the wave number k can be calculated using Equation (5.38) for different values of the time step ($\omega_r = 10^6$ rad/s and the exact solution $k_e = 1$ mm^{-1}). The results, summarised in Figure 5.7 confirm the results shown in Table 5.1. In Sluys *et al.* (1995) results can also be found on the HHT α-method. It is noted that dispersion analyses can also be used to study the effect of the spatial discretisation, including the effects of mass lumping (Huerta and Pijaudier-Cabot 1994; Sluys and de Borst 1994; Sluys *et al.* 1995). It is emphasised that, although the above results confirm some tendencies observed in computational dynamics analyses of linear systems, they are case-dependent, and cannot be generalised. When carrying out dynamic analysis of non-linear systems the

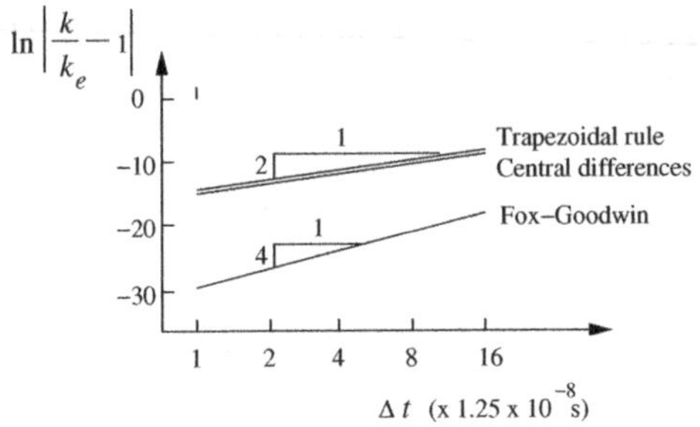

Figure 5.7 Accuracy analysis for different Newmark schemes for a one-dimensional problem with negative stiffness

results should be carefully scrutinised with respect to stability and accuracy, which cannot be guaranteed.

5.6 Energy-conserving Algorithms

In a finite element context, energy-conserving algorithms seem to have been first considered by Hughes *et al.* (1978), who used Lagrangian multipliers to enforce energy conservation as a constraint. While in most structural systems there will be some form of damping or dissipation, and therefore energy will not be conserved, it can nevertheless make sense to start with an algorithm that conserves energy in the absence of damping or dissipation, and to use time integrators that are energy conserving (Crisfield and Shi 1994; Galvanetto and Crisfield 1996; Simo and Tarnow 1992, 1994; Simo *et al.* 1995).

The basic idea of energy-conserving time integrators is to search for a 'mid-point dynamic equilibrium', with associated 'mid-point stresses', which should be the average of those stresses at the beginning and at the end of the time step. We explore this idea by adopting a Total Lagrange formulation, and extend Equation (3.86) by including the inertia term:

$$\int_{V_0} \rho_0 \delta \mathbf{u}^{\mathrm{T}} \ddot{\mathbf{u}} \mathrm{d}V_0 + \int_{V_0} \delta \boldsymbol{\gamma}^{\mathrm{T}} \boldsymbol{\tau} \mathrm{d}V_0 = \int_{S_0} \delta \mathbf{u}^{\mathrm{T}} \mathbf{t}_0 \mathrm{d}S_0 + \int_{V_0} \rho_0 \delta \mathbf{u}^{\mathrm{T}} \mathbf{g} \mathrm{d}V_0 \tag{5.39}$$

which must hold at $t + \alpha\Delta t$, $0 \leq \alpha \leq 1$. With a standard spatial discretisation, and integrating from t till $t + \Delta t$ we transform this identity into:

$$\int_t^{t+\Delta t} \delta \mathbf{a}^{\mathrm{T}} \mathbf{M} \ddot{\mathbf{a}} \mathrm{d}t + \int_t^{t+\Delta t} \delta \mathbf{a}^{\mathrm{T}} \int_{V_0} \mathbf{B}_L^{\mathrm{T}} \boldsymbol{\tau} \mathrm{d}V_0 \mathrm{d}t = \int_t^{t+\Delta t} \delta \mathbf{a}^{\mathrm{T}} \mathbf{f}_{\mathrm{ext}} \mathrm{d}t \tag{5.40}$$

with $\mathbf{B}_L$ as defined in Equation (3.21). We now integrate in time by a combination of trapezoidal and mid-point approximations to obtain:

$$\delta \bar{\mathbf{a}}^{\mathrm{T}} \mathbf{M} \left(\frac{\mathbf{a}^{t+\Delta t} - \mathbf{a}^t}{\Delta t} \right) + \delta \bar{\mathbf{a}}^{\mathrm{T}} \int_{V_0} \bar{\mathbf{B}}_L^{\mathrm{T}} \bar{\boldsymbol{\tau}} \mathrm{d}V_0 = \delta \bar{\mathbf{a}}^{\mathrm{T}} \bar{\mathbf{f}}_{\mathrm{ext}} \tag{5.41}$$

where the bar signifies that a quantity is evaluated at the mid-point,

$$\bar{\mathbf{B}} = \mathbf{B} \left(\frac{\bar{\mathbf{a}}^t + \bar{\mathbf{a}}^{t+\Delta t}}{2} \right)$$

and the Second Piola–Kirchhoff stress tensor averaged over the time step Δt is defined as:

$$\bar{\boldsymbol{\tau}} = \frac{\boldsymbol{\tau}(\mathbf{a}^t) + \boldsymbol{\tau}(\mathbf{a}^{t+\Delta t})}{2}$$

In a compact format Equation (5.41) can be rewritten as:

$$\delta \bar{\mathbf{a}}^{\mathrm{T}} \bar{\mathbf{r}} = 0 \tag{5.42}$$

where the residual vector at the mid-point is defined as:

$$\begin{aligned}\bar{\mathbf{r}} &= \mathbf{M}\left(\frac{\mathbf{a}^{t+\Delta t} - \mathbf{a}^t}{\Delta t}\right) + \int_{V_0} \bar{\mathbf{B}}_L^{\mathrm{T}} \bar{\tau} \mathrm{d}V_0 - \bar{\mathbf{f}}_{\mathrm{ext}} \\ &= \mathbf{M}\left(\frac{\mathbf{a}^{t+\Delta t} - \mathbf{a}^t}{\Delta t}\right) + \bar{\mathbf{f}}_{\mathrm{int}} - \bar{\mathbf{f}}_{\mathrm{ext}}\end{aligned} \tag{5.43}$$

The particular form of the integration that leads to Equation (5.41) relates to energy conservation (Simo and Tarnow 1992). We consider the change in the strain energy over the time step, which can be written as:

$$\Delta\mathcal{E} = \int_{V_0} \Delta\bar{\gamma}^{\mathrm{T}} \bar{\tau} \mathrm{d}V_0 = \Delta\bar{\mathbf{a}}^{\mathrm{T}} \int_{V_0} \bar{\mathbf{B}}_L^{\mathrm{T}} \bar{\tau} \mathrm{d}V_0 = \Delta\bar{\mathbf{a}}^{\mathrm{T}} \bar{\mathbf{f}}_{\mathrm{int}} \tag{5.44}$$

The change in kinetic energy over the time step reads:

$$\begin{aligned}\Delta\mathcal{K} &= \frac{1}{2}(\mathbf{a}^{t+\Delta t})^{\mathrm{T}} \mathbf{M} \mathbf{a}^{t+\Delta t} - \frac{1}{2}(\mathbf{a}^t)^{\mathrm{T}} \mathbf{M} \mathbf{a}^t \\ &= \frac{1}{2}(\mathbf{a}^{t+\Delta t} + \mathbf{a}^t)^{\mathrm{T}} \mathbf{M}(\mathbf{a}^{t+\Delta t} - \mathbf{a}^t) \\ &= \Delta\mathbf{a}^{\mathrm{T}} \mathbf{M}\left(\frac{\mathbf{a}^{t+\Delta t} - \mathbf{a}^t}{\Delta t}\right)\end{aligned} \tag{5.45}$$

Assuming a fixed external load – as in a gravity field – the total energy change is:

$$\Delta(\mathcal{K} + \mathcal{E} + \mathcal{U}) = \Delta\mathbf{a}^{\mathrm{T}}\left(\mathbf{M}\left(\frac{\mathbf{a}^{t+\Delta t} - \mathbf{a}^t}{\Delta t}\right) + \bar{\mathbf{f}}_{\mathrm{int}} - \bar{\mathbf{f}}_{\mathrm{ext}}\right) = \Delta\mathbf{a}^{\mathrm{T}} \bar{\mathbf{r}} \tag{5.46}$$

which will tend to zero upon reaching dynamic equilibrium, i.e. when $\bar{\mathbf{r}}$ vanishes. This class of algorithms thus preserves energy when there are no dissipative processes, as for instance in large rotations, and/or large deflections of slender members. When energy dissipation is present in the problem, as in plasticity, or damage, this class of algorithms can still be used, but of course, there is no conservation of energy. The fact that there is no energy dissipated by the algorithm as such, however, may also be favourable for the stability properties of the algorithm for dissipative non-linear problems.

When we assume that the semi-discrete balance of momentum has been satisfied at $t - \frac{1}{2}\Delta t$, use of a Taylor series gives:

$$\begin{aligned}\mathbf{f}_{\mathrm{int}}^{t+\frac{1}{2}\Delta t} &= \mathbf{f}_{\mathrm{int}}^{t-\frac{1}{2}\Delta t} + \frac{\partial \mathbf{f}_{\mathrm{int}}^{t-\frac{1}{2}\Delta t}}{\partial \mathbf{a}}(\mathbf{a}^{t+\Delta t} - \mathbf{a}^t) \\ &= \mathbf{f}_{\mathrm{int}}^{t-\frac{1}{2}\Delta t} + \mathbf{K}^{t-\frac{1}{2}\Delta t} \Delta\mathbf{a}\end{aligned} \tag{5.47}$$

with the tangential stiffness matrix derived from Equation (5.43):

$$\mathbf{K}^{t-\frac{1}{2}\Delta t} = \int_{V_0} \left(\frac{\bar{\mathbf{B}}_L^{t-\Delta t} + \bar{\mathbf{B}}_L^t}{4}\right)^{\mathrm{T}} \mathbf{D}^t \mathbf{B}_L^t \mathrm{d}V_0 + \int_{V_0} \mathbf{B}_{NL}^{\mathrm{T}} \left(\frac{\mathcal{T}^{t-\Delta t} + \mathcal{T}^t}{4}\right) \mathbf{B}_{NL} \mathrm{d}V_0 \tag{5.48}$$

with $\mathbf{N}_{NL}$ and $\mathcal{T}$ as defined in Equations (3.104) and (3.103). The first contribution to the tangential stiffness matrix becomes non-symmetric. However, for a vanishing time step, this non-symmetry disappears. A similar form of non-symmetry of the algorithmic tangential stiffness matrix will be encountered in computational plasticity (Chapter 7).

We now define the velocity update similar to that for the Newmark scheme, Equation (5.20a):

$$\dot{\mathbf{a}}^{t+\Delta t} = \frac{2(\mathbf{a}^{t+\Delta t} - \mathbf{a}^t)}{\Delta t} - \dot{\mathbf{a}}^t \tag{5.49}$$

Substitution of Equation (5.49) for $\dot{\mathbf{a}}^{t+\Delta t}$ and of Equation (5.47) for $\mathbf{f}_{\text{int}}^{t+\frac{1}{2}\Delta t}$ into the residual vector $\bar{\mathbf{r}}^{t+\frac{1}{2}\Delta t}$ as defined through Equation (5.43) yields

$$(\mathbf{K}^{t-\frac{1}{2}\Delta t})^* \Delta\mathbf{a} = \mathbf{f}^* \tag{5.50}$$

where

$$(\mathbf{K}^{t-\frac{1}{2}\Delta t})^* = \mathbf{K}^{t-\frac{1}{2}\Delta t} + \frac{2}{\Delta t^2}\mathbf{M} \tag{5.51}$$

and

$$\mathbf{f}^* = \mathbf{f}_{\text{ext}}^{t+\frac{1}{2}\Delta t} - \mathbf{f}_{\text{int}}^{t-\frac{1}{2}\Delta t} + \frac{2}{\Delta t}\mathbf{M}\dot{\mathbf{a}}^t \tag{5.52}$$

Having computed $\Delta\mathbf{a}$ using Equation (5.50), the displacements $\mathbf{a}^{t+\Delta t} = \mathbf{a}^t + \Delta\mathbf{a}$ can be computed and, using Equation (5.49), the velocity can be updated. Using the constitutive relation the Second Piola–Kirchhoff stress $\boldsymbol{\tau}^{t+\Delta t}$ can be calculated and, using Equation (5.43), the residual force vector $\bar{\mathbf{r}}^{t+\frac{1}{2}\Delta t}$ can be computed,

$$\bar{\mathbf{r}}^{t+\frac{1}{2}\Delta t} = \mathbf{M}\left(\frac{\dot{\mathbf{a}}^{t+\Delta t} - \dot{\mathbf{a}}^t}{\Delta t}\right) + \bar{\mathbf{f}}_{\text{int}}^{t+\frac{1}{2}\Delta t} - \bar{\mathbf{f}}_{\text{ext}}^{t+\frac{1}{2}\Delta t} \tag{5.53}$$

which will generally not be equal to zero. Again applying a Taylor series to this identity gives the tangential stiffness matrix

$$\mathbf{K}^{t+\frac{1}{2}\Delta t} = \int_{V_0}\left(\frac{\bar{\mathbf{B}}_L^t + \bar{\mathbf{B}}_L^{t+\Delta t}}{4}\right)^{\mathrm{T}}\mathbf{D}^{t+\Delta t}\mathbf{B}_L^{t+\Delta t}\mathrm{d}V_0 + \int_{V_0}\mathbf{B}_{NL}^{\mathrm{T}}\left(\frac{\mathcal{T}^t + \mathcal{T}^{t+\Delta t}}{4}\right)\mathbf{B}_{NL}\mathrm{d}V_0$$

which takes the same form as before. The process is completed by taking the variation of the velocity update, Equation (5.49):

$$\delta\dot{\mathbf{a}}^{t+\Delta t} = \frac{2\delta\mathbf{a}^{t+\Delta t}}{\Delta t} \tag{5.54}$$

and combine this with the linearised form of $\bar{\mathbf{r}}^{t+\frac{1}{2}\Delta t}$ to give the correction to the displacement vector:

$$\mathrm{d}\mathbf{a} = -[(\mathbf{K}^{t+\frac{1}{2}\Delta t})^*]^{-1}\bar{\mathbf{r}}^{t+\frac{1}{2}\Delta t} \tag{5.55}$$

where $(\mathbf{K}^{t+\frac{1}{2}\Delta t})^*$ relates to $\mathbf{K}^{t+\frac{1}{2}\Delta t}$ in a similar manner as $(\mathbf{K}^{t-\frac{1}{2}\Delta t})^*$ relates to $\mathbf{K}^{t-\frac{1}{2}\Delta t}$, Equation (5.51).

5.7 Time Step Size Control and Element Technology

A proper time step should strike a balance between computational costs on one hand, and accuracy, algorithmic stability, and convergence of the solver on the other hand. For instance, for single degree of freedom problems as in Figure 5.4 at least 10 steps would be necessary per period of oscillation already in the linear regime, see Wood (1990) for a comprehensive overview of step size control. The requirements that are imposed by accuracy and convergence of the iterative process that is necessary to solve the non-linear system of equations within each time step, can lead to time steps that are even (much) smaller (Kuhl and Ramm 1999). Also for time integrators that are only conditionally stable in the linear regime, the time step that stems from accuracy considerations and convergence requirements can be much smaller than the critical time step that is derived from the CFL criterion, up to a factor five or ten smaller. Chapter 14 includes some studies in non-linear fracture dynamics, which confirm this.

In view of the high computational demand of (non-linear) dynamics calculations, methods have been sought, which can reduce the computational burden. Among these we find the implicit–explicit methods, in which different parts of the domain can be integrated either using an implicit algorithm, or using an explicit algorithm, depending on the corresponding critical time step. For instance, the part of the domain in which a fine discretisation is applied, and which would therefore lead to a small critical time step, can then be integrated using an implicit method, while the remainder of the domain is integrated in an explicit manner. Typically, a consistent parameterisation is adopted, that encapsulates both an implicit and an explicit time integration algorithm (Daniel 2003a; Hughes and Liu 1978b,a; Miranda *et al.* 1989). It is noted that in the non-linear regime, the efficiency gain of implicit–explicit methods may be less pronounced.

Another way to improve efficiency is to adopt subcycling (Daniel 1998, 2003b; Wu and Smolinski 2000). In this class of methods the spatial domain is partitioned in a manner that depends on the critical time step in a subdomain. Each subdomain is then solved using a time step that is smaller than the critical time step in the subdomain. Synchronisation must take place after a certain time interval, but this can be (much) larger than the most critical time step. The effect is similar to in implicit–explicit methods: the subdomain that has to be integrated with a time step that is below the most critical time step is significantly smaller than the total domain, thus giving a significant efficiency gain. Crucial issues in this class of methods are stability and accuracy, and the synchronisation across subdomains. This holds a fortiori for non-linear analyses.

While the above methods set out to achieve a higher efficiency via optimising the temporal integration, savings can also be made by reducing the number of integration points. In particular when a mass lumping is adopted, the time spent for updating the stress–strain relation in the integration points becomes a considerable part of the total computational effort. This can be achieved by employing reduced integration. However, as will also be discussed in Chapter 7, a uniformly reduced integration can cause the emergence of spurious kinematic modes, or zero-energy modes, such as the hour-glass modes in four-noded elements. Measures to suppress these modes have been proposed (Belytschko *et al.* 1984; Flanagan and Belytschko 1981; Kosloff and Frazier 1978), but for material non-linearities these remedies can embody the danger that they artificially stiffen the solution, since it may no longer be known which spurious modes have to be suppressed.

References

Bathe KJ 1982 *Finite Element Procedures in Engineering Analysis*. Prentice Hall, Inc.

Belytschko T 1983 An overview of semidiscretization and time integration procedures, in *Computational Methods for Transient Analysis* (eds. Belytschko T and Hughes TJR), vol. 1 of *Computational Methods in Mechanics*, pp. 1–65. North-Holland.

Belytschko T, Chiapetta RL and Bartel HD 1976 Efficient large scale non-linear transient analysis by finite elements. *International Journal for Numerical Methods in Engineering* **10**, 579–596.

Belytschko T, Ong SJ, Liu WK and Kennedy JM 1984 Hourglass control in linear and nonlinear problems. *Computer Methods in Applied Mechanics and Engineering* **43**, 251–276.

Bergan PG and Mollestad E 1985 An automatic time-stepping algorithm for dynamic problems. *Computer Methods in Applied Mechanics and Engineering* **49**, 299–318.

Cho SS, Huh H and Park KC 2011 A time-discontinuous implicit variational integrator for stress wave propagation in solids. *Computer Methods in Applied Mechanics and Engineering* **200**, 649–664.

Courant R, Friedrichs K and Lewy H 1928 Über die partiellen Differenzgleichungen der mathematischen Physik (On the partial difference equations of mathematical physics). *Mathematischen Annalen* **100**, 32–74.

Crisfield MA and Shi J 1994 A co-rotational element/time integration strategy for non-linear dynamics. *International Journal for Numerical Methods in Engineering* **37**, 1897–1913.

Daniel WJT 1998 A study of the stability of subcycling algorithms in structural dynamics. *Computer Methods in Applied Mechanics and Engineering* **156**, 1–13.

Daniel WJT 2003a Explicit/implicit partitioning and a new explicit form of the generalized alpha method. *Communications in Numerical Methods in Engineering* **19**, 909–920.

Daniel WJT 2003b A partial velocity approach to subcycling structural dynamics. *Computer Methods in Applied Mechanics and Engineering* **192**, 375–394.

Flanagan DP and Belytschko T 1981 A uniform strain hexahedron and quadrilateral with orthogonal hourglass control. *International Journal for Numerical Methods in Engineering* **17**, 679–706.

Freund LB 1998 *Dynamic Fracture Mechanics*. Cambridge University Press.

Galvanetto U and Crisfield MA 1996 An energy conserving co-rotational procedure for the dynamics of planar structures. *International Journal for Numerical Methods in Engineering* **39**, 2265–2287.

Hilber HM, Hughes TJR and Taylor RL 1977 Improved numerical dissipation for time integration algorithms. *Earthquake Engineering and Structural Dynamics* **5**, 283–292.

Hinton E, Rock A and Zienkiewicz OC 1970 A note on mass lumping and related processes in the finite-element method. *International Journal for Earthquake Engineering and Structural Dynamics* **5**, 245–249.

Houbolt JC 1950 A recurrence matrix solution for the dynamic response of elastic aircraft. *Journal of the Aeronautical Sciences* **17**, 540–550.

Huerta A and Pijaudier-Cabot G 1994 Discretization influence on regularization by two localization limiters. *ASCE Journal of Engineering Mechanics* **120**, 1198–1218.

Hughes TJR 1983 Analysis of transient algorithms with particular reference to stability behavior, in *Computational Methods for Transient Analysis* (eds. Belytschko T and Hughes TJR), vol. 1 of *Computational Methods in Mechanics*, pp. 67–155. North-Holland.

Hughes TJR 1987 *The Finite Element Methods: Linear Static and Dynamic Finite Element Analysis*. Prentice Hall, Inc.

Hughes TJR and Hulbert GM 1988 Space-time finite element methods for elastodynamics: formulations and error estimates. *Computer Methods in Applied Mechanics and Engineering* **66**, 339–363.

Hughes TJR and Liu WK 1978a Implicit-explicit finite elements in transient analysis: implementation and numerical examples. *Journal of Applied Mechanics* **45**, 375–378.

Hughes TJR and Liu WK 1978b Implicit-explicit finite elements in transient analysis: stability theory. *Journal of Applied Mechanics* **45**, 371–374.

Hughes TJR, Liu WK and Caughey TK 1978 Finite element methods for nonlinear elastodynamics which conserve energy. *Journal of Applied Mechanics* **45**, 366–370.

Hulbert GM 2004 Computational structural dynamics, in *The Encyclopedia of Computational Mechanics* (eds. Stein E, de Borst R and Hughes TJR), vol. II, pp. 169–193. John Wiley & Sons, Ltd.

Kosloff D and Frazier GA 1978 Treatment of hourglass patterns in low order finite element codes. *International Journal for Numerical and Analytical Methods in Geomechanics* **2**, 57–72.

Kuhl D and Ramm E 1999 Generalized energy-momentum method for non-linear adaptive shell dynamics. *Computer Methods in Applied Mechanics and Engineering* **178**, 343–366.

Kulkarni M, Belytschko T and Bayliss A 1995 Stability and error analysis for time integrators applied to strain-softening materials. *Computer Methods in Applied Mechanics and Engineering* **124**, 335–363.

Miranda I, Ferencz RM and Hughes TJR 1989 Improved implicit-explicit time integration method for structural dynamics. *Earthquake Engineering and Structural Dynamics* **18**, 643–653.

Newmark NM 1959 A method of computation for structural dynamics. *ASCE Journal of the Engineering Mechanics Division* **85**, 67–94.

Park KC 1975 An improved stiffly stable method for direct integration of nonlinear structural dynamic equations. *Journal of Applied Mechanics* **42**, 464–470.

Park KC and Underwood PG 1980 A variable central difference method for structural dynamic analysis. Part 1: Theoretical aspects. *Computer Methods in Applied Mechanics and Engineering* **22**, 241–258.

Simo JC and Tarnow N 1992 The discrete energy-momentum method. Conserving algorithms for nonlinear elastodynamics. *Zeitschrift für Angewandte Mathematic und Physik* **43**, 757–792.

Simo JC and Tarnow N 1994 A new energy conserving algorithm for the nonlinear dynamics of shells. *International Journal for Numerical Methods in Engineering* **37**, 2527–2549.

Simo JC, Tarnow N and Doblaré M 1995 Nonlinear dynamics of 3-D rods: exact energy and momentum conserving algorithms. *International Journal for Numerical Methods in Engineering* **38**, 1431–1474.

Sluys LJ and de Borst R 1994 Dispersive properties of gradient-dependent and rate-dependent media. *Mechanics of Materials* **18**, 131–149.

Sluys LJ, Cauvern M and de Borst R 1995 Discretization influence in strain-softening problems. *Engineering Computations* **12**, 209–228.

Underwood PG and Park KC 1980 A variable central difference method for structural dynamic analysis. Part 2: Implementation and performance evaluation. *Computer Methods in Applied Mechanics and Engineering* **23**, 259–279.

Whitham GB 1974 *Linear and Nonlinear Waves*. John Wiley & Sons, Ltd.

Wood WL 1990 *Practical Time-Stepping Schemes*. Clarendon Press.

Wu YS and Smolinski P 2000 A multi-time step integration algorithm for structural dynamics based on the modified trapezoid rule. *Computer Methods in Applied Mechanics and Engineering* **187**, 641–660.

Xie YM and Wood WL 1993 On the accuracy of time-stepping schemes for dynamic problems with negative stiffness. *Communications in Numerical Methods in Engineering* **9**, 131–137.

Xu XP and Needleman A 1994 Numerical simulations of fast crack growth in brittle solids. *Journal of the Mechanics and Physics of Solids* **42**, 1397–1434.

Part II

Material Non-linearities

6

Damage Mechanics

Damage mechanics is a branch of continuum mechanics that incorporates changes at the microstructural level in the continuum model via a finite number of scalar or tensor-valued internal variables. In this sense it is very much related to plasticity theory (Chapter 7) where the influence of the history on the stress evolution is also incorporated in the continuum theory via a number of internal variables.

6.1 The Concept of Damage

We consider the system of Figure 6.1, which is composed of m parallel bars. The bars all have the same stiffness k, but the strength of each bar i is different. After the tensile strength in a particular bar is exceeded, we assume a perfectly brittle behaviour, that is the force drops to zero at no additional straining (Figure 6.1). In each individual, unbroken bar, there is a force

$$f_i = ku \tag{6.1}$$

where it is emphasised, that because of the parallel arrangement of the bars, they all experience the same displacement u. We subsequently assume that for a given displacement u, n bars are broken. The total force that can be transmitted is then equal to

$$F = \sum_{i=1}^{m-n} ku = (m-n)ku \tag{6.2}$$

We define the individual stiffness of the bars as $k = EA_0/m\ell$, with E the total stiffness of the system and A_0 and ℓ the total cross-sectional area and the length of the bars, respectively. Then, we can write Equation (6.2) as

$$F = (1 - n/m)EA_0\frac{u}{\ell} \tag{6.3}$$

Non-linear Finite Element Analysis of Solids and Structures, Second Edition.
René de Borst, Mike A. Crisfield, Joris J.C. Remmers and Clemens V. Verhoosel.

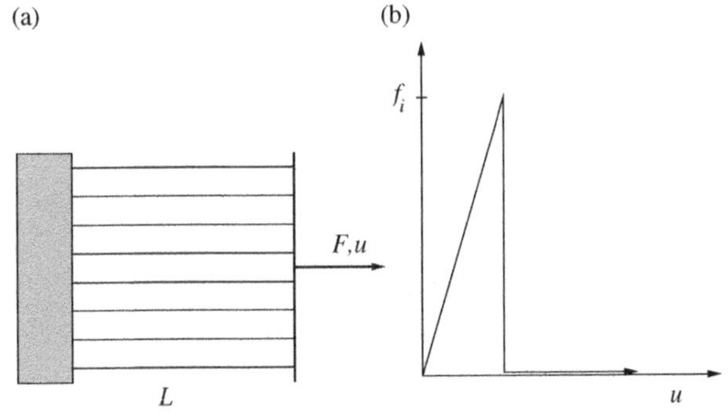

Figure 6.1 Simple damage model composed of m parallel bars (a), each with an elastic perfectly brittle behaviour (b)

When making the transition from a discrete to a continuous model, with $\omega = n/m$ the fraction of broken bars, which starts at 0 and ends at 1, and with the strain in the bars $\epsilon = u/\ell$, we can write:

$$F = (1 - \omega)EA_0\epsilon \tag{6.4}$$

Obviously ω is a function of the end displacement of the bars u, and characterises the state of the system. This so-called damage variable is an internal variable, which signifies how much of the system is still intact. It increases monotonically, since the number of broken bars can only grow or remain constant (during unloading).

Equation (6.4) is a total, non-linear stress–strain relation. For solving the non-linear set of equations that arises when using a non-linear force–displacement relation, we must differentiate Equation (6.4) to obtain:

$$\dot{F} = \left[(1 - \omega) - \frac{d\omega}{d\epsilon}\epsilon\right] EA_0\dot{\epsilon} \tag{6.5}$$

so that the tangential stiffness modulus $E_{\tan}$ reads

$$E_{\tan} = \left[(1 - \omega) - \frac{d\omega}{d\epsilon}\epsilon\right] EA_0 \tag{6.6}$$

During unloading, ω remains constant, so that the second term in Equation (6.6) vanishes, and we have

$$E_{\tan} = (1 - \omega)EA_0$$

Evidently, unloading goes along a secant branch to the stress and strain-free origin of the stress–strain diagram.

Equation (6.4) is also useful to introduce the effective stress concept. F is the force that is instantaneously transmitted by the $m-n$ intact bars, and accordingly, the macroscopically observed stress can be defined as:

$$\sigma = \frac{F}{A_0} = (1-\omega)E\epsilon \tag{6.7}$$

Designating the instanteneous area of unbroken bars by A, we can define the effective stress σ_{eff} that exists in the unbroken bars, as

$$\sigma_{\text{eff}} = \frac{F}{A} = (1-\omega)\frac{EA_0\epsilon}{A} \tag{6.8}$$

It is obvious that the relation $\sigma_{\text{eff}} = E\epsilon$ holds in each individual bar. Accordingly, we obtain the identity

$$A = (1-\omega)A_0 \tag{6.9}$$

between the virgin area A_0 and the still intact load-carrying area A.

6.2 Isotropic Elasticity-based Damage

The basic structure of constitutive models that are set up in the spirit of damage mechanics is simple. We have a total stress–strain relation (Lemaitre and Chaboche 1990):

$$\boldsymbol{\sigma} = \mathbf{D}^{\text{s}}(\omega, \boldsymbol{\omega}, \boldsymbol{\Omega}) : \boldsymbol{\epsilon} \tag{6.10}$$

where $\boldsymbol{\sigma}$ is the stress tensor, $\boldsymbol{\epsilon}$ is the strain tensor and $\mathbf{D}^{\text{s}}$ is a secant, fourth-order stiffness tensor, which can depend on a number of internal variables, like scalar-valued variables ω, second-order tensors $\boldsymbol{\omega}$ and fourth-order tensors $\boldsymbol{\Omega}$. Equation (6.10) differs from non-linear elasticity in the sense that a history dependence is incorporated via a loading–unloading function, f, which vanishes upon loading and is negative otherwise. For damage growth, f must remain zero for an infinitesimal period, so that we have the additional requirement that $\dot{f} = 0$ upon damage growth. The theory is completed by specifying the appropriate (material-dependent) evolution equations for the internal variables.

For isotropic damage evolution, the secant stiffness tensor of Equation (6.10) becomes (in matrix format):

$$\mathbf{D}^{\text{s}} = \frac{E^{\text{s}}}{(1+\nu^{\text{s}})(1-2\nu^{\text{s}})}\begin{bmatrix} 1-\nu^{\text{s}} & \nu^{\text{s}} & \nu^{\text{s}} & 0 & 0 & 0 \\ \nu^{\text{s}} & 1-\nu^{\text{s}} & \nu^{\text{s}} & 0 & 0 & 0 \\ \nu^{\text{s}} & \nu^{\text{s}} & 1-\nu^{\text{s}} & 0 & 0 & 0 \\ 0 & 0 & 0 & \frac{1-2\nu^{\text{s}}}{2} & 0 & 0 \\ 0 & 0 & 0 & 0 & \frac{1-2\nu^{\text{s}}}{2} & 0 \\ 0 & 0 & 0 & 0 & 0 & \frac{1-2\nu^{\text{s}}}{2} \end{bmatrix} \tag{6.11}$$

with $E^s = E(1-\omega_1)$ the secant stiffness modulus, and $\nu^s = \nu(1-\omega_2)$ the secant value of Poisson's ratio, see also Equation (1.118). ω_1 and ω_2 are scalar-valued damage variables, which grow from zero to one at complete damage. A further simplification can be achieved if it is assumed that the Poisson's ratio remains constant during the damage process, which is equivalent to the assumption that the secant shear stiffness and bulk moduli degrade in the same manner during damage evolution. Equation (6.11) then simplifies to:

$$\mathbf{D}^s = (1-\omega)\mathbf{D}^e \tag{6.12}$$

with ω the single damage variable. Alternatively, Equation (6.12) can be expressed as

$$\boldsymbol{\sigma} = (1-\omega)\hat{\boldsymbol{\sigma}} \tag{6.13}$$

with $\hat{\boldsymbol{\sigma}}$ the effective stress tensor,

$$\hat{\boldsymbol{\sigma}} = \mathbf{D}^e : \boldsymbol{\epsilon} \tag{6.14}$$

which is thought to work on the intact material, i.e. the material between the voids or micro-cracks.

The total stress–strain relation (6.12) is complemented by a damage loading function f, which reads:

$$f = f(\tilde{\epsilon}, \tilde{\sigma}, \kappa) \tag{6.15}$$

with $\tilde{\epsilon}$ and $\tilde{\sigma}$ scalar-valued functions of the strain and stress tensors, respectively, and κ the internal variable. The internal variable κ starts at a damage threshold level κ_i and is updated by the requirement that during damage growth $f = 0$, whereas at unloading $f < 0$ and $\dot{\kappa} = 0$. Damage growth occurs according to an evolution law such that $\omega = \omega(\kappa)$, which can be determined from a uniaxial test. The loading–unloading conditions of inelastic constitutive models are often formalised using the Karush–Kuhn–Tucker conditions:

$$f \leq 0 \quad , \quad \dot{\kappa} \geq 0 \quad , \quad f\dot{\kappa} = 0 \tag{6.16}$$

We here limit the treatment to the case that the damage loading function does not depend on $\tilde{\sigma}$. For such a strain-based, or elasticity-based, damage model we have:

$$f(\tilde{\epsilon}, \kappa) = \tilde{\epsilon} - \kappa \tag{6.17}$$

For metals a common choice for $\tilde{\epsilon}$ is the energy measure:

$$\tilde{\epsilon} = \frac{1}{2}\boldsymbol{\epsilon} : \mathbf{D}^e : \boldsymbol{\epsilon} \tag{6.18}$$

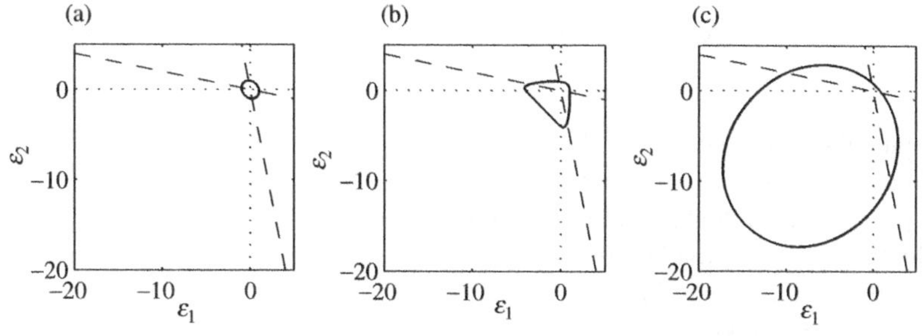

Figure 6.2 Contour plots for $\tilde{\epsilon}$ for (a) the energy-based concept, (b) the Mazars definition (Mazars and Pijaudier-Cabot 1989) and (c) the modified von Mises definition for $k = 10$

Equation (6.18) is less convenient in the sense that it does not reduce to the uniaxial strain for uniaxial stressing. For this reason it is sometimes replaced by the modified expression

$$\tilde{\epsilon} = \sqrt{\frac{1}{E}\,\boldsymbol{\epsilon} : \mathbf{D}^{\mathrm{e}} : \boldsymbol{\epsilon}} \tag{6.19}$$

Expression (6.19) is represented graphically in the principal strain space for plane-stress conditions in Figure 6.2(a). In this figure, a scaling has been applied such that $\tilde{\epsilon} = 1$, while $\nu = 0.2$. The dashed lines are uniaxial stress paths.

The above energy release rate definition for $\tilde{\epsilon}$ gives equal weights to tensile and compressive strain components, which makes it unsuitable to describe the mechanical behaviour of quasi-brittle materials like concrete, rock and ceramics. To remedy this deficiency, Mazars and Pijaudier-Cabot (1989) have suggested the definition

$$\tilde{\epsilon} = \sqrt{\sum_{i=1}^{3}(<\epsilon_i>)^2} \tag{6.20}$$

with ϵ_i the principal strains, with $<\cdot>$ the MacAulay brackets defined such that $<\epsilon_i> = \epsilon_i$ if $\epsilon_i > 0$ and $<\epsilon_i> = 0$ otherwise. A contour plot for $\tilde{\epsilon} = 1$ is given in Figure 6.2(b). A third definition for the equivalent strain $\tilde{\epsilon}$ has been proposed by de Vree *et al.* (1995). This proposition, which has been named a modified von Mises definition, is given by

$$\tilde{\epsilon} = \frac{k-1}{2k(1-\nu)} I_1^{\epsilon} + \frac{1}{2k}\sqrt{\frac{(k-1)^2}{(1-2\nu)^2}(I_1^{\epsilon})^2 + \frac{12k}{(1+\nu)^2} J_2^{\epsilon}} \tag{6.21}$$

with I_1^{ϵ} the first invariant of the strain tensor and J_2^{ϵ} the second invariant of the deviatoric strain tensor. The parameter k governs the sensitivity to the compressive strain components relative to the tensile strain components. The definition of $\tilde{\epsilon}$ is such that a compressive uniaxial stress $k\sigma$ has the same effect as a uniaxial tensile stress σ. k is therefore normally set equal to the ratio of the compressive uniaxial strength and the tensile uniaxial strength. A graphical representation of the modified von Mises definition is given in Figure 6.2(c).

Box 6.1 Algorithmic treatment of isotropic elasticity-based damage model

1. Compute the strain increment: $\Delta\boldsymbol{\epsilon}_{j+1}$
2. Update the total strain: $\boldsymbol{\epsilon}_{j+1} = \boldsymbol{\epsilon}_0 + \Delta\boldsymbol{\epsilon}_{j+1}$
3. Compute the equivalent strain: $\tilde{\epsilon}_{j+1} = \tilde{\epsilon}(\boldsymbol{\epsilon}_{j+1})$
4. Evaluate the damage loading function: $f = \tilde{\epsilon}_{j+1} - \kappa_0$

   if $f \geq 0\,,\ \kappa_{j+1} = \tilde{\epsilon}_{j+1}$

   else $\kappa_{j+1} = \kappa_0$
5. Update the damage variable: $\omega_{j+1} = \omega(\kappa_{j+1})$
6. Compute the new stresses: $\boldsymbol{\sigma}_{j+1} = (1 - \omega_{j+1})\mathbf{D}^{\mathrm{e}} : \boldsymbol{\epsilon}_{j+1}$

From a computational point of view the above elasticity-based damage model is cast easily into a simple and robust algorithm. Indeed, in a displacement-based finite element formulation we can directly compute the strains from the given nodal displacements. The equivalent strain follows in a straightforward fashion, since $\tilde{\epsilon} = \tilde{\epsilon}(\boldsymbol{\epsilon})$. After evaluation of the damage loading function (6.17), the damage variable ω can be updated and the new value for the stress tensor can be computed directly. The simple structure of the algorithm, see Box 6.1 for details, is due to the fact that the stress–strain relation (6.10) is a total stress–strain relation, in the sense that there exists a bijective relation for unloading, and a surjective, but non-injective relation between the stress and strain tensors for loading.

The algorithm described above evaluates the stress from a given strain. To arrive at a computationally efficient procedure that utilises a Newton–Raphson method, it must be complemented by a tangential stiffness tensor, which is derived by a consistent linearisation of the stress–strain relation. Differentiating Equation (6.12) gives:

$$\dot{\boldsymbol{\sigma}} = (1 - \omega)\mathbf{D}^{\mathrm{e}} : \dot{\boldsymbol{\epsilon}} - \dot{\omega}\mathbf{D}^{\mathrm{e}} : \boldsymbol{\epsilon} \tag{6.22}$$

Since $\omega = \omega(\kappa)$, and because the internal variable κ depends on the equivalent strain via $\tilde{\epsilon}$ and the loading function (6.17), we obtain:

$$\dot{\omega} = \frac{\partial \omega}{\partial \kappa}\frac{\partial \kappa}{\partial \tilde{\epsilon}}\dot{\tilde{\epsilon}} \tag{6.23}$$

where $\partial\kappa/\partial\tilde{\epsilon} \equiv 1$ for loading and $\partial\kappa/\partial\tilde{\epsilon} \equiv 0$ for unloading. Considering the dependence $\tilde{\epsilon} = \tilde{\epsilon}(\boldsymbol{\epsilon})$, we can elaborate this relation as:

$$\dot{\omega} = \frac{\partial \omega}{\partial \kappa}\frac{\partial \kappa}{\partial \tilde{\epsilon}}\frac{\partial \tilde{\epsilon}}{\partial \boldsymbol{\epsilon}} : \dot{\boldsymbol{\epsilon}} \tag{6.24}$$

Substitution of Equation (6.24) into the expression for the stress rate yields:

$$\dot{\boldsymbol{\sigma}} = \left((1 - \omega)\mathbf{D}^{\mathrm{e}} - \frac{\partial \omega}{\partial \kappa}\frac{\partial \kappa}{\partial \tilde{\epsilon}}(\mathbf{D}^{\mathrm{e}} : \boldsymbol{\epsilon}) \otimes \frac{\partial \tilde{\epsilon}}{\partial \boldsymbol{\epsilon}}\right) : \dot{\boldsymbol{\epsilon}} \tag{6.25}$$

For unloading the second term in Equation (6.25) cancels and we retrieve the secant stiffness matrix $(1-\omega)\mathbf{D}^e$ as the tangential stiffness matrix for unloading. It is finally noted, cf. Simo and Ju (1987), that the tangential stiffness matrix as defined in (6.25) is generally non-symmetric. For the special choice that the equivalent strain is given by Equation (6.18), symmetry is restored, since then

$$\dot{\sigma} = \left((1-\omega)\mathbf{D}^e - \frac{\partial \omega}{\partial \kappa}\frac{\partial \kappa}{\partial \tilde{\epsilon}}(\mathbf{D}^e : \epsilon) \otimes (\mathbf{D}^e : \epsilon) \right) : \dot{\epsilon} \tag{6.26}$$

6.3 PyFEM: A Plane-strain Damage Model

In this section, we will take a closer look at the implementation of a simple plane-strain damage model in PyFEM. The Python code for this constitutive model can be found in the file `PlaneStrainDamage.py` in the directory `pyfem/materials`. The structure of this file shows many similarities with the elastic plane-strain model, which has been discussed in Section 3.6.

⟨*Plane-strain damage model* ⟩≡

⟨*Initialisation of the plain-strain damage class* 175⟩
⟨*Plane-strain damage class main functions* 176⟩
⟨*Plane-strain damage class utility functions* 177⟩

The plane-strain damage model is implemented as a class, derived from the class `BaseMaterial`:

⟨*Initialisation of the plane-strain damage class* ⟩≡ 175

```
class PlaneStrain( BaseMaterial ):

  def __init__ ( self, props ):

    self.setHistoryParameter( 'kappa', 0. )
    self.commitHistory()

    BaseMaterial.__init__( self, props )
```

The internal variable `kappa` is initialised at zero, and is stored in a similar fashion as the internal variables in, e.g., the truss element discussed in Chapter 3. The material parameters are obtained from the `props` instance in the constructor of the mother class `BaseMaterial`. The main parameters for this model are the Young's modulus, E, and the Poisson's ratio, ν, which are stored as the members E and nu. These parameters are used to construct the elastic stiffness matrix $\mathbf{D}^e$, which is stored as a two-dimensional array De:

⟨*Initialisation of the plane-strain damage class* ⟩+≡ 175

```
self.De = zeros( shape = (3,3) )

self.De[0,0]=self.E*(1.-self.nu)/((1.+self.nu)*(1.-2.*self.nu))
self.De[0,1]=self.De[0,0]*self.nu/(1.-self.nu)
self.De[1,0]=self.De[0,1]
self.De[1,1]=self.De[0,0]
self.De[2,2]=self.De[0,0]*0.5*(1.-2.*self.nu)/(1.-self.nu)

self.a1 = (1./(2.*self.k))
self.a2 = (self.k-1.)/(1.-2.*self.nu)
self.a3 = 12.*self.k/((1.+self.nu)**2)
```

In addition, the constants a1, a2 and a3 are determined and stored for the calculation of the modified von Mises equivalent strain, Equation (6.21). The constants are a function of the parameter k, which is set in the input file and is obtained from the props database in the constructor of BaseMaterial.

The stresses are calculated in the member function getStress. The only input argument of this function is the instance kinematics, which contains the deformation gradient, and the total and the incremental strains at an integration point. In the present model only the total strains are used.

⟨*Plane-strain damage class main functions* ⟩≡ 175

107

```
def getStress( self, kinematics ):

  kappa = self.getHistoryParameter('kappa')

  eps,depsdstrain = self.getEquivStrain( kinematics.strain )
```

177

Having retrieved the internal variable kappa from the database, the equivalent strain eps and its derivative with respect to the total strains, depsdstrain, are calculated. The latter array is needed for the construction of the tangential stiffness matrix.

The current PyFEM implementation uses the modified von Mises formulation of the equivalent strain (de Vree *et al.* 1995). This relation is implemented as a member function of the class and can be easily replaced by an alternative expression. The parameters a1, a2, a3 in this function have been defined in the constructor of the class.

⟨*Plane-strain damage class utility functions* ⟩≡ 175

```
    def getEquivStrain( self , strain ):

      exx = strain[0]
      eyy = strain[1]
      exy = strain[2]
      ezz = self.nu/(self.nu-1.0)*(exx+eyy)

      I1 = exx+eyy+ezz
      J2 = (exx**2+eyy**2+ezz**2-exx*eyy-eyy*ezz-exx*ezz)/3.0+exy**2

      eps = self.a1*(self.a2*I1+sqrt((self.a2*I1)**2+self.a3*J2))
```

Next to the equivalent strain, its derivative is calculated and stored in the array `detadstrain` of length three.

⟨*Plane-strain damage class utility functions* ⟩+≡ 177

```
      depsdstrain = zeros(3)
      ⟨Calculate derivative of equivalent strain with respect to the strains⟩

      return eps , depsdstrain
```

The actual calculation of the terms of this array is not shown in the above fragment. The equivalent strain and the derivatives are the return values of this member function.

In the function `getStress` the equivalent strain `eps` is compared with the internal variable `kappa` to determine whether there is progressive damage or not, see Equation (6.17). In case of progressive damage, the Boolean `progDam` is set to `True`, else `progDam` is set equal to `False`. The updated value of the internal variable `kappa` is subsequently stored in the database.

⟨*Plane-strain damage class main functions* ⟩+≡ 176

```
      if eps > kappa:
        progDam = True
        kappa   = eps
      elif:
        progDam = False

      self.setHistoryParameter( 'kappa', kappa )

      omega , domegadkappa = self.getDamage( kappa )
```

178

Now that the internal variable `kappa` is known, the magnitude of the damage, `omega`, and its derivative with respect to the history, `domegadkappa`, can be calculated. This derivative is needed for the calculation of the tangential stiffness matrix. In this model, we have implemented a linear dependence of the damage parameter omega on the internal variable kappa.

```
⟨Plane-strain damage class utility functions⟩+≡                                  177

    def getDamage( self , kappa ):

      if kappa <= self.kappa0:
        omega          = 0.
        domegadkappa = 0.
      elif self.kappa0 < kappa < self.kappac:
        fac = self.kappac/kappa
        omega          = fac*(kappa-self.kappa0)/(self.kappac-self.kappa0)
        domegadkappa = fac/(self.kappac-self.kappa0)-(omega/kappa)
      else:
        omega          = 1.
        domegadkappa = 0.

      return omega , domegadkappa
```

When kappa is smaller than a threshold value kappa0, the damage omega and its derivative domegadkappa equal zero. When kappa is between kappa0 and kappac, the damage parameter increases linearly from zero to one. The parameters kappa0 and kappac are members of the class and need to be specified in the input file. The scalars omega and domegadkappa are return values of the function.

Upon return in the main member function getStress the stress and the material tangential stiffness matrix in the integration point can be determined:

```
⟨Plane-strain damage class main functions⟩+≡                                      177

      effStress = dot( self.De , kinematics.strain )

      stress    = ( 1. - omega ) * effStress
      tang      = ( 1. - omega ) * self.De

      if progDam:
        tang += -domegadkappa * outer( effStress , detadstrain )

      return stress , tang
```

The stress is calculated according to step 6 in Box 6.1. To this end, we first calculate the effective stress effStress, see also Equation (6.14). Finally, the material tangential stiffness matrix is computed according to Equation (6.25).

The plane-strain damage model has been used in the small example program DamageBar.pro in the directory examples/ch06.

6.4 Stability, Ellipticity and Mesh Sensitivity

A fundamental problem of incorporating damage evolution in standard continuum models is the inherent mesh sensitivity that occurs after reaching a certain damage level. This mesh sensitivity goes beyond the standard discretisation sensitivity of numerical approximation methods for partial differential equations and is not related to deficiencies in the discretisation methods. Instead, the underlying reason for this mesh sensitivity is a local change in character of the governing partial differential equations. This local change of character of the governing set of partial differential equations leads to a loss of well-posedness of the initial boundary value problem and results in an infinite number of possible solutions. After discretisation, a finite number of solutions results. For a finer discretisation, the number of solutions increases, which explains the observed mesh sensitivity.

Since the observed mesh sensitivity is of a fundamental nature, we shall first discuss some basic notions regarding stability and ellipticity. Subsequently, we elucidate the mathematical concepts by simple examples regarding mesh sensitivity.

6.4.1 Stability and Ellipticity

At the continuum level stable material behaviour is usually defined as the scalar product of the stress rate $\dot{\boldsymbol{\sigma}}$ and the strain rate $\dot{\boldsymbol{\epsilon}}$ being positive (Hill 1958; Maier and Hueckel 1979):

$$\dot{\boldsymbol{\epsilon}} : \dot{\boldsymbol{\sigma}} > 0 \tag{6.27}$$

although it can be linked in a rigorous manner to Lyapunov's mathematical definition of stability only for elastic materials (Koiter 1969). In Equation (6.27) restriction is made to geometrical linearity. Extension to geometrical non-linearity is straightforward by replacing $\dot{\boldsymbol{\sigma}}$ by the rate of the First Piola–Kirchhoff stress tensor and $\dot{\boldsymbol{\epsilon}}$ by the velocity gradient, see Box 3.3. Evidently, the scalar product of Equation (6.27) becomes negative when, in a uniaxial tension or compression test, the slope of the homogenised axial stress–strain curve is negative. This phenomenon is named strain softening and is not restricted to a damage mechanics framework, but can also occur in plasticity.

There is a class of material instabilities that can cause the scalar product of stress rate and strain rate to become negative without the occurrence of strain softening in the sense as defined above. These instabilities can arise when the predominant load-carrying mechanism of the material is due to frictional effects such as in sands, rock joints and in pre-cracked concrete. At a phenomenological level such material behaviour typically results in constitutive models which, in a multiaxial context, have a non-symmetric relation between the stress-rate tensor and the strain-rate tensor, e.g. as in Equation (6.25), unless a special choice is made for the equivalent strain $\tilde{\epsilon}$. This lack of symmetry is sufficient to cause loss of material stability, even if the slope of the axial stress–strain curve is still rising (Rudnicki and Rice 1974).

In the above discussion, the terminology 'homogenised' has been used. Here, we refer to the fact that initial flaws and boundary conditions inevitably induce an inhomogeneous stress state in a specimen. During progressive failure of a specimen these flaws and local stress concentrations cause strongly inhomogeneous deformations of the specimen. The procedure that is normally utilised to derive stress–strain relations, i.e. dividing the force by the virgin load-carrying area and dividing the displacement of the end of the specimen by the original length so as to obtain stress and strain, respectively, then no longer reflects what happens at a lower length scale and loses physical significance.

Limiting the discussion to incrementally linear stress–strain relations, that is the relation between the stress rate $\dot{\boldsymbol{\sigma}}$ and the strain rate $\dot{\boldsymbol{\epsilon}}$ can be written as

$$\dot{\boldsymbol{\sigma}} = \mathbf{D} : \dot{\boldsymbol{\epsilon}} \tag{6.28}$$

with $\mathbf{D}$ the material tangential stiffness tensor, inequality (6.27) can be reformulated as

$$\dot{\boldsymbol{\epsilon}} : \mathbf{D} : \dot{\boldsymbol{\epsilon}} > 0 \tag{6.29}$$

The limiting case that the inequality (6.29) is replaced by an equality, marks the onset of unstable material behaviour. Mathematically, this is expressed by the loss of positive definiteness of the material tangential stiffness tensor $\mathbf{D}$:

$$\det(\mathbf{D}^{\mathrm{sym}}) = 0 \tag{6.30}$$

where the superscript *sym* denotes a symmetrised operator. Material instability can lead to structural instability. For a structure that occupies a volume V, Hill's definition (Hill 1958) guarantees structural stability if

$$\int_V \dot{\boldsymbol{\epsilon}} : \dot{\boldsymbol{\sigma}} \, \mathrm{d}V > 0 \tag{6.31}$$

for all kinematically admissible $\dot{\boldsymbol{\epsilon}}$. Obviously, violation of inequality (6.27), i.e. loss of material stability, can lead to violation of Equation (6.31), thus opening the possibility of structural instability. Accordingly, the existence of material instabilities, such as strain softening, can lead to structural instability, even in the absence of geometrically destabilising terms. Of course, there exist many cases where material instabilities and geometrical terms interact and are both (partly) responsible for structural instability.

Yet, the occurrence of unstable material behaviour does not explain the frequently observed discretisation-sensitive behaviour of computations of such solids. Indeed, a crucial consequence of the loss of positive definiteness of the material tangential stiffness tensor $\mathbf{D}$ is that it can result in loss of ellipticity of the governing set of rate equations. Considering quasi-static loading conditions, the governing differential equations – equilibrium equations, kinematic equations and constitutive equations – normally have an elliptic character. Mathematically, this implies that discontinuities in the solution are not possible. Now suppose that within the given context of quasi-static loading conditions, a (possibly curved) plane emerges, say S_d (Figure 6.3), across which the solution can be discontinuous. The difference in the traction rate $\dot{\mathbf{t}}_d$ across this plane reads:

$$[\![\dot{\mathbf{t}}_d]\!] = \mathbf{n}_{S_d} \cdot [\![\dot{\boldsymbol{\sigma}}]\!] \tag{6.32}$$

with $\mathbf{n}_{S_d}$ the normal vector to the discontinuity S_d, or using the tangential stress–strain relation (6.28)

$$[\![\dot{\mathbf{t}}_d]\!] = \mathbf{n}_{S_d} \cdot \mathbf{D} : [\![\dot{\boldsymbol{\epsilon}}]\!] \tag{6.33}$$

where the assumption of a linear comparison solid (Hill 1958) has been introduced, i.e. $\mathbf{D}$ is assumed to have the same value at both sides of the discontinuity S_d. A displacement field $\mathbf{u}$

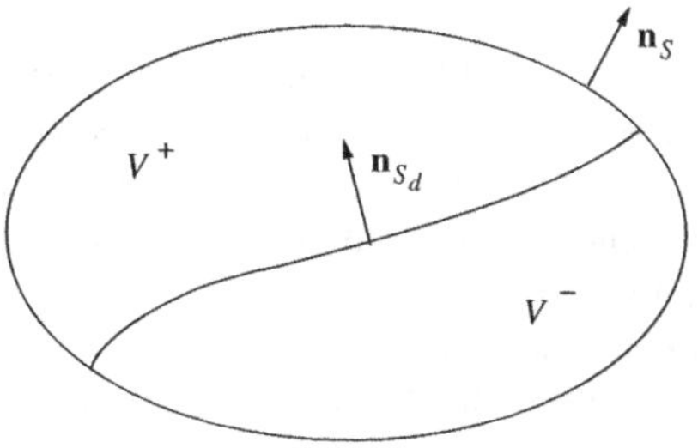

Figure 6.3 Body composed of continuous displacement fields at each side of the discontinuity S_d

that is crossed by a single discontinuity can be represented as:

$$\mathbf{u} = \bar{\mathbf{u}} + \mathcal{H}_{S_d}\tilde{\mathbf{u}} \tag{6.34}$$

with the Heaviside function $\mathcal{H}_{S_d}$ separating the continuous displacement fields $\bar{\mathbf{u}}$ and $\tilde{\mathbf{u}}$. The strain field is subsequently obtained by straightforward differentiation:

$$\boldsymbol{\epsilon} = \nabla^{\mathrm{sym}}\bar{\mathbf{u}} + \mathcal{H}_{S_d}\nabla^{\mathrm{sym}}\tilde{\mathbf{u}} + \delta_{S_d}(\tilde{\mathbf{u}} \otimes \mathbf{n}_{S_d})^{\mathrm{sym}} \tag{6.35}$$

where δ_{S_d} is the Dirac function placed at the discontinuity S_d. For a stationary discontinuity, so that there is no variation of the Heaviside function $\mathcal{H}_{S_d}$ and the Dirac function δ_{S_d}, the strain rate field follows by differentiation with respect to time:

$$\dot{\boldsymbol{\epsilon}} = \nabla^{\mathrm{sym}}\dot{\bar{\mathbf{u}}} + \mathcal{H}_{S_d}\nabla^{\mathrm{sym}}\dot{\tilde{\mathbf{u}}} + \delta_{S_d}(\dot{\tilde{\mathbf{u}}} \otimes \mathbf{n}_{S_d})^{\mathrm{sym}} \tag{6.36}$$

The difference in strain rate fields at S_d is proportional to the unbounded term at the interface:

$$[\![\dot{\boldsymbol{\epsilon}}]\!] = \zeta(\dot{\tilde{\mathbf{u}}} \otimes \mathbf{n}_{S_d})^{\mathrm{sym}} \tag{6.37}$$

also known as the Maxwell compatibility condition and ζ a non-zero scalar. Substitution into Equation (6.33) gives:

$$[\![\dot{\mathbf{t}}_d]\!] = \zeta(\mathbf{n}_{S_d} \cdot \mathbf{D} \cdot \mathbf{n}_{S_d}) \cdot \dot{\tilde{\mathbf{u}}} \tag{6.38}$$

where the minor symmetry of the tangential stiffness tensor has been exploited. A non-trivial solution can exist if and only if the determinant of the acoustic tensor $\mathbf{A} = \mathbf{n}_{S_d} \cdot \mathbf{D} \cdot \mathbf{n}_{S_d}$ vanishes:

$$\det(\mathbf{n}_{S_d} \cdot \mathbf{D} \cdot \mathbf{n}_{S_d}) = 0 \tag{6.39}$$

Thus, if condition (6.39) is met, discontinuous solutions can emerge and loss of ellipticity of the governing differential equations occurs. It is noted that condition (6.39) is coincident with Hill's condition for the propagation of plane acceleration waves in solids (Hill 1962). Analyses that aim at determining the load level at which the determinant of the acoustic tensor vanishes are also denoted as discontinuous bifurcation analyses, cf. Vardoulakis and Sulem (1995).

Ellipticity is a necessary condition for the rate boundary value problem, in the sense that a finite number of linearly independent solutions are admitted, continuously depending on the data and not involving discontinuities, cf. Benallal *et al.* (1988). Loss of ellipticity therefore allows an infinite number of solutions to occur, including those which involve discontinuities. A numerical approximation method will try to capture the discontinuity as good as possible and resolve it in the smallest possible volume which the discretisation allows. Accordingly, mesh refinement will result in a smaller and smaller localisation volume, but obviously, a discontinuity cannot be represented exactly unless special approximation methods are used that can capture a discontinuity rigorously.

For small displacement gradients loss of material stability as expressed by Equation (6.30) is a necessary condition for loss of ellipticity. We show this by substituting the strain field (6.37) into the condition for loss of material stability (6.29):

$$(\tilde{\mathbf{u}} \otimes \mathbf{n}_{S_d}) : \mathbf{D} : (\tilde{\mathbf{u}} \otimes \mathbf{n}_{S_d}) > 0 \tag{6.40}$$

The left-hand side of this inequality vanishes for arbitrary $\tilde{\mathbf{u}}$ if and only if

$$\det(\mathbf{n}_{S_d} \cdot \mathbf{D}^{\mathrm{sym}} \cdot \mathbf{n}_{S_d}) = 0 \tag{6.41}$$

Because the real-valued eigenspectrum of the acoustic tensor $\mathbf{A}$ is bounded by the minimum and maximum eigenvalues of $\mathbf{n}_{S_d} \cdot \mathbf{D}^{\mathrm{sym}} \cdot \mathbf{n}_{S_d}$, Equation (6.41) is always met prior to satisfaction of Equation (6.39). Since Equation (6.41) can only be satisfied if material stability is lost, Equation (6.30), it follows that loss of ellipticity can occur only after loss of material stability. However, when geometrically non-linear terms are included, ellipticity can be lost prior to loss of material stability. This, for instance, can occur at low, but positive values of the plastic hardening modulus, in situations where geometrically non-linear terms have a destabilising effect.

6.4.2 *Mesh Sensitivity*

Mesh sensitivity in a standard continuum equipped with a strain-softening stress–strain relation is conveniently demonstrated by the example of a simple bar loaded in uniaxial tension (Figure 6.4). Let the bar be divided into m elements. Prior to reaching the tensile strength f_t a linear relation is assumed between the normal stress σ and the normal strain ϵ:

$$\sigma = E\epsilon$$

After reaching the peak strength a descending slope is defined in this diagram through an affine transformation from the measured load–displacement curve. The result is given in Figure 6.5(a), where κ_u marks the point where the load-carrying capacity is exhausted. In the post-peak regime

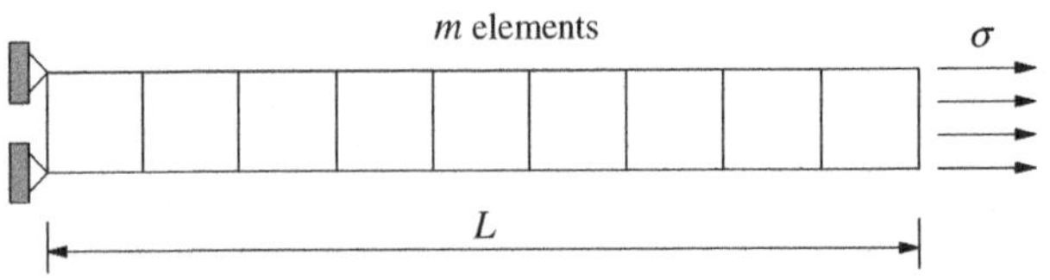

Figure 6.4 Bar of length L subjected to an axial tensile stress σ

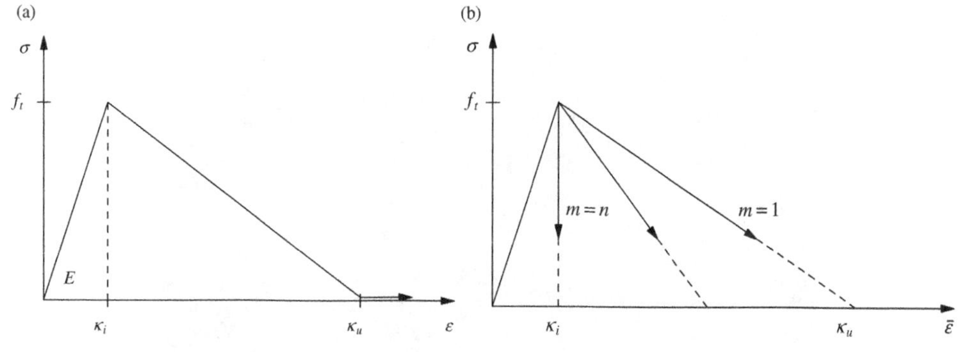

Figure 6.5 (a) Elastic-linear damaging material behaviour. (b) Response of an imperfect bar in terms of a stress-average strain curve

the constitutive model thus reads:

$$\sigma = f_t + h(\epsilon - \kappa_i) \tag{6.42}$$

where, evidently, in case of degrading materials, $h < 0$ and may be termed a softening modulus. For linear strain softening we have

$$h = -\frac{f_t}{\kappa_u - \kappa_i} \tag{6.43}$$

We next suppose that one element has a tensile strength that is marginally below that of the other $m - 1$ elements. Upon reaching the tensile strength of this element, failure will occur. In the other, neighbouring elements the tensile strength is not exceeded and they will unload elastically. Beyond the peak strength the average strain in the bar is thus given by:

$$\bar{\epsilon} = \frac{\sigma}{E} + \frac{E - h}{Eh}\frac{\sigma - f_t}{m} \tag{6.44}$$

Substitution of Equation (6.43) for the softening modulus h and introduction of n as the ratio between the strain κ_u at which the residual load-carrying capacity is exhausted and the threshold damage level κ_i, $n = \kappa_u/\kappa_i$ and $h = -E/(n - 1)$, gives

$$\bar{\epsilon} = \frac{\sigma}{E} + \frac{n(f_t - \sigma)}{mE} \tag{6.45}$$

This result has been plotted in Figure 6.5(b) for different values of m for given n. The computed post-peak curves do not seem to converge to a unique curve. In fact, they do, because the governing equations predict the failure mechanism to be a line crack with zero thickness. The numerical solution simply tries to capture this line crack, which results in localisation in one element, irrespective of the width of the element. The impact on the stress-average strain curve is obvious: for an infinite number of elements ($m \to \infty$) the post-peak curve doubles back on the original loading curve. A major problem is now that, since in continuum mechanics the constitutive model is phrased in terms of a stress–strain relation and not as a force–displacement relation, the energy that is dissipated tends to zero upon mesh refinement, simply because the

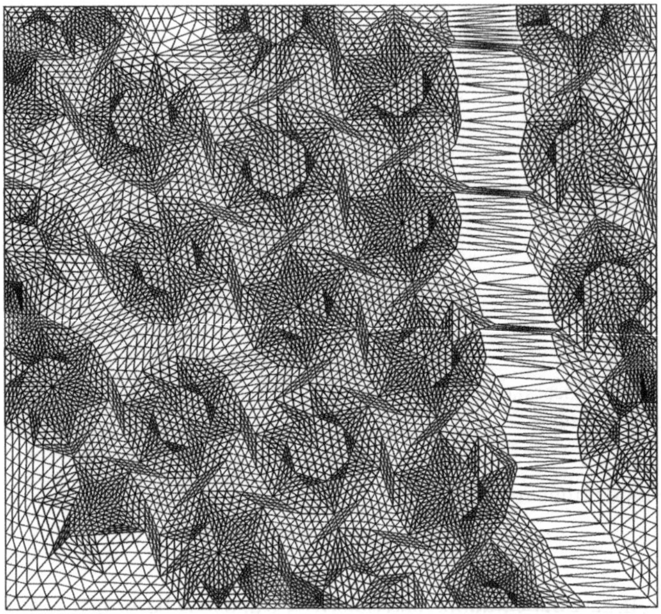

Figure 6.6 Deformed SiC/C specimen beyond the peak load exhibiting a localised failure mode

volume in which the failure process occurs also becomes zero. From a physical point of view this is unacceptable.

The above observations are by no means exclusive to the simple one-dimensional example discussed above. A more complicated boundary value problem is the silicium carbide specimen of Figure 6.6, which is reinforced with carbon fibres (SiC/C composite). The dimensions of the specimen are 30 μm $\times$ 30 μm and a uniform horizontal loading is applied to the vertical sides. The fibres are assumed to remain elastic and also the bond between the fibres and matrix material is assumed to be perfect. A degrading mechanism is only considered for the matrix material, for which a simple softening model has been used.

After the onset of softening a clear localisation zone develops, as is shown in Figure 6.6. This figure shows the fine mesh which consists of 15,568 elements. The computed load–displacement curve has been plotted in Figure 6.7, together with those for the two coarser discretisations, with 3892 and 973 elements, respectively. The same picture arises as for the simple one-dimensional example: a more brittle behaviour is obtained when the mesh is refined and there seems to be convergence towards a solution with zero energy dissipation. In fact, the solution not only becomes more brittle upon mesh refinement, but the peak load is also reduced. Moreover, the solution process becomes very unstable for finer discretisations. This shows through the rather irregular shape of the load–displacement curve for the finest discretisation and by the observation that the solution could not be continued at some stage, no matter how sophisticated the solution techniques employed were. The explanation for this phenomenon is that, as shown in the simple bar problem, a refinement of the discretisation introduces more and more possible equilibrium states. The iterative solution process has to 'choose' between

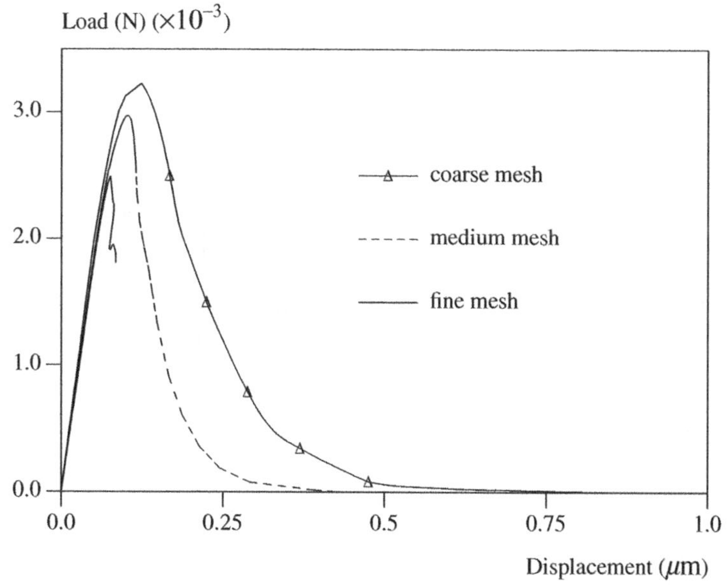

Figure 6.7 Load–displacement curves for a SiC/C specimen obtained with three different discretisations and a softening material model

these equilibrium states and tends to pick another equilibrium state every subsequent iteration. Ultimately, this leads to divergence of the iterative solution procedure.

As discussed, the severe mesh sensitivity is caused by the local loss of ellipticity, or, equivalently, loss of hyperbolicity for dynamic loadings. Since the underlying reason is of a mathematical rather than of a numerical nature, the sensitivity to the discretisation occurs for any discretisation method, including meshfree methods. This is shown in Figures 6.8 and 6.9, which give results of calculations on a one-dimensional tensile bar for different discretisations with the element-free Galerkin method (Belytschko *et al.* 1994) as a prototype meshfree method, see also Pamin *et al.* (2003).

6.5 Cohesive-zone Models

An important issue when considering damage and fracture is the observation that most engineering materials are not perfectly brittle in the Griffith sense, but display some ductility after reaching the strength limit. In fact, there exists a zone in front of the crack tip, in which small-scale yielding, micro-cracking and void initiation, growth and coalescence take place. If this fracture process zone is sufficiently small compared with the structural dimensions, linear-elastic fracture mechanics concepts can apply. However, if this is not the case, the cohesive forces that exist in this fracture process zone must be taken into account. The most powerful and natural way is to use cohesive-zone models, which were introduced by Barenblatt (1962) and Dugdale (1960) for elastic-plastic fracture in ductile metals, and for quasi-brittle materials by Hillerborg *et al.* (1976) in the so-called fictitious crack model.

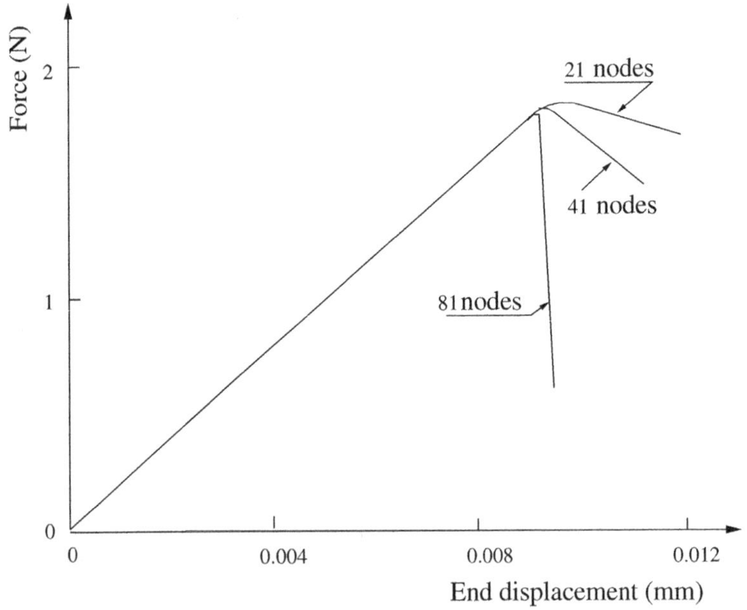

Figure 6.8 Load–displacement curves for a one-dimensional tensile bar composed of a strain-softening elastoplastic material. An element-free Galerkin method has been used with different numbers of nodes along the bar. For each discretisation, the domain of influence equals four times the nodal spacing (Pamin *et al.* 2003)

In cohesive-zone models, the degrading mechanisms in front of the actual crack tip are lumped into a discrete line or plane (Figure 6.10) and a relation between the tractions at the discontinuity $\mathbf{t}_d$ and the relative displacements $\mathbf{v}$ across this line or plane represents the degrading mechanisms in the fracture process zone:

$$\mathbf{t}_d = \mathbf{t}_d(\mathbf{v}, \kappa) \tag{6.46}$$

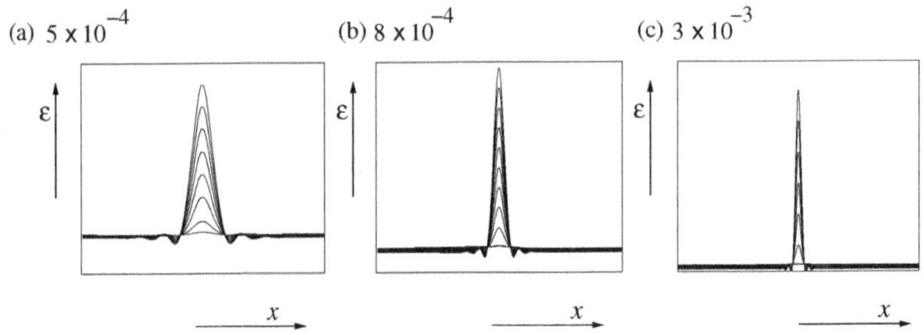

Figure 6.9 Strain evolution in a one-dimensional tensile bar composed of a strain-softening elastoplastic material. Results are shown for discretisations with 21 nodes (a), 41 nodes (b) and 81 nodes (c). Note that the vertical scale is different for the three discretisations (Pamin *et al.* 2003).

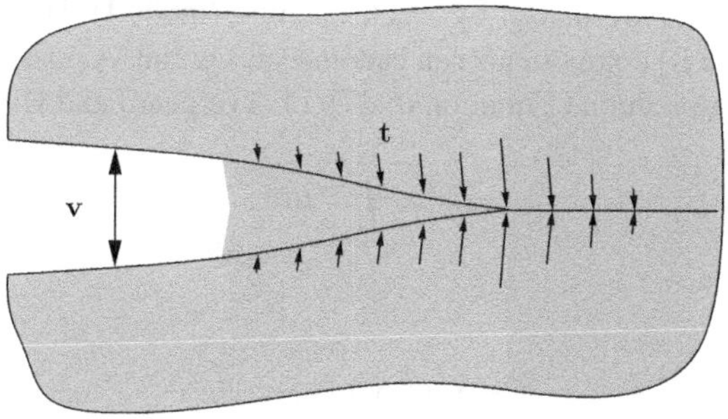

Figure 6.10 Schematic representation of a cohesive zone

with κ an internal variable, which memorises the largest value of a (material-dependent) function of the relative displacements.

Figure 6.11 shows some commonly used decohesion relations, a simple linear relation (a), one for ductile fracture (b) (Tvergaard and Hutchinson 1992) and one for quasi-brittle fracture (c) (Reinhardt and Cornelissen 1984). For ductile fracture, the most important parameters of the cohesive-zone model appear to be the tensile strength f_t and the work of separation or fracture energy $\mathcal{G}_c$ (Hutchinson and Evans 2000), which is the work needed to create a unit area of fully developed crack. It has the dimensions J/m^2 and is formally defined as:

$$\mathcal{G}_c = \int_{v_n=0}^{\infty} t_n \mathrm{d}v_n \tag{6.47}$$

with t_n and v_n the normal traction and the normal relative displacement across the fracture process zone, respectively. For more brittle decohesion relations as shown for instance in Figure 6.11(c), i.e. when the decohesion law stems from micro-cracking as in concrete or ceramics, the shape of the stress–separation relation also plays a role and can be more important than the value of the tensile strength f_t (Chandra *et al.* 2002). In either case, the fracture energy introduces an internal length scale into the model, since the quotient $\mathcal{G}_c/E$ has the dimension of length.

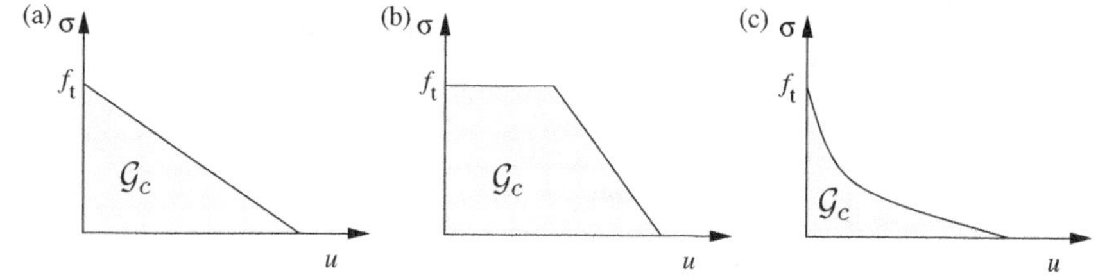

Figure 6.11 Stress–displacement curves for linear decohesion (a), for a ductile solid (b), and for a quasi-brittle solid (c)

Most fracture problems are driven by crack opening (mode I). However, in a number of cases, the sliding (mode II) components can become substantial. A possible way to include the sliding components is to redefine Equation (6.47), cf. Tvergaard and Hutchinson (1993), as:

$$\mathcal{G}_c = \int_{\tilde{v}=0}^{\infty} \tilde{t}\,\mathrm{d}\tilde{v} \tag{6.48}$$

with $\tilde{t} = \tilde{t}(\tilde{v})$, where

$$\tilde{v} = \sqrt{v_n^2 + \alpha(v_s^2 + v_t^2)} \tag{6.49}$$

and v_s and v_t are the sliding components, α being a mode-mixity parameter that sets the ratio between the mode-I and the mode-II components. Alternatively, a mode-II fracture energy can be defined:

$$\mathcal{G}_c^{\mathrm{II}} = \int_{\tilde{v}=0}^{\infty} \tau\,\mathrm{d}\tilde{v} \tag{6.50}$$

with $\tilde{v} = \sqrt{v_s^2 + v_t^2}$ and τ the resolved shear stress, in addition to $\mathcal{G}_c$, which then should be interpreted strictly as a mode-I fracture energy.

Although the cohesive-zone model is essentially a discrete concept, it can be transformed into a continuum formulation by distributing the work of separation or fracture energy $\mathcal{G}_c$ over the thickness w of the volume in which the crack localises (Bažant and Oh 1983). We obtain:

$$\mathcal{G}_c = \int_{n=0}^{w} \int_{\epsilon_{nn}=0}^{\infty} \sigma_{nn}\mathrm{d}\epsilon_{nn}(n)\mathrm{d}n \tag{6.51}$$

with n the coordinate normal to the localisation plane, and σ_{nn} and ϵ_{nn} the normal stress and normal strain in the n-direction, respectively. For linear elements the strains are constant over the width of an element w, so that $\mathcal{G}_c = wg_c$, with g_c the energy dissipated per unit volume of fully damaged material:

$$g_c = \int_{\epsilon_{nn}=0}^{\infty} \sigma_{nn}\mathrm{d}\epsilon_{nn} \tag{6.52}$$

The length scale w which is now introduced into the model is equal or at least proportional to the element size and therefore has a numerical nature.

Assuming a uniform strain distribution over the width of the crack band, carrying out the integration of Equation (6.52) for a linear softening diagram [Figure 6.11(a)] and using the observation that for the bar of Figure 6.3, $w = L/m$, with L the length of the bar and m the number of elements, the softening modulus specialises as:

$$h = \frac{Lf_t^2}{2m\mathcal{G}_c - Lf_t^2/E} \tag{6.53}$$

Evidently, this pseudo-softening modulus is proportional to the structural size and inversely proportional to the number of elements. A model in which the softening modulus was made a function of the element size was first proposed by Pietruszczak and Mróz (1981), but without resorting to an energy concept.

We shall now carry out an analysis for the tension bar of Figure 6.3 and give one element a tensile strength marginally below that of the other elements. As with the stress-based fracture model, the average strain in the post-peak regime is given by Equation (6.44). However, substitution of the fracture-energy-based expression for the pseudo-softening modulus h, Equation (6.53), now results in

$$\bar{\epsilon} = \frac{\sigma}{E} - \frac{2\mathcal{G}_c(\sigma - f_t)}{L f_t^2} \tag{6.54}$$

We observe that, in contrast to the pure stress-based fracture model, Equation (6.45), the number of elements has disappeared from the expression for the ultimate average strain. Therefore, inclusion of the fracture energy $\mathcal{G}_c$ as a material parameter has made the stress–average strain curves, or alternatively, the load–displacement curves, insensitive with respect to mesh refinement. But also the specimen length L has entered the expression for $\bar{\epsilon}$. In other words, the brittleness of the structure now depends on the value of L, and a size effect has been introduced.

When we prescribe the fracture energy $\mathcal{G}_c$ in continuum damage models, the computed load–displacement curves can become reasonably insensitive to the discretisation also for more complicated structures. However, when 'smeared cracks' propagate at lines that are inclined to the grid lay-out, or when quadratic or higher-order finite elements are used, the numerically obtained crack band width normally no longer coincides with the element size. Various formulas have been proposed to estimate the numerical length scale, depending on the interpolation order of the polynomials, the spatial integration scheme and the expected angle between the crack and the grid lines (Feenstra and de Borst 1995b; Oliver 1989). A typical example for quadrilateral elements is:

$$w \approx \alpha_w \sqrt{A_{\text{elem}}} = \alpha_w \sqrt{\sum_{\xi=1}^{n_\xi} \sum_{\eta=1}^{n_\eta} \det(\mathbf{J}) w_\xi w_\eta} \tag{6.55}$$

with A_{elem} the area of the element, and w_ξ and w_η the weight factors of the Gaussian integration rule. The local, isoparametric coordinates of the integration points are given by ξ and η, and $\det(\mathbf{J})$ is the Jacobian of the transformation between the local, isoparametric coordinates and the global coordinate system. The factor α_w is a modification factor which is usually taken equal to one for quadratic elements and equal to $\sqrt{2}$ for linear elements. Experience shows that this approach works reasonably well, but it is of a heuristic nature and can give inaccurate results in particular cases.

When a smeared version of the cohesive-zone model is applied to the SiC/C specimen that was analysed before (Figures 6.6 and 6.7), we obtain load–displacement responses that are fairly independent of the discretisation (Figure 6.12). However, it is emphasised that the number of alternative equilibrium states is not reduced. The numerical procedure is only more stable because the softening branches are more ductile for the finer meshes because of the introduction of a fracture energy. Indeed, if one included the possibility of debonding at the fibre–matrix interface, the number of equilibrium states would increase again, possibly to an extent that divergence would result again.

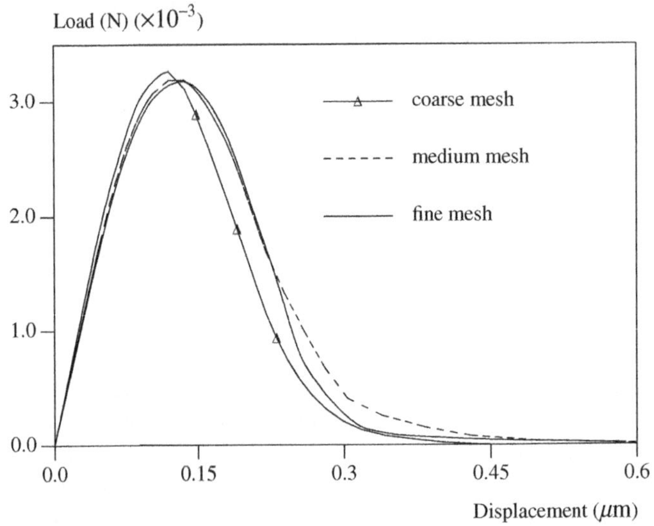

Figure 6.12 Load–displacement curves for SiC/C specimen obtained with a smeared cohesive-zone model

6.6 Element Technology: Embedded Discontinuities

Finite element models with embedded discontinuities provide an elegant way to implement cohesive-zone models in a continuum setting and are probably the most powerful way to analyse cracks, rock faults or shear bands at a macroscopic level. Following the pioneering work of Ortiz *et al.* (1987) and Belytschko *et al.* (1988), who assumed a discontinuity in the displacement gradients, but kept the displacements continuous, many formulations have been published. Starting with the work of Dvorkin *et al.* (1990), Klisinski *et al.* (1991) and Simo *et al.* (1993) discontinuities in the displacements have also been considered; this has been named the 'strong' discontinuity approach, as opposed to the 'weak' discontinuity approach, in which discontinuous displacement gradients are considered. To outline the general framework of embedded discontinuity models, we shall use the kinematics of a strong discontinuity, noting that similar formulations are obtained when adopting the kinematics of a weak discontinuity.

For the derivation of the finite element equations we conventionally take the balance of momentum as a starting point (neglecting body forces for simplicity):

$$\nabla \cdot \boldsymbol{\sigma} = \mathbf{0}$$

Multiplication by a variational field $\delta\bar{\mathbf{u}}$, integrating over the domain V of the body, application of the divergence theorem and use of the appropriate boundary conditions leads to the standard weak formulation:

$$\int_V \nabla\delta\bar{\mathbf{u}} : \boldsymbol{\sigma}\mathrm{d}V = \int_S \delta\mathbf{u} \cdot \mathbf{t}\mathrm{d}S \tag{6.56}$$

which is complemented by the weak forms of the kinematic and the constitutive equations:

$$\int_V \delta\boldsymbol{\sigma} : (\nabla\bar{\mathbf{u}} - \boldsymbol{\epsilon})\mathrm{d}V = 0 \tag{6.57}$$

and

$$\int_V \delta\boldsymbol{\epsilon} : (\boldsymbol{\sigma} - \boldsymbol{\sigma}_\epsilon)\mathrm{d}V = 0 \tag{6.58}$$

with $\delta\boldsymbol{\sigma}$, $\delta\boldsymbol{\epsilon}$ variational fields, $\bar{\mathbf{u}}$ the continuous displacement field and $\boldsymbol{\sigma}_\epsilon$ the stress that is derived from the constitutive relation. Equations (6.56)–(6.58) are the stationarity conditions of the Hu–Washizu variational principle.

We now decompose the strain field into a part $\bar{\boldsymbol{\epsilon}}$ that is derived from $\bar{\mathbf{u}}$ and an additional strain field $\tilde{\boldsymbol{\epsilon}}$:

$$\boldsymbol{\epsilon} = \underbrace{\nabla\bar{\mathbf{u}}}_{\bar{\boldsymbol{\epsilon}}} + \tilde{\boldsymbol{\epsilon}} \tag{6.59}$$

and explicitly allow for the emergence of a discontinuity plane placed at S_d with relative displacements $\mathbf{v}$ and interface tractions $\mathbf{t}_d$. Substitution into Equations (6.56)–(6.58) then results in:

$$\int_V \nabla\delta\bar{\mathbf{u}} : \boldsymbol{\sigma}_\epsilon \mathrm{d}V = \int_S \delta\bar{\mathbf{u}} \cdot \mathbf{t}\mathrm{d}S \tag{6.60}$$

$$\int_{V/S_d} \delta\boldsymbol{\sigma} : \tilde{\boldsymbol{\epsilon}}\mathrm{d}V + \int_{S_d} \delta\mathbf{t}_v \cdot \tilde{\mathbf{v}}\mathrm{d}S = 0 \tag{6.61}$$

and

$$\int_{V/S_d} \delta\tilde{\boldsymbol{\epsilon}} : (\boldsymbol{\sigma} - \boldsymbol{\sigma}_\epsilon)\mathrm{d}V + \int_{S_d} \delta\tilde{\mathbf{v}} \cdot (\mathbf{t}_v - \mathbf{t}_d)\mathrm{d}S = 0 \tag{6.62}$$

with $\delta\mathbf{t}_v$ the variations of $\mathbf{t}_v$, $\delta\mathbf{v}$ those of $\mathbf{v}$ and $\mathbf{t}_v$ the 'static' interface tractions.

In the spirit of the enhanced assumed strain approach (Simo and Rifai 1990), we require that the variations of the stress field and that of the enhanced strain field are orthogonal in an L_2-sense:

$$\int_{V/S_d} \delta\boldsymbol{\sigma} : \delta\tilde{\boldsymbol{\epsilon}}\mathrm{d}V + \int_{S_d} \delta\mathbf{t}_v \cdot \delta\tilde{\mathbf{v}}\mathrm{d}S = 0 \tag{6.63}$$

which implies Equation (6.61) for $\delta\tilde{\boldsymbol{\epsilon}} = \tilde{\boldsymbol{\epsilon}}$ and $\delta\tilde{\mathbf{v}} = \tilde{\mathbf{v}}$ and reduces Equation (6.62) to:

$$\int_{V/S_d} \delta\tilde{\boldsymbol{\epsilon}} : \boldsymbol{\sigma}_\epsilon\mathrm{d}V + \int_{S_d} \delta\tilde{\mathbf{v}} \cdot \mathbf{t}_d\mathrm{d}S = 0 \tag{6.64}$$

if, in addition, $\delta\boldsymbol{\sigma} = \boldsymbol{\sigma}$ and $\delta\mathbf{t}_v = \mathbf{t}_v$. Equation (6.64) can be integrated exactly under the assumptions of a constant strain field and a straight discontinuity S_d – these conditions are automatically satisfied for constant strain elements. Taking into account that in order to pass the patch test, Equation (6.64) must hold for arbitrary constant stress fields and that the orthogonality condition (6.63) must hold for each element:

$$V_{\text{elem}}\delta\tilde{\boldsymbol{\epsilon}} + A_{d,\text{elem}}(\mathbf{n}_{S_d} \otimes \delta\tilde{\mathbf{v}}) = 0 \tag{6.65}$$

with V_{elem} the element volume and $A_{d,\text{elem}}$ the area of S_d in the element. Backsubstituting identity (6.65) into Equation (6.64) then gives:

$$-\int_V \frac{A_{d,\text{elem}}}{V_{\text{elem}}}(\mathbf{n}_{S_d} \otimes \delta\tilde{\mathbf{v}}) : \boldsymbol{\sigma}_\epsilon \mathrm{d}V + \int_{S_d} \delta\tilde{\mathbf{v}} \cdot \mathbf{t}_d \mathrm{d}S = 0 \tag{6.66}$$

We interpolate the continuous part of the displacements in a standard Galerkin manner,

$$\bar{\mathbf{u}} = \mathbf{Ha} \quad \text{and} \quad \delta\bar{\mathbf{u}} = \mathbf{H}\delta\mathbf{a} \tag{6.67}$$

Substituting Equation (6.67) into Equations (6.60) and (6.66), computing the gradients of the variations and the variations of the relative displacements, and requiring that the results hold for all admissible variations, yields the following set of coupled algebraic equations:

$$\int_V \mathbf{B}^{\mathrm{T}} \boldsymbol{\sigma}_\epsilon \mathrm{d}V = \int_S \mathbf{H}^{\mathrm{T}} \mathbf{t} \mathrm{d}S \tag{6.68}$$

and

$$\int_V \mathbf{G}^{\mathrm{T}} \boldsymbol{\sigma}_\epsilon \mathrm{d}V + \int_{S_d} \mathbf{t}_d \mathrm{d}S = \mathbf{0} \tag{6.69}$$

with

$$\mathbf{G} = -\frac{A_{d,\text{elem}}}{V_{\text{elem}}} \begin{bmatrix} n_x & 0 & 0 \\ 0 & n_y & 0 \\ 0 & 0 & n_z \\ n_y & n_x & 0 \\ 0 & n_z & n_y \\ n_z & 0 & n_x \end{bmatrix} \tag{6.70}$$

with n_x, n_y, n_z the x, y, z-components of $\mathbf{n}_{S_d}$. The derivation is completed by relating the stresses to the strains

$$\boldsymbol{\sigma}_\epsilon = \mathbf{D}^{\mathrm{e}} \boldsymbol{\epsilon} \tag{6.71}$$

with $\mathbf{D}^{\mathrm{e}}$ the linear-elastic stiffness matrix, and invoking the discrete relation (6.46) over the interface S_d. In the spirit of a Bubnov–Galerkin approach we interpolate the enhanced strain field by $\mathbf{G}$:

$$\tilde{\boldsymbol{\epsilon}} = \mathbf{G}\boldsymbol{\alpha} \tag{6.72}$$

with $\boldsymbol{\alpha}$ containing discrete parameters at element level. Combining Equations (6.59), (6.71), (6.72) and the kinematic relation then leads to:

$$\boldsymbol{\sigma}_\epsilon = \mathbf{D}^{\mathrm{e}}(\mathbf{Ba} + \mathbf{G}\boldsymbol{\alpha}) \tag{6.73}$$

In order to solve the set of non-linear Equations (6.68) and (6.69) by means of a Newton–Raphson method, linearisation is required. Substituting the linearised form of the constitutive equations (6.46),

$$\mathrm{d}\mathbf{t}_d = \mathbf{D}_d \mathrm{d}\mathbf{v} = \mathbf{D}_d \mathrm{d}\boldsymbol{\alpha} \tag{6.74}$$

with

$$\mathbf{D}_d = \frac{\partial \mathbf{t}_d}{\partial \mathbf{v}} + \frac{\partial \mathbf{t}_d}{\partial \kappa}\frac{\partial \kappa}{\partial \mathbf{v}} \tag{6.75}$$

and the linearised form of Equation (6.73) into the linearised forms of the discrete equations (6.68) and (6.69) yields:

$$\begin{bmatrix} \mathbf{K}_{\mathrm{aa}} & \mathbf{K}_{\mathrm{a}\alpha} \\ \mathbf{K}_{\alpha\mathrm{a}} & \mathbf{K}_{\alpha\alpha} \end{bmatrix} \begin{pmatrix} \mathrm{d}\mathbf{a} \\ \mathrm{d}\boldsymbol{\alpha} \end{pmatrix} = \begin{pmatrix} \mathbf{f}^{\mathrm{a}}_{\mathrm{ext}} - \mathbf{f}^{\mathrm{a}}_{\mathrm{int}} \\ -\mathbf{f}^{\alpha}_{\mathrm{int}} \end{pmatrix} \tag{6.76}$$

with $\mathbf{f}^{\mathrm{a}}_{\mathrm{ext}}$ given by the right-hand side of Equation (6.68) and $\mathbf{f}^{\mathrm{a}}_{\mathrm{int}}$, $\mathbf{f}^{\alpha}_{\mathrm{int}}$ given by the left-hand sides of Equations (6.68) and (6.69). The stiffness matrices are given by:

$$\mathbf{K}_{\mathrm{aa}} = \int_V \mathbf{B}^{\mathrm{T}}\mathbf{D}^{\mathrm{e}}\mathbf{B}\,\mathrm{d}V \tag{6.77}$$

$$\mathbf{K}_{\mathrm{a}\alpha} = \int_V \mathbf{B}^{\mathrm{T}}\mathbf{D}^{\mathrm{e}}\mathbf{G}\,\mathrm{d}V \tag{6.78}$$

$$\mathbf{K}_{\alpha\mathrm{a}} = \int_V \mathbf{G}^{\mathrm{T}}\mathbf{D}^{\mathrm{e}}\mathbf{B}\,\mathrm{d}V \tag{6.79}$$

$$\mathbf{K}_{\alpha\alpha} = \int_V \mathbf{G}^{\mathrm{T}}\mathbf{D}^{\mathrm{e}}\mathbf{G}\,\mathrm{d}V + \int_{S_d} \mathbf{D}_d\,\mathrm{d}S \tag{6.80}$$

The degrees of freedom that correspond to the enhanced strain field can be condensed at element level and, after solution of the global system of equations, be retrieved at element level by an expansion technique, see Box 6.2 for how this method is implemented within the framework of an incremental-iterative procedure. In this manner, the size of the global system is unchanged, and the method can be viewed as a technique that locally improves the element behaviour.

In the above approach, the orthogonality condition in combination with the requirement of traction continuity dictates the form of the enhanced strains $\tilde{\boldsymbol{\epsilon}}$ and the variations thereof, $\delta\tilde{\boldsymbol{\epsilon}}$. Indeed, considering the orthogonality requirement for a single element and piecewise constant stress fields, we obtain from Equation (6.63) for an element:

$$\int_{V_{\mathrm{elem}}} \tilde{\boldsymbol{\epsilon}}\,\mathrm{d}V = \mathbf{0} \tag{6.81}$$

which implies that the enhanced strains make no overall contribution to the element deformations. This requirement makes elements that are enhanced with strains that are obtained from static considerations only, kinematically identical to standard finite elements. Thus, they still suffer relatively strongly from a sensitivity to the direction of the mesh lines. This conclusion holds when the enrichment is done by incorporating weak discontinuities (Sluys and Berends 1998) as well as for strong discontinuities (Wells and Sluys 2001b).

Box 6.2 Condensation and expansion of internal degrees of freedom within an incremental-iterative procedure

A reduction of the global system can be achieved when the internal degrees of freedom are eliminated at element level by a static condensation procedure (Bathe 1982; Hughes 1987). When applying such a process within an incremental-iterative procedure, care must be exercised with respect to the time that the compression of the element stiffness matrix and the expansion of the internal degrees of freedom take place. We will demonstrate this by considering the local compression/expansion process of a static condensation procedure in greater detail. To this end, we first express the internal degrees of freedom in terms of the displacements:

$$\mathrm{d}\boldsymbol{\alpha} = \mathbf{K}_{\alpha\alpha}^{-1}(\mathbf{f}_{\mathrm{int}}^{\alpha} - \mathbf{K}_{\alpha \mathrm{a}}\mathrm{d}\mathbf{a})$$

Inserting this identity into Equation (6.76) yields:

$$(\mathbf{K}_{\mathrm{aa}} - \mathbf{K}_{\mathrm{a}\alpha}\mathbf{K}_{\alpha\alpha}^{-1}\mathbf{K}_{\alpha \mathrm{a}})\mathrm{d}\mathbf{a} = \mathbf{f}_{\mathrm{ext}}^{\mathrm{a}} - \mathbf{f}_{\mathrm{int}}^{\mathrm{a}} - \mathbf{K}_{\mathrm{a}\alpha}\mathbf{K}_{\alpha\alpha}^{-1}\mathbf{f}_{\mathrm{int}}^{\alpha}$$

The stiffness matrix and the right-hand side vector at global level are derived from the second equation of this box. After solving for the corrections to the displacement increments, $\mathrm{d}\mathbf{a}$, the corrections to the increments of the internal degrees of freedom, $\mathrm{d}\boldsymbol{\alpha}$ are computed at element level from the first equation of this box. When using a full Newton–Raphson method, it is essential that the expansion for the internal degrees of freedom is done with the same matrices $\mathbf{K}_{\mathrm{a}\alpha}$, $\mathbf{K}_{\alpha \mathrm{a}}$ and $\mathbf{K}_{\alpha\alpha}$ as have been used for the compression of these degrees of freedom. The danger of using a wrong right-hand side becomes apparent when considering that the force vectors $\mathbf{f}_{\mathrm{int}}^{\mathrm{a}}$ and $\mathbf{f}_{\mathrm{int}}^{\alpha}$ are usually set up at the end of an iteration. The force vector $\mathbf{f}_{\mathrm{ext}}^{\mathrm{a}} - \mathbf{f}_{\mathrm{int}}^{\mathrm{a}} - \mathbf{K}_{\mathrm{a}\alpha}\mathbf{K}_{\alpha\alpha}^{-1}\mathbf{f}_{\mathrm{int}}^{\alpha}$ that is to be used in the next iteration must be computed with the tangential submatrix $\mathbf{K}_{\alpha\alpha}$ which is, however, not available until the beginning of the next iteration. Consequently, the right-hand side vector for the new iteration cannot be set up at the end of the 'old' iteration, but can only be calculated after the tangential stiffness matrices have been set up at the beginning of the new iteration. If this force vector is computed with the 'old' matrix $\mathbf{K}_{\alpha\alpha}$, the quadratic convergence of Newton's method will be lost.

To mitigate this direction sensitivity elements must be enhanced kinematically in the sense that the embedment of a displacement discontinuity results in a relative displacement between nodes on both sides of the discontinuity. As a consequence, the gradient of this enhanced field will be a function of the position of the discontinuity within an element. Generally, a field that contains a single displacement discontinuity, but is continuous otherwise, is given, see Equation (6.34), by:

$$\mathbf{u} = \bar{\mathbf{u}} + \mathcal{H}_{S_d}\tilde{\mathbf{u}}$$

For finite element formulations with embedded displacement discontinuities the effect of the discontinuity is required to vanish at the element boundaries, which can be ensured by modifying Equation (6.34) as follows:

$$\mathbf{u} = \bar{\mathbf{u}} + (\mathcal{H}_{S_d} - \phi)\tilde{\mathbf{u}} \tag{6.82}$$

with $\tilde{\mathbf{u}}$ defined such that $\mathbf{u} = \tilde{\mathbf{u}} \mid_{\mathbf{x} \in S_{\text{elem}}}$ and ϕ a smooth function which vanishes at S^-_{elem} and equals unity at S^+_{elem}. Assuming small strains, it follows that:

$$\boldsymbol{\epsilon} = \mathbf{L}\bar{\mathbf{u}} + (\mathcal{H}_{S_d} - \phi)\mathbf{L}\tilde{\mathbf{u}} + (\tilde{\mathbf{u}} \otimes (\delta_{S_d}\mathbf{n}_{S_d} - \nabla\phi))^{\text{sym}} \tag{6.83}$$

where δ_{S_d} is the Dirac function placed at the discontinuity S_d. Normally, the assumption is made in embedded formulations that $\tilde{\mathbf{u}}$ is elementwise constant, which is rigorously satisfied for constant strain triangles. Then, Equation (6.83) reduces to:

$$\boldsymbol{\epsilon} = \mathbf{L}\bar{\mathbf{u}} + (\tilde{\mathbf{u}} \otimes (\delta_{S_d}\mathbf{n}_{S_d} - \nabla\phi))^{\text{sym}} \tag{6.84}$$

In the interior of an element this formulation captures the kinematics of a strong discontinuity exactly. An element which is purely based on kinematics can now be constructed by replacing $\mathbf{G}$ as defined by Equation (6.70) (Lotfi and Shing 1995):

$$\mathbf{G}^* = -\begin{bmatrix} \frac{\partial\phi}{\partial x} & 0 & 0 \\ 0 & \frac{\partial\phi}{\partial y} & 0 \\ 0 & 0 & \frac{\partial\phi}{\partial z} \\ \frac{\partial\phi}{\partial y} & \frac{\partial\phi}{\partial x} & 0 \\ 0 & \frac{\partial\phi}{\partial z} & \frac{\partial\phi}{\partial y} \\ \frac{\partial\phi}{\partial z} & 0 & \frac{\partial\phi}{\partial x} \end{bmatrix} \tag{6.85}$$

The unbounded term in Equation (6.84) does not enter Equation (6.85), cf. Equation (6.70). In both cases the unbounded term, which acts at S_d only, comes back in the surface integral of Equation (6.80). It is finally noted that this kinematically optimal formulation is defect in the sense that traction continuity is not imposed rigorously.

An optimal element performance can be obtained when traction continuity is imposed following the orthogonality requirement (6.63), but when the kinematics are derived as in Equation (6.84), so that (Armero and Garikipati 1996; Oliver 1996):

$$\boldsymbol{\sigma}_\epsilon = \mathbf{D}^{\text{e}}(\mathbf{Ba} + \mathbf{G}^*\boldsymbol{\alpha}) \tag{6.86}$$

A non-symmetric formulation results, formally still given by the set (6.76), but with the sub-matrices defined as:

$$\mathbf{K}_{\text{aa}} = \int_V \mathbf{B}^{\text{T}}\mathbf{D}^{\text{e}}\mathbf{B}\text{d}V \tag{6.87}$$

$$\mathbf{K}_{\text{a}\alpha} = \int_V \mathbf{B}^{\text{T}}\mathbf{D}^{\text{e}}\mathbf{G}^*\text{d}V \tag{6.88}$$

$$\mathbf{K}_{\alpha\text{a}} = \int_V \mathbf{G}^{\text{T}}\mathbf{D}^{\text{e}}\mathbf{B}\text{d}V \tag{6.89}$$

$$\mathbf{K}_{\alpha\alpha} = \int_V \mathbf{G}^{\text{T}}\mathbf{D}^{\text{e}}\mathbf{G}^*\text{d}V + \int_{S_d} \mathbf{D}_d\text{d}S \tag{6.90}$$

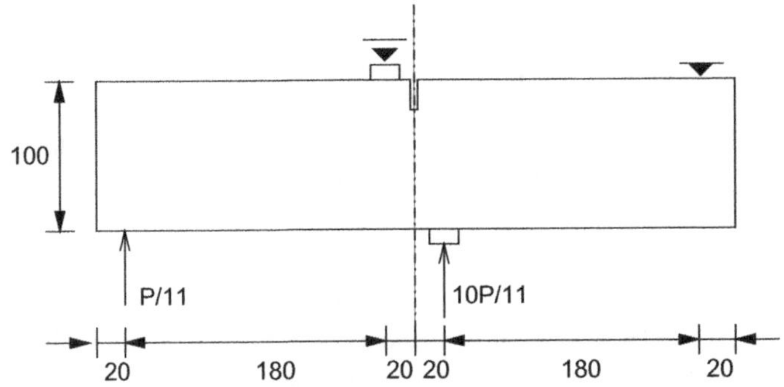

Figure 6.13 Geometry of single-edge notched beam

Numerical experience has shown that this approach is effective in mitigating mesh alignment sensitivity in localisation problems, in spite of the variational inconsistency that is obtained, since, in the spirit of a Petrov–Galerkin approach, the test functions and the trial functions are taken from different spaces (Wells and Sluys 2001b). The relative insensitivity of this statically and kinematically optimal formulation with respect to the mesh orientation is shown in Figure 6.14 for crack propagation in the Single-Edge Notched Beam of Figure 6.13 (Wells and Sluys 2001b). Figure 6.15 shows another three-dimensional calculation – a biaxial test loaded in compression using a strain-softening plasticity model with the aim to simulate shear bands (Wells and Sluys 2001a). The results suggest that the individual discontinuities are correctly oriented, but that the pattern of discontinuities still tends to follow mesh lines, although considerably less than without embedding a discontinuity, or when a statically optimal embedded discontinuity concept is used, see also Gasser and Holzapfel (2003) and Jirasek (2000) for further discussions on embedded discontinuity models.

It is noted that even in the statically and kinematically optimal formulation, a true discontinuity which extends across element boundaries is not obtained. This is because the kinematics of Equation (6.82) are diffused over the element when the governing equations are cast in a weak format. The fact that strain enrichment is discontinuous across element boundaries, makes it possible to resolve the enhanced strain modes at element level by condensation. Restricting the development for simplicity to a case in which $\tilde{\mathbf{u}} = \gamma\mathbf{m}$, with $\mathbf{m}$ giving the shape of the inelastic

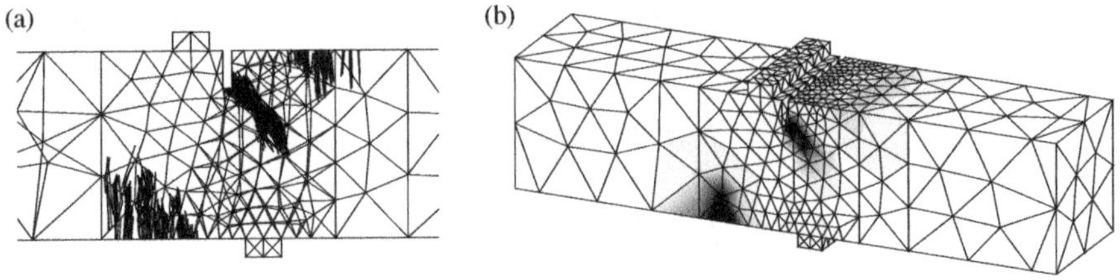

Figure 6.14 Three-dimensional simulation with a statically and kinematically optimal embedded discontinuity model: crack pattern (a) and contours of interface internal variable (b) of the single-edge notched beam at the peak load (Wells and Sluys 2001b)

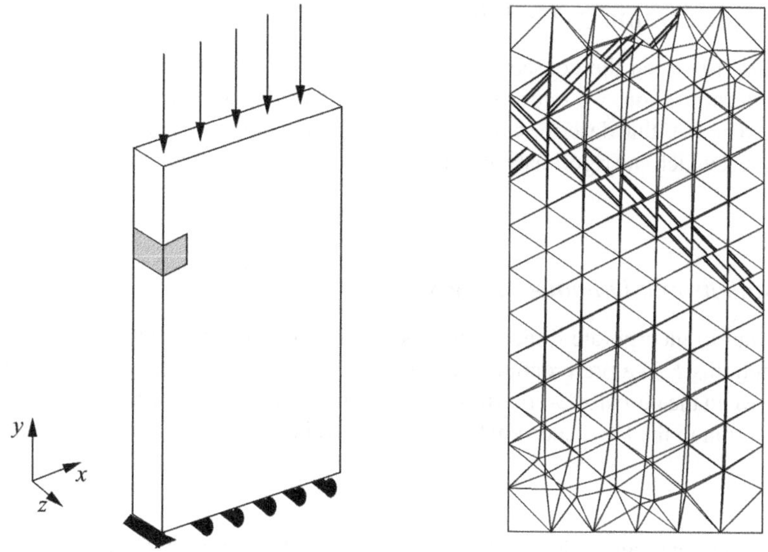

Figure 6.15 Three-dimensional simulation with a statically and kinematically optimal embedded discontinuity model: biaxial test under compressive loading (Wells and Sluys 2001a)

deformation and γ its magnitude – single-surface plasticity models treated in Chapter 7 for instance satisfy this assumption – the second term of the right-hand side of Equation (6.72) can be elaborated as: $\tilde{\epsilon} = \gamma \mathbf{Gm}$. Using this identity, we can resolve the enhanced strain modes by condensation at the element level (Borja 2000):

$$\mathbf{K}_{\text{con}} = \int_V \mathbf{B}^{\mathrm{T}}\mathbf{D}^{\mathrm{e}}\mathbf{B}\mathrm{d}V - \int_V \mathbf{B}^{\mathrm{T}}\mathbf{D}^{\mathrm{e}}\mathbf{G}^*\mathbf{m}\mathrm{d}V \left(\int_{S_d} \mathbf{m}^{\mathrm{T}}\mathbf{D}_d\mathbf{m}\mathrm{d}S + \int_V \mathbf{G}\mathbf{m}^{\mathrm{T}}\mathbf{D}^{\mathrm{e}}\mathbf{G}^*\mathbf{m}\mathrm{d}V \right)^{-1} \int_V \mathbf{G}\mathbf{m}^{\mathrm{T}}\mathbf{D}^{\mathrm{e}}\mathbf{B}\mathrm{d}V \tag{6.91}$$

For constant strain triangles, this expression reduces to:

$$\mathbf{K}_{\text{con}} = V_{\text{elem}}\mathbf{B}^{\mathrm{T}} \left(\mathbf{D}^{\mathrm{e}} - \frac{\mathbf{D}^{\mathrm{e}}(\mathbf{G}^*\mathbf{m})(\mathbf{Gm})^{\mathrm{T}}\mathbf{D}^{\mathrm{e}}}{\underbrace{-\mathbf{m}^{\mathrm{T}}\mathbf{D}_d\mathbf{m}}_{h} + (\mathbf{Gm})^{\mathrm{T}}\mathbf{D}^{\mathrm{e}}(\mathbf{G}^*\mathbf{m})} \right) \mathbf{B} \tag{6.92}$$

This matrix has exactly the format of a tangential stiffness that results from the use of a plasticity model with a non-associated flow rule. For instance, the term $-\mathbf{m}^{\mathrm{T}}\mathbf{Dm}$ can be identified with the conventional softening modulus h. Consequently, under certain assumptions, finite element models with embedded discontinuities can be made identical with a standard finite element model that incorporates softening with a discretisation-dependent softening modulus (Pietruszczak and Mróz 1981).

6.7 Complex Damage Models

After the discussion on the consequences of damage, and in particular strain softening, for the spatial discretisation, we return to the description of damage models, in particular more sophisticated models that describe anisotropic damage, including concrete fracture, and higher-order continua, which can rigorously avoid the ill-posedness caused by the introduction of strain softening.

6.7.1 *Anisotropic Damage Models*

A simple way to incorporate directional dependence of damage evolution is to degrade the Young's modulus E in a preferential direction. When, for plane-stress conditions, distinction is made between the global x, y-coordinate system and a local n, s-coordinate system, a simple loading function in the local coordinate system would be

$$f(\epsilon_{nn}, \kappa) = \epsilon_{nn} - \kappa \tag{6.93}$$

with ϵ_{nn} the normal strain in the local n, s-coordinate system, subject to the standard loading–unloading conditions. The secant stiffness relation (6.10) now becomes (in matrix-vector format):

$$\boldsymbol{\sigma}_{\text{ns}} = \mathbf{D}^{\text{s}}_{\text{ns}} \boldsymbol{\epsilon}_{\text{ns}} \tag{6.94}$$

with $\boldsymbol{\sigma}_{\text{ns}} = [\sigma_{nn}, \sigma_{ss}, \sigma_{ns}]^{\text{T}}$, $\boldsymbol{\epsilon}_{\text{ns}} = [\epsilon_{nn}, \epsilon_{ss}, \gamma_{ns}]^{\text{T}}$, and $\mathbf{D}^{\text{s}}_{\text{ns}}$ given by

$$\mathbf{D}^{\text{s}}_{\text{ns}} = \begin{bmatrix} \frac{(1-\omega)E}{1-(1-\omega)\nu^2} & \frac{(1-\omega)\nu E}{1-(1-\omega)\nu^2} & 0 \\ \frac{(1-\omega)\nu E}{1-(1-\omega)\nu^2} & \frac{E}{1-(1-\omega)\nu^2} & 0 \\ 0 & 0 & \frac{E}{2(1+\nu)} \end{bmatrix} \tag{6.95}$$

with $\omega = \omega(\kappa)$. Using standard transformation rules the components of $\boldsymbol{\epsilon}_{\text{ns}}$ and $\boldsymbol{\sigma}_{\text{ns}}$, can be related to those in the global x, y-coordinate system, Equations (1.102) and (1.105):

$$\boldsymbol{\epsilon}_{\text{ns}} = \mathbf{T}_{\epsilon}\, \boldsymbol{\epsilon}_{\text{xy}} \tag{6.96}$$

and

$$\boldsymbol{\sigma}_{\text{ns}} = \mathbf{T}_{\sigma}\, \boldsymbol{\sigma}_{\text{xy}} \tag{6.97}$$

with $\mathbf{T}_{\epsilon}$ and $\mathbf{T}_{\sigma}$ the standard transformation matrices for the strains and stresses, respectively. Combining Equation (6.94) with Equations (6.96) and (6.97) leads to:

$$\boldsymbol{\sigma}_{\text{xy}} = \mathbf{T}^{\text{T}}_{\sigma} \mathbf{D}^{\text{s}}_{\text{ns}} \mathbf{T}_{\epsilon} \boldsymbol{\epsilon}_{\text{xy}} \tag{6.98}$$

Similar to isotropic damage models, a tangential stiffness matrix can be derived by differentiating the secant stress–strain relation, here Equation (6.98). This results in:

$$\dot{\boldsymbol{\sigma}}_{\text{xy}} = \mathbf{T}^{\text{T}}_{\sigma} \left(\mathbf{D}^{\text{s}}_{\text{ns}} - \Delta \mathbf{D}_{\text{ns}} \right) \mathbf{T}_{\epsilon}\, \dot{\boldsymbol{\epsilon}}_{\text{xy}} \tag{6.99}$$

with $\mathbf{D}^{\text{s}}_{\text{ns}}$ given by Equation (6.95) and

$$\Delta\mathbf{D}_{\text{ns}} = \begin{bmatrix} d_{11} & 0 & 0 \\ \nu d_{11} & 0 & 0 \\ 0 & 0 & 0 \end{bmatrix} \tag{6.100}$$

with

$$d_{11} = \frac{\partial\omega}{\partial\kappa}\frac{\partial\kappa}{\partial\epsilon_{nn}}\frac{E(\epsilon_{nn} + \nu\epsilon_{ss})}{(1-(1-\omega)\nu^2)^2} \tag{6.101}$$

and $\partial\kappa/\partial\epsilon_{nn} \equiv 1$ upon loading and zero otherwise. It is noted that, similar to isotropic damage models, the tangential stiffness matrix normally becomes non-symmetric.

More generally, anisotropic damage can be characterised by a fourth-order damage tensor. This can, for instance, be achieved by broadening the effective stress concept, Equation (6.13) to (Lemaitre and Chaboche 1990):

$$\boldsymbol{\sigma} = (\mathbf{I} - \boldsymbol{\Omega}) : \hat{\boldsymbol{\sigma}} \tag{6.102}$$

or using Equation (6.14),

$$\boldsymbol{\sigma} = (\mathbf{I} - \boldsymbol{\Omega}) : \mathbf{D}^{\text{e}} : \boldsymbol{\epsilon} \tag{6.103}$$

with $\boldsymbol{\Omega}$ a fourth-order damage tensor.

6.7.2 Microplane Models

The framework of anisotropic damage models allows for the incorporation of models that are based on the microplane concept. The microplane concept was originally conceived for metals, where well-defined planes exist in the crystal lattice, along which slip occurs preferentially. This theory, originated by Batdorf and Budiansky (1949), was originally named the slip theory and is now commonly referred to as crystal plasticity (Miehe and Schotte 2004). Later, the concept of preferential slip planes was adapted to damage and fracture in quasi-brittle materials such as concrete and rock and was renamed the microplane model (Bažant and Gambarova 1984) or multilaminate model (Zienkiewicz and Pande 1977). Obviously, the physical basis is now less obvious in the sense that preferential fracture planes cannot be distinguished, but by defining a sufficiently large number of potential fracture planes the damage evolution can be described accurately.

Two major classes of microplane models can be distinguished, namely those based on the kinematic constraint and those based on a static constraint. Herein, we shall only consider a microplane model based on the so-called kinematic constraint, which implies that the normal and tangential strains on a microplane that is labelled α, can be derived by a simple projection of the global strain $\boldsymbol{\epsilon}_{\text{xy}}$ on a microplane with the local n, s-coordinate system, similar to Equation (6.96). In the spirit of Equation (6.10), a secant relation can be set up between the stresses and the strains on each microplane:

$$\boldsymbol{\sigma}^{\alpha}_{\text{ns}} = \mathbf{D}^{\alpha}_{\text{ns}}\boldsymbol{\epsilon}_{\text{ns}} \tag{6.104}$$

with $\mathbf{D}^{\alpha}_{\text{ns}}$ given by

$$\mathbf{D}^{\alpha}_{\text{ns}} = \begin{bmatrix} (1-\omega^{\alpha}_{\text{N}})E_{\text{N}} & 0 & 0 \\ 0 & 0 & 0 \\ 0 & 0 & (1-\omega^{\alpha}_{\text{T}})E_{\text{T}} \end{bmatrix} \tag{6.105}$$

where the initial stiffness moduli E_{N} and E_{T} are functions of the Young's modulus, the Poisson's ratio and a weight parameter (Bažant and Prat 1988). The damage variables $\omega^{\alpha}_{\text{N}}$ and $\omega^{\alpha}_{\text{T}}$ for the normal stiffness and the shear stiffness are functions of the internal variables $\kappa^{\alpha}_{\text{N}}$ and $\kappa^{\alpha}_{\text{T}}$ in a standard fashion: $\omega^{\alpha}_{\text{N}} = \omega^{\alpha}_{\text{N}}(\kappa^{\alpha}_{\text{N}})$ and $\omega^{\alpha}_{\text{T}} = \omega^{\alpha}_{\text{T}}(\kappa^{\alpha}_{\text{T}})$. Evidently, two damage loading functions are required for each microplane α:

$$\begin{aligned} f^{\alpha}_{\text{N}} &= \epsilon^{\alpha}_{nn} - \kappa^{\alpha}_{\text{N}} \\ f^{\alpha}_{\text{T}} &= \gamma^{\alpha}_{ns} - \kappa^{\alpha}_{\text{T}} \end{aligned} \tag{6.106}$$

each subject to the standard loading–unloading conditions. Finally, the stresses in the global x, y-coordinate system are recovered by summing over all the microplanes and by transforming them in a standard fashion according to Equation (6.97):

$$\boldsymbol{\sigma}_{\text{xy}} = \sum_{\alpha=1}^{n} w^{\alpha} (\mathbf{T}^{\alpha}_{\sigma})^{\text{T}} \mathbf{D}^{\alpha}_{\text{ns}} \mathbf{T}^{\alpha}_{\epsilon} \boldsymbol{\epsilon}_{\text{xy}} \tag{6.107}$$

with n the chosen number of microplanes and w^{α} weight factors that stem from numerical integration rules for a sphere.

Attention is drawn to the fact that the second row of $\mathbf{D}^{\alpha}_{\text{ns}}$ consists of zeros. This is because in the microplane concept only the normal stress and the shear stress are resolved on each microplane. The normal stress parallel to this plane is irrelevant. We also note that this is a simple version of the microplane model, namely one in which no splitting in volumetric and deviatoric components is considered. More sophisticated microplane models, e.g. by Bažant and Prat (1988), which incorporate such a split can be captured by the same formalism.

The tangential stiffness matrix of microplane models can be cast in the same format as Equation (6.99). Upon linearisation of Equation (6.107) one obtains:

$$\dot{\boldsymbol{\sigma}}_{\text{xy}} = \sum_{\alpha=1}^{n} w^{\alpha} (\mathbf{T}^{\alpha}_{\sigma})^{\text{T}} (\mathbf{D}^{\alpha}_{\text{ns}} - \Delta\mathbf{D}^{\alpha}_{\text{ns}}) \mathbf{T}^{\alpha}_{\epsilon} \dot{\boldsymbol{\epsilon}}_{\text{xy}} \tag{6.108}$$

with $\mathbf{D}^{\alpha}_{\text{ns}}$ given by Equation (6.105) and

$$\Delta\mathbf{D}^{\alpha}_{\text{ns}} = \begin{bmatrix} d^{\alpha}_{11} & 0 & 0 \\ 0 & 0 & 0 \\ 0 & 0 & d^{\alpha}_{33} \end{bmatrix} \tag{6.109}$$

with

$$d_{11} = \frac{\partial \omega^{\alpha}_{\text{N}}}{\partial \kappa^{\alpha}_{\text{N}}} \frac{\partial \kappa^{\alpha}_{\text{N}}}{\partial \epsilon^{\alpha}_{nn}} E_{\text{N}} \epsilon^{\alpha}_{nn} \tag{6.110}$$

where $\partial\kappa_{\mathrm{N}}^{\alpha}/\partial\epsilon_{nn}^{\alpha} \equiv 1$ if $f_{\mathrm{N}}^{\alpha} = 0$ and zero otherwise, and

$$d_{33} = \frac{\partial\omega_{\mathrm{T}}^{\alpha}}{\partial\kappa_{\mathrm{T}}^{\alpha}} \frac{\partial\kappa_{\mathrm{T}}^{\alpha}}{\partial\gamma_{ns}^{\alpha}} E_{\mathrm{T}} \gamma_{ns}^{\alpha} \tag{6.111}$$

where $\partial\kappa_{\mathrm{T}}^{\alpha}/\partial\gamma_{ns}^{\alpha} \equiv 1$ if $f_{\mathrm{T}}^{\alpha} = 0$ and zero otherwise (Kuhl and Ramm 1998).

6.8 Crack Models for Concrete and Other Quasi-brittle Materials

Two main approaches exist for modelling cracking in concrete, mortar, masonry and rocks, namely discrete crack models (Ingraffea and Saouma 1985; Ngo and Scordelis 1967) and smeared crack models (Rashid 1968).

In the oldest version of the discrete crack model fracture is assumed to occur as soon as the nodal force that is normal to the element boundaries exceeds the maximum tensile force that can be sustained. New degrees of freedom at that node location are created and a geometrical discontinuity is assumed to occur between the 'old' node and the newly created node. Two obvious drawbacks of the method are the continuous change of the topology of the discretisation and the restriction of the crack propagation to follow the mesh lines. Modern, advanced discretisation methods can overcome these limitations.

The counterpart of the discrete crack concept is the smeared crack concept, in which a cracked solid is imagined to be a continuum where the notions of stress and strain continue to apply. The behaviour of cracked concrete is then described in terms of stress–strain relations and, upon cracking, it is sufficient to replace the initial isotropic stress–strain relation by an orthotropic stress–strain relation. As a consequence, the topology of the original finite element mesh remains preserved. This is computationally efficient and it is for this reason that the method has come into widespread use and replaced the early discrete crack models in large-scale computations, especially of reinforced concrete structures.

6.8.1 *Elasticity-based Smeared Crack Models*

In a smeared crack approach, the nucleation of one or more cracks in the volume that is attributed to an integration point is translated into a deterioration of the current stiffness and strength at that integration point. Generally, when the combination of stresses satisfies a specified criterion, e.g. the major principal stress reaching the tensile strength f_t, a crack is initiated. This implies that at the integration point where the stress, the strain and the internal variables are monitored, the isotropic stress–strain relation is replaced by an orthotropic elasticity-type relation with the n, s-axes being axes of orthotropy. In early studies (Rashid 1968), the orthotropic relation was defined by the following secant stiffness matrix:

$$\mathbf{D}_{\mathrm{ns}}^{\mathrm{s}} = \begin{bmatrix} 0 & 0 & 0 \\ 0 & E & 0 \\ 0 & 0 & 0 \end{bmatrix} \tag{6.112}$$

If we introduce ϕ as the angle from the x-axis to the s-axis, we can write Equation (6.98) more explicitly as:

$$\boldsymbol{\sigma}_{\mathrm{xy}} = \mathbf{T}_{\sigma}^{\mathrm{T}}(\phi)\mathbf{D}_{\mathrm{ns}}^{\mathrm{s}}\mathbf{T}_{\epsilon}(\phi)\boldsymbol{\epsilon}_{\mathrm{xy}} \tag{6.113}$$

The approach with ϕ fixed at crack initiation is known as the fixed smeared crack model. Referring to this angle as ϕ_0, we have:

$$\boldsymbol{\sigma}_{\mathrm{xy}} = \mathbf{T}_{\sigma}^{\mathrm{T}}(\phi_0)\mathbf{D}_{\mathrm{ns}}^{\mathrm{s}}\mathbf{T}_{\epsilon}(\phi_0)\boldsymbol{\epsilon}_{\mathrm{xy}} \tag{6.114}$$

Because of ill-conditioning, use of Equation (6.112) can induce convergence difficulties. Also, physically unrealistic and distorted crack patterns may be obtained (Suidan and Schnobrich 1973). For this reason a reduced shear modulus βG, $0 \leq \beta \leq 1$ was introduced into the secant stiffness matrix:

$$\mathbf{D}_{\mathrm{ns}}^{\mathrm{s}} = \begin{bmatrix} 0 & 0 & 0 \\ 0 & E & 0 \\ 0 & 0 & \beta G \end{bmatrix} \tag{6.115}$$

The use of the shear retention factor β not only reduces numerical difficulties, but also improves the capability of fixed smeared crack models to simulate the physics of the cracking process more realistically, because in this way the effects of aggregate interlock, i.e. the locking effect which bigger grains in a crack face have with respect to sliding, and friction in the crack can be represented indirectly.

Setting the stiffness normal to the crack in Equation (6.115) equal to zero gives a sudden stress drop from the tensile strength f_t to zero upon crack initiation. This can cause numerical problems as well. A gradual decrease of the tensile carrying capacity, given by:

$$\mathbf{D}_{\mathrm{ns}}^{\mathrm{s}} = \begin{bmatrix} \mu E & 0 & 0 \\ 0 & E & 0 \\ 0 & 0 & \beta G \end{bmatrix} \tag{6.116}$$

gives results that are physically more appealing and computations that are numerically more stable. In Equation (6.116), μ is a factor which gradually decreases from one to zero as a function of the normal strain ϵ_{nn}, $\mu = \mu(\epsilon_{nn})$. The introduction of the reduced normal stiffness μE was originally motivated by the argument that, in reinforced concrete, the volume attributed to an integration point contains a number of cracks and that due to the bond between concrete and reinforcing steel, the intact concrete between the cracks adds stiffness which would be underestimated by a sudden drop to zero of the tensile strength. Later, servo-controlled experiments on plain concrete have shown that concrete is not a perfectly brittle material in the Griffith sense, but that it has some residual load-carrying capacity after reaching the tensile strength. This experimental observation has led to another interpretation of the reduction factor μ, namely where the descending branch was introduced to model the gradually diminishing tensile strength of plain concrete upon further crack opening (Bažant and Oh 1983). This class of models has been named tension softening models. In fact, such models are a smeared version of cohesive-zone models, in which the fracture energy $\mathcal{G}_c$ is the governing material parameter, and is distributed over the crack band width. In practical finite element computations the crack band width is estimated using formulas like Equation (6.55).

It is nowadays recognised that smeared crack models are, in fact, anisotropic damage models, cf. Equation (6.98). Along the preceding line of development, both the normal stiffness and the shear stiffness are usually reduced, leading to the following definition:

$$\mathbf{D}^{\mathrm{s}}_{\mathrm{ns}} = \begin{bmatrix} \frac{(1-\omega_1)E}{1-(1-\omega_1)\nu^2} & \frac{(1-\omega_1)\nu E}{1-(1-\omega_1)\nu^2} & 0 \\ \frac{(1-\omega_1)\nu E}{1-(1-\omega_1)\nu^2} & \frac{E}{1-(1-\omega_1)\nu^2} & 0 \\ 0 & 0 & \frac{(1-\omega_2)E}{2(1+\nu)} \end{bmatrix} \tag{6.117}$$

instead of Equation (6.95) (Bažant and Oh 1983; de Borst and Nauta 1985; de Borst 1987; Rots 1991). The factor $1-\omega_1$ represents the degradation of the normal stiffness and can be identified with the normal reduction factor μ. The factor $1-\omega_2$ represents the degradation of the shear stiffness and can be identified with the traditional shear retention factor β as introduced by Suidan and Schnobrich (1973). Equation (6.117) differs from Equation (6.116) in the sense that the diminishing of the Poisson effect upon cracking is now properly incorporated. It is noted that the scalar damage variables ω_1 and ω_2 have no relation with the scalar damage variables which enter Equation (6.11). For the fixed crack model the secant stiffness matrix (6.117) takes the place of $\mathbf{D}^{\mathrm{s}}$ in Equation (6.114). Differentiation of Equation (6.114) yields the tangential stress–strain relation needed in an incremental-iterative procedure which utilises the Newton–Raphson method:

$$\dot{\boldsymbol{\sigma}}_{\mathrm{xy}} = \mathbf{T}^{\mathrm{T}}_{\sigma}(\phi_0)\left(\mathbf{D}^{\mathrm{s}}_{\mathrm{ns}} - \Delta\mathbf{D}_{\mathrm{ns}}\right)\mathbf{T}_{\epsilon}(\phi_0)\dot{\boldsymbol{\epsilon}}_{\mathrm{xy}} \tag{6.118}$$

with $\mathbf{D}^{\mathrm{s}}_{\mathrm{ns}}$ given by Equation (6.117) and

$$\Delta\mathbf{D}_{\mathrm{ns}} = \begin{bmatrix} d_{11} & 0 & 0 \\ \nu d_{11} & 0 & 0 \\ d_{31} & 0 & 0 \end{bmatrix} \tag{6.119}$$

with

$$d_{11} = \frac{\partial\omega_1}{\partial\kappa}\frac{\partial\kappa}{\partial\epsilon_{nn}}\frac{E(\epsilon_{nn}+\nu\epsilon_{ss})}{(1-(1-\omega_1)\nu^2)^2} \tag{6.120}$$

and

$$d_{31} = \frac{\partial\omega_2}{\partial\kappa}\frac{\partial\kappa}{\partial\epsilon_{nn}}\frac{E}{2(1+\nu)}\gamma_{ns} \tag{6.121}$$

$\partial\kappa/\partial\epsilon_{nn} = 1$ upon loading and zero otherwise.

The fixed crack model outlined above assumes that, upon violation of the fracture criterion, the direction of the crack plane is fixed. During subsequent loading shear strains can arise along the crack plane, which, in turn, will lead to a build-up of shear stresses over the crack plane. Although the stress normal to the crack plane is reduced gradually, the residual normal stress and the shear stress over the crack can cause principal values of the stress tensor that exceed the tensile strength in a direction different from the normal to the existing crack plane. This problem can be overcome by using a rotating crack model (Cope *et al.* 1980). The rotating crack model is a total stress–strain relation which takes its point of departure in Equation (6.113), similar to the total-strain, elasticity-based fixed crack model. The salient difference with the fixed crack model is that in the rotating crack model the directions of the major principal stress and the normal to the crack are aligned during the entire cracking process. Consequently, the shear stress σ_{ns} is always zero and defining a secant shear stiffness is irrelevant. There is only one remaining damage variable, $\omega = \omega_1$, and we have:

$$\mathbf{D}^{\mathrm{s}}_{\mathrm{ns}} = \begin{bmatrix} \frac{(1-\omega)E}{1-(1-\omega)\nu^2} & \frac{(1-\omega)\nu E}{1-(1-\omega)\nu^2} & 0 \\ \frac{(1-\omega)\nu E}{1-(1-\omega)\nu^2} & \frac{E}{1-(1-\omega)\nu^2} & 0 \\ 0 & 0 & 0 \end{bmatrix} \tag{6.122}$$

Differentiation of Equation (6.113) yields the tangential stress–strain relation needed in an incremental-iterative procedure which utilises the Newton–Raphson method:

$$\dot{\boldsymbol{\sigma}}_{\mathrm{xy}} = \mathbf{T}^{\mathrm{T}}_{\sigma}(\phi)(\mathbf{D}^{\mathrm{s}}_{\mathrm{ns}} - \Delta\mathbf{D}_{\mathrm{ns}})\mathbf{T}_{\epsilon}(\phi)\dot{\boldsymbol{\epsilon}}_{\mathrm{xy}} \tag{6.123}$$

with $\Delta\mathbf{D}_{\mathrm{ns}}$ now given by:

$$\Delta\mathbf{D}_{\mathrm{ns}} = \begin{bmatrix} d_{11} & 0 & 0 \\ \nu d_{11} & 0 & 0 \\ 0 & 0 & -d_{33} \end{bmatrix} \tag{6.124}$$

with

$$d_{11} = \frac{\partial\omega}{\partial\kappa}\frac{\partial\kappa}{\partial\epsilon_{nn}}\frac{E(\epsilon_{nn} + \nu\epsilon_{ss})}{(1-(1-\omega)\nu^2)^2} \tag{6.125}$$

and

$$d_{33} = \frac{\sigma_{nn} - \sigma_{ss}}{2(\epsilon_{nn} - \epsilon_{ss})} \tag{6.126}$$

the tangential shear stiffness which directly follows from the requirement of coaxiality between the stress and strain tensors. It derives from the fact that ϕ is no longer constant, and results from a consistent differentiation of the secant stiffness relation (Box 6.3).

The above smeared crack models are based on total strain concepts. This makes it difficult to properly combine cracking with other non-linear phenomena such as creep, plasticity, or thermal effects. Alternatively, the strain can be decomposed additively into a concrete part $\boldsymbol{\epsilon}^{\mathrm{co}}$ and a cracking part $\boldsymbol{\epsilon}^{\mathrm{cr}}$:

$$\boldsymbol{\epsilon} = \boldsymbol{\epsilon}^{\mathrm{co}} + \boldsymbol{\epsilon}^{\mathrm{cr}} \tag{6.127}$$

The crack strain can be composed of several contributions:

$$\boldsymbol{\epsilon}^{\mathrm{cr}} = \boldsymbol{\epsilon}^{\mathrm{cr}}_1 + \boldsymbol{\epsilon}^{\mathrm{cr}}_2 + \ldots \tag{6.128}$$

with $\boldsymbol{\epsilon}^{\mathrm{cr}}_1$ the strain related to a primary crack, $\boldsymbol{\epsilon}^{\mathrm{cr}}_2$ the strain related to a secondary crack, and so on. The relation between the crack strain rate and the stress rate is conveniently defined in the coordinate system which is aligned with the crack. This necessitates a transformation between the crack strain $\boldsymbol{\epsilon}^{\mathrm{cr}}_{\mathrm{xy},k}$ of crack k in the global x, y-coordinate system and a crack strain $\boldsymbol{\epsilon}^{\mathrm{cr}}_{\mathrm{ns},k}$ expressed in local n, s-coordinates:

$$\boldsymbol{\epsilon}^{\mathrm{cr}}_{\mathrm{ns},k} = \mathbf{T}_{\epsilon}(\phi_k)\boldsymbol{\epsilon}^{\mathrm{cr}}_{\mathrm{xy},k} \tag{6.129}$$

Box 6.3 Derivation of the tangential stiffness matrix for the rotating crack model

From the relation between the stress tensor expressed in the global x, y-coordinate system and that expressed in the local n, s-coordinate system $\boldsymbol{\sigma}_{\mathrm{xy}} = \mathbf{T}_\sigma^{\mathrm{T}}\boldsymbol{\sigma}_{\mathrm{ns}}$, cf. Equation (1.104), we obtain by differentiation: $\dot{\boldsymbol{\sigma}}_{\mathrm{xy}} = \mathbf{T}_\sigma^{\mathrm{T}}\dot{\boldsymbol{\sigma}}_{\mathrm{ns}} + \dot{\mathbf{T}}_\sigma^{\mathrm{T}}\boldsymbol{\sigma}_{\mathrm{ns}}$ and linearise the secant stress–strain relation in the local coordinate system, Equation (6.94):

$$\dot{\boldsymbol{\sigma}}_{\mathrm{ns}} = (\mathbf{D}_{\mathrm{ns}} - \Delta\mathbf{D}_{\mathrm{ns}}^*)\dot{\boldsymbol{\epsilon}}_{\mathrm{ns}} \quad , \quad \Delta\mathbf{D}_{\mathrm{ns}}^* = \begin{bmatrix} d_{11} & 0 & 0 \\ \nu d_{11} & 0 & 0 \\ 0 & 0 & 0 \end{bmatrix}$$

if $\mathbf{D}_{\mathrm{ns}}^{\mathrm{s}}$ is as in Equation (6.122) for the rotating crack model, and d_{11} is given by Equation (6.125). Substitution of the local tangential stress–strain relation into the expression for $\dot{\boldsymbol{\sigma}}_{\mathrm{xy}}$, noting that the transformation matrix $\mathbf{T}_\sigma$ is a function of ϕ, that ϕ is a function of the strains, via $\tan 2\phi = \gamma_{xy}/(\epsilon_{xx} - \epsilon_{yy})$, and using the strain transformation (1.105), we obtain:

$$\dot{\boldsymbol{\sigma}}_{\mathrm{xy}} = \mathbf{T}_\sigma^{\mathrm{T}}(\mathbf{D}_{\mathrm{ns}} - \Delta\mathbf{D}_{\mathrm{ns}}^*)\mathbf{T}_\epsilon\dot{\boldsymbol{\epsilon}}_{\mathrm{xy}} + \left(\frac{\partial\mathbf{T}_\sigma^{\mathrm{T}}}{\partial\phi}\boldsymbol{\sigma}_{\mathrm{ns}}\right)\left(\frac{\partial\phi}{\partial\boldsymbol{\epsilon}_{\mathrm{xy}}}\right)\dot{\boldsymbol{\epsilon}}_{\mathrm{xy}}$$

Noting that for the rotating crack model $\boldsymbol{\sigma}_{\mathrm{ns}}^{\mathrm{T}} = (\sigma_1, \sigma_2, 0)$, the second term on the right-hand side can be elaborated as:

$$\frac{\partial\mathbf{T}_\sigma^{\mathrm{T}}}{\partial\phi}\boldsymbol{\sigma}_{\mathrm{ns}} = (\sigma_1 - \sigma_2)\mathbf{z} \quad , \quad \frac{\partial\phi}{\partial\boldsymbol{\epsilon}_{\mathrm{xy}}} = \frac{\mathbf{z}^{\mathrm{T}}}{2(\epsilon_1 - \epsilon_2)}$$

with $\mathbf{z}^{\mathrm{T}} = (\sin 2\phi, -\sin 2\phi, -\cos 2\phi)$, whence

$$\dot{\boldsymbol{\sigma}}_{\mathrm{xy}} = \mathbf{T}_\sigma^{\mathrm{T}}(\mathbf{D}_{\mathrm{ns}} - \Delta\mathbf{D}_{\mathrm{ns}}^*)\mathbf{T}_\epsilon\dot{\boldsymbol{\epsilon}}_{\mathrm{xy}} + \mathbf{z}\frac{\sigma_1 - \sigma_2}{2(\epsilon_1 - \epsilon_2)}\mathbf{z}^{\mathrm{T}}\dot{\boldsymbol{\epsilon}}_{\mathrm{xy}}$$

which can be further rewritten as:

$$\dot{\boldsymbol{\sigma}}_{\mathrm{xy}} = \mathbf{T}_\sigma^{\mathrm{T}}\left(\mathbf{D}_{\mathrm{ns}} - \Delta\mathbf{D}_{\mathrm{ns}}^* + \frac{\sigma_1 - \sigma_2}{2(\epsilon_1 - \epsilon_2)}\mathbf{Z}\right)\mathbf{T}_\epsilon\dot{\boldsymbol{\epsilon}}_{\mathrm{xy}} \quad , \quad \mathbf{Z} = \begin{bmatrix} 0 & 0 & 0 \\ 0 & 0 & 0 \\ 0 & 0 & 1 \end{bmatrix}$$

This proves Equations (6.123)–(6.126).

with $\mathbf{T}_\epsilon(\phi_k)$ the transformation matrix for crack k, which has an inclination angle ϕ_k between the normal of the crack and the x-axis. In the case of n cracks we have:

$$\boldsymbol{\epsilon}_{\mathrm{xy}}^{\mathrm{cr}} = \sum_{k=1}^{n}\mathbf{T}_\epsilon^{\mathrm{T}}(\phi_k)\boldsymbol{\epsilon}_{\mathrm{ns},k}^{\mathrm{cr}} \tag{6.130}$$

The relation between the stress $\boldsymbol{\sigma}_{\rm xy}$ in the global x, y-coordinate system and the stress $\boldsymbol{\sigma}_{{\rm ns},k}$ in the local coordinate system of crack k can be written as:

$$\boldsymbol{\sigma}_{{\rm ns},k} = \mathbf{T}_\sigma(\phi_k)\boldsymbol{\sigma}_{\rm xy} \tag{6.131}$$

Secant stress–strain relations for the the intact concrete

$$\boldsymbol{\sigma}_{\rm xy} = \mathbf{D}^{\rm co}\boldsymbol{\epsilon}^{\rm co}_{\rm xy} \tag{6.132}$$

and for the smeared cracks

$$\boldsymbol{\sigma}_{{\rm ns},k} = \mathbf{D}^{\rm cr}_k\boldsymbol{\epsilon}^{\rm cr}_{{\rm ns},k} \tag{6.133}$$

with $\mathbf{D}^{\rm cr}_k$ a 2×2 matrix, complete the model. Equations (6.127)–(6.133) allow the derivation of the compliance relation for the cracked concrete:

$$\boldsymbol{\epsilon}_{\rm xy} = [(\mathbf{D}^{\rm co})^{-1} + \sum_{k=1}^{n}\mathbf{T}^{\rm T}_\epsilon(\phi_k)(\mathbf{D}^{\rm cr}_k)^{-1}\mathbf{T}_\sigma(\phi_k)]\boldsymbol{\sigma}_{\rm xy} \tag{6.134}$$

Repeated use of the Sherman–Morrison formula results in the stiffness relation:

$$\boldsymbol{\sigma}_{\rm xy} = \left(\mathbf{D}^{\rm co} - \mathbf{D}^{\rm co}\sum_{k=1}^{n}\mathbf{T}^{\rm T}_\epsilon(\phi_k)(\mathbf{D}^{\rm cr}_k + \mathbf{T}_\sigma(\phi_k)\mathbf{D}^{\rm co}\mathbf{T}^{\rm T}_\epsilon(\phi_k))^{-1}\mathbf{T}_\sigma(\phi_k)\mathbf{D}^{\rm co}\right)\boldsymbol{\epsilon}_{\rm xy} \tag{6.135}$$

In smeared crack analyses of concrete members one experiences a lot of what may be referred to as spurious cracking. Here, we mean that there are quite a number of sampling points which crack, but show only small crack strains. This partly causes the diffuse crack pattern of smeared crack analyses. In fact, we often observe that only a limited number of cracks really open and lead to failure. Yet, these sampling points with small crack strains pose a problem as a number of them show unloading, even close and sometimes open again in a later stage of the loading process. It is important to carefully handle closing and eventually reopening of cracks. In the spirit of damage mechanics a secant approach is usually adopted for the unloading/reloading branch. This assumption neglects the residual strain which we can expect upon crack closing.

6.8.2 *Reinforcement and Tension Stiffening*

The most common way to incorporate reinforcement and prestressing tendons in a computational model of reinforced or prestressed concrete is to adopt the assumption of perfect bond, which states that the reinforcement undergoes the same displacements as the concrete elements in which it is embedded. In this embedded formulation a reinforcing bar is subject to the same state of strain as a concrete 'fibre' that is aligned with the bar.

In the uncracked state the assumption of perfect bond seems natural and will be an accurate model of the reality. This can be different when cracks occur. Then, the assumption of perfect bond is weaker. In consideration of the role of bond in setting the crack width and crack spacing in reinforced concrete structures, it may be no longer possible to adhere to the concept of perfect bond if one strives at obtaining accurate predictions of crack width and crack spacing at a detailed level. For such detailed calculations the use of a discretisation is required which models the displacement discontinuity at the interface between steel and surrounding concrete.

The concept of embedded reinforcement implies that there are two different contributions to the internal virtual work. Starting from the weak form of the equilibrium equations, Equation (2.37), and assuming that the reinforcement only carries normal stresses, we obtain for a composite with a concrete matrix and reinforcement:

$$\int_V \delta\boldsymbol{\epsilon}^{\mathrm{T}}\boldsymbol{\sigma}^{\mathrm{co}}\mathrm{d}V + \int_V \delta\epsilon^{\mathrm{re}}\sigma^{\mathrm{re}}\mathrm{d}V = \int_S \delta\mathbf{u}^{\mathrm{T}}\mathbf{t}\mathrm{d}S \tag{6.136}$$

with $\boldsymbol{\sigma}^{\mathrm{co}}$ and σ^{re} the stresses in the concrete and reinforcement, respectively, and

$$\epsilon^{\mathrm{re}} = \mathbf{B}^{\mathrm{re}}\mathbf{a} \tag{6.137}$$

is the axial strain in the reinforcement. The angle ψ between the local ξ-axis and the axis of the reinforcement bar enters the $\mathbf{B}^{\mathrm{re}}$ matrix. We next specify the stress–strain relationships for the concrete and for the reinforcement. For the concrete, a secant stress–strain relation can be used, as in the foregoing:

$$\boldsymbol{\sigma}^{\mathrm{co}} = \mathbf{D}^{\mathrm{co}}\boldsymbol{\epsilon} \tag{6.138}$$

For the steel an elasto-plastic model is commonly used, which is most conveniently formalised in a rate format:

$$\dot{\sigma}^{\mathrm{re}} = E^{\mathrm{re}}\dot{\epsilon}^{\mathrm{re}} \tag{6.139}$$

where E^{re} is the tangential stiffness modulus of the reinforcement. Substitution of Equations (6.137)–(6.139) into the decomposed form of Equation (6.136), discretising and requiring that the result holds for any admissible kinematic field gives:

$$\begin{aligned}\left[\int_V \mathbf{B}^{\mathrm{T}}\mathbf{D}^{\mathrm{co}}\mathbf{B}\mathrm{d}V + \int_V (\mathbf{B}^{\mathrm{re}})^{\mathrm{T}}E^{\mathrm{re}}\mathbf{B}^{\mathrm{re}}\mathrm{d}V\right]\mathrm{d}\mathbf{a} = \\ \mathbf{f}_{\mathrm{ext}}^{t+\Delta t} - \int_V \mathbf{B}^{\mathrm{T}}\boldsymbol{\sigma}_j^{\mathrm{co}}\mathrm{d}V - \int_V \sigma_j^{\mathrm{re}}(\mathbf{B}^{\mathrm{re}})^{\mathrm{T}}\mathrm{d}V\end{aligned} \tag{6.140}$$

We observe that we have an overlay of a concrete and a reinforcement element which are governed by the same displacement field. The evaluation of the integrals is done separately for the concrete and the steel such that they each have their own integration points.

The limitation of the embedded reinforcement concept – and consequently, of neglecting bond-slip behaviour – can be elucidated by considering an elementary concrete element with a reinforcement bar that is aligned with the global x-axis. When pulling in the x-direction the composite element initially reacts elastically with a stiffness that is the sum of the linear-elastic stiffnesses of the concrete and the reinforcement. This is the linear branch of Figure 6.16 until point A. At point A the concrete cracks. If cracking would occur in a completely brittle manner, the load-carrying capacity would fall back to point B, where the stiffness of the steel bar is picked up until point C, which marks the onset of yielding of the reinforcement. When carrying out an experiment the curve A–C is observed instead of the path A–B–C.

There are two possible explanations why the measured load–elongation curve is stiffer than that which is obtained from the above elementary consideration. A first explanation is that concrete is not a perfectly brittle material, but exhibits a gradual descending branch. This phenomenon can be modelled using a cohesive-zone formulation, for which, as already said,

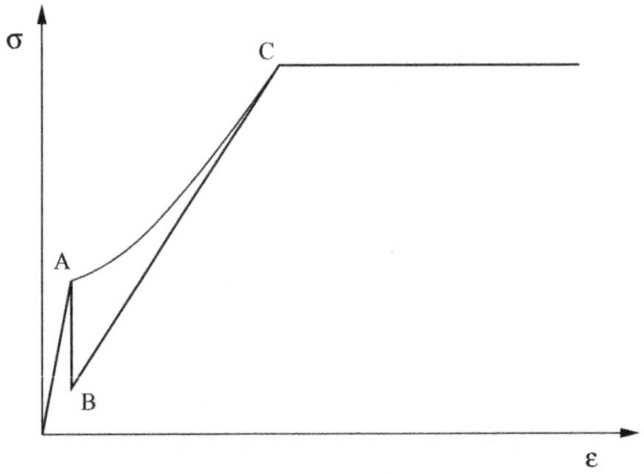

Figure 6.16 Load–displacement curve of a concrete element with an embedded central reinforcing bar

in a smeared format the name 'tension softening' has been coined. But there is an additional effect. After cracking the concrete can still sustain tensile stresses and therefore there can still be shear stress transfer between concrete and the reinforcement along most of the bar. Hence, the concrete between the cracks has a significant residual contribution to the stiffness of the composite element. This is the tension stiffening effect.

A rational approach is to assume that the behaviour of cracked, reinforced concrete can be obtained by superposition of the stiffness of plain concrete, a stiffness of the reinforcement and an additional stiffness due to interaction between concrete and the reinforcement. This leads to the following summation of stress contributions (Feenstra and de Borst 1995a):

$$\boldsymbol{\sigma} = \boldsymbol{\sigma}^{\text{co}} + \boldsymbol{\sigma}^{\text{re}} + \boldsymbol{\sigma}^{\text{ia}} \tag{6.141}$$

with $\boldsymbol{\sigma}^{\text{co}}$ the stress contribution of the plain concrete, $\boldsymbol{\sigma}^{\text{re}}$ the contribution of the reinforcing steel, and $\boldsymbol{\sigma}^{\text{ia}}$ the interaction stress contribution due to tension stiffening [Figure 6.17(a)]. In

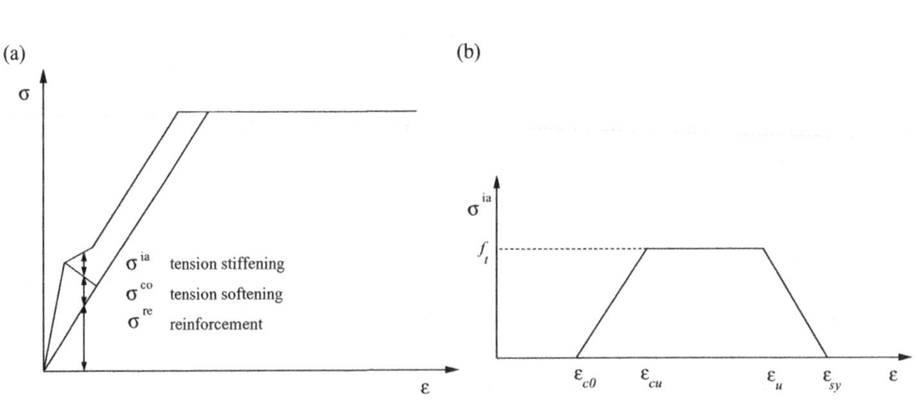

Figure 6.17 (a) Model for separation of tension softening and tension stiffening. (b) Interaction stress between concrete and reinforcement

reinforced concrete structures a number of cracks develop during loading until the cracking process has stabilised and no further cracks develop. The crack spacing at stabilised cracking is determined mainly by the amount of reinforcement. It is assumed that the material model for plain concrete, based on fracture energy, can be applied to reinforced concrete with the total amount of fracture energy dissipated over the equivalent length. In general, the dimensions of the finite elements in simulations of reinforced concrete structures are much larger than the average crack spacing, ℓ_s. Therefore, it is assumed that the released energy can be determined from:

$$\mathcal{G}_c^{\mathrm{rc}} = \min\left[\mathcal{G}_c, \mathcal{G}_c \frac{w}{\ell_s}\right] \tag{6.142}$$

with $\mathcal{G}_c$ the fracture energy of a single crack.

After a stabilised crack pattern has developed, stresses are still transferred from reinforcement to concrete between the cracks due to the bond action which increases the total stiffness of the structure. The additional stress due to tension stiffening is usually assumed to be given as a function of the strain in the direction of the reinforcement. A trilinear function is often adopted for the interaction stress [Figure 6.17(b)]. The interaction stress is only active if the strain in the reinforcement is larger than:

$$\epsilon_{c0} = \frac{f_t}{E_c} \cos^2 \alpha \tag{6.143}$$

with α the angle between the direction of the reinforcement and the direction of the principal stress at incipient cracking. The factor ϵ_{cu} is determined by the crack spacing, the equivalent length of the element and the fracture energy of the concrete and reads:

$$\epsilon_{cu} = 2\cos^2 \alpha \frac{\mathcal{G}_c^{\mathrm{rc}}}{w f_t} \tag{6.144}$$

The constant part of the diagram can be approximated to equal the tensile strength of the concrete f_t. Close to the yield strain of the reinforcement, ϵ_{sy}, the tension-stiffening component is reduced to avoid an artificial increase of the yield stress of the reinforcement. The strain at which the tension-stiffening component is reduced is given by:

$$\epsilon_u = \epsilon_{sy} - \frac{f_t}{\rho_s E_s} \tag{6.145}$$

with ϵ_{sy} the yield strength of the steel, and ρ_s the reinforcement ratio. The local stiffness matrix that incorporates the interaction stiffness then reads:

$$\mathbf{D}^{\mathrm{ia}} = \begin{bmatrix} E_b & 0 & 0 \\ 0 & 0 & 0 \\ 0 & 0 & 0 \end{bmatrix} \tag{6.146}$$

where E_b is the bond stiffness. The transformation to the global coordinate system follows in a standard manner from:

$$\boldsymbol{\sigma}^{\mathrm{ia}} = \left(\mathbf{T}_\sigma^{\mathrm{T}}(\psi)\mathbf{D}^{\mathrm{ia}}\mathbf{T}_\epsilon(\psi)\right)\boldsymbol{\epsilon}$$

6.9 Regularised Damage Models

Loss of well-posedness of the rate boundary value problem causes a complete dependence of the numerical results on the discretisation, not only with respect to mesh refinement, but also, and especially, with respect to mesh alignment, since failure zones exhibit a strong tendency to propagate along lines of discretisation. To avoid loss of ellipticity, the standard, rate-independent continuum must be enhanced. Several possibilities for a regularisation exist: (i) spatial averaging; (ii) introducing a dependence of the stress evolution on strain gradients; (iii) adding couple stresses and conjugate kinematic quantities like micro-curvatures to the continuum description, the so-called Cosserat continuum (de Borst 1991, 1993); or (iv) by adding viscosity (Needleman 1987; Sluys and de Borst 1992). Not all enhancements are as effective in eliminating discretisation sensitivity. Adding viscosity can only be effective when the material under consideration exhibits a sufficiently high rate sensitivity for the given loading rate. Evidently, discretisation dependence is recovered in the rate-independent limit. The Cosserat continuum is usually only effective when there exist a physical motivation for adding couple stresses and micro-curvatures, as is the case in granular materials. Because of the restrictions of the latter two approaches, the non-local and gradient-enhanced models have become widely used in computational analyses.

6.9.1 *Non-local Damage Models*

In a non-local generalisation the equivalent strain $\tilde{\epsilon}$ is normally replaced by a spatially averaged quantity in the damage loading function (Pijaudier-Cabot and Bažant 1987):

$$f(\bar{\epsilon}, \kappa) = \bar{\epsilon} - \kappa \tag{6.147}$$

where the non-local strain $\bar{\epsilon}$ is computed from:

$$\bar{\epsilon}(\mathbf{x}) = \frac{1}{\Psi(\mathbf{x})} \int_V \psi(\mathbf{y}, \mathbf{x}) \tilde{\epsilon}(\mathbf{y}) \mathrm{d}V \quad , \quad \Psi(\mathbf{x}) = \int_V \psi(\mathbf{y}, \mathbf{x}) \mathrm{d}V \tag{6.148}$$

with $\psi(\mathbf{y}, \mathbf{x})$ a weight function. Often, the weight function is assumed to be homogeneous and isotropic, so that it only depends on the norm $s = \| \mathbf{x} - \mathbf{y} \|$. In this formulation all the other relations remain local: the local stress–strain relation (6.12), the loading–unloading conditions (6.16) and the dependence of the damage variable ω on the internal variable κ: $\omega = \omega(\kappa)$. As an alternative to Equation (6.148), the locally defined internal variable κ may be replaced in the damage loading function f by a spatially averaged quantity $\bar{\kappa}$:

$$\bar{\kappa}(\mathbf{x}) = \frac{1}{\Psi(\mathbf{x})} \int_V \psi(\mathbf{y}, \mathbf{x}) \kappa(\mathbf{y}) \mathrm{d}V \tag{6.149}$$

The fact that in elasticity-based damage models the stress can be computed directly from the given strain means that a straightforward algorithm can be set up for non-local damage models. For the non-local damage model defined by Equation (6.148) the algorithm of Box 6.4 applies. Although conceptually straightforward, the tangential stiffness matrix entails some inconvenient properties. Due to the non-local character of the constitutive relation the tangential stiffness matrix is full, i.e. the bandedness is lost. The introduction of a cut-off on the averaging

Box 6.4 Algorithmic treatment of non-local elasticity-based damage model

1. Compute the strain increment: $\Delta\boldsymbol{\epsilon}_{j+1}$
2. Update the total strain: $\boldsymbol{\epsilon}_{j+1} = \boldsymbol{\epsilon}_j + \Delta\boldsymbol{\epsilon}_{j+1}$
3. Compute the equivalent strain: $\tilde{\epsilon}_{j+1} = \tilde{\epsilon}(\boldsymbol{\epsilon}_{j+1})$
4. Compute the non-local equivalent strain:

$$\bar{\epsilon}_{j+1}(\mathbf{x}) = \sum_i w_i \psi(\mathbf{y}_i, \mathbf{x})\tilde{\epsilon}_{j+1}(\mathbf{y}_i) V_{\text{elem}}$$

5. Evaluate the damage loading function: $f = \bar{\epsilon}_{j+1} - \kappa_0$
   if $f \geq 0$, $\kappa_{j+1} = \bar{\epsilon}_{j+1}$
   else $\kappa_{j+1} = \kappa_0$
6. Update the damage variable: $\omega_{j+1} = \omega(\kappa_{j+1})$
7. Compute the new stresses: $\boldsymbol{\sigma}_{j+1} = (1 - \omega_{j+1})\mathbf{D}^{\text{e}} : \boldsymbol{\epsilon}_{j+1}$

function partially remedies this disadvantage, but an increased band width will nevertheless result. Secondly, symmetry can be lost (Pijaudier-Cabot and Huerta 1991).

6.9.2 Gradient Damage Models

Non-local constitutive relations can be considered as a point of departure for constructing gradient models, although we wish to emphasise that the latter class of models can also be defined directly by supplying higher-order gradients in the damage loading function. Yet, we will follow the first-mentioned route to underline the connection between integral and differential type non-local models. This is done either by expanding the kernel $\tilde{\epsilon}$ of the integral in Equation (6.148) in a Taylor series, or by expanding the internal variable κ in Equation (6.149) in a Taylor series. We will first consider the expansion of $\tilde{\epsilon}$ and then we will do the same for κ. If we truncate after the second-order terms and carry out the integration implied in Equation (6.148) under the assumption of isotropy, the following relation ensues:

$$\bar{\epsilon} = \tilde{\epsilon} + c\nabla^2\tilde{\epsilon} \tag{6.150}$$

where c is a gradient parameter of the dimension length squared. It can be related to the averaging volume and then becomes dependent on the precise form of the weight function ψ. For instance, for a one-dimensional continuum and taking

$$\psi(s) = \frac{1}{\sqrt{2\pi}\ell}\mathrm{e}^{-s^2/2\ell^2} \tag{6.151}$$

we obtain $c = 1/2\ell^2$. Here, we adopt the phenomenological view that $\ell = \sqrt{2c}$ reflects the internal length scale of the failure process which we wish to describe macroscopically.

Formulation (6.150), known as the explicit gradient damage model, has a disadvantage when applied in a finite element context, namely that it requires computation of second-order gradients of the local equivalent strain $\tilde{\epsilon}$. Since this quantity is a function of the strain tensor, and since the strain tensor involves first-order derivatives of the displacements, third-order derivatives of the displacements have to be computed, which would necessitate $\mathcal{C}^2$-continuity of the shape functions. To obviate this problem, Equation (6.150) is differentiated twice and the result is substituted again into Equation (6.150). Again neglecting fourth-order terms leads to:

$$\bar{\epsilon} - c\nabla^2\bar{\epsilon} = \tilde{\epsilon} \tag{6.152}$$

In Peerlings *et al.* (2001) it has been shown that the implicit gradient damage model of Equation (6.152) becomes formally identical to a fully non-local formulation for a specific choice of the weighting function ψ in Equation (6.148), which underlines that this formulation has a truly non-local character, in contrast to the explicit gradient formulation of Equation (6.150).

Higher-order continua require additional boundary conditions. With Equation (6.152) governing the damage process, either the averaged equivalent strain $\bar{\epsilon}$ itself or its normal derivative must be specified on the boundary S of the body:

$$\bar{\epsilon} = \bar{\epsilon}_{\mathrm{s}} \quad \text{or} \quad \mathbf{n}_S \cdot \nabla\bar{\epsilon} = \bar{\epsilon}_{\mathrm{ns}} \tag{6.153}$$

In most example calculations in the literature the natural boundary condition $\mathbf{n}_S \cdot \nabla\bar{\epsilon} = 0$ has been adopted.

In a fashion similar to the derivation of the gradient damage models based on the averaging of the equivalent strain $\tilde{\epsilon}$, we can elaborate a gradient approximation of Equation (6.149), i.e. by developing κ into a Taylor series. For an isotropic, infinite medium and truncating after the second term we have (de Borst *et al.* 1996):

$$\bar{\kappa} = \kappa + c\nabla^2\kappa \tag{6.154}$$

Since the weight functions for the different gradient formulations may be quite different, the gradient parameter c may also be very different for the various formulations. For instance, the gradient parameter c of Equation (6.154) may differ considerably from those in Equation (6.150) or (6.152). The additional boundary conditions now apply to κ. Although formally similar to those of Equation (6.153), namely

$$\kappa = \kappa_{\mathrm{s}} \quad \text{or} \quad \mathbf{n}_S \cdot \nabla\kappa = \kappa_{\mathrm{ns}} \tag{6.155}$$

they have a different character, since they apply to an internal variable instead of to a kinematic quantity, which seems somewhat suspect. On the other hand, the physical interpretation that can be given to the boundary condition $(6.155)_2$ is rather clear. Since the damage variable ω is a function of the internal variable κ, and therefore, the differential equation (6.154) and the boundary conditions (6.155) can be replaced by (de Borst *et al.* 1996):

$$\bar{\omega} = \omega + c\nabla^2\omega \tag{6.156}$$

where $\bar{\omega}$ is a spatially averaged damage field, similar to $\bar{\epsilon}$ or $\bar{\kappa}$, and the corresponding boundary conditions

$$\omega = \omega_{\mathrm{s}} \quad \text{or} \quad \mathbf{n}_S \cdot \nabla\omega = \omega_{\mathrm{ns}} \tag{6.157}$$

Equation (6.157) with $\omega_{\mathrm{ns}} = 0$ can be identified as a condition of no damage flux through the boundary S of the body.

Numerical schemes for gradient-enhanced continua typically have the character of a coupled problem and depart from the weak form of the balance of momentum (6.56) and a weak form of the averaging equation, e.g. Equation (6.152):

$$\int_V \delta\bar{\epsilon}(\bar{\epsilon} - c\nabla^2\bar{\epsilon} - \tilde{\epsilon})\mathrm{d}V = 0 \tag{6.158}$$

with $\delta\bar{\epsilon}$ the variational field of the non-local strain $\bar{\epsilon}$. Transforming Equation (6.158), using the divergence theorem and the natural boundary condition $\mathbf{n}_S \cdot \nabla\bar{\epsilon} = 0$ yields:

$$\int_V (\delta\bar{\epsilon}\bar{\epsilon} + c\nabla\delta\bar{\epsilon} \cdot \nabla\bar{\epsilon})\mathrm{d}V = \int_V \delta\bar{\epsilon}\,\tilde{\epsilon}\mathrm{d}V \tag{6.159}$$

From Equation (6.159) it becomes clear that in this formulation a C^0-interpolation for $\bar{\epsilon}$ suffices. Accordingly, we can discretise the displacements $\mathbf{u}$ and the non-local strains

$$\mathbf{u} = \mathbf{Ha} \quad \text{and} \quad \bar{\epsilon} = \bar{\mathbf{H}}\mathbf{e} \tag{6.160}$$

where $\mathbf{H}$ and $\bar{\mathbf{H}}$ contain C^0-interpolation polynomials which can have a different order. Similarly, for the variations

$$\delta\mathbf{u} = \mathbf{H}\delta\mathbf{a} \quad \text{and} \quad \delta\bar{\epsilon} = \bar{\mathbf{H}}\delta\mathbf{e} \tag{6.161}$$

Substitution into Equations (6.56), (6.159) and requiring that the result holds for arbitrary $(\delta\mathbf{a}, \delta\mathbf{e})$, yields the discrete formats of the equilibrium equation (6.68):

$$\int_V \mathbf{B}^{\mathrm{T}}\boldsymbol{\sigma}\mathrm{d}V = \int_S \mathbf{H}^{\mathrm{T}}\mathbf{t}\mathrm{d}S$$

and the averaging equation:

$$\int_V (\bar{\mathbf{H}}^{\mathrm{T}}\bar{\mathbf{H}} + c\bar{\mathbf{B}}^{\mathrm{T}}\bar{\mathbf{B}})\mathrm{d}V = \int_V \bar{\mathbf{H}}^{\mathrm{T}}\tilde{\epsilon}\mathrm{d}V \tag{6.162}$$

where $\bar{\mathbf{B}}$ contains the spatial derivatives of $\bar{\mathbf{H}}$. An algorithm for computing the right-hand side of this model is given in Box 6.5.

The tangential stiffness matrix needed for an iterative solution via the Newton–Raphson method reads (Peerlings *et al.* 1996):

$$\begin{bmatrix} \mathbf{K}_{\mathrm{aa}} & \mathbf{K}_{\mathrm{ae}} \\ \mathbf{K}_{\mathrm{ea}} & \mathbf{K}_{\mathrm{ee}} \end{bmatrix} \begin{pmatrix} \mathrm{d}\mathbf{a} \\ \mathrm{d}\mathbf{e} \end{pmatrix} = \begin{pmatrix} \mathbf{f}^{\mathrm{a}}_{\mathrm{ext}} - \mathbf{f}^{\mathrm{a}}_{\mathrm{int}} \\ \mathbf{f}^{\mathrm{e}}_{\mathrm{int}} - \mathbf{K}_{\mathrm{ee}}\mathbf{e} \end{pmatrix} \tag{6.163}$$

Box 6.5 Algorithmic treatment of second-order implicit gradient damage model

1. Compute the strain increment: $\Delta\boldsymbol{\epsilon}_{j+1}$ and the non-local strain increment $\Delta\bar{\epsilon}_{j+1}$
2. Update the total strain: $\boldsymbol{\epsilon}_{j+1} = \boldsymbol{\epsilon}_j + \Delta\boldsymbol{\epsilon}_{j+1}$ and the non-local strain $\bar{\epsilon}_{j+1} = \bar{\epsilon}_j + \Delta\bar{\epsilon}_{j+1}$
3. Evaluate the damage loading function: $f = \bar{\epsilon}_{j+1} - \kappa_0$

   if $f \geq 0$, $\kappa_{j+1} = \bar{\epsilon}_{j+1}$

   else $\kappa_{j+1} = \kappa_0$
4. Update the damage variable: $\omega_{j+1} = \omega(\kappa_{j+1})$
5. Compute the new stresses: $\boldsymbol{\sigma}_{j+1} = (1 - \omega_{j+1})\mathbf{D}^{\mathrm{e}} : \boldsymbol{\epsilon}_{j+1}$

with $\mathbf{f}^{\mathrm{e}}_{\mathrm{int}}$ given by the right-hand side of Equation (6.162). The stiffness matrices are given by:

$$\mathbf{K}_{\mathrm{aa}} = \int_V (1-\omega)\mathbf{B}^{\mathrm{T}}\mathbf{D}^{\mathrm{e}}\mathbf{B}\,\mathrm{d}V \tag{6.164}$$

$$\mathbf{K}_{\mathrm{ae}} = \int_V q\mathbf{B}^{\mathrm{T}}\mathbf{D}^{\mathrm{e}}\boldsymbol{\epsilon}\bar{\mathbf{H}}\,\mathrm{d}V \tag{6.165}$$

$$\mathbf{K}_{\mathrm{ea}} = \int_V \bar{\mathbf{H}}^{\mathrm{T}}\left(\frac{\partial\tilde{\epsilon}}{\partial\boldsymbol{\epsilon}}\right)\mathbf{B}\,\mathrm{d}V \tag{6.166}$$

$$\mathbf{K}_{\mathrm{ee}} = \int_V \left(\bar{\mathbf{H}}^{\mathrm{T}}\bar{\mathbf{H}} + c\bar{\mathbf{B}}^{\mathrm{T}}\bar{\mathbf{B}}\right)\mathrm{d}V \tag{6.167}$$

where $q = \partial\omega/\partial\kappa$ for loading and vanishes if otherwise. The expressions for $\mathbf{K}_{\mathrm{ae}}$ and $\mathbf{K}_{\mathrm{ea}}$ exhibit a non-symmetry. This non-symmetry is caused by the damage formalism and not by the gradient enhancement, cf. Equation (6.25).

A one-dimensional bar with an imperfection in the centre is shown in Figure 6.18. Results of a computation with this model are shown in Figure 6.19. Figure 6.19 shows that the load–displacement curves converge upon mesh refinement. We finally show the results of a calculation using a gradient damage model for the single-edge notched beam of Figure 6.13. The specimen has been modelled with 1362 elements with an eight-noded quadratic displacement interpolation and a bilinear interpolation for the equivalent non-local strain. A path-following

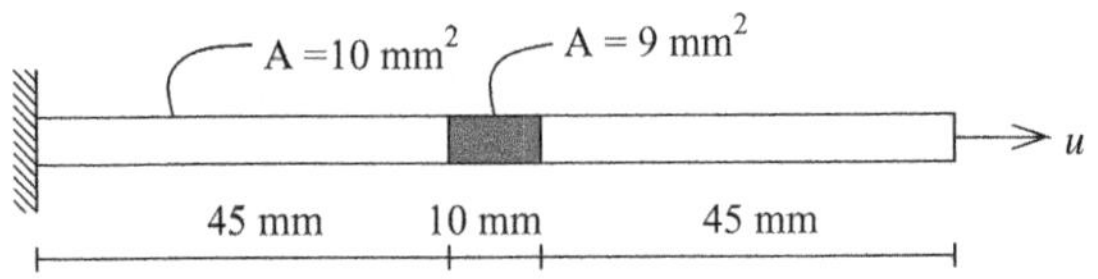

Figure 6.18 Bar with an imperfection subjected to an axial load

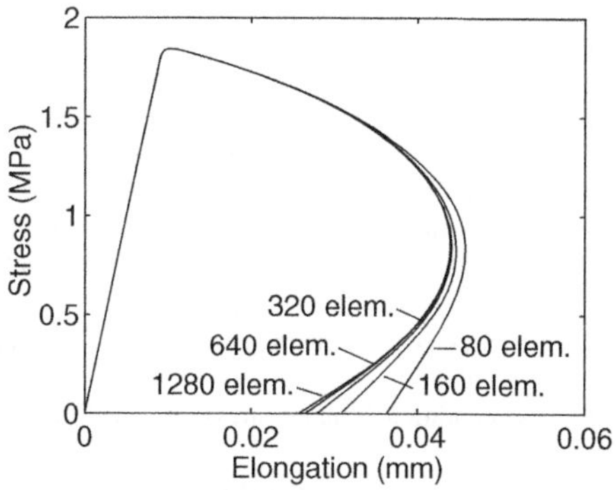

Figure 6.19 Load–displacement curves for a second-order implicit gradient model upon refinement of the finite element discretisation (Peerlings *et al.* 1996)

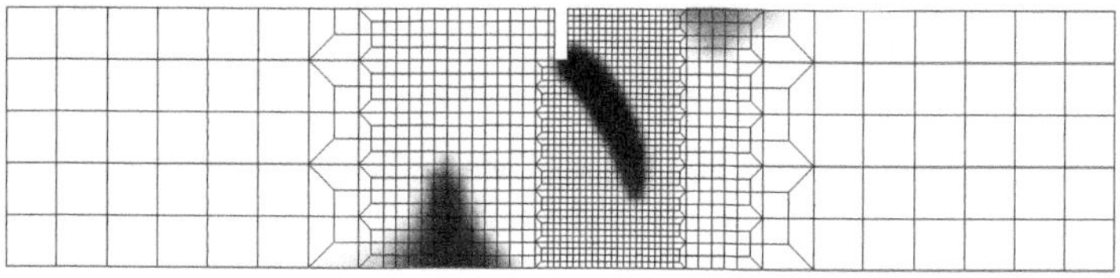

Figure 6.20 Damage contours for the single-edge notched beam of Figure 6.13

procedure has been used to control the loading process, with the crack mouth sliding displacement as the controlling parameter, see Peerlings *et al.* (1998) for details. The damage distribution at impending failure has been plotted in Figure 6.20.

References

Armero F and Garikipati K 1996 An analysis of strong discontinuities in multiplicative finite strain plasticity and their relation with the numerical simulation of strain localization in solids. *International Journal of Solids and Structures* **33**, 2863–2885.

Barenblatt GI 1962 The mathematical theory of equilibrium cracks in brittle fracture. *Advances in Applied Mechanics* **7**, 55–129.

Batdorf SB and Budiansky B 1949 A mathematical theory of plasticity based on the concept of slip. Technical report, National Advisory Committee for Aeronautics, Washington, DC.

Bathe KJ 1982 *Finite Element Procedures in Engineering Analysis*. Prentice Hall, Inc.

Bažant ZP and Gambarova P 1984 Crack shear in concrete: Crack band microplane model. *ASCE Journal of Structural Engineering* **110**, 2015–2036.

Bažant ZP and Oh B 1983 Crack band theory for fracture of concrete. *RILEM Materials and Structures* **16**, 155–177.

Bažant ZP and Prat P 1988 Microplane model for brittle plastic material. Part i: Theory. Part ii: Verification. *ASCE Journal of Engineering Mechanics* **114**, 1672–1702.

Belytschko T, Fish J and Engelman BE 1988 A finite element with embedded localization zones. *Computer Methods in Applied Mechanics and Engineering* **70**, 59–89.

Belytschko T, Lu YY and Gu L 1994 Element-free Galerkin methods. *International Journal for Numerical Methods in Engineering* **37**, 229–256.

Benallal A, Billardon R and Geymonat G 1988 Some mathematical aspects of the damage softening rate problem, in *Cracking and Damage* (eds Mazars J and Bažant ZP), pp. 247–258. Elsevier.

Borja RI 2000 A finite element model for strain localization analysis of strongly discontinuous fields based on standard Galerkin approximation. *Computer Methods in Applied Mechanics and Engineering* **190**, 1529–1549.

Chandra N, Li H, Shet C and Ghonem H 2002 Some issues in the application of cohesive zone models for metal-ceramic interfaces. *International Journal of Solids and Structures* **39**, 2827–2855.

Cope RJ, Rao PV, Clark LA and Norris P 1980 Modelling of reinforced concrete behaviour for finite element analysis of bridge slabs, in *Numerical Methods for Non-linear Problems* (eds Taylor C, Hinton E and Owen DRJ), pp. 457–470, Pineridge Press.

de Borst R 1987 Smeared cracking, plasticity, creep, and thermal loading – a unified approach. *Computer Methods in Applied Mechanics and Engineering* **62**, 89–110.

de Borst R 1991 The zero-normal-stress condition in plane-stress and shell elasto-plasticity. *Communications in Applied Numerical Methods* **7**, 29–33.

de Borst R 1993 A generalization of j_2-flow theory for polar continua. *Computer Methods in Applied Mechanics and Engineering* **103**, 347–362.

de Borst R and Nauta P 1985 Non-orthogonal cracks in a smeared finite element model. *Engineering Computations* **2**, 35–46.

de Borst R, Benallal A and Heeres OM 1996 A gradient-enhanced damage approach to fracture. *Journal de Physique IV* **C6**, 491–502.

de Vree HPJ, Brekelmans WAM and van Gils MAJ 1995 Comparison of nonlocal approaches in continuum damage mechanics. *Computers and Structures* **55**, 581–588.

Dugdale DS 1960 Yielding of steel sheets containing slits. *Journal of the Mechanics and Physics of Solids* **8**, 100–108.

Dvorkin EN, Cuitino AM and Gioia G 1990 Finite elements with dispacement interpolated embedded localization lines insensitive to mesh size and distortions. *International Journal for Numerical Methods in Engineering* **30**, 541–564.

Feenstra PH and de Borst R 1995a Constitutive model for reinforced concrete. *ASCE Journal of Engineering Mechanics* **121**, 587–595.

Feenstra PH and de Borst R 1995b A plasticity model for mode 1 cracking in concrete. *International Journal for Numerical Methods in Engineering* **38**, 2509–2529.

Gasser TC and Holzapfel GA 2003 Geometrically non-linear and consistently embedded strong discontinuity models for 3d problems with an application to the dissection analysis of soft biological tissues. *Computer Methods in Applied Mechanics and Engineering* **192**, 5059–5098.

Hill R 1958 A general theory of uniqueness and stability in elastic-plastic solids. *Journal of the Mechanics and Physics of Solids* **6**, 236–249.

Hill R 1962 Acceleration waves in solid. *Journal of the Mechanics and Physics of Solids* **10**, 1–16.

Hillerborg A, Modeér M and Petersson PE 1976 Analysis of crack formation and crack growth in concrete by means of fracture mechanics and finite elements. *Cement and Concrete Research* **6**, 773–782.

Hughes TJR 1987 *The Finite Element Methods: Linear Static and Dynamic Finite Element Analysis*. Prentice Hall, Inc.

Hutchinson JW and Evans AG 2000 Mechanics of materials: top–down approaches to fracture. *Acta Materialia* **48**, 125–135.

Ingraffea AR and Saouma V 1985 Numerical modelling of discrete crack propagation in reinforced and plain concrete, in *Fracture Mechanics of Concrete* (eds Sih GC and DiTommaso A), pp. 171–225. Martinus Nijhoff.

Jirasek M 2000 Comparative study on finite elements with embedded discontinuities. *Computer Methods in Applied Mechanics and Engineering* **188**, 307–330.

Klisinski M, Runesson K and Sture S 1991 Finite element with inner softening band. *ASCE Journal of Engineering Mechanics* **117**, 575–587.

Koiter WT 1969 On the thermodynamic background of elastic stability theory, in *Problems of Hydrodynamics and Continuum Mechanics* (ed. Radok JRM), pp. 423–433. SIAM.

Kuhl E and Ramm E 1998 On the linearization of the microplane model. *Mechanics of Cohesive-frictional Materials* **3**, 343–364.

Lemaitre J and Chaboche JL 1990 *Mechanics of Solid Materials*. Cambridge University Press.

Lotfi HR and Shing PB 1995 Embedded representation of fracture in concrete with mixed elements. *International Journal for Numerical Methods in Engineering* **38**, 1307–1325.

Maier G and Hueckel T 1979 Nonassociated and coupled flow rules of elastoplasticity for rock-like materials.. *International Journal of Rock Mechanics and Mining Sciences & Geomechanical Abstracts* **16**, 77–92.

Mazars J and Pijaudier-Cabot G 1989 Continuum damage theory – application to concrete. *ASCE Journal of Engineering Mechanics* **115**, 345–365.

Miehe C and Schotte J 2004 Crystal plasticity and the evolution of polycrystalline microstructure, in *The Encyclopedia of Computational Mechanics* (eds Stein E, de Borst R and Hughes TJR), vol. II, pp. 267–289. John Wiley & Sons, Ltd.

Needleman A 1987 Material rate dependence and mesh sensitivity in localization problems. *Computer Methods in Applied Mechanics and Engineering* **28**, 859–878.

Ngo D and Scordelis AC 1967 Finite element analysis of reinforced concrete beams. *Journal of the American Concrete Institute* **64**, 152–163.

Oliver J 1989 A consistent characteristic length for smeared cracking models. *International Journal for Numerical Methods in Engineering* **28**, 461–474.

Oliver J 1996 Modelling strong discontinuities in solid mechanics via strain softening constitutive relations. Part i: Fundamentals. Part ii: Numerical simulation. *International Journal for Numerical Methods in Engineering* **39**, 3575–2724.

Ortiz M, Leroy Y and Needleman A 1987 A finite element method for localized failure analysis. *Computer Methods in Applied Mechanics and Engineering* **61**, 189–214.

Pamin J, Askes H and de Borst R 2003 Two gradient plasticity theories discretized with the element-free Galerkin method. *Computer Methods in Applied Mechanics and Engineering* **192**, 2377–2407.

Peerlings RHJ, de Borst R, Brekelmans WAM and de Vree HPJ 1996 Gradient-enhanced damage for quasi-brittle materials. *International Journal for Numerical Methods in Engineering* **39**, 3391–3403.

Peerlings RHJ, de Borst R, Brekelmans WAM and Geers MGD 1998 Gradient-enhanced modelling of concrete fracture. *Mechanics of Cohesive-frictional Materials* **3**, 323–342.

Peerlings RHJ, Geers MGD, de Borst R and Brekelmans WAM 2001 A critical comparison of nonlocal and gradient-enhanced softening continua. *International Journal of Solids and Structures* **38**, 7723–7746.

Pietruszczak S and Mróz Z 1981 Finite element analysis of deformation of strain softening materials. *International Journal for Numerical Methods in Engineering* **17**, 337–334.

Pijaudier-Cabot G and Bažant ZP 1987 Nonlocal damage theory. *ASCE Journal of Engineering Mechanics* **113**, 1512–1533.

Pijaudier-Cabot G and Huerta A 1991 Finite element analysis of bifurcation in nonlocal strain softening solids. *Computer Methods in Applied Mechanics and Engineering* **90**, 905–919.

Rashid YR 1968 Analysis of reinforced concrete pressure vessels. *Nuclear Engineering and Design* **7**, 334–344.

Reinhardt HW and Cornelissen HAW 1984 Post-peak cyclic behaviour of concrete in uniaxial and alternating tensile and compressive loading. *Cement and Concrete Research* **14**, 263–270.

Rots JG 1991 Smeared and discrete representations of localized fracture. *International Journal of Fracture* **51**, 45–59.

Rudnicki JW and Rice JR 1974 Conditions for the localization of deformation in pressure sensitive dilatant materials. *Journal of the Mechanics and Physics of Solids* **23**, 371–394.

Simo JC and Ju JW 1987 Strain- and stress-based continuum damage formulations. Part I: Formulation. Part II: Computational aspects. *International Journal of Solids and Structures* **23**, 821–869.

Simo JC and Rifai MS 1990 A class of mixed assumed strain methods and the method of incompatible modes. *International Journal for Numerical Methods in Engineering* **29**, 1595–1638.

Simo JC, Oliver J and Armero F 1993 An analysis of strong discontinuities induced by softening relations in rate-independent solids. *Computational Mechanics* **12**, 277–296.

Sluys LJ and Berends AH 1998 Discontinuous failure analysis for mode 1 and mode 2 localization problems. *International Journal of Solids and Structures* **35**, 4257–4274.

Sluys LJ and de Borst R 1992 Wave propagation and localisation in a rate-dependent cracked medium – model formulation and one-dimensional examples. *International Journal of Solids and Structures* **29**, 2945–2958.

Suidan M and Schnobrich WC 1973 Finite element analysis of reinforced concrete. *ASCE Journal of the Structural Division* **99**, 2109–2122.

Tvergaard V and Hutchinson JW 1992 The influence of plasticity on mixed mode interface toughness. *Journal of the Mechanics and Physics of Solids* **40**, 1377–1397.

Tvergaard V and Hutchinson JW 1993 The relation between crack growth resistance and fracture process parameters in elastic-plastic solids. *Journal of the Mechanics and Physics of Solids* **41**, 1119–1135.

Vardoulakis I and Sulem J 1995 *Bifurcation Analysis in Geomechanics*. Blackie.

Wells GN and Sluys LJ 2001a Analysis of slip planes in three-dimensional solids. *Computer Methods in Applied Mechanics and Engineering* **190**, 3591–3606.

Wells GN and Sluys LJ 2001b Three-dimensional embedded discontinuity model for brittle fracture. *International Journal of Solids and Structures* **38**, 897–913.

Zienkiewicz OC and Pande GN 1977 Time-dependent multi-laminate model for rocks – a numerical study of deformation and failure of rock masses. *International Journal for Numerical and Analytical Methods in Geomechanics* **1**, 219–247.

7

Plasticity

One of the most well-developed theories for describing material non-linearity is the theory of plasticity. Its development goes back to Coulomb who postulated the dependence of the sliding resistance on a plane between two bodies to be a function of the adhesion and the frictional properties (Coulomb 1776). Significant developments took place just before and after the Second World War, including the establishment of upper and lower bound theorems and the shakedown theorems. For classical treatises the reader is referred to Hill (1950) or Koiter (1960). More modern treatises are those by Lubliner (1990) and Khan and Huang (1995), while comprehensive treatments of computational aspects have been presented in Simo and Hughes (1998) and in de Souza Neto *et al.* (2008). In this chapter we shall first review some basic notions of standard elasto-plastic theories. Subsequently, we shall focus on the computational setting of continuum plasticity.

7.1 A Simple Slip Model

The simplest 'plasticity' model is probably the spring–sliding system of Figure 7.1. In this formulation the entire horizontal displacement of point A is initially caused by the deformation (elongation) of the spring, since, for low force levels, the adhesion and the friction between the block and the floor prevent any sliding of the block. Only when the maximum shear force that can be exerted by adhesion and friction is exhausted will the block start sliding. From that moment onwards the total horizontal displacement of point A is composed of a contribution of the spring and a contribution of the sliding between block and floor. If u represents the horizontal displacement of point A and if the superscripts e and p are used to denote the contributions due to the deformation in the spring and the sliding of the block, respectively, the total displacement after the onset of sliding is given by the following additive decomposition:

$$u = u^{\mathrm{e}} + u^{\mathrm{p}} \tag{7.1}$$

The first component is called elastic because, upon removal of the force, the deformation in the spring also disappears. The ensuing displacement of point A is recoverable. However, the

Non-linear Finite Element Analysis of Solids and Structures, Second Edition.
René de Borst, Mike A. Crisfield, Joris J.C. Remmers and Clemens V. Verhoosel.

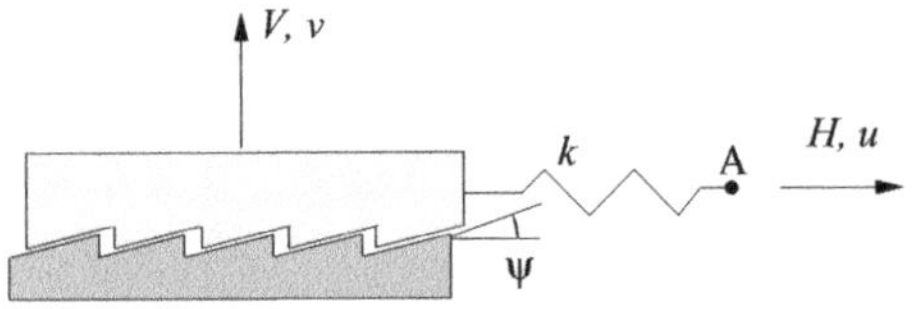

Figure 7.1 Simple spring–sliding system with two degrees of freedom

displacements of the block do not disappear. Any sliding that has occurred is permanent: such deformations cannot be recovered by unloading the system and are named inelastic or plastic.

If the surface between the floor and the sliding block is not perfectly smooth, but microscopically rough, any horizontal sliding will generally also entail a vertical displacement of the block, either an uplift or a downward movement. Let this plastic component be denoted by v^{p}, then both plastic displacement components can be assembled in a vector $\mathbf{u}^{\mathrm{p}}$:

$$\mathbf{u}^{\mathrm{p}} = \begin{pmatrix} u^{\mathrm{p}} \\ v^{\mathrm{p}} \end{pmatrix} \tag{7.2}$$

In a similar fashion the elastic displacements can be assembled in a vector

$$\mathbf{u}^{\mathrm{e}} = \begin{pmatrix} u^{\mathrm{e}} \\ v^{\mathrm{e}} \end{pmatrix} \tag{7.3}$$

where, for the present example, v^{e} cannot be given a physical meaning and is zero. Extension of Equation (7.1) to incorporate the vertical displacements yields

$$\mathbf{u} = \mathbf{u}^{\mathrm{e}} + \mathbf{u}^{\mathrm{p}} \tag{7.4}$$

Another important distinction between the deformations attributable to the spring and those that take place in the sliding block–floor system is the uniqueness between stresses and strains, or in terms of the present discrete mechanical system, between forces and displacements. Between the elastic displacement u^{e} and the horizontal force H a unique relation

$$H = ku^{\mathrm{e}} \tag{7.5}$$

with k the spring constant, can be established. Clearly, if the force H vanishes after reaching a non-zero value, the 'elastic' displacement also reduces to zero. For the inelastic displacement u^{p} it is physically not plausible that such a one-to-one relation can be established, in which the inelastic strain is directly related to the instantaneous value of the stress, here the force (note that in the present section the terms 'forces' and 'stresses', but also 'displacements' and 'strains', are used interchangably). All we can say is that during plastic deformation the rate of the inelastic deformation can be determined. For the moment we assume that the ratio between the horizontal 'plastic' velocity $\dot{u}^{\mathrm{p}}$ and the vertical 'plastic' velocity $\dot{v}^{\mathrm{p}}$ can be obtained from measurements and is governed by a dilatancy angle ψ (Figure 7.1):

$$\tan\psi = \frac{\dot{v}^{\mathrm{p}}}{\dot{u}^{\mathrm{p}}} \tag{7.6}$$

which may be a function of the total accumulated 'plastic' displacements, but for the present, such a dependence will not be considered. With this proposition the direction of the plastic flow is fully determined, since we may write

$$\dot{\mathbf{u}}^{\mathrm{p}} = \dot{\lambda}\mathbf{m} \tag{7.7}$$

with

$$\mathbf{m} = \begin{pmatrix} 1 \\ \tan\psi \end{pmatrix} \tag{7.8}$$

governing the relative magnitudes of the plastic flow components, often named the direction of the plastic flow. The value of the plastic multiplier $\dot{\lambda}$ can be determined from the requirement that during plastic flow the stresses remain bounded. To date, this flow theory of plasticity is the most widely used plasticity theory. It is physically appealing and the transition between elastic and plastic states can also be defined in a straightforward fashion for multi-dimensional stress states. This is not so for the deformation theory of plasticity which is rooted in the assumption that the plastic strain, rather than the plastic strain *rate*, is determined by the instantaneous values of the stresses.

If we define the force vector $\mathbf{f}$ in a similar fashion as the displacement vector, i.e.

$$\mathbf{f} = \begin{pmatrix} H \\ V \end{pmatrix} \tag{7.9}$$

the 'elastic' displacement vector can be related to the force vector symbolically by the matrix-vector relation

$$\mathbf{f} = \mathbf{D}^{\mathrm{e}}\mathbf{u}^{\mathrm{e}} \tag{7.10}$$

where $\mathbf{D}^{\mathrm{e}}$ signifies the elastic stiffness matrix:

$$\mathbf{D}^{\mathrm{e}} = \begin{pmatrix} k & 0 \\ 0 & 0 \end{pmatrix} \tag{7.11}$$

We next differentiate the fundamental decomposition (7.4), so that

$$\dot{\mathbf{u}} = \dot{\mathbf{u}}^{\mathrm{e}} + \dot{\mathbf{u}}^{\mathrm{p}} \tag{7.12}$$

and the elastic relation for the spring (7.10), $\dot{\mathbf{f}} = \mathbf{D}^{\mathrm{e}}\dot{\mathbf{u}}^{\mathrm{e}}$, and combine these results with the relation that sets the direction of the 'plastic' velocity (7.7) for the sliding block. Then, the following relation ensues:

$$\dot{\mathbf{f}} = \mathbf{D}^{\mathrm{e}}(\dot{\mathbf{u}} - \dot{\lambda}\mathbf{m}) \tag{7.13}$$

So far, we have loosely used the terminology 'maximum shear force exhausted' to describe the onset of permanent displacements. Here again, we must have a criterion that sets the borderline between purely 'elastic' displacements and the moment when the block starts sliding, i.e. when 'plastic' displacements occur. For our present system we only have two force components, namely, H and V. The simplest assumption is that sliding starts when the Coulomb friction augmented with some adhesion is fully mobilised, i.e.

$$H + V \tan\varphi - c = 0 \tag{7.14}$$

with $\tan\varphi$ a friction coefficient and c the adhesion. To mobilise friction between the block and the surface the force V must act downwards, and therefore $V < 0$. The second term in Equation (7.14) thus gives a negative contribution. If

$$H + V \tan\varphi - c < 0 \tag{7.15}$$

only elastic deformations (within the spring) will take place. A combination of forces such that

$$H + V \tan\varphi - c > 0 \tag{7.16}$$

is physically impossible, since the maximum horizontal force is bounded by the restriction (7.14). We now assume that φ and c are constants, an assumption that will be dropped later. Use of this assumption leads to

$$\dot{H} + \dot{V} \tan\varphi = 0 \tag{7.17}$$

after differentiation of Equation (7.14). Upon introduction of the vector

$$\mathbf{n} = \begin{pmatrix} 1 \\ \tan\varphi \end{pmatrix} \tag{7.18}$$

Equation (7.17) can be written symbolically as:

$$\mathbf{n}^{\mathrm{T}}\dot{\mathbf{f}} = 0 \tag{7.19}$$

Premultiplying Equation (7.13) with $\mathbf{n}^{\mathrm{T}}$ and utilising the fact that during plastic flow Equation (7.19) must hold, the following explicit expression is obtained for the plastic multiplier $\dot{\lambda}$:

$$\dot{\lambda} = \frac{\mathbf{n}^{\mathrm{T}}\mathbf{D}^{\mathrm{e}}\dot{\mathbf{u}}}{\mathbf{n}^{\mathrm{T}}\mathbf{D}^{\mathrm{e}}\mathbf{m}} \tag{7.20}$$

which can be inserted in Equation (7.13) to yield an explicit relation between velocity $\dot{\mathbf{u}}$ and the rate of the force vector $\dot{\mathbf{f}}$:

$$\dot{\mathbf{f}} = \left(\mathbf{D}^{\mathrm{e}} - \frac{\mathbf{D}^{\mathrm{e}}\mathbf{m}\mathbf{n}^{\mathrm{T}}\mathbf{D}^{\mathrm{e}}}{\mathbf{n}^{\mathrm{T}}\mathbf{D}^{\mathrm{e}}\mathbf{m}}\right)\dot{\mathbf{u}} \tag{7.21}$$

The rate equations (7.21) are in general non-symmetric. This is because in general $\varphi \neq \psi$, so that $\mathbf{n} \neq \mathbf{m}$. In that case the matrix formed by the outer product $\mathbf{m}\mathbf{n}^{\mathrm{T}}$ will be non-symmetric, thus rendering the tangential stiffness matrix that sets the incremental relation between $\dot{\mathbf{f}}$ and $\dot{\mathbf{u}}$ also non-symmetric. It is noted that the present case is perhaps less illustrative, since elaboration shows that the off-diagonal terms are zero, but it can be seen as a precursor to continuum plasticity in the next section, where the off-diagonal terms do not vanish.

By its very nature the flow theory of plasticity does not provide a direct relation between the force (stress) and the displacement (strain). Only a relation between the rate of force (stress rate) and the velocity (strain rate) can be established and even this relation is not

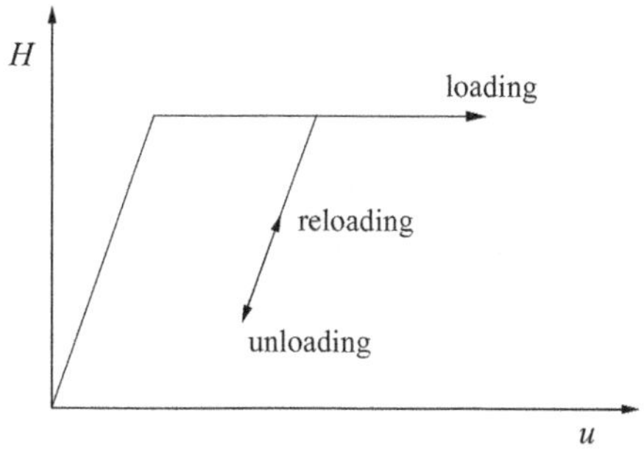

Figure 7.2 Loading, unloading and reloading for the simple slip model of Figure 7.1. The theory of plasticity for continua follows the same principle

necessarily unique, although for most practical applications it is. It is of utmost importance that the incremental equations are integrated accurately, in particular if multi-dimensional stress states are considered. Implicit integration schemes are best suited to carry out this task, as will be discussed later.

A salient feature of this simple slip model, and also of the theory of plasticity for continuous media, is that when the condition for continued sliding as represented in Equation (7.14) is no longer satisfied, in other words the strict inequality (7.15) is again valid, we again have purely elastic behaviour. So, unloading is a purely elastic process. The same statement holds true for reloading, of course subject to inequality (7.15). If Equation (7.14) is again satisfied, permanent contributions to the displacement increment again occur. A graphical representation of this behaviour is given in Figure 7.2.

7.2 Flow Theory of Plasticity

7.2.1 Yield Function

In the preceding section a simple model has been constructed for frictional/adhesive sliding along a fixed plane. This concepts lends itself well for extension to continua in which we deal with stresses rather than with forces. If we extend Coulomb's assumption, i.e. sliding occurs when the shear force on a plane exceeds the normal force multiplied by some friction coefficient plus some adhesion, to continua, we must search for the plane on which the combination of normal stress σ and shear stress τ is critical in the sense that the condition

$$\tau + \sigma \tan \varphi - c = 0 \tag{7.22}$$

is satisfied for the normal stress and the shear stress on that plane. φ now signifies the internal friction angle of the material. For a material such as sand the physical meaning is obvious, since it is related to the friction between the particles. For a continuous medium c represents the cohesion of the material. For metals it is equal to half of the yield strength as we will derive below, while for dry sand the cohesion is almost zero.

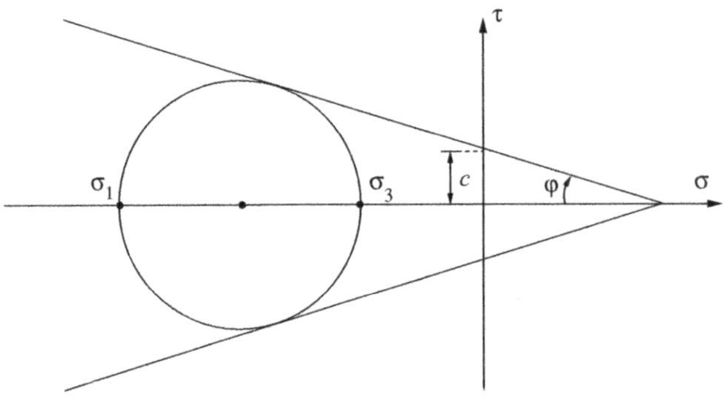

Figure 7.3 Mohr's stress circle and the envelopes that bound all possible stress states for the Mohr–Coulomb yield criterion

Considering a two-dimensional stress state first, we can relate σ and τ to the principal stresses with the aid of Mohr's circle (Figure 7.3):

$$\sigma = \frac{1}{2}(\sigma_3 + \sigma_1) + \frac{1}{2}(\sigma_3 - \sigma_1)\sin\varphi$$

$$\tau = \frac{1}{2}(\sigma_3 - \sigma_1)\cos\varphi$$

with σ_1 and σ_3 the smallest and the largest principal stress, respectively ($\sigma_1 \leq \sigma_3$). Inserting the expressions for σ and τ into Equation (7.22), and multiplying by $\cos\varphi$ gives after rearranging:

$$\frac{1}{2}(\sigma_3 - \sigma_1) + \frac{1}{2}(\sigma_3 + \sigma_1)\sin\varphi - c\cos\varphi = 0 \tag{7.23}$$

Equation (7.23) is the two-dimensional representation of the Mohr–Coulomb yield criterion (Mohr 1900). It is described by two material parameters, namely the angle of internal friction φ and the cohesion c. The model is well suited to describe the strength characteristics of soils, rocks, concrete, and ceramics.

We next extend the Mohr–Coulomb yield condition to fully three-dimensional stress states. Working again in the principal stress space, we observe that Equation (7.23) is valid as long as $\sigma_1 \leq \sigma_2 \leq \sigma_3$. If this is not the case, for instance if $\sigma_2 \leq \sigma_3 \leq \sigma_1$, the shear stress on the plane where the combination of shear stress and normal stress becomes critical is given by $\tau = \frac{1}{2}(\sigma_1 - \sigma_2)\cos\varphi$, while the corresponding normal stress is given by: $\sigma = \frac{1}{2}(\sigma_1 + \sigma_2) + \frac{1}{2}(\sigma_1 - \sigma_2)\sin\varphi$. The yield condition is now formulated in terms of σ_1 and σ_2 instead of σ_3 and σ_1, and reads:

$$\frac{1}{2}(\sigma_1 - \sigma_2) + \frac{1}{2}(\sigma_1 + \sigma_2)\sin\varphi - c\cos\varphi = 0$$

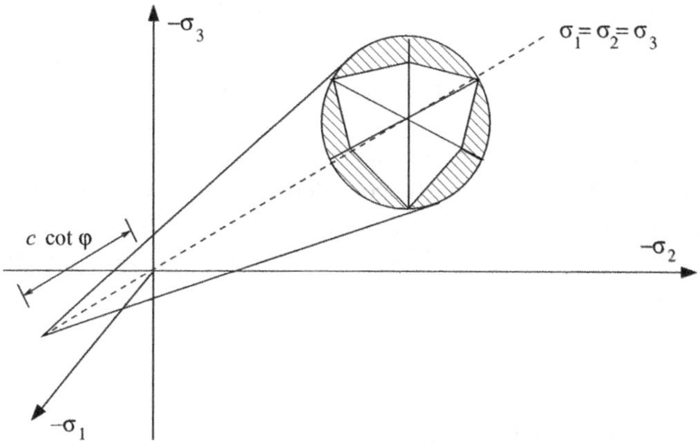

Figure 7.4 Representation of Mohr–Coulomb and Drucker–Prager yield criteria in the three-dimensional principal stress space

instead of by Equation (7.23). By cyclic permutation we successively obtain that the Mohr–Coulomb yield criterion in the three-dimensional principal stress space is complemented by the four conditions:

$$\frac{1}{2}(\sigma_2 - \sigma_3) + \frac{1}{2}(\sigma_2 + \sigma_3)\sin\varphi - c\cos\varphi = 0$$
$$\frac{1}{2}(\sigma_1 - \sigma_3) + \frac{1}{2}(\sigma_1 + \sigma_3)\sin\varphi - c\cos\varphi = 0$$
$$\frac{1}{2}(\sigma_2 - \sigma_1) + \frac{1}{2}(\sigma_2 + \sigma_1)\sin\varphi - c\cos\varphi = 0$$
$$\frac{1}{2}(\sigma_3 - \sigma_2) + \frac{1}{2}(\sigma_3 + \sigma_2)\sin\varphi - c\cos\varphi = 0$$

When represented graphically in the three-dimensional principal stress space, each of the permutations of Equation (7.23) represents a plane. Together they form a cone with six facets which meet in the apex of the yield surface at $\sigma_1 = \sigma_2 = \sigma_3 = c\cot\varphi$ and which opens up in the negative direction of the space diagonal $\sigma_1 = \sigma_2 = \sigma_3$ (Figure 7.4). The six permutations of Equation (7.23) bound all possible stress combinations. In other words, this six-faceted surface can be considered as a limit surface, which acts as an envelope of all possible stress states. Stress states inside this contour cause elastic deformations, while stress states on this yield surface can give rise to elasto-plastic deformations. By definition, stress states outside this yield contour are not possible.

We now introduce the more abstract notion of a yield function. In analogy with Equation (7.23) such a loading function bounds all possible stress states:

$$f(\boldsymbol{\sigma}) \leq 0 \tag{7.24}$$

where the strict equality sign holds for stress states on the yield contour and where the inequality sign is valid whenever stresses are inside the yield contour and cause only elastic deformations.

Considering the Mohr–Coulomb yield function, f attains the form:

$$f(\boldsymbol{\sigma}) = \frac{1}{2}(\sigma_3 - \sigma_1) + \frac{1}{2}(\sigma_3 + \sigma_1)\sin\varphi - c\cos\varphi \tag{7.25}$$

while the other five yield functions are obtained by cyclic permutation.

Many well-known yield criteria can be considered as approximations or refinements of the Mohr–Coulomb yield criterion. A straightforward example is the Tresca yield criterion. Tresca postulated that metals start to yield when in some direction the shear strength has been exhausted (Tresca 1868):

$$\frac{1}{2}(\sigma_3 - \sigma_1) = \tau_{\max} \tag{7.26}$$

From Equation (7.23) it is obvious that this is a special case of the Mohr–Coulomb yield function in which the angle of internal friction is zero. Comparison with Equation (7.26) shows that the Tresca yield function reads:

$$f(\boldsymbol{\sigma}) = (\sigma_3 - \sigma_1) - \bar{\sigma} \tag{7.27}$$

with the yield strength in uniaxial tension, $\bar{\sigma}$, equal to twice the cohesion: $\bar{\sigma} = 2c$. Evidently, the full expression for three-dimensional stress states is given by six yield functions that can be obtained from Equation (7.27) by cyclic permutation. The result is a six-faceted cylinder in the principal stress space, which extends infinitely along the axis $\sigma_1 = \sigma_2 = \sigma_3$ in both the positive and negative directions. Such a yield function, which in shape as well as in size has the same cross section in each plane that is orthogonal to the space diagonal $\sigma_1 = \sigma_2 = \sigma_3$, is called pressure-insensitive. Metals satisfy this property as long as the hydrostatic stress level does not become too high. Yield functions are often represented in the π-plane, which is perpendicular to the space diagonal $\sigma_1 = \sigma_2 = \sigma_3$ and passes through the origin (Figure 7.5). For pressure-insensitive yield functions, such a representation is complete.

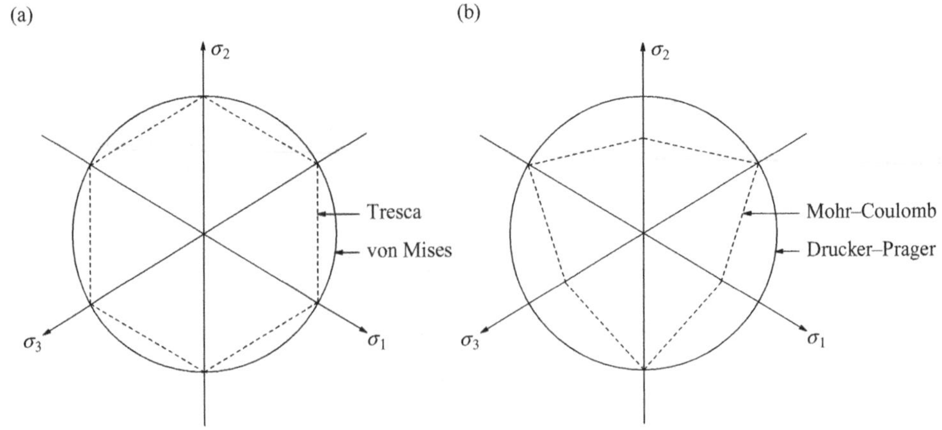

Figure 7.5 Representation of yield surfaces in the π-plane: (a) Tresca and von Mises; (b) Mohr–Coulomb and Drucker–Prager

An inconvenient property of the Tresca and Mohr–Coulomb yield surfaces is that the facets that compose the entire yield contour form corners at the places where they meet. At such places the gradient to the yield surface is no longer uniquely defined which, as we will see, gives rise to difficulties when formulating the incremental stress–strain relations. Therefore, smooth approximations have been proposed to these angular yield surfaces. The smooth approximation to the Tresca yield contour is the von Mises yield criterion, which is a circular cylinder in the principal stress space, and a circle in the π-plane [Figure 7.5(a)] (von Mises 1913). Expressed in principal stresses, the von Mises yield function reads:

$$f(\boldsymbol{\sigma}) = \sqrt{\frac{1}{2}\left[(\sigma_1 - \sigma_2)^2 + (\sigma_2 - \sigma_3)^2 + (\sigma_3 - \sigma_1)^2\right]} - \bar{\sigma} \tag{7.28}$$

where the factor $\frac{1}{2}$ has been chosen in order that in pure uniaxial stressing the yield function reduces to the uniaxial yield strength $\bar{\sigma}$. In terms of normal and shear stresses the yield function can be rewritten as:

$$f(\boldsymbol{\sigma}) = \sqrt{\frac{1}{2}\left[(\sigma_{xx} - \sigma_{yy})^2 + (\sigma_{yy} - \sigma_{zz})^2 + (\sigma_{zz} - \sigma_{xx})^2\right] + 3\sigma_{xy}^2 + 3\sigma_{xy}^2 + 3\sigma_{xy}^2} - \bar{\sigma} \tag{7.29}$$

The expression under the square root is proportional to the second invariant of the deviatoric stresses, Equation (1.88). Introducing the modified stress invariant

$$q = \sqrt{3J_2} \tag{7.30}$$

we can also write:

$$f(\boldsymbol{\sigma}) = q - \bar{\sigma} \tag{7.31}$$

The theory of plasticity that utilises von Mises' yield contour is often called J_2-plasticity. For later use, we also express the von Mises yield function using Voigt notation:

$$f(\boldsymbol{\sigma}) = \sqrt{\frac{3}{2}\boldsymbol{\sigma}^{\mathrm{T}}\mathbf{P}\boldsymbol{\sigma}} - \bar{\sigma} \tag{7.32}$$

with the projection matrix $\mathbf{P}$ defined as:

$$\mathbf{P} = \begin{bmatrix} \frac{2}{3} & -\frac{1}{3} & -\frac{1}{3} & 0 & 0 & 0 \\ -\frac{1}{3} & \frac{2}{3} & -\frac{1}{3} & 0 & 0 & 0 \\ -\frac{1}{3} & -\frac{1}{3} & \frac{2}{3} & 0 & 0 & 0 \\ 0 & 0 & 0 & 2 & 0 & 0 \\ 0 & 0 & 0 & 0 & 2 & 0 \\ 0 & 0 & 0 & 0 & 0 & 2 \end{bmatrix} \tag{7.33}$$

In a similar fashion as von Mises has introduced a yield contour as an approximation to the non-smooth Tresca yield contour, Drucker and Prager (1952) have approximated the Mohr–Coulomb yield contour by a circular cone. While it maintains the linear dependence on the hydrostatic stress level, just as the Mohr–Coulomb contour, the angular shape in the π-plane

– dashed lines in Figure 7.5(b) – is replaced by a circle. The Drucker–Prager yield criterion is defined by:

$$f(\sigma) = q + \alpha p - k \tag{7.34}$$

with the pressure p defined in Equation (1.79), and α and k material constants. Because of the pressure dependence the circle cannot pass through all corners. Often the approximation is chosen that passes through the three outermost corners, and the constants α and k are then related to c and φ by:

$$\alpha = \frac{6 \sin \varphi}{3 - \sin \varphi} \quad \text{and} \quad k = \frac{6c \cos \varphi}{3 - \sin \varphi}$$

Using the projection matrix $\mathbf{P}$, Equation (7.33), and rewriting the pressure as

$$p = \boldsymbol{\pi}^{\mathrm{T}} \boldsymbol{\sigma} \tag{7.35}$$

with the projection vector

$$\boldsymbol{\pi}^{\mathrm{T}} = \left[\frac{1}{3}, \frac{1}{3}, \frac{1}{3}, 0, 0, 0 \right] \tag{7.36}$$

the Drucker–Prager yield function can be rewritten as:

$$f(\boldsymbol{\sigma}) = \sqrt{\frac{3}{2} \boldsymbol{\sigma}^{\mathrm{T}} \mathbf{P} \boldsymbol{\sigma}} + \alpha \boldsymbol{\pi}^{\mathrm{T}} \boldsymbol{\sigma} - k \tag{7.37}$$

7.2.2 Flow Rule

To obtain plastic deformations, the stress point must not only be on the yield contour, it must also remain there for a 'short period'. When the stress point only touches the yield contour and immediately moves inward again, plastic flow will not occur. Plastic straining will take place if and only if the yield function f vanishes:

$$f = 0 \tag{7.38}$$

as well as its 'time derivative':

$$\dot{f} = 0 \tag{7.39}$$

The latter equation is called Prager's consistency condition and expresses that the yield function f must remain zero for at least a small 'time increment' in order that plastic flow can occur.

Within the elastic domain the injective relation $\boldsymbol{\sigma} = \mathbf{D}^{\mathrm{e}} : \boldsymbol{\epsilon}$, with $\mathbf{D}^{\mathrm{e}}$ the continuum elastic stiffness tensor, sets the dependence of the stress $\boldsymbol{\sigma}$ on the strain $\boldsymbol{\epsilon}$. However, such an injective relationship can only be established between the stress $\boldsymbol{\sigma}$ and the elastic strain $\boldsymbol{\epsilon}^{\mathrm{e}}$,

$$\boldsymbol{\sigma} = \mathbf{D}^{\mathrm{e}} : \boldsymbol{\epsilon}^{\mathrm{e}} \tag{7.40}$$

when plastic straining occurs, i.e. when Equations (7.38) and (7.39) hold. The remaining part of the strain is permanent, or plastic, and together with the elastic contribution $\boldsymbol{\epsilon}^{\mathrm{e}}$ forms the total strain $\boldsymbol{\epsilon}$:

$$\boldsymbol{\epsilon} = \boldsymbol{\epsilon}^{e} + \boldsymbol{\epsilon}^{p} \tag{7.41}$$

Using Equation (7.40) this results in:

$$\boldsymbol{\sigma} = \mathbf{D}^{e} : (\boldsymbol{\epsilon} - \boldsymbol{\epsilon}^{p}) \tag{7.42}$$

In a three-dimensional stress space Equation (7.42) constitutes a set of six equations with twelve unknowns, namely the six components of $\boldsymbol{\sigma}$ and those of $\boldsymbol{\epsilon}^{p}$. Note that $\boldsymbol{\epsilon}$ is known, as it can be computed directly from the (known) displacement field. From the missing six equations five must be supplied by measurements, while one equation follows from the consistency requirement, Equation (7.39). To separate the equations that must be supplied from experimental data from the equation that follows from the mathematical structure of the theory of plasticity, the plastic strain rate is written as the product of a scalar $\dot{\lambda}$ and an array $\mathbf{m}$:

$$\dot{\boldsymbol{\epsilon}}^{p} = \dot{\lambda}\mathbf{m} \tag{7.43}$$

cf. Equation (7.7). In Equation (7.43), $\dot{\lambda}$ determines the magnitude of the plastic flow, while $\mathbf{m}$ sets the relative magnitude of the components of the plastic flow, often named the direction. Sometimes $\mathbf{m}$ is normalised, such that $\|\mathbf{m}\|^2 = 1$, but this is not necessary and will not be done here. Since the yield function f has hitherto been assumed to be solely a function of the stress tensor, $f = f(\boldsymbol{\sigma})$, the consistency condition (7.39) can be elaborated as

$$\mathbf{n} : \dot{\boldsymbol{\sigma}} = 0 \tag{7.44}$$

with $\mathbf{n}$ the gradient of the yield function, that is perpendicular to the yield surface at the current stress point $\boldsymbol{\sigma}$:

$$\mathbf{n} = \frac{\partial f}{\partial \boldsymbol{\sigma}} \tag{7.45}$$

Differentiation of Equation (7.42) with respect to a virtual time and combination of the result with Equations (7.43) and (7.44) yields an explicit expression for the magnitude of the plastic flow:

$$\dot{\lambda} = \frac{\mathbf{n} : \mathbf{D}^{e} : \dot{\boldsymbol{\epsilon}}}{\mathbf{n} : \mathbf{D}^{e} : \mathbf{m}} \tag{7.46}$$

Similar to the derivation of the simple slip model a linear relation can now be deduced between the stress rate $\dot{\boldsymbol{\sigma}}$ and the strain rate $\dot{\boldsymbol{\epsilon}}$:

$$\dot{\boldsymbol{\sigma}} = \left(\mathbf{D}^{e} - \frac{(\mathbf{D}^{e} : \mathbf{m}) \otimes (\mathbf{D}^{e} : \mathbf{n})}{\mathbf{n} : \mathbf{D}^{e} : \mathbf{m}} \right) : \dot{\boldsymbol{\epsilon}} \tag{7.47}$$

For a number of materials, the assumption that the plastic flow direction $\mathbf{m}$ is co-linear with the gradient to the yield surface $\mathbf{n}$, is reasonable, and is corroborated by experimental evidence, in particular for metals. Then,

$$\mathbf{m} = \gamma \mathbf{n}$$

with γ an undetermined scalar quantity, and the plastic strain rate can be computed using the associated flow rule of plasticity:

$$\dot{\boldsymbol{\epsilon}}^{\mathrm{p}} = \dot{\lambda}\mathbf{n} \tag{7.48}$$

or, in view of Equation (7.45),

$$\dot{\boldsymbol{\epsilon}}^{\mathrm{p}} = \dot{\lambda}\frac{\partial f}{\partial \boldsymbol{\sigma}} \tag{7.49}$$

The plastic flow direction is thus normal to the yield surface, and the associated flow rule is often referred to as 'normality rule', although the terminology orthogonality rule would be more correct. A major (computational) advantage of the use of an associated flow rule is that the tangential stress–strain relation (7.47), which is non-symmetric for the general case, becomes symmetric:

$$\dot{\boldsymbol{\sigma}} = \left(\mathbf{D}^{\mathrm{e}} - \frac{(\mathbf{D}^{\mathrm{e}} : \mathbf{n}) \otimes \mathbf{D}^{\mathrm{e}} : \mathbf{n})}{\mathbf{n} : \mathbf{D}^{\mathrm{e}} : \mathbf{n}}\right) : \dot{\boldsymbol{\epsilon}} \tag{7.50}$$

An elegant way for deriving the normality rule is to adopt Drucker's Postulate, see Box 7.1. However, Drucker's Postulate is not a law of mechanics, or of thermodynamics, as is sometimes suggested. It is an assumption. Whether this assumption holds, can only be assessed by comparison with experimental data. For metals the agreement is very reasonable, which makes Drucker's Postulate and the normality rule applicable to such materials. For soils, rocks and concrete, the agreement is less good (Vermeer and de Borst 1984).

Flow rules that do not obey identity (7.49) are called non-associated. An important subclass of non-associated flow rules, which covers by far the majority of the applications, are those for which a function of the stresses g, often named the plastic potential function, exists, such that

$$\mathbf{m} = \frac{\partial g}{\partial \boldsymbol{\sigma}} \tag{7.51}$$

According to this identity the plastic flow vector is now coaxial with the gradient to the surface $g = 0$:

$$\dot{\boldsymbol{\epsilon}}^{\mathrm{p}} = \dot{\lambda}\frac{\partial g}{\partial \boldsymbol{\sigma}} \tag{7.52}$$

We now investigate the consequences of the normality rule and take the Mohr–Coulomb criterion as an example. From definition (7.25) and invoking the concept of an associated flow rule we can derive that the plastic volumetric strain rate

$$\dot{\epsilon}^{\mathrm{p}}_{\mathrm{vol}} = \dot{\epsilon}^{\mathrm{p}}_1 + \dot{\epsilon}^{\mathrm{p}}_2 + \dot{\epsilon}^{\mathrm{p}}_3$$

and the rate of plastic shear deformation

$$\dot{\gamma}^{\mathrm{p}} = \dot{\epsilon}^{\mathrm{p}}_3 - \dot{\epsilon}^{\mathrm{p}}_1$$

Box 7.1 Drucker's Postulate and the direction of plastic flow

Instead of determining the plastic flow direction $\mathbf{m}$ experimentally, one can make an assumption, which is then verified for its correctness. A widely used assumption, especially for the plastic behaviour of metals, is to postulate:

During a complete cycle of loading and unloading of an additional stress increment the work that is exerted by external forces is non-negative.

This postulate is usually referred to as Drucker's Postulate. Mathematically, it states that:

$$\int (\boldsymbol{\sigma} - \boldsymbol{\sigma}_0) : \dot{\boldsymbol{\epsilon}} \mathrm{d}t \geq 0 \quad , \quad f(\boldsymbol{\sigma}_0) \leq 0$$

By definition, any cycle in which only elastic deformations occur, results in a zero net work dissipation, so that the preceding expression is equivalent to:

$$\int (\boldsymbol{\sigma} - \boldsymbol{\sigma}_0) : \dot{\boldsymbol{\epsilon}}^{\mathrm{p}} \mathrm{d}t \geq 0 \quad , \quad f(\boldsymbol{\sigma}_0) \leq 0$$

This inequality will certainly be satisfied if the integrand is non-negative. If the current stress is written as $\boldsymbol{\sigma} = \boldsymbol{\sigma}_0 + \dot{\boldsymbol{\sigma}} \mathrm{d}t$, we obtain:

$$\dot{\boldsymbol{\sigma}} : \dot{\boldsymbol{\epsilon}}^{\mathrm{p}} \geq 0$$

In the case of ideal plasticity as considered here, i.e. the yield function is only a function of the current stress state and not of the loading history, $\boldsymbol{\sigma}$ must be on or inside the original yield surface: $f(\boldsymbol{\sigma}) = f(\boldsymbol{\sigma}_0 + \dot{\boldsymbol{\sigma}} \mathrm{d}t) \leq 0$. Accordingly, the stress states $\boldsymbol{\sigma}$ and $\boldsymbol{\sigma}_0$ can be interchanged, which, upon substitution into Drucker's Postulate, yields $\dot{\boldsymbol{\sigma}} : \dot{\boldsymbol{\epsilon}}^{\mathrm{p}} \leq 0$, so that in the case of ideal plasticity the following identity must hold rigorously:

$$\dot{\boldsymbol{\sigma}} : \dot{\boldsymbol{\epsilon}}^{\mathrm{p}} = 0$$

Using Equation (7.43) and considering that $\dot{\lambda}$ is non-negative the latter identity directly leads to:

$$\mathbf{m} : \dot{\boldsymbol{\sigma}} = 0$$

Comparing this result with the consistency condition (7.44) for ideally plastic materials shows that $\mathbf{m}$ must be co-linear with the gradient to the yield surface $\mathbf{n}$, and that the associated flow rule (7.49) holds.

are related by:

$$\dot{\epsilon}^{\mathrm{p}}_{\mathrm{vol}} = \dot{\gamma}^{\mathrm{p}} \sin \varphi \tag{7.53}$$

Experimental evidence shows that Equation (7.53) significantly overpredicts the amount of plastic volume change, also called plastic dilatancy. Introduction of the Mohr–Coulomb like plastic potential function

$$g = \frac{1}{2}(\sigma_3 - \sigma_1) + \frac{1}{2}(\sigma_3 + \sigma_1)\sin\psi + \text{constant} \tag{7.54}$$

results in a much better prediction of the plastic volume change, since the dilatancy angle ψ is now an independent parameter. Use of Equation (7.54) as plastic potential function results in a relation between plastic volume change and plastic shear intensity that is given by:

$$\dot{\epsilon}^{\mathrm{p}}_{\mathrm{vol}} = \dot{\gamma}^{\mathrm{p}}\sin\psi \tag{7.55}$$

which makes it possible to match experimental data. For completeness we also list a common potential function for the Drucker–Prager yield criterion,

$$g(\boldsymbol{\sigma}) = q + \beta p + \text{constant} \tag{7.56}$$

with p and q defined in Equations (1.79) and (7.30). For $\alpha \neq \beta$ this definition renders the Drucker–Prager plasticity model non-associated.

We note that a special case occurs when $\psi = 0$ in the Mohr–Coulomb plastic potential function, Equation (7.54), since then, according to Equation (7.55), the plastic flow is isochoric, i.e. volume-preserving. This implies that for the Tresca yield function with an associated flow rule, the plastic straining is also isochoric. The same holds for the Drucker–Prager yield criterion when $\beta = 0$ in Equation (7.56), and for the von Mises yield function with an associated flow rule.

7.2.3 Hardening Behaviour

In the preceding section it has been assumed that the yield function only depends on the stress tensor. Also in the example of the simple slip model the assumption was made that the friction coefficient is a constant and does not depend upon the loading history. Such a dependence, however, can be envisaged, e.g. due to breaking-off of the asperities between the block and the surface. Making the friction coefficient a descending function of the total amount of sliding would then be a logical choice.

The simplest extension beyond the model of ideal plasticity is to make the yield function dependent on a scalar measure of the plastic strain tensor as well:

$$f = f(\boldsymbol{\sigma}, \kappa) \tag{7.57}$$

The scalar-valued hardening parameter κ is typically dependent on the strain history through invariants of the plastic strain tensor $\boldsymbol{\epsilon}^{\mathrm{p}}$. Another suitable, frame-invariant choice would be the dissipated plastic work. A host of possibilities now emerges, for any function of the invariants of $\boldsymbol{\epsilon}^{\mathrm{p}}$ and/or the plastic work could serve the purpose of defining the hardening parameter such that it is frame-invariant, but three choices have gained popularity. Classically, we have the work-hardening hypothesis,

$$\dot{\kappa} = \boldsymbol{\sigma} : \dot{\boldsymbol{\epsilon}}^{\mathrm{p}} \tag{7.58}$$

which is sometimes, especially for metal plasticity, also formulated as:

$$\dot{\kappa} = \frac{1}{\bar{\sigma}}\boldsymbol{\sigma} : \dot{\boldsymbol{\epsilon}}^{\mathrm{p}} \tag{7.59}$$

and the strain-hardening hypothesis,

$$\dot{\kappa} = \sqrt{\frac{2}{3}\dot{\mathbf{e}}^{\mathrm{p}} : \dot{\mathbf{e}}^{\mathrm{p}}} \tag{7.60}$$

For future use we also list the strain-hardening hypothesis in matrix-vector format:

$$\dot{\kappa} = \sqrt{\frac{2}{3}(\dot{\boldsymbol{\epsilon}}^{\mathrm{p}})^{\mathrm{T}}\mathbf{Q}\dot{\boldsymbol{\epsilon}}^{\mathrm{p}}} \tag{7.61}$$

where the matrix

$$\mathbf{Q} = \begin{bmatrix} \frac{2}{3} & -\frac{1}{3} & -\frac{1}{3} & 0 & 0 & 0 \\ -\frac{1}{3} & \frac{2}{3} & -\frac{1}{3} & 0 & 0 & 0 \\ -\frac{1}{3} & -\frac{1}{3} & \frac{2}{3} & 0 & 0 & 0 \\ 0 & 0 & 0 & \frac{1}{2} & 0 & 0 \\ 0 & 0 & 0 & 0 & \frac{1}{2} & 0 \\ 0 & 0 & 0 & 0 & 0 & \frac{1}{2} \end{bmatrix} \tag{7.62}$$

accounts for the fact that normally only the deviatoric strains are included in the strain-hardening hypothesis and that the double, shear strain components are incorporated in $\boldsymbol{\epsilon}$. For applications in granular materials, the following hypothesis has become popular, as it postulates a history dependence on the plastic volumetric strain:

$$\dot{\kappa} = -\dot{\epsilon}^{\mathrm{p}}_{\mathrm{vol}} \tag{7.63}$$

where the minus sign has been introduced to have a positive hardening parameter for compaction. Using the projection vector (7.36) this hardening hypothesis can also be cast in matrix-vector format:

$$\dot{\kappa} = -3\boldsymbol{\pi}^{\mathrm{T}}\dot{\boldsymbol{\epsilon}}^{\mathrm{p}} \tag{7.64}$$

In all cases the hardening parameter is integrated along the loading path:

$$\kappa = \int \dot{\kappa}\mathrm{d}t \tag{7.65}$$

and, because of the dependence of the yield function on the loading history only through a scalar-valued hardening parameter, the yield surface can only expand or shrink, but not translate or rotate in the stress space (Figure 7.6). Because of the latter property this type of hardening is named isotropic hardening.

Isotropic hardening models cannot accurately capture effects that typically occur in cyclic loading, for instance the Bauschinger effect in metals. An isotropic hardening model would predict the dashed line in Figure 7.7 upon reverse loading, whereas experimental evidence results in the solid line. Kinematic hardening is better suited to model this phenomenon. In this

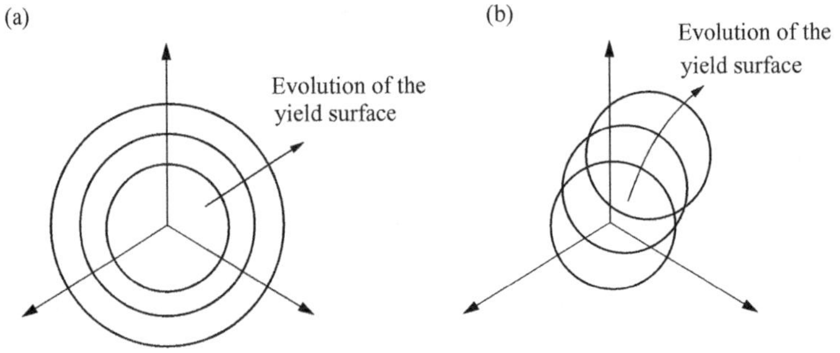

Figure 7.6 Different types of hardening. (a) Isotropic hardening. (b) Kinematic hardening

class of hardening models the yield surface translates (Figure 7.6). A second-order tensor, $\boldsymbol{\alpha}$, often called the back stress tensor, describes the movement of the origin of the yield surface. In its original form, where the yield surface only translates, but does not change size, the yield function only depends on the difference of the stress tensor and the back stress tensor:

$$f = f(\boldsymbol{\sigma} - \boldsymbol{\alpha}) \tag{7.66}$$

For von Mises plasticity evolution equations for the back stress tensor were formulated by Prager (1955), who assumed that the translation of the back stress was in the direction of the plastic strain rate:

$$\dot{\boldsymbol{\alpha}} = a(\boldsymbol{\alpha})\dot{\boldsymbol{\epsilon}}^{\mathrm{p}} \tag{7.67}$$

with $a(\boldsymbol{\alpha})$ a scalar-valued function of the back stress tensor. To remedy some inconsistencies that can be encountered with Prager's hardening rule – especially when working in subspaces such as plane stress – Ziegler (1959) proposed the following hardening rule:

$$\dot{\boldsymbol{\alpha}} = \dot{\lambda} a(\boldsymbol{\alpha})(\boldsymbol{\sigma} - \boldsymbol{\alpha}) \tag{7.68}$$

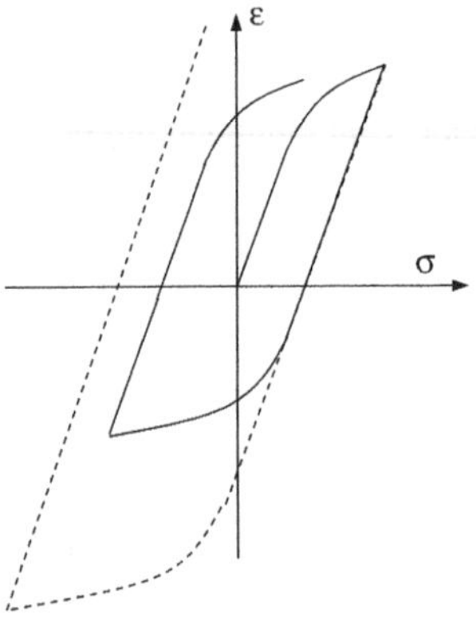

Figure 7.7 Cyclic loading in metals: Bauschinger effect

Box 7.2 Alternative plasticity models

The evolution equations for the back stress tensor can be hard to formulate. An early alternative approach is the sublayer model (Besseling 1958), a parallel arrangement of spring–slider elements (Figure 7.8). In the same figure a related model has been plotted, the so-called nested yield surfaces model (Mróz 1967). In this model hardening starts when the stress point touches the innermost yield surface and ideal-plastic yielding occurs when the outermost circle has been reached. The bounding surface plasticity model (Dafalias and Popov 1975; Krieg 1975) can be considered as a limiting case of the latter model, and consists of two surfaces only, a yield surface (the innermost circle), and a bounding surface. The hardening rate is determined from the distance between both surfaces, measured in a suitable norm.

In classical plasticity, the hardening modulus h and the vectors $\mathbf{n}$ and $\mathbf{m}$ are fully determined through the definition of the yield surface and the evolution of the plastic strains. This is not so for plasticity models in which a yield surface has not been defined explicitly. For instance, in generalised plasticity (Pastor *et al.* 1990), they can be chosen without recourse to a yield or loading surface. Different expressions can be selected, depending on whether the material is loading or unloading, thus making the approach suitable to cyclic loading conditions. In the absence of an explicitly defined loading surface, plastic loading/unloading is determined by comparing the direction of the elastic stress increment with $\mathbf{n}$,

$$\begin{aligned} \mathbf{n} : \mathbf{D}^{\mathrm{e}} : \dot{\boldsymbol{\epsilon}} > 0 \quad &\rightarrow \quad \text{loading} \\ \mathbf{n} : \mathbf{D}^{\mathrm{e}} : \dot{\boldsymbol{\epsilon}} = 0 \quad &\rightarrow \quad \text{neutral loading} \\ \mathbf{n} : \mathbf{D}^{\mathrm{e}} : \dot{\boldsymbol{\epsilon}} < 0 \quad &\rightarrow \quad \text{unloading} \end{aligned}$$

Since consistency cannot be enforced, the rate of plastic flow is simply defined as:

$$\dot{\lambda} \equiv \frac{\mathbf{n} : \mathbf{D}^{\mathrm{e}} : \dot{\boldsymbol{\epsilon}}}{h_{L/U} + \mathbf{n} : \mathbf{D}^{\mathrm{e}} : \mathbf{m}_{L/U}}$$

where the subscripts used are L for loading and U for unloading. When $\mathbf{n} : \mathbf{D}^{\mathrm{e}} : \dot{\boldsymbol{\epsilon}} = 0$ (neutral loading), both predict $\dot{\lambda} = 0$, and continuity between loading and unloading is guaranteed.

which corresponds to a radial motion. Combinations of isotropic and kinematic hardening models are possible as well. In these mixed hardening models the yield function translates and changes size:

$$f = f(\boldsymbol{\sigma} - \boldsymbol{\alpha}, \kappa) \tag{7.69}$$

Some alternative plasticity models for cyclic loading are discussed in Box 7.2.

When using matrix-vector notation, one can assemble all internal variables, either components of a second-order tensor as $\boldsymbol{\alpha}$, or scalars like κ in an array $\boldsymbol{\kappa}$. Then, we have the following

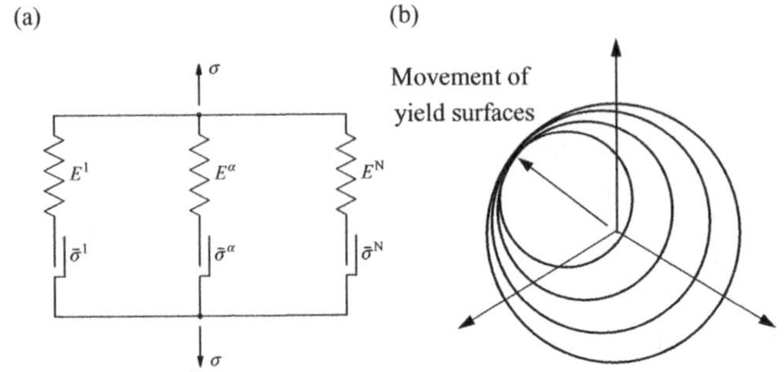

Figure 7.8 Alternative plasticity models. (a) Sublayer model (Besseling 1958). (b) Nested yield surfaces model (Mróz 1967)

general format for a yield function that includes isotropic and kinematic hardening:

$$f = f(\boldsymbol{\sigma}, \boldsymbol{\kappa}) \tag{7.70}$$

Owing to the presence of hardening the mathematical structure of the tangential constitutive relation between stress rate and strain rate has changed. The consistency condition (7.39) is now elaborated as:

$$\mathbf{n}^{\mathrm{T}}\dot{\boldsymbol{\sigma}} + \left(\frac{\partial f}{\partial \boldsymbol{\kappa}}\right)^{\mathrm{T}} \dot{\boldsymbol{\kappa}} = 0 \tag{7.71}$$

From the definitions (7.58), (7.60) and (7.63) it can be inferred that the hardening variables are proportional to the consistency parameter $\dot{\lambda}$, so that

$$\dot{\boldsymbol{\kappa}} = \dot{\lambda}\mathbf{p}(\boldsymbol{\sigma}, \boldsymbol{\kappa}) \tag{7.72}$$

with $\mathbf{p}$ a vector function of the stress $\boldsymbol{\sigma}$ and the hardening variables collected in $\boldsymbol{\kappa}$. Equation (7.71) can then be rewritten as:

$$\mathbf{n}^{\mathrm{T}}\dot{\boldsymbol{\sigma}} - h\dot{\lambda} = 0 \tag{7.73}$$

with

$$h = -\left(\frac{\partial f}{\partial \boldsymbol{\kappa}}\right)^{\mathrm{T}} \mathbf{p}(\boldsymbol{\sigma}, \boldsymbol{\kappa}) \tag{7.74}$$

the hardening modulus. Equation (7.74) shows that the hardening modulus cannot be defined independently. When the dependence of the yield function f on the array of hardening parameters $\boldsymbol{\kappa}$, on the hardening hypothesis and on the evolution of the plastic strains have been set, the hardening modulus is fully determined.

For the derivation of the tangential stiffness relation for hardening plasticity we could proceed along the same lines as with the derivation for the simple slip model and for ideal plasticity. For future reference it is more convenient to first set up the tangential compliance relation. To this end we combine the time derivative of Equation (7.42) and the flow rule (7.43) with the consistency condition for hardening plasticity (7.73). This results in:

$$\dot{\boldsymbol{\epsilon}} = \left((\mathbf{D}^{\mathrm{e}})^{-1} + \frac{1}{h}\mathbf{m}\mathbf{n}^{\mathrm{T}} \right) \dot{\boldsymbol{\sigma}} \tag{7.75}$$

Note that a compliance relation cannot be derived for ideal plasticity, since then $h = 0$. Equation (7.75) can be inverted with the aid of the Sherman–Morrison formula. The result is similar to Equation (7.47):

$$\dot{\boldsymbol{\sigma}} = \left(\mathbf{D}^{\mathrm{e}} - \frac{\mathbf{D}^{\mathrm{e}}\mathbf{m}\mathbf{n}^{\mathrm{T}}\mathbf{D}^{\mathrm{e}}}{h + \mathbf{n}^{\mathrm{T}}\mathbf{D}^{\mathrm{e}}\mathbf{m}} \right) \dot{\boldsymbol{\epsilon}} \tag{7.76}$$

As noted before, the work-hardening hypothesis in metal plasticity is often cast in the modified format of Equation (7.59). The von Mises yield function that incorporates hardening reads:

$$f(\boldsymbol{\sigma}, \kappa) = q - \bar{\sigma}(\kappa) \tag{7.77}$$

while for the Tresca yield function one obtains:

$$f(\boldsymbol{\sigma}, \kappa) = (\sigma_3 - \sigma_1) - \bar{\sigma}(\kappa) \tag{7.78}$$

For both yield functions, the rate of the hardening parameter $\dot{\kappa}$ and the plastic multiplier $\dot{\lambda}$ then become identical, $\dot{\kappa} = \dot{\lambda}$, and, using Equation (7.59), the hardening modulus reduces to:

$$h = -\frac{\partial f}{\partial \kappa} \tag{7.79}$$

which can also be expressed as:

$$h = \frac{\partial \bar{\sigma}}{\partial \kappa} \tag{7.80}$$

Consequently, for this hardening hypothesis and these yield functions, the hardening modulus simply equals the slope of the hardening diagram.

For the Tresca and von Mises yield functions an appealing interpretation of Equation (7.80) can be made. For uniaxial stressing, $\bar{\sigma} = \sigma_3$ and the associated flow rule states that $\dot{\lambda} = \dot{\epsilon}_3^{\mathrm{p}}$, or, equivalently: $\dot{\kappa} = \dot{\epsilon}_3^{\mathrm{p}}$. Note that in keeping with Equation (7.23), the axial stress σ_3 is the major principal stress in uniaxial tension. The uniaxial plastic strain ϵ_3^{p} in the loading direction thus coincides with the hardening parameter and can directly be used to plot the hardening diagram as: ϵ_3^{p} *vs* $\bar{\sigma}$ (Figure 7.9). The hardening modulus simply reads:

$$h = \frac{\partial \bar{\sigma}}{\partial \epsilon_3^{\mathrm{p}}} \tag{7.81}$$

which shows the analogy between the Young's modulus E and the hardening modulus h, since the latter governs the plastic strain rate in a manner similar to the way Young's modulus determines the elastic strain rate. The rationale for preferring Equation (7.59) over Equation (7.58) now becomes apparent. If Equation (7.58) had been used for the work-hardening hypothesis, the uniaxial plastic strain could not have served as the x-coordinate in the hardening diagram.

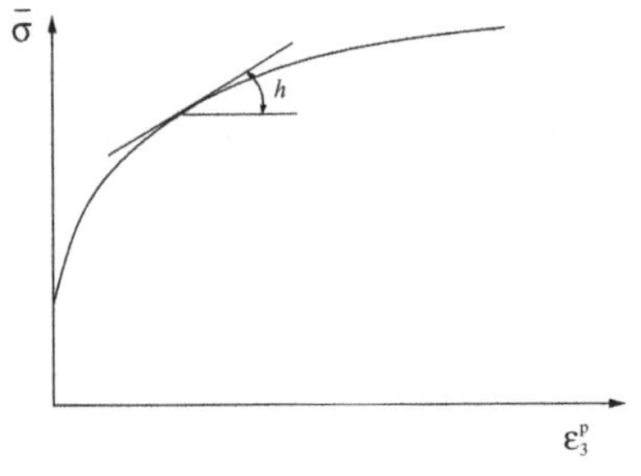

Figure 7.9 Hardening diagram for metal plasticity

Instead, we would have been forced to plot the plastic work on the horizontal axis of the hardening diagram. Finally, it is observed from this identity that the hardening modulus will be positive if the slope of the curve in the hardening diagram is also positive. The latter property is shared by all yield functions and hardening hypotheses.

For other plasticity models the relation between the rate of the hardening parameter $\dot{\kappa}$ and the rate of plastic flow $\dot{\lambda}$ is more complicated. For instance, the strain-hardening hypothesis in combination with the Mohr–Coulomb yield function results in the following expression for the hardening modulus h:

$$h = -\sqrt{\frac{1}{3}(1 + \sin^2 \varphi)}\frac{\partial f}{\partial \kappa} \tag{7.82}$$

or when a non-associated flow rule with a plastic potential as defined in Equation (7.54) is employed:

$$h = -\sqrt{\frac{1}{3}(1 + \sin^2 \psi)}\frac{\partial f}{\partial \kappa} \tag{7.83}$$

When more strength parameters than the uniaxial yield strength $\bar{\sigma}$ enter the yield function each of these parameters can, in principle, be a function of the amount of hardening. For instance, in the Mohr–Coulomb yield function (7.23) both the cohesion c and the angle of internal friction φ can be made a function of the plastic strain through the hardening parameter:

$$f(\sigma, \kappa) = \frac{1}{2}(\sigma_3 - \sigma_1) + \frac{1}{2}(\sigma_3 + \sigma_1)\sin\varphi(\kappa) - c(\kappa)\cos\varphi(\kappa) \tag{7.84}$$

The derivative of the yield function f with respect to the hardening parameter κ now involves the stress level:

$$\frac{\partial f}{\partial \kappa} = \frac{1}{2}(\sigma_3 + \sigma_1)\frac{\partial \sin\varphi}{\partial \kappa} - \frac{\partial (c\cos\varphi)}{\partial \kappa}$$

and two hardening diagrams are needed to completely define the hardening process. Clearly, for the Mohr–Coulomb and Drucker–Prager yield criteria the hardening behaviour cannot be inferred from a simple tension test. The plastic strains in all three directions must be measured in a triaxial device in order to construct a proper hardening diagram in which one or more strength measures are plotted as a function of the hardening parameter κ.

7.3 Integration of the Stress–strain Relation

Computational models for elasto-plasticity were first developed in the early 1970s, e.g., in the seminal paper of Nayak and Zienkiewicz (1972). An account of the developments in these early years has been presented in the book by Owen and Hinton (1980). A firm basis of algorithms in plasticity was established in the 1980s, with the introduction of important notions like return-mapping algorithms and consistent tangent operators, see for instance Simo and Taylor (1985), Ortiz and Popov (1985) and Runesson *et al.* (1986). With the introduction of these concepts, large-scale computations have become feasible.

To obtain the strains and stresses in a structure that relate to a generic loading stage Equation (7.76) must be integrated along the loading path. The most straightforward way is to use a one-point Euler forward integration rule. Such a scheme is fully explicit: the stresses and the value of the hardening modulus h are known at the beginning of the strain increment so that the tangential stiffness matrix can be evaluated directly. If the initial stress point $\boldsymbol{\sigma}_0$ is on the yield contour the stress increment can be computed as:

$$\Delta\boldsymbol{\sigma} = \left(\mathbf{D}^{\mathrm{e}} - \frac{\mathbf{D}^{\mathrm{e}}\mathbf{m}_0\mathbf{n}_0^{\mathrm{T}}\mathbf{D}^{\mathrm{e}}}{h_0 + \mathbf{n}_0^{\mathrm{T}}\mathbf{D}^{\mathrm{e}}\mathbf{m}_0}\right)\Delta\boldsymbol{\epsilon} \tag{7.85}$$

where the subscript 0 refers to the fact that quantities are evaluated at the beginning of the load increment. The estimate in iteration $j+1$ for the stress at the end of the loading step, $\boldsymbol{\sigma}_{j+1}$, then follows from:

$$\boldsymbol{\sigma}_{j+1} = \boldsymbol{\sigma}_0 + \Delta\boldsymbol{\sigma} \tag{7.86}$$

If the stress point is initially inside the yield contour the total strain increment must be subdivided into a purely elastic part, i.e. a part that is needed to make the stress point reach the yield surface, $\Delta\boldsymbol{\epsilon}_A$ in Figure 7.10, and a part that involves elasto-plastic straining. Now, the stress increment is computed as:

$$\Delta\boldsymbol{\sigma} = \mathbf{D}^{\mathrm{e}}\Delta\boldsymbol{\epsilon}_A + \left(\mathbf{D}^{\mathrm{e}} - \frac{\mathbf{D}^{\mathrm{e}}\mathbf{m}_c\mathbf{n}_c^{\mathrm{T}}\mathbf{D}^{\mathrm{e}}}{h_c + \mathbf{n}_c^{\mathrm{T}}\mathbf{D}^{\mathrm{e}}\mathbf{m}_c}\right)\Delta\boldsymbol{\epsilon}_B \tag{7.87}$$

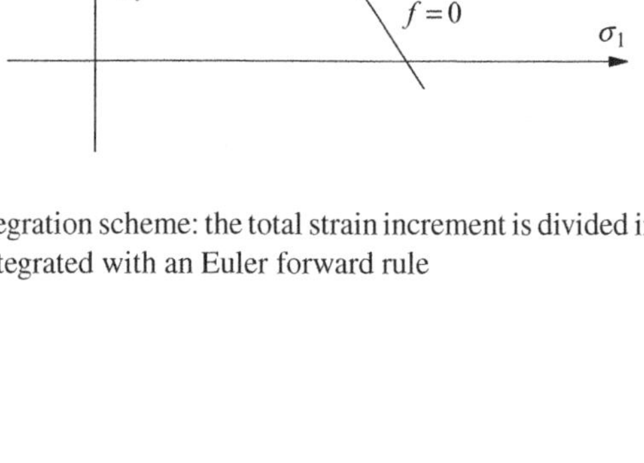

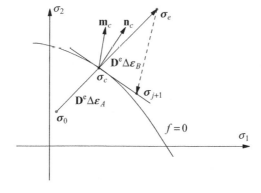

Figure 7.10 Explicit integration scheme: the total strain increment is divided into an elastic and a plastic part. The plastic part is integrated with an Euler forward rule

where the subscript c denotes that the respective quantities are evaluated at $\boldsymbol{\sigma} = \boldsymbol{\sigma}_c$. A major disadvantage of the explicit Euler method now becomes apparent: the contact stress must be calculated explicitly. The procedure can also be viewed as follows. First, the 'elastic' stress increment

$$\Delta\boldsymbol{\sigma}_e = \mathbf{D}^e \Delta\boldsymbol{\epsilon} \tag{7.88}$$

is calculated. For this calculation it is irrelevant whether the initial location of the stress is inside or on the current yield surface. This stress increment may be conceived as a trial stress increment, which rests upon the assumption of purely elastic behaviour during the load increment. Possible plastic straining is not considered during this trial step. The total trial stress $\boldsymbol{\sigma}_e$ is set up as the sum of the stress at the beginning of the loading step, $\boldsymbol{\sigma}_0$, and the 'elastic' stress increment $\Delta\boldsymbol{\sigma}_e$:

$$\boldsymbol{\sigma}_e = \boldsymbol{\sigma}_0 + \mathbf{D}^e \Delta\boldsymbol{\epsilon} \tag{7.89}$$

If the trial stress $\boldsymbol{\sigma}_e$ violates the yield condition, $f(\boldsymbol{\sigma}_e, \kappa_0) > 0$, a correction is applied. The direction and the magnitude of this correction are inferred from Equation (7.87):

$$\boldsymbol{\sigma}_{j+1} - \boldsymbol{\sigma}_e = -\frac{\mathbf{n}_c^{\mathrm{T}} \mathbf{D}^e \Delta\boldsymbol{\epsilon}_B}{h_c + \mathbf{n}_c^{\mathrm{T}} \mathbf{D}^e \mathbf{m}_c} \mathbf{D}^e \mathbf{m}_c \tag{7.90}$$

for the most general case that the initial stress point is located within the yield surface. Equation (7.90) shows that in the explicit Euler method, the flow direction $\mathbf{m}$, the gradient to the yield surface $\mathbf{n}$ and the hardening modulus h are computed either at the initial stress state $\boldsymbol{\sigma}_0$, or at the stresses at the contact or intersection point of the 'elastic' stress path with the yield contour $\boldsymbol{\sigma}_c$ if the initial stress state is within the yield contour. Combining Equations (7.88)–(7.90) shows that the elastic predictor–plastic corrector process has the following format:

$$\boldsymbol{\sigma}_{j+1} = \boldsymbol{\sigma}_e - \Delta\lambda \mathbf{D}^e \mathbf{m}_c \tag{7.91}$$

where $\Delta\lambda$ is the amount of plastic flow within this loading step:

$$\Delta\lambda = \frac{\mathbf{n}_c^{\mathrm{T}} \mathbf{D}^e \Delta\boldsymbol{\epsilon}_B}{h_c + \mathbf{n}_c^{\mathrm{T}} \mathbf{D}^e \mathbf{m}_c} \tag{7.92}$$

Equation (7.91) can be interpreted as follows. First, a trial stress is computed assuming fully elastic behaviour. Then, the trial stress is mapped back, i.e. projected in the direction of the yield surface. Therefore, the name return-mapping algorithm has become popular for this type of integration method.

Figure 7.10 shows that, in the absence of hardening, the stress $\boldsymbol{\sigma}_{j+1}$ is found at the intersection of the hyperplane that is a tangent to the yield surface at $\boldsymbol{\sigma}_c$ and the return direction $\mathbf{m}_c$. Apparently, the forward Euler method does not guarantee a rigorous return to the yield surface. An error is committed with a magnitude which depends upon the local curvature of the yield surface. A strongly curved yield surface gives rise to larger errors than an almost flat yield contour. Especially when relatively large loading steps are used, the accumulation of errors can become significant, which decreases the accuracy, and may lead to numerical instability of the algorithm. This is shown in Figure 7.11, where the projection from the trial stress onto the yield surface fails. Possible corrections to improve the estimate (7.85) for the stress increment

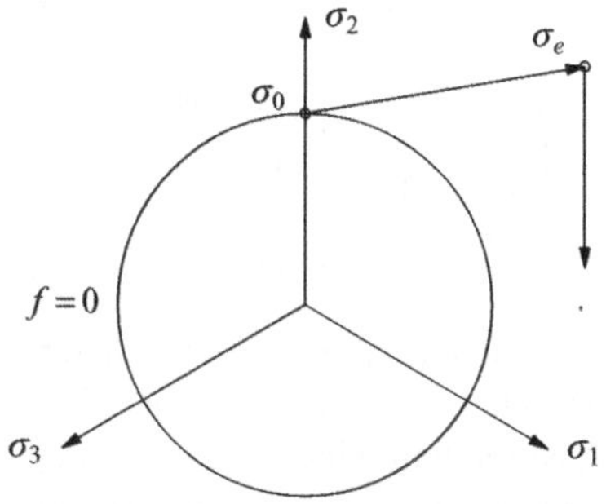

Figure 7.11 Failure of the Euler forward integration rule.The return direction is such that the yield contour will be missed

such that the plastic part of the strain increment is computed more accurately and the drifting error is made smaller, are not of much help in this situation. The direction of the correction for plastic flow is simply such that the yield surface will never be reached. For the Euler forward algorithm stability of the integration scheme is only ensured for small loading steps.

The latter property of the Euler forward method is at variance with the wish to use large loading steps. Therefore, the conceptual simplicity of the Euler forward method is normally sacrificed to a more robust algorithm which warrants numerical stability irrespective of the step size. A good and relatively simple algorithm is the implicit Euler backward method. Formally, this algorithm is also given by Equation (7.91), but all quantities are now evaluated at $(\boldsymbol{\sigma}_{j+1}, \boldsymbol{\kappa}_{j+1})$:

$$\boldsymbol{\sigma}_{j+1} = \boldsymbol{\sigma}_e - \Delta\lambda \mathbf{D}^{\mathrm{e}}\mathbf{m}(\boldsymbol{\sigma}_{j+1}, \boldsymbol{\kappa}_{j+1}) \tag{7.93}$$

Comparing Equations (7.91) and (7.93), we observe that in the fully implicit Euler backward algorithm there is no need to determine the contact stress $\boldsymbol{\sigma}_c$. However, the equations are now implicit in the sense that neither $\Delta\lambda$, $\boldsymbol{\sigma}_{j+1}$ or $\boldsymbol{\kappa}_{j+1}$ can be computed directly. Moreover, the six equations of (7.93) contain fourteen unknowns, namely the six stress components of $\boldsymbol{\sigma}_{j+1}$, the seven components of $\boldsymbol{\kappa}_{j+1}$, and $\Delta\lambda$. We must therefore augment Equations (7.93) by the integrated form of the hardening rule (7.72):

$$\boldsymbol{\kappa}_{j+1} = \boldsymbol{\kappa}_0 + \Delta\lambda \mathbf{p}(\boldsymbol{\sigma}_{j+1}, \boldsymbol{\kappa}_{j+1}) \tag{7.94}$$

with $\boldsymbol{\kappa}_0$ containing the values of the hardening variables at the beginning of the load increment, and the requirement that the yield condition is complied with at the end of the loading step:

$$f(\boldsymbol{\sigma}_{j+1}, \boldsymbol{\kappa}_{j+1}) = 0 \tag{7.95}$$

The set of non-linear equations (7.93)–(7.95) is subsequently cast in a format of local residuals (in the sense that the residuals are defined at integration point level):

$$\begin{cases} \mathbf{r}_\sigma = \boldsymbol{\sigma}_{j+1} - \boldsymbol{\sigma}_e + \Delta\lambda \mathbf{D}^{\mathrm{e}}\mathbf{m}(\boldsymbol{\sigma}_{j+1}, \boldsymbol{\kappa}_{j+1}) \\ \mathbf{r}_\kappa = \boldsymbol{\kappa}_{j+1} - \boldsymbol{\kappa}_0 - \Delta\lambda \mathbf{p}(\boldsymbol{\sigma}_{j+1}, \boldsymbol{\kappa}_{j+1}) \\ r_f = f(\boldsymbol{\sigma}_{j+1}, \boldsymbol{\kappa}_{j+1}) \end{cases} \tag{7.96}$$

This system can be solved using an iterative procedure, e.g. a Newton–Raphson method:

$$\begin{pmatrix} \boldsymbol{\sigma}_{j+1}^{k+1} \\ \kappa_{j+1}^{k+1} \\ \lambda_{j+1}^{k+1} \end{pmatrix} = \begin{pmatrix} \boldsymbol{\sigma}_{j+1}^{k} \\ \kappa_{j+1}^{k} \\ \lambda_{j+1}^{k} \end{pmatrix} - \begin{bmatrix} \frac{\partial \mathbf{r}_\sigma}{\partial \boldsymbol{\sigma}} & \frac{\partial \mathbf{r}_\sigma}{\partial \kappa} & \frac{\partial \mathbf{r}_\sigma}{\partial \lambda} \\ \frac{\partial \mathbf{r}_\kappa}{\partial \boldsymbol{\sigma}} & \frac{\partial \mathbf{r}_\kappa}{\partial \kappa} & \frac{\partial \mathbf{r}_\kappa}{\partial \lambda} \\ \frac{\partial r_f}{\partial \boldsymbol{\sigma}} & \frac{\partial r_f}{\partial \kappa} & 0 \end{bmatrix}^{-1} \begin{pmatrix} \mathbf{r}_\sigma^k \\ \mathbf{r}_\kappa^k \\ r_f^k \end{pmatrix} \tag{7.97}$$

Within the global iteration $j+1$, k denotes the iteration counter of this local Newton–Raphson method at integration point level. The equations (7.96) and (7.97) are elaborated in Box 7.3 for the case of a von Mises yield function with kinematic hardening. Box 7.4 shows that the same methodology can also be applied to non-standard plasticity models like generalised plasticity or bounding surface plasticity.

For isotropic hardening, the evolution equations of the hardening variable (7.72) reduce to a quasi-linear relation between the rate of the hardening variable $\dot{\kappa}$ and the consistency parameter $\dot{\lambda}$:

$$\dot{\kappa} = \dot{\lambda} p(\boldsymbol{\sigma}, \kappa) \tag{7.98}$$

We can exploit this proportionality by reducing the system of residuals

$$\begin{cases} \mathbf{r}_\sigma = \boldsymbol{\sigma}_{j+1} - \boldsymbol{\sigma}_e + \Delta\lambda \mathbf{D}^{\mathrm{e}} \mathbf{m}(\boldsymbol{\sigma}_{j+1}, \kappa_{j+1}) \\ r_\kappa = \kappa_{j+1} - \kappa_0 - \Delta\lambda p(\boldsymbol{\sigma}_{j+1}, \kappa_{j+1}) \\ r_f = f(\boldsymbol{\sigma}_{j+1}, \kappa_{j+1}) \end{cases}$$

to:

$$\begin{cases} \mathbf{r}_\sigma = \boldsymbol{\sigma}_{j+1} - \boldsymbol{\sigma}_e + \Delta\lambda \mathbf{D}^{\mathrm{e}} \mathbf{m}(\boldsymbol{\sigma}_{j+1}, \lambda_{j+1}) \\ r_f = f(\boldsymbol{\sigma}_{j+1}, \lambda_{j+1}) \end{cases} \tag{7.99}$$

and κ_{j+1} is then computed from

$$\kappa_{j+1} = \kappa_0 + \Delta\lambda p(\boldsymbol{\sigma}_{j+1}, \lambda_{j+1}) \tag{7.100}$$

after convergence of the local Newton–Raphson iterative process, now described by:

$$\begin{pmatrix} \boldsymbol{\sigma}_{j+1}^{k+1} \\ \lambda_{j+1}^{k+1} \end{pmatrix} = \begin{pmatrix} \boldsymbol{\sigma}_{j+1}^{k} \\ \lambda_{j+1}^{k} \end{pmatrix} - \begin{bmatrix} \frac{\partial \mathbf{r}_\sigma}{\partial \boldsymbol{\sigma}} & \frac{\partial \mathbf{r}_\sigma}{\partial \lambda} \\ \frac{\partial r_f}{\partial \boldsymbol{\sigma}} & \frac{\partial r_f}{\partial \lambda} \end{bmatrix}^{-1} \begin{pmatrix} \mathbf{r}_\sigma^k \\ r_f^\lambda \end{pmatrix} \tag{7.101}$$

The differentials in Equation (7.101) can be elaborated as:

$$\frac{\partial \mathbf{r}_\sigma}{\partial \boldsymbol{\sigma}} = \mathbf{I} + \Delta\lambda \mathbf{D}^{\mathrm{e}} \frac{\partial \mathbf{m}}{\partial \boldsymbol{\sigma}} \equiv \mathbf{A} \tag{7.102}$$

$$\frac{\partial \mathbf{r}_\sigma}{\partial \lambda} = \mathbf{D}^{\mathrm{e}} \bar{\mathbf{m}} \tag{7.103}$$

$$\frac{\partial r_f}{\partial \boldsymbol{\sigma}} = \frac{\partial f}{\partial \boldsymbol{\sigma}} = \mathbf{n}^{\mathrm{T}} \tag{7.104}$$

Box 7.3 Von Mises plasticity with kinematic hardening

The von Mises yield function equipped with kinematic hardening reads:

$$f(\boldsymbol{\sigma} - \boldsymbol{\alpha}) = \sqrt{\frac{3}{2}(\boldsymbol{\sigma} - \boldsymbol{\alpha})^{\mathrm{T}}\mathbf{P}(\boldsymbol{\sigma} - \boldsymbol{\alpha})} - \bar{\sigma}$$

With an associated flow rule the flow direction becomes

$$\mathbf{m} = \frac{3\mathbf{P}(\boldsymbol{\sigma} - \boldsymbol{\alpha})}{2\sqrt{\frac{3}{2}(\boldsymbol{\sigma} - \boldsymbol{\alpha})^{\mathrm{T}}\mathbf{P}(\boldsymbol{\sigma} - \boldsymbol{\alpha})}}$$

or, using the yield function $f = 0$:

$$\mathbf{m} = \frac{3}{2\bar{\sigma}}\mathbf{P}(\boldsymbol{\sigma} - \boldsymbol{\alpha})$$

Using Ziegler's kinematic hardening rule (7.68) we have: $\mathbf{p} = a(\boldsymbol{\alpha})(\boldsymbol{\sigma} - \boldsymbol{\alpha})$ and the set (7.96) can be elaborated as:

$$\begin{cases} \mathbf{r}_\sigma = \boldsymbol{\sigma}_{j+1} - \boldsymbol{\sigma}_e + \frac{3\Delta\lambda}{2\bar{\sigma}}\mathbf{D}^{\mathrm{e}}\mathbf{P}(\boldsymbol{\sigma}_{j+1} - \boldsymbol{\alpha}_{j+1}) \\ \mathbf{r}_\alpha = \boldsymbol{\alpha}_{j+1} - \boldsymbol{\alpha}_0 - \Delta\lambda a(\boldsymbol{\alpha}_{j+1})(\boldsymbol{\sigma}_{j+1} - \boldsymbol{\alpha}_{j+1}) \\ r_f = \sqrt{\frac{3}{2}(\boldsymbol{\sigma}_{j+1} - \boldsymbol{\alpha}_{j+1})^{\mathrm{T}}\mathbf{P}(\boldsymbol{\sigma}_{j+1} - \boldsymbol{\alpha}_{j+1})} - \bar{\sigma} \end{cases}$$

whence the derivates needed in the local iteration (7.97) become:

$$\begin{cases} \frac{\partial \mathbf{r}_\sigma}{\partial \boldsymbol{\sigma}} = \mathbf{I} + \frac{3\Delta\lambda}{2\bar{\sigma}}\mathbf{D}^{\mathrm{e}}\mathbf{P} \\ \frac{\partial \mathbf{r}_\sigma}{\partial \boldsymbol{\alpha}} = -\frac{3\Delta\lambda}{2\bar{\sigma}}\mathbf{D}^{\mathrm{e}}\mathbf{P} \\ \frac{\partial \mathbf{r}_\sigma}{\partial \lambda} = \frac{3}{2\bar{\sigma}}\mathbf{D}^{\mathrm{e}}\mathbf{P}(\boldsymbol{\sigma}_{j+1} - \boldsymbol{\alpha}_{j+1}) \\ \frac{\partial \mathbf{r}_\alpha}{\partial \boldsymbol{\sigma}} = -\Delta\lambda a(\boldsymbol{\alpha}_{j+1})\mathbf{I} \\ \frac{\partial \mathbf{r}_\alpha}{\partial \boldsymbol{\alpha}} = \left(1 + \Delta\lambda a(\boldsymbol{\alpha}_{j+1})\right)\mathbf{I} - \Delta\lambda(\boldsymbol{\sigma}_{j+1} - \boldsymbol{\alpha}_{j+1})\left(\frac{\partial a}{\partial \boldsymbol{\alpha}}\right)^{\mathrm{T}} \\ \frac{\partial \mathbf{r}_\alpha}{\partial \lambda} = -a(\boldsymbol{\alpha}_{j+1})(\boldsymbol{\sigma}_{j+1} - \boldsymbol{\alpha}_{j+1}) \\ \frac{\partial r_f}{\partial \boldsymbol{\sigma}} = \frac{3}{2\bar{\sigma}}\mathbf{P}(\boldsymbol{\sigma}_{j+1} - \boldsymbol{\alpha}_{j+1}) \\ \frac{\partial r_f}{\partial \boldsymbol{\alpha}} = -\frac{3}{2\bar{\sigma}}\mathbf{P}(\boldsymbol{\sigma}_{j+1} - \boldsymbol{\alpha}_{j+1}) \end{cases}$$

and, using Equations (7.74) and (7.98):

$$\frac{\partial r_f}{\partial \lambda} = \frac{\partial f}{\partial \kappa}\frac{\partial \kappa}{\partial \lambda} = -h \tag{7.105}$$

In Equation (7.103) the vector $\bar{\mathbf{m}}$ has been introduced, which is defined as:

$$\bar{\mathbf{m}} = \mathbf{m} + \Delta\lambda\frac{\partial \mathbf{m}}{\partial \lambda} \tag{7.106}$$

Box 7.4 Integration of alternative plasticity models

For generalised plasticity, the stress update and the internal variable update remain valid. The difference is that the yield condition is replaced by an evolution equation for the consistency parameter. If the discretised loading conditions for generalised plasticity are violated, the governing equations read:

$$\begin{aligned} \mathbf{r}_\sigma &= \boldsymbol{\sigma}_{j+1} - \boldsymbol{\sigma}_e + \Delta\lambda \mathbf{D}^{\mathrm{e}}(\mathbf{m}_L)_{j+1} \\ \mathbf{r}_\kappa &= \boldsymbol{\kappa}_{j+1} - \boldsymbol{\kappa}_0 - \Delta\lambda \mathbf{p}_{j+1} \\ r_f &= \Delta\lambda((h_L)_{j+1} + \mathbf{n}_{j+1}^{\mathrm{T}} \mathbf{D}^{\mathrm{e}}(\mathbf{m}_L)_{j+1}) - \mathbf{n}_{j+1}^{\mathrm{T}} \mathbf{D}^{\mathrm{e}} \Delta\boldsymbol{\epsilon} \end{aligned}$$

In bounding surface plasticity the hardening modulus h depends on the distance between the current state and the bounding surface, and the vectors **m** and **n** depend on the bounding surface. Since the equations do not require the specification of a yield function, the algorithm also includes the integration of bounding surface plasticity, see also Auricchio and Taylor (1995).

As we will observe in the next section, **A**, $\bar{\mathbf{m}}$, $\mathbf{n}^{\mathrm{T}}$ and $(-h)$ are reused when assembling the tangential stiffness matrix at global level.

A number of commonly used yield functions permit expressing $\boldsymbol{\sigma}_{j+1}$ explicitly in terms of $\boldsymbol{\sigma}_e$, simplifying the procedure to the solution of a single non-linear equation. An example is the Drucker–Prager yield function with isotropic hardening:

$$f(\boldsymbol{\sigma}, \kappa) = \sqrt{\frac{3}{2}\boldsymbol{\sigma}^{\mathrm{T}}\mathbf{P}\boldsymbol{\sigma}} + \alpha\boldsymbol{\pi}^{\mathrm{T}}\boldsymbol{\sigma} - k(\kappa) \tag{7.107}$$

Considering that the Drucker–Prager yield function is often used for granular materials, strain hardening is a reasonable hypothesis for this yield function. Substitution of the (non-associated) flow rule (7.52) with the Drucker–Prager plastic potential (7.56) into Equation (7.60), and considering that for the projection matrices we have:

$$\mathbf{PQP} = \mathbf{P} \tag{7.108}$$

cf. Equations (7.33) and (7.62), we have $\dot{\kappa} = \dot{\lambda}$, and the following expression for $\boldsymbol{\sigma}_{j+1}$ ensues upon substitution into Equation (7.93):

$$\boldsymbol{\sigma}_{j+1} = \boldsymbol{\sigma}_e - \Delta\lambda \left(\frac{3\mathbf{D}^{\mathrm{e}}\mathbf{P}\boldsymbol{\sigma}_{j+1}}{2\sqrt{\frac{3}{2}\boldsymbol{\sigma}_{j+1}^{\mathrm{T}}\mathbf{P}\boldsymbol{\sigma}_{j+1}}} + \beta\mathbf{D}^{\mathrm{e}}\boldsymbol{\pi} \right) \tag{7.109}$$

We next use the yield condition, $f(\boldsymbol{\sigma}_{j+1}, \lambda_{j+1}) = 0$ to transform this equation into:

$$\boldsymbol{\sigma}_{j+1} = \boldsymbol{\sigma}_e - \Delta\lambda \left(\frac{3\mathbf{D}^{\mathrm{e}}\mathbf{P}\boldsymbol{\sigma}_{j+1}}{2(k(\lambda_{j+1}) - \alpha\boldsymbol{\pi}^{\mathrm{T}}\boldsymbol{\sigma}_{j+1})} + \beta\mathbf{D}^{\mathrm{e}}\boldsymbol{\pi} \right) \tag{7.110}$$

The unknown stress $\boldsymbol{\sigma}_{j+1}$ appears in both sides of the equation. This problem can be solved by the following projection. We premultiply both sides by $\boldsymbol{\pi}$, note that $\mathbf{P}\boldsymbol{\pi} = \mathbf{0}$, and obtain:

$$\boldsymbol{\pi}^{\mathrm{T}}\boldsymbol{\sigma}_{j+1} = \boldsymbol{\pi}^{\mathrm{T}}\boldsymbol{\sigma}_e - \Delta\lambda\beta K \tag{7.111}$$

with $K = \boldsymbol{\pi}^{\mathrm{T}}\mathbf{D}^{\mathrm{e}}\boldsymbol{\pi}$ the bulk modulus, Equation (1.114). Substitution of this result into Equation (7.110) yields:

$$\boldsymbol{\sigma}_{j+1} = \mathbf{A}^{-1}(\boldsymbol{\sigma}_e - \Delta\lambda\beta\mathbf{D}^{\mathrm{e}}\boldsymbol{\pi}) \tag{7.112}$$

with

$$\mathbf{A} = \mathbf{I} + \frac{3\Delta\lambda\mathbf{D}^{\mathrm{e}}\mathbf{P}}{2k(\lambda_{j+1}) + \Delta\lambda\alpha\beta K - \alpha\boldsymbol{\pi}^{\mathrm{T}}\boldsymbol{\sigma}_e} \tag{7.113}$$

Enforcing the yield condition $f(\boldsymbol{\sigma}_{j+1}, \lambda_{j+1}) = 0$ results in a single non-linear equation in terms of $\Delta\lambda$:

$$\begin{aligned} f(\Delta\lambda) = & \sqrt{\frac{3}{2}(\boldsymbol{\sigma}_e - \Delta\lambda\beta\mathbf{D}^{\mathrm{e}}\boldsymbol{\pi})^{\mathrm{T}}\mathbf{A}^{-\mathrm{T}}\mathbf{P}\mathbf{A}^{-1}(\boldsymbol{\sigma}_e - \Delta\lambda\beta\mathbf{D}^{\mathrm{e}}\boldsymbol{\pi})} \\ & + \alpha\boldsymbol{\pi}^{\mathrm{T}}\mathbf{A}^{-1}(\boldsymbol{\sigma}_e - \Delta\lambda\beta\mathbf{D}^{\mathrm{e}}\boldsymbol{\pi}) - k(\lambda_{j+1}) = 0 \end{aligned} \tag{7.114}$$

The solution of this non-linear equation can be accomplished using a Newton–Raphson method or a secant method, or to apply a spectral decomposition to $\mathbf{P}$ and $\mathbf{D}^{\mathrm{e}}$ (Matthies 1989), which is convenient in the present case because of the isotropic character of Drucker–Prager plasticity and the assumed isotropy of the elastic part. Most anisotropic yield functions allow a similar reduction to a single non-linear equation expressed in terms of the increment of the plastic multiplier $\Delta\lambda$ (Box 7.5).

Slope stability problems are often used to illustrate pressure-dependent plasticity, e.g. using the Drucker–Prager yield function. Typically, and this assumption has also been made in the example shown here, the slope is modelled as a plane strain configuration. A Drucker–Prager yield function with a non-associated, non-dilatant flow rule ($\psi = 0$) has been used. The slope has been is discretised using 2854 six-noded triangular elements with a three-point Gauss integration scheme. Further details regarding the material data and geometry are given in Verhoosel *et al.* (2009).

Figure 7.12 shows the slip plane that develops in the final stages of such a computation. The load application for such an embankment is not trivial, since the self-weight that causes failure cannot be increased further after the critical value has been reached. A force-controlled simulation is therefore not capable of tracing the entire equilibrium path, and a path-following method, see Chapter 4, must be used. Since the problem shows a localised failure mode which involves energy dissipation, an energy release constraint that is based on Equation (4.32) has been used, for which a more detailed formulation in the case of plasticity has been given by Verhoosel *et al.* (2009). In Figure 7.13 the self-weight $|\mathbf{b}|$ is plotted as a function of the downward vertical displacement of the crest of the slope.

Algorithms in which the flow direction and the state variables are not evaluated at the beginning or at the end of the loading step have also been proposed. More precisely, we can distinguish between the generalised trapezoidal return-mapping scheme and the generalised

Box 7.5 Anisotropic plasticity: the Hoffman and Hill yield criteria

The Hill (1950) and the Hoffman (1967) failure criteria have been developed to describe the fracture behaviour of anisotropic (rolled) metals and of composites. Later, they have been adapted to serve as yield criteria for anisotropic materials in a general sense (de Borst and Feenstra 1990; Hashagen and de Borst 2001; Schellekens and de Borst 1990). The Hoffman failure criterion is a quadratic function of the stresses:

$$f(\sigma) = \alpha_{23}(\sigma_{22} - \sigma_{33})^2 + \alpha_{31}(\sigma_{33} - \sigma_{11})^2 + \alpha_{12}(\sigma_{11} - \sigma_{22})^2 + \\ \alpha_{11}\sigma_{11} + \alpha_{22}\sigma_{22} + \alpha_{33}\sigma_{33} + 3\alpha_{44}\sigma_{23}^2 + 3\alpha_{55}\sigma_{31}^2 + 3\alpha_{66}\sigma_{12}^2 - \bar{\sigma}^2$$

In the principal stress space the criterion forms an elliptic paraboloid. The intersections of the yield surface with planes parallel to the deviatoric plane are ellipses, with shapes which are determined by the quadratic part of the function. The expansion of the function along its space diagonal is governed by the terms that are linear in the stress. Setting $\alpha_{11} = \alpha_{22} = \alpha_{33} = 0$, the dependence on the hydrostatic stress is eliminated and the Hill criterion results. For finite element implementation the yield function is reformulated as:

$$f(\boldsymbol{\sigma}) = \frac{1}{2}\boldsymbol{\sigma}^{\mathrm{T}}\mathbf{P}_{\alpha}\boldsymbol{\sigma} + \boldsymbol{\sigma}^{\mathrm{T}}\boldsymbol{\pi}_{\alpha} - \bar{\sigma}^2$$

with

$$\mathbf{P}_{\alpha} = \begin{bmatrix} 2(\alpha_{31}+\alpha_{12}) & -2\alpha_{12} & -2\alpha_{31} & 0 & 0 & 0 \\ -2\alpha_{12} & 2(\alpha_{23}+\alpha_{12}) & -2\alpha_{23} & 0 & 0 & 0 \\ -2\alpha_{31} & -2\alpha_{23} & 2(\alpha_{31}+\alpha_{23}) & 0 & 0 & 0 \\ 0 & 0 & 0 & 6\alpha_{44} & 0 & 0 \\ 0 & 0 & 0 & 0 & 6\alpha_{55} & 0 \\ 0 & 0 & 0 & 0 & 0 & 6\alpha_{66} \end{bmatrix}$$

and $\boldsymbol{\pi}_{\alpha}^{\mathrm{T}} = (\alpha_{11}, \alpha_{22}, \alpha_{33}, 0, 0, 0)$. Use of an associated flow rule gives $\mathbf{m} = \mathbf{P}_{\alpha}\boldsymbol{\sigma} + \boldsymbol{\pi}_{\alpha}$ and the Euler backward algorithm, Equation (7.93), then gives:

$$\boldsymbol{\sigma}_{j+1} = (\mathbf{I} + \Delta\lambda\mathbf{D}^{\mathrm{e}}\mathbf{P}_{\alpha})^{-1}(\boldsymbol{\sigma}_e - \Delta\lambda\mathbf{D}^{\mathrm{e}}\boldsymbol{\pi}_{\alpha})$$

Subsequent substition into the yield function $f(\boldsymbol{\sigma})$ results in a non-linear equation in $\Delta\lambda$.

midpoint return-mapping scheme. In the first class the residuals are defined as:

$$\begin{cases} \mathbf{r}_{\sigma} = \boldsymbol{\sigma}_{j+1} - \boldsymbol{\sigma}_e + \Delta\lambda\mathbf{D}^{\mathrm{e}}\left((1-\theta)\mathbf{m}(\boldsymbol{\sigma}_j, \boldsymbol{\kappa}_j) + \theta\mathbf{m}(\boldsymbol{\sigma}_{j+1}, \boldsymbol{\kappa}_{j+1})\right) \\ \mathbf{r}_{\kappa} = \boldsymbol{\kappa}_{j+1} - \boldsymbol{\kappa}_0 - \Delta\lambda\left((1-\theta)\mathbf{p}(\boldsymbol{\sigma}_j, \boldsymbol{\kappa}_j) + \theta\mathbf{p}(\boldsymbol{\sigma}_{j+1}, \boldsymbol{\kappa}_{j+1})\right) \\ r_f = f(\boldsymbol{\sigma}_{j+1}, \boldsymbol{\kappa}_{j+1}) \end{cases} \tag{7.115}$$

where $0 \leq \theta \leq 1$, so that the flow direction and the hardening variables are determined as a weighted average of the values at the beginning and at the end of the step. In the midpoint

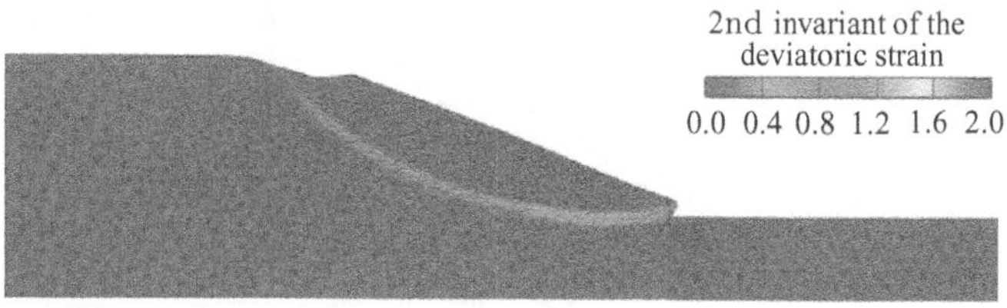

Figure 7.12 Failure mode of the slope stability problem

return-mapping scheme the stress and the hardening variables are evaluated at an intermediate stage, i.e. somewhere between the beginning and the end of the load increment, and these are taken as input for the computation of the local residuals:

$$\begin{cases} \mathbf{r}_\sigma = \boldsymbol{\sigma}_{j+1} - \boldsymbol{\sigma}_e + \Delta\lambda \mathbf{D}^e \mathbf{m}(\boldsymbol{\sigma}_{j+\theta}, \boldsymbol{\kappa}_{j+\theta}) \\ \mathbf{r}_\kappa = \boldsymbol{\kappa}_{j+1} - \boldsymbol{\kappa}_0 - \Delta\lambda \mathbf{p}(\boldsymbol{\sigma}_{j+\theta}, \boldsymbol{\kappa}_{j+\theta}) \\ r_f = f(\boldsymbol{\sigma}_{j+1}, \boldsymbol{\kappa}_{j+1}) \end{cases} \tag{7.116}$$

with $\boldsymbol{\sigma}_{j+\theta} = (1-\theta)\boldsymbol{\sigma}_j + \theta\boldsymbol{\sigma}_{j+1}$ and $\boldsymbol{\kappa}_{j+\theta} = (1-\theta)\boldsymbol{\kappa}_j + \theta\boldsymbol{\kappa}_{j+1}$. Clearly, for both schemes the fully explicit Euler forward method is retrieved when $\theta = 0$, while the (implicit) Euler backward method is obtained for $\theta = 1$.

Ortiz and Simo (1986) have proposed the tangent cutting plane (TCP) algorithm as a successive application of a number of Euler forward steps (Figure 7.14). As with the Euler forward and backward algorithms a trial stress $\boldsymbol{\sigma}_e$ is computed assuming purely elastic behaviour. Then, a yield surface f_0 is constructed through this stress point, and a flow direction $\mathbf{m}_0 = \mathbf{m}(\boldsymbol{\sigma}_e, \boldsymbol{\kappa}_0)$ is computed, with $\boldsymbol{\kappa}_0$ the hardening variables at the end of the previous loading step. When the new stress $\boldsymbol{\sigma}_1$ is not compliant with the yield condition (7.95), the process is repeated until the yield condition is met within a certain tolerance. This process can be summarised as:

$$\boldsymbol{\sigma}_{j+1} = \boldsymbol{\sigma}_j - \Delta\lambda \mathbf{D}^e \mathbf{m}_j \quad , \quad \mathbf{m}_j = \mathbf{m}(\boldsymbol{\sigma}_j, \boldsymbol{\kappa}_j) \tag{7.117}$$

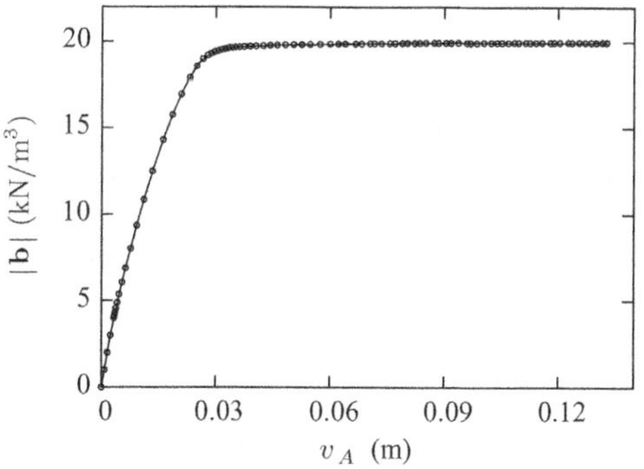

Figure 7.13 Self-weight as a function of the vertical displacement of the crest of the slope

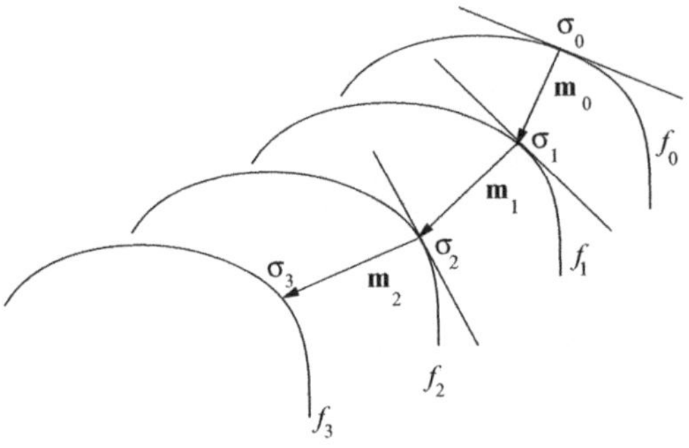

Figure 7.14 The tangent cutting plane algorithm (Ortiz and Simo 1986)

with the initial flow direction $\mathbf{m}_0 = \mathbf{m}(\boldsymbol{\sigma}_e, \kappa_0)$. The plastic flow intensity $\Delta\lambda$ is obtained from a linearisation of the yield condition (7.95):

$$f(\boldsymbol{\sigma}_{j+1}, \kappa_{j+1}) = f(\boldsymbol{\sigma}_j, \kappa_j) - \left(h_j + \mathbf{n}_j^{\mathrm{T}}\mathbf{D}^{\mathrm{e}}\mathbf{m}_j\right)\Delta\lambda \tag{7.118}$$

Requiring the yield condition to hold at $j+1$ provides the explicit expression:

$$\Delta\lambda = \frac{f(\boldsymbol{\sigma}_j, \kappa_j)}{h_j + \mathbf{n}_j^{\mathrm{T}}\mathbf{D}^{\mathrm{e}}\mathbf{m}_j} \tag{7.119}$$

so that we have the following explicit update for the stress at iteration $j+1$:

$$\boldsymbol{\sigma}_{j+1} = \boldsymbol{\sigma}_j - \frac{f(\boldsymbol{\sigma}_j, \kappa_j)}{h_j + \mathbf{n}_j^{\mathrm{T}}\mathbf{D}^{\mathrm{e}}\mathbf{m}_j}\mathbf{D}^{\mathrm{e}}\mathbf{m}_j \tag{7.120}$$

Although the TCP algorithm is conceptually simple and accurate, it is not unconditionally stable because of its explicit character.

For a single iteration, Equation (7.120) specialises as:

$$\boldsymbol{\sigma}_1 = \boldsymbol{\sigma}_e - \frac{f(\boldsymbol{\sigma}_e, \kappa_0)}{h_0 + \mathbf{n}_0^{\mathrm{T}}\mathbf{D}^{\mathrm{e}}\mathbf{m}_0}\mathbf{D}^{\mathrm{e}}\mathbf{m}_0 \tag{7.121}$$

For linear, isotropic hardening $h = h_0$ is constant. Furthermore, for (piecewise) linear yield functions such as Tresca and Mohr–Coulomb, higher-order terms in the expansion of Equation (7.118) vanish and the gradients to the yield surface are the same for the trial stress $\boldsymbol{\sigma}_e$ and for the stress that follows after the return map, $\boldsymbol{\sigma}_1$ – we will return to this property in Section 7.5. The single-iteration TCP algorithm then coincides with an Euler backward algorithm. Indeed, a rigorous return to the yield surface is obtained in a single iteration. The circular shape of the yield surface in the π-plane in combination with the linearity in terms of the pressure p warrant the same properties for the von Mises yield function (7.31) and for the Drucker–Prager yield function (7.34), and it can be shown that higher-order terms cancel (de Borst and Feenstra 1990). For the von Mises yield function the return map attains a particularly simple format:

$$\boldsymbol{\sigma}_1 - \boldsymbol{\sigma}_e = \frac{3\mu}{h + 3\mu}\left(1 - \frac{\bar{\sigma}(\kappa_0)}{\sqrt{3(J_2)_e}}\right)\mathbf{s}_e \tag{7.122}$$

The name radial return method has been coined for this case, since the stress return direction is radially in the π-plane (Wilkins 1964).

As regards the important issue of stability of the integration scheme, Ortiz and Popov (1985) have shown that both for the generalised trapezoidal rule and for the generalised midpoint rule, unconditional stability is obtained for $\theta \geq \frac{1}{2}$, irrespective of the magnitude of the loading step. For the case of the von Mises yield criterion thorough accuracy studies have been carried out for the generalised midpoint rule for $\theta = 0, \theta = \frac{1}{2}$ – the mean-normal method (Rice and Tracey 1971) – and $\theta = 1$ (Krieg and Krieg 1977; Schreyer *et al.* 1979). It was concluded that for this particular yield criterion the Euler backward method is especially accurate for larger loading steps, while the mean-normal method is the most accurate for smaller loading steps.

It is finally noted that Bushnell (1977) has proposed the substepping technique, where the strain increment $\Delta\boldsymbol{\epsilon}$ is divided in a number of, say n, subincrements, see also Pérez-Foguet *et al.* (2001). The elasto-plastic stress computation is then carried out for each subincrement. The substepping technique is mostly used in connection with an Euler forward scheme, but can be applied equally well to the Euler backward schemes. Substepping still involves the linearisation of the strain path over the loading step. This approximation can only be improved when smaller loading steps are taken.

7.4 Tangent Stiffness Operators

Equation (7.93), which sets the dependence of the stress increment on the prescribed strain increment can be conceived as a total stress–strain relation within the loading step. Accordingly, we have a deformation theory of plasticity within a finite loading step rather than a flow theory of plasticity when a return-mapping algorithm is used. For this reason the tangential stiffness relation between stress rate and strain rate that is required when a Newton–Raphson method is used at a global (structural) level, bears resemblance to the tangential operators that result from a deformation theory of plasticity.

For the derivation of the consistent tangent operator the set (7.96) is differentiated to give:

$$\begin{bmatrix} \left(\mathbf{I} + \Delta\lambda \mathbf{D}^{\mathrm{e}} \frac{\partial \mathbf{m}}{\partial \boldsymbol{\sigma}}\right) & \Delta\lambda \mathbf{D}^{\mathrm{e}} \frac{\partial \mathbf{m}}{\partial \boldsymbol{\kappa}} & \mathbf{D}^{\mathrm{e}}\mathbf{m} \\ -\Delta\lambda \frac{\partial \mathbf{p}}{\partial \boldsymbol{\sigma}} & \left(\mathbf{I} - \Delta\lambda \frac{\partial \mathbf{p}}{\partial \boldsymbol{\kappa}}\right) & -\mathbf{p} \\ \mathbf{n}^{\mathrm{T}} & \left(\frac{\partial f}{\partial \boldsymbol{\kappa}}\right)^{\mathrm{T}} & 0 \end{bmatrix} \begin{pmatrix} \dot{\boldsymbol{\sigma}} \\ \dot{\boldsymbol{\kappa}} \\ \dot{\lambda} \end{pmatrix} = \begin{pmatrix} \mathbf{D}^{\mathrm{e}}\dot{\boldsymbol{\epsilon}} \\ \mathbf{0} \\ 0 \end{pmatrix} \tag{7.123}$$

Inversion of this system formally gives:

$$\begin{pmatrix} \dot{\boldsymbol{\sigma}} \\ \dot{\boldsymbol{\kappa}} \\ \dot{\lambda} \end{pmatrix} = \begin{bmatrix} \mathbf{D}_{11} & \mathbf{D}_{12} & \mathbf{D}_{13} \\ \mathbf{D}_{21} & \mathbf{D}_{22} & \mathbf{D}_{23} \\ \mathbf{D}_{31} & \mathbf{D}_{32} & D_{33} \end{bmatrix} \begin{pmatrix} \mathbf{D}^{\mathrm{e}}\dot{\boldsymbol{\epsilon}} \\ \mathbf{0} \\ 0 \end{pmatrix} \tag{7.124}$$

so that

$$\dot{\boldsymbol{\sigma}} = \mathbf{D}_{11}\mathbf{D}^{\mathrm{e}}\dot{\boldsymbol{\epsilon}} \tag{7.125}$$

and $\mathbf{D}_{11}\mathbf{D}^{\mathrm{e}}$ is the consistent tangent operator as it sets the relation between the stress rate $\dot{\boldsymbol{\sigma}}$ and the strain rate $\dot{\boldsymbol{\epsilon}}$.

The simpler structure of the reduced system (7.99) permits the construction of the consistent tangential stiffness operator in an explicit format. Differentiation of Equations (7.99) results in

$$\begin{cases} \dot{\boldsymbol{\sigma}} = \mathbf{D}^{\mathrm{e}}\dot{\boldsymbol{\epsilon}} - \dot{\lambda}\mathbf{D}^{\mathrm{e}}\mathbf{m} - \Delta\lambda\mathbf{D}^{\mathrm{e}}\frac{\partial \mathbf{m}}{\partial \boldsymbol{\sigma}}\dot{\boldsymbol{\sigma}} - \Delta\lambda\mathbf{D}^{\mathrm{e}}\frac{\partial \mathbf{m}}{\partial \lambda}\dot{\lambda} \\ \mathbf{n}^{\mathrm{T}}\dot{\boldsymbol{\sigma}} - h\dot{\lambda} = 0 \end{cases} \tag{7.126}$$

Using Equations (7.102) and (7.106) we rewrite Equation (7.126) as:

$$\begin{cases} \mathbf{A}\dot{\boldsymbol{\sigma}} = \mathbf{D}^{\mathrm{e}}\dot{\boldsymbol{\epsilon}} - \dot{\lambda}\mathbf{D}^{\mathrm{e}}\bar{\mathbf{m}} \\ \mathbf{n}^{\mathrm{T}}\dot{\boldsymbol{\sigma}} - h\dot{\lambda} = 0 \end{cases} \tag{7.127}$$

Introduction of the 'pseudo-elastic' stiffness matrix

$$\mathbf{H} = \mathbf{A}^{-1}\mathbf{D}^{\mathrm{e}} \tag{7.128}$$

permits the rewriting of Equation (7.126) in the following format:

$$\begin{cases} \dot{\boldsymbol{\sigma}} = \mathbf{H}(\dot{\boldsymbol{\epsilon}} - \dot{\lambda}\bar{\mathbf{m}}) \\ \mathbf{n}^{\mathrm{T}}\dot{\boldsymbol{\sigma}} - h\dot{\lambda} = 0 \end{cases} \tag{7.129}$$

Invoking the second equation of the above set, the algorithmic tangential stiffness relation between stress rate and strain rate can be derived as:

$$\dot{\boldsymbol{\sigma}} = \left(\mathbf{H} - \frac{\mathbf{H}\bar{\mathbf{m}}\mathbf{n}^{\mathrm{T}}\mathbf{H}}{h + \mathbf{n}^{\mathrm{T}}\mathbf{H}\bar{\mathbf{m}}}\right)\dot{\boldsymbol{\epsilon}} \tag{7.130}$$

or, invoking Equation (7.128) again:

$$\dot{\boldsymbol{\sigma}} = \underbrace{\left(\mathbf{A}^{-1} - \frac{\mathbf{H}\bar{\mathbf{m}}\mathbf{n}^{\mathrm{T}}\mathbf{A}^{-1}}{h + \mathbf{n}^{\mathrm{T}}\mathbf{H}\bar{\mathbf{m}}}\right)\mathbf{D}^{\mathrm{e}}}_{\mathbf{D}_{11}}\dot{\boldsymbol{\epsilon}} \tag{7.131}$$

which better brings out the similarity with the consistent tangent operator derived for the full set, cf. Equation (7.125). For von Mises plasticity with isotropic hardening, a particularly simple form of the tangential stiffness matrix can be constructed, which consists of a simple modification of the elastic and plastic stiffness parameters, see Box 7.6 for details. For infinitesimally small load steps $\Delta\lambda \to 0$, and consequently, $\mathbf{A} \to \mathbf{I}$, cf. Equation (7.102). Then, $\mathbf{H} \to \mathbf{D}^{\mathrm{e}}$ and the tangential stiffness matrix for continuum plasticity is recovered, Equation (7.76). However, for large increments of plastic strain, $\mathbf{H}$ can differ significantly from $\mathbf{D}^{\mathrm{e}}$.

Normally, the matrix $\mathbf{H}$ will be symmetric. This can be seen most easily by considering its inverse:

$$\mathbf{H}^{-1} = (\mathbf{D}^{\mathrm{e}})^{-1}\mathbf{A} \tag{7.132}$$

or, with $\mathbf{A}$ according to Equation (7.102),

$$\mathbf{H}^{-1} = (\mathbf{D}^{\mathrm{e}})^{-1} + \Delta\lambda\frac{\partial \mathbf{m}}{\partial \boldsymbol{\sigma}} \tag{7.133}$$

Box 7.6 Von Mises (J_2) plasticity with isotropic hardening

For von Mises plasticity the 'pseudo-elastic' stiffness relation (7.128) particularises as (de Borst 1989; Simo and Taylor 1985):

$$\mathbf{H} = (\mathbf{D}^e)^*$$

where $(\mathbf{D}^e)^*$ has the same structure as $\mathbf{D}^e$, but with a modified shear modulus

$$\mu^* = \frac{E}{2(1+\nu) + 3E\Delta\lambda/\sqrt{3J_2}}$$

and a modified Poisson ratio

$$\nu^* = \frac{\nu + E\Delta\lambda/2\sqrt{3J_2}}{1 + E\Delta\lambda/\sqrt{3J_2}}$$

which reduce to their continuum counterparts when plastic flow is not included ($\Delta\lambda \equiv 0$). Furthermore, a modified value for the plastic hardening modulus

$$h^* = \frac{h}{1 - E\Delta\lambda/\sqrt{3J_2}}$$

has to be inserted in the consistent tangential stiffness matrix, Equation (7.130). The above expressions provide a simple modification to the continuum elasto-plastic stiffness matrix.

For coaxial flow rules, i.e. when Equation (7.51) holds, we can further reduce $\mathbf{H}^{-1}$ to:

$$\mathbf{H}^{-1} = (\mathbf{D}^e)^{-1} + \Delta\lambda \frac{\partial^2 g}{\partial \sigma^2} \tag{7.134}$$

which is clearly symmetric, and therefore also $\mathbf{H}$. The symmetry of $\mathbf{H}$ is, however, a necessary, but not a sufficient condition for the symmetry of the consistent tangential stiffness matrix as expressed through Equation (7.130). Clearly, a non-associated flow rule ($\mathbf{m} \neq \mathbf{n}$) causes the consistent tangential stiffness matrix to become non-symmetric, but this also happens when the plastic flow direction $\mathbf{m}$ is dependent on the plastic strain history through the hardening parameter κ. Then, $\bar{\mathbf{m}}$ differs from the plastic flow direction $\mathbf{m}$, Equation (7.106), and therefore from the gradient to the yield function $\mathbf{n}$. This source of non-symmetry, which is called non-associated hardening, is not present in the conventional expression for the elasto-plastic tangential stiffness relation. Furthermore, the scheme that is used to integrate the rate equations influences the tangential stiffness matrix and can cause the matrix to become non-symmetric, even when the tangential stiffness matrix for the rate problem is symmetric (Box 7.7). It is finally noted that use of a consistent tangential stiffness matrix is meaningful only when a full Newton–Raphson procedure is employed to solve the set of non-linear equations at a global (structural) level. This is because the magnitude of the plastic strain within the loading step enters the tangential stiffness matrix.

Box 7.7 Symmetry *vs* non-symmetry of consistently linearised tangent operators

Among the class of implicit integration schemes, the Euler backward scheme ($\theta = 1$) is the only scheme that can preserve symmetry (Ortiz and Martin 1989). For the von Mises yield criterion with an associated flow rule and no hardening this can be shown straightforwardly. Then, **m** and **n** are co-linear, and we have instead of Equation (7.126):

$$\dot{\boldsymbol{\sigma}} = \mathbf{D}^{\mathrm{e}}\dot{\boldsymbol{\epsilon}} - \dot{\lambda}\mathbf{D}^{\mathrm{e}}\mathbf{n}_{j+\theta} - \Delta\lambda\mathbf{D}^{\mathrm{e}}\frac{\partial \mathbf{n}_{j+\theta}}{\partial \boldsymbol{\sigma}_{j+1}}\dot{\boldsymbol{\sigma}}$$

Defining **A** now as:

$$\mathbf{A} = \mathbf{I} + \Delta\lambda\mathbf{D}^{\mathrm{e}}\frac{\partial \mathbf{n}_{j+\theta}}{\partial \boldsymbol{\sigma}_{j+1}}$$

one obtains:

$$\dot{\boldsymbol{\sigma}} = \mathbf{A}^{-1}\mathbf{D}^{\mathrm{e}}(\dot{\boldsymbol{\epsilon}} - \dot{\lambda}\mathbf{n}_{j+\theta})$$

Since compliance with the yield function is enforced at the end of the loading step, cf. Equation (7.99), the consistency condition gives:

$$\mathbf{n}^{\mathrm{T}}_{j+1}\dot{\boldsymbol{\sigma}} = 0$$

and the consistent tangential stiffness relation reads:

$$\dot{\boldsymbol{\sigma}} = \left(\mathbf{A}^{-1} - \frac{\mathbf{A}^{-1}\mathbf{D}^{\mathrm{e}}\mathbf{n}_{j+\theta}\mathbf{n}^{\mathrm{T}}_{j+1}\mathbf{A}^{-1}}{\mathbf{n}^{\mathrm{T}}_{j+1}\mathbf{A}^{-1}\mathbf{D}^{\mathrm{e}}\mathbf{n}_{j+\theta}}\right)\mathbf{D}^{\mathrm{e}}\dot{\boldsymbol{\epsilon}}$$

Thus, symmetry is lost even for associated von Mises plasticity as a pure consequence of the update algorithm.

7.5 Multi-surface Plasticity

7.5.1 Koiter's Generalisation

A complication occurs when the plastic potential function, or in the case of an associated flow rule the yield function, is not continously differentiable along the entire yield surface. This for instance occurs for the Tresca and Mohr–Coulomb yield surfaces at places where two planes intersect. At such vertices a unique flow direction cannot be defined. At most, it can be stated that the plastic flow direction is within the wedge spanned by the flow direction vectors that belong to either of the plastic potential functions g_1 and g_2 that are adjacent to the corner of Figure 7.15. Mathematically, we obtain that

$$\dot{\boldsymbol{\epsilon}}^{\mathrm{p}} = \dot{\lambda}_1\mathbf{m}_1 + \dot{\lambda}_2\mathbf{m}_2 \tag{7.135}$$

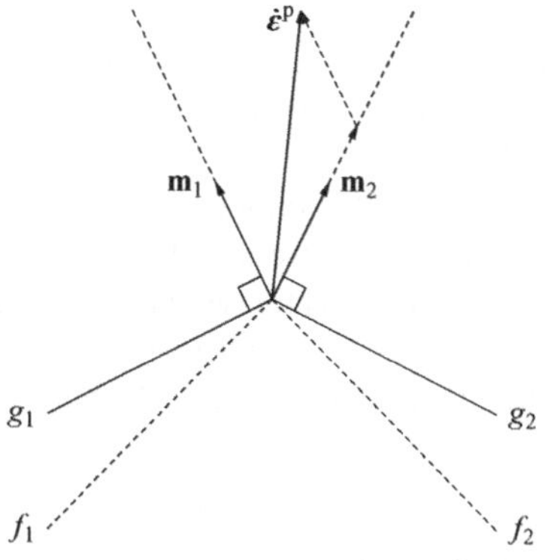

Figure 7.15 Corner in the yield surface and/or plastic potential function

where

$$\mathbf{m}_1 = \frac{\partial g_1}{\partial \boldsymbol{\sigma}} \quad , \quad \mathbf{m}_2 = \frac{\partial g_2}{\partial \boldsymbol{\sigma}}$$

The constraints are that we cannot have a negative magnitude of the plastic flow in either direction $\mathbf{m}_1$ or $\mathbf{m}_2$:

$$\dot{\lambda}_1 \geq 0 \quad \wedge \quad \dot{\lambda}_2 \geq 0$$

Thus, Koiter's generalisation for singularities in the yield surface states that the plastic strain-rate vector can be written as a linear combination of both flow vectors $\mathbf{m}_i, i = 1, 2$, with non-negative coefficients (Koiter 1953). Application of the Euler backward algorithm results in the following discrete set of equations:

$$\begin{cases} \boldsymbol{\sigma}_{j+1} = \boldsymbol{\sigma}_e - \sum_{i=1}^{2} \Delta\lambda_i \mathbf{D}^e \mathbf{m}_i \\ \boldsymbol{\kappa}_{j+1} = \boldsymbol{\kappa}_0 + \sum_{i=1}^{2} \Delta\lambda_i \mathbf{p}_i \end{cases} \tag{7.136}$$

subject to the discrete Karush–Kuhn–Tucker conditions:

$$\Delta\lambda_i \geq 0 \quad , \quad f_i \leq 0 \quad , \quad \Delta\lambda_i f_i = 0 \tag{7.137}$$

The relative magnitudes of the plastic multipliers $\dot{\lambda}_1$ and $\dot{\lambda}_2$ depend on the external kinematic constraints that are imposed on the deformation. Taking the Tresca yield function with an associated flow rule as an example, we find for pure uniaxial tension that the plastic strain rate is given by:

$$\begin{pmatrix} \dot{\epsilon}_1^p \\ \dot{\epsilon}_2^p \\ \dot{\epsilon}_3^p \end{pmatrix} = \dot{\lambda}_1 \begin{pmatrix} 1 \\ 0 \\ -1 \end{pmatrix} + \dot{\lambda}_2 \begin{pmatrix} 1 \\ -1 \\ 0 \end{pmatrix} \tag{7.138}$$

Suppose now that the boundary conditions are such that while loading is in the 1-direction, the displacements in the 2- and 3-directions are not restrained. Then, symmetry dictates that the displacements in the 2- and 3-directions will be equal, which results in $\dot{\lambda}_1 = \dot{\lambda}_2 = \frac{1}{2}\dot{\lambda}$.

Next, suppose that the boundary conditions are such that the displacements in the 3-direction are prohibited, and that only those in the 2-direction are free (plane-strain conditions). Now, one of the flow mechanisms will not become active, so that $\dot{\lambda}_1 = 0$, but $\dot{\lambda}_2 = \dot{\lambda}$. Note that under this kinematic condition there is no plastic straining in the 3-direction. Since also the total strain rate in this direction is zero, this implies that there is also no elastic strain rate in this direction, and consequently, there is no stress build-up in the 3-direction. Here, we have a salient difference with the von Mises criterion for which an associated flow rule predicts plastic straining in the intermediate direction, and as an ultimate consequence thereof, a change in the stress level in the direction of the intermediate principal stress.

7.5.2 *Rankine Plasticity for Concrete*

The proper modelling of tension-compression biaxial stress states in plain and reinforced concrete is of major practical importance, since such stress states occur in critical regions and crack initiation under such stress conditions often acts as a precursor to progressive collapse of concrete structures. Examples include shear-critical reinforced concrete beams and splitting failure in plain concrete structures. The modelling involves different inelastic mechanisms which play a role under such stress conditions. Often, smeared crack or damage models are used for tensile cracking and a plasticity formalism is employed for the failure of concrete in (biaxial) compression. The simultaneous satisfaction of a fracture/damage criterion and a yield function is not trivial. In de Borst and Nauta (1985) this issue is solved through a local procedure at integration point level in which cracking and plasticity were treated in an iterative fashion, such that during the computation of the plastic flow the cracking strain increment was treated as an initial strain increment, and subsequently the fracture strain increment was updated while freezing the computed plastic flow increment etc. Although this procedure has been applied successfully in a number of calculations, numerical difficulties with state changes have been reported (Crisfield and Wills 1989), especially for tension–compression stress states.

An alternative, robust option is to model both the fracture behaviour under tensile stresses and the 'crushing' of concrete under compressive stresses via a plasticity formalism. For this purpose a composite yield function can be constructed (Figure 7.16), in which a Rankine (principal stress) yield function is used to limit the tensile stresses (Rankine 1858) and a Drucker–Prager yield criterion is employed to model the compression–compression regime and part of the tension–compression regime. This composite yield contour appears to closely match classical experimental data (Kupfer and Gerstle 1973). Since both tensile and compressive failures are now modelled using plasticity, the 'corner' regime, where both yield contours intersect, can be handled using Koiter's generalisation for plastic flow at a corner in the yield contour (Koiter 1953), and an Euler backward integration algorithm can be developed, including a consistently linearised tangential stiffness matrix. A drawback of the plasticity-based approach is that the stiffness degradation due to progressive damage cannot be modelled, which is disadvantageous especially under cyclic loadings.

The composite yield surface consists of a Rankine yield function, f_1, and a Drucker–Prager yield function, f_2. Both yield functions will subsequently be formulated for plane-strain conditions. Plane-stress conditions can be derived either by direct reduction, or indirectly by enforcing the plane stress condition, $\sigma_{zz} = 0$, via a compression/expansion algorithm (de Borst 1991). In this algorithm, the strain vector is expanded by including the normal strain in the

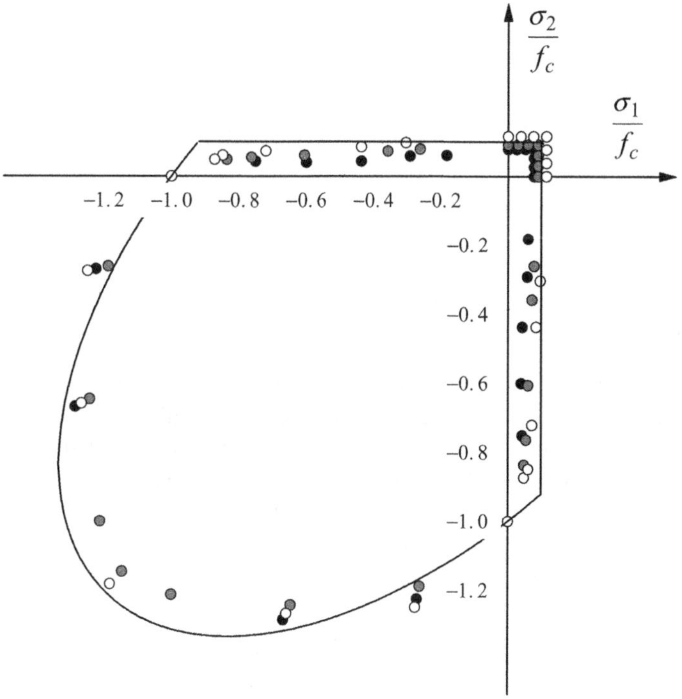

Figure 7.16 Comparison of a composite Rankine/Drucker–Prager yield surface with experimental data for biaxially loaded concrete (Kupfer and Gerstle 1973)

constrained direction in the beginning of an iteration. Then, the constitutive model is evaluated in the expanded, or plane-strain stress space. Finally, the stress vector is compressed such that the condition $\sigma_{zz} = 0$ is enforced rigorously.

The composite yield contour is thus expressed by the following yield functions:

$$\begin{cases} f_1 = \sqrt{\frac{1}{2}\boldsymbol{\sigma}^{\mathrm{T}}\mathbf{P}_1\boldsymbol{\sigma}} + \frac{1}{2}\boldsymbol{\pi}_1^{\mathrm{T}}\boldsymbol{\sigma} - \bar{\sigma}_1(\kappa_1) \\ f_2 = \sqrt{\frac{1}{2}\boldsymbol{\sigma}^{\mathrm{T}}\mathbf{P}_2\boldsymbol{\sigma}} + \alpha_f\boldsymbol{\pi}_2^{\mathrm{T}}\boldsymbol{\sigma} - \beta\bar{\sigma}_2(\kappa_2) \end{cases} \tag{7.139}$$

with $\boldsymbol{\sigma}^{\mathrm{T}} = (\sigma_{xx}, \sigma_{yy}, \sigma_{zz}, \sigma_{xy})$ and the projection matrices $\mathbf{P}_1$ and $\mathbf{P}_2$ are given by:

$$\mathbf{P}_1 = \begin{bmatrix} \frac{1}{2} & -\frac{1}{2} & 0 & 0 \\ -\frac{1}{2} & \frac{1}{2} & 0 & 0 \\ 0 & 0 & 0 & 0 \\ 0 & 0 & 0 & 2 \end{bmatrix} \tag{7.140}$$

and

$$\mathbf{P}_2 = \begin{bmatrix} 2 & -1 & -1 & 0 \\ -1 & 2 & -1 & 0 \\ -1 & -1 & 2 & 0 \\ 0 & 0 & 0 & 6 \end{bmatrix} \tag{7.141}$$

respectively. The projection vectors read:

$$\begin{cases} \boldsymbol{\pi}_1^{\mathrm{T}} = (1, 1, 0, 0) \\ \boldsymbol{\pi}_2^{\mathrm{T}} = (1, 1, 1, 0) \end{cases} \tag{7.142}$$

The equivalent stress $\bar{\sigma}_1$ is the uniaxial tensile strength, and is a function of an internal variable, κ_1. The equivalent stress $\bar{\sigma}_2$ is the uniaxial compressive strength, which is also expressed as a function of an internal variable, κ_2. The factors α_f and β can be related to the uniaxial and biaxial compressive strengths. In the tensile regime an associated flow rule suffices, but in the compressive regime an excessive plastic volume increase is predicted by an associated flow rule, hence the following potential functions are used:

$$\begin{cases} g_1 = f_1 \\ g_2 = \sqrt{\frac{1}{2}\boldsymbol{\sigma}^{\mathrm{T}}\mathbf{P}_2\boldsymbol{\sigma}} + \alpha_g \boldsymbol{\pi}_2^{\mathrm{T}}\boldsymbol{\sigma} - \beta\bar{\sigma}_2(\kappa_2) \end{cases} \tag{7.143}$$

with typically $\alpha_g < \alpha_f$.

For the choice of g_i formulated in Equations (7.143) the Euler backward procedure (7.136) evolves as:

$$\begin{cases} \boldsymbol{\sigma}_{j+1} = \mathbf{A}^{-1}\left(\boldsymbol{\sigma}_e - \frac{1}{2}\Delta\lambda_1\mathbf{D}^{\mathrm{e}}\boldsymbol{\pi}_1 - \alpha_g\Delta\lambda_2\mathbf{D}^{\mathrm{e}}\boldsymbol{\pi}_2\right) \\ \boldsymbol{\kappa}_{j+1} = \boldsymbol{\kappa}_0 + \sum_{i=1}^{2}\Delta\lambda_i\mathbf{p}_i \end{cases} \tag{7.144}$$

with the matrix $\mathbf{A}$ given by:

$$\mathbf{A} = \mathbf{I} + \frac{\Delta\lambda_1}{2\Psi_1}\mathbf{D}^{\mathrm{e}}\mathbf{P}_1 + \frac{\Delta\lambda_2}{2\Psi_2}\mathbf{D}^{\mathrm{e}}\mathbf{P}_2 \tag{7.145}$$

where the denominators read:

$$\begin{cases} \Psi_1 = \sqrt{\frac{1}{2}\boldsymbol{\sigma}_{j+1}^{\mathrm{T}}\mathbf{P}_1\boldsymbol{\sigma}_{j+1}} \\ \Psi_2 = \sqrt{\frac{1}{2}\boldsymbol{\sigma}_{j+1}^{\mathrm{T}}\mathbf{P}_2\boldsymbol{\sigma}_{j+1}} \end{cases} \tag{7.146}$$

These expressions are not convenient because they involve the updated stress $\boldsymbol{\sigma}_{j+1}$. Using projections as outlined for the Drucker–Prager yield function, expressions can be obtained for Ψ_1 and Ψ_2 which are only functions of the variables in the trial state and of the plastic multipliers (Feenstra and de Borst 1996).

The singular point at the apex in the Rankine and the Drucker–Prager yield surfaces can cause numerical problems because the denominators Ψ_1 as well as Ψ_2 can become zero at the apex of these yield surfaces, see Equation (7.146). Because of the assumption of isotropic elasticity, the elastic stiffness matrix $\mathbf{D}^e$ and the projection matrices $\mathbf{P}_1$ and $\mathbf{P}_2$ have the same eigenvector space. This means that the spectral decomposition is given by the same transformation, according to:

$$\begin{cases} \mathbf{D}^e = \mathbf{W}\boldsymbol{\Lambda}_D\mathbf{W}^T \\ \mathbf{P}_1 = \mathbf{W}\boldsymbol{\Lambda}_{P1}\mathbf{W}^T \\ \mathbf{P}_2 = \mathbf{W}\boldsymbol{\Lambda}_{P2}\mathbf{W}^T \end{cases} \tag{7.147}$$

with the diagonal matrices

$$\begin{cases} \boldsymbol{\Lambda}_D = \text{diag}\left[\frac{E}{1+\nu}, \frac{E}{1+\nu}, \frac{E}{1-2\nu}, \frac{E}{2(1+\nu)}\right] \\ \boldsymbol{\Lambda}_{P1} = \text{diag}\,[0, 1, 0, 2] \\ \boldsymbol{\Lambda}_{P2} = \text{diag}\,[3, 3, 0, 6] \end{cases} \tag{7.148}$$

and the orthogonal matrix

$$\mathbf{W} = \begin{bmatrix} \frac{1}{\sqrt{6}} & -\frac{1}{\sqrt{2}} & \frac{1}{\sqrt{3}} & 0 \\ \frac{1}{\sqrt{6}} & \frac{1}{\sqrt{2}} & \frac{1}{\sqrt{3}} & 0 \\ -\frac{2}{\sqrt{6}} & 0 & \frac{1}{\sqrt{3}} & 0 \\ 0 & 0 & 0 & 1 \end{bmatrix} \tag{7.149}$$

The matrix $\mathbf{A}$ now simplifies to:

$$\mathbf{A} = \mathbf{W}\left[\mathbf{I} + \frac{\Delta\lambda_1}{2\Psi_1}\boldsymbol{\Lambda}_D\boldsymbol{\Lambda}_{P1} + \frac{\Delta\lambda_2}{2\Psi_2}\boldsymbol{\Lambda}_D\boldsymbol{\Lambda}_{P2}\right]\mathbf{W}^T \tag{7.150}$$

which can be inverted to give:

$$\mathbf{A}^{-1} = \mathbf{W}\,\text{diag}\left[\frac{\Psi_2}{\Psi_2 + 3\Delta\lambda_2 G}, \frac{\Psi_1\Psi_2}{\Psi_1\Psi_2 + (\Psi_2\Delta\lambda_1 + 3\Psi_1\Delta\lambda_2)G}, 1, \frac{\Psi_1\Psi_2}{\Psi_1\Psi_2 + (\Psi_2\Delta\lambda_1 + 3\Psi_1\Delta\lambda_2)G}\right]\mathbf{W}^T \tag{7.151}$$

When the stress is at the apex of the Rankine yield function, i.e. when $\boldsymbol{\sigma}^T = [\bar{\sigma}_1, \bar{\sigma}_1, 0, 0]$, we have $\Psi_1 = 0$. Since the Drucker–Prager yield function is not active at this point, $\Delta\lambda_2$ equals zero, and the limit of the mapping matrix is given by

$$\lim_{\Psi_1 \to 0} \mathbf{A}^{-1} = \begin{bmatrix} \frac{1}{2} & \frac{1}{2} & 0 & 0 \\ \frac{1}{2} & \frac{1}{2} & 0 & 0 \\ 0 & 0 & 1 & 0 \\ 0 & 0 & 0 & 0 \end{bmatrix} \tag{7.152}$$

so that the return map reads:

$$\boldsymbol{\sigma}_{j+1} = \lim_{\Psi_1 \to 0} \mathbf{A}^{-1}(\boldsymbol{\sigma}_e - \frac{1}{2}\Delta\lambda_1 \mathbf{D}^{\mathrm{e}} \boldsymbol{\pi}_1) \tag{7.153}$$

Another situation which could cause numerical problems is when the final stress is at the apex of the Drucker–Prager yield surface, $\Psi_2 = 0$. Because the tensile stresses are bounded by the tensile strength via the Rankine criterion, this situation cannot occur.

When the stress is at an intersection of two yield surfaces, both yield functions $f_i(\boldsymbol{\sigma}_{j+1}, \boldsymbol{\kappa}_{j+1})$ must vanish. Upon substitution of $\boldsymbol{\sigma}_{j+1}$ and $\boldsymbol{\kappa}_{j+1}$, as given by Equations (7.136), the yield functions f_i can be expressed in terms of the plastic multipliers $\Delta\lambda_i$, and reduce to a coupled system of scalar equations:

$$\begin{cases} f_1(\Delta\lambda_1, \Delta\lambda_2) = 0 \\ f_2(\Delta\lambda_1, \Delta\lambda_2) = 0 \end{cases} \tag{7.154}$$

which have to be solved for $(\Delta\lambda_1, \Delta\lambda_2)$ to obtain the final stress state which is subject to the constraints of the discrete Karush–Kuhn–Tucker conditions, Equation (7.137). The location of the intersection between two yield surfaces is unknown at the beginning of a step and the initial configuration cannot provide sufficient information for determining which surface is active at the end of the load step. Simo *et al.* (1988) have proposed an algorithm in which the assumption is made that the number of active yield surfaces in the final stress state is less than or equal to the number of active yield surfaces in the trial stress state. This implies that it is not possible for a yield function which is inactive in the trial state to become active during the return map. As explained by Pramono and Willam (1989) this assumption is not valid for softening plasticity. For this reason Feenstra and de Borst (1996) have modified the approach of Simo *et al.* (1988) to account for the fact that a yield surface can become active during the return map. Additional constraints c_j are introduced, which indicate the status of the yield functions. Initially, the constraints are determined by the violation of the yield criterion in the trial state:

$$c_i = \begin{cases} 1 & \text{if} \quad f_i(\boldsymbol{\sigma}_e, \boldsymbol{\kappa}_0) > 0 \\ 0 & \text{if} \quad f_i(\boldsymbol{\sigma}_e, \boldsymbol{\kappa}_0) \leq 0 \end{cases} \tag{7.155}$$

During the return map the active yield functions are determined through the conditions

$$c_i = \begin{cases} 1 & \text{if} \quad \Delta\lambda_i > 0 \vee f_i > 0 \\ 0 & \text{if} \quad \Delta\lambda_i < 0 \wedge f_i < 0 \end{cases} \tag{7.156}$$

The additional constraints are introduced for numerical convenience, since the non-linear constraint equations, Equations (7.154), are now expressed as

$$\begin{cases} c_1 f_1(\Delta\lambda_1, \Delta\lambda_2) + (1 - c_1)\Delta\lambda_1 = 0 \\ c_2 f_2(\Delta\lambda_1, \Delta\lambda_2) + (1 - c_2)\Delta\lambda_2 = 0 \end{cases} \tag{7.157}$$

so that the conditions (7.137) are enforced simultaneously. The solution of this system of equations is obtained through a local iterative procedure, for instance with a Quasi Newton update of the Jacobian (Dennis and Schnabel 1983). The success of this approach depends on

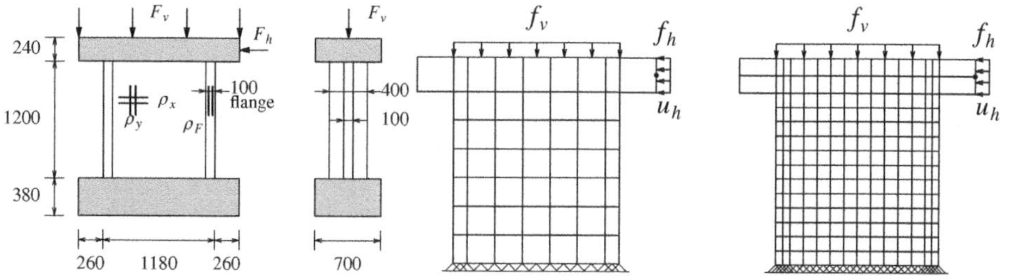

Figure 7.17 Experimental set-up and finite element discretisation of shear wall [panel S2 of Maier and Thürlimann (1985), all dimensions are in millimetres]

the initial Jacobian, which is typically determined from the linearisation of the yield functions in the trial state. With this algorithm it is usually possible to arrive at the correct number of active yield surfaces within ten iterations, also when more yield conditions are violated in the trial state.

The analysis of shear wall panels is an example of the application of a composite plasticity model to reinforced concrete. The panel which is shown in Figure 7.17 has been tested by Maier and Thürlimann (1985). The panel is initially loaded by a vertical compressive force, and then by a horizontal force until the experiment becomes unstable and failure occurs. In the experimental set-up, the panels were supported on a base block and loaded through a thick top slab. Linear elasticity has been assumed for the top slab and no reinforcement has been added in the top slab. The supporting block has been replaced by fixed supports in the x- and y-directions. Figure 7.17 shows that the horizontal and vertical loads have been applied as uniformly distributed loads. The figure also shows the finite element discretisation, with quadratic plane-stress elements and nine-point Gaussian integration for the concrete as well as for the reinforcement. The reinforcement consists of reinforcing grids in two directions with a diameter of 8 mm and a cover of 10 mm. The reinforcement ratios in the web of the panel are equal to 0.0103 and 0.0116 for the x- and y-directions, respectively. The reinforcement ratio in the flange of the panel is equal to 0.0116. The material properties have been averaged from the experimental data, but with a reduced compressive strength $f_c = 27.5$ MPa. The following material parameters have been used: Young's modulus $E_c = 30$ GPa, Poisson's ratio $\nu = 0.15$, tensile strength $f_t = 2.2$ MPa and a tensile fracture energy $\mathcal{G}_c = 0.07$ N/mm. To strive for mesh objectivity in the compressive region as well, a compressive fracture energy $\mathcal{G}_c^{\mathrm{II}} = 50$ N/mm has been included in the model (Feenstra and de Borst 1996).

The panel is subjected to an initial vertical load of 1653 kN, which results in an initial horizontal displacement of 0.29 mm in the experiment. The calculated initial displacement is equal to -54×10^{-6} mm which indicates a possible eccentricity in the experimental set-up. After the initial vertical load, the horizontal load is applied using a path-following method with the horizontal direction u_h as the active degree of freedom and load steps of approximately 0.1 mm. Using a Newton–Raphson iterative procedure, converged solutions could be obtained in the entire loading regime. The load–displacement diagram for u_h is shown in Figure 7.18, which shows a fair agreement between the experimental and calculated response for both discretisations. The failure mechanism was rather explosive and caused a complete loss of

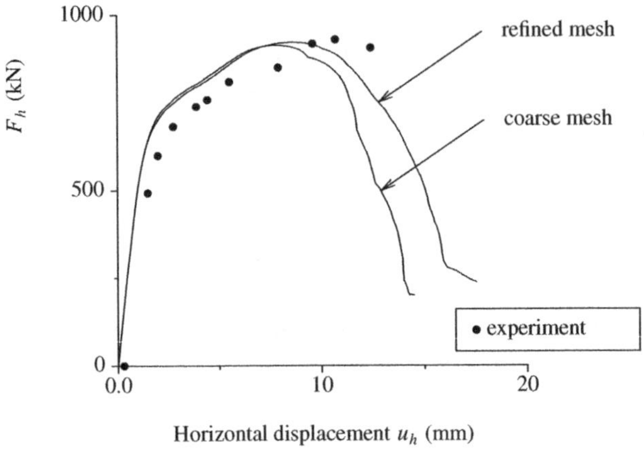

Figure 7.18 Experimental and computed load–displacement curves for shear wall [panel S2 of Maier and Thürlimann (1985)]

load-carrying capacity (Maier and Thürlimann 1985), which is well simulated by the computed brittle behaviour of the panel after maximum load.

7.5.3 Tresca and Mohr–Coulomb Plasticity

Early numerical work on yield surfaces with corners focused on the Tresca and Mohr–Coulomb criteria. As a first solution, it was proposed to locally round-off the corners (Nayak and Zienkiewicz 1972; Owen and Hinton 1980). However, the early approach in which the Mohr–Coulomb yield criterion was replaced by the Drucker–Prager yield criterion in the vicinity of the corner effectively introduces new corners at the intersection of the Mohr–Coulomb and the Drucker–Prager yield surfaces. A better approach is to place a small cone in the corner such that a smooth transition between the regular part of the yield surface and this cone is obtained. A disadvantage of this procedure is that the introduction of such a strongly curved part in the yield surface causes the accuracy and stability properties to deteriorate (Ortiz and Popov 1985).

A more robust approach is to adopt Koiter's generalisation (de Borst 1987). At the corner of the Mohr–Coulomb or Tresca yield surfaces, both yield functions must be satisfied identically:

$$\begin{cases} f_1(\boldsymbol{\sigma}_{j+1}, \kappa_{j+1}) = 0 \\ f_2(\boldsymbol{\sigma}_{j+1}, \kappa_{j+1}) = 0 \end{cases} \tag{7.158}$$

Substitution of Equation $(7.136)_1$ leads to:

$$\begin{cases} f_1(\boldsymbol{\sigma}_e - \Delta\lambda_1 \mathbf{D}^e \mathbf{m}_1 - \Delta\lambda_2 \mathbf{D}^e \mathbf{m}_2, \kappa_{j+1}) = 0 \\ f_2(\boldsymbol{\sigma}_e - \Delta\lambda_1 \mathbf{D}^e \mathbf{m}_1 - \Delta\lambda_2 \mathbf{D}^e \mathbf{m}_2, \kappa_{j+1}) = 0 \end{cases} \tag{7.159}$$

We recall that the Mohr–Coulomb and Tresca yield functions are linear in the principal stress space. Since for a single hardening mechanism, $\kappa_{j+1} = \kappa_0 + h(\Delta\lambda_1 + \Delta\lambda_2)$, cf.

Equation (7.136), and assuming linear hardening with respect to the cohesion, so that the hardening modulus h becomes a constant, a first-order Taylor series expansion results in a stress $\boldsymbol{\sigma}_1$ that exactly satisfies both yield functions f_1 and f_2, and one obtains:

$$\begin{cases} \mathcal{H}_1 \Delta\lambda_1 + \mathcal{H}_2 \Delta\lambda_2 = f_1(\boldsymbol{\sigma}_e, \kappa_0) \\ \mathcal{H}_2 \Delta\lambda_1 + \mathcal{H}_3 \Delta\lambda_2 = f_2(\boldsymbol{\sigma}_e, \kappa_0) \end{cases} \tag{7.160}$$

with

$$\begin{cases} \mathcal{H}_1 = h + \mathbf{n}_1^{\mathrm{T}} \mathbf{D}^{\mathrm{e}} \mathbf{m}_1 \\ \mathcal{H}_2 = h + \mathbf{n}_1^{\mathrm{T}} \mathbf{D}^{\mathrm{e}} \mathbf{m}_2 = h + \mathbf{n}_2^{\mathrm{T}} \mathbf{D}^{\mathrm{e}} \mathbf{m}_1 \\ \mathcal{H}_3 = h + \mathbf{n}_2^{\mathrm{T}} \mathbf{D}^{\mathrm{e}} \mathbf{m}_2 \end{cases} \tag{7.161}$$

The set $(\Delta\lambda_1, \Delta\lambda_2)$ can be be solved in closed form as:

$$\begin{cases} \Delta\lambda_1 = \frac{\mathcal{H}_3 f_1(\boldsymbol{\sigma}_e,\kappa_0) - \mathcal{H}_2 f_2(\boldsymbol{\sigma}_e,\kappa_0)}{\mathcal{H}_1\mathcal{H}_3 - \mathcal{H}_2^2} \\ \Delta\lambda_2 = \frac{\mathcal{H}_1 f_2(\boldsymbol{\sigma}_e,\kappa_0) - \mathcal{H}_2 f_1(\boldsymbol{\sigma}_e,\kappa_0)}{\mathcal{H}_1\mathcal{H}_3 - \mathcal{H}_2^2} \end{cases} \tag{7.162}$$

and we have for the final stress:

$$\boldsymbol{\sigma}_1 = \boldsymbol{\sigma}_e - \Delta\lambda_1 \mathbf{D}^{\mathrm{e}} \mathbf{m}_1 - \Delta\lambda_2 \mathbf{D}^{\mathrm{e}} \mathbf{m}_2 \tag{7.163}$$

The Mohr–Coulomb yield function (7.25) is expressed in terms of principal stresses. It can be expressed in terms of invariants by first solving the characteristic equation for the deviatoric stresses, Equation (1.87):

$$s^3 - J_2 s - J_3 = 0$$

with J_2 and J_3 the deviatoric stress invariants. Using Cardano's formula, the principal values, or eigenvalues, of the deviatoric stress tensor can be derived, and addition of the hydrostatic component yields the principal values of the stress tensor:

$$\begin{pmatrix} \sigma_1 \\ \sigma_2 \\ \sigma_3 \end{pmatrix} - 2\sqrt{J_2/3} \begin{pmatrix} \sin(\theta - \frac{2}{3}\pi) \\ \sin(\theta) \\ \sin(\theta + \frac{2}{3}\pi) \end{pmatrix} + p \begin{pmatrix} 1 \\ 1 \\ 1 \end{pmatrix} \tag{7.164}$$

where Lode's angle $-\pi/6 \leq \theta \leq \pi/6$ follows from (Figure 7.19):

$$\sin(3\theta) = -\frac{J_2}{2(J_3/3)^{3/2}} \tag{7.165}$$

For $\sigma_1 \leq \sigma_2 \leq \sigma_3$ the Mohr–Coulomb yield function is given by Equation (7.25). Substitution of the expressions for σ_1 and σ_3 into this identity results in an expression in terms of stress invariants:

$$f = \sqrt{J_2} \cos\theta - \left(2\sqrt{J_2/3} - p\right) \sin\varphi - c \tag{7.166}$$

and we have for the gradient to the yield surface:

$$\mathbf{n} = \sin\varphi \frac{\partial p}{\partial \boldsymbol{\sigma}} + a\frac{\partial J_2}{\partial \boldsymbol{\sigma}} + b\frac{\partial \theta}{\partial \boldsymbol{\sigma}} \tag{7.167}$$

with the scalars a and b given by

$$\begin{cases} a = \frac{1}{2\sqrt{J_2}}(\cos\theta - \sin\theta\sin\varphi/\sqrt{3}) \\ b = -\sqrt{J_2}(\sin\theta + \cos\theta\sin\varphi/\sqrt{3}) \end{cases} \tag{7.168}$$

and

$$\frac{\partial \theta}{\partial \boldsymbol{\sigma}} = \frac{\sqrt{3}}{2\cos 3\theta}\left(\frac{3}{2}J_2^{-5/2}J_3\frac{\partial J_2}{\partial \boldsymbol{\sigma}} - J_2^{-3/2}\frac{\partial J_3}{\partial \boldsymbol{\sigma}}\right) \tag{7.169}$$

The derivatives of the invariants read:

$$\begin{cases} \left(\frac{\partial p}{\partial \boldsymbol{\sigma}}\right)^{\mathrm{T}} = (1\ ,\ 1\ ,\ 1\ ,\ 0\ ,\ 0\ ,\ 0) \\ \left(\frac{\partial J_2}{\partial \boldsymbol{\sigma}}\right)^{\mathrm{T}} = (s_{xx}\ ,\ s_{yy}\ ,\ s_{zz}\ ,\ 2s_{xy}\ ,\ 2s_{yz}\ ,\ 2s_{zx}) \\ \left(\frac{\partial J_3}{\partial \boldsymbol{\sigma}}\right)^{\mathrm{T}} = \Big(s_{yy}s_{zz} - s_{yz}^2 + J_2/3\ ,\ s_{zz}s_{xx} - s_{zx}^2 + J_2/3\ ,\ s_{xx}s_{yy} - s_{xy}^2 + J_2/3\ , \\ \qquad 2(s_{yz}s_{zx} - s_{zz}s_{xy})\ ,\ 2(s_{zx}s_{xy} - s_{xx}s_{yz})\ ,\ 2(s_{xy}s_{yz} - s_{yy}s_{zx})\Big) \end{cases} \tag{7.170}$$

The expression for the flow direction $\mathbf{m}$ is identical to that for the gradient $\mathbf{n}$ except for the replacement of the friction angle φ by the dilatancy angle ψ.

When the strict inequality signs of $\sigma_1 \leq \sigma_2 \leq \sigma_3$ do not hold, i.e. when the stress point is in a corner of the Mohr–Coulomb yield surface, the plastic strain rate is determined via Koiter's generalisation. For the Mohr–Coulomb surface, we essentially have two yield corners for the present ordering of the principal stresses (Figure 7.19). We first consider the case for which $\sigma_1 = \sigma_2$, so that the yield function

$$f = \frac{1}{2}(\sigma_3 - \sigma_2) + \frac{1}{2}(\sigma_3 + \sigma_2)\sin\varphi - c$$

is also active. Carrying out a similar operation as for the yield function (7.25) the following values for the scalars a and b are obtained:

$$\begin{cases} a = \frac{1}{4\sqrt{J_2}}\left(\cos\theta - \sqrt{3}\sin\theta + (\cos\theta + \sin\theta/\sqrt{3})\sin\varphi\right) \\ b = -\frac{1}{2}\sqrt{J_2}\left(\sqrt{3}\cos\theta + \sin\theta + (\cos\theta/\sqrt{3} - \sin\theta)\sin\varphi\right) \end{cases} \tag{7.171}$$

Similarly, for $\sigma_2 = \sigma_3$, we obtain:

$$\begin{cases} a = \frac{1}{4\sqrt{J_2}}\left(\sin\theta + \sqrt{3}\cos\theta + (\sin\theta/\sqrt{3} - \cos\theta)\sin\varphi\right) \\ b = -\frac{1}{2}\sqrt{J_2}\left(\sin\theta - \sqrt{3}\cos\theta - (\sin\theta + \cos\theta/\sqrt{3})\sin\varphi\right) \end{cases} \tag{7.172}$$

A problem can arise if two principal stresses of the trial stress become exactly equal, since the derivative $\frac{\partial \theta}{\partial \sigma}$ then becomes indeterminate. However, even when two principal stresses are

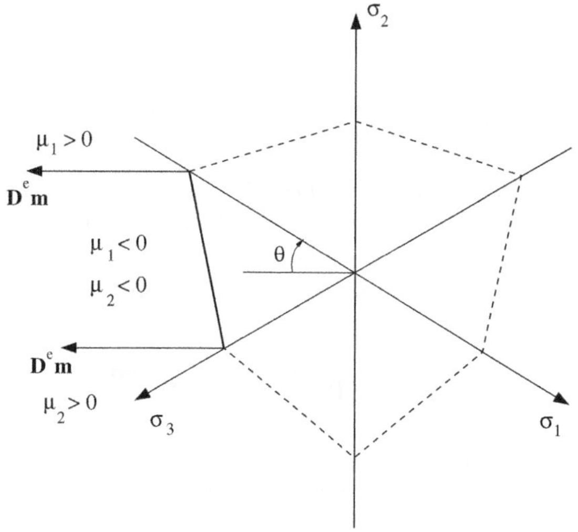

Figure 7.19 Active part of the Mohr–Coulomb yield surface in the π-plane, with Lode's angle $-\pi/6 \leq \theta \leq \pi/6$

exactly equal, the expression for the gradient **n** remains bounded. Indeed, for this limiting case the gradient to the Mohr–Coulomb yield surface becomes identical to that of the Drucker–Prager yield surface. As an example we consider the case that $\theta = \frac{1}{6}\pi$. Then:

$$\begin{pmatrix} \sigma_1 \\ \sigma_2 \\ \sigma_3 \end{pmatrix} = \sqrt{J_2/3} \begin{pmatrix} -2 \\ 1 \\ 1 \end{pmatrix} + p \begin{pmatrix} 1 \\ 1 \\ 1 \end{pmatrix} \tag{7.173}$$

which upon substitution in the yield function (7.25) gives

$$f = (3 - \sin\varphi)\sqrt{J_2/12} + p \sin\varphi \tag{7.174}$$

Differentation results in:

$$\mathbf{n} = \frac{3 - \sin\varphi}{4\sqrt{3J_2}} \frac{\partial J_2}{\partial \boldsymbol{\sigma}} + \sin\varphi \frac{\partial p}{\partial \boldsymbol{\sigma}} \tag{7.175}$$

which is precisely the gradient which is obtained when we differentiate the Drucker–Prager yield function.

Another singularity in the Mohr–Coulomb yield surface occurs at the apex of the yield surface. The algorithm has to check whether the stress is beyond the apex. If this happens to be the case, an additional correction should be applied to bring the stress point back to the apex of the yield cone. This problem only arises for cohesionless materials such as sand, because for cohesive materials, a fracture criterion bounds the tensile stresses.

Equation (7.163) can be differentiated to yield the consistent tangent stiffness matrix for the corner regime:

$$\left(\mathbf{I} + \Delta\lambda_1 \mathbf{D}^{\mathrm{e}} \frac{\partial \mathbf{m}_1}{\partial \boldsymbol{\sigma}} + \Delta\lambda_2 \mathbf{D}^{\mathrm{e}} \frac{\partial \mathbf{m}_2}{\partial \boldsymbol{\sigma}}\right) \dot{\boldsymbol{\sigma}} = \mathbf{D}^{\mathrm{e}} \dot{\boldsymbol{\epsilon}} - \dot{\lambda}_1 \mathbf{D}^{\mathrm{e}} \mathbf{m}_1 - \dot{\lambda}_2 \mathbf{D}^{\mathrm{e}} \mathbf{m}_2 \tag{7.176}$$

assuming that the flow directions do not depend on the plastic strain, i.e. that the dilatancy angle ψ is constant. This assumption certainly does not hold for many frictional materials, including soils, rocks, and concrete. The assumption has merely been made here to not overcomplicate the derivation. A matrix $\mathbf{A}$ can now be defined that derives from the sum of both plastic flow mechanisms, so that:

$$\mathbf{A} = \mathbf{I} + \Delta\lambda_1 \mathbf{D}^{\mathrm{e}} \frac{\partial \mathbf{m}_1}{\partial \boldsymbol{\sigma}} + \Delta\lambda_2 \mathbf{D}^{\mathrm{e}} \frac{\partial \mathbf{m}_2}{\partial \boldsymbol{\sigma}} \tag{7.177}$$

and the 'pseudo-elastic' stiffness matrix $\mathbf{H}$ can be defined according to Equation (7.128). Equation (7.176) can subsequently be written in a format that resembles Equation (7.129):

$$\dot{\boldsymbol{\sigma}} = \mathbf{H}\dot{\boldsymbol{\epsilon}} - \dot{\lambda}_1 \mathbf{H}\mathbf{m}_1 - \dot{\lambda}_2 \mathbf{H}\mathbf{m}_2 \tag{7.178}$$

In the corner regime not only yield functions $f_1(\boldsymbol{\sigma}, \kappa)$ and $f_2(\boldsymbol{\sigma}, \kappa)$ must be satisfied, but also their 'time' derivatives (consistency condition). This leads to:

$$\mathbf{n}_1^{\mathrm{T}} \dot{\boldsymbol{\sigma}} - h(\dot{\lambda}_1 + \dot{\lambda}_2) = 0 \quad \wedge \quad \mathbf{n}_2^{\mathrm{T}} \dot{\boldsymbol{\sigma}} - h(\dot{\lambda}_1 + \dot{\lambda}_2) = 0 \tag{7.179}$$

cf. Equation (7.129). Combining Equations (7.178) and (7.179) and defining

$$\begin{cases} \tilde{\mathcal{H}}_1 = h + \mathbf{n}_1^{\mathrm{T}} \mathbf{H}\mathbf{m}_1 \\ \tilde{\mathcal{H}}_2 = h + \mathbf{n}_1^{\mathrm{T}} \mathbf{H}\mathbf{m}_2 = h + \mathbf{n}_2^{\mathrm{T}} \mathbf{H}\mathbf{m}_1 \\ \tilde{\mathcal{H}}_3 = h + \mathbf{n}_2^{\mathrm{T}} \mathbf{H}\mathbf{m}_2 \end{cases} \tag{7.180}$$

results in the consistent tangential stiffness matrix in the corner regime for a Mohr–Coulomb yield function:

$$\dot{\boldsymbol{\sigma}} = \left(\mathbf{H} - \frac{1}{\tilde{\mathcal{H}}_1 \tilde{\mathcal{H}}_3 - \tilde{\mathcal{H}}_2^2} \mathbf{H} \left(\tilde{\mathcal{H}}_3 \mathbf{m}_1 \mathbf{n}_1^{\mathrm{T}} - \tilde{\mathcal{H}}_2 \mathbf{m}_1 \mathbf{n}_2^{\mathrm{T}} - \tilde{\mathcal{H}}_2 \mathbf{m}_2 \mathbf{n}_1^{\mathrm{T}} + \tilde{\mathcal{H}}_1 \mathbf{m}_2 \mathbf{n}_2^{\mathrm{T}}\right) \mathbf{H}\right) \dot{\boldsymbol{\epsilon}} \tag{7.181}$$

This expression becomes singular if two principal stresses are equal, which is at the corner. As observed by Crisfield (1987) this problem can be circumvented by observing that in view of the linearity of the Mohr–Coulomb yield criterion in the principal stress space, the gradients $\mathbf{n}$ and $\mathbf{m}$ at the trial stress $\boldsymbol{\sigma}_e$ and those for the final stress $\boldsymbol{\sigma}_1$ are coaxial. Therefore, the gradients at the trial stress $\boldsymbol{\sigma}_e$ can be used for the evaluation of the consistent tangential stiffness matrix (7.181). In practical computations, no singularities are then encountered, and also not when two principal stresses are equal in the trial stress state, since round-off errors then usually prevent a singularity.

The observation that, for the Mohr–Coulomb yield function, the gradients at the trial stress $\boldsymbol{\sigma}_e$ and at the final stress $\boldsymbol{\sigma}_1$ are coaxial can be exploited to derive a return mapping that operates in principal stress space (Larsson and Runesson 1996; Perić and de Souza Neto 1999). It is based on the observation that the gradient of a scalar function with respect to a second-order tensor is coaxial with the tensor (Ogden 1984). This implies that both at the trial stress state and at the final stress state the gradient $\mathbf{n}$ is coaxial with the stress tensor. Since both gradients are coaxial in this particular case, the tensors $\boldsymbol{\sigma}_e$ and $\boldsymbol{\sigma}_1$ are also coaxial. This implies that they have the same eigenvectors, and therefore, the transformation to the principal stress axes can be carried out with the same transformation matrix $\mathbf{T}_\sigma$. Using Equation (1.102) we can transform the Euler backward method, Equation (7.93), to the coordinate system of the principal stresses:

$$\bar{\boldsymbol{\sigma}}_1 = \mathbf{T}_\sigma \boldsymbol{\sigma}_1 = \mathbf{T}_\sigma \left(\boldsymbol{\sigma}_e - \Delta\lambda \mathbf{D}^{\mathrm{e}} \mathbf{m}_1\right) = \bar{\boldsymbol{\sigma}}_e - \Delta\lambda \mathbf{D}^{\mathrm{e}} \bar{\mathbf{m}}_1 \tag{7.182}$$

where the bar above a symbol denotes a quantity in the coordinate system of the principal stresses. For simplicity we now restrict the discussion to a regular part of the Mohr–Coulomb yield surface. The derivation for the corner regime is identical and straightforward, but involves a more lengthy derivation. Using Equations (7.25) and (7.54) the following explicit expressions in the principal directions ensue:

$$\begin{pmatrix} \sigma_1 \\ \sigma_2 \\ \sigma_3 \end{pmatrix}_1 = \begin{pmatrix} \sigma_1 \\ \sigma_2 \\ \sigma_3 \end{pmatrix}_e - \frac{\mu f(\boldsymbol{\sigma}_e, \kappa_0)}{h + \mu(1 + \frac{\sin\varphi \sin\psi}{1-2\nu})} \begin{pmatrix} -1 + \frac{\sin\psi}{1-2\nu} \\ \frac{2\nu \sin\psi}{1-2\nu} \\ 1 + \frac{\sin\psi}{1-2\nu} \end{pmatrix} \tag{7.183}$$

The updated stresses in the x, y, z-coordinate system can subsequently be obtained in a standard manner via, cf. Equation (1.104):

$$\boldsymbol{\sigma}_1 = \mathbf{T}_\sigma^{\mathrm{T}} \bar{\boldsymbol{\sigma}}_1 \tag{7.184}$$

The consistent tangential stiffness matrix can be obtained in a manner similar to that for the stiffness matrix of the rotating crack model, Equation (6.123). From Equation (7.184) we have by differentiation:

$$\dot{\boldsymbol{\sigma}}_1 = \mathbf{T}_\sigma^{\mathrm{T}} \dot{\bar{\boldsymbol{\sigma}}}_1 + \dot{\mathbf{T}}_\sigma^{\mathrm{T}} \bar{\boldsymbol{\sigma}}_1 \tag{7.185}$$

Using the derivative of Equation (7.183), observing that the transformation matrix $\mathbf{T}_\sigma$ is a function of the angle ϕ between the global coordinate system and the directions of the principal stresses, and noting that ϕ is a function of the strains, cf. Box 6.3, we can rewrite this identity as:

$$\dot{\boldsymbol{\sigma}} = \mathbf{T}_\sigma^{\mathrm{T}} \left(\mathbf{D}^{\mathrm{e}} \dot{\bar{\boldsymbol{\epsilon}}} - \dot{\lambda} \mathbf{D}^{\mathrm{e}} \bar{\mathbf{m}}_1\right) + \left(\frac{\partial \mathbf{T}_\sigma^{\mathrm{T}}}{\partial \phi} \bar{\boldsymbol{\sigma}}\right) \left(\frac{\partial \phi}{\partial \boldsymbol{\epsilon}}\right) \dot{\boldsymbol{\epsilon}} \tag{7.186}$$

with

$$\dot{\lambda} = \frac{\bar{\mathbf{n}}^{\mathrm{T}} \dot{\bar{\boldsymbol{\sigma}}}_e}{h + \mu(1 + \frac{\sin\varphi \sin\psi}{1-2\nu})}$$

Noting that $\bar{\mathbf{n}}^{\mathrm{T}} \dot{\bar{\boldsymbol{\sigma}}}_e = \bar{\mathbf{n}}^{\mathrm{T}} \mathbf{D}^{\mathrm{e}} \dot{\bar{\boldsymbol{\epsilon}}} = \mu (\mathbf{n}^*)^{\mathrm{T}} \dot{\bar{\boldsymbol{\epsilon}}}$ with

$$(\mathbf{n}^*)^{\mathrm{T}} = \mu(-1 + \frac{\sin\varphi}{1-2\nu}, \frac{2\nu \sin\varphi}{1-2\nu}, 1 + \frac{\sin\varphi}{1-2\nu})$$

and using the transformation of the strain components, Equation (1.105), the tangential stiffness relation in the global coordinate system $\dot{\boldsymbol{\sigma}} = \mathbf{D}\dot{\boldsymbol{\epsilon}}$ for a smooth part of the Mohr–Coulomb yield

Box 7.8 Construction of the tangential stiffness matrix for the Mohr–Coulomb yield function using eigenprojections

1. Given $\boldsymbol{\sigma}_e$, compute the principal values $(\sigma_i)_e \quad , \quad i = 1, 2$
2. Compute the principal stresses $(\sigma_i)_1$, via a return map in the principal stress space, Equation (7.183)
3. Compute the derivatives in the principal stress space, $\frac{\partial(\sigma_i)_1}{\partial(\sigma_j)_e}, i = 1, 2, j = 1, 2$, from Equation (7.183)
4. Compute the eigenprojections: $(\sigma_m)_e = (\sigma_1)_e + (\sigma_2)_e$
   (a) $\boldsymbol{E}_i = \frac{1}{2(\sigma_i)_e - (\sigma_m)_e}[\boldsymbol{\sigma}_e - ((\sigma_i)_e - (\sigma_m)_e)\mathbf{I}] \quad , \quad (\sigma_1)_e \neq (\sigma_2)_e$
   (b) $\boldsymbol{E}_i = \mathbf{I} \quad , \quad (\sigma_1)_e = (\sigma_2)_e$
5. Compute the derivatives:
   (a) $\mathbf{D} = \frac{(\sigma_1)_1 - (\sigma_2)_1}{(\sigma_1)_e - (\sigma_2)_e}[\mathbf{I}^{\rm sym} - \boldsymbol{E}_1 \otimes \boldsymbol{E}_1 - \boldsymbol{E}_2 \otimes \boldsymbol{E}_2] + \sum_{i=1}^{2}\sum_{j=1}^{2} \frac{\partial(\sigma_i)_1}{\partial(\sigma_j)_e} \boldsymbol{E}_i \otimes \boldsymbol{E}_j \quad , \quad (\sigma_1)_e \neq (\sigma_2)_e$
   (b) $\mathbf{D} = \left(\frac{\partial(\sigma_1)_1}{\partial(\sigma_1)_e} - \frac{\partial(\sigma_1)_1}{\partial(\sigma_2)_e}\right)\mathbf{I}^{\rm sym} + \frac{\partial(\sigma_1)_1}{\partial(\sigma_2)_e}\mathbf{I} \otimes \mathbf{I} \quad , \quad (\sigma_1)_e = (\sigma_2)_e$

surface is obtained, with the tangential stiffness matrix:

$$\mathbf{D} = \mathbf{T}_\sigma^{\rm T}\left(\mathbf{D}^{\rm e} - \frac{\mathbf{m}^*(\mathbf{n}^*)^{\rm T}}{h + \mu(1 + \frac{\sin\varphi \sin\psi}{1-2\nu})}\right)\mathbf{T}_\epsilon + \left(\frac{\partial \mathbf{T}_\sigma^{\rm T}}{\partial \phi}\bar{\boldsymbol{\sigma}}\right)\left(\frac{\partial \phi}{\partial \boldsymbol{\epsilon}}\right) \tag{7.187}$$

where $\mathbf{m}^*$ and $\mathbf{n}^*$ are identical, except for the replacement of the friction angle φ by the dilatancy angle ψ. In a manner similar to the derivation of the tangential stiffness matrix for orthotropic damage (Box 6.3), the second term on the right-hand side can be elaborated to give an explicit form of the tangential stiffness matrix in two dimensions:

$$\mathbf{D} = \mathbf{T}_\sigma^{\rm T}\left(\mathbf{D}^{\rm e} - \frac{\mathbf{m}^*(\mathbf{n}^*)^{\rm T}}{h + \mu(1 + \frac{\sin\varphi \sin\psi}{1-2\nu})} + \frac{\sigma_1 - \sigma_2}{2(\epsilon_1 - \epsilon_2)}\mathbf{Z}\right)\mathbf{T}_\epsilon \tag{7.188}$$

Alternatively, the tangential stiffness matrix can be constructed using eigenprojections (Box 7.8), from which the similarity with the expression derived in Equation (7.188), as well as with the tangential stiffness for the rotating crack model, Equation (6.123), becomes apparent. For the construction of the tangential stiffness matrix at corners of the yield surface, or for three-dimensional configurations, eigenvalues and eigenprojections are more suited, see Box 7.8 for the elaboration in two dimensions and de Souza Neto *et al.* (2008) for the three-dimensional case.

In the preceding subsection a general procedure has been outlined to determine which yield functions have been violated, and therefore, which return mapping should be applied. For yield functions that are linear in the principal stress space, like those of Mohr–Coulomb or

Tresca, a simple indicator can be derived whether the stress point is in a corner regime or not. The procedure is rigorous for ideal plasticity and a good first approximation for hardening or softening plasticity. The idea is to construct a plane which is spanned by the vector that points along the intersection of two planes of the Mohr–Coulomb yield surface and the direction vector $\mathbf{D}^e\mathbf{m}$. This plane distinguishes between trial stresses that are mapped back onto a regular part of the yield surface and those that are in the corner regime. The trial stress is mapped onto a regular part of the yield surface if

$$\mu_1 \leq 0 \quad \wedge \quad \mu_2 \leq 0 \tag{7.189}$$

with

$$\begin{cases} \mu_1 = f + \frac{1-2\nu+\sin\varphi\sin\psi}{(1-2\nu)(1+\sin\psi)}(\sigma_2 - \sigma_3) \\ \mu_2 = f + \frac{1-2\nu+\sin\varphi\sin\psi}{(1-2\nu)(1-\sin\psi)}(\sigma_2 - \sigma_1) \end{cases} \tag{7.190}$$

see also Figure7.19. If $\mu_1 > 0$ we are in the upper corner ($\sigma_2 = \sigma_3$) and for $\mu_2 > 0$ the stress return must be towards the lower corner ($\sigma_1 = \sigma_2$) (de Borst *et al.* 1991). For hardening or softening plasticity, this procedure cannot rigorously predict the correct regime, because the position of the yield surface is unknown. An a posteriori check must then be done to verify the prediction, and if it turns out that the prediction is falsified, a return map assuming the other regime must be carried out.

7.6 Soil Plasticity: Cam-clay Model

When shearing a clay sample either of the stress–strain curves of Figure 7.20 will be obtained, depending on the initial stress state. When the clay layer has been deposited in a normal manner, the monotonically rising curve, labelled 'normally consolidated' will be obtained. However, when there has been a (vertical) prestress in the past, for instance resulting from an overburden such as a thick layer of ice, there can be large horizontal initial stresses present, and the hardening part of the curve will be followed by softening (Wood 1990). Since the plastic compaction is an important measure for the loading history, it is natural to take the volumetric plastic compaction $\dot{\kappa} = -\dot{\epsilon}^{\mathrm{p}}_{\mathrm{vol}}$ as an internal variable, Equation (7.63). Within the framework

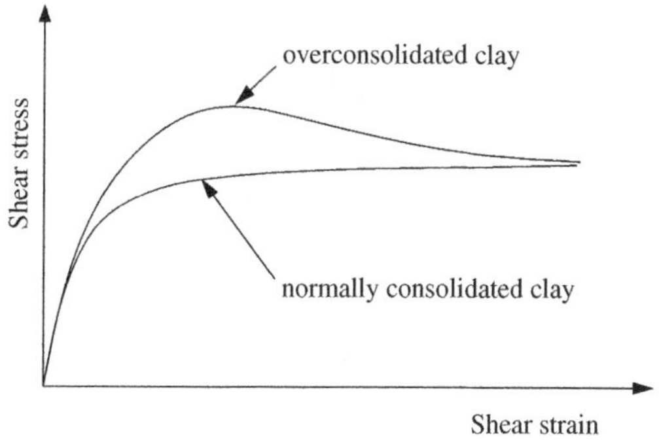

Figure 7.20 Shearing of normally consolidated and overconsolidated clay

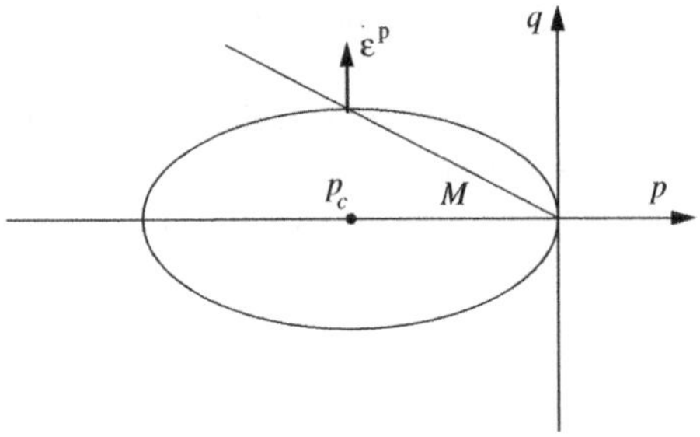

Figure 7.21 Elliptical yield surface of the Cam-clay model in the p, q-space

of single surface plasticity, the (modified) Cam-clay model is able to capture this phenomenon. In its simplest form, it uses an ellipse in the p, q-space as yield surface, and the yield function reads:

$$f = \left(\frac{q}{M}\right)^2 + p(p - 2p_c) \tag{7.191}$$

see Figure 7.21, where, for simplicity, restriction has been made to a cohesionless soil. In Equation (7.191) M is the stress ratio q/p at the critical state, i.e. the state for which continued plastic shearing will occur at a zero plastic volume change, and p_c is the preconsolidation pressure. Defining $(p_c)_0$ as the initial preconsolidation pressure and p_0 being the initial value of the pressure, normally consolidated clays are characterised by $p_0 < (p_c)_0$, and an associated flow rule $\dot{\boldsymbol{\epsilon}}^{\text{p}} = \dot{\lambda}\mathbf{n}$ predicts plastic volume compaction and an expansion of the yield surface (hardening). For overconsolidated clays, $p_0 > (p_c)_0$, and a plastic volume expansion occurs, accompanied by softening. A hardening rule that can accommodate this behaviour reads:

$$\dot{p}_c = \left(\frac{p_c}{\kappa^* - \lambda^*}\right)\dot{\epsilon}^{\text{p}}_{\text{vol}} \tag{7.192}$$

with κ^* and λ^* the modified swelling index and the modified compression index, respectively. This relation can be integrated to give:

$$\Delta p_c = \exp\left(\frac{\Delta\epsilon^{\text{p}}_{\text{vol}}}{\kappa^* - \lambda^*}\right) \tag{7.193}$$

Unlike metals, concrete and most other materials, it is not possible to distinguish an initial linear-elastic branch in the stress–strain curve for soils, and stress-dependent elastic moduli are required for a proper description. For clays, this non-linear elasticity primarily applies to the volumetric behaviour, while in shear linearity is often assumed. Then, the shear modulus μ can be taken as constant, while for the volumetric elastic behaviour the following relation is postulated:

$$\dot{p} = \left(\frac{p}{\kappa^*}\right)\dot{\epsilon}^{\text{e}}_{\text{vol}} \tag{7.194}$$

instead of the linear relation (1.113). Integrating Equation (7.194) for a finite strain increment yields:

$$\Delta p = \exp\left(\frac{\Delta\epsilon_{\text{vol}}^{\text{e}}}{\kappa^*}\right) \tag{7.195}$$

By definition the secant bulk modulus $K^{\text{s}} = \frac{\Delta p}{\Delta\epsilon_{\text{vol}}^{\text{e}}}$, so that

$$K^{\text{s}} = \frac{\exp\left(\frac{\Delta\epsilon_{\text{vol}}^{\text{e}}}{\kappa^*}\right)}{\Delta\epsilon_{\text{vol}}^{\text{e}}} \tag{7.196}$$

has to be inserted in the elastic secant stiffness matrix $\mathbf{D}^{\text{s}}$ that is used in the algorithm to compute the updated stresses for finite strain increments.

Using the projection matrix (7.36) and the projection matrix (7.33) the yield function can be rewritten as:

$$f = \frac{3\boldsymbol{\sigma}^{\text{T}}\mathbf{P}\boldsymbol{\sigma}}{2M^2} + \boldsymbol{\pi}^{\text{T}}\boldsymbol{\sigma}(\boldsymbol{\pi}^{\text{T}}\boldsymbol{\sigma} - 2p_c) \tag{7.197}$$

and the flow rule becomes:

$$\dot{\boldsymbol{\epsilon}}^{\text{p}} = \dot{\lambda}\mathbf{n} = \dot{\lambda}\left(\frac{3\mathbf{P}\boldsymbol{\sigma}}{M^2} + 2p(1 - p_c)\boldsymbol{\pi}\right) \tag{7.198}$$

Note that the Cam-clay model involves a dependence of the plastic flow on the hardening (through p_c), and therefore, exhibits non-associated hardening behaviour, leading to a non-symmetric consistent tangent stiffness matrix. Using the flow rule, Equation (7.198), the evolution of the hardening variable κ becomes:

$$\dot{\kappa} = -\dot{\epsilon}_{\text{vol}}^{\text{p}} = -3\boldsymbol{\pi}^{\text{T}}\dot{\boldsymbol{\epsilon}}^{\text{p}} = -3\dot{\lambda}\boldsymbol{\pi}^{\text{T}}\mathbf{n} = -2\dot{\lambda}p(1 - p_c) \tag{7.199}$$

Since $(p_c)_{j+1} = p_c(\kappa_{j+1}) = p_c(\kappa_0 + \Delta\kappa)$ and since, according to Equation (7.199), $\Delta\kappa$ is a function of the preconsolidation pressure p_c, it is not possible to explicitly resolve p_c, and therefore, Equations (7.99) cannot be used. The non-linear dependence of the elastic volumetric strain on the pressure is another reason that the framework of Equations (7.99) is insufficient. Indeed, this non-linear dependence means that a further residual equation has to be added. For a proper return-mapping algorithm for Cam-clay plasticity, the set of local residuals (7.99) has to be expanded to include a variable elastic stiffness (de Borst and Heeres 2002; Rouainia and Wood 2000):

$$\begin{cases} \mathbf{r}_\sigma = \boldsymbol{\sigma}_{j+1} - (\boldsymbol{\sigma}_0 + \mathbf{D}^{\text{s}}\Delta\boldsymbol{\epsilon}) + \Delta\lambda\mathbf{D}^{\text{s}}\mathbf{n}(\boldsymbol{\sigma}_{j+1}, (p_c)_{j+1}) \\ r_p = p_{j+1} - p_0 \exp\left(\frac{\Delta\epsilon_{\text{vol}} - 2\Delta\lambda p_{j+1}[1-(p_c)_{j+1}]}{\kappa^*}\right) \\ r_{p_c} = (p_c)_{j+1} - (p_c)_0 \exp\left(\frac{\Delta\epsilon_{\text{vol}}^{\text{p}} + 2\Delta\lambda p_{j+1}[1-(p_c)_{j+1}]}{\kappa^* - \lambda^*}\right) \\ r_f = f(\boldsymbol{\sigma}_{j+1}, (p_c)_{j+1}) \end{cases} \tag{7.200}$$

This set of non-linear equations in the unknowns ($\boldsymbol{\sigma}$, p, p_c, $\Delta\lambda$) can be solved in a standard manner, via a local Newton–Raphson procedure. Simplifications can be made by exploiting the fact that the Cam-clay model is formulated in terms of the stress invariants p and q only. Using a procedure similar to those for Drucker–Prager or Hoffman plasticity, the residuals can be reduced to three equations (Rouainia and Wood 2000). When non-linear elasticity is left out of consideration a reduction to two equations is possible (de Souza Neto *et al.* 2008).

7.7 Coupled Damage–Plasticity Models

A possible way to combine plasticity and damage in a rational and physically meaningful manner is to adopt a damage-like stress–strain relation:

$$\boldsymbol{\sigma} = (1 - \omega)\mathbf{D}^{\mathrm{e}} : \boldsymbol{\epsilon}^{\mathrm{e}} \tag{7.201}$$

and to postulate that plasticity only applies to the intact matrix material and not to the voids of the micro-cracks. Using the effective stress concept, Equation (6.13), the effective stresses in a coupled damage–plasticity model can therefore be defined as:

$$\hat{\boldsymbol{\sigma}} = \mathbf{D}^{\mathrm{e}} : \boldsymbol{\epsilon}^{\mathrm{e}} \tag{7.202}$$

which enter the yield function (7.24) and the flow rule,

$$\dot{\boldsymbol{\epsilon}}^{\mathrm{p}} = \dot{\lambda}\frac{\partial f}{\partial \hat{\boldsymbol{\sigma}}} \tag{7.203}$$

An algorithm for this coupled damage–plasticity model is given in Box 7.9.

Box 7.9 Algorithm for coupled damage–plasticity model

1. Compute the strain increment: $\Delta\boldsymbol{\epsilon}$
2. Update the strain: $\boldsymbol{\epsilon}_{j+1} = \boldsymbol{\epsilon}_j + \Delta\boldsymbol{\epsilon}$
3. Compute the trial stress in the effective stress space:
   $\hat{\boldsymbol{\sigma}}_0 = \boldsymbol{\sigma}_0/(1 - \omega_0)$
   $\hat{\boldsymbol{\sigma}}_e = \hat{\boldsymbol{\sigma}}_0 + \mathbf{D}^{\mathrm{e}}\Delta\boldsymbol{\epsilon}$
   $\tilde{\sigma}_e = \tilde{\sigma}(\hat{\boldsymbol{\sigma}}_e)$
4. Evaluate the plastic loading function: $f^{\mathrm{p}} = \tilde{\sigma}_e - \bar{\sigma}(\kappa_j^{\mathrm{p}})$
   if $f^{\mathrm{p}} \geq 0$, $\hat{\boldsymbol{\sigma}}_{j+1} = \hat{\boldsymbol{\sigma}}_e - \Delta\lambda\mathbf{D}^{\mathrm{e}}\left(\frac{\partial f^{\mathrm{p}}}{\partial \hat{\boldsymbol{\sigma}}}\right)_{j+1}$
   else $\Delta\lambda = 0 \rightarrow \hat{\boldsymbol{\sigma}}_{j+1} = \hat{\boldsymbol{\sigma}}_e$
5. Evaluate the damage loading function: $f^{\mathrm{d}} = \tilde{\epsilon}(\boldsymbol{\epsilon}_{j+1}) - \kappa_j^{\mathrm{d}}$
   if $f^{\mathrm{d}} \geq 0$, $\kappa_{j+1}^{\mathrm{d}} = \tilde{\epsilon}(\boldsymbol{\epsilon}_{j+1})$
   else $\kappa_{j+1}^{\mathrm{d}} = \kappa_j^{\mathrm{d}}$
6. Update the damage paramater: $\omega_{j+1} = \omega(\kappa_{j+1}^{\mathrm{d}})$
7. Compute the new stresses: $\boldsymbol{\sigma}_{j+1} = (1 - \omega_{j+1})\hat{\boldsymbol{\sigma}}_{j+1}$

An alternative way for coupling damage and plasticity is provided by the modified Gurson model, which has been widely used for the analysis of void nucleation and growth in porous metals. Essentially, this model is a plasticity model, equipped with the usual decomposition of the strain into elastic and plastic components, Equation (7.41), the injective relation between the elastic part of the strain and the stress, Equation (7.40), the flow rule (7.43) and the loading–unloading conditions, but with the yield function incorporating a damage parameter ω:

$$f(I_1, J_2, \kappa, \phi^*) = \frac{3J_2}{2\bar{\sigma}^2(\kappa)} + 2q_1\omega(\phi^*)\cosh\left(\frac{q_2 I_1}{2\bar{\sigma}(\kappa)}\right) - q_3\omega^2(\phi^*) - 1 \tag{7.204}$$

with σ_0 the uniaxial yield strength of the material without voids, which depends on the hardening parameter κ. q_1, q_2 and q_3 are material parameters, which, for many conditions, can be taken as: $q_1 = 1.5$, $q_2 = 1$ and $q_3 = q_1^2$. The coupling is provided by the inclusion of the damage parameter ω in the yield function f. The damage parameter ω is specified as:

$$\omega = \begin{cases} \phi^* & \text{if } \phi^* \leq \phi_c^* \\ \phi_c^* + \frac{q_1^{-1} - \phi_c^*}{\phi_f^* - \phi_c^*}(\phi^* - \phi_c^*) & \text{if } \phi^* > \phi_c^* \end{cases} \tag{7.205}$$

with ϕ_c^* a critical value of the void volume fraction ϕ^* at which coalescence begins and ϕ_f^* its final value. The model is completed by an evolution equation for ϕ^*, which consists of two terms, one due to void nucleation, $\dot{\phi}^*_{\text{nucl}}$ and a second term due to void growth $\dot{\phi}^*_{\text{growth}}$: $\dot{\phi}^* = \dot{\phi}^*_{\text{nucl}} + \dot{\phi}^*_{\text{growth}}$, and starts at $\phi^* = \phi_0^*$. Detailed expressions are given by Needleman and Tvergaard (1987). In the absence of voids, $\phi^* = 0$ and the yield function reduces to the von Mises contour.

The coupling of damage to plasticity causes a degradation of the strength at a certain level of deformation. A softening behaviour is then observed, material stability is lost, and the possibility of loss of ellipticity exists in quasi-static calculations. As a consequence, an excessive dependence on the discretisation can be obtained in numerical simulations, as discussed in Chapter 6. Regularisation strategies, for instance the addition of spatial gradients, must then be adopted in order to restore ellipticity, so that physically meaningful results are obtained, see de Borst *et al.* (1999) for a gradient-enhancement of coupled damage–plasticity models. A degradation mechanism can also be introduced in plasticity alone by making the yield strength a descending function of the hardening variable. Also in this case regularisation must be adopted to restore ellipticity in quasi-static calculations (de Borst and Mühlhaus 1992).

7.8 Element Technology: Volumetric Locking

An important issue in computational elasto-plasticity is the phenomenon of mesh locking at fully developed plastic flow, which can lead to large errors when computing collapse loads (Nagtegaal *et al.* 1974). Much effort has therefore been put in the development of simple elements that alleviate or even remove this problem.

For a proper discussion of locking of finite elements under kinematic constraints induced by the constitutive relation, we take our point of departure at the rate form of the principle of virtual work (2.37):

$$\int_V \delta\boldsymbol{\epsilon}^{\mathrm{T}}\dot{\boldsymbol{\sigma}}\mathrm{d}V = \int_V \rho\delta\mathbf{u}^{\mathrm{T}}\dot{\mathbf{g}}\mathrm{d}V + \int_S \delta\mathbf{u}^{\mathrm{T}}\dot{\mathbf{t}}\mathrm{d}S \tag{7.206}$$

When a limit point is attained the external loads become stationary, $\dot{\mathbf{g}} = \mathbf{0}$ and $\dot{\mathbf{t}} = \mathbf{0}$, and Equation (7.206) reduces to:

$$\int_V \delta\boldsymbol{\epsilon}^{\mathrm{T}}\dot{\boldsymbol{\sigma}}\mathrm{d}V = 0 \tag{7.207}$$

This equation is satisfied for all possible $\delta\boldsymbol{\epsilon}$ if and only if $\dot{\boldsymbol{\sigma}} = \mathbf{0}$. Considering linear elasticity, we have $\dot{\boldsymbol{\sigma}} = \mathbf{D}^{\mathrm{e}}\dot{\boldsymbol{\epsilon}}^{\mathrm{e}}$. Clearly, the requirement that the stress becomes stationary implies $\dot{\boldsymbol{\epsilon}}^{\mathrm{e}} = \mathbf{0}$. This observation, together with the fact that the plastic flow components are interdependent, effectively imposes a kinematic constraint upon the velocity field $\dot{\mathbf{u}}$ for certain types of constitutive operators and configurations, notably plane-strain, axisymmetric and three-dimensional conditions. The issue is most conveniently illustrated for a Mohr–Coulomb yield function, Equation (7.23), where a non-associated flow rule with a plastic potential of the form (7.54) results in the following relation between the plastic volume change and the plastic shear strain, see also Equation (7.55):

$$\dot{\epsilon}^{\mathrm{p}}_{\mathrm{vol}} = \dot{\gamma}^{\mathrm{p}}\sin\psi$$

Since the elastic strain rates have been shown to vanish at a limit point, we can replace this identity by:

$$\dot{\epsilon}_{\mathrm{vol}} = \dot{\gamma}\sin\psi \tag{7.208}$$

which states that at collapse any amount of shear strain imposed upon the element is necessarily accompanied with an amount of volumetric strain, governed by the dilatancy angle ψ. The consequences for the element performance can be illustrated simply for the two constant strain triangles of Figure 7.22, which are representative for the whole mesh. According to the constraint (7.208) the right-upper node can only move along the dashed lines, which effectively means that the node must remain on its place, thus preventing the possibility of a collapse mechanism. This phenomenon of volumetric locking is known for incompressible elasticity and isochoric plasticity ($\psi = 0$ for Mohr–Coulomb plasticity), but is present whenever the constitutive relation imposes a kinematic constraint on the strain rate field.

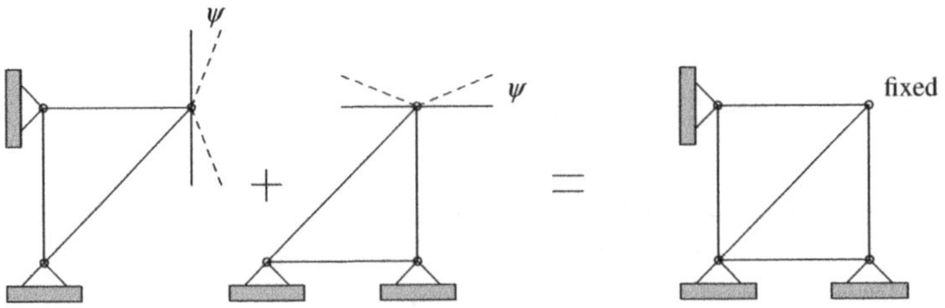

Figure 7.22 Locking of two three-noded triangles for fully developed plastic flow

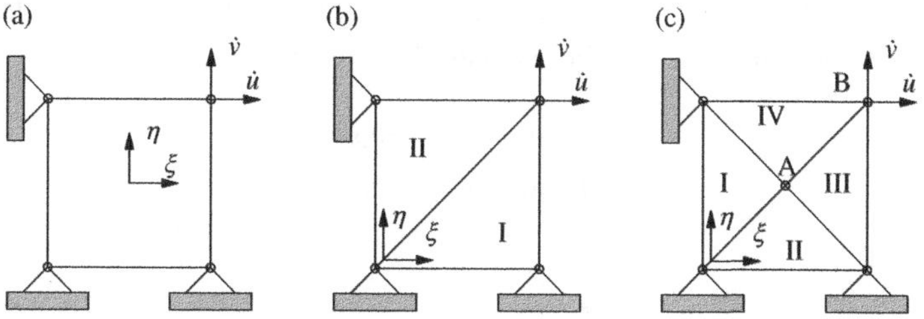

Figure 7.23 (a) Four-noded quadrilateral element; (b) patch of two three-noded triangular elements; (c) patch of four three-noded triangular elements

To present a more quantitative treatment of the phenomenon and of some of the remedies that have been proposed, we consider the quadrilateral element of Figure 7.23. We choose the principal axes of the strain rate tensor to coincide with the local ξ, η-coordinate system of an element. This choice is permissible, since under planar deformations $\dot{\epsilon}_{\text{vol}}$ and $\dot{\gamma}$ are invariant. For the Mohr–Coulomb yield function resembling plastic potential g, the kinematic constraint (7.208) then specialises as

$$(1 - \sin\psi)\dot{\epsilon}_{\xi\xi} + (1 + \sin\psi)\dot{\epsilon}_{\eta\eta} + \dot{\epsilon}_{\zeta\zeta} = 0 \tag{7.209}$$

The velocities within the element are interpolated in an isoparametric manner, so that

$$\begin{cases} \dot{u}(\xi,\eta) = \frac{1}{4}(1+\xi)(1+\eta)\dot{u} \\ \dot{v}(\xi,\eta) = \frac{1}{4}(1+\xi)(1+\eta)\dot{v} \end{cases} \tag{7.210}$$

with $\dot{u}$, $\dot{v}$ the horizontal and vertical velocities of the right upper node of the element. The normal strain rates within the element are obtained by differentiation as:

$$\begin{pmatrix} \dot{\epsilon}_{\xi\xi} \\ \dot{\epsilon}_{\eta\eta} \\ \dot{\epsilon}_{\zeta\zeta} \end{pmatrix} = \frac{1}{4}\begin{bmatrix} 1+\eta & 0 \\ 0 & 0 \\ 0 & 1+\xi \end{bmatrix}\begin{pmatrix} \dot{u} \\ \dot{v} \end{pmatrix} \tag{7.211}$$

Upon substitution of these expressions into the kinematic constraint (7.209), the following restriction upon the velocity field ensues:

$$[(1 - \sin\psi)\dot{u} + (1 + \sin\psi)\dot{v}] + (1 - \sin\psi)\dot{u}\eta + (1 + \sin\psi)\dot{v}\xi = 0 \tag{7.212}$$

The term between brackets sets the ratio between the horizontal velocity $\dot{u}$ and the vertical velocity $\dot{v}$ of the right upper node of the element. It vanishes, which is a direct reflection of the kinematic constraint imposed on the possible velocity field by the constitutive relation. Since this term must be zero, disappearance of the entire identity can only be achieved for arbitrary pairs ξ, η if $\dot{u}$ and $\dot{v}$ are both zero. This implies that the element is not able to deform and locks. It is emphasised that this observation holds for all values of ψ, including the isochoric case ($\psi = 0$). Using the same methodology it can be shown that a patch of two three-noded triangular elements [Figure 7.23(b)], also exhibits volumetric locking.

As first observed by Nagtegaal *et al.* (1974) a cross-diagonal patch of four three-noded triangular element performs well under incompressibility. This statement also holds true for dilatant/contractant plasticity. We demonstrate this as follows. First, we consider elements I and II of the patch of Figure 7.23(c). For element I we have:

$$\begin{cases} \dot{u}_{\rm I}(\xi,\eta) = 2\xi\dot{u}_{\rm A} \\ \dot{v}_{\rm I}(\xi,\eta) = 2\xi\dot{v}_{\rm A} \end{cases}$$

and for element II:

$$\begin{cases} \dot{u}_{\rm II}(\xi,\eta) = 2\eta\dot{u}_{\rm A} \\ \dot{v}_{\rm II}(\xi,\eta) = 2\eta\dot{v}_{\rm A} \end{cases}$$

Differentiation of both velocity fields and substitution of the results into the kinematic constraint (7.209) then results in

$$\begin{cases} (1+\sin\psi)\dot{v}_{\rm A} = 0 \\ (1-\sin\psi)\dot{u}_{\rm A} = 0 \end{cases}$$

which obviously yields $\dot{u}_{\rm A} = \dot{v}_{\rm A} = 0$. Using this result we can derive for elements III and IV that:

$$\begin{cases} \dot{u}_{\rm III/IV}(\xi,\eta) = (\xi+\eta-1)\dot{u}_{\rm B} \\ \dot{v}_{\rm III/IV}(\xi,\eta) = (\xi+\eta-1)\dot{v}_{\rm B} \end{cases}$$

Substitution of these expressions in the kinematic constraint (7.209) gives:

$$(1-\sin\psi)\dot{u} + (1+\sin\psi)\dot{v} = 0 \tag{7.213}$$

which is by definition satisfied for arbitrary pairs ξ, η.

A trick that is often adopted to alleviate mesh locking is to underintegrate the stiffness matrix and the internal force vector, i.e. is to apply an integration scheme that, even in linear elasticity and for rectangular elements, would not evaluate the integrals exactly. Typically, an integration scheme is used that is one order lower than a scheme that would yield an exact integration. For instance, for the four-noded quadrilateral element reduced Gauss integration would use a single integration point in the centre of the element instead of the consistent two-by-two integration scheme. Obviously, the strain rate distribution over the element is now constant, so that:

$$\begin{pmatrix} \dot{\epsilon}_{\xi\xi} \\ \dot{\epsilon}_{\eta\eta} \\ \dot{\epsilon}_{\xi\zeta} \end{pmatrix} = \frac{1}{4}\begin{bmatrix} 1 & 0 \\ 0 & 1 \\ 0 & 0 \end{bmatrix}\begin{pmatrix} \dot{u} \\ \dot{v} \end{pmatrix} \tag{7.214}$$

Substitution of this expression in the kinematic constraint (7.209) again gives Equation (7.213), which excludes volumetric locking. However, the four-noded quadrilateral element with a uniform, one-point integration has two spurious kinematic modes, which can occur without an increase of the internal energy. These modes can propagate into neighbouring elements, which makes practical computations not feasible without proper stabilisation (Belytschko *et al.* 1984; Flanagan and Belytschko 1981; Kosloff and Frazier 1978). The situation seems less severe for eight-noded quadrilateral elements with reduced, four-point Gauss integration. This element also has a spurious mode, but unlike the four-noded quadrilateral elements, it cannot propagate into neighbouring elements. However, this statement only holds for linear elasticity, and for non-linear constitutive relations, propagation of spurious kinematic modes into neighbouring elements and an ensuing lack of reliability of the solution is still possible (de Borst and Vermeer 1984).

Considering that uniform reduced integration is impractical for the four-noded element, it has been proposed to apply selective integration, such that four Gauss integration points are used for the shear strains, and only one, the centre point, for the volumetric strain. This approach can be cast within the $\bar{\mathbf{B}}$-concept (Hughes 1980) and the normal strain rates are redefined as:

$$\begin{pmatrix} \dot{\epsilon}_{\xi\xi} \\ \dot{\epsilon}_{\eta\eta} \\ \dot{\epsilon}_{\zeta\zeta} \end{pmatrix} = \frac{1}{4} \begin{bmatrix} 1+\frac{2}{3}\eta & -\frac{1}{3}\xi \\ -\frac{1}{3}\eta & 1+\frac{2}{3}\xi \\ -\frac{1}{3}\eta & -\frac{1}{3}\xi \end{bmatrix} \begin{pmatrix} \dot{u} \\ \dot{v} \end{pmatrix} \tag{7.215}$$

Note that the normal strain rate in the third direction, $\dot{\epsilon}_{\zeta\zeta}$, does not vanish pointwise, but only in an average sense. Substitution of this strain rate field in the kinematic constraint (7.209) leads to

$$[(1-\sin\psi)\dot{u} + (1+\sin\psi)\dot{v}] - \sin\psi(\dot{u}\eta - \dot{v}\xi) = 0 \tag{7.216}$$

Obviously, this condition can only be satisfied for arbitrary pairs ξ, η when $\psi = 0$, the case of plastically volume-preserving flow. For arbitrary values of ψ, $\dot{u}$ and $\dot{v}$ must vanish identically, which means that the $\bar{\mathbf{B}}$ element locks for the general case of $\psi \neq 0$, but is effective for isochoric plastic flow.

Another possible way to avoid volumetric locking is to adopt higher-order interpolation of pure displacement-based finite elements. This solution is expensive in terms of computer time, but is robust. It suffers less from unreliable element behaviour such as the emergence of spurious modes than mixed/hybrid approaches. The favourable properties of higher-order elements regarding volumetric locking are now demonstrated for the nine-noded Lagrangian element of Figure 7.24(b). This element has four 'free' nodes, labelled A, B, C and D, for

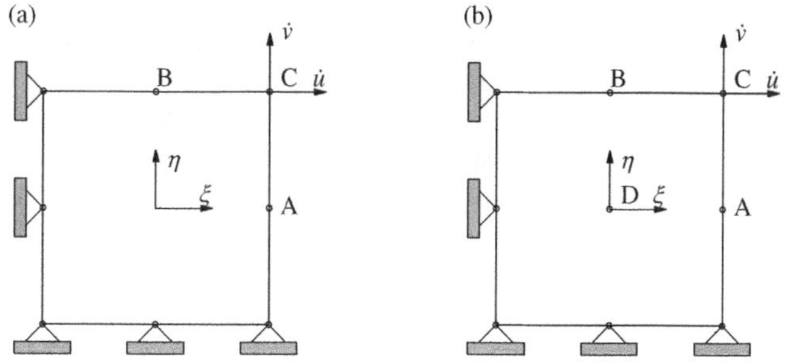

Figure 7.24 (a) Eight-noded quadrilateral element; (b) nine-noded quadrilateral element

which the shape functions read:

$$\begin{cases} h_A = \frac{1}{2}\xi(1+\xi)(1-\eta^2) \\ h_B = \frac{1}{2}\eta(1+\eta)(1-\xi^2) \\ h_C = \frac{1}{4}\xi\eta(1+\xi)(1+\eta) \\ h_D = (1-\xi^2)(1-\eta^2) \end{cases}$$

For the normal strain rates we obtain by differentiation:

$$\begin{cases} \dot{\epsilon}_{\xi\xi} = \frac{1}{2}(1+2\xi)(1-\eta^2)\dot{u}_A - \xi\eta(1+\eta)\dot{u}_B + \frac{1}{4}\eta(1+2\xi)(1+\eta)\dot{u}_C - 2\xi(1-\eta^2)\dot{u}_D \\ \dot{\epsilon}_{\eta\eta} = -\xi\eta(1+\xi)\dot{v}_A + \frac{1}{2}(1+2\eta)(1-\xi^2)\dot{v}_B + \frac{1}{4}\xi(1+\xi)(1+2\eta)\dot{v}_C - 2\eta(1-\xi^2)\dot{v}_D \end{cases}$$

which, upon substitution into the kinematic constraint (7.209), results in an equation with a constant term, and with terms η, ξ, η^2, $\xi\eta$, η^2, $\xi\eta^2$ and $\xi^2\eta$. Since this equation must hold for arbitrary pairs (ξ, η), we obtain the following homogeneous system of equations:

$$\begin{cases} (1-\sin\psi)\dot{u}_A + (1+\sin\psi)\dot{v}_B = 0 \\ (1-\sin\psi)(4\dot{u}_A - 8\dot{u}_D) + (1+\sin\psi)\dot{v}_C = 0 \\ (1-\sin\psi)\dot{u}_C + (1+\sin\psi)(4\dot{v}_B - 8\dot{v}_D) = 0 \\ \dot{v}_C - 2\dot{v}_B = 0 \\ (1-\sin\psi)(\dot{u}_C - 2\dot{u}_B) + (1+\sin\psi)(\dot{v}_C - 2\dot{v}_A) = 0 \\ \dot{u}_C - 2\dot{u}_A = 0 \\ -2\dot{u}_A - 2\dot{u}_B + \dot{u}_C + 4\dot{u}_D = 0 \\ -2\dot{v}_A - 2\dot{v}_B + \dot{v}_C + 4\dot{v}_D = 0 \end{cases}$$

This system can be shown to be singular, so that there exists a solution for non-zero $\dot{u}_A, \ldots, \dot{v}_D$, and, therefore, volumetric locking does not occur for this element. For the eight-noded quadrilateral element of Figure 7.24(a), there are only six degrees of freedom to satisfy eight constraints. A slightly different homogeneous system of eight equations arises, which can only be satisfied if $\dot{u}_A = \dot{v}_A = \dot{u}_B = \dot{v}_B = \dot{u}_C = \dot{v}_C = 0$. Accordingly, volumetric locking occurs for this element, although numerical experience indicates that it is less severe than for the four-noded quadrilateral element or for the three-noded triangular element.

The favourable properties of higher-order interpolations with respect to volumetric locking can be preserved if the higher-order displacement modes are eliminated at element level by static condensation. In this manner incompatible displacement modes are obtained, e.g. the modified incompatible modes element by Taylor *et al.* (1976), in which the displacement field is locally enriched by quadratic polynomials:

$$\begin{cases} \tilde{u}(\xi,\eta) = \frac{1}{2}(\xi^2-1)\alpha_1 + \frac{1}{2}(\eta^2-1)\alpha_4 \\ \tilde{v}(\xi,\eta) = \frac{1}{2}(\xi^2-1)\alpha_3 + \frac{1}{2}(\eta^2-1)\alpha_2 \end{cases} \tag{7.217}$$

with $\alpha_1, \ldots, \alpha_4$ displacement parameters which are defined locally at element level. For the normal strain rates we thus find that:

$$\begin{pmatrix} \dot{\epsilon}_{\xi\xi} \\ \dot{\epsilon}_{\eta\eta} \\ \dot{\epsilon}_{\zeta\zeta} \end{pmatrix} = \frac{1}{4}\begin{pmatrix} 1+\eta & 0 \\ 0 & 1+\xi \\ 0 & 0 \end{pmatrix}\begin{pmatrix} \dot{u} \\ \dot{v} \end{pmatrix} + \begin{pmatrix} \xi & 0 \\ 0 & \eta \\ 0 & 0 \end{pmatrix}\begin{pmatrix} \dot{\alpha}_1 \\ \dot{\alpha}_2 \end{pmatrix} \tag{7.218}$$

Substitution in the kinematic constraint (7.209) gives:

$$[(1-\sin\psi)\dot{u} + (1+\sin\psi)\dot{v}] + [(1+\sin\psi)\dot{v} + 4(1-\sin\psi)\dot{\alpha}_1]\xi + \\ [(1-\sin\psi)\dot{u} + 4(1+\sin\psi)\dot{\alpha}_2]\eta = 0$$

The first term between brackets vanishes and so do the other two terms, because we can always find values for $\dot{\alpha}_1$ and $\dot{\alpha}_2$ which nullify them for non-zero $\dot{u}$ and $\dot{v}$. Accordingly, a locking-free element is obtained. This element has been shown to belong to a more general class of mixed elements that follow the enhanced assumed strain approach (Simo and Rifai 1990), in which the strain field that stems from the continuous displacement field is augmented by a strain field that is defined locally per element. This class of elements was introduced in Chapter 6 where the enrichment consisted of a discontinuous field. The fact that only the strains are augmented, makes it possible to straightforwardly utilise standard methods for the stress integration. In general, the elements are accurate, but as all mixed formulations, they can exhibit spurious modes when plasticity or damage is introduced.

It is noted that the incompatible displacement element is a somewhat exceptional case in the sense that the augmented strain field can be derived from a displacement field at element level. Since this element has been used widely, both in the original format and cast in the enhanced assumed strain framework, we list the full, two-dimensional strain field for completeness:

$$\begin{pmatrix} \dot{\epsilon}_{\xi\xi} \\ \dot{\epsilon}_{\eta\eta} \\ \dot{\gamma}_{\xi\eta} \end{pmatrix} = \frac{1}{4}\begin{pmatrix} 1+\eta & 0 \\ 0 & 1+\xi \\ 1+\xi & 1+\eta \end{pmatrix}\begin{pmatrix} \dot{u} \\ \dot{v} \end{pmatrix} + \begin{pmatrix} \xi & 0 & 0 & 0 \\ 0 & \eta & 0 & 0 \\ 0 & 0 & \xi & \eta \end{pmatrix}\begin{pmatrix} \dot{\alpha}_1 \\ \dot{\alpha}_2 \\ \dot{\alpha}_3 \\ \dot{\alpha}_4 \end{pmatrix} \tag{7.219}$$

which shows also the enrichment for the shear strain, which considerably improves the behaviour when shear stresses are (locally) important.

Finally, it is noted that also mixed formulations have been proposed (Sussmann and Bathe 1987) in which the displacements are interpolated as well as the pressures (u/p-formulation). Typically, the pressure degrees of freedom are eliminated at element level by static condensation, so that only displacements enter the global system of equations. While generally effective, they bear the disadvantage that the presence of stress-like quantities as degrees of freedom makes it less straightforward to apply the stress integration algorithms outlined before, since these are strain-driven.

References

Auricchio F and Taylor RL 1995 Two material models for cyclic plasticity: Nonlinear kinematic hardening and generalized plasticity. *International Journal of Plasticity* **11**, 65–98.

Belytschko T, Ong SJ, Liu WK and Kennedy JM 1984 Hourglass control in linear and nonlinear problems. *Computer Methods in Applied Mechanics and Engineering* **43**, 251–276.

Besseling JF 1958 A theory of elastic, plastic and creep deformations of an initially isotropic material showing anisotropic strain-hardening, creep and recovery, and secondary creep. *Journal of Applied Mechanics* **24**, 529–536.

Bushnell D 1977 A strategy for the solution of problems involving large deflections, plasticity and creep. *International Journal for Numerical Methods in Engineering* **11**, 682–708.

Coulomb CA 1776 Essai sur une application des règles de maximis et minimis à quelques problèmes de statique, relatifs à l'architecture. *Mémoires de Mathématique et de Physique présentés à l'Académie Royale des Sciences par divers Savans* **7**, 343–382.

Crisfield MA 1987 Plasticity computations using the Mohr-Coulomb yield criterion. *Engineering Computations* **4**, 300–308.

Crisfield MA and Wills J 1989 Analysis of r/c panels using different concrete models. *ASCE Journal of Engineering Mechanics* **115**, 578–597.

Dafalias YF and Popov EP 1975 Plastic internal variables formalism of cyclic plasticity. *Journal of Applied Mechanics* **32**, 645–651.

de Borst R 1987 Computation of post-bifurcation and post-failure behaviour of strain-softening solids. *Computers & Structures* **25**, 211–224.

de Borst R 1989 Numerical methods for bifurcation analysis in geomechanics. *Ingenieur-Archiv* **59**, 160–174.

de Borst R 1991 The zero-normal-stress condition in plane-stress and shell elasto-plasticity. *Communications in Applied Numerical Methods* **7**, 29–33.

de Borst R and Feenstra PH 1990 Studies in anisotropic plasticity with reference. *International Journal for Numerical Methods in Engineering* **29**, 315–336.

de Borst R and Heeres OM 2002 A unified approach to the implicit integration of standard, non-standard and viscous plasticity models. *International Journal for Numerical and Analytical Methods in Geomechanics* **26**, 1059–1070.

de Borst R and Mühlhaus HB 1992 Gradient-dependent plasticity: formulation and algorithmic aspects. *International Journal for Numerical Methods in Engineering* **35**, 521–539.

de Borst R and Nauta P 1985 Non-orthogonal cracks in a smeared finite element model. *Engineering Computations* **2**, 35–46.

de Borst R and Vermeer PA 1984 Possibilities and limitations of finite elements for limit analysis. *Geotechnique* **34**, 199–210.

de Borst R, Pamin J and Geers MGD 1999 On coupled gradient-dependent plasticity and damage theories with a view to localization analysis. *European Journal of Mechanics: A/Solids* **18**, 939–962.

de Borst R, Pankaj and Bićanić N 1991 A note on singularity indicators for Mohr-Coulomb type yield criteria. *Computers & Structures* **39**, 219–220.

de Souza Neto EA, Perić D and Owen DRJ 2008 *Computational Methods for Plasticity: Theory and Applications*. John Wiley & Sons, Ltd.

Dennis JE and Schnabel RB 1983 *Numerical Methods for Unconstrained Optimization and Nonlinear Equations*. Prentice-Hall.

Drucker DC and Prager W 1952 Soil mechanics and plastic analysis for limit design. *Quarterly of Applied Mathematics* **10**, 157–165.

Feenstra PH and de Borst R 1996 A composite plasticity model for concrete. *International Journal of Solids and Structures* **33**, 707–730.

Flanagan DP and Belytschko T 1981 A uniform strain hexahedron and quadrilateral with orthogonal hourglass control. *International Journal for Numerical Methods in Engineering* **17**, 679–706.

Hashagen F and de Borst R 2001 Enhancement of the Hoffman yield criterion with an anisotropic hardening model. *Composite Structures* **79**, 637–651.

Hill R 1950 *The Mathematical Theory of Plasticity*. Oxford University Press.

Hoffman O 1967 The brittle strength of orthotropic materials. *Journal of Composite Materials* **1**, 200–206.

Hughes TJR 1980 Generalization of selective integration procedures to anisotropic and nonlinear media. *International Journal for Numerical Methods in Engineering* **15**, 1413–1418.

Khan AS and Huang S 1995 *Continuum Theory of Plasticity*. John Wiley & Sons, Ltd.

Koiter WT 1953 Stress-strain relations, uniqueness and variational theorems for elastic-plastic materials with a singular yield surface. *Quarterly of Applied Mathematics* **11**, 350–354.

Koiter WT 1960 General theorems of elastic-plastic solids, in *Progress in Solid Mechanics* (eds Sneddon IN and Hill R), vol. 1, pp. 165–221. North-Holland.

Kosloff D and Frazier GA 1978 Treatment of hourglass patterns in low order finite element codes. *International Journal for Numerical and Analytical Methods in Geomechanics* **2**, 57–72.

Krieg RD 1975 A practical two-surface plasticity theory. *Journal of Applied Mechanics* **31**, 641–646.

Krieg RD and Krieg DB 1977 Accuracies of numerical solution methods for the elastic-perfectly plastic model. *Journal of Pressure Vessel Technology* **99**, 510–515.

Kupfer HB and Gerstle KH 1973 Behavior of concrete under biaxial stresses. *ASCE Journal of Engineering Mechanics* **99**, 853–866.

Larsson R and Runesson K 1996 Implicit integration and consistent linearization for yield criteria of the Mohr-Coulomb type. *Mechanics of Cohesive-frictional Materials* **1**, 367–383.

Lubliner J 1990 *Plasticity Theory*. Macmillan.

Maier J and Thürlimann B 1985 Bruchversuche an Stahlbetonscheiben. Technical report, Eidgenössische Technische Hochschule, Zürich.

Matthies HG 1989 A decomposition method for the integration of the elastic-plastic rate problem. *International Journal for Numerical Methods in Engineering* **28**, 1–11.

Mohr O 1900 Welche Umstände bedingen die Elastizitätsgrenze und den Bruch eines Materials. *Civilingenieur: Zeitschrift des Vereins deutscher Ingenieure* **44-45**, 1524–1530, 1572–1577.

Mróz Z 1967 On the description of anisotropic workhardening. *Journal of the Mechanics and Physics of Solids* **15**, 163–175.

Nagtegaal JC, Parks DM and Rice JR 1974 On numerically accurate finite element solutions in the fully plastic range. *Computer Methods in Applied Mechanics and Engineering* **4**, 113–135.

Nayak GC and Zienkiewicz OC 1972 Elasto-plastic stress analysis. a generalization for various constitutive relations including strain softening. *International Journal for Numerical Methods in Engineering* **5**, 113–135.

Needleman A and Tvergaard V 1987 An analysis of ductile rupture modes at a crack tip. *Journal of the Mechanics and Physics of Solids* **35**, 151–183.

Ogden RW 1984 *Non-linear Elastic Deformations*. Ellis Horwood.

Ortiz M and Martin JB 1989 Symmetry-preserving return mapping algorithms and incrementally extremal paths: a unification of concepts. *International Journal for Numerical Methods in Engineering* **28**, 1839–1854.

Ortiz M and Popov EP 1985 Accuracy and stability of integration algorithms for elastoplastic constitutive relations. *International Journal for Numerical Methods in Engineering* **21**, 1561–1576.

Ortiz M and Simo JC 1986 An analysis of a new class of integration algorithms for elastoplastic constitutive relations. *International Journal for Numerical Methods in Engineering* **23**, 353–366.

Owen DRJ and Hinton E 1980 *Finite Elements in Plasticity: Theory and Practice*. Pineridge Press.

Pastor M, Zienkiewicz OC and Chan AC 1990 Generalized plasticity and the modeling of soil behavior. *International Journal for Numerical and Analytical Methods in Geomechanics* **14**, 151–190.

Pérez-Foguet A, Rodríguez-Ferran A and Huerta A 2001 Consistent tangent matrices for substepping schemes. *Computer Methods in Applied Mechanics and Engineering* **190**, 4627–4647.

Perić D and de Souza Neto EA 1999 A new computational model for tresca plasticity at finite strains with optimal parametrization in the principal space. *Computer Methods in Applied Mechanics and Engineering* **171**, 463–489.

Prager W 1955 The theory of plasticity: a survey of recent achievements. *Institution of Mechanical Engineers* **169**, 41–57.

Pramono E and Willam KJ 1989 Implicit integration of composite yield surfaces with corners. *Engineering Computations* **7**, 186–197.

Rankine WJM 1858 *Manual of Applied Mechanics*. R. Griffin and Co.

Rice JR and Tracey DM 1971 Computational fracture mechanics, in *Proceedings of the Symposium on Numerical and Computational Methods in Structural Mechanics* (ed. Fenves SJ), pp. 585–624. Academic Press.

Rouainia M and Wood DM 2000 An implicit constitutive algorithm for finite strain cam-clay elasto-plastic model. *Mechanics of Cohesive-frictional Material* **5**, 469–489.

Runesson K, Samuelsson A and Bernspang L 1986 Numerical technique in plasticity including solution advancement control. *International Journal for Numerical Methods in Engineering* **22**, 769–788.

Schellekens JCJ and de Borst R 1990 The use of the Hoffman yield criterion in finite element analysis of anisotropic composites. *Computers & Structures* **37**, 1087–1096.

Schreyer HL, Kulak RF and Kramer JM 1979 Accurate numerical solutions for elastic-plastic models. *Journal of Pressure Vessel Technology* **101**, 226–234.

Simo JC and Hughes TJR 1998 *Computational Inelasticity: Theory and Applications*. Springer.

Simo JC and Rifai MS 1990 A class of mixed assumed strain methods and the method of incompatible modes. *International Journal for Numerical Methods in Engineering* **29**, 1595–1638.

Simo JC and Taylor RL 1985 Consistent tangent operators for rate-independent plasticity. *Computer Methods in Applied Mechanics and Engineering* **48**, 101–118.

Simo JC, Kennedy JG and Govindjee S 1988 Non-smooth multisurface plasticity and viscoplasticity. loading/unloading conditions and numerical algorithms. *International Journal for Numerical Methods in Engineering* **26**, 2161–2185.

Sussmann T and Bathe KJ 1987 A finite element formulation for nonlinear incompressible elastic and inelastic analysis. *Computers & Structures* **29**, 357–409.

Taylor RL, Beresford PJ and Wilson EL 1976 A non-conforming element for stress analysis. *International Journal for Numerical Methods in Engineering* **10**, 1211–1219.

Tresca H 1868 Mémoire sur l' écoulement des corps solides. *Mémoires Savants de l' Académie des Sciences Paris* **18**, 733–799.

Verhoosel CV, Remmers JJC and Gutiérrez MA 2009 A dissipation-based arc-length method for robust simulation of brittle and ductile failure. *International Journal for Numerical Methods in Engineering* **77**, 1290–1321.

Vermeer PA and de Borst R 1984 Non-associated plasticity for soils, concrete and rock. *Heron* **29(1)**, 1–64.

von Mises R 1913 Mechanik der festen Körper im plastisch deformablen Zustand. *Göttinger Nachrichten, mathematisch-physische Klasse* **1**, 582–592.

Wilkins ML 1964 Calculation of elastic-plastic flow, in *Methods in Computational Physics* (eds Alder B, Fernbach S and Rotenberg M), vol. 3, pp. 211–263. Academic Press.

Wood DM 1990 *Soil Behaviour and Critical State Soil Mechanics*. Cambridge University Press.

Ziegler H 1959 A modification of Prager's hardening rule. *Quarterly of Applied Mathematics* **17**, 55–65.

8

Time-dependent Material Models

Time-dependent material models like visco-elasticity, creep, and visco-plasticity, differ from rate-independent material models. In rate-independent models, the relation between the stress and the strain can be linearised to give

$$\Delta\boldsymbol{\sigma}_1 = \mathbf{D}_1 \Delta\boldsymbol{\epsilon}_1 \tag{8.1}$$

where the subscript '1' denotes quantities that are related to the first iteration, and $\mathbf{D}$ is the material tangential stiffness matrix, while in the ensuing iterations one obtains:

$$\mathrm{d}\boldsymbol{\sigma}_j = \mathbf{D}_j \mathrm{d}\boldsymbol{\epsilon}_j \tag{8.2}$$

for the iterative improvements. As we will see in this chapter, rate-dependent material models result in an incremental relation that attains the following format

$$\Delta\boldsymbol{\sigma}_1 = \mathbf{D}_1 \Delta\boldsymbol{\epsilon}_1 + \mathbf{q} \tag{8.3}$$

where the vector $\mathbf{q}$, which can depend on the stress at the beginning of the time step, on a set of internal variables, or on non-mechanical quantities like thermal or hygral strains, and often, on a parameter from the time integration scheme. Clearly, $\mathbf{q}$ has to be added to the right-hand side of the discretised balance of momentum, and can be conceived as a pseudo-load vector. However, this only has to be done in the first iteration, since $\mathbf{q}$ does not vary from iteration to iteration, so that in the next iterations Equation (8.2) still holds.

8.1 Linear Visco-elasticity

In this section we show how visco-elastic models can be used to analyse the time-dependent behaviour of materials and structures. We introduce the theory of linear visco-elasticity with the help of a simple one-dimensional model consisting of a linear spring and a linear dashpot. Next, the extension to three-dimensional models is made and aspects concerning the numerical implementation are discussed. The formulation is generalised to incorporate the possible effects of aging.

Non-linear Finite Element Analysis of Solids and Structures, Second Edition.
René de Borst, Mike A. Crisfield, Joris J.C. Remmers and Clemens V. Verhoosel.

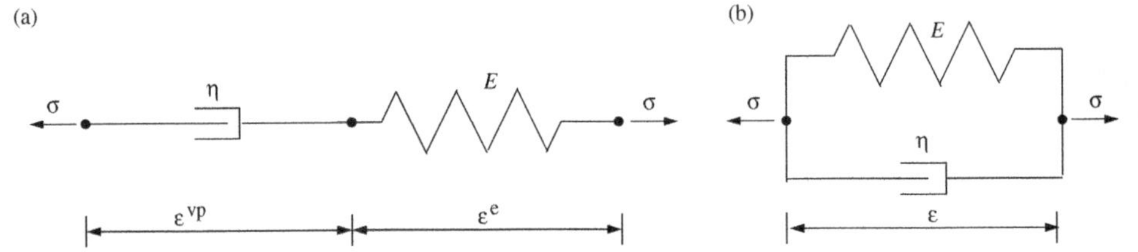

Figure 8.1 Maxwell element (a) and Kelvin element (b)

8.1.1 One-dimensional Linear Visco-elasticity

The two simplest linear visco-elastic models are the Maxwell element and the Kelvin element (Figure 8.1), where the term 'element' is here used for a constitutive element, and not for a finite element. The former model arises when a linear spring is coupled in series with a linear dashpot, while the latter model ensues when these two rheological elements are connected in parallel.

For the Maxwell element the total strain rate $\dot{\epsilon}$ is obtained as the sum of the elastic strain rate that stems from the spring $\dot{\epsilon}^{\mathrm{e}}$ and the viscous strain rate $\dot{\epsilon}^{\mathrm{v}}$ that results from the mechanical action of the dashpot:

$$\dot{\epsilon} = \dot{\epsilon}^{\mathrm{e}} + \dot{\epsilon}^{\mathrm{v}} \tag{8.4}$$

When E denotes the Young's modulus of the spring and η is the viscosity of the dashpot, we can set up the following constitutive relations

$$\dot{\epsilon}^{\mathrm{e}} = \frac{\dot{\sigma}}{E} \tag{8.5}$$

$$\dot{\epsilon}^{\mathrm{v}} = \frac{\sigma}{\eta} \tag{8.6}$$

where the observation that the stress σ is equal for the spring and dashpot has been utilised. Combination of Equations (8.4)–(8.6) gives the differential equation for the Maxwell element:

$$\dot{\epsilon} = \frac{\dot{\sigma}}{E} + \frac{\sigma}{\eta} \tag{8.7}$$

Now suppose that at time $t = t^0$ a constant strain ϵ^0 is imposed on the system. The system will then react according to

$$\sigma(t) = E\epsilon^0 \exp\left(-\frac{t - t^0}{\tau}\right) \tag{8.8}$$

which follows from the solution of the differential equation (8.7) with the initial condition, and

$$\tau = \frac{\eta}{E} \tag{8.9}$$

the so called relaxation time, i.e. the time that is needed to bring the stress back to $\exp(-1)$ of its value immediately after imposing the strain $[\sigma(0) = E\epsilon^0]$.

Although the response function of a Maxwell element on some imposed strain is now known, this is not so for the response of this element to an arbitrary strain history. A fundamental property of linear visco-elasticity enters at this point: the response of the system on two individual strain histories is equal to the sum of the responses on each of those strain histories. This assumption is named the superposition principle and states that the total response on a strain ϵ^0 that is applied at t^0 and a strain ϵ^1 that is applied at t^1 is given by

$$\sigma(t) = E\epsilon^0 \exp\left(-\frac{t - t^0}{\tau}\right) + E\epsilon^1 \exp\left(-\frac{t - t^1}{\tau}\right)$$

More generally we have for n excitations:

$$\sigma(t) = \sum_{i=1}^{n} E\epsilon^i \exp\left(-\frac{t - t^i}{\tau}\right) \tag{8.10}$$

For the limiting case that ϵ^i becomes infinitesimally small we obtain the integral

$$\sigma(t) = \int E \exp\left(-\frac{t - \tilde{t}}{\tau}\right) \mathrm{d}\epsilon \tag{8.11}$$

or replacing this so-called Stieltjes integral by a Riemann integral

$$\sigma(t) = \int_0^t E \exp\left(-\frac{t - \tilde{t}}{\tau}\right) \dot{\epsilon}(\tilde{t})\mathrm{d}\tilde{t} \tag{8.12}$$

The relaxation function

$$E(t - \tilde{t}) = E \exp\left(-\frac{t - \tilde{t}}{\tau}\right) \tag{8.13}$$

is characteristic for a Maxwell element that is subjected to a relaxation experiment, since another combination of springs and dashpots results in a different response function $E(t - \tilde{t})$. Upon introduction of the relaxation function we can set up a general expression for the stress at time t as a function of the strain history:

$$\sigma(t) = \int_0^t E(t - \tilde{t})\dot{\epsilon}(\tilde{t})\mathrm{d}\tilde{t} \tag{8.14}$$

For each combination of springs and dashpots a specific expression of $E(t - \tilde{t})$ can be derived. We can even postulate relaxation functions for Equation (8.14) that do not have a direct mechanical interpretation in terms of springs and dashpots.

Equation (8.14) gives the current stress as a function of the strain. Creep experiments on the other hand give the current strain as a function of the applied stress history. This necessitates an inversion of Equation (8.14), which results in the compliance formulation

$$\epsilon(t) = \int_0^t J(t - \tilde{t})\dot{\sigma}(\tilde{t})\mathrm{d}\tilde{t} \tag{8.15}$$

with $J(t - \tilde{t})$ the creep function. The creep function gives the current value of the strain for a given unit stress increment applied at $\tilde{t}$, i.e. $\sigma(\tilde{t}) = \mathcal{H}(\tilde{t})$ with $\mathcal{H}$ the Heaviside function, as can be inferred since then $\dot{\sigma}(\tilde{t}) = \delta(\tilde{t})$, with δ the Dirac function, so that substitution into Equation (8.15) renders: $\epsilon(t) = J(t - \tilde{t})$.

8.1.2 Three-dimensional Visco-elasticity

We will now generalise Equations (8.14) and (8.15) to three dimensions. In addition we will generalise the formulation to include the phenomenon of aging – the increase of the stiffness of a material over the course of time, which has been observed experimentally for many materials including concrete and polymers.

Generalising Equation (8.14) to three dimensions and taking into account the possible influence of aging and of an initial strain $\boldsymbol{\epsilon}(0)$ we obtain

$$\boldsymbol{\sigma}(t) = \mathbf{D}(t)\boldsymbol{\epsilon}(0) + \int_0^t \mathbf{D}(t - \tilde{t}, \tilde{t})\dot{\boldsymbol{\epsilon}}(\tilde{t})\mathrm{d}\tilde{t} \tag{8.16}$$

The matrix $\mathbf{D}(t - \tilde{t}, \tilde{t})$ is the three-dimensional generalisation of the relaxation function $E(t - \tilde{t}, \tilde{t})$. Aging is incorporated because $\mathbf{D}$ is not only a function of the time difference $t - \tilde{t}$ as in Equation (8.14), but also of $\tilde{t}$. The first term on the right-hand side of Equation (8.16) represents the possible presence of initial strains.

Similar to Equation (8.16) the strain ϵ at time t can be expressed as a function of the stress history:

$$\boldsymbol{\epsilon}(t) = \mathbf{C}(t)\boldsymbol{\sigma}(0) + \int_0^t \mathbf{C}(t - \tilde{t}, \tilde{t})\dot{\boldsymbol{\sigma}}(\tilde{t})\mathrm{d}\tilde{t} \tag{8.17}$$

The matrix $\mathbf{C}(t - \tilde{t}, \tilde{t})$ is the three-dimensional generalisation of the creep function $J(t - \tilde{t}, \tilde{t})$. As with the three-dimensional generalisation of the relaxation function the time of loading $\tilde{t}$ has entered the formulation as an independent quantity to account for possible effects of aging of the material. The first term on the right-hand size represents the possible presence of initial stresses. We shall continue the treatment of three-dimensional, linear visco-elasticity by taking Equation (8.16) as a point of departure. In a displacement-based finite element formulation this has the advantage that the stresses are directly expressed as a function of the strain history.

An important simplification of Equation (8.16) can be achieved if it is assumed that the material remains isotropic during the entire loading history. Then, $\mathbf{D}$ reads:

$$\mathbf{D}(t - \tilde{t}, \tilde{t}) = \begin{bmatrix} \tilde{\lambda} + 2\tilde{\mu} & \tilde{\lambda} & \tilde{\lambda} & 0 & 0 & 0 \\ \tilde{\lambda} & \tilde{\lambda} + 2\tilde{\mu} & \tilde{\lambda} & 0 & 0 & 0 \\ \tilde{\lambda} & \tilde{\lambda} & \tilde{\lambda} + 2\tilde{\mu} & 0 & 0 & 0 \\ 0 & 0 & 0 & \tilde{\mu} & 0 & 0 \\ 0 & 0 & 0 & 0 & \tilde{\mu} & 0 \\ 0 & 0 & 0 & 0 & 0 & \tilde{\mu} \end{bmatrix} \tag{8.18}$$

with $\tilde{\lambda} = \lambda(t - \tilde{t}, \tilde{t})$ and $\tilde{\mu} = \mu(t - \tilde{t}, \tilde{t})$ the Lamé constants, which are now time- and age-dependent. In many applications it is more convenient to work with the Young's modulus and the Poisson's ratio, especially since the variation of Poisson's ratio with time is often of minor importance compared with that of the change of the Young's modulus. Making this assumption and using the definitions (1.116) we obtain:

$$\sigma(t) = E(t)\bar{\mathbf{D}}\epsilon(0) + \int_0^t E(t-\tilde{t},\tilde{t})\bar{\mathbf{D}}\dot{\epsilon}(\tilde{t})\mathrm{d}\tilde{t} \tag{8.19}$$

with $\bar{\mathbf{D}}$ the dimensionless matrix

$$\bar{\mathbf{D}} = \frac{1}{(1+\nu)(1-2\nu)}\begin{bmatrix} 1-\nu & \nu & \nu & 0 & 0 & 0 \\ \nu & 1-\nu & \nu & 0 & 0 & 0 \\ \nu & \nu & 1-\nu & 0 & 0 & 0 \\ 0 & 0 & 0 & \frac{1}{2}-\nu & 0 & 0 \\ 0 & 0 & 0 & 0 & \frac{1}{2}-\nu & 0 \\ 0 & 0 & 0 & 0 & 0 & \frac{1}{2}-\nu \end{bmatrix} \tag{8.20}$$

8.1.3 Algorithmic Aspects

Equations (8.16) and (8.19) require that we memorise the entire strain history in order to calculate the value of the stress after the new time step. Put differently, we have to store all previous strain increments in order to compute the new stress increment $\Delta\sigma$ with the aid of Equation (8.19). This is inconvenient for large-scale computations. Starting with the pioneering work of Zienkiewicz *et al.* (1968) and of Taylor *et al.* (1970), which was later extended to aging visco-elasticity by Bažant and Wu (1974), most algorithms in finite element programs are based on an expansion of the relaxation function $E(t-\tilde{t},\tilde{t})$, which is known as the kernel of the hereditary integral. After an expansion of the original kernel in a series of polynomials or negative exponential powers a so-called degenerated kernel arises. For instance, if we expand $E(t-\tilde{t},\tilde{t})$ in a series of negative exponential powers, we obtain:

$$E(t-\tilde{t},\tilde{t}) = E^0(\tilde{t}) + \sum_{\alpha=1}^{N} E^{\alpha}(\tilde{t})\exp\left(-\frac{t-\tilde{t}}{\tau^{\alpha}}\right) \tag{8.21}$$

In Equation (8.21) E^{α} is a stiffness and τ^{α} has the dimension of time. Neglecting the possible effect of initial strains, substitution of Equation (8.21) in (8.19) gives:

$$\sigma(t) = \int_0^t \left(E^0(\tilde{t}) + \sum_{\alpha=1}^{N} E^{\alpha}(\tilde{t})\exp\left(-\frac{t-\tilde{t}}{\tau^{\alpha}}\right)\right)\bar{\mathbf{D}}\dot{\epsilon}(\tilde{t})\mathrm{d}\tilde{t} \tag{8.22}$$

We now compare Equations (8.22) and (8.12), which is the response function for a single Maxwell element. We observe that the response of (8.22) is exactly the same as that obtained when a parallel arrangement of N Maxwell elements, each with its own relaxation time τ^{α} and spring stiffness E^{α}, and one spring element with stiffness E^0, is loaded by the same strain history $\epsilon(\tilde{t})$. The parallel chain of Figure 8.2 is named a Maxwell chain, and can be derived formally by first differentiating both sides of Equation (8.22). This gives:

$$\dot{\sigma}(t) = \frac{\mathrm{d}}{\mathrm{d}t}\left(\int_0^t \left(E^0(\tilde{t}) + \sum_{\alpha=1}^{N} E^{\alpha}(\tilde{t})\exp\left(-\frac{t-\tilde{t}}{\tau^{\alpha}}\right)\right)\bar{\mathbf{D}}\dot{\epsilon}(\tilde{t})\mathrm{d}\tilde{t}\right) \tag{8.23}$$

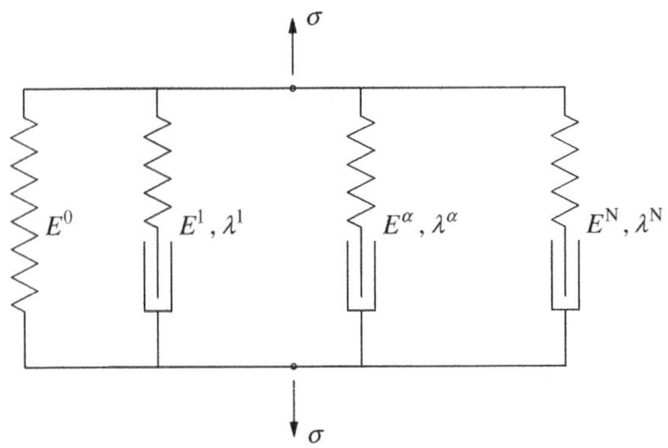

Figure 8.2 Maxwell chain

When we define the stress within each Maxwell element as

$$\boldsymbol{\sigma}^{\alpha}(t) = \int_0^t E^{\alpha}(\tilde{t}) \exp\left(-\frac{t-\tilde{t}}{\tau^{\alpha}}\right) \bar{\mathbf{D}}\dot{\boldsymbol{\epsilon}}(\tilde{t})\mathrm{d}\tilde{t} \tag{8.24}$$

and subsequently interchange the order of integration and summation we obtain

$$\dot{\boldsymbol{\sigma}}(t) = E^0\bar{\mathbf{D}}\dot{\boldsymbol{\epsilon}}(t) + \sum_{\alpha=1}^{N} \dot{\boldsymbol{\sigma}}^{\alpha}(t) \tag{8.25}$$

Finally, Leibnitz' rule can be invoked to derive from Equation (8.24) that

$$\dot{\boldsymbol{\sigma}}^{\alpha}(t) = E^{\alpha}\bar{\mathbf{D}}\dot{\boldsymbol{\epsilon}}(t) - \frac{1}{\tau^{\alpha}}\boldsymbol{\sigma}^{\alpha}(t) \tag{8.26}$$

which is the three-dimensional generalisation of the differential equation for a single Maxwell element, cf. Equation (8.7). The above derivation shows that the expansion in negative exponential powers – also called a Dirichlet series – of a relaxation function as in Equation (8.21) can be interpreted as a Maxwell chain. In a manner similar to this derivation one can show that the expansion of a creep function in a Dirichlet series mechanically results in a Kelvin chain, i.e. a series arrangement of Kelvin elements.

Equation (8.22) is a useful point of departure for the development of an algorithm that is suitable for large-scale computations, in the sense that storage of the entire strain history is not required for the computation of a new stress increment. Instead, the stress at time t can be calculated on the basis of the strain increment and a finite number of state variables, all of which are known at the current time $t - \Delta t$. Information of previous time steps is not needed.

When we bring the dimensionless matrix $\bar{\mathbf{D}}$ outside the integral and interchange the order of integration and summation in Equation (8.22) we obtain:

$$\boldsymbol{\sigma}(t) = \bar{\mathbf{D}}\left(\int_0^t E^0(\tilde{t})\dot{\boldsymbol{\epsilon}}(\tilde{t})\mathrm{d}\tilde{t} + \sum_{\alpha=1}^{N}\int_0^t E^{\alpha}(\tilde{t}) \exp\left(-\frac{t-\tilde{t}}{\tau^{\alpha}}\right)\dot{\boldsymbol{\epsilon}}(\tilde{t})\mathrm{d}\tilde{t}\right) \tag{8.27}$$

Next, the time interval is divided into two parts, one from $\tilde{t} = 0$ to $\tilde{t} = t - \Delta t$, and one from $\tilde{t} = t - \Delta t$ to $\tilde{t} = t$. When we subtract

$$\sigma(t - \Delta t) = \bar{\mathbf{D}} \left(\int_0^{t-\Delta t} E^0(\tilde{t})\dot{\epsilon}(\tilde{t})\mathrm{d}\tilde{t} + \sum_{\alpha=1}^{N} \int_0^{t-\Delta t} E^\alpha(\tilde{t}) \exp\left(-\frac{t - \Delta t - \tilde{t}}{\tau^\alpha}\right) \dot{\epsilon}(\tilde{t})\mathrm{d}\tilde{t} \right) \tag{8.28}$$

from both sides of the previous equation, the following incremental stress–strain relation results:

$$\Delta\sigma = \bar{\mathbf{D}}E^0(\tilde{t})\Delta\epsilon + \sum_{\alpha=1}^{N} \bar{\mathbf{D}} \int_{t-\Delta t}^{t} E^\alpha(\tilde{t}) \exp\left(-\frac{t - \tilde{t}}{\tau^\alpha}\right) \dot{\epsilon}(\tilde{t})\mathrm{d}\tilde{t} - \sum_{\alpha=1}^{N} \left(1 - \exp\left(-\frac{\Delta t}{\tau^\alpha}\right)\right) \sigma^\alpha(t - \Delta t) \tag{8.29}$$

Using the assumption that the strain rate is constant over the time step,

$$\dot{\epsilon} \approx \frac{\Delta\epsilon}{\Delta t} \tag{8.30}$$

the integral can be elaborated in a semi-analytical manner, as follows:

$$\Delta\sigma = \bar{\mathbf{D}}E^0(\tilde{t})\Delta\epsilon + \sum_{\alpha=1}^{N} \left(1 - \exp\left(-\frac{\Delta t}{\tau^\alpha}\right)\right) \left(\frac{E^\alpha(\tilde{t})}{\Delta t/\tau^\alpha}\bar{\mathbf{D}}\Delta\epsilon - \sigma^\alpha(t - \Delta t)\right) \tag{8.31}$$

For non-aging materials, E^α does not depend on $\tilde{t}$, and the integration is exact. Clearly, Equation (8.31) can be brought in the format of Equation (8.3), with

$$\mathbf{D}_1 = \bar{\mathbf{D}}E^0(\tilde{t}) + \sum_{\alpha=1}^{N} \left(1 - \exp\left(-\frac{\Delta t}{\tau^\alpha}\right)\right) \frac{E^\alpha(\tilde{t})}{\Delta t/\tau^\alpha}\bar{\mathbf{D}} \tag{8.32}$$

and

$$\mathbf{q} = -\sum_{\alpha=1}^{N} \left(1 - \exp\left(-\frac{\Delta t}{\tau^\alpha}\right)\right) \sigma^\alpha(t - \Delta t) \tag{8.33}$$

8.2 Creep Models

Creep models are often used to describe the time-dependent behaviour of metals, and can be considered as a generalisation of visco-elasticity. To elucidate this, we note that, similar to the strain-rate decomposition in visco-elasticity, Equation (8.4), we have a strain-rate decomposition into an elastic and a creep strain:

$$\dot{\epsilon} = \dot{\epsilon}^{\mathrm{e}} + \dot{\epsilon}^{\mathrm{c}} \tag{8.34}$$

and, considering the Bailey–Norton power law, which is often used to characterise the creep behaviour of metals (Hult 1966):

$$\dot{\epsilon}^{\mathrm{c}} = \frac{1}{\varphi(t)}\left(\frac{\sigma}{\sigma_n}\right)^n \tag{8.35}$$

we observe that this is a non-linear generalisation of the linear dashpot of Equation (8.6). In Equation (8.35), σ_n and n are (temperature-dependent) material parameters, and $\varphi(t)$ is a (monotonically increasing) function of time. Equation (8.35) is a function of the applied stress σ, the temperature T and the time t:

$$\dot{\epsilon}^{\mathrm{c}} = \dot{\epsilon}^{\mathrm{c}}(\sigma, T, t) \tag{8.36}$$

Creep laws of this kind are called time hardening and are convenient for simplified design calculations. For a constant stress σ and temperature T, we can equivalently write:

$$\dot{\epsilon}^{\mathrm{c}} = \dot{\epsilon}^{\mathrm{c}}(\sigma, T, \epsilon^{\mathrm{c}}) \tag{8.37}$$

where the creep strain takes the role of an internal variable, similar to the plastic strain ϵ^{p} in elasto-plasticity. Obviously, for non-constant stressing or a varying temperature, time-hardening creep, Equation (8.36), and strain-hardening creep, Equation (8.37), differ and will not give the same result. Note also that by replacing the time by the creep strain in the expression for the strain rate, the framework of (non-linear) visco-elasticity no longer applies, and that the total creep strain cannot be obtained via the computation of hereditary integrals. Instead, the creep strain increment is directly calculated as:

$$\Delta\epsilon^{\mathrm{c}} = \Delta t(\dot{\epsilon}^{\mathrm{c}})^{t+\theta\Delta t} \tag{8.38}$$

for the time increment Δt, with $(\dot{\epsilon}^{\mathrm{c}})^{t+\theta\Delta t}$ evaluated via a generalised midpoint rule. Alternatively, a trapezoidal rule can be used. The new stress is computed in a manner similar to elasto-plasticity (Zienkiewicz and Taylor 1991), namely via:

$$\sigma^{t+\Delta t} = \sigma^t + E(\Delta\epsilon - \Delta\epsilon^{\mathrm{c}}) \tag{8.39}$$

From Equations (8.38) and (8.39) we can derive the tangential relation needed for the iterative solution of the resulting non-linear equations at structural level using a Newton–Raphson method. Linearisation of both equations gives:

$$\dot{\sigma} = \left(E^{-1} + \Delta t\frac{\partial\epsilon^{\mathrm{c}}}{\partial\sigma}\right)^{-1}\dot{\epsilon} \tag{8.40}$$

which necessitates the assembly and the factorisation of the tangential stiffness matrix at every time step, except for $\theta = 0$, i.e. when explicit time integration is used. However, explicit integration schemes suffer from limited accuracy and stability, see also Chapter 5 in the context of the integration of the balance of momentum equation.

The generalisation of the above one-dimensional constitutive relation to the three-dimensional case is straightforward. Since creep models as formulated above are typically used in computations for metals, this is usually done via a creep potential function, similar to plasticity, so that for the multiaxial generalisation of the creep strain rate one obtains:

$$\Delta\boldsymbol{\epsilon}^{c} = \Delta t(\dot{\epsilon}^{c})^{t+\theta\Delta t}\frac{\partial f}{\partial\boldsymbol{\sigma}} \tag{8.41}$$

with $f = f(\boldsymbol{\sigma})$ a potential function and $\dot{\epsilon}^{c} = \dot{\epsilon}^{c}(\boldsymbol{\epsilon}^{c})$ the equivalent creep strain rate. The generalisation of Equations (8.39) and (8.40) is straightforward.

8.3 Visco-plasticity

Visco-elasticity and creep models as described in the preceding sections predict time-dependent deformations for all stress levels. This may not always be realistic, and for certain materials time-dependent strains only become noticeable above a threshold stress level. This can be well modelled using visco-plasticity. Similar to inviscid plasticity (Chapter 7), there exists a yield surface, and when the stress remains inside the yield surface, no time-dependent strains will develop.

8.3.1 One-dimensional Visco-plasticity

The visco-plastic extension of the rate-independent equations is demonstrated with the rheological model in Figure 8.3. The model consists of an elastic element with stiffness E, which is connected in series to an inelastic element, which consists of a dashpot with a viscosity η in parallel with a plastic slider with a current yield strength $\bar{\sigma}$. As in rate-independent small-strain plasticity an additive decomposition is assumed with respect to the total strain:

$$\epsilon = \epsilon^{e} + \epsilon^{vp} \tag{8.42}$$

with ϵ^{e} the strain in the spring and ϵ^{vp} the visco-plastic strain in the inelastic element (Figure 8.4). When we denote the applied stress by σ we have

$$\sigma = E\epsilon^{e} \tag{8.43}$$

which can be combined with Equation (8.42) to give:

$$\sigma = E(\epsilon - \epsilon^{vp}) \tag{8.44}$$

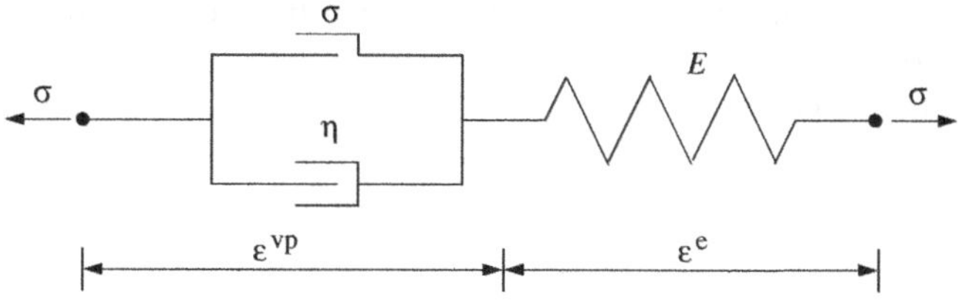

Figure 8.3 One-dimensional representation of a visco-plasticity model

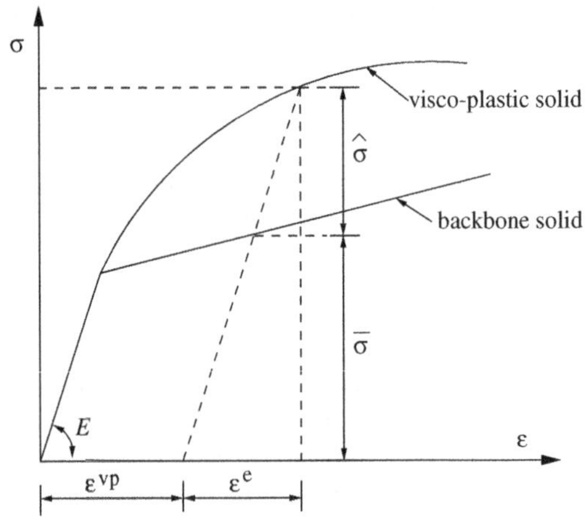

Figure 8.4 Schematic representation of the stress response for visco-plasticity

As in rate-independent plasticity the solid remains elastic when the stress σ is smaller than the yield stress $\bar{\sigma}$ (assuming that $\sigma > 0$), and the yield function reads:

$$f(\sigma) = \sigma - \bar{\sigma} \tag{8.45}$$

In rate-independent plasticity, stresses that are larger than the yield strength $\bar{\sigma}$ are impossible, and the strict requirement that $f \leq 0$ is imposed, which is formalised through the Karush–Kuhn–Tucker conditions, Equation (6.16). This requirement is relaxed in visco-plasticity, where, during yielding, an additional stress, sometimes called the overstress, can be carried by the dashpot:

$$\hat{\sigma} = f(\sigma) > 0 \tag{8.46}$$

which is assumed to react according to a viscous relation,

$$\hat{\sigma} = \eta \dot{\epsilon}^{\mathrm{vp}} \tag{8.47}$$

with the viscosity parameter η. Combination of the latter two equations yields:

$$\dot{\epsilon}^{\mathrm{vp}} = \frac{1}{\eta} f(\sigma) \quad \text{if} \quad f(\sigma) \geq 0 \tag{8.48}$$

which represents a one-dimensional visco-plastic constitutive equation of the Perzyna type (Perzyna 1966).

Viscosity effectively introduces a time scale in the boundery-value problem. This can be demonstrated most straightforwardly by defining

$$\tau = \frac{\eta}{E}$$

as in Equation (8.9). As in visco-elasticity, τ can be interpreted as the relaxation time of the model. To show this, we subject the model of Figure 8.3 to a relaxation test, where an

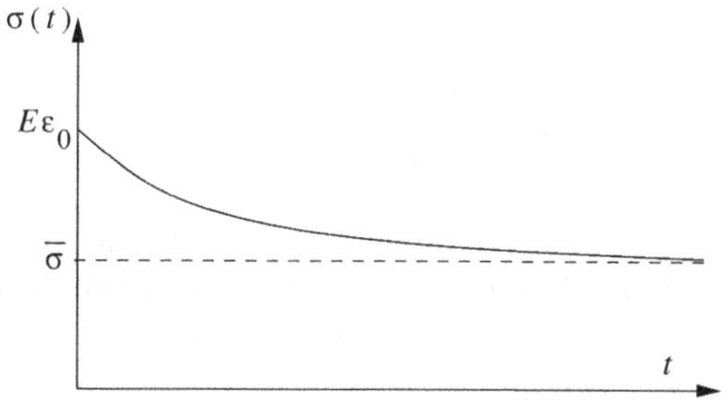

Figure 8.5 One-dimensional relaxation test on a visco-plastic element

instantaneous strain $\epsilon_0 > \frac{\bar{\sigma}}{E}$ is applied, hence inducing visco-plastic straining. We first substitute the definition of the yield function into Equation (8.48). Assuming continued loading we then have

$$\dot{\epsilon}^{\rm vp} = \frac{1}{\tau} E^{-1}(\sigma - \bar{\sigma}) \tag{8.49}$$

Substitution of this expression into Equation (8.44) and using the definition for τ yields:

$$\dot{\sigma} + \frac{1}{\tau}\sigma = E\dot{\epsilon} + \frac{1}{\tau}\bar{\sigma} \tag{8.50}$$

Since $\dot{\epsilon} = 0$ for a relaxation experiment the following closed-form solution is obtained:

$$\sigma(t) = (E\epsilon_0 - \bar{\sigma}) \exp\left(-\frac{t}{\tau}\right) + \bar{\sigma} \tag{8.51}$$

which is shown in Figure 8.5.

8.3.2 *Integration of the Rate Equations*

As for rate-independent plasticity, the algorithms for integrating the rate equations of visco-plasticity are strain driven. This implies that we depart from a converged state at time t, with known quantities $\boldsymbol{\epsilon}^t$, $(\boldsymbol{\epsilon}^{\rm vp})^t$, $\boldsymbol{\sigma}^t$, κ^t in a three-dimensional context. Here, κ is the hardening parameter, cf. Equation (7.57). The new stress $\boldsymbol{\sigma}^{t+\Delta t}$ is then computed from

$$\boldsymbol{\sigma}^{t+\Delta t} = \boldsymbol{\sigma}^t + \mathbf{D}^{\rm e}(\Delta\boldsymbol{\epsilon} - \Delta\boldsymbol{\epsilon}^{\rm vp}) \tag{8.52}$$

where the strain increment $\Delta\boldsymbol{\epsilon}$ follows from the displacement increments collected in $\Delta\mathbf{a}$ using standard kinematic operators (Chapters 2 and 3) and where the visco-plastic strain increment is computed either using a generalised trapezoidal rule

$$\begin{cases} \Delta\boldsymbol{\epsilon}^{\rm vp} = \left((1-\theta)(\dot{\boldsymbol{\epsilon}}^{\rm vp})^t + \theta(\dot{\boldsymbol{\epsilon}}^{\rm vp})^{t+\Delta t}\right)\Delta t \\ \Delta\kappa = \left((1-\theta)\dot{\kappa}^t + \theta\dot{\kappa}^{t+\Delta t}\right)\Delta t \end{cases} \tag{8.53}$$

or using a generalised midpoint rule

$$\begin{cases} \Delta\boldsymbol{\epsilon}^{\text{vp}} = (\dot{\boldsymbol{\epsilon}}^{\text{vp}})^{t+\theta\Delta t}\Delta t \\ \Delta\kappa = \dot{\kappa}^{t+\theta\Delta t}\Delta t \end{cases} \tag{8.54}$$

similar to rate-independent plasticity (Chapter 7). The following subsections elaborate the generalised trapezoidal rule for three different classes of visco-plasticity, namely the theory proposed by Perzyna (1966), the theory of Duvaut and Lions (1972), and the consistency visco-plasticity model (Wang *et al.* 1996, 1997).

8.3.3 Perzyna Visco-plasticity

The Perzyna theory of visco-plasticity is the oldest visco-plasticity theory. In it, the visco-plastic strain rate is defined as:

$$\dot{\boldsymbol{\epsilon}}^{\text{vp}} = \eta < \varphi(f) > \mathbf{m} \tag{8.55}$$

with $\mathbf{m}$ the visco-plastic flow direction, Equation (7.43), and $< \cdot >$ the MacAulay brackets. $\varphi(f)$ is an arbitrary function of f, for which a power law is commonly utilised:

$$\varphi(f) = \left(\frac{f}{\bar{\sigma}_0}\right)^n \tag{8.56}$$

with n a constant and $\bar{\sigma}_0$ the initial yield stress.

The visco-plastic strain rate at $t + \Delta t$ can be approximated using a truncated Taylor series as:

$$\begin{aligned} (\dot{\boldsymbol{\epsilon}}^{\text{vp}})^{t+\Delta t} &= (\dot{\boldsymbol{\epsilon}}^{\text{vp}})^t + \left(\frac{\partial\dot{\boldsymbol{\epsilon}}^{\text{vp}}}{\partial\boldsymbol{\sigma}}\right)^t \Delta\boldsymbol{\sigma} + \left(\frac{\partial\dot{\boldsymbol{\epsilon}}^{\text{vp}}}{\partial\kappa}\right)^t \Delta\kappa \\ &= (\dot{\boldsymbol{\epsilon}}^{\text{vp}})^t + \mathbf{G}^t\Delta\boldsymbol{\sigma} + \mathbf{h}^t\Delta\kappa \end{aligned} \tag{8.57}$$

where Equation (8.55) has been substituted to give:

$$\begin{aligned} \mathbf{G}^t &= \eta\left(\frac{\partial\varphi}{\partial\boldsymbol{\sigma}}\mathbf{m}^{\text{T}} + \varphi\frac{\partial\mathbf{m}}{\partial\boldsymbol{\sigma}}\right)^t \\ \mathbf{h}^t &= \eta\left(\frac{\partial\varphi}{\partial\kappa}\mathbf{m} + \varphi\frac{\partial\mathbf{m}}{\partial\kappa}\right)^t \end{aligned} \tag{8.58}$$

and $\Delta\kappa$ stems from the previous time step or from the previous equilibrium iteration. Substitution of Equation (8.57) into Equation (8.53) yields:

$$\Delta\boldsymbol{\epsilon}^{\text{vp}} = \left((\dot{\boldsymbol{\epsilon}}^{\text{vp}})^t + \theta\mathbf{G}^t\Delta\boldsymbol{\sigma} + \theta\mathbf{h}^t\Delta\kappa\right)\Delta t \tag{8.59}$$

which can be substituted into the expression for the updated stress, Equation (8.52), leading to:

$$\boldsymbol{\sigma}^{t+\Delta t} = \boldsymbol{\sigma}^t + \mathbf{D}\Delta\boldsymbol{\epsilon} - \mathbf{q} \tag{8.60}$$

where

$$\mathbf{D} = \left((\mathbf{D}^{\rm e})^{-1} + \theta \Delta t \mathbf{G}^t \right)^{-1}$$
$$\mathbf{q} = \mathbf{D} \left((\dot{\boldsymbol{\epsilon}}^{\rm vp})^t + \theta \mathbf{h}^t \Delta\kappa \right) \Delta t \tag{8.61}$$

are the tangential stiffness matrix and the pseudo-load vector, which arises due to the time-dependency, respectively. Box 8.1 summarises this single-step Euler algorithm for Perzyna visco-plasticity.

The single-step integration algorithm for the Perzyna visco-plasticity model may suffer from a limited stability and/or accuracy, especially when $\theta \leq \frac{1}{2}$, or when no global equilibrium iterations are added after the first estimate for the stress increment. A more rigorous approach

Box 8.1 Perzyna visco-plasticity: single-step Euler algorithm for $t \to t + \Delta t$

1. Begin time step. Initialise: $\Delta \mathbf{a} = \mathbf{0}$, $\mathbf{f}_{\rm ext}^{t+\Delta t}$
2. *For each integration point i:*
   - Compute: $\mathbf{G}_{i,0} = \eta \left(\frac{\partial \varphi}{\partial \boldsymbol{\sigma}} \mathbf{m}^{\rm T} + \varphi \frac{\partial \mathbf{m}}{\partial \boldsymbol{\sigma}} \right)_{i,0}$, $\mathbf{h}_{i,0} = \eta \left(\frac{\partial \varphi}{\partial \kappa} \mathbf{m} + \varphi \frac{\partial \mathbf{m}}{\partial \kappa} \right)_{i,0}$
   - Compute material tangential stiffness matrix: $\mathbf{D}_{i,0} = \left((\mathbf{D}_i^{\rm e})^{-1} + \theta \Delta t \mathbf{G}_{i,0} \right)^{-1}$
   - Compute pseudo-load vector: $\mathbf{q}_i = \mathbf{D}_{i,0} \left((\dot{\boldsymbol{\epsilon}}^{\rm vp})_i^t + \theta \mathbf{h}_{i,0} \Delta \kappa_i^t \right) \Delta t$
3. Compute the internal force vector: $\mathbf{f}_{{\rm int},0} = \mathbf{f}_{\rm int}^t + \int_V \mathbf{B}^{\rm T} \mathbf{q} {\rm d}V$
4. Iterations $j = 0, \ldots$ for finding equilibrium within the time step:
   - *For each integration point i:* compute $\mathbf{D}_{i,j} = \left((\mathbf{D}_i^{\rm e})^{-1} + \theta \Delta t \mathbf{G}_{i,j} \right)^{-1}$
   - Compute tangential stiffness matrix: $\mathbf{K}_j = \int_V \mathbf{B}^{\rm T} \mathbf{D}_j \mathbf{B} {\rm d}V$
   - Solve the linear system: $\mathbf{da}_{j+1} = (\mathbf{K}_j)^{-1} (\mathbf{f}_{\rm ext}^{t+\Delta t} - \mathbf{f}_{{\rm int},j})$
   - Update the displacement increments: $\Delta \mathbf{a}_{j+1} = \Delta \mathbf{a}_j + \mathbf{da}_{j+1}$
   - *For each integration point i:*
     - Compute the strain increment: $\Delta \boldsymbol{\epsilon}_{i,j+1} = \mathbf{B}_i \Delta \mathbf{a}_{j+1}$
     - Compute the trial stress: $(\boldsymbol{\sigma}_e)_{i,j+1} = \boldsymbol{\sigma}_i^t + \mathbf{D}_i^{\rm e} \Delta \boldsymbol{\epsilon}_{i,j+1}$
     - Evaluate the loading function: $f = f((\boldsymbol{\sigma}_e)_{i,j+1}, \kappa_i^t)$.
     - If $f \geq 0$: $\Delta \boldsymbol{\epsilon}_{i,j+1}^{\rm vp} = \left((\dot{\boldsymbol{\epsilon}}^{\rm vp})_i^t + \theta \mathbf{G}_{i,j} \Delta \boldsymbol{\sigma}_{i,j} + \theta \mathbf{h}_{i,j} \Delta \kappa_{i,j} \right) \Delta t$
       Else: $\Delta \boldsymbol{\epsilon}_{i,j+1}^{\rm vp} = \mathbf{0}$
     - Compute the stress increment: $\Delta \boldsymbol{\sigma}_{i,j+1} = \mathbf{D}_i^{\rm e} (\Delta \boldsymbol{\epsilon}_{i,j+1} - \Delta \boldsymbol{\epsilon}_{i,j+1}^{\rm vp})$
     - Compute the total stress: $\boldsymbol{\sigma}_{i,j+1} = \boldsymbol{\sigma}_i^t + \Delta \boldsymbol{\sigma}_{i,j+1}$
     - Update the hardening parameter increment: $\Delta \kappa_{i,j+1} = \Delta \kappa (\Delta \boldsymbol{\epsilon}_{i,j+1}^{\rm vp})$
   - Compute internal force: $\mathbf{f}_{{\rm int},j} = \int_V \mathbf{B}^{\rm T} \boldsymbol{\sigma}_{j+1} {\rm d}V$
   - Check convergence: if $\|\mathbf{f}_{\rm ext}^{t+\Delta t} - \mathbf{f}_{{\rm int},j+1}\| < \eta$, continue, else go to 4
5. *For each integration point i:* $\kappa_i^{t+\Delta t} = \kappa_i^t + \Delta \kappa_{i,j+1}$
6. End time step

is to rewrite the Perzyna visco-plasticity model using residuals. Similar to inviscid plasticity, Equations (7.99), we obtain:

$$\begin{cases} \mathbf{r}_\sigma = \boldsymbol{\sigma}_{j+1} - \boldsymbol{\sigma}_e + \Delta\lambda \mathbf{D}^{\mathrm{e}} \mathbf{m}(\boldsymbol{\sigma}_{j+1}, \lambda_{j+1}) \\ r_f = \varphi\left(f(\boldsymbol{\sigma}_{j+1}, \lambda_{j+1})\right) - \frac{\Delta\lambda}{\eta\Delta t} \end{cases} \tag{8.62}$$

where the second term in Equation $(8.62)_2$ satisfies the visco-plastic relation (8.55) in an incremental sense, as was also done for the generalised and bounding surface plasticity models in Box 7.4 (de Borst and Heeres 2002). A local Newton–Raphson iterative process can now be carried out, cf. Equation (7.101):

$$\begin{pmatrix} \boldsymbol{\sigma}_{j+1}^{k+1} \\ \lambda_{j+1}^{k+1} \end{pmatrix} = \begin{pmatrix} \boldsymbol{\sigma}_{j+1}^{k} \\ \lambda_{j+1}^{k} \end{pmatrix} - \begin{bmatrix} \frac{\partial \mathbf{r}_\sigma}{\partial \boldsymbol{\sigma}} & \frac{\partial \mathbf{r}_\sigma}{\partial \lambda} \\ \frac{\partial r_f}{\partial \boldsymbol{\sigma}} & \frac{\partial r_f}{\partial \lambda} \end{bmatrix}^{-1} \begin{pmatrix} \mathbf{r}_\sigma^k \\ r_f^k \end{pmatrix}$$

where the derivatives $\frac{\partial \mathbf{r}_\sigma}{\partial \boldsymbol{\sigma}}$, $\frac{\partial \mathbf{r}_\sigma}{\partial \lambda}$ and $\frac{\partial r_f}{\partial \boldsymbol{\sigma}}$ are given by Equations (7.102), (7.103) and (7.104), while, using Equations (7.74) and (7.98), $\frac{\partial r_f}{\partial \lambda}$ becomes:

$$\frac{\partial r_f}{\partial \lambda} = -\left(h\frac{\partial \varphi}{\partial f} + \frac{1}{\eta\Delta t}\right) \tag{8.63}$$

instead of Equation (7.105), which holds for inviscid plasticity. For the tangential stiffness matrix that is associated with this integration algorithm we take the variations of Equations (8.62) to give:

$$\begin{cases} \delta\boldsymbol{\sigma} = \mathbf{D}^{\mathrm{e}}\delta\boldsymbol{\epsilon} - \mathbf{D}^{\mathrm{e}}\mathbf{m}\delta\lambda - \Delta\lambda\mathbf{D}^{\mathrm{e}}\frac{\partial \mathbf{m}}{\partial \boldsymbol{\sigma}}\delta\boldsymbol{\sigma} - \Delta\lambda\mathbf{D}^{\mathrm{e}}\frac{\partial \mathbf{m}}{\partial \lambda}\delta\lambda \\ \mathbf{n}^{\mathrm{T}}\delta\boldsymbol{\sigma} - \left(h + \frac{1}{\eta\Delta t}\right)\delta\lambda = 0 \end{cases} \tag{8.64}$$

Using Equation (7.102), (7.106) and (7.128) we rewrite Equation (8.64) as:

$$\begin{cases} \delta\boldsymbol{\sigma} = \mathbf{H}(\delta\boldsymbol{\epsilon} - \bar{\mathbf{m}}\delta\lambda) \\ \mathbf{n}^{\mathrm{T}}\delta\boldsymbol{\sigma} - \left(h + \frac{1}{\eta\Delta t}\right)\delta\lambda = 0 \end{cases} \tag{8.65}$$

and, using arguments as in Chapter 7, the algorithmic tangential stiffness relation between stress rate and strain rate can be derived as:

$$\delta\boldsymbol{\sigma} = \left(\mathbf{H} - \frac{\mathbf{H}\bar{\mathbf{m}}\mathbf{n}^{\mathrm{T}}\mathbf{H}}{h + \frac{1}{\eta\Delta t} + \mathbf{n}^{\mathrm{T}}\mathbf{H}\bar{\mathbf{m}}}\right)\delta\boldsymbol{\epsilon} \tag{8.66}$$

8.3.4 Duvaut–Lions Visco-plasticity

An alternative approach, which in its elaboration more closely connects to rate-independent plasticity, has been proposed by Duvaut and Lions (1972), and is based on the difference in response between the visco-plastic model and the underlying, rate-independent plasticity model. The visco-plastic strain rate and the hardening law are now defined as:

$$\dot{\boldsymbol{\epsilon}}^{\mathrm{vp}} = \frac{1}{\tau}(\mathbf{D}^{\mathrm{e}})^{-1}(\boldsymbol{\sigma} - \bar{\boldsymbol{\sigma}}) \tag{8.67a}$$

$$\dot{\kappa} = -\frac{1}{\tau}(\kappa - \bar{\kappa}) \tag{8.67b}$$

with τ the relaxation time, and $\bar{\boldsymbol{\sigma}}$ the rate-independent material response. Quantities that relate to the inviscid plasticity model or back bone model, denoted by $\bar{\cdot}$, can be viewed as a projection of the current stress on the yield surface. The visco-plastic strain rate is determined by the difference between the total stress and the stress in the inviscid back-bone model, which marks a difference with the Perzyna model, but it is noted that under certain conditions, both visco-plasticity formulations can be made to coincide (Runesson *et al.* 1999). The Duvaut–Lions visco-plasticity model has a marked advantage, namely that it can be used for yield surfaces for which the gradient is discontinuous at some point (Simo *et al.* 1988).

In the Duvaut–Lions visco-plastic model, the stress update is carried out in two steps. First, the inviscid back bone stress $\bar{\boldsymbol{\sigma}}$ is updated using a standard Euler backward return-mapping algorithm for inviscid plasticity, see Chapter 7. Subsequently, the visco-plastic response at $t + \Delta t$ is computed according to Equation (8.67):

$$(\dot{\boldsymbol{\epsilon}}^{\mathrm{vp}})^{t+\Delta t} = \frac{1}{\tau}(\mathbf{D}^{\mathrm{e}})^{-1}\left(\boldsymbol{\sigma}^{t+\Delta t} - \bar{\boldsymbol{\sigma}}^{t+\Delta t}\right) \tag{8.68}$$

Substitution into the generalised trapezoidal rule, Equation (8.53), gives the visco-plastic strain increment:

$$\Delta\boldsymbol{\epsilon}^{\mathrm{vp}} = \left((1-\theta)(\dot{\boldsymbol{\epsilon}}^{\mathrm{vp}})^{t} + \frac{\theta}{\tau}(\mathbf{D}^{\mathrm{e}})^{-1}(\boldsymbol{\sigma}^{t+\Delta t} - \bar{\boldsymbol{\sigma}}^{t+\Delta t})\right)\Delta t \tag{8.69}$$

Using this expression the new stress $\boldsymbol{\sigma}^{t+\Delta t}$ can be computed by substitution into Equation (8.52), while the new value of the hardening parameter follows from

$$\kappa^{t+\Delta t} = \kappa^{t} + \Delta\kappa$$

with $\Delta\kappa = \Delta\kappa(\Delta\boldsymbol{\epsilon}^{\mathrm{vp}})$.

The tangential stiffness matrix can be derived by substitution of Equation (8.69) into Equation (8.52), which yields Equation (8.60), with

$$\begin{cases} \mathbf{D} = \frac{\tau}{\tau+\theta\Delta t}\left(\mathbf{D}^{\mathrm{e}} + \frac{\theta\Delta t}{\tau}\bar{\mathbf{D}}\right) \\ \mathbf{q} = \frac{\tau\Delta t}{\tau+\theta\Delta t}\left((1-\theta)\mathbf{D}^{\mathrm{e}}(\dot{\boldsymbol{\epsilon}}^{\mathrm{vp}})^{t} + \frac{\theta}{\tau}(\boldsymbol{\sigma}^{t} - \bar{\boldsymbol{\sigma}}^{t})\right) \end{cases} \tag{8.70}$$

and $\bar{\mathbf{D}}$ the tangential stiffness matrix that is computed for the inviscid back bone plasticity model, and relates the stress rate in the back bone inviscid plasticity model to the strain rate: $\dot{\bar{\boldsymbol{\sigma}}} = \bar{\mathbf{D}}\dot{\boldsymbol{\epsilon}}$. It is noted that for $\theta = 1$ the tangential stiffness matrix $\mathbf{D}$ for the backward Euler

Box 8.2 Duvaut–Lions visco-plasticity: stress update for $t \rightarrow t + \Delta t$

1. Begin time step. Initialise: $\Delta\mathbf{a} = \mathbf{0}$, $\mathbf{f}_{\text{ext}}^{t+\Delta t}$
2. *For each integration point i:*
   - Compute pseudo-load vector: $\mathbf{q}_i = \frac{\tau\Delta t}{\tau+\theta\Delta t}\left((1-\theta)\mathbf{D}_i^{\text{e}}(\dot{\boldsymbol{\epsilon}}^{\text{vp}})_i^t + \frac{\theta}{\tau}\boldsymbol{\sigma}_i^t - \bar{\boldsymbol{\sigma}}_i^t)\right)$
3. Compute the internal force vector: $\mathbf{f}_{\text{int},0} = \mathbf{f}_{\text{int}}^t + \int_V \mathbf{B}^{\text{T}}\mathbf{q}\text{d}V$
4. Iterations $j = 0, \ldots$ for finding equilibrium within the time step:
   - *For each integration point i:* compute $\mathbf{D}_{i,j} = \frac{\tau}{\tau+\theta\Delta t}\left(\mathbf{D}_i^{\text{e}} + \frac{\theta\Delta t}{\tau}\bar{\mathbf{D}}_{i,j}\right)$
   - Compute tangential stiffness matrix: $\mathbf{K}_j = \int_V \mathbf{B}^{\text{T}}\mathbf{D}_j\mathbf{B}\text{d}V$
   - Solve the linear system: $\text{d}\mathbf{a}_{j+1} = (\mathbf{K}_j)^{-1}(\mathbf{f}_{\text{ext}}^{t+\Delta t} - \mathbf{f}_{\text{int},j})$
   - Update the displacement increments: $\Delta\mathbf{a}_{j+1} = \Delta\mathbf{a}_j + \text{d}\mathbf{a}_{j+1}$
   - *For each integration point i:*
     - Compute the strain increment: $\Delta\boldsymbol{\epsilon}_{i,j+1} = \mathbf{B}_i\Delta\mathbf{a}_{j+1}$
     - Compute the trial stress: $(\boldsymbol{\sigma}_e)_{i,j+1} = \boldsymbol{\sigma}_i^t + \mathbf{D}_i^{\text{e}}\Delta\boldsymbol{\epsilon}_{i,j+1}$
     - Compute the backbone stress $\bar{\boldsymbol{\sigma}}_{i,j+1}$ using a return-mapping algorithm
     - Compute the visco-plastic strain increment:

$$\Delta\boldsymbol{\epsilon}_{i,j+1}^{\text{vp}} = \left((1-\theta)(\dot{\boldsymbol{\epsilon}}^{\text{vp}})_i^t + \frac{\theta}{\tau}(\mathbf{D}_i^{\text{e}})^{-1}(\boldsymbol{\sigma}_{i,j+1} - \bar{\boldsymbol{\sigma}}_{i,j+1})\right)\Delta t$$

     - Compute the stress increment: $\Delta\boldsymbol{\sigma}_{i,j+1} = \mathbf{D}_i^{\text{e}}(\Delta\boldsymbol{\epsilon}_{i,j+1} - \Delta\boldsymbol{\epsilon}_{i,j+1}^{\text{vp}})$
     - Compute the total stress: $\boldsymbol{\sigma}_{i,j+1} = \boldsymbol{\sigma}_i^t + \Delta\boldsymbol{\sigma}_{i,j+1}$
   - Compute internal force: $\mathbf{f}_{\text{int},j+1} = \int_V \mathbf{B}^{\text{T}}\boldsymbol{\sigma}_{j+1}\text{d}V$
   - Check convergence: if $\|\mathbf{f}_{\text{ext}}^{t+\Delta t} - \mathbf{f}_{\text{int},j+1}\| < \eta$, continue, else go to 4.
5. *For each integration point i:*
   - Update: $\Delta\kappa_{i,j+1} = \Delta\kappa(\Delta\boldsymbol{\epsilon}_{i,j+1}^{\text{vp}})$, $\kappa_i^{t+\Delta t} = \kappa_i^t + \Delta\kappa_{i,j+1}$
   - Update the visco-plastic strain rate: $(\dot{\boldsymbol{\epsilon}}^{\text{vp}})_i^{t+\Delta t} = \frac{1}{\tau}(\mathbf{D}_i^{\text{e}})^{-1}\left(\boldsymbol{\sigma}_{i,j+1} - \bar{\boldsymbol{\sigma}}_{i,j+1}\right)$
6. End time step

algorithm as derived by Ju (1990) is recovered. In Box 8.2 the algorithm for Duvaut–Lions visco-plasticity is summarised.

8.3.5 Consistency Model

The Perzyna and Duvaut–Lions visco-plasticity theories differ from the inviscid plasticity theory in that the current stress may violate the yield criterion, so that the Karush–Kuhn–Tucker conditions, Equation (6.16), do not apply. In the stress space this has the implication that the stress point can be outside the yield surface, which is the reason that the terminology overstress plasticity models has been coined for these classes of visco-plasticity models. For a constant external loading the stresses return, or relax, to the yield surface in the course of time.

However, visco-plasticity models have been proposed where the rate effect is introduced via a rate-dependent yield surface (Ristinmaa and Ottosen 2000; Wang *et al.* 1996, 1997). The Karush–Kuhn–Tucker conditions are then enforced, hence they are known by the name consistency visco-plasticity models. For isotropic hardening/softening the yield function then attains the format:

$$f(\boldsymbol{\sigma}, \kappa, \dot{\kappa}) = 0 \tag{8.71}$$

so that the vanishing of its variation, $\delta f = 0$, which is an alternative expression for the consistency condition, Equation (7.39), can be elaborated to give:

$$\mathbf{n}^{\mathrm{T}}\delta\boldsymbol{\sigma} - h\delta\lambda - s\delta\dot{\lambda} = 0 \tag{8.72}$$

where $\mathbf{n}$ is the gradient to the yield surface, Equation (7.45), h follows the standard definition of a hardening modulus, Equation (7.74), and s signifies the rate sensitivity parameter:

$$s = -\frac{\partial f}{\partial \dot{\kappa}}\frac{\partial \dot{\kappa}}{\partial \dot{\lambda}} \tag{8.73}$$

A notable advantage of the consistency visco-plastic model is that the current yield strength is not only dependent on the accumulated plastic strain, but can also account for strain-rate softening or hardening, which occurs in certain alloys. Like the Duvaut–Lions visco-plasticity model, the consistency visco-plasticity model can be made to coincide with the Perzyna visco-plasticity model (Heeres *et al.* 2002). For continued plastic loading and in the absence of strain-rate effects, this is enforced by choosing:

$$s = \eta\left(\frac{\mathrm{d}\varphi(f)}{\mathrm{d}f}\right)^{-1} \tag{8.74}$$

Since the consistency model of visco-plasticity is equipped with a yield surface, a standard return-mapping algorithm (Chapter 7), can be applied to integrate the rate equations of elasto-visco-plasticity. The tangential stiffness matrix follows by taking the variation of the updated stress, Equation (8.52), with the generalised trapezoidal rule for the visco-plastic strain increment, Equation $(8.53)_1$:

$$\delta\boldsymbol{\sigma} = \mathbf{D}^{\mathrm{e}}\delta\boldsymbol{\epsilon} - \theta\Delta t\mathbf{D}^{\mathrm{e}}\delta(\dot{\boldsymbol{\epsilon}}^{\mathrm{vp}})^{t+\Delta t} \tag{8.75}$$

Following Chapter 7 we consider the wide class of plasticity models for which the rate of the hardening parameter $\dot{\kappa}$ is proportional to the consistency parameter $\dot{\lambda}$, Equation (7.98), so that the yield function can also be written as: $f = f(\boldsymbol{\sigma}, \lambda, \dot{\lambda})$. Accordingly,

$$\dot{\boldsymbol{\epsilon}}^{\mathrm{vp}} = \dot{\lambda}\,\mathbf{m}(\boldsymbol{\sigma}, \lambda, \dot{\lambda}) \tag{8.76}$$

and the variation of the stress, $\delta\boldsymbol{\sigma}$ can be elaborated as:

$$\delta\boldsymbol{\sigma} = \mathbf{D}^{\mathrm{e}}\delta\boldsymbol{\epsilon} - \mathbf{D}^{\mathrm{e}}\theta\Delta t\left(\dot{\lambda}\left(\frac{\partial\mathbf{m}}{\partial\boldsymbol{\sigma}}\delta\boldsymbol{\sigma} + \frac{\partial\mathbf{m}}{\partial\lambda}\delta\lambda + \frac{\partial\mathbf{m}}{\partial\dot{\lambda}}\delta\dot{\lambda}\right) + \mathbf{m}\delta\dot{\lambda}\right)$$

In consideration of Equation (8.53) we have

$$\delta\lambda = \theta\Delta t\delta\dot{\lambda} \tag{8.77}$$

so that the latter identity can be re-expressed as:

$$\delta\sigma = \mathbf{H}\delta\epsilon - \mathbf{H}\bar{\mathbf{m}}\delta\lambda \tag{8.78}$$

At variance with Chapter 7, $\mathbf{H}$ and $\bar{\mathbf{m}}$ are now defined as:

$$\mathbf{H} = \left((\mathbf{D}^{\mathrm{e}})^{-1} + \theta\Delta t\dot{\lambda}\frac{\partial\mathbf{m}}{\partial\sigma}\right)^{-1}$$
$$\bar{\mathbf{m}} = \mathbf{m} + \theta\Delta t\dot{\lambda}\frac{\partial\mathbf{m}}{\partial\lambda} + \dot{\lambda}\frac{\partial\mathbf{m}}{\partial\dot{\lambda}} \tag{8.79}$$

where, according to Equation $(8.53)_2$

$$\dot{\lambda} \equiv \dot{\lambda}^{t+\Delta t} = \frac{\Delta\lambda - (1-\theta)\dot{\lambda}^t}{\theta\Delta t} \tag{8.80}$$

Inserting the expression of Equation (8.77) for $\delta\dot{\lambda}$ into the variation of the yield function, Equation (8.72), yields:

$$\mathbf{n}^{\mathrm{T}}\delta\sigma = \left(h + \frac{s}{\theta\Delta t}\right)\delta\lambda \tag{8.81}$$

from which $\delta\lambda$ can be resolved, and Equation (8.78) can be rewritten as:

$$\delta\sigma = \mathbf{D}\delta\epsilon \tag{8.82}$$

with

$$\mathbf{D} = \mathbf{H} - \frac{\mathbf{H}\bar{\mathbf{m}}\mathbf{n}^{\mathrm{T}}\mathbf{H}}{\left(h + \frac{s}{\theta\Delta t}\right) + \mathbf{n}^{\mathrm{T}}\mathbf{H}\bar{\mathbf{m}}} \tag{8.83}$$

the tangential stiffness matrix for the consistency model of visco-plasticity.

8.3.6 *Propagative or Dynamic Instabilities*

As discussed in Chapter 6, localisation of deformation followed by failure can be caused by descending branches in the equivalent stress–strain diagram and by loss of the major symmetry in the tangential stiffness tensor, while geometrical non-linearities may either be stabilising or destabilising. The instabilities discussed in Chapter 6 are called static instabilities, because the localised strain mode remains confined to a certain part in the body. However, experimental observations have revealed other types of instabilities, like patterning in rock masses and salt formations and propagative instabilities, also named dynamic instabilities, like Lüders bands and Portevin–Le Chatelier (PLC) bands in metals and alloys, where a shear band propagates through the body. The latter types of instabilities are caused by rehardening and by strain-rate softening, respectively.

For a classification of instability problems we shall consider the simple problem of a uniaxially stressed tensile bar subject to a dynamic loading. In this case, the equation of motion and the continuity equation can be expressed in a rate format as:

$$\begin{cases} \dfrac{\partial \dot{\sigma}}{\partial x} = \rho \dfrac{\partial^2 \dot{u}}{\partial t^2} \\ \dot{\epsilon} = \dfrac{\partial \dot{u}}{\partial x} \end{cases}$$

where ρ is the mass density. Using the format of classical small-strain plasticity, the strain rate $\dot{\epsilon}$ is additively decomposed into an elastic contribution $\dot{\epsilon}^{\mathrm{e}}$ and a plastic contribution $\dot{\epsilon}^{\mathrm{p}}$, cf. Equation (7.41):

$$\dot{\epsilon} = \dot{\epsilon}^{\mathrm{e}} + \dot{\epsilon}^{\mathrm{p}} \tag{8.84}$$

Assuming linear elasticity, the elastic contribution is related to the stress rate $\dot{\sigma}$ according to Equation (7.40), which yields for one-dimensional conditions:

$$\dot{\sigma} = E\dot{\epsilon}^{\mathrm{e}}$$

Differentation of the one-dimensional equation of motion with respect to the spatial coordinate x and substitution of the kinematic equation, the strain decomposition and the linear relation between the stress rate and the elastic strain rate yields:

$$\frac{\partial^2 \dot{\sigma}}{\partial x^2} - \frac{\rho}{E}\frac{\partial^2 \dot{\sigma}}{\partial t^2} = \rho \frac{\partial^2 \dot{\epsilon}^{\mathrm{p}}}{\partial t^2} \tag{8.85}$$

We now postulate the stress to be dependent on the plastic strain, the plastic strain rate and, following arguments advocated by Aifantis (1984), on the second spatial gradient of the plastic strain

$$\sigma = \sigma\left(\epsilon^{\mathrm{p}}, \dot{\epsilon}^{\mathrm{p}}, \frac{\partial^2 \epsilon^{\mathrm{p}}}{\partial x^2}\right) \tag{8.86}$$

which constitutes the simplest possible, symmetric extension of a standard, rate-independent plasticity model, where $\sigma = \sigma(\epsilon^{\mathrm{p}})$. In a rate format we obtain

$$\dot{\sigma} = h\dot{\epsilon}^{\mathrm{p}} + s\ddot{\epsilon}^{\mathrm{p}} + c\frac{\partial^2 \dot{\epsilon}^{\mathrm{p}}}{\partial x^2} \tag{8.87}$$

with

$$h = \frac{\partial \sigma}{\partial \epsilon^{\mathrm{p}}}, \quad s = \frac{\partial \sigma}{\partial \dot{\epsilon}^{\mathrm{p}}}, \quad c = \frac{\partial \sigma}{\partial(\partial^2 \epsilon^{\mathrm{p}}/\partial x^2)} \tag{8.88}$$

Herein, h, s and c refer to the hardening/softening modulus, the strain-rate sensitivity and to the gradient parameter, respectively. In general, they can be strain and strain-rate dependent. We now combine Equation (8.85) with the constitutive equation in rate format, Equation (8.87), to obtain:

$$h\tilde{\nabla}\dot{\epsilon}^{\mathrm{p}} + s\tilde{\nabla}\ddot{\epsilon}^{\mathrm{p}} + c\tilde{\nabla}\left(\frac{\partial^2 \dot{\epsilon}^{\mathrm{p}}}{\partial x^2}\right) = \rho\frac{\partial^2 \dot{\epsilon}^{\mathrm{p}}}{\partial t^2} \tag{8.89}$$

with

$$\tilde{\nabla} = \frac{\partial^2}{\partial x^2} - \frac{\rho}{E}\frac{\partial^2}{\partial t^2} \tag{8.90}$$

To investigate the stability of an equilibrium state, we assume a harmonic perturbation $\dot{\epsilon}^{\mathrm{p}}$ starting from a homogeneous deformation state

$$\dot{\epsilon}^{\mathrm{p}} = A\mathrm{e}^{\mathrm{i}(kx+\lambda t)} \tag{8.91}$$

where A is the amplitude, k is the wave number and λ is the eigenvalue. Substitution of Equation (8.91) into Equation (8.89) gives the characteristic equation:

$$\lambda^3 + a\lambda^2 + b\lambda + d = 0 \tag{8.92}$$

with

$$a = \frac{h - ck^2 + E}{s}\,, \quad b = \frac{Ek^2}{\rho}\,, \quad d = \frac{Ek^2}{\rho}\frac{h - ck^2}{s} \tag{8.93}$$

According to the Routh–Hurwitz stability theorem all solutions $\lambda(k)$ have a negative real part when the following conditions are fulfilled simultaneously:

$$a = \frac{h - ck^2 + E}{s} > 0\,, \quad d = \frac{Ek^2}{\rho}\frac{h - ck^2}{s} > 0\,, \quad ab - d = \frac{E^2k^2}{\rho s} > 0 \tag{8.94}$$

If the Routh–Hurwitz criterion fails to hold, an eigenvalue with a real, positive part will exist, which implies that the homogeneous state is unstable and a small perturbation can grow into, for instance, a shear band instability. Here, we consider two possible types of instabilities:

1. An h-type instability which is associated with the formation of a stationary localisation band

$$s > 0\,, \quad h - ck^2 < 0 \tag{8.95}$$

It is noted that this type of instability is dependent on k, the wave number, which has a cut-off value

$$k = \sqrt{\frac{h}{c}} \tag{8.96}$$

Only waves with a wave length smaller than $L = 2\pi/k$ can propagate in the localisation band. This sets an internal length scale in this gradient-enhanced continuum

$$\ell = \frac{L}{2\pi} = \sqrt{\frac{c}{h}} \tag{8.97}$$

since the size of the localisation band in the one-dimensional case coincides with the largest possible wavelength.

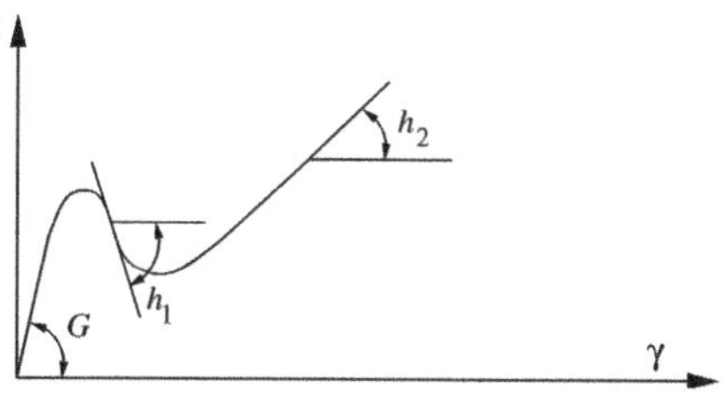

Figure 8.6 Simplified softening–rehardening model for Lüders band propagation

2. An s-type instability, which can be associated with the occurrence of travelling PLC bands

$$s < 0\,, \quad h > 0 \tag{8.98}$$

The consistency model of visco-plasticity has the advantage that it is well suited to describe s-type instability phenomena, i.e. that are are caused by strain-rate softening.

As an example of Lüders band propagation we consider a one-dimensional bar, but now subjected to a shear force that is applied instantaneously (Wang *et al.* 1997). The boundary conditions are such that the bar is loaded purely in shear (no bending effects). Eight-noded elements with a nine-point Gaussian integration scheme have been used. The elastic parameters are such that the shear wave speed is $c_s = 1000$ m/s. The time step has been chosen such that $\Delta t = \Delta l / c_s$, with Δl the finite element size, and c_s the shear wave velocity. The time integration has been done using a Newmark scheme, with parameters $\beta = 0.9$ and $\gamma = 0.49$ to introduce some numerical damping (Chapter 5).

The propagation of the instability has been investigated for the softening–rehardening model sketched in Figure 8.6, using two different meshes, with 20 and 40 elements, respectively. In Figure 8.7 analytical and numerical results are shown for strain distributions when the reflected wave front has travelled to $x_r = 15$ mm and the propagative instability has reached $x_s = 8.6$ mm.

For both discretisations, the numerical results well capture the reflected wave front at x_r. Strain softening occurs as soon as the shear stress exceeds the initial yield strength. Due to the rehardening, another discontinuity emerges at x_s which propagates with a different (lower) velocity c_s and stress fluctuations arise (Figure 8.8). For strain softening followed by ideal

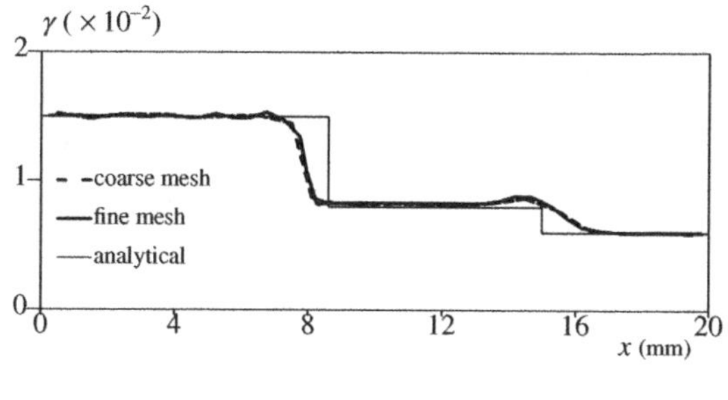

Figure 8.7 Shear strain distribution for the softening–rehardening model after wave reflection (Wang *et al.* 1997)

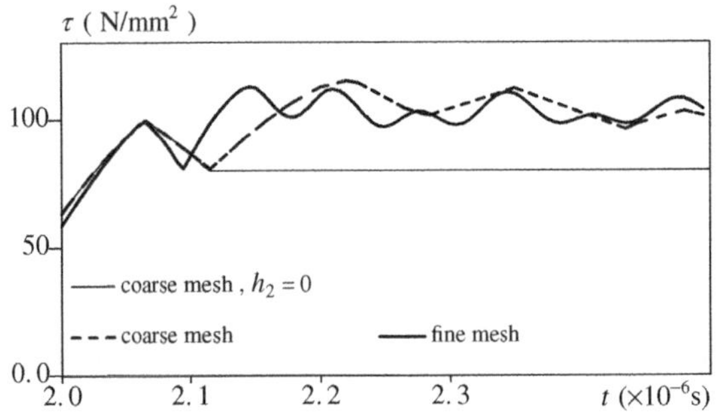

Figure 8.8 Stress evolution at the left-most integration point in the bar for ideal and softening plasticity, considering two different discretisations (Wang *et al.* 1997)

plasticity, $h_2 = 0$ in Figure 8.8, these fluctuations do not occur, since then the wave speed becomes imaginary, so that the instability cannot propagate and is trapped at this position. However, when rehardening takes place after softening, the wave speed becomes real again and the wave front propagates to the next point, where this sequence of events is repeated: softening, trapping of the wave, followed by rehardening, which causes wave propagation. Due to the instantaneous drop at the onset of softening, a reflected wave propagates to the centre region and a fluctuation propagates to the left. Hence, the number of fluctuations increases with the number of elements in the bar, which is confirmed in Figure 8.8. Therefore, conventional strain-softening models are mesh sensitive even if rehardening is introduced. However, mesh sensitivity must now be interpreted in the sense that, upon mesh refinement, the stress distribution will show more fluctuations.

The bar loaded in shear is now reanalysed with a material model, which incorporates strain hardening ($h > 0$) and strain-rate softening ($s < 0$) (Figure 8.9) (Wang *et al.* 1997). For this s-type instability a propagative shear band is observed, which commonly is referred to as a PLC band. The competition between strain hardening ($h > 0$) and strain-rate softening ($s < 0$) determines the propagation velocity of this shear band in the absence of temperature effects.

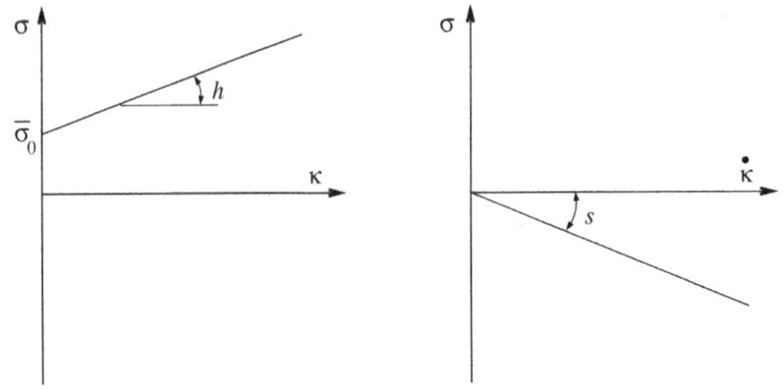

Figure 8.9 Simplified model for an s-type instability such as the Portevin–Le Chatelier effect

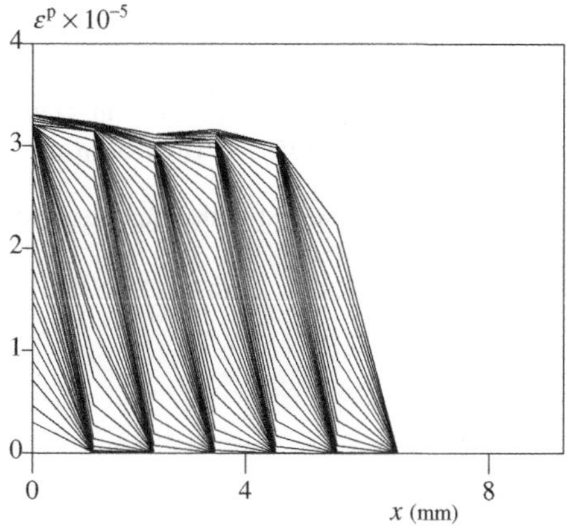

Figure 8.10 Stroboscopic picture of the plastic strain distribution during wave propagation in a bar due to an s-type instability (Wang *et al.* 1997)

When the deformation reaches a critical value, the local PLC band is arrested due to the contribution of the strain hardening. Subsequently, a new PLC band will be initiated at the adjacent element, so that the PLC effect progresses from one end of the bar to the other in a discontinuous, but orderly and periodic fashion (Figure 8.10).

References

Aifantis EC 1984 On the microstructural origin of certain inelastic models. *ASME Journal of Engineering Materials and Technology* **106**, 326–330.

Bažant ZP and Wu ST 1974 Rate-type creep law for aging concrete based on Maxwell chain. *RILEM Materials and Structures* **7**, 45–60.

de Borst R and Heeres OM 2002 A unified approach to the implicit integration of standard, non-standard and viscous plasticity models. *International Journal for Numerical and Analytical Methods in Geomechanics* **26**, 1059–1070.

Duvaut G and Lions JL 1972 *Les Inequations en Mécanique et en Physique*. Dunod.

Heeres OM, Suiker ASJ and de Borst R 2002 A comparison between the Perzyna viscoplastic model and the consistency viscoplastic model. *European Journal of Mechanics: A/Solids* **21**, 1–12.

Hult J 1966 *Creep in Engineering Structures*. Blaisdell.

Ju JW 1990 Consistent tangent moduli for a class of viscoplasticity. *ASCE Journal of Engineering Mechanics* **116**, 1764–1779.

Perzyna P 1966 Fundamental problems in viscoplasticity, in *Advances in Applied Mechanics* (ed. Chernyi GG), vol. 9, pp. 243–377. Academic Press.

Ristinmaa M and Ottosen NS 2000 Consequences of dynamic yield surfaces in viscoplasticity. *International Journal of Solids and Structures* **37**, 4601–4622.

Runesson K, Ristinmaa M and Mähler L 1999 A comparison of viscoplasticity formats and algorithms. *Mechanics of Cohesive-frictional Materials* **4**, 75–98.

Simo JC, Kennedy JG and Govindjee S 1988 Non-smooth multisurface plasticity and viscoplasticity. loading/unloading conditions and numerical algorithms. *International Journal for Numerical Methods in Engineering* **26**, 2161–2185.

Taylor RL, Pister KS and Goudreau GL 1970 Thermomechanical analysis of viscoelastic solids. *International Journal for Numerical Methods in Engineering* **2**, 45–59.

Wang WM, Sluys LJ and de Borst R 1996 Interaction between material length scale and imperfection size for localisation in viscoplastic media. *European Journal of Mechanics: A/Solids* **15**, 447–464.

Wang WM, Sluys LJ and de Borst R 1997 Viscoplasticity for instabilities due to strain softening and strain-rate softening. *International Journal for Numerical Methods in Engineering* **40**, 3839–3864.

Zienkiewicz OC and Taylor RL 1991 *The Finite Element Method, Vol. 2, Solid and Fluid Mechanics, Dynamics and Non-linearity*. McGraw-Hill.

Zienkiewicz OC, Watson M and King IP 1968 A numerical method for viscoelastic stress analysis. *International Journal of Engineering Science* **10**, 807–827.

Part III

Structural Elements

9

Beams and Arches

The majority of this chapter is devoted to two-dimensional beam elements, while in the final part of the chapter the extension to three-dimensional formulations is made. Beam elements for two-dimensional analysis are of interest in their own right, but also have a didactic role. Much more so than for three-dimensional beam formulations, or for other structural elements such as plates and shells, the mathematical complications remain limited, and transparency is preserved. Moreover, it allows us to easily connect with the developments in Part I. We therefore start with a shallow-arch formulation. Subsequently, a corotational approach for two-dimensional beam elements is introduced, which can be considered as an extension of the corotational formulation for truss elements in Chapter 3. Before entering a three-dimensional formulation, we will then consider a degenerated continuum beam element using the Total Lagrange formulation, which connects with the discussion on continuum elements in Chapter 3.

9.1 A Shallow Arch

9.1.1 Kirchhoff Formulation

We will start the discussion by departing from an initially flat element (Figure 9.1). Using a degenerated form of the Green–Lagrange strain tensor – Equation (3.70) with $\left(\frac{\mathrm{d}u}{\mathrm{d}x}\right)^2 \ll \left(\frac{\mathrm{d}w}{\mathrm{d}x}\right)^2$ – the axial strain can be expressed as:

$$\epsilon = \frac{\mathrm{d}u}{\mathrm{d}x} + \frac{1}{2}\left(\frac{\mathrm{d}w}{\mathrm{d}x}\right)^2 \tag{9.1}$$

Note that we have adopted the short-hand notation $\epsilon = \epsilon_{xx}$, $u = u_x$ and $w = u_z$ in conformity with the majority of the literature on beam elements. With u_ℓ the axial displacement at the centre line and assuming that plane cross-sections remain plane [Figure 9.2(a)], the displacement in

Non-linear Finite Element Analysis of Solids and Structures, Second Edition.
René de Borst, Mike A. Crisfield, Joris J.C. Remmers and Clemens V. Verhoosel.

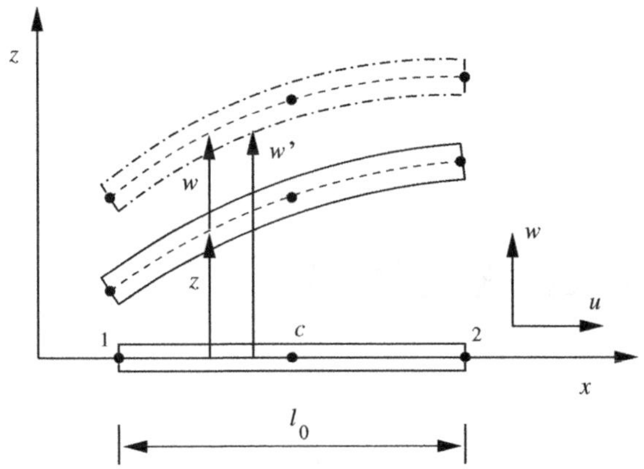

Figure 9.1 A shallow-arch element

the x-direction is given by:

$$u = u_\ell - z_\ell \frac{dw}{dx} \tag{9.2}$$

with z_ℓ measured with respect to the centreline of the element. Substitution of Equation (9.2) into Equation (9.1) results in:

$$\epsilon = \frac{du_\ell}{dx} + \frac{1}{2}\left(\frac{dw}{dx}\right)^2 + z_\ell \chi \tag{9.3}$$

with the curvature

$$\chi = -\frac{d^2 w}{dx^2} \tag{9.4}$$

For an initially curved element (Figure 9.1), Equation (9.3) must be modified to become:

$$\epsilon = \epsilon_\ell + z_\ell \chi \tag{9.5}$$

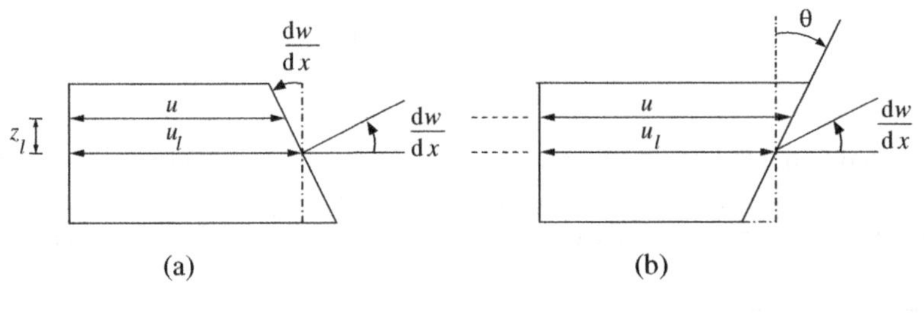

Figure 9.2 Detail for shallow-arch element: (a) using the Kirchhoff assumptions (no shear deformation); (b) for a Timoshenko beam (including shear deformation)

with the membrane strain

$$\epsilon_\ell = \frac{\mathrm{d}u_\ell}{\mathrm{d}x} + \frac{1}{2}\left[\left(\frac{\mathrm{d}w'}{\mathrm{d}x}\right)^2 - \left(\frac{\mathrm{d}z}{\mathrm{d}x}\right)^2\right] \tag{9.6}$$

and

$$w' = w + z \tag{9.7}$$

Note that for $z = 0$, Equations (9.5)–(9.7) reduce to Equation (9.3). The virtual strain follows in a straightforward manner as:

$$\delta\epsilon = \delta\epsilon_\ell + z_\ell \delta\chi \tag{9.8}$$

with

$$\delta\epsilon_\ell = \frac{\mathrm{d}\delta u_\ell}{\mathrm{d}x} + \frac{\mathrm{d}w'}{\mathrm{d}x}\frac{\mathrm{d}\delta w}{\mathrm{d}x} \tag{9.9}$$

since $\delta w' = \delta w$.

As in truss elements the internal virtual work involves only the axial stress and the axial strain, Equation (3.3):

$$\int_{V_0} \sigma\delta\epsilon \mathrm{d}V_0 = \delta\mathbf{u}^\mathrm{T}\mathbf{f}_\mathrm{ext}$$

where the superscripts $t + \Delta t$ have been dropped for notational convenience. Substitution of Equation (9.8) and integration through the depth yields:

$$\int_{\ell_0} (N\delta\epsilon_\ell + M\delta\chi)\,\mathrm{d}x = \delta\mathbf{u}^\mathrm{T}\mathbf{f}_\mathrm{ext} \tag{9.10}$$

with

$$N = \int_{-h/2}^{+h/2} b(z_\ell)\sigma \mathrm{d}z_\ell \tag{9.11}$$

the normal force, and

$$M = \int_{-h/2}^{+h/2} b(z_\ell)\sigma z_\ell \mathrm{d}z_\ell \tag{9.12}$$

the bending moment, being the stress resultants. h is the height of the beam and $b(z_\ell)$ is its width at z_ℓ.

The finite element shape functions are now formally introduced as:

$$u_\ell = \mathbf{h}_u^\mathrm{T}\mathbf{a}\,, \quad w = \mathbf{h}_w^\mathrm{T}\mathbf{w} \tag{9.13}$$

In this section, we assume that the beam is thin, and that the Kirchhoff assumption holds. A possible, matching interpolation is then a quadratic, hierarchical interpolation for u_ℓ, and a cubic, Hermitian interpolation for w:

$$\mathbf{h}_u^{\mathrm{T}} = \frac{1}{2}\left(1-\xi, 1+\xi, 2(1-\xi^2)\right) \tag{9.14a}$$

$$\mathbf{h}_w^{\mathrm{T}} = \frac{1}{8}\left(4-6\xi+2\xi^3, \ell_0(\xi^2-1)(\xi-1), 4+6\xi-2\xi^3, \ell_0(\xi^2-1)(\xi+1)\right) \tag{9.14b}$$

with ξ the isoparametric coordinate along the axis, and the nodal arrays:

$$\mathbf{a}^{\mathrm{T}} = (a_1, a_2, \Delta a_c)$$
$$\mathbf{w}^{\mathrm{T}} = (w_1, \theta_1, w_2, \theta_2)$$

with, according to the Kirchhoff assumption:

$$\theta = \frac{\mathrm{d}w}{\mathrm{d}x} \tag{9.15}$$

and Δa_c the relative displacement at the hierarchical mid-side node c. Differentiation of Equation (9.13) leads to:

$$\begin{aligned}\frac{\mathrm{d}u_\ell}{\mathrm{d}x} &= \mathbf{b}_u^{\mathrm{T}}\mathbf{a}\\ \frac{\mathrm{d}w}{\mathrm{d}x} &= \mathbf{b}_w^{\mathrm{T}}\mathbf{w}\\ \chi = -\frac{\mathrm{d}^2w}{\mathrm{d}x^2} &= \mathbf{c}^{\mathrm{T}}\mathbf{w}\end{aligned} \tag{9.16}$$

with

$$\begin{aligned}\mathbf{b}_u &= \frac{1}{\ell_0}(-1, +1, -4\xi)\\ \mathbf{b}_w &= \frac{1}{4\ell_0}\left(6(\xi^2-1), \ell_0(3\xi^2-2\xi-1), -6(\xi^2-1), \ell_0(3\xi^2+2\xi-1)\right)\\ \mathbf{c} &= -\frac{1}{\ell_0^2}\left(6\xi, \ell_0(3\xi-1), -6\xi, \ell_0(3\xi+1)\right)\end{aligned} \tag{9.17}$$

so that the strain, Equation (9.5), becomes:

$$\epsilon = \underbrace{\mathbf{b}_u^{\mathrm{T}}\mathbf{a} + \frac{1}{2}(\mathbf{b}_w^{\mathrm{T}}\mathbf{w}')^2 - \frac{1}{2}\left(\frac{\mathrm{d}z}{\mathrm{d}x}\right)^2}_{\epsilon_\ell} + z_\ell \underbrace{\mathbf{c}^{\mathrm{T}}\mathbf{w}}_{\chi} \tag{9.18}$$

where $\mathbf{w}' = \mathbf{w} + \mathbf{z}$. Assuming a Bubnov–Galerkin approach, the virtual displacements are interpolated in the same manner as the displacements, and the expression for the virtual strain, Equation (9.8), attains the following discrete format:

$$\delta\epsilon = \underbrace{\mathbf{b}_u^{\mathrm{T}}\delta\mathbf{a} + (\mathbf{b}_w^{\mathrm{T}}\mathbf{w}')\mathbf{b}_w^{\mathrm{T}}\delta\mathbf{w}}_{\delta\epsilon_\ell} + z_\ell \underbrace{\mathbf{c}^{\mathrm{T}}\delta\mathbf{w}}_{\delta\chi} \tag{9.19}$$

For the present we will assume linear elasticity, so that the stress resultants, Equations (9.11) and (9.12), can be integrated explicitly to give:

$$N = EA\epsilon_\ell \tag{9.20a}$$

$$M = EI\chi \tag{9.20b}$$

with I the moment of inertia of the beam. It has been assumed implicitly that the strains remain small, so that for the cross-sectional area of the beam we have: $A \approx A_0$. Substitution of these expressions and Equation (9.19) into the virtual work expression for the beam, Equation (9.10), leads to:

$$\int_{\ell_0} \left[\left(N\delta\mathbf{a}^{\mathrm{T}}\mathbf{b}_u + N(\mathbf{b}_w^{\mathrm{T}}\mathbf{w}')\delta\mathbf{w}^{\mathrm{T}}\mathbf{b}_w \right) + M\delta\mathbf{w}^{\mathrm{T}}\mathbf{c} \right] \mathrm{d}x = \delta\mathbf{a}^{\mathrm{T}}\mathbf{f}_{\mathrm{ext}}^{\mathrm{a}} + \delta\mathbf{w}^{\mathrm{T}}\mathbf{f}_{\mathrm{ext}}^{\mathrm{w}} \tag{9.21}$$

Since this identity must hold for arbitrary $\delta\mathbf{a}$, $\delta\mathbf{w}$, the following set of equations results:

$$\begin{pmatrix} \mathbf{f}_{\mathrm{int}}^{\mathrm{a}} \\ \mathbf{f}_{\mathrm{int}}^{\mathrm{w}} \end{pmatrix} = \begin{pmatrix} \mathbf{f}_{\mathrm{ext}}^{\mathrm{a}} \\ \mathbf{f}_{\mathrm{ext}}^{\mathrm{w}} \end{pmatrix} \tag{9.22}$$

with the internal force vectors defined as:

$$\mathbf{f}_{\mathrm{int}}^{\mathrm{a}} = \int_{\ell_0} N\mathbf{b}_u \mathrm{d}x \tag{9.23a}$$

$$\mathbf{f}_{\mathrm{int}}^{\mathrm{w}} = \int_{\ell_0} \left(N(\mathbf{b}_w^{\mathrm{T}}\mathbf{w}')\mathbf{b}_w + M\mathbf{c} \right) \mathrm{d}x \tag{9.23b}$$

The set of equations (9.22) is non-linear and must be solved using an iterative procedure, e.g. a Newton–Raphson method. For this purpose a truncated Taylor series is applied

$$\begin{cases} (\mathbf{f}_{\mathrm{int}}^{\mathrm{a}})_{j+1} = (\mathbf{f}_{\mathrm{int}}^{\mathrm{a}})_j + \left(\frac{\partial \mathbf{f}_{\mathrm{int}}^{\mathrm{a}}}{\partial \mathbf{a}} \right)_j \mathrm{d}\mathbf{a} + \left(\frac{\partial \mathbf{f}_{\mathrm{int}}^{\mathrm{a}}}{\partial \mathbf{w}} \right)_j \mathrm{d}\mathbf{w} \\ (\mathbf{f}_{\mathrm{int}}^{\mathrm{w}})_{j+1} = (\mathbf{f}_{\mathrm{int}}^{\mathrm{w}})_j + \left(\frac{\partial \mathbf{f}_{\mathrm{int}}^{\mathrm{w}}}{\partial \mathbf{a}} \right)_j \mathrm{d}\mathbf{a} + \left(\frac{\partial \mathbf{f}_{\mathrm{int}}^{\mathrm{w}}}{\partial \mathbf{w}} \right)_j \mathrm{d}\mathbf{w} \end{cases}$$

and the residual is forced to zero, so that the following matrix-vector equation results:

$$\begin{bmatrix} \mathbf{K}_{\mathrm{aa}} & \mathbf{K}_{\mathrm{aw}} \\ \mathbf{K}_{\mathrm{aw}}^{\mathrm{T}} & \mathbf{K}_{\mathrm{ww}} \end{bmatrix} \begin{pmatrix} \mathrm{d}\mathbf{a} \\ \mathrm{d}\mathbf{w} \end{pmatrix} = \begin{pmatrix} \mathbf{f}_{\mathrm{ext}}^{\mathrm{a}} - \mathbf{f}_{\mathrm{int}}^{\mathrm{a}} \\ \mathbf{f}_{\mathrm{ext}}^{\mathrm{w}} - \mathbf{f}_{\mathrm{int}}^{\mathrm{w}} \end{pmatrix} \tag{9.24}$$

with the submatrices of the tangential stiffness matrix defined as:

$$\begin{aligned} \mathbf{K}_{\mathrm{aa}} &= \frac{\partial \mathbf{f}_{\mathrm{int}}^{\mathrm{a}}}{\partial \mathbf{a}} = \int_{\ell_0} EA\mathbf{b}_u\mathbf{b}_u^{\mathrm{T}}\mathrm{d}x \\ \mathbf{K}_{\mathrm{aw}} &= \frac{\partial \mathbf{f}_{\mathrm{int}}^{\mathrm{a}}}{\partial \mathbf{w}} = \int_{\ell_0} EA(\mathbf{b}_w^{\mathrm{T}}\mathbf{w}')\mathbf{b}_u\mathbf{b}_w^{\mathrm{T}}\mathrm{d}x \\ \mathbf{K}_{\mathrm{ww}} &= \frac{\partial \mathbf{f}_{\mathrm{int}}^{\mathrm{w}}}{\partial \mathbf{w}} = \int_{\ell_0} \left(EI\mathbf{c}\mathbf{c}^{\mathrm{T}} + EA(\mathbf{b}_w^{\mathrm{T}}\mathbf{w}')^2\mathbf{b}_w\mathbf{b}_w^{\mathrm{T}} + N\mathbf{b}_w\mathbf{b}_w^{\mathrm{T}} \right) \mathrm{d}x \end{aligned} \tag{9.25}$$

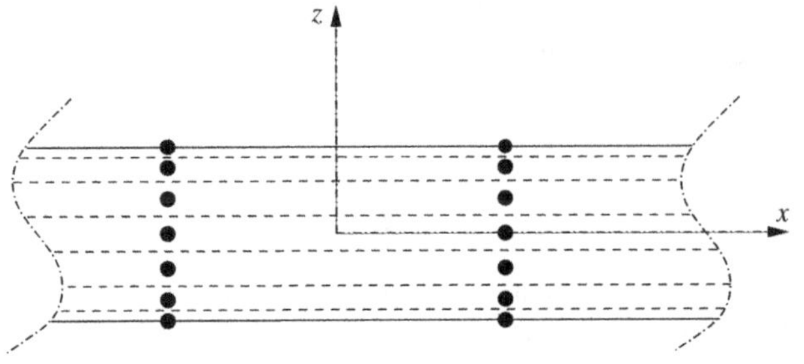

Figure 9.3 Gauss integration along the axis and Newton–Cotes, Simpson or Lobatto integration through the depth for a layered beam element

where the last term of $\mathbf{K}_{ww}$ can be identified as the geometric contribution to the tangential stiffness matrix.

When material non-linear behaviour is to be included in the analysis, e.g. plasticity or damage, the beam must be divided into a number of layers, in order to be able to monitor the spread of plasticity or damage through the depth of the beam. The inelasticity is, like along the x-axis, sampled at integration points. At variance with the integration along the x-axis which is typically done using Gauss integration, Newton–Cotes, Simpson and Lobatto integration schemes are more suitable for through-the-depth integration, since the outermost fibres of the beam, which are the most severely strained, and therefore the starting point of yielding or damage evolution, are then explicitly monitored (Figure 9.3) (Burgoyne and Crisfield 1990). For most practical purposes, using five to nine integration points through the depth suffices. Considering that plasticity and damage can be cast in an incrementally linear relation between the axial stress σ and the strain ϵ,

$$\dot{\sigma} = E_{\tan}(z_\ell)\dot{\epsilon} \tag{9.26}$$

with $E_{\tan}$ the tangential stiffness modulus which, for a given x, only depends on the position with respect to the centreline, closed-form expressions for the normal force and the bending moment can no longer be derived. Instead, we obtain for the rate of the normal force and the rate of the bending moment,

$$\dot{N} = \underbrace{\left(\int_{-h/2}^{+h/2} b(z_\ell)E_{\tan}(z_\ell)\mathrm{d}z_\ell\right)}_{\overline{EA}}\dot{\epsilon}_\ell + \underbrace{\left(\int_{-h/2}^{+h/2} b(z_\ell)E_{\tan}(z_\ell)z_\ell\mathrm{d}z_\ell\right)}_{\overline{EX}}\dot{\chi} \tag{9.27}$$

and

$$\dot{M} = \underbrace{\left(\int_{-h/2}^{+h/2} b(z_\ell)E_{\tan}(z_\ell)z_\ell\mathrm{d}z_\ell\right)}_{\overline{EX}}\dot{\epsilon}_\ell + \underbrace{\left(\int_{-h/2}^{+h/2} b(z_\ell)E_{\tan}(z_\ell)z_\ell^2\mathrm{d}z_\ell\right)}_{\overline{EI}}\dot{\chi} \tag{9.28}$$

respectively. For a layered beam the stiffness matrices thus become:

$$\mathbf{K}_{\mathrm{aa}} = \int_{\ell_0} \overline{EA}\mathbf{b}_u\mathbf{b}_u^{\mathrm{T}}\mathrm{d}x$$

$$\mathbf{K}_{\mathrm{aw}} = \int_{\ell_0} \left(\overline{EA}(\mathbf{b}_w^{\mathrm{T}}\mathbf{w}')\mathbf{b}_u\mathbf{b}_w^{\mathrm{T}} + \overline{EX}\mathbf{b}_u\mathbf{c}^{\mathrm{T}}\right)\mathrm{d}x \tag{9.29}$$

$$\mathbf{K}_{\mathrm{ww}} = \int_{\ell_0} \Big(\overline{EI}\mathbf{c}\mathbf{c}^{\mathrm{T}} + \overline{EA}(\mathbf{b}_w^{\mathrm{T}}\mathbf{w}')^2\mathbf{b}_w\mathbf{b}_w^{\mathrm{T}} + N\mathbf{b}_w\mathbf{b}_w^{\mathrm{T}}$$
$$+\overline{EX}(\mathbf{b}_w^{\mathrm{T}}\mathbf{w}')(\mathbf{b}_w\mathbf{c}^{\mathrm{T}} + \mathbf{c}\mathbf{b}_w^{\mathrm{T}})\Big)\,\mathrm{d}x \tag{9.30}$$

For the special case that z_ℓ is measured from the centreline, that $b(z_\ell)$ is symmetric with respect to the centreline, and that the material behaves identically in tension and in compression (so that E_{tan} is symmetric with respect to z_ℓ), the second integral of Equation (9.27) and the first integral of Equation (9.28) – the coupling terms – vanish.

With a Kirchhoff bending theory we cannot use an interpolation for the transverse displacement w that is lower than cubic polynomials. By contrast, it is well possible, with respect to all continuity requirements, to adopt any function, from linear functions onwards, for the axial displacement at the centreline, u_ℓ. However, with a cubic interpolation for w, an interpolation for u_ℓ would be required that involves quintic polynomials in order that the interpolations for w and u_ℓ are balanced, and that we can represent a constant membrane strain, and, in particular, the zero membrane strain associated with inextensional bending. The inability to represent such a constant membrane strain can lead to overstiff solutions, commonly denoted as membrane locking. However, a quintic interpolation for the axial strain would be extremely cumbersome, particularly when extended to plate or shell elements.

Instead of resorting to such a complicated element, a number of techniques have been proposed to remove, or at least to ameliorate, membrane locking (Stolarski and Belytschko 1982; Crisfield 1986; Bischoff *et al.* 2004). When adopting a linear interpolation function for u_ℓ, a possible solution is to use a single point, selective-reduced integration for the membrane strain $\bar{\epsilon}$. The problem remains, however, that a linear interpolation for u_ℓ in conjunction with a cubic interpolation for w leads to terms in the expression for the displacement u, Equation (9.2), that do not match, since the spatial derivative of w is then quadratic. Ignoring the non-linear terms we then have a solution that even for bending-dominant problems depends on the reference plane, i.e. where u_ℓ acts (Crisfield 1991). Hence, for eccentricity, for which the coupling terms that involve the $\overline{EX}$-terms are non-zero, an overstiff solution may be induced. It is noted that eccentricity can stem from the initial geometry, but also from induced material non-linearity when there is a difference in tensile and compressive behaviour. As an alternative solution, one can use different interpolations for w, e.g. a linear interpolation with respect to the membrane strain, but a cubic interpolation for the curvature, χ.

Methods to remove or reduce membrane locking can be put on a more rigorous footing using the Hu–Washizu variational principle (Crisfield 1986; Washizu 1975; Wempner 1969). This involves the membrane strain ϵ_ℓ being replaced by an effective membrane strain, ϵ_{eff}, such that

$$\epsilon_{\mathrm{eff}} = \frac{1}{\ell_0}\int_{\ell_0} \epsilon_\ell \mathrm{d}x \tag{9.31}$$

Using a linear interpolation for u_ℓ (so that $\Delta a_c = 0$), we can elaborate this identity for the shape functions (9.14) as:

$$\bar{\epsilon}_{\text{eff}} = \frac{a_2 - a_1}{\ell_0} + \frac{1}{2\ell_0}(\mathbf{w}')^{\mathrm{T}} \left(\int_{\ell_0} \mathbf{b}_w \mathbf{b}_w^{\mathrm{T}} \mathrm{d}x \right) \mathbf{w}' - \frac{1}{2\ell_0}(\mathbf{z})^{\mathrm{T}} \left(\int_{\ell_0} \mathbf{b}_w \mathbf{b}_w^{\mathrm{T}} \mathrm{d}x \right) \mathbf{z} \tag{9.32}$$

If the original expression for $\bar{\epsilon}$ is used instead, it is essential to use at least a quadratic polynomial for u_ℓ and to include the variable Δa_c to limit the self-straining. With a two-point Gauss integration along the beam axis, reasonable solutions are obtained. Without this quadratic term, overstiff solutions can result (Crisfield 1986).

9.1.2 *Including Shear Deformation: Timoshenko Beam*

As an alternative to using the Kirchhoff hypothesis, Equation (9.15), we can adopt a Timoshenko beam formulation, which includes shear deformation (Timoshenko 1921). As a consequence, θ, the rotation of the normal to the centreline, becomes an independent variable [Figure 9.2(b)], and the curvature is given by:

$$\chi = \frac{\mathrm{d}\theta}{\mathrm{d}x} \tag{9.33}$$

Defining

$$Q = \int_{-h/2}^{+h/2} b(z_\ell)\tau \mathrm{d}z_\ell \tag{9.34}$$

as the shear force, with $\tau = \sigma_{xz}$ the short-hand notation for the shear stress, the virtual work contribution

$$\int_{\ell_0} Q\delta\gamma \mathrm{d}x$$

with the shear strain γ defined as:

$$\gamma = \theta + \frac{\mathrm{d}w}{\mathrm{d}x} \tag{9.35}$$

must be added to Equation (9.10) to give:

$$\int_{\ell_0} (N\delta\bar{\epsilon} + M\delta\chi + Q\delta\gamma)\,\mathrm{d}x = \delta\mathbf{u}^{\mathrm{T}}\mathbf{f}_{\text{ext}} \tag{9.36}$$

The interpolation now formally follows from:

$$u_\ell = \mathbf{h}_u^{\mathrm{T}}\mathbf{a}\,, \quad w = \mathbf{h}_w^{\mathrm{T}}\mathbf{w}\,, \quad \theta = \mathbf{h}_\theta^{\mathrm{T}}\boldsymbol{\theta} \tag{9.37}$$

and the spatial derivatives follow by straightforward differentiation:

$$\frac{\mathrm{d}u_\ell}{\mathrm{d}x} = \mathbf{b}_u^{\mathrm{T}}\mathbf{a}\,, \quad \frac{\mathrm{d}w}{\mathrm{d}x} = \mathbf{b}_w^{\mathrm{T}}\mathbf{w}\,, \quad \frac{\mathrm{d}\theta}{\mathrm{d}x} = \mathbf{b}_\theta^{\mathrm{T}}\boldsymbol{\theta} \tag{9.38}$$

Using quadratic, hierarchical shape functions for the interpolation of u_ℓ, w and θ gives $\mathbf{h}_u = \mathbf{h}_w = \mathbf{h}_\theta$, with $\mathbf{h}_u$ as defined in Equation (9.14), and the nodal variables:

$$\begin{aligned}\mathbf{a}^{\mathrm{T}} &= (a_1, a_2, \Delta a_c)\\ \mathbf{w}^{\mathrm{T}} &= (w_1, w_2, \Delta w_c)\\ \boldsymbol{\theta}^{\mathrm{T}} &= (\theta_1, \theta_2, \Delta\theta_c)\end{aligned}$$

Note that the nodal rotational variables have a sign that is opposite of that in the Kirchhoff formulation, since they no longer follow the slope of the centreline ($\frac{\mathrm{d}w}{\mathrm{d}x}$), but that of the rotation of the normal. This slight notational anomaly could have been remedied by defining the z-axis in the downward direction (Hartsuijker and Welleman 2007). From the spatial derivates and Equations (9.33) and (9.35) we obtain the following expressions for the shear strain:

$$\gamma = \mathbf{h}_\theta^{\mathrm{T}}\boldsymbol{\theta} + \mathbf{b}_w^{\mathrm{T}}\mathbf{w} \tag{9.39}$$

and the curvature:

$$\chi = \mathbf{b}_\theta^{\mathrm{T}}\boldsymbol{\theta} \tag{9.40}$$

Substitution of the interpolations (9.37) into Equation (9.36) and noting that the discrete nodal variables now consist of the set $(\mathbf{a}, \mathbf{w}, \boldsymbol{\theta})$ yields:

$$\int_{\ell_0}\left[N\left(\delta\mathbf{a}^{\mathrm{T}}\mathbf{b}_u + (\mathbf{b}_w^{\mathrm{T}}\mathbf{w}')\delta\mathbf{w}^{\mathrm{T}}\mathbf{b}_w\right) + M\delta\boldsymbol{\theta}^{\mathrm{T}}\mathbf{b}_\theta + Q\left(\delta\boldsymbol{\theta}^{\mathrm{T}}\mathbf{h}_\theta + \delta\mathbf{w}^{\mathrm{T}}\mathbf{b}_w\right)\right]\mathrm{d}x = \delta\mathbf{a}^{\mathrm{T}}\mathbf{f}_{\mathrm{ext}}^{\mathrm{a}} + \delta\mathbf{w}^{\mathrm{T}}\mathbf{f}_{\mathrm{ext}}^{\mathrm{w}} + \delta\boldsymbol{\theta}^{\mathrm{T}}\mathbf{f}_{\mathrm{ext}}^{\theta} \tag{9.41}$$

Since this identity must hold for arbitrary $(\delta\mathbf{a}, \delta\mathbf{w}, \delta\boldsymbol{\theta})$, a set of equations results that has the familiar appearance of the balance of the external and the internal forces:

$$\begin{pmatrix}\mathbf{f}_{\mathrm{int}}^{\mathrm{a}}\\ \mathbf{f}_{\mathrm{int}}^{\mathrm{w}}\\ \mathbf{f}_{\mathrm{int}}^{\theta}\end{pmatrix} = \begin{pmatrix}\mathbf{f}_{\mathrm{ext}}^{\mathrm{a}}\\ \mathbf{f}_{\mathrm{ext}}^{\mathrm{w}}\\ \mathbf{f}_{\mathrm{ext}}^{\theta}\end{pmatrix}$$

with the internal force vectors now defined as:

$$\begin{aligned}\mathbf{f}_{\mathrm{int}}^{\mathrm{a}} &= \int_{\ell_0} N\mathbf{b}_u\mathrm{d}x\\ \mathbf{f}_{\mathrm{int}}^{\mathrm{w}} &= \int_{\ell_0}\left(N(\mathbf{b}_w^{\mathrm{T}}\mathbf{w}')\mathbf{b}_w + Q\mathbf{b}_w\right)\mathrm{d}x\\ \mathbf{f}_{\mathrm{int}}^{\theta} &= \int_{\ell_0}(M\mathbf{b}_\theta + Q\mathbf{h}_\theta)\,\mathrm{d}x\end{aligned} \tag{9.42}$$

As with the beam based on the Kirchhoff hypothesis, a Newton–Raphson method is used to solve the resulting set of non-linear algebraic equations, and the linearisation process results in:

$$\begin{bmatrix} \mathbf{K}_{\mathrm{aa}} & \mathbf{K}_{\mathrm{aw}} & \mathbf{K}_{\mathrm{a}\theta} \\ \mathbf{K}_{\mathrm{aw}}^{\mathrm{T}} & \mathbf{K}_{\mathrm{ww}} & \mathbf{K}_{\mathrm{w}\theta} \\ \mathbf{K}_{\mathrm{a}\theta}^{\mathrm{T}} & \mathbf{K}_{\mathrm{w}\theta}^{\mathrm{T}} & \mathbf{K}_{\theta\theta} \end{bmatrix} \begin{pmatrix} \mathrm{d}\mathbf{a} \\ \mathrm{d}\mathbf{w} \\ \mathrm{d}\boldsymbol{\theta} \end{pmatrix} = \begin{pmatrix} \mathbf{f}_{\mathrm{ext}}^{\mathrm{a}} - \mathbf{f}_{\mathrm{int}}^{\mathrm{a}} \\ \mathbf{f}_{\mathrm{ext}}^{\mathrm{w}} - \mathbf{f}_{\mathrm{int}}^{\mathrm{w}} \\ \mathbf{f}_{\mathrm{ext}}^{\theta} - \mathbf{f}_{\mathrm{int}}^{\theta} \end{pmatrix} \tag{9.43}$$

with the submatrices of the tangential stiffness matrix defined as:

$$\begin{aligned} \mathbf{K}_{\mathrm{aa}} &= \int_{\ell_0} \overline{EA}\mathbf{b}_u\mathbf{b}_u^{\mathrm{T}}\mathrm{d}x \\ \mathbf{K}_{\mathrm{aw}} &= \int_{\ell_0} \overline{EA}(\mathbf{b}_w^{\mathrm{T}}\mathbf{w}')\mathbf{b}_u\mathbf{b}_w^{\mathrm{T}}\mathrm{d}x \\ \mathbf{K}_{\mathrm{a}\theta} &= \int_{\ell_0} \overline{EX}\mathbf{b}_u\mathbf{b}_\theta^{\mathrm{T}}\mathrm{d}x \\ \mathbf{K}_{\mathrm{ww}} &= \int_{\ell_0} \left(\overline{EA}(\mathbf{b}_w^{\mathrm{T}}\mathbf{w}')^2 + \overline{GA} + N\right)\mathbf{b}_w\mathbf{b}_w^{\mathrm{T}}\mathrm{d}x \\ \mathbf{K}_{\mathrm{w}\theta} &= \int_{\ell_0} \left(\overline{EX}(\mathbf{b}_w^{\mathrm{T}}\mathbf{w}')\mathbf{b}_w\mathbf{b}_\theta^{\mathrm{T}} + \overline{GA}\mathbf{b}_w\mathbf{h}_\theta^{\mathrm{T}}\right)\mathrm{d}x \\ \mathbf{K}_{\theta\theta} &= \int_{\ell_0} \left(\overline{EI}\mathbf{b}_\theta\mathbf{b}_\theta^{\mathrm{T}} + \overline{GA}\mathbf{h}_\theta\mathbf{h}_\theta^{\mathrm{T}}\right)\mathrm{d}x \end{aligned} \tag{9.44}$$

where the last term of $\mathbf{K}_{\mathrm{ww}}$ can again be identified as the geometric contribution to the tangential stiffness matrix.

Similar to $\overline{EI}$ etc., the expression $\overline{GA}$ can be derived by considering the rate of the shear force, as defined in Equation (9.34)

$$\dot{Q} = \int_{-h/2}^{+h/2} b(z_\ell)\dot{\tau}\mathrm{d}z_\ell = \left(\int_{-h/2}^{+h/2} b(z_\ell)G_{\mathrm{tan}}(z_\ell)\mathrm{d}z_\ell\right)\dot{\gamma} \tag{9.45}$$

with G_{tan} the tangential shear modulus, which depends on the position with respect to the centreline. It follows that

$$\overline{GA} = \int_{-h/2}^{+h/2} b(z_\ell)G_{\mathrm{tan}}(z_\ell)\mathrm{d}z_\ell \tag{9.46}$$

and for a homogeneous cross section $\overline{GA} = GA$. It has been shown by Cowper (1966) that even for a homogeneous cross section and linear elasticity, a shear correction factor k should be applied:

$$\overline{GA} = kGA \tag{9.47}$$

with $k = \frac{5}{6}$ for rectangular cross sections.

Very acceptable solutions can be obtained when quadratic, hierarchic shape functions are used for all variables, in conjunction with a two-point Gauss integration. However, for slender beams, i.e. if the length to thickness ratio becomes high, so called shear locking can occur (Bischoff *et al.* 2004; Crisfield 1986). This can be overcome by forcing the shear strain to be effectively constant along the beam axis, by adopting the constraint (Crisfield 1986):

$$\Delta w_c = \frac{\ell}{8}(\theta_2 - \theta_1) \tag{9.48}$$

and the set of nodal variables becomes:

$$\begin{aligned}\mathbf{a}^{\mathrm{T}} &= (a_1, a_2, \Delta a_c)\\ \mathbf{w}^{\mathrm{T}} &= (w_1, w_2)\\ \boldsymbol{\theta}^{\mathrm{T}} &= (\theta_1, \theta_2, \Delta\theta_c)\end{aligned}$$

The constraint can be imposed after the stiffness matrix has been formed, but it is usually advantageous to directly modify the shape function derivatives in $\mathbf{b}_w$.

9.2 PyFEM: A Kirchhoff Beam Element

The Kirchhoff and the Timoshenko beam elements have been implemented in PyFEM in the files `KirchhoffBeam.py` and `TimoshenkoBeam.py`, which are located in the directory `pyfem/elements`. In this section, we will take a closer look at the implementation of the Kirchhoff beam element.

In general, each node of a continuum element has the same number of degrees of freedom. In structural elements however, this is not always the case. The Kirchhoff and the Timoshenko beam elements both contain additional degrees of freedom, which do not represent the displacement or the rotation of a node, but are used to construct a higher-order field. In the Kirchhoff beam model Δa_c, introduced in Equation (9.15), is an example of such a degree of freedom.

In most finite element codes, including PyFEM, it is assumed that each node in an element contains the same degrees of freedom. In order to implement a structural element with this restriction, the following trick can be used. We define the beam element as a three-noded element, where nodes 1 and 3 are the nodes which define the position of the element. Both nodes have three degrees of freedom $[u, w, \theta]$. Node 2 is a so-called 'dummy' node, which is only used to introduce an element specific degree of freedom, Δa_c. Since the node automatically contains the same set of degrees of freedom as the other nodes, the first degree of freedom u represents Δa_c. The other two degrees of freedom of this node, w and θ, are constrained.

The file `KirchhoffBeam.py` is organised as follows:

⟨*Kirchhoff beam element* ⟩≡

⟨*Kirchhoff beam class definition* 318⟩
⟨*Kirchhoff beam class main functions* 318⟩
⟨*Kirchhoff beam class utility functions* 319⟩

In the class definition, the nodal degrees of freedom of the element are specified:

⟨*Kirchhoff beam class definition* ⟩≡ 317

```
class KirchhoffBeam ( Element ):

  dofTypes = [ 'u' , 'w' , 'theta' ]

  def __init__ ( self, elnodes , props ):
    Element.__init__( self, elnodes , props )

    self.EA = self.E * self.A
    self.EI = self.E * self.I
    ⟨Construct arrays with sample points and integration weights⟩
```

The list `dofTypes` is a member of the class, and is used to construct the solution space of the problem. The items in this list represent the three nodal degrees of freedom u, w and θ. In the constructor of the base class the other parameters of the model are obtained: the Young's modulus `E`, the cross-sectional area `A` and the second moment of inertia `I`. They are used to compute the parameters `EA` and `EI`, which are stored as members of the class. Finally, two one-dimensional arrays are created that contain the sample point positions `intpoints` and integration weights `weights` of a third-order Gauss integration scheme.

The most important member function is `getTangentStiffness`:

⟨*Kirchhoff beam class main functions* ⟩≡ 317

```
  def getTangentStiffness ( self, elemdat ):

    l0  = norm( elemdat.coords[2]-elemdat.coords[0] )
    jac = 0.5 * l0

    a_bar = self.glob2Elem( elemdat.state , elemdat.coords )

    fint  = zeros(9)
    stiff = zeros( elemdat.stiff.shape )
```

First, the length of the current element is calculated and stored as `l0`. It is noted that the length is set by the positions of the first and the third node. The determinant of the Jacobian matrix, which is needed for the numerical integration, is calculated next. In this case, it is equal to half the length of the element. Then, the total state vector is rotated to the local element coordinate system. The function `glob2Elem` is also a member of this class. Finally, the new, empty arrays `fint` and `stiff` are created and are used to compute the internal force vector and the stiffness matrix in the local element coordinate system.

The numerical integration of the element is carried out by looping over the integration points:

```
⟨Kirchhoff beam class main functions ⟩+≡                                   318
    for xi,alpha in zip( self.intpoints , self.weights ):

      bu = self.getBu( l0 , xi )                                           319
      bw = self.getBw( l0 , xi )
      c  = self.getC ( l0 , xi )

      epsl = dot( bu , a_bar ) + 0.5 *( dot( bw , a_bar ) )**2
      chi  = dot( c  , a_bar )
```

The operators bu, bw and c that map the local state variables onto the derivatives, are first calculated for a given integration point xi. These vectors have a length 9 (the total number of degrees of freedom for this element), and are relatively sparse. This causes some numerical overhead, but in this manner, the mapping from the degree of freedom can be done automatically. As an example, the implementation of the function getBu is shown here:

```
⟨Kirchhoff beam class utility functions ⟩≡                                 317
  def getBu( self , l0 , xi ):

    Bu = zeros( 9 )

    Bu[0] = -1.0/l0
    Bu[3] = -4.0*xi/l0
    Bu[6] =  1.0/l0

    return Bu
```

When the axial strain and the curvature are known, the normal force N and bending moment M can be computed according to Equation (9.20):

```
⟨Kirchhoff beam class main functions ⟩+≡                                   319
      N = self.EA * epsl
      M = self.EI * chi

      wght = jac * alpha

      fint  += N * bu * wght
      fint  += ( N * dot( bw , a_bar ) * bw + M * c ) * wght
```

The weight factor is equal to the integration weight `alpha` times the determinant of the Jacobian matrix `jac`. The internal force vector is computed according to Equation (9.23). Because of the sparse structure of the arrays `bu`, `bw` and `c`, the terms of the internal force vector end up in the correct position automatically. The stiffness matrix is constructed according to Equation (9.25):

⟨*Kirchhoff beam class main functions* ⟩+≡ 319

```
stiff += self.EA * outer( bu , bu ) * wght
stiff += self.EA * dot( bw , a_bar ) * outer( bu , bw ) * wght
stiff += self.EA * dot( bw , a_bar ) * outer( bw , bu ) * wght
stiff += ( self.EI * outer( c , c ) + \
           self.EA * (dot( bw , a_bar ))**2 * outer( bw , bw ) + \
           N  * outer( bw , bw ) ) * wght
```

Note that the third line in the fragment has been added to calculate the term $\mathbf{K}_{wa} = \mathbf{K}_{aw}^{T}$.

When the loop over the integration points has been completed, the internal force vector and stiffness matrix have been calculated. However, the fourth and fifth terms in them belong to a dummy degree of freedom, and contain only zeros. To prevent ill-conditioning, these terms must be constrained manually. As a simple solution a unit value has been put on the diagonal of these terms:

⟨*Kirchhoff beam class main functions* ⟩+≡ 320

```
stiff[4:5,4:5]  = eye(2)

elemdat.fint   = self.elem2Glob( fint   , elemdat.coords )
elemdat.stiff  = self.elem2Glob( stiff  , elemdat.coords )
```

The function is completed by transforming the internal force vector and the stiffness matrix back to the global coordinate system and storing the values in the `elemdat` container.

The performance of the Kirchhoff beam element is now demonstrated through the simulation of the buckling of a simply supported, slender column shown in Figure 9.4(a). The column has length $L = 200$ and a uniform cross section with an area $A = 6$ and a moment of inertia $I = 2$. The Young's modulus is $E = 7.2 \times 10^6$. Both ends of the column are simply supported and the column is loaded by a point load P. The input file is `KirchhoffEuler.pro` and can be found in the directory `examples/ch09`. The beam has been discretised by ten Kirchhoff beam elements. The equilibrium path is obtained using Riks' arc-length method, see Chapter 4. In order to trigger the correct buckling mode, a small sinusoidal imperfection has been assumed, with an amplitude h that varies from 0.001 to 0.1.

The buckling load P_{cr}, at which loss of stability occurs, follows from the classical analytical relation:

$$P_{cr} = \frac{\pi^2 EI}{L^2} \tag{9.49}$$

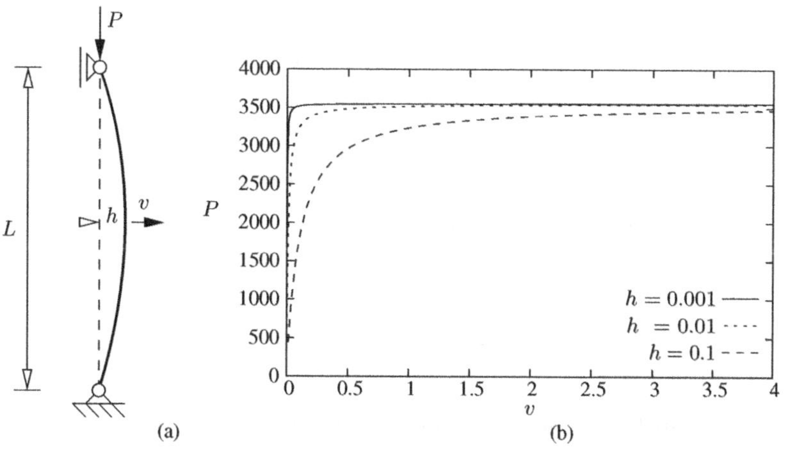

Figure 9.4 Euler buckling of a slender beam under compression. (a) Geometry and boundary conditions of the specimen. The beam has a sinusoidal imperfection with an amplitude h. (b) The applied load P *vs* the lateral displacement v of the beam for different amplitudes of the geometric imperfection.

For the given dimensions and material parameters, $P_{cr} = 3553$. From the graph in Figure 9.4(b) it is observed that all the simulations converge to this buckling load.

9.3 Corotational Elements

Corotational formulations have been addressed in relation to truss elements and to continuum elements in Chapter 3. The origins of the corotational formulation can be traced back to Belytschko and Glaum (1979), Belytschko and Hsieh (1973) and Wempner (1969), and also to the 'natural approach' (Argyris *et al.* 1979), which has much in common with corotational formulations. The terminology 'corotational' has been used in different contexts, but will be reserved here to describe the situation of a single element frame that continuously rotates with the element. In this coordinate frame the hypothesis of small strains is normally applied. In Chapter 11 we will show how large strains can be accommodated within the corotational framework in an approximate manner.

Since much of the early work was directed towards high-speed dynamics calculations, tangential stiffness matrices were often not derived explicitly. As also has become clear in Chapter 3 when deriving corotational formulations for truss elements and (two-dimensional) continuum elements, the key to a consistent linearisation in a corotational framework is the proper introduction of the variation of the local-global transformation matrices (Oran 1973; Oran and Kassimali 1976).

9.3.1 Kirchhoff Theory

Herein we will first describe a two-dimensional corotational beam element based on the Kirchhoff assumption. Simple engineering concepts will be used without recourse to shape functions. Throughout, an overbar will be used to denote a quantity that is referred to local coordinates.

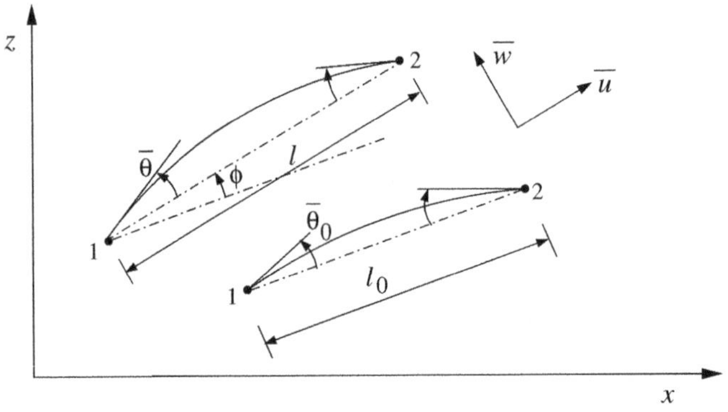

Figure 9.5 Stretch and local slopes in the corotational formulation

The elongation of an element follows that of the truss elements in Chapter 3 and the local axial strain reads, cf. Box 3.1:

$$\bar{u} = \ell - \ell_0 = \sqrt{(\ell_0 + \bar{u}_{21})^2 + \bar{w}_{21}^2} - \ell_0 \tag{9.50}$$

where, as in Chapter 3, the abbreviated notation $\bar{u}_{21} = \bar{u}_2 - \bar{u}_1$ etc. has been used. Assuming a homogeneous cross section – i.e. no layers – for the present, the normal force follows from Equation (9.20a):

$$N = \frac{EA}{\ell_0}\bar{u} \tag{9.51}$$

With respect to bending the standard engineering beam theory relations are assumed to apply in the local coordinate system, so that with the local transverse displacements being zero (Figure 9.5):

$$\begin{pmatrix} \bar{M}_1 \\ \bar{M}_2 \end{pmatrix} = \frac{2EI}{\ell_0}\begin{bmatrix} 2 & 1 \\ 1 & 2 \end{bmatrix}\begin{pmatrix} \bar{\theta}_1 \\ \bar{\theta}_2 \end{pmatrix} \tag{9.52}$$

where the local slopes are according to the Kirchhoff hypothesis given by:

$$\bar{\theta} = \frac{\mathrm{d}\bar{w}}{\mathrm{d}\bar{x}}$$

and are related to the slope in the global system, θ, the rigid rotation of the bar, ϕ, and the initial slope in the local system, $\bar{\theta}_0$, through:

$$\bar{\theta} = \theta - \phi - \bar{\theta}_0 \tag{9.53}$$

Expanding Equation (9.52) to include the linear-elastic relation in the axial direction, $\bar{N} = \frac{EA}{\ell_0}$, yields:

$$\begin{pmatrix} \bar{N} \\ \bar{M}_1 \\ \bar{M}_2 \end{pmatrix} = \begin{bmatrix} \frac{EA}{\ell_0} & 0 & 0 \\ 0 & \frac{4EI}{\ell_0} & \frac{2EI}{\ell_0} \\ 0 & \frac{2EI}{\ell_0} & \frac{4EI}{\ell_0} \end{bmatrix} \begin{pmatrix} \bar{u} \\ \bar{\theta}_1 \\ \bar{\theta}_2 \end{pmatrix} \tag{9.54}$$

Assembling the internal forces in the local coordinate system as:

$$\bar{\mathbf{f}}_{\text{int}}^{\text{T}} = \left(\bar{N}, \bar{M}_1, \bar{M}_2\right) \tag{9.55}$$

and the generalised displacements in the local coordate system as:

$$\bar{\mathbf{a}}^{\text{T}} = (\bar{u}, \bar{\theta}_1, \bar{\theta}_2) \tag{9.56}$$

we can write Equation (9.54) in a compact manner as:

$$\bar{\mathbf{f}}_{\text{int}} = \bar{\mathbf{D}}^{\text{e}}\bar{\mathbf{a}} \tag{9.57}$$

with

$$\bar{\mathbf{D}}^{\text{e}} = \begin{bmatrix} \frac{EA}{\ell_0} & 0 & 0 \\ 0 & \frac{4EI}{\ell_0} & \frac{2EI}{\ell_0} \\ 0 & \frac{2EI}{\ell_0} & \frac{4EI}{\ell_0} \end{bmatrix} \tag{9.58}$$

the elastic relation between the generalised internal forces and generalised displacements expressed in the local coordinate system.

In principle, the virtual strains can be derived from Equation (3.11). Alternatively, an approach can be followed as in Chapter 3. Referring to Figure 3.3, see also Box 3.2, we have:

$$\delta\bar{u} = \begin{pmatrix} \cos\phi \\ \sin\phi \end{pmatrix}^{\text{T}} \delta\mathbf{u}_{21} = \boldsymbol{e}_1^{\text{T}}\delta\mathbf{u}_{21} \tag{9.59}$$

where, as in Chapter 1, the unit vector $\boldsymbol{e}_1^{\text{T}} = (\cos\phi, \sin\phi)$ is directed along the axis of the beam in the current configuration. Assembling the unknowns in the global coordinate frame as:

$$\mathbf{a}^{\text{T}} = (u_1, w_1, \theta_1, u_2, w_2, \theta_2) \tag{9.60}$$

we can rewrite Equation (9.59) as:

$$\delta\bar{u} = \mathbf{v}^{\text{T}}\delta\mathbf{a} \tag{9.61}$$

with

$$\mathbf{v}^{\text{T}} = (-\cos\phi, -\sin\phi, 0, \cos\phi, \sin\phi, 0) \tag{9.62}$$

Next, we use the unit vector $\boldsymbol{e}_2^{\text{T}} = (-\sin\phi, \cos\phi)$, which is normal to the beam axis in the current configuration, to express the variation of the rigid rotation of the beam, see also Box 3.2, as:

$$\delta\phi = \frac{1}{\ell}\boldsymbol{e}_2^{\mathrm{T}}\delta\mathbf{u}_{21} \tag{9.63}$$

or, using **a**,

$$\delta\phi = \frac{1}{\ell}\mathbf{z}^{\mathrm{T}}\delta\mathbf{a} \tag{9.64}$$

with

$$\mathbf{z}^{\mathrm{T}} = (\sin\phi, -\cos\phi, 0, -\sin\phi, \cos\phi, 0) \tag{9.65}$$

Combining this identity with Equation (9.53) leads to:

$$\begin{pmatrix} \bar{\theta}_1 \\ \bar{\theta}_2 \end{pmatrix} = \left(\begin{bmatrix} 0 & 0 & 1 & 0 & 0 & 0 \\ 0 & 0 & 0 & 0 & 0 & 1 \end{bmatrix} - \frac{1}{\ell} \begin{bmatrix} \mathbf{z}^{\mathrm{T}} \\ \mathbf{z}^{\mathrm{T}} \end{bmatrix} \right) \delta\mathbf{a} \tag{9.66}$$

In view of Equation (9.56) the latter equation can be combined with Equation (9.61) to give:

$$\delta\bar{\mathbf{a}} = \mathbf{B}\delta\mathbf{a} \tag{9.67}$$

with

$$\mathbf{B} = \begin{bmatrix} 0 & 0 & 0 & 0 & 0 & 0 \\ 0 & 0 & 1 & 0 & 0 & 0 \\ 0 & 0 & 0 & 0 & 0 & 1 \end{bmatrix} + \begin{bmatrix} \mathbf{v}^{\mathrm{T}} \\ -\frac{1}{\ell}\mathbf{z}^{\mathrm{T}} \\ -\frac{1}{\ell}\mathbf{z}^{\mathrm{T}} \end{bmatrix} \tag{9.68}$$

Equating the internal virtual work in the local and the global coordinate systems gives:

$$\delta\bar{\mathbf{a}}^{\mathrm{T}}\bar{\mathbf{f}}_{\mathrm{int}} = \delta\mathbf{a}^{\mathrm{T}}\mathbf{f}_{\mathrm{int}} \tag{9.69}$$

while making use of Equation (9.67) results in:

$$\delta\mathbf{a}^{\mathrm{T}}\mathbf{f}_{\mathrm{int}} = \delta\mathbf{a}^{\mathrm{T}}\mathbf{B}^{\mathrm{T}}\bar{\mathbf{f}}_{\mathrm{int}} \tag{9.70}$$

Equating both right-hand sides, and considering that the result must hold for arbitrary $\delta\mathbf{a}$, we derive for the internal force vector:

$$\mathbf{f}_{\mathrm{int}} = \mathbf{B}^{\mathrm{T}}\bar{\mathbf{f}}_{\mathrm{int}} \tag{9.71}$$

The tangential stiffness matrix can be derived directly from Equation (9.71), by taking the variation:

$$\delta\mathbf{f}_{\mathrm{int}} = \mathbf{B}^{\mathrm{T}}\delta\bar{\mathbf{f}}_{\mathrm{int}} + \delta\mathbf{B}^{\mathrm{T}}\bar{\mathbf{f}}_{\mathrm{int}} \tag{9.72}$$

Using Equations (9.57) and (9.67), the first term can be elaborated as:

$$\mathbf{B}^{\mathrm{T}}\delta\bar{\mathbf{f}}_{\mathrm{int}} = \mathbf{B}^{\mathrm{T}}\bar{\mathbf{D}}^{\mathrm{e}}\mathbf{B}\delta\mathbf{a} \tag{9.73}$$

For the second term we can write:

$$\begin{aligned}\delta\mathbf{B}^{\mathrm{T}}\bar{\mathbf{f}}_{\mathrm{int}} &= \bar{N}\delta\mathbf{v} - \frac{\bar{M}_1 + \bar{M}_2}{\ell}\delta\mathbf{z} + \frac{\bar{M}_1 + \bar{M}_2}{\ell^2}\mathbf{z}\delta\ell \\ &= \bar{N}\mathbf{z}\delta\phi + \frac{\bar{M}_1 + \bar{M}_2}{\ell}\mathbf{v}\delta\phi + \frac{\bar{M}_1 + \bar{M}_2}{\ell^2}\mathbf{z}\delta\bar{u}\end{aligned} \tag{9.74}$$

which, using Equations (9.61) and (9.64), gives

$$\delta\mathbf{B}^{\mathrm{T}}\bar{\mathbf{f}}_{\mathrm{int}} = \left(\frac{\bar{N}}{\ell}\mathbf{z}\mathbf{z}^{\mathrm{T}} + \frac{\bar{M}_1 + \bar{M}_2}{\ell^2}\mathbf{v}\mathbf{z}^{\mathrm{T}} + \frac{\bar{M}_1 + \bar{M}_2}{\ell^2}\mathbf{z}\mathbf{v}^{\mathrm{T}}\right)\delta\mathbf{a} \tag{9.75}$$

and the tangential stiffness matrix has been derived as:

$$\mathbf{K} = \mathbf{B}^{\mathrm{T}}\bar{\mathbf{D}}\mathbf{B} + \frac{\bar{N}}{\ell}\mathbf{z}\mathbf{z}^{\mathrm{T}} + \frac{\bar{M}_1 + \bar{M}_2}{\ell^2}\left(\mathbf{v}\mathbf{z}^{\mathrm{T}} + \mathbf{z}\mathbf{v}^{\mathrm{T}}\right) \tag{9.76}$$

Whereas a direct, engineering approach has been used in the preceding derivation, a more formal, conventional finite element formulation that utilises shape functions, is also possible. To this end, the local displacement, $\bar{u}(\xi)$, is expressed as:

$$\bar{u}(\xi) = \frac{1}{2}(1+\xi)(\bar{u}_2 - \bar{u}_1) \tag{9.77}$$

and

$$\bar{x}(\xi) = \frac{1}{2}(1+\xi)\ell_0 \tag{9.78}$$

so that the local, axial strain is obtained as:

$$\bar{\epsilon} = \frac{\mathrm{d}\bar{u}}{\mathrm{d}\bar{x}} = \frac{\mathrm{d}\bar{u}}{\mathrm{d}\xi}\frac{\mathrm{d}\xi}{\mathrm{d}\bar{x}} = \frac{\bar{u}_2 - \bar{u}_1}{\ell_0} \tag{9.79}$$

and the axial force is given by substituting this expresion into Equation (9.20a), resulting in Equation (9.51). The local transverse displacement is assumed to be given by a conventional cubic interpolation

$$\bar{w}(\xi) = \frac{\ell_0}{8}\left((\xi^2-1)(\xi-1), (\xi^2-1)(\xi+1)\right)\begin{pmatrix}\bar{\theta}_1 \\ \bar{\theta}_2\end{pmatrix} \tag{9.80}$$

which is such that $\bar{w}$ vanishes at both ends (Figure 9.5). Straightforward differentation leads to the local rotation:

$$\bar{\theta}(\xi) = \frac{\mathrm{d}\bar{w}}{\mathrm{d}\bar{x}} = \frac{1}{4}\left(-1-2\xi+3\xi^2, -1+2\xi+3\xi^2\right)\begin{pmatrix}\bar{\theta}_1 \\ \bar{\theta}_2\end{pmatrix} \tag{9.81}$$

while further differentation results in the curvature:

$$\bar{\chi}(\xi) = \frac{\mathrm{d}\bar{\theta}}{\mathrm{d}\bar{x}} = \mathbf{b}^{\mathrm{T}}\bar{\boldsymbol{\theta}} \tag{9.82}$$

with

$$\mathbf{b} = \frac{1}{\ell_0}(-1 + 3\xi, 1 + 3\xi) \tag{9.83}$$

and $\bar{\boldsymbol{\theta}}^{\mathrm{T}} = (\bar{\theta}_1, \bar{\theta}_2)$. From Equation (9.20b) we then obtain the expression for the bending moment:

$$M = EI\chi = EI\mathbf{b}^{\mathrm{T}}\bar{\boldsymbol{\theta}} \tag{9.84}$$

The virtual work equation can now readily be expressed as:

$$\delta\bar{\mathbf{a}}^{\mathrm{T}}\bar{\mathbf{f}}_{\mathrm{int}} = \int_{\ell_0} \left(N\frac{\delta\bar{u}}{\ell_0} + M\delta\chi \right) \mathrm{d}x \tag{9.85}$$

$$= N\delta\bar{u} + \bar{\boldsymbol{\theta}}^{\mathrm{T}} \left(\int_{\ell_0} EI\mathbf{b}\mathbf{b}^{\mathrm{T}}\mathrm{d}x \right) \delta\bar{\boldsymbol{\theta}} \tag{9.86}$$

or after integration, and noting that the result must hold for arbitrary $\delta\bar{\mathbf{a}}$:

$$\bar{\mathbf{f}}_{\mathrm{int}} = \bar{\mathbf{D}}^{\mathrm{e}}\bar{\mathbf{a}} \tag{9.87}$$

with $\bar{\mathbf{D}}^{\mathrm{e}}$ as in Equation (9.58). The latter equation coincides with Equation (9.57), and the internal force vector is identical for both formulations. It follows that the tangential stiffness matrices also coincide.

Equation (9.51) is based on the approximation that the axial strain in the beam is equal to the relative axial deformation of the ends divided by the original length. This assumption does not allow for straining caused by the beam shape departing from a straight line. Such an effect can be incorporated by introducing the local shallow-arch terms (Belytschko and Glaum 1979). A general approach is offered by invoking the Green–Lagrange strain relative to the rotating coordinate system. When $\bar{\theta}_0$ denotes the effect of initial slopes for a shallow arch, the strain is obtained as:

$$\bar{\epsilon} = \frac{\mathrm{d}\bar{u}}{\mathrm{d}\bar{x}} + \frac{1}{2}\left(\frac{\mathrm{d}\bar{u}}{\mathrm{d}\bar{x}}\right)^2 + \frac{1}{2}\bar{\theta}^2 - \frac{1}{2}\bar{\theta}_0^2 \tag{9.88}$$

instead of Equation (9.79).

9.3.2 *Timoshenko Beam Theory*

The preceding derivation for a corotational beam element based on Kirchhoff's assumptions can be modified easily to incorporate shear deformation, hence to arrive at a corotational Timoshenko beam theory. While the description of the stretching of the beam is unaffected, the expressions for the bending moment and the shear force change into:

$$\bar{M} = EI\bar{\chi} = -\frac{EI}{\ell_0}(\bar{\theta}_2 - \bar{\theta}_1) = -\frac{EI}{\ell_0}(\theta_2 - \theta_1) \tag{9.89}$$

and

$$\bar{Q} = GA\gamma = -GA\left(\frac{\bar{\theta}_2 + \bar{\theta}_1}{2}\right) = -GA\left(\frac{\theta_2 + \theta_1}{2} - \phi\right) \tag{9.90}$$

The minus signs are required in order to maintain the sign convention adopted for the shallow arch in the previous section. Note that in Equations (9.89) and (9.90) possible initial curvatures have been omitted.

From Equations (9.89) and (9.90) and using Equations (9.61) and (9.64) we can formally relate the variations of the generalised displacements in the local and in the global reference frames as in Equation (9.67),

$$\delta\bar{\mathbf{a}} = \mathbf{B}\delta\mathbf{a}$$

but with the generalised displacements given by:

$$\mathbf{a}^{\mathrm{T}} = (\bar{u}, \ell_0\chi, \ell_0\gamma) \tag{9.91}$$

and the $\mathbf{B}$ matrix given by:

$$\mathbf{B} = \begin{bmatrix} 0 & 0 & 0 & 0 & 0 & 0 \\ 0 & 0 & 1 & 0 & 0 & -1 \\ 0 & 0 & -\frac{\ell_0}{2} & 0 & 0 & -\frac{\ell_0}{2} \end{bmatrix} + \begin{bmatrix} \mathbf{v}^{\mathrm{T}} \\ \mathbf{0}^{\mathrm{T}} \\ -\frac{\ell_0}{\ell}\mathbf{z}^{\mathrm{T}} \end{bmatrix} \tag{9.92}$$

in lieu of Equation (9.68). Equating the internal virtual work in the local and in the global reference frame yields again Equation (9.69), but the local internal force vector now reads:

$$\bar{\mathbf{f}}_{\mathrm{int}}^{\mathrm{T}} = (\bar{N}, \bar{M}, \bar{Q}) \tag{9.93}$$

On account of Equation (9.69) the relation between the internal force vectors expressed in the local and in the global coordinate systems formally still reads as in Equation (9.71), but with the $\mathbf{B}$ matrix now as in Equation (9.92).

Taking the variation of the internal force formally leads to Equation (9.72):

$$\delta\mathbf{f}_{\mathrm{int}} = \mathbf{B}^{\mathrm{T}}\delta\bar{\mathbf{f}}_{\mathrm{int}} + \delta\mathbf{B}^{\mathrm{T}}\bar{\mathbf{f}}_{\mathrm{int}}$$

For an elastic Timoshenko beam, the constitutive relation becomes:

$$\begin{pmatrix} \bar{N} \\ \bar{M} \\ \bar{Q} \end{pmatrix} = \begin{bmatrix} EA & 0 & 0 \\ 0 & EI & 0 \\ 0 & 0 & GA \end{bmatrix} \begin{pmatrix} \bar{u} \\ \ell_0\chi \\ \ell_0\gamma \end{pmatrix} \tag{9.94}$$

so that the elastic constitutive matrix expressed in the local coordinate system reads:

$$\bar{\mathbf{D}}^{\mathrm{e}} = \begin{bmatrix} EA & 0 & 0 \\ 0 & EI & 0 \\ 0 & 0 & GA \end{bmatrix} \tag{9.95}$$

The first term of Equation (9.72) leads to the 'material' contribution in the tangential stiffness matrix, while the geometric contribution can be elaborated in the same manner as for the

Kirchhoff beam, leading to:

$$\mathbf{K} = \frac{1}{\ell_0}\mathbf{B}^{\mathrm{T}}\bar{\mathbf{D}}^{\mathrm{e}}\mathbf{B} + \frac{\bar{N}}{\ell}\mathbf{z}\mathbf{z}^{\mathrm{T}} - \frac{\bar{Q}\ell_0}{\ell^2}\left(\mathbf{v}\mathbf{z}^{\mathrm{T}} + \mathbf{z}\mathbf{v}^{\mathrm{T}}\right) \tag{9.96}$$

9.4 A Two-dimensional Isoparametric Degenerate Continuum Beam Element

Figure 9.6 shows a two-dimensional three-noded isoparametric degenerate continuum beam element. For the linear theory the reader is referred to, for example, Bathe (1982) and Crisfield (1986). The non-linear formulation that follows, relates to a general non-linear isoparametric beam element (Bathe and Bolourchi 1975; Surana 1983b; Wood and Zienkiewicz 1977). Note that the element is formulated in the x, y-plane, which follows the two-dimensional continuum formulations of Chapter 3, but marks a departure from the beam formulations in the above, which are formulated in the x, z-plane.

Following standard degenerate-continuum techniques, the displacements can be interpolated as:

$$u(\xi) = \sum_{k=1}^{n} h_k(\xi)u_k + \frac{\zeta}{2}\sum_{k=1}^{n} h_k(\xi)t_k \cos\phi_k \tag{9.97a}$$

$$v(\xi) = \sum_{k=1}^{n} h_k(\xi)v_k + \frac{\zeta}{2}\sum_{k=1}^{n} h_k(\xi)t_k \sin\phi_k \tag{9.97b}$$

with $-1 \leq \xi \leq +1, -1 \leq \zeta \leq +1$ the isoparametric coordinates along the beam axis and perpendicular to the axis, respectively, t_k is the thickness at node k, and u_k and v_k are the nodal displacements of the centreline. Furthermore, $\cos\phi_k$ and $\sin\phi_k$ are the components of the normalised director that connects the positions of node k at the top and the bottom of the beam:

$$\mathbf{d}_k = \frac{(\mathbf{x}_k)_{\mathrm{top}} - (\mathbf{x}_k)_{\mathrm{bottom}}}{\left\|(\mathbf{x}_k)_{\mathrm{top}} - (\mathbf{x}_k)_{\mathrm{bottom}}\right\|} \tag{9.98}$$

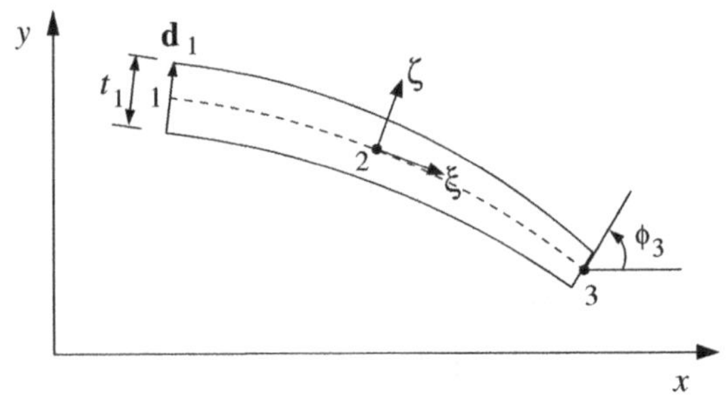

Figure 9.6 Three-noded degenerate continuum arch element

In the isoparametric concept, the geometry is interpolated in the same manner, thus:

$$x(\xi) = \sum_{k=1}^{n} h_k(\xi)x_k + \frac{\zeta}{2}\sum_{k=1}^{n} h_k(\xi)t_k \cos\phi_k \tag{9.99a}$$

$$y(\xi) = \sum_{k=1}^{n} h_k(\xi)y_k + \frac{\zeta}{2}\sum_{k=1}^{n} h_k(\xi)t_k \sin\phi_k \tag{9.99b}$$

where x_k and y_k are the nodal coordinates at the centreline. The derivatives with respect to the isoparametric coordinates, $\frac{\partial u}{\partial \xi}$ etc., can be obtained by straightforward differentiation, and the derivatives with respect to the global coordinates x, y then follow, cf. Chapter 2, from:

$$\begin{pmatrix} \frac{\partial u}{\partial x} \\ \frac{\partial u}{\partial y} \end{pmatrix} = \mathbf{J}^{-1} \begin{pmatrix} \frac{\partial u}{\partial \xi} \\ \frac{\partial u}{\partial \zeta} \end{pmatrix}, \quad \begin{pmatrix} \frac{\partial v}{\partial x} \\ \frac{\partial v}{\partial y} \end{pmatrix} = \mathbf{J}^{-1} \begin{pmatrix} \frac{\partial v}{\partial \xi} \\ \frac{\partial v}{\partial \zeta} \end{pmatrix}, \tag{9.100}$$

with $\mathbf{J}$ the standard Jacobian, Equation (2.22). We define

$$\mathbf{a}^{\mathrm{T}} = (u_1, v_1, u_2, v_2, \ldots, u_n, v_n) \tag{9.101}$$

as the array that contains the nodal displacements,

$$\mathbf{H} = \begin{bmatrix} h_1 & 0 & h_2 & 0 & \ldots & \ldots & h_n & 0 \\ 0 & h_1 & 0 & h_2 & \ldots & \ldots & 0 & h_n \end{bmatrix} \tag{9.102}$$

as the matrix that contains the shape functions h_k, and

$$\mathbf{v} = \left(\frac{1}{2}t_1 \cos\phi_1, \frac{1}{2}t_1 \sin\phi_1, \frac{1}{2}t_2 \cos\phi_2, \frac{1}{2}t_2 \sin\phi_2, \ldots, \frac{1}{2}t_n \cos\phi_n, \frac{1}{2}t_n \sin\phi_n\right) \tag{9.103}$$

as the array that contains the thickness at the nodes and the components of the directors, $\cos\phi_k$ and $\sin\phi_k$. Equation (9.97) can then be rewritten in matrix-vector format:

$$\mathbf{u} = \mathbf{H}(\mathbf{a} + \zeta\mathbf{v}) \tag{9.104}$$

and the virtual displacements, needed in the subsequent derivations, follow directly as:

$$\delta\mathbf{u} = \mathbf{H}(\delta\mathbf{a} + \zeta\mathbf{V}\delta\boldsymbol{\phi}) \tag{9.105}$$

with

$$\mathbf{V} = \begin{bmatrix} -\frac{1}{2}t_1 \sin\phi_1 & 0 & \ldots & 0 \\ +\frac{1}{2}t_1 \cos\phi_1 & 0 & \ldots & 0 \\ 0 & -\frac{1}{2}t_2 \sin\phi_2 & \ldots & 0 \\ 0 & +\frac{1}{2}t_2 \cos\phi_2 & \ldots & 0 \\ \vdots & \vdots & \ddots & \vdots \\ 0 & 0 & \ldots & -\frac{1}{2}t_n \sin\phi_n \\ 0 & 0 & \ldots & +\frac{1}{2}t_n \sin\phi_n \end{bmatrix} \tag{9.106}$$

and

$$\boldsymbol{\phi}^{\mathrm{T}} = (\phi_1, \phi_2, \ldots, \phi_n) \tag{9.107}$$

Defining the array $\hat{\mathbf{a}}$ that contains all nodal variables of the beam element,

$$\hat{\mathbf{a}} = \begin{pmatrix} \mathbf{a} \\ \boldsymbol{\phi} \end{pmatrix} \tag{9.108}$$

we can rewrite Equation (9.105) as:

$$\delta\mathbf{u} = [\mathbf{H}, \zeta\mathbf{H}\mathbf{V}]\,\delta\hat{\mathbf{a}} \tag{9.109}$$

For a Lagrange formulation we depart from Equation (3.86)

$$\int_{V_0} \delta\boldsymbol{\gamma}^{\mathrm{T}}\boldsymbol{\tau}^{t+\Delta t}\mathrm{d}V_0 = \int_{S_0} \delta\mathbf{u}^{\mathrm{T}}\mathbf{t}_0\mathrm{d}S_0 + \int_{V_0} \rho_0\delta\mathbf{u}^{\mathrm{T}}\mathbf{g}\mathrm{d}V_0$$

where, because of the beam assumption the normal stress perpendicular to the centreline must vanish, τ_{yy}, so that τ_{xx} , τ_{xy}, referred to a local coordinate system attached to the centreline, are the only non-vanishing components of the Second Piola–Kirchhoff stress tensor. The relevant components of the Green–Lagrange strain tensor are therefore, cf. Equation (3.61):

$$\gamma_{xx} = \frac{\partial u}{\partial \xi_1} + \frac{1}{2}\left(\left(\frac{\partial u}{\partial \xi_1}\right)^2 + \left(\frac{\partial v}{\partial \xi_1}\right)^2\right) \tag{9.110a}$$

$$\gamma_{xy} = \frac{\partial u}{\partial \xi_2} + \frac{\partial v}{\partial \xi_1} + \frac{\partial u}{\partial \xi_1}\frac{\partial u}{\partial \xi_2} + \frac{\partial v}{\partial \xi_1}\frac{\partial v}{\partial \xi_2} \tag{9.110b}$$

where it is noted that the engineering definition of the shear strain component has been used. Attention is drawn to the fact that in Equations (9.110), ξ_1 and ξ_2 are the local material coordinates, pointing along and perpendicular to the centreline, respectively, and are not to be confused with ξ, the isoparametric coordinate. Because of the assumption of small strains an incrementally linear relation holds between the increments of the Green–Lagrange strain tensor, $\Delta\boldsymbol{\gamma}$, and the increment of the Second Piola–Kirchhoff stress tensor:

$$\Delta\boldsymbol{\tau} = \mathbf{D}\Delta\boldsymbol{\gamma}$$

Because of the beam assumption the normal stress perpendicular to the centreline must vanish, and $\mathbf{D}$ attains the following format for linear elasticity:

$$\mathbf{D} = \begin{bmatrix} E & 0 \\ 0 & kG \end{bmatrix} \tag{9.111}$$

with k, as before, the shear stress correction factor.

As in the treatment for continuum elements in Chapter 3, the Second Piola–Kirchhoff stress at $t + \Delta t$ can be decomposed into its value at t and an increment $\Delta\boldsymbol{\tau}$. Use of the linear relation between the increments of the Second Piola–Kirchhoff stress tensor and the Green–Lagrange strain tensor, and decomposing the increment of the Green–Lagrange strain tensor into a part that is linear in the generalised displacement increments, $\Delta\mathbf{e}$, and a part that is non-linear in the generalised displacement increments, $\Delta\boldsymbol{\eta}$:

$$\Delta\boldsymbol{\gamma} = \Delta\mathbf{e} + \Delta\boldsymbol{\eta}$$

permits a linearisation of the virtual work equation, Equation (3.86), to give:

$$\begin{gathered}\int_{V_0} \delta\mathbf{e}^{\mathrm{T}}\mathbf{D}\Delta\mathbf{e}\mathrm{d}V_0 + \int_{V_0} \delta\boldsymbol{\eta}^{\mathrm{T}}\boldsymbol{\tau}^t\mathrm{d}V_0 = \\ \int_{S_0} \delta\mathbf{u}^{\mathrm{T}}\mathbf{t}_0\mathrm{d}S_0 + \int_{V_0} \rho_0\delta\mathbf{u}^{\mathrm{T}}\mathbf{g}\mathrm{d}V_0 - \int_{V_0} \delta\mathbf{e}^{\mathrm{T}}\boldsymbol{\tau}^t\mathrm{d}V_0\end{gathered} \tag{9.112}$$

which is identical to Equation (3.92).

From Equations (9.110) the increments of the strain components can be derived:

$$\Delta\gamma_{xx} = F_{11}\frac{\partial\Delta u}{\partial\xi_1} + F_{21}\frac{\partial\Delta v}{\partial\xi_1} + \frac{1}{2}\left(\left(\frac{\partial\Delta u}{\partial\xi_1}\right)^2 + \left(\frac{\partial\Delta v}{\partial\xi_1}\right)^2\right) \tag{9.113a}$$

$$\Delta\gamma_{xy} = F_{11}\frac{\partial\Delta u}{\partial\xi_2} + F_{12}\frac{\partial\Delta u}{\partial\xi_1} + F_{21}\frac{\partial\Delta v}{\partial\xi_2} + F_{22}\frac{\partial\Delta v}{\partial\xi_1} + \frac{\partial\Delta u}{\partial\xi_1}\frac{\partial\Delta u}{\partial\xi_2} + \frac{\partial\Delta v}{\partial\xi_1}\frac{\partial\Delta v}{\partial\xi_2} \tag{9.113b}$$

with F_{ij} the components of the deformation gradient, Equation (3.54). The part of the strain increment that is linear in the displacement increments thus becomes:

$$\Delta\mathbf{e} = \mathbf{L}\Delta\mathbf{u}$$

with

$$\mathbf{L} = \begin{bmatrix} F_{11}\frac{\partial}{\partial\xi_1} & F_{21}\frac{\partial}{\partial\xi_1} \\ F_{11}\frac{\partial}{\partial\xi_2} + F_{12}\frac{\partial}{\partial\xi_1} & F_{21}\frac{\partial}{\partial\xi_2} + F_{22}\frac{\partial}{\partial\xi_1} \end{bmatrix} \tag{9.114}$$

By straightforward differentiation a similar expression results for the variation of the linear part of the strain increment:

$$\delta\mathbf{e} = \mathbf{L}\delta\mathbf{u}$$

Substitution of Equation (9.109) yields

$$\delta\mathbf{e} = \widehat{\mathbf{B}}_L\delta\hat{\mathbf{a}} \tag{9.115}$$

with

$$\widehat{\mathbf{B}}_L = [\mathbf{B}_L, \zeta\mathbf{B}_L\mathbf{V}] \tag{9.116}$$

where

$$\mathbf{B}_L = \begin{bmatrix} F_{11}\frac{\partial h_1}{\partial\xi_1} & F_{21}\frac{\partial h_1}{\partial\xi_1} & \dots \\ F_{11}\frac{\partial h_1}{\partial\xi_2} + F_{12}\frac{\partial h_1}{\partial\xi_1} & F_{21}\frac{\partial h_1}{\partial\xi_2} + F_{22}\frac{\partial h_1}{\partial\xi_1} & \dots \end{bmatrix} \tag{9.117}$$

Finally, the non-linear parts of the strain increments read according to Equations (9.113a and b):

$$\Delta\eta_{xx} = \frac{1}{2}\left(\left(\frac{\partial\Delta u}{\partial\xi_1}\right)^2 + \left(\frac{\partial\Delta v}{\partial\xi_1}\right)^2\right) \tag{9.118a}$$

$$\Delta\eta_{xy} = \frac{\partial\Delta u}{\partial\xi_1}\frac{\partial\Delta u}{\partial\xi_2} + \frac{\partial\Delta v}{\partial\xi_1}\frac{\partial\Delta v}{\partial\xi_2} \tag{9.118b}$$

so that we obtain for the variations of the non-linear parts:

$$\delta\eta_{xx} = \frac{\partial\delta u}{\partial\xi_1}\frac{\partial\Delta u}{\partial\xi_1} + \frac{\partial\delta v}{\partial\xi_1}\frac{\partial\Delta v}{\partial\xi_1} \tag{9.119a}$$

$$\delta\eta_{xy} = \frac{\partial\delta u}{\partial\xi_1}\frac{\partial\Delta u}{\partial\xi_2} + \frac{\partial\delta u}{\partial\xi_2}\frac{\partial\Delta u}{\partial\xi_1} + \frac{\partial\delta v}{\partial\xi_1}\frac{\partial\Delta v}{\partial\xi_2} + \frac{\partial\delta v}{\partial\xi_2}\frac{\partial\Delta v}{\partial\xi_1} \tag{9.119b}$$

Substition of the expressions for $\delta\mathbf{e}$ and $\Delta\mathbf{e}$ into the first term on the left-hand side of the linearised virtual work equation, Equation (9.112), and into the last term on the right-hand side of this equation yields the first part of the tangential stiffness matrix

$$\mathbf{K}_L = \int_{V_0} \widehat{\mathbf{B}}_L^{\mathrm{T}}\mathbf{D}\widehat{\mathbf{B}}_L \mathrm{d}V_0 \tag{9.120}$$

and the internal force vector

$$\mathbf{f}_{\mathrm{int}}^t = \int_{V_0} \widehat{\mathbf{B}}_L^{\mathrm{T}}\boldsymbol{\tau}^t \mathrm{d}V_0 \tag{9.121}$$

respectively.

Similar to Chapter 3 the second term of Equation (3.92) can formally be rewritten as:

$$\int_{V_0} (\delta\boldsymbol{\eta})^{\mathrm{T}}\boldsymbol{\tau}^t \mathrm{d}V_0 = (\delta\hat{\mathbf{a}})^{\mathrm{T}}\mathbf{K}_{NL}\Delta\hat{\mathbf{a}} \tag{9.122}$$

where the geometric part of the tangential stiffness matrix is now given by:

$$\mathbf{K}_{NL} = \int_{V_0} \widehat{\mathbf{B}}_{NL}^{\mathrm{T}}\boldsymbol{\mathcal{T}}^t\widehat{\mathbf{B}}_{NL}\mathrm{d}V_0 \tag{9.123}$$

The matrix form of the Second Piola–Kirchhoff stress, $\boldsymbol{\mathcal{T}}$, is identical to the expression of Equation (3.103), except for the vanishing of τ_{yy}:

$$\boldsymbol{\mathcal{T}} = \begin{bmatrix} \tau_{xx} & \tau_{xy} & 0 & 0 \\ \tau_{xy} & 0 & 0 & 0 \\ 0 & 0 & \tau_{xx} & \tau_{xy} \\ 0 & 0 & \tau_{xy} & 0 \end{bmatrix} \tag{9.124}$$

In consideration of Equation (9.109) the matrix $\widehat{\mathbf{B}}_{NL}$ is defined such that:

$$\widehat{\mathbf{B}}_{NL} = [\mathbf{B}_{NL}, \zeta\mathbf{B}_{NL}\mathbf{V}] \tag{9.125}$$

with $\mathbf{B}_{NL}$ as in Equation (3.104).

Similar to the truss and continuum elements we derive that

$$(\delta\hat{\mathbf{a}})^{\mathrm{T}}(\mathbf{K}_L + \mathbf{K}_{NL})\Delta\hat{\mathbf{a}} = (\delta\hat{\mathbf{a}})^{\mathrm{T}}\left(\mathbf{f}_{\mathrm{ext}}^{t+\Delta t} - \mathbf{f}_{\mathrm{int}}^{t}\right) \tag{9.126}$$

with the external force vector defined as:

$$\mathbf{f}_{\mathrm{ext}}^{t+\Delta t} = \int_{S_0} \begin{pmatrix} \mathbf{H}^{\mathrm{T}}\mathbf{t}_0 \\ \mathbf{0} \end{pmatrix} \mathrm{d}S_0 + \int_{V_0} \begin{pmatrix} \rho_0\mathbf{H}^{\mathrm{T}}\mathbf{g} \\ \mathbf{0} \end{pmatrix} \mathrm{d}V_0 \tag{9.127}$$

when it is assumed that no external loads act on the rotational degrees of freedom ϕ_k. Identity (9.126) must hold for any virtual displacement increment $\delta\hat{\mathbf{a}}$, whence

$$(\mathbf{K}_L + \mathbf{K}_{NL})\Delta\hat{\mathbf{a}} = \mathbf{f}_{\mathrm{ext}}^{t+\Delta t} - \mathbf{f}_{\mathrm{int}}^{t} \tag{9.128}$$

Although the above derivation is for a homogeneous cross section, different material behaviour through the depth of the beam can be accommodated in a straightforward manner, similar to the treatment of the shallow arch. Also, enhancements in terms of element technology can be made, e.g. in order to improve the performance with respect to membrane locking and shear locking (Bischoff *et al.* 2004).

9.5 A Three-dimensional Isoparametric Degenerate Continuum Beam Element

The derivation of a three-dimensional degenerate continuum beam element shows similarities to that of the two-dimensional beam element discussed in the preceding section. A complication arises, however, from the fact that we now have two directors at each node, $\mathbf{v}_k$ and $\mathbf{w}_k$ (Figure 9.7). The motion of these directors can be chacterised by rotations. However, finite rotations are of a non-vectorial nature, and they cannot be added like displacements. When the rotation increments are small, the error that is committed, can be acceptable, but this is not so for large rotation increments. Moreover, when the latter effect is properly accounted for, a consistent linearisation introduces additional terms in the geometric part of the tangential stiffness. Neglecting these terms normally causes a loss of the quadratic convergence of the Newton–Raphson method.

For a three-dimensional isoparametric beam element the displacements are interpolated as (Bathe and Bolourchi 1975; Dvorkin *et al.* 1988):

$$\mathbf{u}(\xi) = \sum_{k=1}^{n} h_k(\xi)\mathbf{a}_k + +\frac{\eta}{2}\sum_{k=1}^{n} h_k(\xi)b_k\left(\mathbf{v}_k - \mathbf{v}_k^0\right) + \frac{\zeta}{2}\sum_{k=1}^{n} h_k(\xi)t_k\left(\mathbf{w}_k - \mathbf{w}_k^0\right) \tag{9.129}$$

with $-1 \le \xi \le +1, -1 \le \eta \le +1, -1 \le \zeta \le +1$ the isoparametric coordinates along the beam axis and perpendicular to the axis, respectively, while the width and the thickness of the beam at node k are b_k and t_k, respectively. The array $\mathbf{a}_k^{\mathrm{T}} = ((u_x)_k, (u_y)_k, (u_z)_k)$ assembles

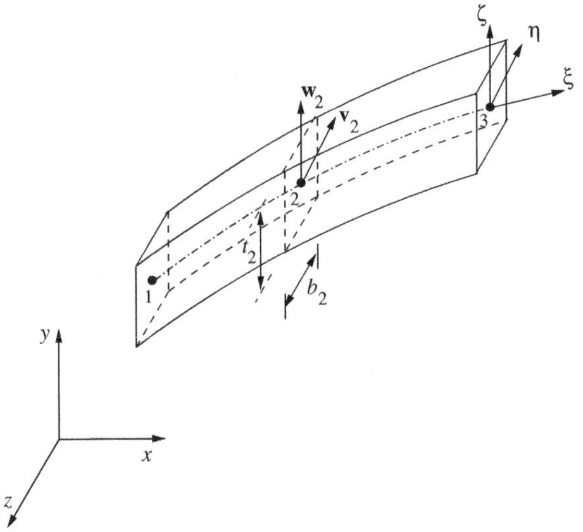

Figure 9.7 Three-dimensional degenerate continuum beam element

the displacements of node k at the centreline, and $\mathbf{v}_k$ and $\mathbf{w}_k$ are the normalised directors, which are measured with respect to their initial direction, denoted by a superscript 0. In an incremental format Equation (9.129) becomes:

$$\Delta\mathbf{u}(\xi) = \sum_{k=1}^{n} h_k(\xi)\Delta\mathbf{a}_k + +\frac{\eta}{2}\sum_{k=1}^{n} h_k(\xi)b_k\Delta\mathbf{v}_k + \frac{\zeta}{2}\sum_{k=1}^{n} h_k(\xi)t_k\Delta\mathbf{w}_k \tag{9.130}$$

The directors $\mathbf{v}$ and $\mathbf{w}$ are assumed to have a unit length throughout the deformation process. Together with a unit vector at node k that points in the axial direction of the beam, they form an orthonormal triad, that rotates without deforming. As a consequence their motion can be described by:

$$\begin{cases} \mathbf{v}_k = \mathbf{R}_k\mathbf{v}_k^0 \\ \mathbf{w}_k = \mathbf{R}_k\mathbf{w}_k^0 \end{cases} \tag{9.131}$$

and an identical relation holds for the update of $\mathbf{v}_k$, $\mathbf{w}_k$ from time t to $t + \Delta t$. The rotation matrix $\mathbf{R}$ can be expanded as follows (Argyris 1982; Atluri and Cazzani 1995):

$$\mathbf{R} = \mathbf{I} + \mathbf{\Theta} + \frac{1}{2!}\mathbf{\Theta}^2 + \frac{1}{3!}\mathbf{\Theta}^3 + \ldots \tag{9.132}$$

with

$$\mathbf{\Theta} = \begin{bmatrix} 0 & -\theta_3 & \theta_2 \\ \theta_3 & 0 & -\theta_1 \\ -\theta_2 & \theta_1 & 0 \end{bmatrix} \tag{9.133}$$

where the axial vector $\boldsymbol{\theta}$, which points in the direction of the axis of rotation, contains the components $\theta_1, \theta_2, \theta_3$:

$$\boldsymbol{\theta}^{\mathrm{T}} = (\theta_1, \theta_2, \theta_3) \tag{9.134}$$

Since two angles are sufficient to fully characterise a vector in the three-dimensional space, it follows that the components of $\boldsymbol{\theta}$ are not independent. Limiting the expansion of Equation (9.132) to the linear and the quadratic contributions, the increments of the directors for a finite step can be written as:

$$\begin{cases} \Delta \mathbf{v}_k = \Delta \boldsymbol{\Theta}_k \mathbf{v}_k^t + \frac{1}{2}(\Delta \boldsymbol{\Theta}_k)^2 \mathbf{v}_k^t \\ \Delta \mathbf{w}_k = \Delta \boldsymbol{\Theta}_k \mathbf{w}_k^t + \frac{1}{2}(\Delta \boldsymbol{\Theta}_k)^2 \mathbf{w}_k^t \end{cases} \tag{9.135}$$

where $\Delta\boldsymbol{\Theta}_k$ contains the incremental rotations $(\theta_1, \theta_2, \theta_3)$ at node k. Using the axial vector $\boldsymbol{\theta}_k$ we can alternatively write:

$$\begin{cases} \Delta \mathbf{v}_k = \Delta \boldsymbol{\theta}_k \times \mathbf{v}_k^t + \frac{1}{2}\Delta \boldsymbol{\theta}_k \times (\Delta \boldsymbol{\theta}_k \times \mathbf{v}_k^t) \\ \qquad = -\mathbf{v}_k^t \times \Delta \boldsymbol{\theta}_k + \frac{1}{2}(\mathbf{v}_k^t \times \Delta \boldsymbol{\theta}_k) \times \Delta \boldsymbol{\theta}_k \\ \Delta \mathbf{w}_k = \Delta \boldsymbol{\theta}_k \times \mathbf{w}_k^t + \frac{1}{2}\Delta \boldsymbol{\theta}_k \times (\Delta \boldsymbol{\theta}_k \times \mathbf{w}_k^t) \\ \qquad = -\mathbf{w}_k^t \times \Delta \boldsymbol{\theta}_k + \frac{1}{2}(\mathbf{w}_k^t \times \Delta \boldsymbol{\theta}_k) \times \Delta \boldsymbol{\theta}_k \end{cases} \tag{9.136}$$

We next substitute Equation (9.136) into Equation (9.130) and decompose the incremental displacements into a contribution that depends linearly on the rotation increments, $\Delta\mathbf{u}_L$, and a contribution $\Delta\mathbf{u}_{NL}$, that is quadratic in the rotation increments:

$$\Delta \mathbf{u}(\xi) = \Delta \mathbf{u}_L(\xi) + \Delta \mathbf{u}_{NL}(\xi) \tag{9.137}$$

where

$$\Delta \mathbf{u}_L(\xi) = \sum_{k=1}^{n} h_k(\xi) \Delta \mathbf{a}_k - \frac{\eta}{2} \sum_{k=1}^{n} h_k(\xi) b_k \mathbf{v}_k^t \times \Delta \boldsymbol{\theta}_k - \frac{\zeta}{2} \sum_{k=1}^{n} h_k(\xi) t_k \mathbf{w}_k^t \times \Delta \boldsymbol{\theta}_k \tag{9.138}$$

and

$$\Delta \mathbf{u}_{NL}(\xi) = \frac{\eta}{4} \sum_{k=1}^{n} h_k(\xi) b_k (\mathbf{v}_k^t \times \Delta \boldsymbol{\theta}_k) \times \Delta \boldsymbol{\theta}_k + \frac{\zeta}{4} \sum_{k=1}^{n} h_k(\xi) t_k (\mathbf{w}_k^t \times \Delta \boldsymbol{\theta}_k) \times \Delta \boldsymbol{\theta}_k \tag{9.139}$$

We next define the array $\hat{\mathbf{a}}$ that contains all nodal variables of the beam element,

$$\hat{\mathbf{a}} = \begin{pmatrix} \mathbf{a} \\ \boldsymbol{\theta} \end{pmatrix} \tag{9.140}$$

where $\mathbf{a}$ assembles the translational degrees of freedom,

$$\mathbf{a}^{\mathrm{T}} = \big((u_x)_1, (u_y)_1, (u_z)_1, \ldots, (u_x)_n, (u_y)_n, (u_z)_n\big) \tag{9.141}$$

and $\boldsymbol{\theta}$ assembles the rotation components,

$$\boldsymbol{\theta} = ((\theta_1)_1, (\theta_2)_1, (\theta_3)_1, \ldots, (\theta_1)_n, (\theta_2)_n, (\theta_3)_n) \tag{9.142}$$

Further,

$$\mathbf{H} = \begin{bmatrix} h_1 & 0 & 0 & h_2 & 0 & 0 & \dots & \dots & h_n & 0 & 0 \\ 0 & h_1 & 0 & 0 & h_2 & 0 & \dots & \dots & 0 & h_n & 0 \\ 0 & 0 & h_1 & 0 & 0 & h_2 & \dots & \dots & 0 & 0 & h_n \end{bmatrix}$$

cf. Equation (2.11), is the matrix that contains the $h_k(\xi)$. With the block-diagonal matrix

$$\mathbf{V} = \begin{bmatrix} \mathbf{V}_1 & \dots & \mathbf{0} \\ \vdots & \ddots & \vdots \\ \mathbf{0} & \dots & \mathbf{V}_n \end{bmatrix} \tag{9.143}$$

where

$$\mathbf{V}_k = \begin{bmatrix} 0 & -\frac{\eta b_k}{2}(v_3)_k - \frac{\zeta t_k}{2}(w_3)_k & \frac{\eta b_k}{2}(v_2)_k + \frac{\zeta t_k}{2}(w_2)_k \\ \frac{\eta b_k}{2}(v_3)_k + \frac{\zeta t_k}{2}(w_3)_k & 0 & -\frac{\eta b_k}{2}(v_1)_k - \frac{\zeta t_k}{1}(w_2)_k \\ -\frac{\eta b_k}{2}(v_2)_k - \frac{\zeta t_k}{2}(w_2)_k & \frac{\eta b_k}{2}(v_1)_k + \frac{\zeta t_k}{1}(w_1)_k & 0 \end{bmatrix} \tag{9.144}$$

Equation (9.138) can be rewritten in matrix-vector format:

$$\Delta \mathbf{u}_L = [\mathbf{H}, \mathbf{HV}]\, \Delta \hat{\mathbf{a}} \tag{9.145}$$

Taking the variation gives directly:

$$\delta \mathbf{u}_L = [\mathbf{H}, \mathbf{HV}]\, \delta \hat{\mathbf{a}} \tag{9.146}$$

since $\delta\mathbf{V}$ vanishes.

Because of the beam assumption the normal stresses perpendicular to the centreline must vanish, τ_{yy} and τ_{zz}, as well as the shear stress τ_{yz}, so that τ_{xx} , τ_{xy} and τ_{xz}, referred to the local coordinate system attached to the centreline, are the non-vanishing components of the Second Piola–Kirchhoff stress tensor. The relevant components of the Green–Lagrange strain tensor are therefore, cf. Equation (3.61):

$$\gamma_{xx} = \frac{\partial u}{\partial \xi_1} + \frac{1}{2}\left(\left(\frac{\partial u}{\partial \xi_1}\right)^2 + \left(\frac{\partial v}{\partial \xi_1}\right)^2 + \left(\frac{\partial w}{\partial \xi_1}\right)^2\right) \tag{9.147a}$$

$$\gamma_{xy} = \frac{\partial u}{\partial \xi_2} + \frac{\partial v}{\partial \xi_1} + \frac{\partial u}{\partial \xi_1}\frac{\partial u}{\partial \xi_2} + \frac{\partial v}{\partial \xi_1}\frac{\partial v}{\partial \xi_2} + \frac{\partial w}{\partial \xi_1}\frac{\partial w}{\partial \xi_2} \tag{9.147b}$$

$$\gamma_{xz} = \frac{\partial u}{\partial \xi_3} + \frac{\partial w}{\partial \xi_1} + \frac{\partial u}{\partial \xi_1}\frac{\partial u}{\partial \xi_3} + \frac{\partial v}{\partial \xi_1}\frac{\partial v}{\partial \xi_3} + \frac{\partial w}{\partial \xi_1}\frac{\partial w}{\partial \xi_3} \tag{9.147c}$$

where the engineering definition of the shear strain component has again been used, and material coordinates ξ_i should not be confused with ξ, the isoparametric coordinate along the beam axis. The components of the strain increments can be derived from Equations (9.147):

$$\Delta\gamma_{xx} = F_{11}\frac{\partial \Delta u}{\partial \xi_1} + F_{21}\frac{\partial \Delta v}{\partial \xi_1} + F_{31}\frac{\partial \Delta w}{\partial \xi_1} + \frac{1}{2}\left(\left(\frac{\partial \Delta u}{\partial \xi_1}\right)^2 + \left(\frac{\partial \Delta v}{\partial \xi_1}\right)^2 + \left(\frac{\partial \Delta w}{\partial \xi_1}\right)^2\right) \tag{9.148a}$$

$$\Delta\gamma_{xy} = F_{11}\frac{\partial \Delta u}{\partial \xi_2} + F_{12}\frac{\partial \Delta u}{\partial \xi_1} + F_{21}\frac{\partial \Delta v}{\partial \xi_2} + F_{22}\frac{\partial \Delta v}{\partial \xi_1} + F_{31}\frac{\partial \Delta w}{\partial \xi_2} + F_{32}\frac{\partial \Delta w}{\partial \xi_1} + \frac{\partial \Delta u}{\partial \xi_1}\frac{\partial \Delta u}{\partial \xi_2} + \frac{\partial \Delta v}{\partial \xi_1}\frac{\partial \Delta v}{\partial \xi_2} + \frac{\partial \Delta w}{\partial \xi_1}\frac{\partial \Delta w}{\partial \xi_2} \tag{9.148b}$$

$$\Delta\gamma_{xz} = F_{11}\frac{\partial \Delta u}{\partial \xi_3} + F_{13}\frac{\partial \Delta u}{\partial \xi_1} + F_{21}\frac{\partial \Delta v}{\partial \xi_3} + F_{23}\frac{\partial \Delta v}{\partial \xi_1} + F_{31}\frac{\partial \Delta w}{\partial \xi_3} + F_{33}\frac{\partial \Delta w}{\partial \xi_1} + \frac{\partial \Delta u}{\partial \xi_1}\frac{\partial \Delta u}{\partial \xi_3} + \frac{\partial \Delta v}{\partial \xi_1}\frac{\partial \Delta v}{\partial \xi_3} + \frac{\partial \Delta w}{\partial \xi_1}\frac{\partial \Delta w}{\partial \xi_3} \tag{9.148c}$$

so that the part of the strain increment that is linear in the generalised displacement increments attains the format:

$$\Delta \mathbf{e} = \mathbf{L}\Delta \mathbf{u}_L$$

with $\mathbf{L}$ defined as:

$$\mathbf{L} = \begin{bmatrix} F_{11}\frac{\partial}{\partial \xi_1} & F_{21}\frac{\partial}{\partial \xi_1} & F_{31}\frac{\partial}{\partial \xi_1} \\ F_{11}\frac{\partial}{\partial \xi_2} + F_{12}\frac{\partial}{\partial \xi_1} & F_{21}\frac{\partial}{\partial \xi_2} + F_{22}\frac{\partial}{\partial \xi_1} & F_{31}\frac{\partial}{\partial \xi_2} + F_{32}\frac{\partial}{\partial \xi_1} \\ F_{13}\frac{\partial}{\partial \xi_1} + F_{11}\frac{\partial}{\partial \xi_3} & F_{23}\frac{\partial}{\partial \xi_1} + F_{21}\frac{\partial}{\partial \xi_3} & F_{33}\frac{\partial}{\partial \xi_1} + F_{31}\frac{\partial}{\partial \xi_3} \end{bmatrix} \tag{9.149}$$

Substitution of Equation (9.145) yields:

$$\Delta \mathbf{e} = \widehat{\mathbf{B}}_L \Delta \hat{\mathbf{a}} \tag{9.150}$$

with

$$\widehat{\mathbf{B}}_L = [\mathbf{B}_L, \mathbf{B}_L \mathbf{V}] \tag{9.151}$$

where

$$\mathbf{B}_L = \begin{bmatrix} F_{11}\frac{\partial h_1}{\partial \xi_1} & F_{21}\frac{\partial h_1}{\partial \xi_1} & F_{31}\frac{\partial h_1}{\partial \xi_1} & \dots \\ F_{11}\frac{\partial h_1}{\partial \xi_2} + F_{12}\frac{\partial h_1}{\partial \xi_1} & F_{21}\frac{\partial h_1}{\partial \xi_2} + F_{22}\frac{\partial h_1}{\partial \xi_1} & F_{31}\frac{\partial h_1}{\partial \xi_2} + F_{32}\frac{\partial h_1}{\partial \xi_1} & \dots \\ F_{13}\frac{\partial h_1}{\partial \xi_1} + F_{11}\frac{\partial h_1}{\partial \xi_3} & F_{23}\frac{\partial h_1}{\partial \xi_1} + F_{21}\frac{\partial h_1}{\partial \xi_3} & F_{33}\frac{\partial h_1}{\partial \xi_1} + F_{31}\frac{\partial h_1}{\partial \xi_3} & \dots \end{bmatrix} \tag{9.152}$$

and the same relation holds for the variation $\delta\mathbf{e}$, cf. Equation (9.146).

The non-linear part of the strain increment that results from the terms in Equation (9.148a) that are quadratic in the gradients of the displacement increments reads:

$$\begin{aligned}
\Delta\eta_{xx} &= \frac{1}{2}\left(\left(\frac{\partial\Delta u}{\partial\xi_1}\right)^2 + \left(\frac{\partial\Delta v}{\partial\xi_1}\right)^2 + \left(\frac{\partial\Delta w}{\partial\xi_1}\right)^2\right) \\
\Delta\eta_{xy} &= \frac{\partial\Delta u}{\partial\xi_1}\frac{\partial\Delta u}{\partial\xi_2} + \frac{\partial\Delta v}{\partial\xi_1}\frac{\partial\Delta v}{\partial\xi_2} + \frac{\partial\Delta w}{\partial\xi_1}\frac{\partial\Delta w}{\partial\xi_2} \\
\Delta\eta_{xz} &= \frac{\partial\Delta u}{\partial\xi_1}\frac{\partial\Delta u}{\partial\xi_3} + \frac{\partial\Delta v}{\partial\xi_1}\frac{\partial\Delta v}{\partial\xi_3} + \frac{\partial\Delta w}{\partial\xi_1}\frac{\partial\Delta w}{\partial\xi_3}
\end{aligned} \tag{9.153}$$

so that we obtain for the variations:

$$\begin{aligned}
\delta\eta_{xx} &= \frac{\partial\delta u}{\partial\xi_1}\frac{\partial\Delta u}{\partial\xi_1} + \frac{\partial\delta v}{\partial\xi_1}\frac{\partial\Delta v}{\partial\xi_1} + \frac{\partial\delta w}{\partial\xi_1}\frac{\partial\Delta w}{\partial\xi_1} \\
\delta\eta_{xy} &= \frac{\partial\delta u}{\partial\xi_1}\frac{\partial\Delta u}{\partial\xi_2} + \frac{\partial\delta u}{\partial\xi_2}\frac{\partial\Delta u}{\partial\xi_1} + \frac{\partial\delta v}{\partial\xi_1}\frac{\partial\Delta v}{\partial\xi_2} + \frac{\partial\delta v}{\partial\xi_2}\frac{\partial\Delta v}{\partial\xi_1} \\
&\quad + \frac{\partial\delta w}{\partial\xi_1}\frac{\partial\Delta w}{\partial\xi_2} + \frac{\partial\delta w}{\partial\xi_2}\frac{\partial\Delta w}{\partial\xi_1} \\
\delta\eta_{xz} &= \frac{\partial\delta u}{\partial\xi_1}\frac{\partial\Delta u}{\partial\xi_3} + \frac{\partial\delta u}{\partial\xi_3}\frac{\partial\Delta u}{\partial\xi_1} + \frac{\partial\delta v}{\partial\xi_1}\frac{\partial\Delta v}{\partial\xi_3} + \frac{\partial\delta v}{\partial\xi_3}\frac{\partial\Delta v}{\partial\xi_1} \\
&\quad + \frac{\partial\delta w}{\partial\xi_1}\frac{\partial\Delta w}{\partial\xi_3} + \frac{\partial\delta w}{\partial\xi_3}\frac{\partial\Delta w}{\partial\xi_1}
\end{aligned} \tag{9.154}$$

However, there is another contribution that is non-linear in the generalised displacement increments. It stems from the contribution to the displacement increment that is quadratic in terms of the rotation increments, $\Delta\mathbf{u}_{NL}$, defined in Equation (9.139). When substituted into the Green–Lagrange strain increment, this contribution also gives rise to non-linear terms.

As in the previous section we take the virtual work balance, Equation (3.86), as the point of departure for the derivation of the internal force and the tangential stiffness matrix:

$$\int_{V_0} \delta\boldsymbol{\gamma}^{\mathrm{T}}\boldsymbol{\tau}^{t+\Delta t}\mathrm{d}V_0 = \int_{S_0} \delta\mathbf{u}^{\mathrm{T}}\mathbf{t}_0\mathrm{d}S_0 + \int_{V_0} \rho_0\delta\mathbf{u}^{\mathrm{T}}\mathbf{g}\mathrm{d}V_0$$

We decompose the Second Piola–Kirchhoff stress tensor $\boldsymbol{\tau}^{t+\Delta t}$ into $\boldsymbol{\tau}^t$ and $\Delta\boldsymbol{\tau}$, and relate the stress increment to the increment of the Green–Lagrange strain tensor using the small-strain assumption: $\Delta\boldsymbol{\tau} = \mathbf{D}\Delta\boldsymbol{\gamma}$, where, for linear elasticity:

$$\mathbf{D} = \begin{bmatrix} E & 0 & 0 \\ 0 & kG & 0 \\ 0 & 0 & kG \end{bmatrix} \tag{9.155}$$

with k the shear stress correction factor. Subsequently, we linearise and write the right-hand side in a compact format to arrive at:

$$\int_{V_0} \delta\mathbf{e}^{\mathrm{T}}\mathbf{D}\Delta\mathbf{e}\mathrm{d}V_0 + \int_{V_0} \delta\boldsymbol{\eta}^{\mathrm{T}}\boldsymbol{\tau}^t\mathrm{d}V_0 + \int_{V_0} \delta\tilde{\boldsymbol{\eta}}^{\mathrm{T}}\boldsymbol{\tau}^t\mathrm{d}V_0 = \delta\hat{\mathbf{a}}^{\mathrm{T}}\mathbf{f}_{\mathrm{ext}}^{t+\Delta t} - \int_{V_0} \delta\mathbf{e}^{\mathrm{T}}\boldsymbol{\tau}^t\mathrm{d}V_0 \tag{9.156}$$

where the external force vector is as defined in Equation (9.127). Elaboration of the first term on the left-hand side and of the last term on the right-hand side of Equation (9.156) results in the first part of the tangential stiffness matrix, Equation (9.120), and the internal force vector Equation (9.121), respectively. They are formally similar to the two-dimensional case, except for the redefinition of $\widehat{\mathbf{B}}_L$ and $\mathbf{D}$, which are now given by Equations (9.151) and (9.155), respectively.

The geometric stiffness matrix is now composed of two contributions. The first part is obtained by elaborating the second term on the left-hand side of Equation (9.156), and is similar to the geometric part of the tangential stiffness matrix that has been derived in the preceding section, Equation (9.123). Similar to Equation (9.125) the matrix $\widehat{\mathbf{B}}_{NL}$ is defined such that:

$$\widehat{\mathbf{B}}_{NL} = [\mathbf{B}_{NL}, \mathbf{B}_{NL}\mathbf{V}] \tag{9.157}$$

with $\mathbf{B}_{NL}$ as in Equation (3.106). Similarly, the matrix form of the Second Piola–Kirchhoff stress, $\mathcal{T}$, is identical to the expression of Equation (3.105), except for the fact that τ_{yy}, τ_{zz} and τ_{yz} vanish:

$$\mathcal{T} = \begin{bmatrix} \tau_{xx} & \tau_{xy} & \tau_{xz} & 0 & 0 & 0 & 0 & 0 & 0 \\ \tau_{xy} & 0 & 0 & 0 & 0 & 0 & 0 & 0 & 0 \\ \tau_{xz} & 0 & 0 & 0 & 0 & 0 & 0 & 0 & 0 \\ 0 & 0 & 0 & \tau_{xx} & \tau_{xy} & \tau_{xz} & 0 & 0 & 0 \\ 0 & 0 & 0 & \tau_{xy} & 0 & 0 & 0 & 0 & 0 \\ 0 & 0 & 0 & \tau_{xz} & 0 & 0 & 0 & 0 & 0 \\ 0 & 0 & 0 & 0 & 0 & 0 & \tau_{xx} & \tau_{xy} & \tau_{xz} \\ 0 & 0 & 0 & 0 & 0 & 0 & \tau_{xy} & 0 & 0 \\ 0 & 0 & 0 & 0 & 0 & 0 & \tau_{xz} & 0 & 0 \end{bmatrix} \tag{9.158}$$

The second contribution to the geometric stiffness matrix stems from the variation of the linear part of the Green–Lagrange strain tensor which operates on the part of the displacement increment that is quadratic in terms of the rotation increments:

$$\delta\tilde{\eta} = \mathbf{L}\delta\mathbf{u}_{NL} \tag{9.159}$$

where, in consideration of Equation (9.139),

$$\delta\mathbf{u}_{NL} = [\mathbf{0}, \mathbf{HW}]\delta\hat{\mathbf{a}} \tag{9.160}$$

with $\mathbf{0}$ a $3 \times 3n$ matrix and

$$\mathbf{W} = \begin{bmatrix} \mathbf{W}_1 & \dots & \mathbf{0} \\ \vdots & \ddots & \vdots \\ \mathbf{0} & \dots & \mathbf{W}_n \end{bmatrix} \tag{9.161}$$

a block-diagonal matrix with the submatrices $\mathbf{W}_k$ defined as:

$$\mathbf{W}_k = \frac{\eta b_k}{4}\begin{bmatrix} (v_2)_k(\Delta\theta_2)_k + (v_3)_k(\Delta\theta_3)_k & (v_2)_k(\Delta\theta_1)_k - 2(v_1)_k(\Delta\theta_2)_k & (v_3)_k(\Delta\theta_1)_k - 2(v_1)_k(\Delta\theta_3)_k \\ (v_1)_k(\Delta\theta_2)_k - 2(v_2)_k(\Delta\theta_1)_k & (v_3)_k(\Delta\theta_3)_k + (v_1)_k(\Delta\theta_1)_k & (v_3)_k(\Delta\theta_2)_k - 2(v_2)_k(\Delta\theta_3)_k \\ (v_1)_k(\Delta\theta_3)_k - 2(v_3)_k(\Delta\theta_1)_k & (v_2)_k(\Delta\theta_3)_k - 2(v_3)_k(\Delta\theta_2)_k & (v_1)_k(\Delta\theta_1)_k + (v_2)_k(\Delta\theta_2)_k \end{bmatrix} + \frac{\zeta t_k}{4}\begin{bmatrix} (w_2)_k(\Delta\theta_2)_k + (w_3)_k(\Delta\theta_3)_k & (w_2)_k(\Delta\theta_1)_k - 2(w_1)_k(\Delta\theta_2)_k & (w_3)_k(\Delta\theta_1)_k - 2(w_1)_k(\Delta\theta_3)_k \\ (w_1)_k(\Delta\theta_2)_k - 2(w_2)_k(\Delta\theta_1)_k & (w_3)_k(\Delta\theta_3)_k + (w_1)_k(\Delta\theta_1)_k & (w_3)_k(\Delta\theta_2)_k - 2(w_2)_k(\Delta\theta_3)_k \\ (w_1)_k(\Delta\theta_3)_k - 2(w_3)_k(\Delta\theta_1)_k & (w_2)_k(\Delta\theta_3)_k - 2(w_3)_k(\Delta\theta_2)_k & (w_1)_k(\Delta\theta_1)_k + (w_2)_k(\Delta\theta_2)_k \end{bmatrix} \tag{9.162}$$

Substitution of Equation (9.160) into Equation (9.159) gives:

$$\delta\tilde{\boldsymbol{\eta}} = [\mathbf{0}, \mathbf{B}_L\mathbf{W}]\delta\hat{\mathbf{a}} \tag{9.163}$$

This expression for $\delta\tilde{\boldsymbol{\eta}}$ can be substituted into the third term of Equation (9.156), which then becomes:

$$\int_{V_0} \delta\tilde{\boldsymbol{\eta}}^{\mathrm{T}}\boldsymbol{\tau}^t \mathrm{d}V_0 = \int_{V_0}\left(\delta\tilde{\eta}_{xx}\tau^t_{xx} + \delta\tilde{\eta}_{xy}\tau^t_{xy} + \delta\tilde{\eta}_{xz}\tau^t_{xz}\right)\mathrm{d}V_0 = \delta\hat{\mathbf{a}}^{\mathrm{T}}\mathbf{K}^*_{NL}\Delta\hat{\mathbf{a}} \tag{9.164}$$

with

$$\mathbf{K}^*_{NL} = \begin{bmatrix} \mathbf{0} & \mathbf{0} \\ \mathbf{0} & \tau_{xx}\mathbf{K}^*_{xx} + \tau_{xy}\mathbf{K}^*_{xy} + \tau_{xz}\mathbf{K}^*_{xz} \end{bmatrix} \tag{9.165}$$

the second contribution to the geometric stiffness matrix, which results from the multiplication of $\mathbf{B}_L$ and $\mathbf{W}$, followed by factoring out $\delta\boldsymbol{\theta}$ from the product, and $\mathbf{0}$ are $3n \times 3n$ matrices. As an example we elaborate the submatrix $\mathbf{K}^*_{xx}$, which, like $\mathbf{W}$, has a block-diagonal structure:

$$\mathbf{K}^*_{xx} = \begin{bmatrix} (\mathbf{K}^*_{xx})_1 & \dots & \mathbf{0} \\ \vdots & \ddots & \vdots \\ \mathbf{0} & \dots & (\mathbf{K}^*_{xx})_n \end{bmatrix} \tag{9.166}$$

with the submatrices:

$$(\mathbf{K}^*_{xx})_k = \int_{V_0} \frac{\eta b_k}{4} \frac{\partial h_k}{\partial \xi_1} \begin{bmatrix} -2F_{21}(v_2)_k - 2F_{31}(v_3)_k & F_{11}(v_2)_k + F_{21}(v_1)_k & F_{11}(v_3)_k + F_{31}(v_1)_k \\ F_{11}(v_2)_k + F_{21}(v_1)_k & -2F_{11}(v_2)_k - 2F_{31}(v_3)_k & F_{21}(v_3)_k + F_{31}(v_2)_k \\ F_{11}(v_3)_k + F_{31}(v_1)_k & F_{21}(v_3)_k + F_{31}(v_2)_k & -2F_{11}(v_1)_k - 2F_{21}(v_2)_k \end{bmatrix} \mathrm{d}V_0$$
$$+ \int_{V_0} \frac{\zeta t_k}{4} \frac{\partial h_k}{\partial \xi_1} \begin{bmatrix} -2F_{21}(w_2)_k - 2F_{31}(w_3)_k & F_{11}(w_2)_k + F_{21}(w_1)_k & F_{11}(w_3)_k + F_{31}(w_1)_k \\ F_{11}(w_2)_k + F_{21}(w_1)_k & -2F_{11}(w_2)_k - 2F_{31}(w_3)_k & F_{21}(w_3)_k + F_{31}(w_2)_k \\ F_{11}(w_3)_k + F_{31}(w_1)_k & F_{21}(w_3)_k + F_{31}(w_2)_k & -2F_{11}(w_1)_k - 2F_{21}(w_2)_k \end{bmatrix} \mathrm{d}V_0 \tag{9.167}$$

which is symmetric. The above derivation resembles that of Dvorkin *et al.* (1988), but other derivations exist that take into account the effect of finite rotation increments (Surana 1983; Simo 1985; Simo and Quoc 1986). They result in different formulations for the tangential stiffness matrix. It is remarked that, while the displacements are updated in a standard vectorial manner,

$$\mathbf{a}^{t+\Delta t} = \mathbf{a}^t + \Delta\mathbf{a}$$

the update of the rotations follows in a multiplicative sense:

$$\mathbf{R}^{t+\Delta t} = \Delta\mathbf{R}\mathbf{R}^t$$

It is finally noted that the above derivation is for a rectangular and homogeneous cross section, and for linear elasticity, but other cross-sectional shapes and different material behaviour can be accommodated by subdividing the cross section of the beam in a number of fibres, which is the equivalent of the layered structure of the two-dimensional beam. Evidently, Simpson, Lobatto or Newton–Cotes integration now has to be carried out in both cross-sectional directions of the beam.

References

Argyris JH 1982 An excursion into large rotations. *Computer Methods in Applied Mechanics and Engineering* **32**, 85–155.

Argyris JH, Balmer H, Doltsinis JS, Dunne PC, Haase M, Kleiber M, Malejannakis GA, Mlejnek HP, Muller M and Scharpf DW 1979 Finite element method – the natural approach. *Computer Methods in Applied Mechanics and Engineering* **17/18**, 1–106.

Atluri SN and Cazzani A 1995 Rotations in computational solid mechanics. *Archives of Computational Methods in Engineering* **2**, 49–138.

Bathe KJ 1982 *Finite Element Procedures in Engineering Analysis*. Prentice Hall, Inc.

Bathe KJ and Bolourchi S 1975 Large displacement analysis of three-dimensional beam structures. *International Journal for Numerical Methods in Engineering* **14**, 961–986.

Belytschko T and Glaum LW 1979 Application of higher-order corotational stretch theories to nonlinear finite element analysis. *Computers & Structures* **10**, 175–182.

Belytschko T and Hsieh BJ 1973 Non-linear transient finite element analysis with convected co-ordinates. *International Journal for Numerical Methods in Engineering* **7**, 255–271.

Bischoff M, Wall WA, Bletzinger KU and Ramm E 2004 Models and finite elements for thin-walled structures, in *The Encyclopedia of Computational Mechanics* (eds Stein E, de Borst R and Hughes TJR), vol. II, pp. 59–137. John Wiley & Sons, Ltd.

Burgoyne CJ and Crisfield MA 1990 Numerical integration strategy for plates and shells. *International Journal for Numerical Methods in Engineering* **29**, 105–121.

Cowper GR 1966 The shear coefficient in Timoshenko's beam theory. *Journal of Applied Mechanics* **33**, 335–340.

Crisfield MA 1986 *Finite Elements and Solution Procedures for Structural Analysis: Linear Analysis*. Pineridge Press.

Crisfield MA 1991 The 'eccentricity' issue in the design of beam, plate and shell elements. *Communications in Applied Numerical Methods* **7**, 47–56.

Dvorkin EN, Onate E and Oliver J 1988 On a non-linear formulation for curved Timoshenko beam elements considering large displacement/rotation increments. *International Journal for Numerical Methods in Engineering* **26**, 1597–1613.

Hartsuijker C and Welleman JW 2007 *Engineering Mechanics: Stresses, Strains, Displacements*. Springer.

Oran C 1973 Tangent stiffness in space frames. *ASCE Journal of the Structural Division* **99**, 973–985.

Oran C and Kassimali A 1976 Large deformation of framed structures under static and dynamic loads. *Computers & Structures* **6**, 539–547.

Simo JC 1985 A finite strain beam formulation. The three-dimensional dynamic problem. Part I. *Computer Methods in Applied Mechanics and Engineering* **49**, 55–70.

Simo JC and Quoc LV 1986 A three-dimensional finite strain rod model. Part II: computational aspects. *Computer Methods in Applied Mechanics and Engineering* **58**, 79–116.

Stolarski H and Belytschko T 1982 Membrane locking and reduced integration for curved elements. *Journal of Applied Mechanics* **49**, 172–176.

Surana KS 1983a Geometrically non-linear formulation for the curved shell elements. *International Journal for Numerical Methods in Engineering* **19**, 353–386.

Surana KS 1983b Geometrically non-linear formulation for two-dimensional curved beam elements. *Computers & Structures* **17**, 105–114.

Timoshenko SP 1921 On the correction for shear of the differential equation for transverse vibration of prismatic bars. *Philosophical Magazine* **41**, 744–746.

Washizu K 1975 *Variational Methods in Elasticity and Plasticity*, 2nd edn. Pergamon Press.

Wempner G 1969 Finite elements, finite rotations and small strains of flexible shells. *International Journal of Solids and Structures* **5**, 117–153.

Wood RD and Zienkiewicz OC 1977 Geometrically nonlinear finite element analysis of beams, frames, arches and axisymmetric shells. *Computers & Structures* **7**, 725–735.

10

Plates and Shells

Probably, the majority of the work on non-linear shell elements has followed the seminal paper of Ahmad *et al.* (1970), in which the linear, degenerate continuum shell element was introduced (Bathe and Bolourchi 1980; Büchter and Ramm 1992; Dvorkin and Bathe 1984; Hughes and Liu 1981; Parisch 1981; Stander *et al.* 1989; Surana 1983). As noted in Chapter 9, structural elements can suffer from various forms of locking, such as membrane locking and shear locking. An early remedy has been the use of reduced integration (Zienkiewicz *et al.* 1971), while assumed strain approaches form a more recent solution. For a comprehensive overview the reader is referred to Bischoff *et al.* (2004) and references therein.

The isoparametric degenerate continuum concept adopts shape functions for the components of the displacement in a fixed rectangular Cartesian system. Consequently, it allows for the exact satisfaction of rigid-body modes, even when the plane sections remain plane constraint is applied in the thickness direction (Crisfield 1986).

Considerable savings in computer time can be gained by resorting to shell elements where the through-the-depth integration is done directly instead of through numerical integration. As discussed in Chapter 9 for beams and shallow arches, the ensuing formulations employ the membrane strain, $\boldsymbol{\epsilon}_\ell$, and the curvature, $\boldsymbol{\chi}$, along with the corresponding stress resultants, $\mathbf{N}$ and $\mathbf{M}$ (Milford and Schnobrich 1986). For materially non-linear behaviour, the constitutive relations must then also be phrased in terms of stress resultants, which can be non-trivial (Burgoyne and Brennan 1993; Ilyushin 1956), or sometimes even impossible, for example when the behaviour in compression differs from that in tension, since the neutral line then shifts upon subsequent loadings. But even when a direct approach is feasible, the accuracy is considerably lower than that of a layered approach, because the yielding of the outermost fibres is not represented at occurrence, but delayed, since the yielding of the entire cross section will take place later. Nevertheless, considerable work has been done in developing stress-resultant shell elements (Simo and Fox 1989; Simo and Kennedy 1992; Simo *et al.* 1989, 1990).

An anomaly with traditional shell elements, including those based on the degenerate continuum concept, is that the normal stress perpendicular to the shell mid-surface is assumed to vanish. For materially non-linear analysis this poses the problem that the full three-dimensional constitutive equations cannot be used, and have to be reduced somehow. Such a reduced set

Non-linear Finite Element Analysis of Solids and Structures, Second Edition.
René de Borst, Mike A. Crisfield, Joris J.C. Remmers and Clemens V. Verhoosel.

of constitutive equations can be less trivial to handle numerically. For instance, the return-mapping algorithms that have been described in Chapter 7 for handling elasto-plasticity may have to be modified in order to properly account for this constraint. Alternatively, a standard algorithm for a three-dimensional stress state can be used, whereafter a compression is applied to enforce the zero normal-stress condition. The expansion of the reduced stress state to a full three-dimensional stress state at the next iteration then follows using the linearised tangent moduli (de Borst 1991). A rigorous approach is to develop shell elements that incorporate a non-zero normal stress in the thickness direction. Key to these solid-like shell elements, or solid-shell elements, is the inclusion of the stretch in the thickness direction (Braun *et al.* 1994; Büchter *et al.* 1994; Hauptmann and Schweizerhof 1998; Kühhorn and Schoop 1992; Parisch 1995; Sansour 1995). An important feature is that only translational degrees of freedom can be employed, which is advantageous when coupling such a rotation-free shell element to solid elements, or when employing them in a stacked manner, e.g. for layered structures where a single element with a through-the-thickness integration no longer suffices.

10.1 Shallow-shell Formulations

Similar to the previous chapter, where first a shallow-arch formulation was outlined, we will start this chapter with a shallow-shell formulation. At variance with the previous chapter, we will not start with a shallow-shell formulation based on the Kirchhoff assumption, but will directly adopt a Reissner–Mindlin formulation (Crisfield 1986). This formulation includes the case without shear deformation when the so-called discrete Kirchhoff constraints are enforced (Crisfield 1983, 1984, 1986). As we have seen in the previous chapter, the Reissner–Mindlin formulation is favourable in the context of finite element formulations, since no higher-order continuity is required. In Chapter 15 we will revisit the Kirchhoff–Love formulation in the framework of isogeometric analysis, an element technology that creates higher-order continuous shape functions in a rigorous manner.

Assuming that plane sections remain plane, Equation (9.5) can be extended to give:

$$\boldsymbol{\epsilon} = \boldsymbol{\epsilon}_\ell + z_\ell \boldsymbol{\chi} \tag{10.1}$$

where the subscript ℓ refers to the reference plane, which can but need not be the centre plane of the shell. The array $\boldsymbol{\epsilon}$ contains the in-plane strains

$$\boldsymbol{\epsilon}^{\mathrm{T}} = (\epsilon_{xx}, \epsilon_{yy}, \epsilon_{xy})$$

so that the $\boldsymbol{\epsilon}_\ell$ is given by:

$$\boldsymbol{\epsilon}_\ell = \begin{pmatrix} \frac{\partial u_\ell}{\partial x} \\ \frac{\partial v_\ell}{\partial y} \\ \frac{\partial u_\ell}{\partial y} + \frac{\partial v_\ell}{\partial x} \end{pmatrix} + \begin{pmatrix} \frac{1}{2}\left(\frac{\partial w'}{\partial x}\right)^2 - \frac{1}{2}\left(\frac{\partial z}{\partial x}\right)^2 \\ \frac{1}{2}\left(\frac{\partial w'}{\partial y}\right)^2 - \frac{1}{2}\left(\frac{\partial z}{\partial y}\right)^2 \\ \frac{\partial w'}{\partial x}\frac{\partial w'}{\partial y} - \frac{\partial z}{\partial x}\frac{\partial z}{\partial y} \end{pmatrix} \tag{10.2}$$

where z is the initial vertical coordinate of the shell reference plane, w is the displacement measured with respect to the reference plane, and $w' = w + z$, cf. Figure 9.1. Note that it has been assumed that the in-plane strains remain small: $\left(\frac{\partial u}{\partial x}\right)^2 \ll \left(\frac{\partial w}{\partial x}\right)^2$ etc., and that the

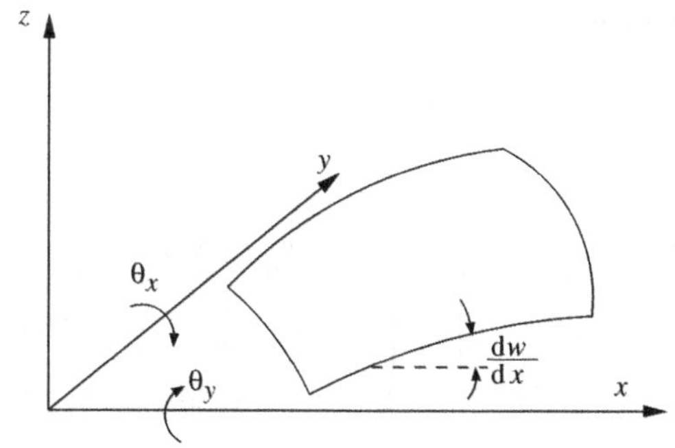

Figure 10.1 Coordinate system for a shallow shell

out-of-plane normal strain, ϵ_{zz}, has been assumed to vanish. Using a Reissner–Mindlin approach, the curvatures are given by:

$$\boldsymbol{\chi} = \begin{pmatrix} \chi_x \\ \chi_y \\ \chi_{xy} \end{pmatrix} = \begin{pmatrix} \frac{\partial \theta_x}{\partial x} \\ \frac{\partial \theta_y}{\partial y} \\ \frac{\partial \theta_x}{\partial y} + \frac{\partial \theta_y}{\partial x} \end{pmatrix} \tag{10.3}$$

where θ_x and θ_y are the rotations of the normal (Figure 10.1). The out-of-plane shear strains follow a standard format, cf. Equation (9.35):

$$\boldsymbol{\gamma} = \begin{pmatrix} \gamma_{xz} \\ \gamma_{yz} \end{pmatrix} = \begin{pmatrix} \theta_x \\ \theta_y \end{pmatrix} + \begin{pmatrix} \frac{\partial w}{\partial x} \\ \frac{\partial w}{\partial y} \end{pmatrix} \tag{10.4}$$

where it is emphasised that $\boldsymbol{\gamma}$ as in Equation (10.4) should not be confused with the Green–Lagrange strain tensor.

The variations of the strains and the strain increments follow in a standard manner. For the variations of the in-plane strains we have:

$$\delta\boldsymbol{\epsilon} = \delta\boldsymbol{\epsilon}_\ell + z_\ell \delta\boldsymbol{\chi} \tag{10.5}$$

where

$$\delta\boldsymbol{\epsilon}_\ell = \begin{pmatrix} \frac{\partial \delta u_\ell}{\partial x} \\ \frac{\partial \delta v_\ell}{\partial y} \\ \frac{\partial \delta u_\ell}{\partial y} + \frac{\partial \delta v_\ell}{\partial x} \end{pmatrix} + \begin{pmatrix} \frac{\partial w'}{\partial x}\frac{\partial \delta w}{\partial x} \\ \frac{\partial w'}{\partial y}\frac{\partial \delta w}{\partial y} \\ \frac{\partial w'}{\partial x}\frac{\partial \delta w}{\partial y} + \frac{\partial w'}{\partial y}\frac{\partial \delta w}{\partial x} \end{pmatrix} \tag{10.6}$$

and

$$\delta\boldsymbol{\chi} = \begin{pmatrix} \frac{\partial \delta\theta_x}{\partial x} \\ \frac{\partial \delta\theta_x}{\partial y} \\ \frac{\partial \delta\theta_x}{\partial y} + \frac{\partial \delta\theta_y}{\partial x} \end{pmatrix} \tag{10.7}$$

while the strain increment reads:

$$\Delta\boldsymbol{\epsilon} = \Delta\boldsymbol{\epsilon}_\ell + z_\ell \Delta\boldsymbol{\chi} \tag{10.8}$$

where

$$\Delta\boldsymbol{\epsilon}_\ell = \begin{pmatrix} \frac{\partial \Delta u_\ell}{\partial x} \\ \frac{\partial \Delta v_\ell}{\partial y} \\ \frac{\partial \Delta u_\ell}{\partial y} + \frac{\partial \Delta v_\ell}{\partial x} \end{pmatrix} + \begin{pmatrix} \frac{\partial w'}{\partial x}\frac{\partial \Delta w}{\partial x} + \frac{1}{2}\left(\frac{\partial \Delta w}{\partial x}\right)^2 \\ \frac{\partial w'}{\partial y}\frac{\partial \Delta w}{\partial y} + \frac{1}{2}\left(\frac{\partial \Delta w}{\partial y}\right)^2 \\ \frac{\partial w'}{\partial x}\frac{\partial \Delta w}{\partial y} + \frac{\partial w'}{\partial y}\frac{\partial \Delta w}{\partial x} + \frac{\partial \Delta w}{\partial x}\frac{\partial \Delta w}{\partial y} \end{pmatrix} \tag{10.9}$$

and $\Delta\boldsymbol{\chi}$ has a format similar to $\delta\boldsymbol{\chi}$, Equation (10.7). The variations of the out-of-plane shear strains become, Equation (10.4):

$$\delta\boldsymbol{\gamma} = \begin{pmatrix} \delta\theta_x \\ \delta\theta_y \end{pmatrix} + \begin{pmatrix} \frac{\partial \delta w}{\partial x} \\ \frac{\partial \delta w}{\partial y} \end{pmatrix} \tag{10.10}$$

while a similar equation ensues for $\Delta\boldsymbol{\gamma}$, the out-of-plane shear strain increment.

In a shell we have five non-vanishing stress components, the in-plane normal stresses, σ_{xx} and σ_{yy}, the in-plane shear stress σ_{xy}, and the out-of-plane shear stresses σ_{xz} and σ_{yz}. It is noted that under the assumption for the strain in the thickness direction, the use of a full three-dimensional isotropic elasticity relation as in Equation (1.115) does not result in this assumed stress state. This inconsistency can be remedied in an ad-hoc manner by modifying the constitutive relation.

These can be integrated through the depth to form generalised stresses, in particular the normal forces:

$$\mathbf{N} = \begin{pmatrix} N_x \\ N_y \\ N_{xy} \end{pmatrix} = \int_{-h/2}^{+h/2} \begin{pmatrix} \sigma_{xx}(z_\ell) \\ \sigma_{yy}(z_\ell) \\ \sigma_{xy}(z_\ell) \end{pmatrix} \mathrm{d}z_\ell \tag{10.11}$$

the bending moments,

$$\mathbf{M} = \begin{pmatrix} M_x \\ M_y \\ M_{xy} \end{pmatrix} = \int_{-h/2}^{+h/2} \begin{pmatrix} \sigma_{xx}(z_\ell) \\ \sigma_{yy}(z_\ell) \\ \sigma_{xy}(z_\ell) \end{pmatrix} z_\ell \mathrm{d}z_\ell \tag{10.12}$$

and the out-of-plane shear forces:

$$\mathbf{Q} = \begin{pmatrix} Q_x \\ Q_y \end{pmatrix} = \int_{-h/2}^{+h/2} \begin{pmatrix} \sigma_{xz}(z_\ell) \\ \sigma_{yz}(z_\ell) \end{pmatrix} \mathrm{d}z_\ell \tag{10.13}$$

Figure 10.2 shows the normal forces and bending moments that act on a structural element.

As in the shallow-arch formulation we allow for material non-linearity, and model the spread through the depth via a layered approach. Using the assumption that the strains remain small, we can relate the rate (or increment) of the in-plane components of the stress tensor $\boldsymbol{\sigma}$ to the strain tensor $\boldsymbol{\epsilon}$ (which can be conceived as a degenerated version of the Green–Lagrange strain

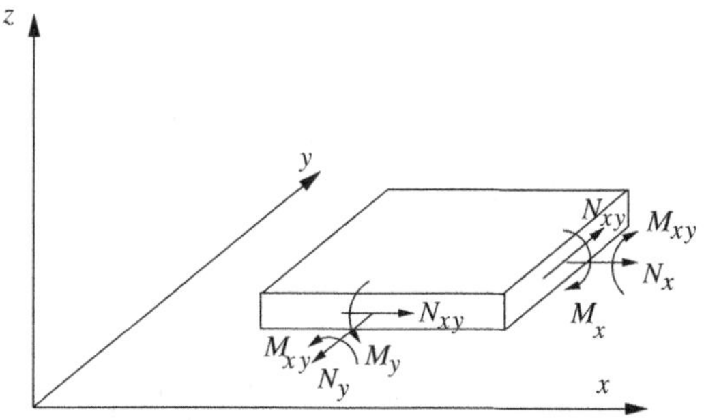

Figure 10.2 Normal forces and bending moments that act on a shell element

tensor), according to:

$$\begin{pmatrix} \dot{\sigma}_{xx} \\ \dot{\sigma}_{yy} \\ \dot{\sigma}_{xy} \end{pmatrix} = \mathbf{D} \begin{pmatrix} \dot{\epsilon}_{xx} \\ \dot{\epsilon}_{yy} \\ \dot{\epsilon}_{xy} \end{pmatrix} \tag{10.14}$$

where, for linear elasticity, $\mathbf{D}$ specialises to:

$$\mathbf{D} = \mathbf{D}^{\mathrm{e}} = \frac{E}{1-\nu^2} \begin{bmatrix} 1 & \nu & 0 \\ \nu & 1 & 0 \\ 0 & 0 & \frac{1-\nu}{2} \end{bmatrix} \tag{10.15}$$

In a manner similar to that employed in Chapter 9, cf. Equations (9.27) and (9.28), we obtain for the rates of the in-plane forces:

$$\dot{\mathbf{N}} = \underbrace{\left(\int_{-h/2}^{+h/2} \mathbf{D}(z_\ell) \mathrm{d}z_\ell \right)}_{\mathbf{D}_{\mathrm{m}}} \dot{\boldsymbol{\epsilon}}_\ell + \underbrace{\left(\int_{-h/2}^{+h/2} \mathbf{D}(z_\ell) z_\ell \mathrm{d}z_\ell \right)}_{\mathbf{D}_{\mathrm{c}}} \dot{\boldsymbol{\chi}} \tag{10.16}$$

and for the rates of the bending moments:

$$\dot{\mathbf{M}} = \underbrace{\left(\int_{-h/2}^{+h/2} \mathbf{D}(z_\ell) z_\ell \mathrm{d}z_\ell \right)}_{\mathbf{D}_{\mathrm{c}}} \dot{\boldsymbol{\epsilon}}_\ell + \underbrace{\left(\int_{-h/2}^{+h/2} \mathbf{D}(z_\ell) z_\ell^2 \mathrm{d}z_\ell \right)}_{\mathbf{D}_{\mathrm{b}}} \dot{\boldsymbol{\chi}} \tag{10.17}$$

with $\mathbf{D}_{\mathrm{m}}$ and $\mathbf{D}_{\mathrm{b}}$ the tangential constitutive matrices for membrane action and for bending, respectively, and $\mathbf{D}_{\mathrm{c}}$ the tangential cross-coupling matrix. To supplement these tangential relations for the stress resultants $\mathbf{N}$ and $\mathbf{M}$ we take for the rate of the shear forces

$$\dot{\mathbf{Q}} = \underbrace{\left(\int_{-h/2}^{+h/2} G_{\tan}(z_\ell) \mathbf{I} \mathrm{d}z_\ell \right)}_{\mathbf{D}_{\mathrm{s}}} \dot{\boldsymbol{\gamma}} \tag{10.18}$$

with $\mathbf{I}$ the 2×2 identity matrix, and $G_{\tan}$ a tangential shear modulus. For linear elasticity this equation reduces to:

$$\begin{pmatrix} \dot{Q}_x \\ \dot{Q}_y \end{pmatrix} = \begin{bmatrix} \mu & 0 \\ 0 & \mu \end{bmatrix} \begin{pmatrix} \dot{\gamma}_{xz} \\ \dot{\gamma}_{yz} \end{pmatrix} \tag{10.19}$$

with $G_{\tan} = \mu$ the elastic shear modulus. The expressions for rates of the the stress resultants can be combined to give:

$$\begin{pmatrix} \dot{\mathbf{N}} \\ \dot{\mathbf{M}} \\ \dot{\mathbf{Q}} \end{pmatrix} = \underbrace{\begin{bmatrix} \mathbf{D}_{\mathrm{m}} & \mathbf{D}_{\mathrm{c}} & \mathbf{0} \\ \mathbf{D}_{\mathrm{c}} & \mathbf{D}_{\mathrm{b}} & \mathbf{0} \\ \mathbf{0} & \mathbf{0} & \mathbf{D}_{\mathrm{s}} \end{bmatrix}}_{\mathbf{D}_{\mathrm{shell}}} \begin{pmatrix} \dot{\boldsymbol{\epsilon}}_{\ell} \\ \dot{\boldsymbol{\chi}} \\ \dot{\boldsymbol{\gamma}} \end{pmatrix} \tag{10.20}$$

The shallow shell theory of this section can be applied to a range of shell formulations, but herein we will adhere to a standard Reissner–Mindlin formulation with isoparametric interpolation functions. As in the Timoshenko beam element of Chapter 9, the basic variables, $u_\ell, v_\ell, w, \theta_x, \theta_y$, are expanded using the same shape functions $h_1, \ldots, h_n$, collected in the array $\mathbf{h}$. Defining

$$\mathbf{H} = \begin{bmatrix} h_1 & 0 & h_2 & 0 & \ldots & \ldots & h_n & 0 \\ 0 & h_1 & 0 & h_2 & \ldots & \ldots & 0 & h_n \end{bmatrix}$$

as in Equation (9.102), and assembling the in-plane nodal displacements in $\mathbf{a}$ and the nodal rotations in $\boldsymbol{\theta}$, as follows:

$$\mathbf{a} = \begin{pmatrix} (u_\ell)_1 \\ (v_\ell)_1 \\ \ldots \\ \ldots \\ (u_\ell)_n \\ (v_\ell)_n \end{pmatrix}, \quad \boldsymbol{\theta} = \begin{pmatrix} (\theta_x)_1 \\ (\theta_y)_1 \\ \ldots \\ \ldots \\ (\theta_x)_n \\ (\theta_y)_n \end{pmatrix} \tag{10.21}$$

we then have:

$$\begin{pmatrix} u_\ell \\ v_\ell \end{pmatrix} = \mathbf{H}\mathbf{a}\,, \quad w = \mathbf{h}^{\mathrm{T}}\mathbf{w}\,, \quad \begin{pmatrix} \theta_x \\ \theta_y \end{pmatrix} = \mathbf{H}\boldsymbol{\theta} \tag{10.22}$$

The shape functions $h_1, \ldots, h_n$ either belong to the serendipity family, or to the Lagrange family of interpolants, cf. Bathe (1982), Crisfield (1986) and Hughes (1987). It can be advantageous to formulate the shape function in an hierarchic manner, since this facilitates the implementation of constraints, for instance to remedy shear locking (Crisfield 1983, 1984).

We next invoke the definition of the variation of the in-plane strains, Equation (10.6), to derive the discretised expression:

$$\delta\boldsymbol{\epsilon}_\ell = \mathbf{B}\delta\mathbf{a} + (\mathbf{B}\mathcal{W}')\mathbf{B}_w\delta\mathbf{w} \tag{10.23}$$

with **B** the conventional strain-nodal displacement matrix for small displacement gradients:

$$\mathbf{B} = \begin{bmatrix} \frac{\partial h_1}{\partial x} & 0 & \dots & \dots & \frac{\partial h_n}{\partial x} & 0 \\ 0 & \frac{\partial h_1}{\partial y} & \dots & \dots & 0 & \frac{\partial h_n}{\partial y} \\ \frac{\partial h_1}{\partial y} & \frac{\partial h_1}{\partial x} & \dots & \dots & \frac{\partial h_n}{\partial y} & \frac{\partial h_n}{\partial x} \end{bmatrix}$$

cf. Equation (3.128),

$$\mathcal{W}' = \begin{bmatrix} w_1' & 0 \\ 0 & w_1' \\ \vdots & \vdots \\ w_n' & 0 \\ 0 & w_n' \end{bmatrix} \tag{10.24}$$

and

$$\mathbf{B}_w = \begin{bmatrix} \frac{\partial h_1}{\partial x} & \dots & \frac{\partial h_n}{\partial x} \\ \frac{\partial h_1}{\partial y} & \dots & \frac{\partial h_n}{\partial y} \end{bmatrix} \tag{10.25}$$

The discretised expressions for the variation of the curvature, $\delta\boldsymbol{\chi}$, and for the variation of the out-of-plane shear strains, $\delta\boldsymbol{\gamma}$, become

$$\delta\boldsymbol{\chi} = \mathbf{B}\delta\boldsymbol{\theta} \tag{10.26}$$

and

$$\delta\boldsymbol{\gamma} = \mathbf{H}\delta\boldsymbol{\theta} + \mathbf{B}_w\delta\mathbf{w} \tag{10.27}$$

The starting point for the derivation of the internal force vector and the tangential stiffness matrix is the virtual work equation, Equation (2.37), at $t + \Delta t$, which for the case of a shallow shell can be written as:

$$\int_V (\delta\boldsymbol{\epsilon}^{t+\Delta t})^{\mathrm{T}}\boldsymbol{\sigma}^{t+\Delta t}\mathrm{d}V = (\delta\mathbf{u}^{t+\Delta t})^{\mathrm{T}}\mathbf{f}_{\mathrm{ext}}^{t+\Delta t}$$

Using the kinematic assumption (10.1), considering that σ_{zz} vanishes, adopting that for the shell surface $A \approx A_0$, and integrating through the depth, this equation can be modified to give:

$$\int_A \left((\delta\boldsymbol{\epsilon}_\ell^{t+\Delta t})^{\mathrm{T}}\mathbf{N}^{t+\Delta t} + (\delta\boldsymbol{\chi}^{t+\Delta t})^{\mathrm{T}}\mathbf{M}^{t+\Delta t} + (\delta\boldsymbol{\gamma}^{t+\Delta t})^{\mathrm{T}}\mathbf{Q}^{t+\Delta t}\right)\mathrm{d}A = (\delta\mathbf{u}^{t+\Delta t})^{\mathrm{T}}\mathbf{f}_{\mathrm{ext}}^{t+\Delta t} \tag{10.28}$$

Decomposing the generalised stresses **N**, **M** and **Q** into their values after iteration j and the iterative correction during iteration $j + 1$, substituting the expressions for the variations, Equations (10.23), (10.26) and (10.27), and linearising gives:

$$\begin{aligned} \int_A & \left[\delta\mathbf{a}^{\mathrm{T}}\mathbf{B}^{\mathrm{T}}\mathrm{d}\mathbf{N} + \delta\boldsymbol{\theta}^{\mathrm{T}}\mathbf{B}^{\mathrm{T}}\mathrm{d}\mathbf{M} + \left(\delta\boldsymbol{\theta}^{\mathrm{T}}\mathbf{H}^{\mathrm{T}} + \delta\mathbf{w}^{\mathrm{T}}\mathbf{B}_w^{\mathrm{T}}\right)\mathrm{d}\mathbf{Q} + \delta\mathbf{w}^{\mathrm{T}}\mathbf{B}_w^{\mathrm{T}}(\mathbf{B}\mathrm{d}\mathcal{W}')^{\mathrm{T}}\mathbf{N}\right]\mathrm{d}A \\ &= \delta\mathbf{a}^{\mathrm{T}}\left(\mathbf{f}_{\mathrm{ext}}^{\mathrm{a}} - \mathbf{f}_{\mathrm{int}}^{\mathrm{a}}\right) + \delta\mathbf{w}^{\mathrm{T}}\left(\mathbf{f}_{\mathrm{ext}}^{\mathrm{w}} - \mathbf{f}_{\mathrm{int}}^{\mathrm{w}}\right) + \delta\boldsymbol{\theta}^{\mathrm{T}}\left(\mathbf{f}_{\mathrm{ext}}^{\theta} - \mathbf{f}_{\mathrm{int}}^{\theta}\right) \end{aligned} \tag{10.29}$$

with the internal force vectors defined as:

$$\begin{aligned}
\mathbf{f}_{\mathrm{int}}^{\mathrm{a}} &= \int_A \mathbf{B}^{\mathrm{T}}\mathbf{N}\mathrm{d}A \\
\mathbf{f}_{\mathrm{int}}^{\mathrm{w}} &= \int_A \mathbf{B}_w^{\mathrm{T}}\left((\mathbf{B}\mathcal{W}')^{\mathrm{T}}\mathbf{N}+\mathbf{Q}\right)\mathrm{d}A \\
\mathbf{f}_{\mathrm{int}}^{\theta} &= \int_A \left(\mathbf{B}^{\mathrm{T}}\mathbf{M}+\mathbf{H}^{\mathrm{T}}\mathbf{Q}\right)\mathrm{d}A
\end{aligned} \tag{10.30}$$

where the superscripts $t+\Delta t$ to the external force vectors and the subscript j to the internal force vectors have been dropped for notational simplicity. Inserting the constitutive relation for the shallow shell, Equation (10.20), and requiring that Equation (10.29) holds for arbitrary $(\delta\mathbf{a}, \delta\mathbf{w}, \delta\boldsymbol{\theta})$ yields the following set of equations:

$$\begin{bmatrix} \mathbf{K}_{\mathrm{aa}} & \mathbf{K}_{\mathrm{aw}} & \mathbf{K}_{\mathrm{a}\theta} \\ \mathbf{K}_{\mathrm{aw}}^{\mathrm{T}} & \mathbf{K}_{\mathrm{ww}} & \mathbf{K}_{\mathrm{w}\theta} \\ \mathbf{K}_{\mathrm{a}\theta}^{\mathrm{T}} & \mathbf{K}_{\mathrm{w}\theta}^{\mathrm{T}} & \mathbf{K}_{\theta\theta} \end{bmatrix} \begin{pmatrix} \mathrm{d}\mathbf{a} \\ \mathrm{d}\mathbf{w} \\ \mathrm{d}\boldsymbol{\theta} \end{pmatrix} = \begin{pmatrix} \mathbf{f}_{\mathrm{ext}}^{\mathrm{a}} - \mathbf{f}_{\mathrm{int}}^{\mathrm{a}} \\ \mathbf{f}_{\mathrm{ext}}^{\mathrm{w}} - \mathbf{f}_{\mathrm{int}}^{\mathrm{w}} \\ \mathbf{f}_{\mathrm{ext}}^{\theta} - \mathbf{f}_{\mathrm{int}}^{\theta} \end{pmatrix}$$

which formally resembles those for the shallow arch using the Timoshenko assumption, Equation (9.43), but now with the submatrices of the tangential stiffness matrix defined as:

$$\begin{aligned}
\mathbf{K}_{\mathrm{aa}} &= \int_A \mathbf{B}^{\mathrm{T}}\mathbf{D}_{\mathrm{m}}\mathbf{B}\mathrm{d}A \\
\mathbf{K}_{\mathrm{aw}} &= \int_A \mathbf{B}^{\mathrm{T}}\mathbf{D}_{\mathrm{m}}(\mathbf{B}\mathcal{W}')\mathbf{B}_w\mathrm{d}A \\
\mathbf{K}_{\mathrm{a}\theta} &= \int_A \mathbf{B}^{\mathrm{T}}\mathbf{D}_{\mathrm{c}}\mathbf{B}\mathrm{d}A \\
\mathbf{K}_{\mathrm{ww}} &= \int_A \mathbf{B}_w^{\mathrm{T}}\mathbf{D}_{\mathrm{s}}\mathbf{B}_w\mathrm{d}A + \int_A \mathbf{B}_w^{\mathrm{T}}(\mathbf{B}\mathcal{W}')^{\mathrm{T}}\mathbf{D}_{\mathrm{m}}(\mathbf{B}\mathcal{W}')\mathbf{B}_w\mathrm{d}A \\
&\quad + \int_A \mathbf{B}_w^{\mathrm{T}}\mathcal{N}\mathbf{B}_w\mathrm{d}A \\
\mathbf{K}_{\mathrm{w}\theta} &= \int_A \mathbf{B}_w^{\mathrm{T}}\mathbf{D}_{\mathrm{s}}\mathbf{H}\mathrm{d}A + \int_A \mathbf{B}_w^{\mathrm{T}}(\mathbf{B}\mathcal{W}')^{\mathrm{T}}\mathbf{D}_{\mathrm{c}}\mathbf{B}\mathrm{d}A \\
\mathbf{K}_{\theta\theta} &= \int_A \mathbf{B}^{\mathrm{T}}\mathbf{D}_{\mathrm{b}}\mathbf{B}\mathrm{d}A + \int_A \mathbf{H}^{\mathrm{T}}\mathbf{D}_{\mathrm{s}}\mathbf{H}\mathrm{d}A
\end{aligned} \tag{10.31}$$

The third contribution to $\mathbf{K}_{\mathrm{ww}}$ can be identified as the geometric contribution to the tangential stiffness matrix, and stems from the last term on the left-hand side of Equation (10.29), which can be rewritten as:

$$\int_A \delta\mathbf{w}^{\mathrm{T}}\mathbf{B}_w^{\mathrm{T}}(\mathbf{B}\mathrm{d}\mathcal{W}')^{\mathrm{T}}\mathbf{N}\mathrm{d}A = \int_A \delta\mathbf{w}^{\mathrm{T}}\mathbf{B}_w^{\mathrm{T}} \begin{pmatrix} N_x \dfrac{\partial \mathrm{d}w}{\partial x} + N_{xy}\dfrac{\partial \mathrm{d}w}{\partial y} \\ N_y \dfrac{\partial \mathrm{d}w}{\partial y} + N_{xy}\dfrac{\partial \mathrm{d}w}{\partial x} \end{pmatrix} \mathrm{d}A \tag{10.32}$$

$$= \delta\mathbf{w}^{\mathrm{T}}\left(\int_A \mathbf{B}_w^{\mathrm{T}}\mathcal{N}\mathbf{B}_w\mathrm{d}A\right)\mathrm{d}\mathbf{w} \tag{10.33}$$

where $\mathcal{N}$ assembles the normal forces in a matrix:

$$\mathcal{N} = \begin{pmatrix} N_x & N_{xy} \\ N_{xy} & N_y \end{pmatrix} \tag{10.34}$$

10.2 An Isoparametric Degenerate Continuum Shell Element

The derivation of an isoparametric degenerate continuum shell element runs along similar lines as those for the degenerate continuum beam elements in Chapter 9. For a k-noded degenerate continuum element, the displacements can be interpolated as:

$$\begin{aligned} u(\xi,\eta) &= \sum_{k=1}^{n} h_k(\xi,\eta)u_k + \frac{\zeta}{2}\sum_{k=1}^{n} h_k(\xi,\eta)t_k\cos\phi_k \\ v(\xi,\eta) &= \sum_{k=1}^{n} h_k(\xi,\eta)v_k + \frac{\zeta}{2}\sum_{k=1}^{n} h_k(\xi,\eta)t_k\sin\phi_k\cos\psi_k \\ w(\xi,\eta) &= \sum_{k=1}^{n} h_k(\xi,\eta)w_k + \frac{\zeta}{2}\sum_{k=1}^{n} h_k(\xi,\eta)t_k\sin\phi_k\sin\psi_k \end{aligned} \tag{10.35}$$

with $-1 \leq \xi \leq +1$, $-1 \leq \eta \leq +1$ the isoparametric coordinates tangential to the shell surface, and $-1 \leq \zeta \leq +1$ the isoparametric coordinate perpendicular to the surface (Figure 10.3). Figure 10.3 shows that t_k is the thickness at node k, while u_k, v_k and w_k are the nodal displacements of the centre surface. The angles ϕ_k and ψ_k (Figure 10.4), describe the position of the normalised director vector $\mathbf{d}_k$ that connects the positions of node k at the top and the bottom of the shell, Equation (9.98). In the isoparametric concept, the geometry is interpolated in the

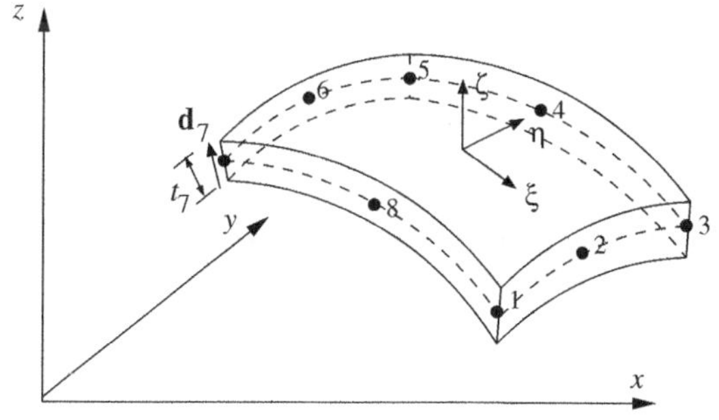

Figure 10.3 An eight-noded degenerate continuum shell

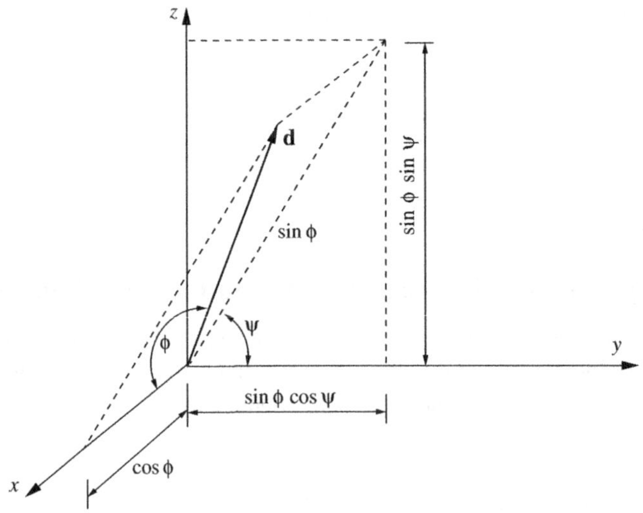

Figure 10.4 Spatial angles ϕ_k and ψ_k that define the director vector $\mathbf{d}_k$. Note that the subscript k has been omitted in the figure

same manner, thus:

$$\begin{aligned} x(\xi,\eta) &= \sum_{k=1}^{n} h_k(\xi,\eta)x_k + \frac{\zeta}{2}\sum_{k=1}^{n} h_k(\xi,\eta)t_k\cos\phi_k \\ y(\xi,\eta) &= \sum_{k=1}^{n} h_k(\xi,\eta)y_k + \frac{\zeta}{2}\sum_{k=1}^{n} h_k(\xi,\eta)t_k\sin\phi_k\cos\psi_k \\ z(\xi,\eta) &= \sum_{k=1}^{n} h_k(\xi,\eta)z_k + \frac{\zeta}{2}\sum_{k=1}^{n} h_k(\xi,\eta)t_k\sin\phi_k\sin\psi_k \end{aligned} \tag{10.36}$$

where x_k, y_k and z_k are the nodal coordinates at the centre surface. The derivatives with respect to the isoparametric coordinates, $\frac{\partial u}{\partial \xi}$ etc. can be obtained by straightforward differentiation, and the derivatives with respect to the global coordinates x, y, z then follow in a standard manner (Chapters 2 and 9).

We define

$$\mathbf{a}^{\mathrm{T}} = (u_1, v_1, w_1, \ldots, \ldots, \ldots, u_n, v_n, w_n)$$

as the array that contains the nodal displacements and $\mathbf{H}$ as the matrix that contains the shape functions h_k, cf. Equation (2.11). The array

$$\begin{aligned} \mathbf{v} = \; & (\frac{1}{2}t_1\cos\phi_1, \frac{1}{2}t_1\sin\phi_1\cos\psi_1, \frac{1}{2}t_1\sin\phi_1\sin\psi_1, \ldots, \\ & \frac{1}{2}t_n\cos\phi_n, \frac{1}{2}t_n\sin\phi_n\cos\psi_n, \frac{1}{2}t_n\sin\phi_n\sin\psi_n) \end{aligned} \tag{10.37}$$

contains the thickness at the nodes and the components of the directors, $\cos\phi_k$, $\sin\phi_k\cos\psi_k$ and $\sin\phi_k\sin\psi_k$. Equation (10.35) can then be rewritten in matrix-vector format:

$$\mathbf{u} = \mathbf{H}\left(\mathbf{a} + \zeta\mathbf{v}\right)$$

which is formally identical to that of the degenerate continuum beams, Equation (9.104), and the virtual displacements, needed in the subsequent derivations, follow directly as:

$$\delta\mathbf{u} = \mathbf{H}\left(\delta\mathbf{a} + \zeta\mathbf{V}\delta\boldsymbol{\theta}\right)$$

This identity resembles Equation (9.105), but with $\mathbf{V}$ defined as:

$$\mathbf{V} = \begin{bmatrix} -\frac{1}{2}t_1\sin\phi_1 & 0 & \dots & 0 & 0 \\ \frac{1}{2}t_1\cos\phi_1\cos\psi_1 & -\frac{1}{2}t_1\sin\phi_1\sin\psi_1 & \dots & 0 & 0 \\ \frac{1}{2}t_1\cos\phi_1\sin\psi_1 & -\frac{1}{2}t_1\sin\phi_1\sin\psi_1 & \dots & 0 & 0 \\ \vdots & \vdots & \ddots & \vdots & \vdots \\ 0 & 0 & \dots & -\frac{1}{2}t_n\sin\phi_n & 0 \\ 0 & 0 & \dots & \frac{1}{2}t_n\cos\phi_n\cos\psi_n & -\frac{1}{2}t_n\sin\phi_n\sin\psi_n \\ 0 & 0 & \dots & \frac{1}{2}t_n\cos\phi_n\sin\psi_n & -\frac{1}{2}t_n\sin\phi_n\sin\psi_n \end{bmatrix} \tag{10.38}$$

and

$$\boldsymbol{\theta}^{\mathrm{T}} = (\phi_1, \psi_1, \dots, \phi_n, \psi_n) \tag{10.39}$$

Defining $\hat{\mathbf{a}}$ as the array that contains all the nodal variables of the shell element:

$$\hat{\mathbf{a}} = \begin{pmatrix} \mathbf{a} \\ \boldsymbol{\theta} \end{pmatrix} \tag{10.40}$$

we obtain:

$$\delta\mathbf{u} = [\mathbf{H}, \zeta\mathbf{H}\mathbf{V}]\,\delta\hat{\mathbf{a}} \tag{10.41}$$

cf. Equation (9.109).

For deriving the discretised equations we depart, as usual, from Equation (3.86)

$$\int_{V_0} \delta\boldsymbol{\gamma}^{\mathrm{T}}\boldsymbol{\tau}^{t+\Delta t}\mathrm{d}V_0 = \int_{S_0} \delta\mathbf{u}^{\mathrm{T}}\mathbf{t}_0\mathrm{d}S_0 + \int_{V_0} \rho_0\delta\mathbf{u}^{\mathrm{T}}\mathbf{g}\mathrm{d}V_0$$

where, because of the shell assumption the normal stress orthogonal to the shell surface must vanish, τ_{zz}, and $\gamma_{xx}, \gamma_{yy}, \gamma_{xy}, \gamma_{yz}, \gamma_{zx}$ are the relevant components of the Green–Lagrange strain tensor, Equation (3.61). Because of the assumption of small strains an incrementally linear relation holds between the increments of the Green–Lagrange strain tensor, $\mathrm{d}\boldsymbol{\gamma}$, and the increment of the Second Piola–Kirchhoff stress tensor:

$$\mathrm{d}\boldsymbol{\tau} = \tilde{\mathbf{D}}\mathrm{d}\boldsymbol{\gamma}$$

where, for linear elasticity, $\tilde{\mathbf{D}}$ attains the following format:

$$\tilde{\mathbf{D}} = \frac{E}{1-\nu^2}\begin{bmatrix} 1 & \nu & 0 & 0 & 0 \\ \nu & 1 & 0 & 0 & 0 \\ 0 & 0 & \frac{1}{2}k(1-\nu) & 0 & 0 \\ 0 & 0 & 0 & \frac{1}{2}k(1-\nu) & 0 \\ 0 & 0 & 0 & 0 & \frac{1}{2}k(1-\nu) \end{bmatrix} \tag{10.42}$$

with, as for the Timoshenko beam, k the shear correction factor.

As in the preceding chapters, the Second Piola–Kirchhoff stress is decomposed into its value at iteration j and a correction $\mathrm{d}\boldsymbol{\tau}$. Use of the linear relation between the increments of the Second Piola–Kirchhoff stress tensor and the Green–Lagrange strain tensor, and decomposing the increment of the Green–Lagrange strain tensor into a part that is linear in the generalised displacement increments, $\mathrm{d}\mathbf{e}$, and a part that is non-linear in the generalised displacement increments, $\mathrm{d}\boldsymbol{\eta}$:

$$\mathrm{d}\boldsymbol{\gamma} = \mathrm{d}\mathbf{e} + \mathrm{d}\boldsymbol{\eta}$$

permits a linearisation of the virtual work equation, Equation (3.86), to give:

$$\int_{V_0} \delta\mathbf{e}^{\mathrm{T}}\mathbf{D}\mathrm{d}\mathbf{e}\mathrm{d}V_0 + \int_{V_0} \delta\boldsymbol{\eta}^{\mathrm{T}}\boldsymbol{\tau}_j\mathrm{d}V_0 =$$
$$\int_{S_0} \delta\mathbf{u}^{\mathrm{T}}\mathbf{t}_0\mathrm{d}S_0 + \int_{V_0} \rho_0\delta\mathbf{u}^{\mathrm{T}}\mathbf{g}\mathrm{d}V_0 - \int_{V_0} \delta\mathbf{e}^{\mathrm{T}}\boldsymbol{\tau}_j\mathrm{d}V_0$$

which is identical to Equation (3.92).

By straightforward differentiation the variation of the linear part of the strain increment can be derived:

$$\delta\mathbf{e} = \mathbf{L}\delta\mathbf{u}$$

with, for the degenerate continuum shell:

$$\mathbf{L} = \begin{bmatrix} F_{11}\frac{\partial}{\partial\xi_1} & F_{21}\frac{\partial}{\partial\xi_1} & F_{31}\frac{\partial}{\partial\xi_1} \\ F_{12}\frac{\partial}{\partial\xi_2} & F_{22}\frac{\partial}{\partial\xi_2} & F_{32}\frac{\partial}{\partial\xi_2} \\ F_{11}\frac{\partial}{\partial\xi_2} + F_{12}\frac{\partial}{\partial\xi_1} & F_{21}\frac{\partial}{\partial\xi_2} + F_{22}\frac{\partial}{\partial\xi_1} & F_{31}\frac{\partial}{\partial\xi_2} + F_{32}\frac{\partial}{\partial\xi_1} \\ F_{12}\frac{\partial}{\partial\xi_3} + F_{13}\frac{\partial}{\partial\xi_2} & F_{22}\frac{\partial}{\partial\xi_3} + F_{23}\frac{\partial}{\partial\xi_2} & F_{32}\frac{\partial}{\partial\xi_3} + F_{33}\frac{\partial}{\partial\xi_2} \\ F_{13}\frac{\partial}{\partial\xi_1} + F_{11}\frac{\partial}{\partial\xi_3} & F_{23}\frac{\partial}{\partial\xi_1} + F_{21}\frac{\partial}{\partial\xi_3} & F_{33}\frac{\partial}{\partial\xi_1} + F_{31}\frac{\partial}{\partial\xi_3} \end{bmatrix} \tag{10.43}$$

and F_{ij} the components of the deformation gradient. Substitution of Equation (10.41) yields

$$\delta\mathbf{e} = \widehat{\mathbf{B}}_L\delta\hat{\mathbf{a}} \tag{10.44}$$

with

$$\widehat{\mathbf{B}}_L = [\mathbf{B}_L, \zeta\mathbf{B}_L\mathbf{V}] \tag{10.45}$$

where

$$\mathbf{B}_L = \begin{bmatrix} F_{11}\frac{\partial h_1}{\partial \xi_1} & F_{21}\frac{\partial h_1}{\partial \xi_1} & F_{31}\frac{\partial h_1}{\partial \xi_1} & \cdots \\ F_{12}\frac{\partial h_1}{\partial \xi_2} & F_{22}\frac{\partial h_1}{\partial \xi_2} & F_{32}\frac{\partial h_1}{\partial \xi_2} & \cdots \\ F_{11}\frac{\partial h_1}{\partial \xi_2} + F_{12}\frac{\partial h_1}{\partial \xi_1} & F_{21}\frac{\partial h_1}{\partial \xi_2} + F_{22}\frac{\partial h_1}{\partial \xi_1} & F_{31}\frac{\partial h_1}{\partial \xi_2} + F_{32}\frac{\partial h_1}{\partial \xi_1} & \cdots \\ F_{12}\frac{\partial h_1}{\partial \xi_3} + F_{13}\frac{\partial h_1}{\partial \xi_2} & F_{22}\frac{\partial h_1}{\partial \xi_3} + F_{23}\frac{\partial h_1}{\partial \xi_2} & F_{32}\frac{\partial h_1}{\partial \xi_3} + F_{33}\frac{\partial h_1}{\partial \xi_2} & \cdots \\ F_{13}\frac{\partial h_1}{\partial \xi_1} + F_{11}\frac{\partial h_1}{\partial \xi_3} & F_{23}\frac{\partial h_1}{\partial \xi_1} + F_{21}\frac{\partial h_1}{\partial \xi_3} & F_{33}\frac{\partial h_1}{\partial \xi_1} + F_{31}\frac{\partial h_1}{\partial \xi_3} & \cdots \end{bmatrix} \tag{10.46}$$

which is, of course, a degenerated form of Equation (3.98). Substition of the expressions for $\delta\mathbf{e}$ and $\mathrm{d}\mathbf{e}$ into the first term on the left-hand side of the linearised virtual work equation, and into the last term on the right-hand side of this equation yields the first part of the tangential stiffness matrix, which, with the appropriate definitions for $\mathbf{B}_L$ and $\mathbf{D}$ is given by Equation (9.120), and the internal force vector by Equation (9.121), but, again with a redefinition of $\mathbf{B}_L$ and $\boldsymbol{\tau}$. Furthermore, the second term of Equation (3.92) can be rewritten as:

$$\int_{V_0} (\delta\boldsymbol{\eta})^{\mathrm{T}}\boldsymbol{\tau}^t \mathrm{d}V_0 = (\delta\hat{\mathbf{a}})^{\mathrm{T}}\mathbf{K}_{NL}\mathrm{d}\hat{\mathbf{a}}$$

where the geometric part of the tangential stiffness matrix is now given by:

$$\mathbf{K}_{NL} = \int_{V_0} \widehat{\mathbf{B}}_{NL}^{\mathrm{T}}\widehat{\mathcal{T}}_j\widehat{\mathbf{B}}_{NL}\mathrm{d}V_0 \tag{10.47}$$

The matrix form of the Second Piola–Kirchhoff stress, $\widehat{\mathcal{T}}$, is identical to the expression of Equation (3.105), except for the fact that $\tau_{zz} = 0$:

$$\widehat{\mathcal{T}} = \begin{bmatrix} \tau_{xx} & \tau_{xy} & \tau_{zx} & 0 & 0 & 0 & 0 & 0 & 0 \\ \tau_{xy} & \tau_{yy} & \tau_{yz} & 0 & 0 & 0 & 0 & 0 & 0 \\ \tau_{zx} & \tau_{yz} & 0 & 0 & 0 & 0 & 0 & 0 & 0 \\ 0 & 0 & 0 & \tau_{xx} & \tau_{xy} & \tau_{zx} & 0 & 0 & 0 \\ 0 & 0 & 0 & \tau_{xy} & \tau_{yy} & \tau_{yz} & 0 & 0 & 0 \\ 0 & 0 & 0 & \tau_{zx} & \tau_{yz} & 0 & 0 & 0 & 0 \\ 0 & 0 & 0 & 0 & 0 & 0 & \tau_{xx} & \tau_{xy} & \tau_{zx} \\ 0 & 0 & 0 & 0 & 0 & 0 & \tau_{xy} & \tau_{yy} & \tau_{yz} \\ 0 & 0 & 0 & 0 & 0 & 0 & \tau_{zx} & \tau_{yz} & 0 \end{bmatrix} \tag{10.48}$$

In consideration of Equation (10.41) the matrix $\widehat{\mathbf{B}}_{NL}$ is defined such that:

$$\widehat{\mathbf{B}}_{NL} = [\mathbf{B}_{NL}, \zeta\mathbf{B}_{NL}\mathbf{V}]$$

with $\mathbf{B}_{NL}$ given by Equation (3.106).

As with the truss, continuum and beam elements we have for the discretised weak form:

$$(\delta\hat{\mathbf{a}})^{\mathrm{T}}(\mathbf{K}_L + \mathbf{K}_{NL})\mathrm{d}\hat{\mathbf{a}} = (\delta\hat{\mathbf{a}})^{\mathrm{T}}\left(\mathbf{f}_{\mathrm{ext}}^{t+\Delta t} - \mathbf{f}_{\mathrm{int},j}\right)$$

with the external force vector defined as in Equation (9.127). The above weak form must hold for any virtual displacement increment $\delta\hat{\mathbf{a}}$, whence

$$(\mathbf{K}_L + \mathbf{K}_{NL})\mathrm{d}\hat{\mathbf{a}} = \mathbf{f}_{\mathrm{ext}}^{t+\Delta t} - \mathbf{f}_{\mathrm{int},j} \tag{10.49}$$

Although the above derivation is for a shell element that has homogeneous properties through the depth, different material behaviour, or material non-linearity, can be accommodated in a straightforward manner by adopting a layered approach, similar to the case of the degenerate continuum beam. Similarly, improvements can be made in terms of element technology, e.g. with respect to membrane locking and shear locking (Bischoff *et al.* 2004).

10.3 Solid-like Shell Elements

The degenerate continuum shell elements are widely used in linear and non-linear finite element analysis, but suffer from the drawback that the normal strain in the thickness direction is zero. In solid-like shell elements, or solid-shell elements, an additional set of internal degrees of freedom is supplied which provides a quadratic term in the displacement field in the thickness direction, which leads to a stretching in the thickness direction. Since the strain normal to the shell surface is now non-zero, a normal stress that is perpendicular to the shell surface can now meaningfully be computed via the full three-dimensional stress–strain relation. In particular for material non-linearities this is an advantage. Moreover, the conditioning of the stiffness matrix tends to be superior to that of degenerate continuum shell elements.

We consider the shell shown in Figure 10.5. The position of a material point in the shell in the undeformed configuration can be written as a function of the curvilinear coordinates (ξ, η, ζ):

$$\boldsymbol{\xi}(\xi, \eta, \zeta) = \boldsymbol{\xi}_0(\xi, \eta) + \zeta\mathbf{d}(\xi, \eta) \tag{10.50}$$

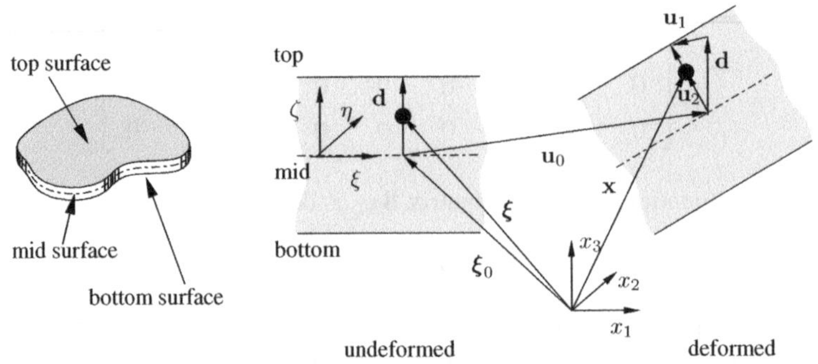

Figure 10.5 Kinematic relations of a solid-like shell element

where $\boldsymbol{\xi}_0(\xi, \eta)$ is the projection of the point on the mid-surface of the shell and $\mathbf{d}(\xi, \eta)$ is the thickness director at this point:

$$\boldsymbol{\xi}_0(\xi, \eta) = \frac{1}{2}\left(\boldsymbol{\xi}_{\text{top}}(\xi, \eta) + \boldsymbol{\xi}_{\text{bottom}}(\xi, \eta)\right) \tag{10.51}$$

$$\mathbf{d}(\xi, \eta) = \frac{1}{2}\left(\boldsymbol{\xi}_{\text{top}}(\xi, \eta) - \boldsymbol{\xi}_{\text{bottom}}(\xi, \eta)\right) \tag{10.52}$$

Note that the thickness director vector $\mathbf{d}$, in contrast to the treatment for the degenerate continuum shell, has not been normalised. The position of a material point in the deformed configuration $\mathbf{x}(\xi, \eta, \zeta)$ is related to $\boldsymbol{\xi}(\xi, \eta, \zeta)$ via the displacement field $\hat{\mathbf{u}}(\xi, \eta, \zeta)$ according to:

$$\mathbf{x}(\xi, \eta, \zeta) = \boldsymbol{\xi}(\xi, \eta, \zeta) + \hat{\mathbf{u}}(\xi, \eta, \zeta) \tag{10.53}$$

where

$$\hat{\mathbf{u}}(\xi, \eta, \zeta) = \mathbf{u}_0(\xi, \eta) + \zeta\mathbf{u}_1(\xi, \eta) + (1 - \zeta^2)\mathbf{u}_2(\xi, \eta) \tag{10.54}$$

Herein, $\mathbf{u}_0$ and $\mathbf{u}_1$ are the displacements of a material point on the shell mid-surface, $\boldsymbol{\xi}_0$, and of the thickness director vector $\mathbf{d}$:

$$\mathbf{u}_0(\xi, \eta) = \frac{1}{2}\left(\mathbf{u}_{\text{top}}(\xi, \eta) + \mathbf{u}_{\text{bottom}}(\xi, \eta)\right) \tag{10.55}$$

and

$$\mathbf{u}_1(\xi, \eta) = \frac{1}{2}\left(\mathbf{u}_{\text{top}}(\xi, \eta) - \mathbf{u}_{\text{bottom}}(\xi, \eta)\right) \tag{10.56}$$

respectively. The displacement field $\mathbf{u}_2(\xi, \eta)$ provides the stretching in the thickness direction of the element, and is co-linear with the thickness director vector in the deformed configuration. It is a function of the stretch parameter w:

$$\mathbf{u}_2(\xi, \eta) = w(\xi, \eta)\left(\mathbf{d}(\xi, \eta) + \mathbf{u}_1(\xi, \eta)\right) \tag{10.57}$$

The displacement field $\hat{\mathbf{u}}$ is thus a function of two types of variables: the ordinary displacement field $\mathbf{u}$, which is split in a displacement of the top and bottom surfaces $\mathbf{u}_{\text{top}}$ and $\mathbf{u}_{\text{bottom}}$, and the internal stretch parameter w: $\hat{\mathbf{u}} = \hat{\mathbf{u}}(\mathbf{u}_{\text{top}}, \mathbf{u}_{\text{bottom}}, w)$. The derivation of the strains is quite lengthy and the reader is referred to Parisch (1995), which also gives details on the finite element implementation, see also Remmers *et al.* (2003b).

10.4 Shell Plasticity: Ilyushin's Criterion

While a layerwise approach is the most accurate way to handle plasticity and other non-linear material models in structural elements such as beams, plates and shells, one can also directly operate on the stress resultants, i.e. the normal forces, the bending moments and the shear forces. Such an approach usually results in an overstiff structural response after the onset of non-linear behaviour, since the outer fibres are assumed to remain elastic in such an approach, while in practice they are already yielding, damaging or showing other kinds of material non-linear behaviour. The advantage is the reduction in computer time, which can be significant, since there is no integration through the depth, and the time spent in evaluating the stress–strain relation and setting up the tangent stiffness matrix in integration points, constitutes the major share of the total computing time in large-scale non-linear computations. With the increase of computer power, this advantage has decreased in importance.

One of the most important and more widely used yield criteria for shell structures that is formulated in terms of stress resultants is due to Ilyushin (1956), and departs from the von Mises yield criterion. Normally, and also herein, the so-called approximate Ilyushin yield function is employed, which is given by:

$$f = \frac{\overline{N}}{N_0^2} + \frac{\overline{M}}{M_0^2} + \frac{s}{\sqrt{3}}\frac{\overline{MN}}{M_0 N_0} - 1 \tag{10.58}$$

with

$$\begin{aligned}\overline{N} &= N_x^2 + N_y^2 - N_x N_y + 3N_{xy}^2 \\ \overline{M} &= M_x^2 + M_y^2 - M_x M_y + 3M_{xy}^2 \\ \overline{MN} &= M_x N_x + M_y N_y - \frac{1}{2}(M_x N_y + M_y N_x) + 3M_{xy} N_{xy}\end{aligned} \tag{10.59}$$

see also Figure 10.2 for the definition of the normal force components and the moment components, and

$$N_0 = \bar{\sigma} t$$

is the uniaxial yield force,

$$M_0 = \frac{1}{4}\bar{\sigma} t^2$$

is the uniaxial yield moment, and

$$s = \frac{\overline{MN}}{|\overline{MN}|} \tag{10.60}$$

The yield criterion of Equation (10.58) is an approximation to the original proposal of Ilyushin, since the latter is rather complicated to implement, while Equation (10.58) gives a yield contour that is close to the original formulation (Figure 10.6), in which both yield contours are visualised in the $\overline{N}, \overline{M}$-space (Burgoyne and Brennan 1993; Skallerud and Haugen 1999).

As with continuum plasticity, it has advantages to reformulate the Ilyushin yield criterion in the following compact format:

$$f = \sqrt{\boldsymbol{\sigma}^{\mathrm{T}} \boldsymbol{\mathcal{P}} \boldsymbol{\sigma}} - \bar{\sigma}(\kappa) \tag{10.61}$$

where the dependence of the yield strength $\bar{\sigma}$ on a history parameter κ has now been made explicit. The vector $\boldsymbol{\sigma}$ contains the generalised stresses:

$$\boldsymbol{\sigma} = \begin{pmatrix} \mathbf{N} \\ \mathbf{M} \end{pmatrix} \tag{10.62}$$

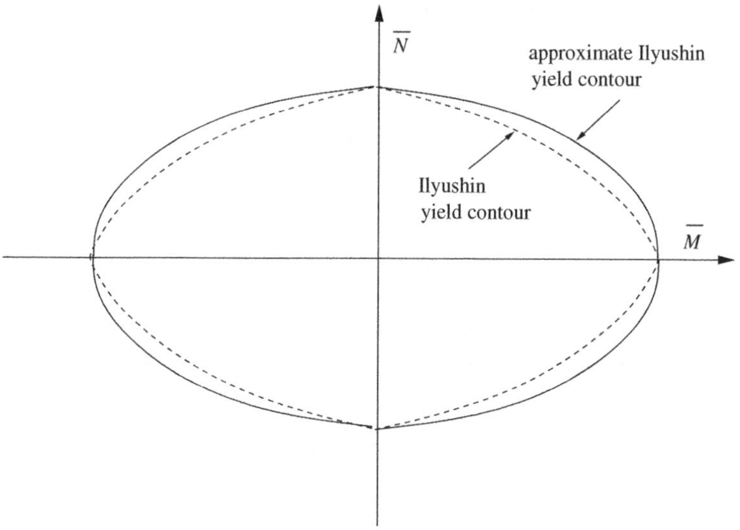

Figure 10.6 Illyushin's yield criterion

and $\mathcal{P}$ is given by:

$$\mathcal{P} = \begin{bmatrix} \frac{3}{2}t^2\bar{\mathbf{P}} & st\sqrt{3}\bar{\mathbf{P}} \\ st\sqrt{3}\bar{\mathbf{P}} & 24\bar{\mathbf{P}} \end{bmatrix} \tag{10.63}$$

where $\bar{\mathbf{P}}$ is composed of the rows and columns that correspond to the 'in-plane' entries of the projection matrix $\mathbf{P}$ introduced in Equation (7.33) for the von Mises yield criterion:

$$\bar{\mathbf{P}} = \begin{bmatrix} \frac{2}{3} & -\frac{1}{3} & 0 \\ -\frac{1}{3} & \frac{2}{3} & 0 \\ 0 & 0 & 2 \end{bmatrix} \tag{10.64}$$

We note that the form of Equation (10.61) for the Ilyushin approximate yield function differs from that given by Skallerud and Haugen (1999), since the present format tends to result in a faster convergence to satisfaction of the yield function at the end of the step, as is required in Euler backward procedures for the integration of the plasticity rate equations (de Borst and Feenstra 1990).

Considering that the Ilyushin yield criterion has been constructed on the basis of the von Mises yield criterion, which is primarily applicable to metals, and therefore usually used in conjunction with an associated flow rule, it is reasonable to now also utilise an associated flow rule:

$$\dot{\boldsymbol{\epsilon}}^{\mathrm{p}} = \dot{\lambda}\mathbf{n}$$

where the plastic strain rate vector contains generalised plastic strains that are conjugate to the generalised stresses in $\boldsymbol{\sigma}$, so that:

$$\boldsymbol{\epsilon} = \begin{pmatrix} \boldsymbol{\epsilon}_\ell \\ \boldsymbol{\chi} \end{pmatrix} \tag{10.65}$$

Elaboration then gives:

$$\dot{\boldsymbol{\epsilon}}^{\mathrm{p}} = \dot{\lambda}\frac{\boldsymbol{\mathcal{P}}\boldsymbol{\sigma}}{\bar{\sigma}(\kappa)} \tag{10.66}$$

where the identity $f = 0$ has been used, with f according to Equation (10.61). An Euler backward integration scheme follows straightforwardly along the lines laid out in Chapter 7, and is given by:

$$\boldsymbol{\sigma}_{j+1} = \mathbf{A}^{-1}\boldsymbol{\sigma}_e \tag{10.67}$$

with $\boldsymbol{\sigma}_e$ the 'elastic' trial stress, and

$$\mathbf{A} = \mathbf{I} + \frac{\Delta\lambda\boldsymbol{\mathcal{D}}^{\mathrm{e}}\boldsymbol{\mathcal{P}}}{\bar{\sigma}(\kappa(\lambda_{j+1}))} \tag{10.68}$$

where the fact has been used that the hardening parameter κ can be written as a function of λ for common hardening hypotheses, cf. Chapter 7. Given that $\boldsymbol{\sigma}$ and $\boldsymbol{\epsilon}$ contain generalised stresses and generalised strains, respectively, $\boldsymbol{\mathcal{D}}^{\mathrm{e}}$ reads:

$$\boldsymbol{\mathcal{D}}^{\mathrm{e}} = \begin{bmatrix} t\mathbf{D}^{\mathrm{e}} & \mathbf{0} \\ \mathbf{0} & \frac{t^3}{12}\mathbf{D}^{\mathrm{e}} \end{bmatrix} \tag{10.69}$$

with the submatrix $\mathbf{D}^{\mathrm{e}}$ defined as in Equation (10.15). Substitution of Equation (10.67) into the yield function, Equation (10.61), and requiring $f = 0$ yields:

$$\sqrt{\boldsymbol{\sigma}_e^{\mathrm{T}}\mathbf{A}^{-\mathrm{T}}\boldsymbol{\mathcal{P}}\mathbf{A}^{-1}\boldsymbol{\sigma}_e} - \bar{\sigma}(\kappa(\lambda_0 + \Delta\lambda)) = 0 \tag{10.70}$$

This non-linear algebraic equation in $\Delta\lambda$ can be solved using standard procedures, e.g. a local Newton–Raphson or a local Quasi-Newton method.

The consistent tangent stiffness matrix is subsequently derived by a standard procedure, cf. Chapter 7, through differentiation of the Euler backward integration scheme, and reads:

$$\boldsymbol{\mathcal{D}} = \mathbf{H} - \frac{\mathbf{H}\mathbf{n}\mathbf{n}^{\mathrm{T}}\mathbf{H}}{h + \mathbf{n}^{\mathrm{T}}\mathbf{H}\mathbf{n}} \tag{10.71}$$

cf. Equation (7.130), where for structural plasticity,

$$\mathbf{H} = \mathbf{A}^{-1}\boldsymbol{\mathcal{D}}^{\mathrm{e}} \tag{10.72}$$

with $\mathbf{A}$ and $\boldsymbol{\mathcal{D}}^{\mathrm{e}}$ defined in Equations (10.68) and (10.69), respectively.

We observe from Equation (10.66) with $\boldsymbol{\mathcal{P}}$ defined as in Equation (10.63) that the approximate Ilyushin yield contour is non-smooth in the six-dimensional space spanned by $(N_x, N_y, N_{xy}, M_x, M_y, M_{xy})$, as the gradients to the yield function f are discontinuous when s

switches sign, although it may be smooth in the $\overline{N}$, $\overline{M}$-space (Figure 10.6). At points where the gradient to the yield surface is discontinuous, the procedure must be utilised that is described in Chapter 7 for treating singularities in the yield surface, where it is noted that the Koiter's generalisation equally holds for yield functions that are expressed in terms of generalised stresses. As with the Tresca, Mohr–Coulomb, and Rankine yield functions, the singularity of the Ilyushin yield surface requires distinction between the smooth parts of the yield surface and the corner regimes. This is most conveniently done by first assuming that the stress point is on a smooth part of the yield surface and to apply the procedure for a single active yield mechanism. Subsequently, it is checked whether s has the same value for the trial stress $\boldsymbol{\sigma}_e$ and for the final stress $\boldsymbol{\sigma}_{j+1}$ at this iteration, i.e. $j+1$. If this is the case, the assumption has been correct, otherwise the return-mapping procedure has to be repeated under the assumption that two plastic mechanisms are active.

References

Ahmad S, Irons BM and Zienkiewicz OC 1970 Analysis of thick and thin shell structures by curved finite elements. *International Journal for Numerical Methods in Engineering* **2**, 419–451.

Bathe KJ 1982 *Finite Element Procedures in Engineering Analysis*. Prentice Hall, Inc.

Bathe KJ and Bolourchi S 1980 A geometric and material nonlinear plate and shell element. *Computers & Structures* **11**, 23–48.

Bischoff M, Wall WA, Bletzinger KU and Ramm E 2004 Models and finite elements for thin-walled structures, in *The Encyclopedia of Computational Mechanics* (eds Stein E, de Borst R and Hughes TJR), vol. II, pp. 59–137. John Wiley & Sons, Ltd.

Braun M, Bischoff M and Ramm E 1994 Nonlinear shell formulations for complete three-dimensional constitutive laws including composites and laminates. *Computational Mechanics* **15**, 1–18.

Büchter N and Ramm E 1992 Shell theory versus degeneration. A comparison in large rotation shell theory. *International Journal for Numerical Methods in Engineering* **34**, 39–59.

Büchter N, Ramm E and Roehl D 1994 Three-dimensional extension of nonlinear shell formulation based on the enhanced assumed strain concept. *International Journal for Numerical Methods in Engineering* **37**, 2551–2568.

Burgoyne CJ and Brennan MG 1993 Exact Ilyushin yield surface. *International Journal of Solids and Structures* **30**, 1113–1131.

Crisfield MA 1983 A four-noded thin-plate bending element using shear constraints – a modified version of Lyons' element. *Computer Methods in Applied Mechanics and Engineering* **38**, 93–120.

Crisfield MA 1984 A quadratic mindlin element using shear constraints. *Computers & Structures* **18**, 833–852.

Crisfield MA 1986 *Finite Elements and Solution Procedures for Structural Analysis: Linear Analysis*. Pineridge Press.

de Borst R 1991 The zero-normal-stress condition in plane-stress and shell elasto-plasticity. *Communications in Applied Numerical Methods* **7**, 29–33.

de Borst R and Feenstra PH 1990 Studies in anisotropic plasticity with reference. *International Journal for Numerical Methods in Engineering* **29**, 315–336.

Dvorkin EN and Bathe KJ 1984 A continuum mechanics based four-node shell element for general non-linear analysis. *Engineering Computations* **1**, 77–88.

Hauptmann R and Schweizerhof K 1998 A systematic development of 'solid-shell' element formulations for linear and non-linear analyses employing only displacement degrees of freedom. *International Journal for Numerical Methods in Engineering* **42**, 49–69.

Hughes TJR 1987 *The Finite Element Methods: Linear Static and Dynamic Finite Element Analysis*. Prentice Hall, Inc.

Hughes TJR and Liu WK 1981 Nonlinear finite element analysis of shells. Part 1: Three-dimensional shells. *Computer Methods in Applied Mechanics and Engineering* **26**, 331–362.

Ilyushin AA 1956 *Plasticité (Deformation elasto-plastiques)*. Editions Eyrolles.

Kühhorn A and Schoop H 1992 A nonlinear theory for sandwich shells including the wrinkling phenomenon. *Archive of Applied Mechanics* **62**, 413–427.

Milford RV and Schnobrich WC 1986 Degenerated isoparametric elements using explicit integration. *International Journal for Numerical Methods in Engineering* **23**, 133–154.

Parisch H 1981 Large displacements of shells including material nonlinearities. *Computer Methods in Applied Mechanics and Engineering* **27**, 183–214.

Parisch H 1995 A continuum-based shell theory for non-linear applications. *International Journal for Numerical Methods in Engineering* **38**, 1855–1883.

Remmers JJC, Wells GN and de Borst R 2003 A discontinuous solid-like shell element for arbitrary delaminations. *International Journal for Numerical Methods in Engineering* **58**, 2013–2040.

Sansour C 1995 A theory and finite element formulation of shells at finite deformations involving thickness change. *Archive of Applied Mechanics* **65**, 194–216.

Simo JC and Fox DD 1989 On a stress resultant geometrically exact shell model. Part I: Formulation and optimal representation. *Computer Methods in Applied Mechanics and Engineering* **72**, 267–304.

Simo JC and Kennedy JG 1992 On a stress resultant geometrically exact shell model. Part V: Nonlinear plasticity: formulation and integration algorithms. *Computer Methods in Applied Mechanics and Engineering* **96**, 133–171.

Simo JC, Fox DD and Rifai MS 1989 On a stress resultant geometrically exact shell model. Part II: The linear theory: computational aspects. *Computer Methods in Applied Mechanics and Engineering* **73**, 53–92.

Simo JC, Fox DD and Rifai MS 1990 On a stress resultant geometrically exact shell model. Part III: Computational aspects of the nonlinear theory. *Computer Methods in Applied Mechanics and Engineering* **79**, 21–70.

Skallerud B and Haugen B 1999 Collapse of thin shell structures – stress resultant plasticity modelling within a co-rotated ANDES finite element formulation. *International Journal for Numerical Methods in Engineering* **46**, 1961–1986.

Stander N, Matzenmiller A and Ramm E 1989 An assessment of assumed strain methods in finite element rotation shell analysis. *Engineering Computations* **6**, 58–66.

Surana KS 1983 Geometrically non-linear formulation for the curved shell elements. *International Journal for Numerical Methods in Engineering* **19**, 353–386.

Zienkiewicz OC, Taylor RL and Too JM 1971 Reduced integration techniques in general analysis of plates and shells. *International Journal for Numerical Methods in Engineering* **3**, 275–290.

Part IV
Large Strains

11

Hyperelasticity

In the first chapter we have recapitulated some basic notions of continuum mechanics, while in Chapter 3 an extension has been made to non-linear kinematics. However, in that chapter restriction was made to small strains. This limitation will be relaxed in the present chapter, in which hyperelastic material models will be discussed, which are often used to describe the mechanical behaviour of rubberlike materials. To provide a proper setting, we will commence this chapter by a presentation of additional topics from continuum mechanics, which have not been discussed before, but which are necessary for a proper discussion of large elastic strains – the topic of this chapter – and of large elasto-plastic strains (Chapter 12).

11.1 More Continuum Mechanics

11.1.1 Momentum Balance and Stress Tensors

As point of departure we take the balance of momentum in the current configuration, Equation (2.4). Since the discussion in this chapter and in Chapter 12 will be limited to quasi-static deformations, it is recalled here without the inertia term:

$$\nabla \cdot \boldsymbol{\sigma} + \rho \mathbf{g} = \mathbf{0} \tag{11.1}$$

with $\boldsymbol{\sigma}$ the Cauchy stress tensor and ρ the mass density in the current configuration. In component form Equation (11.1) reads:

$$\frac{\partial \sigma_{ij}}{\partial x_i} + \rho g_j = 0$$

The weak form of the equilibrium equation (11.1) can be obtained through multiplication by a test function, $\delta \mathbf{u}$, which can be interpreted as a virtual displacement field. After integration over the domain V currently occupied by the body, so that,

$$\int_V \delta \mathbf{u} \cdot (\nabla \cdot \boldsymbol{\sigma} + \rho \mathbf{g}) \mathrm{d}V = 0 \tag{11.2}$$

Non-linear Finite Element Analysis of Solids and Structures, Second Edition.
René de Borst, Mike A. Crisfield, Joris J.C. Remmers and Clemens V. Verhoosel.

and utilising the divergence theorem, one obtains:

$$\int_V \nabla(\delta\mathbf{u}) : \boldsymbol{\sigma}\mathrm{d}V = \int_V \delta\mathbf{u} \cdot \rho\mathbf{g}\mathrm{d}V + \int_S \delta\mathbf{u} \cdot \mathbf{t}\mathrm{d}S \tag{11.3}$$

with $\mathbf{t} = \mathbf{n} \cdot \boldsymbol{\sigma}$ the stress vector in the current configuration, cf. Equation (1.75).

From Equation (11.3) we infer that the Cauchy stress tensor and the gradient of the virtual displacement field are energetically conjugate. This is a property that is shared by several pairs of stress tensors and deformation measures. To show this, we will transform Equation (11.3) to the undeformed configuration, denoted by the subscript 0. The transformation of an elementary volume $\mathrm{d}V$ to that in the undeformed configuration, $\mathrm{d}V_0$, is straightforward using conservation of mass, Equation (3.74). Such a transformation is less easy for the transformation of an elementary surface $\mathrm{d}S_0$. Denoting the normal vectors of the elementary surfaces $\mathrm{d}S_0$ and $\mathrm{d}S$ by $\mathbf{n}_0$ and $\mathbf{n}$, respectively, and defining an arbitrary vector $\mathrm{d}\boldsymbol{\ell}_0$, that transforms into $\mathrm{d}\boldsymbol{\ell}$ and is not orthogonal to $\mathbf{n}_0$, then there exists an elementary volume $\mathrm{d}V_0 = \mathrm{d}\boldsymbol{\ell}_0 \cdot \mathbf{n}_0\mathrm{d}S_0$ which transforms into $\mathrm{d}V = \mathrm{d}\boldsymbol{\ell} \cdot \mathbf{n}\mathrm{d}S$. Since $\mathrm{d}\boldsymbol{\ell} = \mathbf{F} \cdot \mathrm{d}\boldsymbol{\ell}_0$ and $\rho\mathrm{d}V = \rho_0\mathrm{d}V_0$, cf. Equation (3.74), one obtains:

$$\rho\mathbf{n} \cdot \mathbf{F} \cdot \mathrm{d}\boldsymbol{\ell}_0\mathrm{d}S = \rho_0\mathbf{n}_0 \cdot \mathrm{d}\boldsymbol{\ell}_0\mathrm{d}S_0$$

This identity must hold for arbitrary $\mathrm{d}\boldsymbol{\ell}_0$, which results in Nanson's formula for the transformation of surface elements:

$$\mathbf{n}\mathrm{d}S = \frac{\rho_0}{\rho}\mathbf{n}_0 \cdot \mathbf{F}^{-1}\mathrm{d}S_0 \tag{11.4}$$

or in component form:

$$n_i\mathrm{d}S = \frac{\rho_0}{\rho}(n_0)_j(F^{-1})_{ji}\mathrm{d}S_0$$

Considering Equation (3.76) we can also write Equation (11.4) as:

$$\mathbf{n}\mathrm{d}S = \det\mathbf{F}\ \mathbf{n}_0 \cdot \mathbf{F}^{-1}\mathrm{d}S_0 \tag{11.5}$$

Using Equations (1.75), (3.74) and (11.4) we can rewrite Equation (11.3) as:

$$\int_{V_0} \frac{\rho_0}{\rho}\nabla(\delta\mathbf{u}) : \boldsymbol{\sigma}\mathrm{d}V = \int_{V_0} \delta\mathbf{u} \cdot \rho_0\mathbf{g}\mathrm{d}V + \int_{S_0} \delta\mathbf{u} \cdot \left(\frac{\rho_0}{\rho}\mathbf{n}_0 \cdot \mathbf{F}^{-1} \cdot \boldsymbol{\sigma}\right)\mathrm{d}S \tag{11.6}$$

From the surface integral we can define the stress tensor $\mathbf{p}$ that relates the nominal traction $\mathbf{t}_0$, which is the force per surface area in the undeformed configuration, to the normal $\mathbf{n}_0$ of the surface in the undeformed configuration:

$$\mathbf{t}_0 = \mathbf{n}_0 \cdot \mathbf{p} \tag{11.7}$$

where the nominal stress tensor $\mathbf{p}$ relates to the Cauchy stress tensor $\boldsymbol{\sigma}$ through:

$$\mathbf{p} = \frac{\rho_0}{\rho}\mathbf{F}^{-1} \cdot \boldsymbol{\sigma} \tag{11.8}$$

or inversely:

$$\boldsymbol{\sigma} = \frac{\rho}{\rho_0}\mathbf{F} \cdot \mathbf{p} \tag{11.9}$$

The components of the nominal stress tensor can hence be interpreted as the stresses that result from a force that acts on a surface in the undeformed configuration. The nominal stress tensor is the transpose of the First Piola–Kirchhoff stress tensor, $\mathbf{p}^*$, defined in Box 3.3, which can be shown as follows (Ogden 1984):

$$\mathbf{p}^{\mathrm{T}} = \frac{\rho_0}{\rho}\left(\mathbf{F}^{-1}\cdot\boldsymbol{\sigma}\right)^{\mathrm{T}} = \frac{\rho_0}{\rho}\boldsymbol{\sigma}^{\mathrm{T}}\cdot(\mathbf{F}^{-1})^{\mathrm{T}} = \frac{\rho_0}{\rho}\boldsymbol{\sigma}\cdot(\mathbf{F}^{-1})^{\mathrm{T}} = \mathbf{p}^*$$

Using Equation (11.8) the left-hand side of the weak form of the equilibrium equation (11.6) can be recast as:

$$\int_{V_0}\frac{\rho_0}{\rho}\nabla(\delta\mathbf{u}):\boldsymbol{\sigma}\mathrm{d}V = \int_{V_0}(\nabla(\delta\mathbf{u})\cdot\mathbf{F}):\mathbf{p}\mathrm{d}V = \int_{V_0}\nabla_0(\delta\mathbf{u}):\mathbf{p}\mathrm{d}V = \int_{V_0}\delta\mathbf{F}:\mathbf{p}\mathrm{d}V \quad (11.10)$$

so that the nominal stress tensor is energetically conjugate to the variation of the deformation gradient. In deriving Equation (11.10) use has been made of the identity:

$$\nabla_0\mathbf{u} = \nabla\mathbf{u}\cdot\mathbf{F}$$

Using Equations (11.7) and (11.8) we subsequently can write the weak form of the equilibrium equation in the undeformed configuration as:

$$\int_{V_0}\nabla_0(\delta\mathbf{u}):\mathbf{p}\mathrm{d}V = \int_{V_0}\delta\mathbf{u}\cdot\rho_0\mathbf{g}\mathrm{d}V + \int_{S_0}\delta\mathbf{u}\cdot\mathbf{t}_0\mathrm{d}S \quad (11.11)$$

This identity must hold for any variation of the displacement field $\delta\mathbf{u}$. Use of the divergence theorem then yields the equilibrium equation in the original, undeformed configuration:

$$\nabla_0\cdot\mathbf{p} + \rho_0\mathbf{g} = \mathbf{0} \quad (11.12)$$

A disadvantage of the nominal stress tensor is its asymmetry, which makes it less suitable for computations. For this reason the Second Piola–Kirchhoff stress tensor is more frequently used in finite element analysis, in particular when using a Lagrangian framework. We recall the definition of the Second Piola–Kirchhoff stress tensor, Equation (3.73):

$$\boldsymbol{\sigma} = \frac{\rho}{\rho_0}\mathbf{F}\cdot\boldsymbol{\tau}\cdot\mathbf{F}^{\mathrm{T}}$$

or in its inverse form:

$$\boldsymbol{\tau} = \frac{\rho_0}{\rho}\mathbf{F}^{-1}\cdot\boldsymbol{\sigma}\cdot(\mathbf{F}^{-1})^{\mathrm{T}} \quad (11.13)$$

Using this identity, the left-hand side of the equilibrium equation in the reference configuration, Equation (11.6) can be recast as:

$$\int_{V_0}\frac{\rho_0}{\rho}\nabla(\delta\mathbf{u}):\boldsymbol{\sigma}\mathrm{d}V = \int_{V_0}\mathrm{tr}(\nabla(\delta\mathbf{u})\cdot\mathbf{F}\cdot\boldsymbol{\tau}\cdot\mathbf{F}^{\mathrm{T}})\mathrm{d}V = \int_{V_0}\mathrm{tr}(\delta\mathbf{F}\cdot\boldsymbol{\tau}\cdot\mathbf{F}^{\mathrm{T}})\mathrm{d}V$$

and further, using the expression for the variation of the Green–Lagrange strain tensor $\delta\boldsymbol{\gamma}$, cf. Equation (3.79), and the symmetry of the Second Piola–Kirchhoff stress tensor,

$$\int_{V_0}\frac{\rho_0}{\rho}\nabla(\delta\mathbf{u}):\boldsymbol{\sigma}\mathrm{d}V = \int_{V_0}\delta\boldsymbol{\gamma}:\boldsymbol{\tau}\mathrm{d}V$$

Clearly, the Second Piola–Kirchhoff stress tensor is energetically conjugate to the variation of the Green–Lagrange strain tensor. The weak form of the equilibrium equation in the reference configuration expressed with aid of the Second Piola–Kirchhoff stress tensor thus reads:

$$\int_{V_0} \delta\boldsymbol{\gamma} : \boldsymbol{\tau} \mathrm{d}V = \int_{V_0} \delta\mathbf{u} \cdot \rho_0 \mathbf{g} \mathrm{d}V_0 + \int_{S_0} \delta\mathbf{u} \cdot \mathbf{t}_0 \mathrm{d}S \tag{11.14}$$

As shown in Chapter 3, the Second Piola–Kirchhoff stress tensor is a suitable point of departure for finite element formulations.

A further stress measure that is encountered in the literature is the Kirchhoff stress tensor, defined as

$$\boldsymbol{\kappa} = \frac{\rho_0}{\rho}\boldsymbol{\sigma} = \det\mathbf{F}\,\boldsymbol{\sigma} \tag{11.15}$$

which is, in view of Equation (11.6), energetically conjugate to the gradient of the deformation rate with respect to the initial volume. Finally, we mention the Biot stress tensor $\boldsymbol{T}$ which is energetically conjugate to the variation of the right stretch tensor $\mathbf{U}$ with respect to the initial volume (Bonet and Wood 1997). This can be seen by departing from the last identity of Equation (11.10). Inserting the polar decomposition, Equation (3.58) gives:

$$\int_{V_0} \delta\mathbf{F} : \mathbf{p} \mathrm{d}V = \int_{V_0} \mathbf{p} : (\mathbf{R} \cdot \delta\mathbf{U} + \delta\mathbf{R} \cdot \mathbf{U}) \mathrm{d}V = \int_{V_0} (\mathbf{p} \cdot \mathbf{R}) : (\delta\mathbf{U} + \mathbf{R}^{\mathrm{T}} \cdot \delta\mathbf{R} \cdot \mathbf{U}) \mathrm{d}V$$

Noting that by taking the variation of the identity $\mathbf{R}^{\mathrm{T}} \cdot \mathbf{R} = \mathbf{I}$ it can be shown that the tensor $\mathbf{R}^{\mathrm{T}} \cdot \delta\mathbf{R}$ is antisymmetric, whereas the right stretch tensor $\mathbf{U}$ is symmetric, we observe that the second term cancels, so that:

$$\int_{V_0} \delta\mathbf{F} : \mathbf{p} \mathrm{d}V = \int_{V_0} \delta\mathbf{U} : (\mathbf{p} \cdot \mathbf{R}) \mathrm{d}V \tag{11.16}$$

from which the Biot stress tensor

$$\boldsymbol{T} = \mathbf{p} \cdot \mathbf{R} \tag{11.17}$$

can be identified as the work conjugate to the variation of the right stretch tensor $\mathbf{U}$. In general, the Biot stress tensor is unsymmetric, like the nominal stress tensor. Using Equations (11.8) and (11.13), and using the identity $\mathbf{R}^{\mathrm{T}} \cdot \mathbf{R} = \mathbf{I}$, the Biot stress can also be written as:

$$\boldsymbol{T} = \boldsymbol{\tau} \cdot \mathbf{U} \tag{11.18}$$

For isotropic material behaviour the Second Piola–Kirchhoff stress tensor $\boldsymbol{\tau}$ and the right stretch tensor $\mathbf{U}$ are coaxial, and therefore, commute. Consequently, under these circumstances, their product $\boldsymbol{T}$ is symmetric. The Biot stress tensor will be used later in this chapter within the context of corotational formulations for continuum elements subjected to large strains.

11.1.2 Objective Stress Rates

The mechanical behaviour of rubbers, which will be the focus of the next section, is characterised by a unique relation between the stress and the strain state: upon unloading the stress response is virtually identical to that in loading. In other words, rubbers show almost no history dependence. Indeed, for most practical applications rubbers can be considered as instantly and fully recoverable upon unloading. Such a behaviour can be captured with hyperelastic models, where the stress at any state can be derived directly by differentiating a strain energy function with respect to a suitable strain measure.

Most materials, however, show a history dependence, and in Chapters 6–8 we have discussed a host of damage, plasticity and time-dependent models that can accommodate this behaviour. A common and essential denominator of these models is that the stress cannot be derived solely from the current strain state, but is also dependent on one or more history variables, e.g. the plastic strain tensor. This can be accommodated by casting the constitutive relation in a rate format, i.e. a linear relation is postulated between the strain rate tensor and the stress rate tensor, and this relation is integrated along the loading path to obtain the current stress state. When formulating such relations for materials that undergo large deformations, as we will do in Chapter 12 for elasto-plasticity, the issue arises whether the stress rate that is used in the constitutive relation is objective.

Any constitutive relation must satisfy the principle of objectivity, i.e. the mechanical response of a system must be the same irrespective of the frame of reference that is being used. We consider two coordinate systems, which for the sake of simplicity, are both Cartesian, but one is fixed in space, while the other coordinate system is rotating. A vector **n** will then have components n_i in the fixed x, y, z-coordinate frame, and components $\bar{n}_i$ in the $\bar{x}$, $\bar{y}$, $\bar{z}$-coordinate system. According to Equation (1.50), the components of the vector represented in the rotating coordinate system can be obtained from those in the fixed coordinate system by:

$$\bar{\mathbf{n}} = \mathbf{R} \cdot \mathbf{n}$$

with **R** the rotation tensor. Because **R** is an orthogonal tensor, $\mathbf{R}^{-1} = \mathbf{R}^{\mathrm{T}}$ and the inverse relation becomes:

$$\mathbf{n} = \mathbf{R}^{\mathrm{T}} \cdot \bar{\mathbf{n}}$$

Differentiation of Equation (1.50) with respect to time gives:

$$\bar{\mathbf{n}} = \mathbf{R} \cdot \mathbf{n} + \mathbf{R} \cdot \mathbf{n} \tag{11.19}$$

Back-transformation to the fixed reference frame through pre-multiplication with $\mathbf{R}^{\mathrm{T}}$, using the property of orthogonal tensors and defining:

$$\boldsymbol{\Omega} = \dot{\mathbf{R}}^{\mathrm{T}} \cdot \mathbf{R} = -\mathbf{R}^{\mathrm{T}} \cdot \dot{\mathbf{R}} \tag{11.20}$$

then results in the following, objective derivative of a vector:

$$\overset{\diamond}{\mathbf{n}} = \mathbf{R}^{\mathrm{T}} \cdot \bar{\mathbf{n}} = \dot{\mathbf{n}} - \boldsymbol{\Omega} \cdot \mathbf{n} \tag{11.21}$$

which represents the temporal change of the vector **n** in the rotating coordinate system, but with its components expressed in the fixed reference frame.

The concept of an objective derivative straightforwardly generalises to second and higher-order tensors. Differentiating Equation (1.55),

$$\bar{\mathbf{C}} = \mathbf{R} \cdot \mathbf{C} \cdot \mathbf{R}^{\mathrm{T}}$$

with $\mathbf{C}$ an arbitrary second-order tensor with respect to time yields:

$$\bar{\mathbf{C}} = \mathbf{R} \cdot \mathbf{C} \cdot \mathbf{R}^{\mathrm{T}} + \mathbf{R} \cdot \mathbf{C} \cdot \mathbf{R}^{\mathrm{T}} + \mathbf{R} \cdot \mathbf{C} \cdot \mathbf{R}^{\mathrm{T}} \tag{11.22}$$

Back-transformation of $\bar{\mathbf{C}}$ to the fixed reference frame, i.e. pre-multiplying this expression by $\mathbf{R}^{\mathrm{T}}$ and post-multiplication by $\mathbf{R}$, exploiting the property of orthogonal tensors and using definition (11.20) results in:

$$\overset{\diamond}{\mathbf{C}} = \mathbf{R}^{\mathrm{T}} \cdot \bar{\mathbf{C}} \cdot \mathbf{R} = \mathbf{C} - \boldsymbol{\Omega} \cdot \mathbf{C} + \mathbf{C} \cdot \boldsymbol{\Omega} \tag{11.23}$$

which defines the Green–Naghdi rate of a second-order tensor. Accordingly, the Green–Naghdi derivative of the Cauchy stress tensor reads:

$$\overset{\diamond}{\boldsymbol{\sigma}} = \mathbf{R}^{\mathrm{T}} \cdot \bar{\boldsymbol{\sigma}} \cdot \mathbf{R} = \boldsymbol{\sigma} - \boldsymbol{\Omega} \cdot \boldsymbol{\sigma} + \boldsymbol{\sigma} \cdot \boldsymbol{\Omega} \tag{11.24}$$

The material derivative of the Cauchy stress tensor, $\dot{\boldsymbol{\sigma}}$, is therefore not an objective stress rate, and cannot be used directly in constitutive relations. This also holds for the Kirchhoff stress $\boldsymbol{\kappa}$ for which the Green–Naghdi derivative reads in a similar manner:

$$\overset{\diamond}{\boldsymbol{\kappa}} = \mathbf{R}^{\mathrm{T}} \cdot \bar{\boldsymbol{\kappa}} \cdot \mathbf{R} = \boldsymbol{\kappa} - \boldsymbol{\Omega} \cdot \boldsymbol{\kappa} + \boldsymbol{\kappa} \cdot \boldsymbol{\Omega} \tag{11.25}$$

Another commonly used objective derivative is that proposed by Jaumann (1911). For its derivation we first define the velocity gradient $\boldsymbol{\ell}$ as:

$$\boldsymbol{\ell} = \nabla \mathbf{x} \tag{11.26}$$

For the time derivative of the deformation gradient we have:

$$\mathbf{F} = \nabla \mathbf{x} \cdot \mathbf{F}$$

so that the velocity gradient can alternatively be expressed as:

$$\boldsymbol{\ell} = \mathbf{F} \cdot \mathbf{F}^{-1} \tag{11.27}$$

or, using the time derivative of the identity $\mathbf{F} \cdot \mathbf{F}^{-1} = \mathbf{I}$,

$$\boldsymbol{\ell} = -\mathbf{F} \cdot \dot{\mathbf{F}}^{-1} \tag{11.28}$$

The velocity gradient can be decomposed into the symmetric rate of deformation tensor, or stretching tensor $\boldsymbol{\epsilon}$ and the antisymmetric spin tensor $\boldsymbol{w}$:

$$\boldsymbol{\ell} = \boldsymbol{\epsilon} + \boldsymbol{w} \tag{11.29}$$

where, by definition:

$$\boldsymbol{\epsilon} = \frac{1}{2}(\boldsymbol{\ell} + \boldsymbol{\ell}^{\mathrm{T}}) \tag{11.30}$$

and

$$\boldsymbol{w} = \frac{1}{2}(\boldsymbol{\ell} - \boldsymbol{\ell}^{\mathrm{T}}) \tag{11.31}$$

Accordingly, the spin tensor can be elaborated as:

$$w = \frac{1}{2}\left(\mathbf{F}\mathbf{F}^{-1} - (\mathbf{F}\mathbf{F}^{-1})^{\mathrm{T}}\right) \tag{11.32}$$

Substitution of the polar decomposition of $\mathbf{F}$, Equation (3.58), using the time derivative of the identity $\mathbf{R}^{\mathrm{T}} \cdot \mathbf{R} = \mathbf{I}$, and definition (11.20) finally yields:

$$w = \mathbf{\Omega} + \frac{1}{2}\mathbf{R} \cdot \left(\mathbf{U} \cdot \mathbf{U}^{-1} - \mathbf{U}^{-1} \cdot \mathbf{U}\right) \cdot \mathbf{R}^{\mathrm{T}} \tag{11.33}$$

Evidently, the second term vanishes for rigid body motions. Since the second term in Equation (11.33) is objective, it can be used in the definition of objective stress rates. Hence, $\mathbf{\Omega}$ can be replaced by w, which results in the Jaumann rate of the Cauchy stress:

$$\overset{\circ}{\sigma} = \sigma - w \cdot \sigma + \sigma \cdot w \tag{11.34}$$

and the Jaumann rate of the Kirchhoff stress:

$$\overset{\circ}{\kappa} = \kappa - w \cdot \kappa + \kappa \cdot w \tag{11.35}$$

In contrast to the Cauchy and Kirchhoff stress tensors, the Second Piola–Kirchhoff stress tensor is intrinsically independent of rigid body rotations. Indeed, time differentiation of Equation (11.13) yields:

$$\begin{aligned} \tau = {} & \frac{\rho_0}{\rho}\dot{\mathbf{F}}^{-1} \cdot \sigma \cdot (\mathbf{F}^{-1})^{\mathrm{T}} + \frac{\rho_0}{\rho}\mathbf{F}^{-1} \cdot \dot{\sigma} \cdot (\mathbf{F}^{-1})^{\mathrm{T}} \\ & + \frac{\rho_0}{\rho}\mathbf{F}^{-1} \cdot \sigma \cdot (\dot{\mathbf{F}}^{-1})^{\mathrm{T}} + \frac{\dot{\overline{\rho_0}}}{\rho}\mathbf{F}^{-1} \cdot \sigma \cdot (\mathbf{F}^{-1})^{\mathrm{T}} \end{aligned}$$

Back-transformation then yields the Truesdell rate of the Cauchy stress tensor:

$$\overset{\star}{\sigma} = \frac{\rho}{\rho_0}\mathbf{F} \cdot \tau \cdot \mathbf{F}^{\mathrm{T}} = \sigma - \boldsymbol{\ell} \cdot \sigma - \sigma \cdot \boldsymbol{\ell}^{\mathrm{T}} + (\mathrm{tr}\boldsymbol{\ell})\sigma \tag{11.36}$$

since $\mathrm{tr}\boldsymbol{\ell} = \mathrm{d}\dot{V}/\mathrm{d}V_0 = \rho/\dot{\rho}$. The Truesdell stress rate is objective as it can be shown straightforwardly that $\overset{\star}{\bar{\sigma}} = \mathbf{R} \cdot \overset{\star}{\sigma} \cdot \mathbf{R}^{\mathrm{T}}$ (Bonet and Wood 1997). In a similar manner the Truesdell rate of the Kirchhoff stress can be derived:

$$\overset{\star}{\kappa} = \mathbf{F} \cdot \tau \cdot \mathbf{F}^{\mathrm{T}} = \kappa - \boldsymbol{\ell} \cdot \kappa - \kappa \cdot \boldsymbol{\ell}^{\mathrm{T}} \tag{11.37}$$

Even though all the stress rates discussed in the preceding are objective, they give different responses. This can be shown for instance when a block of material is subjected to pure shear. To bring out the differences most clearly the material is assumed to be elastic. More specifically, a hypoelastic relation is assumed, which implies that there is a linear relation between an objective stress rate, say $\overset{\circ}{\sigma}$, and the rate of deformation tensor ϵ:

$$\overset{\circ}{\sigma} = \mathbf{D}^{\mathrm{e}} : \epsilon \tag{11.38}$$

with $\mathbf{D}^{\mathrm{e}}$ the standard elastic stiffness tensor. Integrating this equation, and using the Green–Naghdi rate, Equation (11.24), the Jaumann rate, Equation (11.34) and the Truesdell rate,

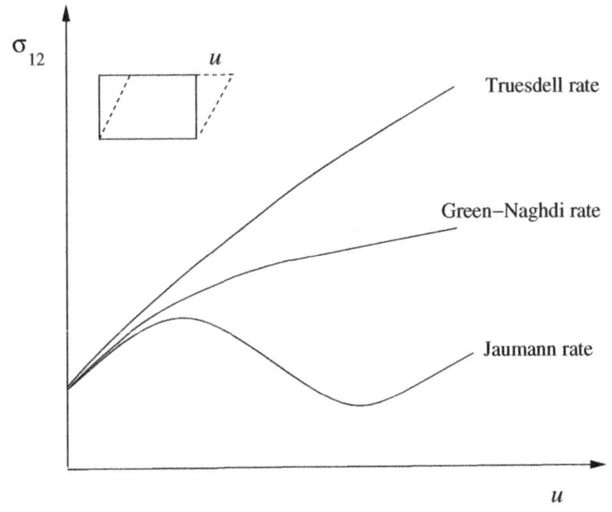

Figure 11.1 Shear stress response of a block of hypoelastic material for various stress rate definitions. *Source:* en.wikiversity.org

Equation (11.36), of the Cauchy stress tensor, yields the shear stress response that is shown in Figure 11.1. The oscillatory, non-physical behaviour of the Jaumann stress rate has been the subject of an intensive debate in the literature (Dienes 1979, 1986; Johnson and Bamman 1984; Nagtegaal and de Jong 1982). It can be remedied, for instance, by reverting to the Green–Naghdi or Truesdell rates (Figure 11.1) but certain objections apply to all hypoelastic formulations when very large strains and rotations occur. For instance, Simo and Pister (1984) have pointed out that for other stress rates energy dissipation can occur during a closed cycle when using a purely elastic constitutive relation, which is at variance with the very notion of elasticity. Although these effects are usually less important in large-scale analyses, which only seldom show such large strains that these anomalous effects become significant, there has been a trend to move away from hypoelastic relations and to use hyperelastic relations instead. Indeed, hyperelasticity will be adopted in the next section, where a number of strain energy functions for modelling rubberlike behaviour will be discussed. But also in large-strain elasto-plasticity, which will be the subject of Chapter 12, there has been a tendency to adopt hyperelasticity for the elastic part of the deformation, and to use a multiplicative decomposition of the deformation gradient into an elastic contribution and a plastic contribution, and thus bypassing the need to use stress rates, rather than using constitutive equations that are framed within a hypoelastic concept a priori.

11.1.3 Principal Stretches and Invariants

Each of the stress tensors introduced in the preceding section can be decomposed into volumetric and deviatoric components, cf. Chapter 1, and invariants and principal stresses can be computed. The principal values C_i of the right Cauchy–Green deformation tensor $\mathbf{C}$ can be computed from:

$$C_i^3 - I_1^C C_i^2 + I_2^C C_i - I_3^C = 0 \tag{11.39}$$

with I_i^C the invariants of the right Cauchy–Green deformation tensor,

$$\begin{aligned} I_1^C &= C_{xx} + C_{yy} + C_{zz} \\ I_2^C &= C_{xx}C_{yy} + C_{yy}C_{zz} + C_{zz}C_{xx} + C_{xy}^2 + C_{yz}^2 + C_{zx}^2 \\ I_3^C &= C_{xx}C_{yy}C_{zz} + 2C_{xy}C_{yz}C_{zx} - C_{xx}C_{yz}^2 - C_{yy}C_{zx}^2 - C_{zz}C_{xy}^2 \end{aligned} \tag{11.40}$$

In view of the definition of the right Cauchy–Green deformation tensor, Equation (3.60), the principal values C_i are the squares of the principal stretches λ_i, which, in turn, are defined as the quotient of the length of an elementary cube in the deformed state ℓ_i and that in the undeformed state ℓ_{i0}:

$$C_i = \lambda_i^2 = \left(\frac{\ell_i}{\ell_{i0}}\right)^2 \tag{11.41}$$

Similar to the small-strain tensor $\boldsymbol{\epsilon}$, the right Cauchy–Green deformation tensor can be decomposed into a volumetric and a deviatoric contribution. For this purpose, it is first noted that the volumetric deformation is fully characterised by the third invariant I_3^C, which, in the principal directions, reduces to:

$$I_3^C = C_1 C_2 C_3 = (\lambda_1 \lambda_2 \lambda_3)^2 \tag{11.42}$$

Since for an elementary volume, $\lambda_1\lambda_2\lambda_3 = \mathrm{d}V/\mathrm{d}V_0$, I_3^C sets the ratio between the volume in the deformed configuration $\mathrm{d}V$ and the volume in the undeformed configuration $\mathrm{d}V_0$:

$$I_3^C = \left(\frac{\mathrm{d}V}{\mathrm{d}V_0}\right)^2 \tag{11.43}$$

or, using the deformation gradient $\mathbf{F}$,

$$I_3^C = (\det \mathbf{F})^2 \tag{11.44}$$

In contrast to I_3^C, the invariants I_1^C and I_2^C depend on the volumetric as well as on the deviatoric deformation. In order to be able to define strain energy functions that are separable into volumetric and deviatoric parts, we decompose the deformation gradient in a multiplicative sense:

$$\mathbf{F} = \mathbf{F}_{\mathrm{vol}} \cdot \tilde{\mathbf{F}} = \tilde{\mathbf{F}} \cdot \mathbf{F}_{\mathrm{vol}} \tag{11.45}$$

where

$$\mathbf{F}_{\mathrm{vol}} = (\det \mathbf{F})^{\frac{1}{3}} \mathbf{I} \tag{11.46}$$

describes the purely volumetric deformation and $\tilde{\mathbf{F}}$ captures the isochoric or volume-preserving deformation (Simo *et al.* 1985; Moran *et al.* 1990), as can be shown by the simple consideration

$$\det \mathbf{F}_{\mathrm{vol}} = \left((\det \mathbf{F})^{\frac{1}{3}}\right)^3 \times \det \mathbf{I} = \det \mathbf{F}$$

and therefore,

$$\det \tilde{\mathbf{F}} = (\det \mathbf{F}_{\text{vol}})^{-1} \times \det \mathbf{F} = (\det \mathbf{F})^{-1} \times \det \mathbf{F} = 1 \tag{11.47}$$

The modified right Cauchy–Green deformation tensor can subsequently be defined as:

$$\tilde{\mathbf{C}} = \tilde{\mathbf{F}}^{\mathrm{T}} \tilde{\mathbf{F}} \tag{11.48}$$

which, in view of Equation (11.47), has a purely deviatoric character. From Equations (11.44), (11.45) and (11.46) we derive that it is related to the right Cauchy–Green deformation tensor via:

$$\tilde{\mathbf{C}} = \left(I_3^C\right)^{-\frac{1}{3}} \mathbf{C} \tag{11.49}$$

and the following set of invariants can be derived for the modified right Cauchy–Green deformation tensor (Penn 1970):

$$\begin{aligned} J_1^C &= I_1^C \left(I_3^C\right)^{-\frac{1}{3}} \\ J_2^C &= I_2^C \left(I_3^C\right)^{-\frac{2}{3}} \\ J_3^C &= \left(I_3^C\right)^{\frac{1}{2}} \end{aligned} \tag{11.50}$$

These invariants are normally used for the definition of strain energy functions of slightly compressible rubberlike materials. In a similar manner the principal values of the modified right Cauchy–Green deformation tensor can be derived as:

$$\tilde{C}_i = C_i \left(I_3^C\right)^{-\frac{1}{3}} \tag{11.51}$$

which satisfy the characteristic equation

$$\tilde{C}_i^3 - J_1^C \tilde{C}_i^2 + J_2^C \tilde{C}_i - 1 = 0 \tag{11.52}$$

and the modified principal stretches become:

$$\tilde{\lambda}_i = \lambda_i \left(I_3^C\right)^{-\frac{1}{6}} \tag{11.53}$$

11.2 Strain Energy Functions

The most notable mechanical properties of rubbers are their ability to undergo very large deformations, up to several hundred percent, without tearing and with almost instantaneously recoverable strains. For these reasons, rubbers are sometimes termed 'ideally elastic' materials. An ideally elastic material is defined by a unique relation between stress and strain, the stress being dependent only on the current strain state and not on the deformation history as is the case in damage or plasticity. Properties like a unique relation between stress and strain and no energy dissipation in a closed cycle of application and removal of stress, can be ensured by requiring the strain energy density e to be a function of the strain tensor $\boldsymbol{\gamma}$ only, which characterises the class of hyperelastic materials. In consideration of the relation between the Green–Lagrange strain tensor and the right Cauchy–Green deformation tensor, Equation (3.61), we can equivalently require e to be a single-valued function of the right Cauchy–Green deformation tensor $\mathbf{C}$:

$$e = e(\mathbf{C}) \tag{11.54}$$

In a purely mechanical theory, i.e. without the consideration of thermal effects, an equilibrium state is characterised by the vanishing of the first variation of the difference of the total deformation energy $\mathcal{E}$ of the body and the potential energy $\mathcal{U}$ of the loads:

$$\delta(\mathcal{E} - \mathcal{U}) = 0 \tag{11.55}$$

In the current configuration $\mathcal{E}$ is given by:

$$\mathcal{E} = \int_V \rho e(\mathbf{C}) \mathrm{d}V \tag{11.56}$$

while for the potential energy $\mathcal{U}$ we have in a standard manner:

$$\mathcal{U} = \int_V \rho \mathbf{x} \cdot \mathbf{g} \mathrm{d}V + \int_S \mathbf{x} \cdot \mathbf{t} \mathrm{d}S \tag{11.57}$$

Since e is a function of the right Cauchy–Green deformation tensor only, the variation of the strain energy density is given by:

$$\delta e = 2\mathrm{tr}\left(\delta \mathbf{F} \cdot \frac{\partial e}{\partial \mathbf{C}} \cdot \mathbf{F}^{\mathrm{T}}\right) \tag{11.58}$$

so that, assuming that the loads are conservative:

$$\int_V 2\rho \mathrm{tr}\left(\delta \mathbf{F} \cdot \frac{\partial e}{\partial \mathbf{C}} \cdot \mathbf{F}^{\mathrm{T}}\right) = \int_V \rho \delta \mathbf{x} \cdot \mathbf{g} \mathrm{d}V + \int_S \delta \mathbf{x} \cdot \mathbf{t} \mathrm{d}S \tag{11.59}$$

Invoking the divergence theorem for the first integral,

$$\begin{aligned} \int_V \mathrm{tr}\left(\delta \mathbf{F} \cdot \frac{\partial e}{\partial \mathbf{C}} \cdot \mathbf{F}^{\mathrm{T}}\right) \mathrm{d}V = \\ \int_S \delta \mathbf{x} \cdot \left(\mathbf{F} \cdot \frac{\partial e}{\partial \mathbf{C}} \cdot \mathbf{F}^{\mathrm{T}}\right) \cdot \mathbf{n} \mathrm{d}S - \int_V \delta \mathbf{x} \cdot \nabla \cdot \left(\mathbf{F} \cdot \frac{\partial e}{\partial \mathbf{C}} \cdot \mathbf{F}^{\mathrm{T}}\right) \mathrm{d}V \end{aligned} \tag{11.60}$$

substitution of this result into Equation (11.59), and noting that the result must hold for all admissible $\delta \mathbf{x}$, then yields

$$\nabla \cdot \left(2\rho \mathbf{F} \cdot \frac{\partial e}{\partial \mathbf{C}} \cdot \mathbf{F}^{\mathrm{T}}\right) + \rho \mathbf{g} = \mathbf{0} \tag{11.61}$$

in each material point within the body. A comparison with the equilibrium equation in the current configuration shows that the expression in parentheses can be identified as the Cauchy stress tensor $\boldsymbol{\sigma}$, whence, use of relation (3.73) between the Cauchy stress tensor $\boldsymbol{\sigma}$ and the Second Piola–Kirchhoff stress tensor $\boldsymbol{\tau}$ yields:

$$\boldsymbol{\tau} = 2\rho_0 \frac{\partial e}{\partial \mathbf{C}}$$

When we define

$$\mathcal{W} = \rho_0 e \tag{11.62}$$

as the strain energy function, the simple relation

$$\boldsymbol{\tau} = 2\frac{\partial \mathcal{W}}{\partial \mathbf{C}} \tag{11.63}$$

ensues. When the material is isotropic, the strain energy function $\mathcal{W}$ becomes only a function of the stretch invariants I_1^C, I_2^C and I_3^C of the right Cauchy–Green deformation tensor $\mathbf{C}$:

$$\mathcal{W} = \mathcal{W}(I_1^C, I_2^C, I_3^C) \tag{11.64}$$

with $\mathcal{W}(3, 3, 1) = 0$ since $\mathcal{W}$ must vanish in the undeformed state, Equation (11.40) with $\mathbf{C} = \mathbf{I}$. For the Second Piola–Kirchhoff stress tensor we then obtain:

$$\boldsymbol{\tau} = 2\left(\frac{\partial \mathcal{W}}{\partial I_1^C}\frac{\partial I_1^C}{\partial \mathbf{C}} + \frac{\partial \mathcal{W}}{\partial I_2^C}\frac{\partial I_2^C}{\partial \mathbf{C}} + \frac{\partial \mathcal{W}}{\partial I_3^C}\frac{\partial I_3^C}{\partial \mathbf{C}}\right) \tag{11.65}$$

11.2.1 Incompressibility and Near-incompressibility

The task of constructing a function $\mathcal{W}$ that accurately captures experimental data, is further alleviated when the assumption is made that the strain energy is separable into a volumetric part that is purely dependent on the volumetric deformations, and a deviatoric part that is a function of the distortion. As a first step in writing $\mathcal{W}$ as the sum of a volumetric and a deviatoric part, we will decompose $\mathcal{W}$ into $\mathcal{W}^*(I_1^C, I_2^C)$ and $f_p(I_3^C)$:

$$\mathcal{W} = \mathcal{W}^*(I_1^C, I_2^C) + f_p(I_3^C)$$

such that $\mathcal{W}^*(3, 3) = 0$ and $f_p(1) = 0$. A problem for the identification of $f_p(I_3^C)$ is that $\mathcal{W}^*$ is also affected by purely volumetric deformations. Using the modified invariants of the right Cauchy–Green deformation tensor, Equation (11.50), we can formulate the strain energy function as (Penn 1970; Peng and Landel 1975):

$$\mathcal{W} = \mathcal{W}^*(J_1^C, J_2^C) + f_p(J_3^C - 1) \tag{11.66}$$

which provides a complete separation of the distortional and the volumetric work. It is noted that by making f_p a function of $J_3^C - 1$ rather than of J_3^C, the argument of f_p vanishes in the undeformed state. When using Equation (11.66) as the strain energy function of a slightly compressible solid instead of Equation (11.64), the Second Piola–Kirchhoff stress becomes:

$$\boldsymbol{\tau} = 2\left(\frac{\partial \mathcal{W}^*(J_1^C, J_2^C)}{\partial \mathbf{C}} + f_p'\frac{\partial J_3^C}{\partial \mathbf{C}}\right) \tag{11.67}$$

where f_p' denotes differentiation of $f_p(J_3^C - 1)$ with respect to J_3^C.

A simple form for f_p, which is convenient for numerical implementation, is obtained by assuming a linear relation between the hydrostatic pressure p and the volume change ΔV:

$$p = K\frac{\Delta V}{V_0} \tag{11.68}$$

with K the bulk modulus. Experimental evidence suggests that K is independent of J_3^C for a wide range of pressures (Peng and Landel 1975; Penn 1970). In view of the definition of J_3^C, Equation (11.68) can be rewritten as:

$$p = K\left(J_3^C - 1\right) \tag{11.69}$$

The hydrostatic pressure is defined as one-third of the trace of the Cauchy stress tensor $\boldsymbol{\sigma}$, cf. Equation (1.79). For purely volumetric deformations, all principal stretches are equal to λ, so that for the deformation gradient we have $\mathbf{F} = \lambda\mathbf{I}$. Accordingly, relation (3.73) between the Cauchy stress tensor and the Second Piola–Kirchhoff stress tensor simplifies to:

$$\boldsymbol{\sigma} = \lambda^{-1}\boldsymbol{\tau} \tag{11.70}$$

and the hydrostatic pressure is given by

$$p = \frac{1}{3}\lambda^{-1}\mathrm{tr}\boldsymbol{\sigma} \tag{11.71}$$

Furthermore, the contribution of the deviatoric part $\mathcal{W}^*$ of the strain energy vanishes under an all-round uniform pressure p, so that in view of Equation (11.67), the Second Piola–Kirchhoff stress tensor reduces to:

$$\boldsymbol{\tau} = 2f_p'\frac{\partial J_3^C}{\partial \mathbf{C}} \tag{11.72}$$

and, under purely volumetric deformations, the following expression for the hydrostatic pressure p ensues:

$$p = f_p' \tag{11.73}$$

Equating expressions (11.69) and (11.73) results in $K(J_3^C - 1) = f_p'$ which, considering that f_p must vanish for $J_3^C - 1 = 0$, can be solved to yield:

$$f_p = \frac{1}{2}K\left(J_3^C - 1\right)^2 \tag{11.74}$$

Substitution of Equation (11.74) into Equation (11.67) results in the following strain energy function:

$$\mathcal{W} = \mathcal{W}^*\left(J_1^C, J_2^C\right) + \frac{1}{2}K\left(J_3^C - 1\right)^2 \tag{11.75}$$

Simo and Taylor (1982) have shown that definition (11.74) produces an instability in the compressive regime, and have proposed to replace it by:

$$f\left(J_3^C - 1\right) = \frac{1}{2}K\left(\left(J_3^C - 1\right)^2 + \ln^2\left(J_3^C\right)\right) \tag{11.76}$$

The full expression for the strain energy function then changes accordingly:

$$\mathcal{W} = \mathcal{W}^*\left(J_1^C, J_2^C\right) + \frac{1}{2}K\left(\left(J_3^C - 1\right)^2 + \ln^2\left(J_3^C\right)\right) \tag{11.77}$$

When the bulk modulus K is set equal to infinity, $J_3^C - 1$ and $\ln(J_3^C)$ vanish, and since $(J_3^C - 1)^2 + \ln^2(J_3^C)$ approaches zero faster than K tends to infinity, the second term in Equation (11.77) also vanishes, and the strain energy function reduces to:

$$\mathcal{W} = \mathcal{W}^*\left(J_1^C, J_2^C\right) \tag{11.78}$$

or, since $I_1^C = J_1^C$, $I_2^C = J_2^C$ for $J_3^C - 1 = 0$,

$$\mathcal{W} = \mathcal{W}\left(I_1^C, I_2^C\right) \tag{11.79}$$

Most early finite element calculations assume the incompressible formulation (11.79) to model rubber behaviour (Scharnhorst and Pian 1978; Cescotto and Fonder 1979; Jankovich *et al.* 1981; Sussmann and Bathe 1987; Glowinski and Tallec 1985). Yet, the assumption of incompressibility may be too crude for industrial rubbers and finite element analyses that a priori assume incompressibility will not always give a realistic prediction of the stresses and especially the deformations of rubber components.

11.2.2 Strain Energy as a Function of Stretch Invariants

The first proposals for strain energy functions that can capture the mechanical behaviour of rubberlike materials used the assumption of incompressibility and were formulated using the strain invariants of the right Cauchy–Green deformation tensor. Mooney (1940) was the first to propose a strain energy function of the form (11.79) and Rivlin (1948) generalised this to include higher-order terms and postulated the finite series:

$$\mathcal{W} = \sum_{i=0}^{l}\sum_{j=0}^{m} K_{ij}\left(I_1^C - 3\right)^i\left(I_2^C - 3\right)^j \tag{11.80}$$

where the model parameters l and m are natural numbers. The constant K_{00} stands for the energy level in the reference configuration and is normally set equal to zero. However, its choice is irrelevant, since it vanishes upon differentiation with respect to the right Cauchy–Green deformation tensor, i.e. when calculating the stress tensor. For the analysis of rubber components often only the linear terms are retained. Setting $l = m = 1$, $K_{11} = 0$, $K_{10} = K_1$, and $K_{01} = K_2$ results in the simple Mooney–Rivlin model:

$$\mathcal{W} = K_1\left(I_1^C - 3\right) + K_2\left(I_2^C - 3\right) \tag{11.81}$$

A further simplification occurs when K_2 is also assumed to vanish. This gives the so-called neo-Hookean model:

$$\mathcal{W} = K_1\left(I_1^C - 3\right) \tag{11.82}$$

Although the Mooney–Rivlin model was originally proposed for strictly incompressible rubbers, it can be used equally well to model the distortional response of a compressible rubber. For this purpose, Equation (11.81) is replaced by

$$\mathcal{W}^* = K_1\left(J_1^C - 3\right) + K_2\left(J_2^C - 3\right) \tag{11.83}$$

Strictly speaking, the constants K_1 and K_2 should be modified when replacing Equation (11.81) by Equation (11.83), but the observation that $J_3^C - 1$ hardly ever exceeds 10^{-4} justifies the assumption to use the same values for K_1 and K_2, irrespective of whether Equation (11.81) or Equation (11.83) is used. Addition of the volumetric part of the strain energy, Equation (11.74), subsequently gives the complete expression for the strain energy:

$$\mathcal{W} = K_1\left(J_1^C - 3\right) + K_2\left(J_2^C - 3\right) + \frac{1}{2}K\left(J_3^C - 1\right)^2 \tag{11.84}$$

so that the Second Piola–Kirchhoff stresses are given by:

$$\boldsymbol{\tau} = 2\left(K_1\frac{\partial J_1^C}{\partial \mathbf{C}} + K_2\frac{\partial J_2^C}{\partial \mathbf{C}} + K\left(J_3^C - 1\right)\frac{\partial J_3^C}{\partial \mathbf{C}}\right) \tag{11.85}$$

When elaborating this equation for pure shear conditions, one obtains that $2(K_1 + K_2) \to \mu$, with μ the ground-state shear modulus, i.e. the shear modulus for small strains (de Borst *et al.* 1988). For a given strain the stress can thus be computed in a straightforward manner, and the internal force vector follows directly from Chapter 3, see Box 11.1.

The tangential stiffness tensor, needed for equilibrium iterations within the Newton–Raphson method, can be obtained in a regular manner by differentiating the Second Piola–Kirchhoff stress tensor with respect to the Green–Lagrange strain tensor:

$$\mathbf{D} = \frac{\partial \boldsymbol{\tau}}{\partial \boldsymbol{\gamma}} = 2\frac{\partial \boldsymbol{\tau}}{\partial \mathbf{C}} \tag{11.86}$$

which results in:

$$\mathbf{D} = 4\left(K_1\frac{\partial^2 J_1^C}{\partial \mathbf{C}^2} + K_2\frac{\partial^2 J_2^C}{\partial \mathbf{C}^2} + K\left(J_3^C - 1\right)\frac{\partial^2 J_3^C}{\partial \mathbf{C}^2} + K\frac{\partial J_3^C}{\partial \mathbf{C}} \otimes \frac{\partial J_3^C}{\partial \mathbf{C}}\right) \tag{11.87}$$

The terms in this equation have been elaborated in Box 11.2.

The Mooney–Rivlin strain energy function has been used frequently for calculations in engineering practice. An example is the application of shock cells in the lock gates of the Eastern Scheldt storm surge barrier in The Netherlands, which was constructed in the early 1980s. These shock cells have been used to prevent damage accumulation between the steel lock gates and the concrete structure due to forces generated by incoming waves. Figure 11.2 shows the function and the position of the shock cell in the structure.

The shock cell consists of two enclosing steel cylinders of different diameters. The cylinders, which can move with respect to each other along the axis of symmetry, are connected by a rubber structure according to Figure 11.3. The inner cylinder is supported by the barrier, while the outer cylinder is loaded by the wave attacks on the gates. This load case has been modelled by applying a prescribed displacement to the outer cylinder parallel to the axis of symmetry.

Box 11.1 Algorithmic treatment for compressible Mooney–Rivlin hyperelasticity

For a given correction to the displacement increment in iteration $j+1$, $\mathbf{da}_{j+1}$:

1. Update the nodal displacements: $\mathbf{a}_{j+1} = \mathbf{a}_j + \mathbf{da}_{j+1}$
2. Compute the deformation gradient in each integration point: $\mathbf{F} = \mathbf{I} + \sum_{i=1}^{n} \mathbf{a}_i \frac{\partial h_i}{\partial \xi}$, with n the number of nodes in the element
3. Compute the Green–Lagrange strain tensor $\mathbf{C} = \mathbf{F}^{\mathrm{T}}\mathbf{F}$
4. Compute the derivatives of the invariants with respect to $\mathbf{C}$ (in Voigt notation):

$$\frac{\partial I_1^C}{\partial \mathbf{C}} = \begin{pmatrix} 1 \\ 1 \\ 1 \\ 0 \\ 0 \\ 0 \end{pmatrix}, \quad \frac{\partial I_2^C}{\partial \mathbf{C}} = \begin{pmatrix} C_{xx} + C_{yy} \\ C_{yy} + C_{zz} \\ C_{zz} + C_{xx} \\ -C_{xy} \\ -C_{yz} \\ -C_{zx} \end{pmatrix}, \quad \frac{\partial I_3^C}{\partial \mathbf{C}} = \begin{pmatrix} C_{yy}C_{zz} - C_{yz}^2 \\ C_{zz}C_{xx} - C_{zx}^2 \\ C_{xx}C_{yy} - C_{xy}^2 \\ C_{yz}C_{zx} - C_{zz}C_{xy} \\ C_{zx}C_{xy} - C_{xx}C_{yz} \\ C_{xy}C_{yz} - C_{yy}C_{zx} \end{pmatrix}$$

5. Compute the derivates of the modified invariants:

$$\begin{cases} \frac{\partial J_1^C}{\partial \mathbf{C}} = (I_3^C)^{-1/3} \frac{\partial I_1^C}{\partial \mathbf{C}} - \frac{1}{3} I_1^C (I_3^C)^{-4/3} \frac{\partial I_3^C}{\partial \mathbf{C}} \\ \frac{\partial J_2^C}{\partial \mathbf{C}} = (I_3^C)^{-2/3} \frac{\partial I_2^C}{\partial \mathbf{C}} - \frac{2}{3} I_2^C (I_3^C)^{-5/3} \frac{\partial I_3^C}{\partial \mathbf{C}} \\ \frac{\partial J_3^C}{\partial \mathbf{C}} = \frac{1}{2} (I_3^C)^{-1/2} \frac{\partial I_3^C}{\partial \mathbf{C}} \end{cases}$$

6. Compute the Second Piola–Kirchhoff stress tensor $\boldsymbol{\tau}_{j+1}$ from Equation (11.85)
7. Update the internal force vector: $\mathbf{f}_{\mathrm{int},j+1} = \int_{V_0} \mathbf{B}_L^{\mathrm{T}} \boldsymbol{\tau}_{j+1} \mathrm{d}V$

Note that the computational flow is somewhat more compact than that of Equation (2.52), which is possible because hyperelasticity is a total stress–strain relation, and involves no path dependency, so that there is no integration of a rate constitutive equation along the strain path.

In the calculations for the shock cell the incompressible Mooney–Rivlin model has been used. The material parameters have been determined for a virgin material as well as for a pre-stressed rubber. A difference is observed between both parameter identifications, but for the type of deformation to which the shock cell is subjected (mainly shear) only the sum of K_1 and K_2 is of interest, and then only a slight difference remains between both identifications (de Borst *et al.* 1988).

First, a finite element analysis has been made for an undamaged shock cell (Figure 11.4). Since the maximum tensile stresses in the left-lower corner exceeded the permissible tensile stress for the used rubber compound, a second calculation was made in which a pre-defined crack was inserted. A comparison of the global load–displacement curves with an experimental measurement is shown in Figure 11.4.

Box 11.2 Tangent stiffness matrix for compressible Mooney–Rivlin hyperelasticity

1. Compute the second derivatives of the invariants with respect to the right Cauchy–Green deformation tensor:

$$\frac{\partial^2 I_1^C}{\partial \mathbf{C}^2} = \mathbf{0}$$

$$\frac{\partial^2 I_2^C}{\partial \mathbf{C}^2} = \begin{bmatrix} 0 & 1 & 1 & 0 & 0 & 0 \\ 1 & 0 & 1 & 0 & 0 & 0 \\ 1 & 1 & 0 & 0 & 0 & 0 \\ 0 & 0 & 0 & -\frac{1}{2} & 0 & 0 \\ 0 & 0 & 0 & 0 & -\frac{1}{2} & 0 \\ 0 & 0 & 0 & 0 & 0 & -\frac{1}{2} \end{bmatrix}$$

$$\frac{\partial^2 I_3^C}{\partial \mathbf{C}^2} = \begin{bmatrix} 0 & C_{zz} & C_{yy} & 0 & -C_{yz} & 0 \\ C_{zz} & 0 & C_{xx} & 0 & 0 & -C_{zx} \\ C_{yy} & C_{xx} & 0 & -C_{xy} & 0 & 0 \\ 0 & 0 & -C_{xy} & -C_{zz}/2 & C_{zx}/2 & C_{yz}/2 \\ -C_{yz} & 0 & 0 & C_{zx}/2 & -C_{xx}/2 & C_{xy}/2 \\ 0 & -C_{yz} & 0 & C_{zx}/2 & C_{xy}/2 & -C_{yy}/2 \end{bmatrix}$$

2. Compute the second derivatives of the modified invariants:

$$\begin{cases} \dfrac{\partial^2 J_1^C}{\partial \mathbf{C}^2} = I_3^{-1/3}\dfrac{\partial^2 I_1^C}{\partial \mathbf{C}^2} + \frac{4}{9} I_1 I_3^{-7/3}\dfrac{\partial I_3^C}{\partial \mathbf{C}}\left(\dfrac{\partial I_3^C}{\partial \mathbf{C}}\right)^{\mathrm{T}} \\ \qquad -\frac{1}{3} I_3^{-4/3}\left[\dfrac{\partial I_1^C}{\partial \mathbf{C}}\left(\dfrac{\partial I_3^C}{\partial \mathbf{C}}\right)^{\mathrm{T}} + I_1\dfrac{\partial^2 I_3^C}{\partial \mathbf{C}^2} + \dfrac{\partial I_3^C}{\partial \mathbf{C}}\left(\dfrac{\partial I_1^C}{\partial \mathbf{C}}\right)^{\mathrm{T}}\right] \\ \dfrac{\partial^2 J_2^C}{\partial \mathbf{C}^2} = I_3^{-2/3}\dfrac{\partial^2 I_2^C}{\partial \mathbf{C}^2} + \frac{10}{9} I_2 I_3^{-8/3}\dfrac{\partial I_3^C}{\partial \mathbf{C}}\left(\dfrac{\partial I_3^C}{\partial \mathbf{C}}\right)^{\mathrm{T}} \\ \qquad -\frac{2}{3} I_3^{-5/3}\left[\dfrac{\partial I_2^C}{\partial \mathbf{C}}\left(\dfrac{\partial I_3^C}{\partial \mathbf{C}}\right)^{\mathrm{T}} + I_2\dfrac{\partial^2 I_3^C}{\partial \mathbf{C}^2} + \dfrac{\partial I_3^C}{\partial \mathbf{C}}\left(\dfrac{\partial I_2^C}{\partial \mathbf{C}}\right)^{\mathrm{T}}\right] \\ \dfrac{\partial^2 J_3^C}{\partial \mathbf{C}^2} = \frac{1}{2} I_3^{-1/2}\dfrac{\partial^2 I_3^C}{\partial \mathbf{C}^2} - \frac{1}{4} I_3^{-3/2}\dfrac{\partial I_3^C}{\partial \mathbf{C}}\left(\dfrac{\partial I_3^C}{\partial \mathbf{C}}\right)^{\mathrm{T}} \end{cases}$$

3. Compute the tangential stiffness matrix **D** according to Equation (11.87).

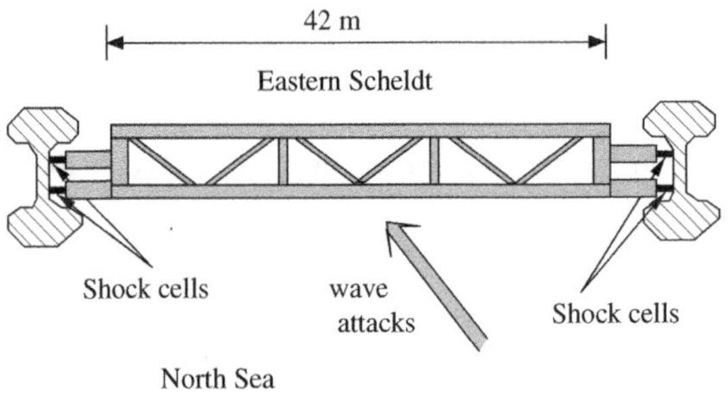

Figure 11.2 Top view of a lock gate of the Easter Scheldt storm surge barrier

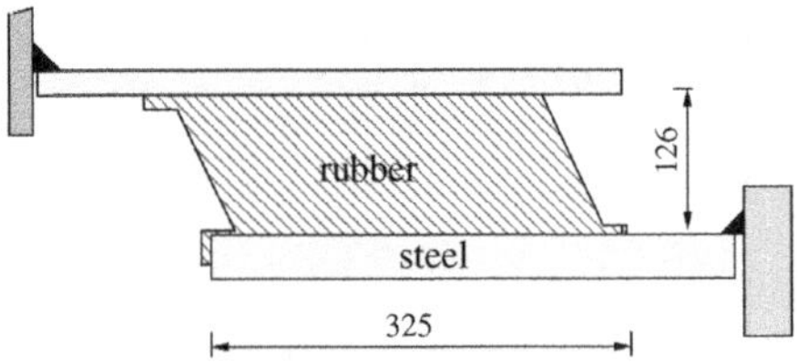

Figure 11.3 Geometry of the shock cell (dimensions in millimetre)

11.2.3 Strain Energy as a Function of Principal Stretches

Strain energy functions that are expressed in terms of stretch invariants can be less accurate for very large stretches. It appears that expressing the strain energy function in terms of principal stretches offers more flexibility and makes it easier to accurately capture the constitutive behaviour for stretches $\lambda \geq 2$ (Figure 11.5). The first proposal along this line was formulated by Varga (1966). This approach was generalised by Valanis and Landel (1967), who formulated the strain energy function for incompressible materials as a symmetric function of the principal

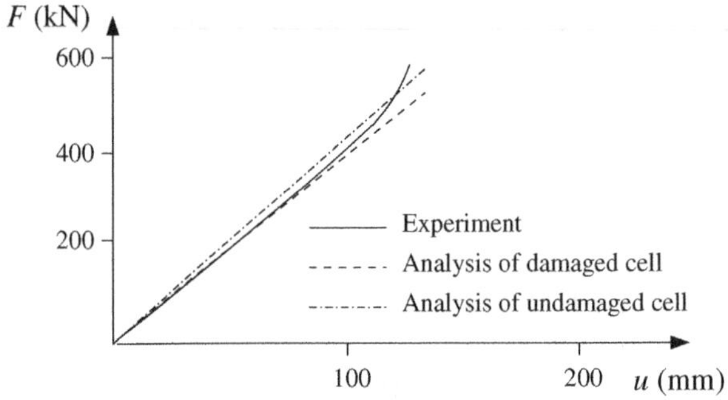

Figure 11.4 Computed and experimentally determined axial force in the shock cell

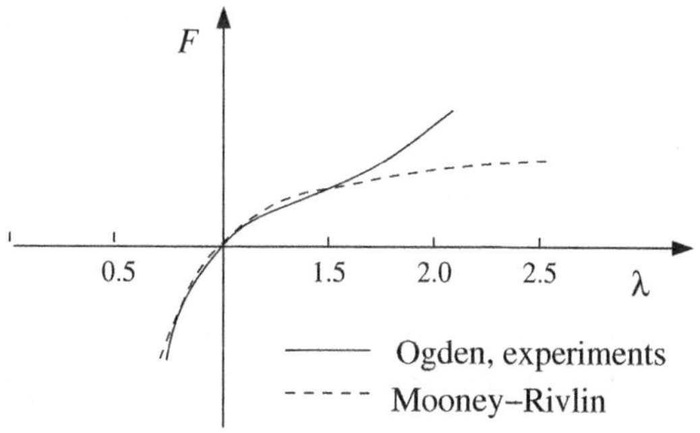

Figure 11.5 Rubber in uniaxial tension: the Mooney–Rivlin and Ogden models *vs* observed rubber behaviour

stretches:

$$\mathcal{W} = w(\lambda_1) + w(\lambda_2) + w(\lambda_3) \tag{11.88}$$

Within this formalism Ogden (1972) postulated the following series of functions in the principal stretches:

$$\mathcal{W} = \sum_{r=1}^{n_r} \frac{\mu_r}{\alpha_r} \left(\lambda_1^{\alpha_r} + \lambda_2^{\alpha_r} + \lambda_3^{\alpha_r} - 3\right) \tag{11.89}$$

in which μ_r and α_r are model parameters, which satisfy the relation

$$\sum_{r=1}^{n_r} \mu_r \alpha_r = 2\mu \tag{11.90}$$

with μ the ground-state shear modulus. Equivalently, the strain energy function can be expressed in terms of the principal values of the right Cauchy–Green deformation tensor:

$$\mathcal{W} = \sum_{r=1}^{n_r} \frac{\mu_r}{\alpha_r} \left(C_1^{\alpha_r/2} + C_2^{\alpha_r/2} + C_3^{\alpha_r/2} - 3\right) \tag{11.91}$$

The Ogden model incorporates a number of other formulations, including the Mooney–Rivlin model, which is obtained by setting $n_r = 2, \alpha_1 = 2$ and $\alpha_2 = -2$, and the neo-Hookean model, which results when $n_r = 1, \alpha_1 = 2$. As with the Mooney–Rivlin model a model for compressible rubbers can be obtained by simply replacing Equation (11.89) by:

$$\mathcal{W}^* = \sum_{r=1}^{n_r} \frac{\mu_r}{\alpha_r} \left(\tilde{\lambda}_1^{\alpha_r} + \tilde{\lambda}_2^{\alpha_r} + \tilde{\lambda}_3^{\alpha_r} - 3\right) \tag{11.92}$$

with $\tilde{\lambda}_i$ the modified principal stretches, Equation (11.53). Again, a reformulation in terms of the principal values of the modified right Cauchy–Green tensor is possible:

$$\mathcal{W}^* = \sum_{r=1}^{n_r} \frac{\mu_r}{\alpha_r} \left(\tilde{C}_1^{\alpha_r/2} + \tilde{C}_2^{\alpha_r/2} + \tilde{C}_3^{\alpha_r/2} - 3 \right) \tag{11.93}$$

Addition of the volumetric contribution, Equation (11.74), gives the complete expression for the strain energy, similar to the Mooney–Rivlin model in Equation (11.84):

$$\mathcal{W} = \sum_{r=1}^{n_r} \frac{\mu_r}{\alpha_r} \left(\tilde{C}_1^{\alpha_r/2} + \tilde{C}_2^{\alpha_r/2} + \tilde{C}_3^{\alpha_r/2} - 3 \right) + \frac{1}{2} K (J_3^C - 1)^2 \tag{11.94}$$

so that the Second Piola–Kirchhoff stress is given by:

$$\boldsymbol{\tau} = 2 \left[\sum_{i=1}^{3} \tilde{\tau}_i \left(\frac{\partial \tilde{C}_i}{\partial J_1^C} \frac{\partial J_1^C}{\partial \mathbf{C}} + \frac{\partial \tilde{C}_i}{\partial J_2^C} \frac{\partial J_2^C}{\partial \mathbf{C}} + \frac{\partial \tilde{C}_i}{\partial J_3^C} \frac{\partial J_3^C}{\partial \mathbf{C}} \right) + K (J_3^C - 1) \frac{\partial J_3^C}{\partial \mathbf{C}} \right] \tag{11.95}$$

with

$$\tilde{\tau}_i = \frac{\partial \mathcal{W}^*}{\partial \tilde{C}_i} = \sum_{r=1}^{n_r} \frac{\mu_r}{2} \tilde{C}_i^{(\alpha_r/2)-1} \tag{11.96}$$

the principal values of the deviatoric part of the Second Piola–Kirchhoff stress tensor. The most convenient method to compute the derivatives of the principal values with respect to the invariants makes use of the implicit function theorem (Peng 1979). For this purpose the characteristic equation (11.52) is denoted by $f = f(J_1^C, J_2^C, J_3^C, \tilde{C}_i(J_1^C, J_2^C, J_3^C))$, and

$$\frac{\partial f}{\partial J_j^C} + \frac{\partial f}{\partial \tilde{C}_i} \frac{\partial \tilde{C}_i}{\partial J_j^C} = 0$$

which results in:

$$\begin{aligned} \frac{\partial \tilde{C}_i}{\partial J_1^C} &= \frac{\tilde{C}_i^2}{3\tilde{C}_i^2 - 2J_1^C \tilde{C}_i + J_2^C} \\ \frac{\partial \tilde{C}_i}{\partial J_2^C} &= \frac{-\tilde{C}_i}{3\tilde{C}_i^2 - 2J_1^C \tilde{C}_i + J_2^C} \\ \frac{\partial \tilde{C}_i}{\partial J_3^C} &= 0 \end{aligned} \tag{11.97}$$

The tangential stiffness matrix can be obtained in a manner similar to that for the Mooney–Rivlin model, cf. Equation (11.87):

$$
\begin{aligned}
\mathbf{D} = 4\sum_{i=1}^{3}\Bigg[& \frac{\partial\tilde{\tau}_i}{\partial\tilde{C}_i}\left(\frac{\partial\tilde{C}_i}{\partial J_1^C}\frac{\partial J_1^C}{\partial\mathbf{C}}+\frac{\partial\tilde{C}_i}{\partial J_2^C}\frac{\partial J_2^C}{\partial\mathbf{C}}+\frac{\partial\tilde{C}_i}{\partial J_3^C}\frac{\partial J_3^C}{\partial\mathbf{C}}\right)\otimes \\
& \left(\frac{\partial\tilde{C}_i}{\partial J_1^C}\frac{\partial J_1^C}{\partial\mathbf{C}}+\frac{\partial\tilde{C}_i}{\partial J_2^C}\frac{\partial J_2^C}{\partial\mathbf{C}}+\frac{\partial\tilde{C}_i}{\partial J_3^C}\frac{\partial J_3^C}{\partial\mathbf{C}}\right) \\
& +\tilde{\tau}_i\left(\frac{\partial^2\tilde{C}_i}{\partial(J_1^C)^2}\frac{\partial J_1^C}{\partial\mathbf{C}}+\frac{\partial^2\tilde{C}_i}{\partial J_1^C\partial J_2^C}\frac{\partial J_2^C}{\partial\mathbf{C}}+\frac{\partial^2\tilde{C}_i}{\partial J_1^C\partial J_3^C}\frac{\partial J_3^C}{\partial\mathbf{C}}\right)\otimes\frac{\partial J_1^C}{\partial\mathbf{C}} \\
& +\tilde{\tau}_i\left(\frac{\partial^2\tilde{C}_i}{\partial J_2^C\partial J_1^C}\frac{\partial J_1^C}{\partial\mathbf{C}}+\frac{\partial^2\tilde{C}_i}{\partial(J_2^C)^2}\frac{\partial J_2^C}{\partial\mathbf{C}}+\frac{\partial^2\tilde{C}_i}{\partial J_2^C\partial J_3^C}\frac{\partial J_3^C}{\partial\mathbf{C}}\right)\otimes\frac{\partial J_2^C}{\partial\mathbf{C}} \\
& +\tilde{\tau}_i\left(\frac{\partial^2\tilde{C}_i}{\partial J_3^C\partial J_1^C}\frac{\partial J_1^C}{\partial\mathbf{C}}+\frac{\partial^2\tilde{C}_i}{\partial J_3^C\partial J_2^C}\frac{\partial J_2^C}{\partial\mathbf{C}}+\frac{\partial^2\tilde{C}_i}{\partial(J_3^C)^2}\frac{\partial J_3^C}{\partial\mathbf{C}}\right)\otimes\frac{\partial J_3^C}{\partial\mathbf{C}} \\
& +\tilde{\tau}_i\left(\frac{\partial\tilde{C}_i}{\partial J_1^C}\frac{\partial^2 J_1^C}{\partial\mathbf{C}^2}+\frac{\partial\tilde{C}_i}{\partial J_2^C}\frac{\partial^2 J_2^C}{\partial\mathbf{C}^2}+\frac{\partial\tilde{C}_i}{\partial J_3^C}\frac{\partial^2 J_3^C}{\partial\mathbf{C}^2}\right)\Bigg] \\
& +4\left(K(J_3^C-1)\frac{\partial^2 J_3^C}{\partial\mathbf{C}^2}+K\frac{\partial J_3^C}{\partial\mathbf{C}}\otimes\frac{\partial J_3^C}{\partial\mathbf{C}}\right)
\end{aligned}
\tag{11.98}
$$

where the derivatives of the principal deviatoric stresses with respect to the principal values of the modified Green–Lagrange deformation tensor are given by:

$$
\frac{\partial\tilde{\tau}_i}{\partial\tilde{C}_i}=\sum_{r=1}^{n_r}\frac{1}{4}(\mu_r\alpha_r-2\mu_r)\tilde{C}_i^{(\alpha_r/2)-2} \tag{11.99}
$$

and the second derivatives of the principal values $\tilde{C}_i$ with respect to the invariants are again computed using the implicit function theorem, resulting in:

$$
\begin{aligned}
\frac{\partial^2\tilde{C}_i}{\partial(J_1^C)^2} &= \frac{2\tilde{C}_i^3(3\tilde{C}_i^2-3J_1^C\tilde{C}_i+2J_2^C)}{(3\tilde{C}_i^2-2J_1^C\tilde{C}_i+J_2^C)^3} \\
\frac{\partial^2\tilde{C}_i}{\partial J_1^C\partial J_2^C} &= -\frac{\tilde{C}_i^2(3\tilde{C}_i^2-4J_1^C\tilde{C}_i+3J_2^C)}{(3\tilde{C}_i^2-2J_1^C\tilde{C}_i+J_2^C)^3} \\
\frac{\partial^2\tilde{C}_i}{\partial(J_2^C)^2} &= -\frac{2\tilde{C}_i(J_c^C\tilde{C}_i-J_2^C)}{(3\tilde{C}_i^2-2J_1^C\tilde{C}_i+J_2^C)^3}
\end{aligned}
\tag{11.100}
$$

The approach in which the strain energy function (11.93) is transformed and expressed in terms of invariants can be less robust, since singularities occur when two or more principal values become equal. Indeed, when $C_1=C_2=C$, $I_1^C=2C+C_3$, $I_2^C=C^2+2CC_3$, and the denominators of Equation (11.97) become zero. Although limits can be computed which render the full expression for $\boldsymbol{\tau}$ finite, a more convenient approach is to directly compute the principal values of the deviatoric part of the Second Piola–Kirchhoff stress tensor according to Equation (11.96), whereafter the stresses in the x, y, z-system are obtained through eigenprojections, as outlined in Box 11.3 (de Souza Neto *et al.* 2008). This algorithm rests upon the property that the eigenvectors of the (modified) right Cauchy–Green deformation tensor and the (deviatoric part of the) Second Piola–Kirchhoff stress tensor are coaxial.

As discussed in Chapter 7 the use of eigenprojections is closely related to an approach in which the strain tensor is expressed in principal directions via (in Voigt notation) $\bar{\mathbf{C}} = \mathbf{T}_{\epsilon}\mathbf{C}$, followed by a computation of the principal values $\tilde{\tau}_i$ according to Equation (11.96). The stresses in the x, y, z-system are subsequently obtained by back-transformation using $\mathbf{T}_{\sigma}$:

$$\boldsymbol{\tau} = \mathbf{T}_{\sigma}\mathbf{D}_{\tilde{\tau}} + 2K(J_3^C - 1)\frac{\partial J_3^C}{\partial \mathbf{C}} \tag{11.101}$$

where

$$\mathbf{D}_{\tilde{\tau}} = \begin{bmatrix} \sum_{r=1}^{n_r} \frac{\mu_r}{2}\tilde{C}_1^{(\alpha_r/2)-1} & 0 & 0 \\ 0 & \sum_{r=1}^{n_r} \frac{\mu_r}{2}\tilde{C}_2^{(\alpha_r/2)-1} & 0 \\ 0 & 0 & \sum_{r=1}^{n_r} \frac{\mu_r}{2}\tilde{C}_3^{(\alpha_r/2)-1} \end{bmatrix} \tag{11.102}$$

The approach is warranted because of the material isotropy, and a similar approach has been followed for Mohr–Coulomb plasticity in Chapter 7. The tangential stiffness matrix is obtained in a manner similar to that for the Mohr–Coulomb yield criterion, Equation (7.187), or for the rotating crack model (Box 6.3).

11.2.4 Logarithmic Extension of Linear Elasticity: Hencky Model

In Equation (11.54) the strain energy density e has been defined as a function of the right Cauchy–Green deformation tensor $\mathbf{C}$. Equivalently, the strain energy density can be made a function of the left Cauchy–Green deformation tensor $\mathbf{B}$. Instead of Equation (11.56) we now have:

$$\mathcal{E} = \int_v \rho e(\mathbf{B})\mathrm{d}V \tag{11.103}$$

and the variation of the strain energy density becomes:

$$\delta e = 2\mathrm{tr}\left(\delta\mathbf{F}^{\mathrm{T}} \cdot \frac{\partial e}{\partial \mathbf{B}} \cdot \mathbf{F}\right) \tag{11.104}$$

Invoking the divergence theorem now yields:

$$\int_V \mathrm{tr}\left(\delta\mathbf{F}^{\mathrm{T}} \cdot \frac{\partial e}{\partial \mathbf{B}} \cdot \mathbf{F}\right)\mathrm{d}V = \int_S \delta\mathbf{x} \cdot \left(\frac{\partial e}{\partial \mathbf{B}} \cdot \mathbf{B}\right) \cdot \mathbf{n}\mathrm{d}S - \int_V \delta\mathbf{x} \cdot \nabla \cdot \left(\frac{\partial e}{\partial \mathbf{B}} \cdot \mathbf{B}\right)\mathrm{d}V \tag{11.105}$$

Box 11.3 The computation of deviatoric stresses and the deviatoric part of the tangential stiffness matrix for strain energy functions that are expressed in terms of principal stretches

1. For $\tilde{\mathbf{C}}$, compute the principal values $\tilde{C}_i$
2. Compute the principal stresses $\tilde{\tau}_i$ using Equation (11.96)
3. Compute the eigenprojections:
   (a) $\boldsymbol{E}_i = \frac{\tilde{C}_i}{2\tilde{C}_i^3 - J_1^C \tilde{C}_i^2 + J_3^C}\left(\tilde{\mathbf{C}}^2 - (J_1^C - \tilde{C}_i)\tilde{\mathbf{C}} + \frac{J_3^C}{\tilde{C}_i}\mathbf{I}\right)$, $\tilde{C}_1 \neq \tilde{C}_2 \neq \tilde{C}_3$
   (b) $\boldsymbol{E}_1$ as under (a), $\boldsymbol{E}_2 = \mathbf{I} - \boldsymbol{E}_i$, $\tilde{C}_1 \neq \tilde{C}_2 = \tilde{C}_3$
   (c) $\boldsymbol{E}_i = \mathbf{I}$, $\tilde{C}_1 = \tilde{C}_2 = \tilde{C}_3$
4. Compute the Second Piola–Kirchhoff stress tensor: $\boldsymbol{\tau} = \sum_{i=1}^{3} \tau_i \boldsymbol{E}_i$
5. Compute the derivatives in the principal stress space $\frac{\partial \tilde{\tau}_i}{\partial \tilde{C}_i}$ from Equation (11.99)
6. Compute, using cyclic permutation for i, the deviatoric tangential stiffness tensor:
   (a) $\tilde{\mathbf{D}} = \sum_{i=1}^{3}\Big[\frac{\tilde{\tau}_i}{(\tilde{C}_i - \tilde{C}_{i+1})(\tilde{C}_i - \tilde{C}_{i+2})}\Big(\frac{\mathrm{d}\tilde{\mathbf{C}}^2}{\mathrm{d}\tilde{\mathbf{C}}} - (\tilde{C}_{i+1} + \tilde{C}_{i+2})\mathbf{I}$
   $-(2\tilde{C}_i - \tilde{C}_{i+1} - \tilde{C}_{i+2})\boldsymbol{E}_i \otimes \boldsymbol{E}_i$
   $-(\tilde{C}_{i+1} - \tilde{C}_{i+2})(\boldsymbol{E}_{i+1} \otimes \boldsymbol{E}_{i+1} - \boldsymbol{E}_{i+2} \otimes \boldsymbol{E}_{i+2})\Big)$
   $+\frac{\partial \tilde{\tau}_i}{\partial \tilde{C}_i}\boldsymbol{E}_i \otimes \boldsymbol{E}_i\Big]$, $\tilde{C}_1 \neq \tilde{C}_2 \neq \tilde{C}_3$
   (b) $\tilde{\mathbf{D}} = s_1 \frac{\mathrm{d}\tilde{\mathbf{C}}^2}{\mathrm{d}\tilde{\mathbf{C}}} - s_2\mathbf{I} - s_3\tilde{\mathbf{C}} \otimes \tilde{\mathbf{C}} + s_4(\tilde{\mathbf{C}} \otimes \mathbf{I})^{\mathrm{sym}} - s_5\mathbf{I} \otimes \mathbf{I}$, $\tilde{C}_1 \neq \tilde{C}_2 = \tilde{C}_3$
   (c) $\tilde{\mathbf{D}} = \frac{\partial \tilde{\tau}_1}{\partial \tilde{C}_1}\mathbf{I}$, $\tilde{C}_1 = \tilde{C}_2 = \tilde{C}_3$

The scalars $s_1, \ldots, s_5$ are defined as:

$$\begin{cases} s_1 = \frac{\tilde{\tau}_i - \tilde{\tau}_{i+2}}{(\tilde{C}_i - \tilde{C}_{i+2})^2} - \frac{1}{\tilde{C}_i - \tilde{C}_{i+2}}\frac{\partial \tilde{\tau}_{i+2}}{\partial \tilde{C}_{i+2}} \\ s_2 = 2\tilde{C}_{i+2}\frac{\tilde{\tau}_i - \tilde{\tau}_{i+2}}{(\tilde{C}_i - \tilde{C}_{i+2})^2} - \frac{\tilde{C}_i + \tilde{C}_{i+2}}{\tilde{C}_i - \tilde{C}_{i+2}}\frac{\partial \tilde{\tau}_{i+2}}{\partial \tilde{C}_{i+2}} \\ s_3 = 2\frac{\tilde{\tau}_i - \tilde{\tau}_{i+2}}{(\tilde{C}_i - \tilde{C}_{i+2})^3} - \frac{1}{(\tilde{C}_i - \tilde{C}_{i+2})^2}\left(\frac{\partial \tilde{\tau}_i}{\partial \tilde{C}_i} + \frac{\partial \tilde{\tau}_{i+2}}{\partial \tilde{C}_{i+2}}\right) \\ s_4 = 2s_3\tilde{C}_{i+2} \\ s_5 = s_3\tilde{C}_{i+2}^2 \end{cases}$$

while

$$\left(\frac{\mathrm{d}\tilde{\mathbf{C}}^2}{\mathrm{d}\tilde{\mathbf{C}}}\right)_{ijkl} = \frac{1}{2}\left(\delta_{ik}\tilde{C}_{lj} + \delta_{il}\tilde{C}_{kj} + \delta_{jl}\tilde{C}_{ik} + \delta_{kj}\tilde{C}_{il}\right)$$

Substitution of this expression into Equation (11.55), use of Equation (11.57), and noting that the result must hold for all admissible $\delta\mathbf{x}$ thus gives:

$$\nabla \cdot \left(2\rho\frac{\partial e}{\partial \mathbf{B}} \cdot \mathbf{B}\right) + \rho\mathbf{g} = \mathbf{0} \qquad (11.106)$$

The expression in parentheses can be identified as the Cauchy stress tensor $\boldsymbol{\sigma}$. Using relation (11.15) between the Cauchy and the Kirchhoff stress tensor, and using definition (11.62) for the strain energy function $\mathcal{W}$, the following relation between the Kirchhoff stress tensor $\boldsymbol{\kappa}$ and the left Cauchy–Green deformation tensor $\mathbf{B}$ can be derived:

$$\boldsymbol{\kappa} = 2\frac{\partial \mathcal{W}}{\partial \mathbf{B}} \cdot \mathbf{B} \tag{11.107}$$

In passing from Equation (11.92) to Equation (11.93), we expressed the strain energy function $\mathcal{W}$ in terms of the principal values of the right Cauchy–Green deformation tensor $\mathbf{C}$ instead of in terms of the right stretch tensor $\mathbf{U}$. Accordingly, the strain energy function, which for the Mooney–Rivlin class of strain energy functions of Equation (11.80) has been expressed in terms of (the invariants of) the right Cauchy–Green deformation tensor, can equally be expressed in terms of (the invariants of) the right stretch tensor, so that $\mathcal{W} = \mathcal{W}(\mathbf{U})$. Indeed, any objective deformation measure can be used to define the strain energy function, but also functions of objective deformation measures can be used for this purpose. An often used function is the logarithm of the left stretch tensor $\mathbf{V}$:

$$\boldsymbol{\varepsilon} = \ln \mathbf{V} \tag{11.108}$$

which is coaxial with the left stretch tensor since taking the logarithm is an isotropic tensor function, and the principal values of $\boldsymbol{\varepsilon}$ are obtained as $\ln \lambda_i$, with λ_i the principal stretches. By virtue of Equation (3.66), this so-called Eulerian logarithmic strain tensor can also be expressed as:

$$\boldsymbol{\varepsilon} = \frac{1}{2} \ln \mathbf{B} \tag{11.109}$$

A particularly convenient expression for the strain energy function is given by:

$$\mathcal{W}(\boldsymbol{\varepsilon}) = \frac{1}{2}\boldsymbol{\varepsilon} : \mathbf{D}^{\mathrm{e}} : \boldsymbol{\varepsilon} \tag{11.110}$$

which is due to Hencky (1933), and can be considered as the finite (logarithmic) strain extension of small-strain elasticity. Indeed, the format of Hencky's strain energy function resembles that of classical, infinitesimal elasticity theory, with $\mathbf{D}^{\mathrm{e}}$ the standard elastic stiffness tensor, Equation (1.109).

Noting that the Hencky strain energy function is defined in terms of the Eulerian logarithmic strain $\boldsymbol{\varepsilon}$, Equation (11.107) can be reworked as:

$$\boldsymbol{\kappa} = \frac{\partial \mathcal{W}}{\partial \boldsymbol{\varepsilon}} \cdot \frac{\partial \ln \mathbf{B}}{\partial \mathbf{B}} \cdot \mathbf{B} \tag{11.111}$$

Since we have material isotropy, and since $\ln \mathbf{B}$ is an isotropic tensor function of $\mathbf{B}$, the operation $\frac{\partial \ln \mathbf{B}}{\partial \mathbf{B}} \cdot \mathbf{B}$ can be carried out in any coordinate system. In the coordinate system with the axes aligned with the eigenvectors, this operation becomes particularly simple. In each of the principal directions we have:

$$\frac{\partial \ln(\lambda_i^2)}{\partial \lambda_i^2}\lambda_i^2 = \frac{1}{\lambda_i^2}\lambda_i^2 = 1$$

whence

$$\frac{\partial \ln \mathbf{B}}{\partial \mathbf{B}} \cdot \mathbf{B} = \mathbf{I}$$

For the Hencky definition of the strain energy function $\mathcal{W}$, Equation (11.110), we obtain a linear relation between the Kirchhoff stress tensor $\boldsymbol{\kappa}$ and the Eulerian logarithmic strain $\boldsymbol{\varepsilon}$:

$$\boldsymbol{\kappa} = \mathbf{D}^{\mathrm{e}} : \boldsymbol{\varepsilon} \tag{11.112}$$

11.3 Element Technology

When the ratio of the bulk modulus over the ground-state shear modulus K/μ tends to infinity, i.e. when the limiting case of incompressibility is approached, volumetric locking can occur, similar to the element behaviour for isochoric/dilatant plasticity, as described in Chapter 7. In principle, the same strategies can be applied to ameliorate the behaviour of finite elements. However, the strategies that have been advocated in small-strain elasto-plasticity cannot always be carried over straightforwardly to incompressible hyperelasticity – and therefore also not to large-strain elasto-plasticity. The classical strategies that use crossed triangular elements (Nagtegaal *et al.* 1974) and reduced integration (Doll *et al.* 2000) apply directly in the large-strain regime. Mixed formulations that use independent interpolations of displacements and pressures, usually with a pressure field that is condensed at the element level, can also be extended in a rather straightforward manner to large strains (Brink and Stein 1996; de Borst *et al.* 1988; Sussmann and Bathe 1987; van den Bogert *et al.* 1991). The same holds, in principle, for enhanced assumed strain methods (Simo and Armero 1992), but instabilities have been detected in the large-strain regime (de Souza Neto *et al.* 1995; Wriggers and Reese 1996). Strategies that can remedy the anomalous behaviour have been proposed by Glaser and Armero (1997), Korelc and Wriggers (1996), Reese and Wriggers (2000), Wall *et al.* (2000). An extension to large deformations of the $\bar{\mathbf{B}}$-concept has been proposed by Moran *et al.* (1990), while de Souza Neto *et al.* (1996, 2005, 2008) have proposed a related concept that utilises a volumetric/deviatoric split of the deformation gradient $\mathbf{F}$. Finally, Crisfield and Moita (1996); Moita and Crisfield (1996) have extended the corotational approach to large strains of two-dimensional and three-dimensional continuum elements, while enhancing the element performance using the incompatible modes technique.

11.3.1 u/p Formulation

In the u/p formulation the pressure field is interpolated in addition to the displacement field. Since the pressure enters as an independent variable, Equation (11.67) is rewritten using expression (11.73), to give:

$$\boldsymbol{\tau} = 2\left(\frac{\partial \mathcal{W}^*(J_1^C, J_2^C)}{\partial \mathbf{C}} + p\frac{\partial J_3^C}{\partial \mathbf{C}}\right) \tag{11.113}$$

which can be differentiated to give the material tangential stiffness relation:

$$\boldsymbol{\tau} = \mathbf{D} : \boldsymbol{\gamma} + 2\frac{\partial J_3^C}{\partial \mathbf{C}}\dot{p} \tag{11.114}$$

with

$$\mathbf{D} = 4\left(\frac{\partial^2 \mathcal{W}^*}{\partial \mathbf{C}^2} + p\frac{\partial^2 J_3^C}{\partial \mathbf{C}^2}\right) \tag{11.115}$$

We next consider the virtual work expression (3.86) at iteration $j+1$:

$$\int_{V_0} \delta\boldsymbol{\gamma}^{\mathrm{T}}\boldsymbol{\tau}_{j+1}\mathrm{d}V \int_{S_0} \delta\mathbf{u}^{\mathrm{T}}\mathbf{t}\mathrm{d}S + \int_{V_0} \rho_0\delta\mathbf{u}^{\mathrm{T}}\mathbf{g}\mathrm{d}V \tag{11.116}$$

and the weak form of the hydrostatic pressure–volume change relation (11.73):

$$\int_{V_0} \delta p\left(p_{j+1} - f_p'\right)\mathrm{d}V = 0 \tag{11.117}$$

Decomposing $\boldsymbol{\tau}_{j+1}$ additively in $\boldsymbol{\tau}_j$ and the correction $\mathrm{d}\boldsymbol{\tau}$ during the iteration, and substitution of Equation (11.114) in an incremental format results in:

$$\begin{aligned}\int_{V_0} \delta\boldsymbol{\gamma}^{\mathrm{T}}\mathbf{D}\mathrm{d}\boldsymbol{\gamma}\mathrm{d}V + \int_{V_0} \delta\boldsymbol{\gamma}^{\mathrm{T}}\boldsymbol{\tau}_j\mathrm{d}V + \int_{V_0} \delta\boldsymbol{\gamma}^{\mathrm{T}}\left(2\frac{\partial J_3^C}{\partial \mathbf{C}}\right)\mathrm{d}p\mathrm{d}V \\ = \int_{S_0} \delta\mathbf{u}^{\mathrm{T}}\mathbf{t}\mathrm{d}S + \int_{V_0} \rho_0\delta\mathbf{u}^{\mathrm{T}}\mathbf{g}\mathrm{d}V\end{aligned} \tag{11.118}$$

By a similar decomposition with respect to the pressure we can rewrite (11.117) as:

$$-\int_{V_0} \delta p\left(2f_p''\frac{\partial J_3^C}{\partial \mathbf{C}}\right)\mathrm{d}\boldsymbol{\gamma}\mathrm{d}V + \int_{V_0} \delta p\mathrm{d}p\mathrm{d}V + \int_{V_0} \delta p\left(p_j - f_p'\right)\mathrm{d}V = 0 \tag{11.119}$$

where for the choice (11.74) obviously $f_p'' = K$. Applying a linearisation of Equation (11.118) similar to that after Equation (3.91), premultiplying Equation (11.119) with $-f_p''$, and rearranging yields:

$$\begin{aligned}\int_{V_0} \delta\mathbf{e}^{\mathrm{T}}\mathbf{D}\mathrm{d}\mathbf{e}\mathrm{d}V + \int_{V_0} \delta\boldsymbol{\eta}^{\mathrm{T}}\boldsymbol{\tau}_j\mathrm{d}V + \int_{V_0} \delta\mathbf{e}^{\mathrm{T}}\left(2\frac{\partial J_3^C}{\partial \mathbf{C}}\right)\mathrm{d}p\mathrm{d}V \\ = \int_{S_0} \delta\mathbf{u}^{\mathrm{T}}\mathbf{t}\mathrm{d}S + \int_{V_0} \rho_0\delta\mathbf{u}^{\mathrm{T}}\mathbf{g}\mathrm{d}V - \int_{V_0} \delta\mathbf{e}^{\mathrm{T}}\boldsymbol{\tau}_j\mathrm{d}V\end{aligned} \tag{11.120}$$

and

$$\int_{V_0} \delta p\left(2\frac{\partial J_3^C}{\partial \mathbf{C}}\right)\mathrm{d}\mathbf{e}\mathrm{d}V - \int_{V_0} (f_p'')^{-1}\delta p\mathrm{d}p\mathrm{d}V = \int_{V_0} (f_p'')^{-1}\delta p\left(p_j - f_p'\right)\mathrm{d}V \tag{11.121}$$

Since the pressure has been assumed to be an independent variable a separate interpolation is necessary for p. Because of its origin as an additional constraint no explicit boundary conditions are required for the pressure field and a wider range of interpolations is permissible than for the displacement field. Furthermore, the pressure degrees of freedom can be introduced either as system or element degrees of freedom. The latter approach, which is more customary, allows the condensation of the pressure degrees of freedom at the element level (Box 6.2), and implies a discontinuous pressure field across the element boundaries. Assembling the interpolation polynomials for the pressure degrees of freedom in a matrix $\mathbf{H}_{\mathrm{p}}$, we have:

$$p = \mathbf{H}_{\mathrm{p}}\mathbf{p} \tag{11.122}$$

where $\mathbf{p}$ contains the values of the nodal pressures. Substitution of the relations (3.96)–(3.104) and (11.122) into Equations (11.120) and (11.121), and considering that the resulting equations must hold for any admissible variation of the displacements and the pressures, gives:

$$\begin{bmatrix} \mathbf{K}_{\mathrm{aa}} & \mathbf{K}_{\mathrm{ap}} \\ \mathbf{K}_{\mathrm{ap}}^{\mathrm{T}} & \mathbf{K}_{\mathrm{pp}} \end{bmatrix} \begin{pmatrix} \mathrm{d}\mathbf{a} \\ \mathrm{d}\mathbf{p} \end{pmatrix} = \begin{pmatrix} \mathbf{f}_{\mathrm{ext}}^{\mathrm{a}} - \mathbf{f}_{\mathrm{int}}^{\mathrm{a}} \\ \mathbf{f}_{\mathrm{int}}^{\mathrm{p}} \end{pmatrix} \tag{11.123}$$

with $\mathbf{f}_{\mathrm{int}}^{\mathrm{a}}$ and $\mathbf{f}_{\mathrm{int}}^{\mathrm{p}}$ given by:

$$\mathbf{f}_{\mathrm{int}}^{\mathrm{a}} = \int_{V_0} \mathbf{B}_L^{\mathrm{T}} \boldsymbol{\tau}_j \mathrm{d}V \tag{11.124}$$

$$\mathbf{f}_{\mathrm{int}}^{\mathrm{p}} = \int_{V_0} (f_p'')^{-1} \mathbf{H}_p^{\mathrm{T}} \left(p_j - f_p' \right) \mathrm{d}V \tag{11.125}$$

while the stiffness matrices are given by:

$$\mathbf{K}_{\mathrm{aa}} = \int_{V_0} \mathbf{B}_L^{\mathrm{T}} \mathbf{D} \mathbf{B}_L \mathrm{d}V + \int_{V_0} \mathbf{B}_{NL}^{\mathrm{T}} \mathcal{T}^t \mathbf{B}_{NL} \mathrm{d}V \tag{11.126}$$

$$\mathbf{K}_{\mathrm{ap}} = \int_{V_0} \mathbf{B}_L^{\mathrm{T}} \left(2 \frac{\partial J_3^C}{\partial \mathbf{C}} \right) \mathbf{H}_p \mathrm{d}V \tag{11.127}$$

$$\mathbf{K}_{\mathrm{pp}} = -\int_{V_0} (f_p'')^{-1} \mathbf{H}_p^{\mathrm{T}} \mathbf{H}_p \mathrm{d}V \tag{11.128}$$

with $\mathcal{T}$ the matrix representation of the Second Piola–Kirchhoff stress tensor, and $\mathbf{B}_L$ and $\mathbf{B}_{NL}$ the 'linear' and 'non-linear' $\mathbf{B}$ matrices, cf. Equations (3.97), (3.103) and (3.104) for the two-dimensional representation.

For the calculation of the element stiffness matrix in Equation (11.123) different interpolation polynomials can be chosen for the displacement and pressure field. An arbitrary combination of interpolation functions, however, may lead to a poor numerical performance. For mixed finite elements a sound mathematical theory is available through the Ladyzenskaya–Babuška–Brezzi (LBB) condition (Hughes 1987). Unfortunately, proof that an element passes the LBB condition is usually rather difficult. Instead, a simple and heuristic method, based on constraint counting (Nagtegaal *et al.* 1974) is often applied to mixed elements with a discontinuous pressure field for a first assessment of their suitability for the analysis of (nearly) incompressible media.

To provide a proper setting, the fundamental difficulty in (nearly) incompressible elasticity is recalled by means of Figure 7.24(a), in which a four-noded, quadrilateral two-dimensional element is plotted with eight displacement degrees of freedom. At the left and bottom edges the displacements have been prescribed to zero, so that the model possesses just two degrees of freedom. We may now envisage a model composed of an arbitrary number of these basic elements. Each element that is added in either direction increases the total number of displacement degrees of freedom by two. The assumption of incompressibility implies a constant surface of the element, and, as a consequence, a restricted movement of the free node along a line with an angle of $-45°$. The remaining degree of freedom is determined by an equation with one pressure degree of freedom as an additional unknown. When two or more pressure points are specified no displacement degrees of freedom are left, which causes volumetric locking. When, on the other hand, no pressure degrees of freedom are introduced, spurious kinematic modes, such as hour-glass modes, see also Chapter 7, can arise during the numerical simulation. For two-dimensional elements the optimal ratio $r = n_{\mathrm{a}}/n_{\mathrm{p}}$ of the number of displacement degrees of freedom n_{a} over the number of pressure degrees of freedom n_{p} therefore equals two, while for three-dimensional elements the optimal ratio $r = 3$. The method outlined above has been applied to determine the optimal number of pressure points in three-dimensional hexahedral elements in Box 11.4.

The method suggests an influence of the boundary conditions on the number of constraint conditions. In fact, the ratio r of the entire finite element model should be considered when calculating the optimal number of pressure degrees of freedom. Ideally, also the number of constraint conditions should be distributed inhomogeneously over the mesh in order to arrive at a locally optimal ratio: fewer pressure degrees of freedom near edges with a prescribed displacement and more pressure degrees of freedom near edges with a prescribed boundary traction.

11.3.2 Enhanced Assumed Strain Elements

Simo and Rifai (1990) have developed a rigorous framework for the enrichment of the kinematics of displacement-based finite elements, the enhanced assumed strain approach. Since the resulting elements are displacement based, the standard methodology for handling inelastic constitutive laws like plasticity and damage, discussed in Chapters 6–8, holds. This is unlike the u/p formulation discussed in Section 11.3.1, and is a major advantage. The methodology was extended to geometrical non-linearity by Simo and Armero (1992). We take the balance of momentum in the current configuration, Equation (11.1), as the starting point, multiply it by the variation of the displacement field, $\delta\mathbf{u}$, utilise the divergence theorem and transform the result to the original configuration to yield Equation (11.6). Using expressions (11.7) and (11.8) we obtain:

$$\int_{V_0} \frac{\rho_0}{\rho} \nabla(\delta\mathbf{u}) : \boldsymbol{\sigma} \mathrm{d}V = \int_{V_0} \delta\mathbf{u} \cdot \rho_0 \mathbf{g} \mathrm{d}V + \int_{S_0} \delta\mathbf{u} \cdot \mathbf{t}_0 \mathrm{d}S \qquad (11.129)$$

which is complemented by the weak forms of the kinematic and the constitutive equations:

$$\int_{V_0} \delta\boldsymbol{\tau} \cdot \left(\frac{1}{2}(\mathbf{F}^{\mathrm{T}} \cdot \mathbf{F} - \mathbf{I}) - \boldsymbol{\gamma} \right) \mathrm{d}V = 0 \qquad (11.130)$$

Box 11.4 Shape functions for three-dimensional u/p elements

Three-dimensional tetrahedral elements which have a constant interpolation polynomial for the pressure field are the 8/1 elements, which have a trilinear displacement interpolation with 8 degrees of freedom, and the 20/1 elements, which have a quadratic, serendipity interpolation for the displacements. For the 8/1 element $n_a = 3$ and $n_p = 1$, so that $r = 3$, which is optimal. For the 20/1 element, however, the ratio $r = 12$, which implies that spurious kinematic modes can arise rather easily. An optimal element is therefore the 20/4 element, for which $r = 12/4 = 3$. In the three-dimensional space it needs a linear interpolation with four parameters, which can be derived by degeneration of the trilinear shape functions, which are for instance used to describe the displacement field within the 8/1 element. Let $\boldsymbol{\xi} = (\xi, \eta, \zeta)$ be the place vector in the isoparametric coordinates and $\mathbf{x} = (x, y, z)$ be the place vector in the model coordinates. Then, the mapping of a point from model to isoparametric coordinates is defined by the shape functions h_i according to

$$\mathbf{x}(\boldsymbol{\xi}) = \sum_{i=1}^{n} h_i(\boldsymbol{\xi})\mathbf{x}_i^e$$

where $\mathbf{x}_i^e$ are the model coordinates of node i of element e. The degeneration process implies the mapping of more than one point in the isoparametric $\boldsymbol{\xi}$-space onto a point in the $\mathbf{x}$-space. A proper linear shape function is obtained when four independent parameters are used to span a three-dimensional space. To meet this requirement and to obtain a symmetric set of polynomials the following degeneration can be employed:

$$h_1^* = h_1 + h_2 = \frac{1}{4}(1 - \eta)(1 - \zeta)$$

$$h_3^* = h_3 + h_4 = \frac{1}{4}(1 + \eta)(1 - \zeta)$$

$$h_6^* = h_6 + h_7 = \frac{1}{4}(1 + \xi)(1 + \zeta)$$

$$h_8^* = h_8 + h_5 = \frac{1}{4}(1 - \xi)(1 + \zeta)$$

This interpolation scheme satisfies the basic convergence requirements with respect to the smoothness and completeness on the element domain (Bathe 1982; Hughes 1987), but continuity across the element boundaries is lost due to the degeneration process. Because of the assumed element-wise condensation of the pressure degrees of freedom this is of no further consequence.

and

$$\int_{V_0} \delta\boldsymbol{\gamma} : (\boldsymbol{\tau} - \boldsymbol{\tau}_\gamma)\mathrm{d}V = 0 \tag{11.131}$$

with $\mathbf{F}$ the deformation gradient that is derived from the displacement field $\mathbf{u}$,

$$\delta\boldsymbol{\tau}_\gamma = \frac{\partial\boldsymbol{\tau}}{\partial\boldsymbol{\gamma}} : \delta\boldsymbol{\gamma} = \mathbf{D} : \delta\boldsymbol{\gamma}$$

the stress that is derived from the constitutive relation, and $\boldsymbol{\tau}$ and $\boldsymbol{\gamma}$ independent stress and strain fields. Equations (11.129)–(11.131) are the stationarity conditions of the three-field Hu–Washizu variational principle, cf. Wall *et al.* (2000). Substitution of the relation between the Cauchy and Second Piola–Kirchhoff stress tensors, Equation (11.13), the variation of the Green–Lagrange strain tensor, Equation (3.79), and exploiting the symmetry of the Second Piola–Kirchhoff stress tensor, transform Equation (11.129) into:

$$\int_{V_0} \mathrm{tr}(\delta \mathbf{F}^{\mathrm{T}} \cdot \boldsymbol{\tau} \cdot \mathbf{F}) \mathrm{d}V = \int_{V_0} \delta \mathbf{u} \cdot \rho_0 \mathbf{g} \mathrm{d}V + \int_{S_0} \delta \mathbf{u} \cdot \mathbf{t}_0 \mathrm{d}S \tag{11.132}$$

The enhanced assumed strain methodology rests on the decomposition of the strain field into a part $\bar{\boldsymbol{\gamma}}$ which is derived from the (continuous) displacement field $\mathbf{u}$ and an additional strain field $\tilde{\boldsymbol{\gamma}}$:

$$\boldsymbol{\gamma} = \underbrace{\frac{1}{2}(\mathbf{F}^{\mathrm{T}} \cdot \mathbf{F} - \mathbf{I})}_{\bar{\boldsymbol{\gamma}}} + \tilde{\boldsymbol{\gamma}} \tag{11.133}$$

Substitution into Equations (11.132), (11.130) and (11.131) results in:

$$\int_{V_0} \mathrm{tr}(\delta \mathbf{F}^{\mathrm{T}} \cdot \boldsymbol{\tau}_\gamma \cdot \mathbf{F}) \mathrm{d}V = \int_{V_0} \delta \mathbf{u} \cdot \rho_0 \mathbf{g} \mathrm{d}V + \int_{S_0} \delta \mathbf{u} \cdot \mathbf{t}_0 \mathrm{d}S \tag{11.134}$$

$$\int_{V_0} \delta \boldsymbol{\tau} : \tilde{\boldsymbol{\gamma}} \mathrm{d}V = 0 \tag{11.135}$$

and

$$\int_{V_0} \delta \tilde{\boldsymbol{\gamma}} : (\boldsymbol{\tau} - \boldsymbol{\tau}_\gamma) \mathrm{d}V = 0 \tag{11.136}$$

In the enhanced assumed strain approach (Simo and Rifai 1990; Simo and Armero 1992) the variations of the stress field and that of the enhanced part of the strain field are assumed to be orthogonal in an L_2-sense:

$$\int_{V_0} \delta \boldsymbol{\tau} : \delta \tilde{\boldsymbol{\gamma}} \mathrm{d}V = 0 \tag{11.137}$$

which can be satisfied for a proper choice of the trial functions for $\tilde{\boldsymbol{\gamma}}$ and $\boldsymbol{\tau}$. Equation (11.137) satisfies Equation (11.135) for $\delta \tilde{\boldsymbol{\gamma}} = \tilde{\boldsymbol{\gamma}}$ and, for $\delta \boldsymbol{\tau} = \boldsymbol{\tau}$, reduces Equation (11.136) to:

$$\int_{V_0} \delta \tilde{\boldsymbol{\gamma}} : \boldsymbol{\tau}_\gamma \mathrm{d}V = 0 \tag{11.138}$$

We interpolate the continuous part of the displacements in a standard manner,

$$\mathbf{u} = \mathbf{H}\mathbf{a}$$

while the enhanced strains can be interpolated via

$$\tilde{\boldsymbol{\gamma}} = \mathbf{G}\boldsymbol{\alpha} \tag{11.139}$$

with $\boldsymbol{\alpha}$ the array that contains the discrete parameters that govern the magnitude of the enhanced strains at the element level. The (non-square) matrix $\mathbf{G}$ depends on how the enhanced strains are formulated. Examples are the strain enrichment to accommodate a discontinuity in Equation (6.70) and the strain enrichment in the modified incompatible modes element of Taylor *et al.* (1976), see Equation (7.219) for the enhanced normal strains of this element. Substitution of these interpolations into Equations (11.134) and (11.138), elaborating the resulting equations in a manner similar to that between Equations (3.86) and (3.104), and requiring that the results hold for all admissible variations, yields the following linearised set of coupled algebraic equations:

$$\begin{bmatrix} \mathbf{K}_{\mathrm{aa}} & \mathbf{K}_{\mathrm{a}\alpha} \\ \mathbf{K}_{\mathrm{a}\alpha}^{\mathrm{T}} & \mathbf{K}_{\alpha\alpha} \end{bmatrix} \begin{pmatrix} \mathrm{d}\mathbf{a} \\ \mathrm{d}\boldsymbol{\alpha} \end{pmatrix} = \begin{pmatrix} \mathbf{f}_{\mathrm{ext}}^{\mathrm{a}} - \mathbf{f}_{\mathrm{int}}^{\mathrm{a}} \\ -\mathbf{f}_{\mathrm{int}}^{\alpha} \end{pmatrix} \tag{11.140}$$

with $\mathbf{f}_{\mathrm{ext}}^{\mathrm{a}}$ the standard external load vector, and the internal force vectors $\mathbf{f}_{\mathrm{int}}^{\mathrm{a}}$, $\mathbf{f}_{\mathrm{int}}^{\alpha}$ given by

$$\mathbf{f}_{\mathrm{int}}^{\mathrm{a}} = \int_{V_0} \mathbf{B}_L^{\mathrm{T}} \boldsymbol{\tau}_\gamma \mathrm{d}V \tag{11.141}$$

$$\mathbf{f}_{\mathrm{int}}^{\alpha} = \int_{V_0} \mathbf{G}^{\mathrm{T}} \boldsymbol{\tau}_\gamma \mathrm{d}V \tag{11.142}$$

with $\mathbf{B}_L$ defined in Equation (3.21). The stiffness matrices are given by:

$$\mathbf{K}_{\mathrm{aa}} = \int_{V_0} \mathbf{B}_L^{\mathrm{T}} \mathbf{D} \mathbf{B}_L \mathrm{d}V + \int_{V_0} \mathbf{B}_{NL}^{\mathrm{T}} \boldsymbol{\mathcal{T}} \mathbf{B}_{NL} \mathrm{d}V \tag{11.143}$$

$$\mathbf{K}_{\mathrm{a}\alpha} = \int_{V_0} \mathbf{B}_L^{\mathrm{T}} \mathbf{D} \mathbf{G} \mathrm{d}V \tag{11.144}$$

$$\mathbf{K}_{\alpha\alpha} = \int_{V_0} \mathbf{G}^{\mathrm{T}} \mathbf{D} \mathbf{G} \mathrm{d}V \tag{11.145}$$

with $\boldsymbol{\mathcal{T}}$ and $\mathbf{B}_{NL}$ defined in Equations (3.103) and (3.104), respectively. Since the enhanced strains $\boldsymbol{\gamma}$ are discontinuous across element boundaries, the array of discrete parameters $\boldsymbol{\alpha}$ resides at the element level, and can be condensed prior to assembling the stiffness matrix at the structural level (Box 6.2).

Several authors (de Souza Neto *et al.* 1995; Wriggers and Reese 1996) have reported that enhanced assumed strain elements are not necessarily stable for large deformations, and remedies have been proposed by Glaser and Armero (1997), Korelc and Wriggers (1996) and Reese and Wriggers (2000). Wall *et al.* (2000) have presented an in-depth analysis of the problem and have proposed a stabilised finite element method based on concepts advocated in computational fluid dynamics.

11.3.3 F-bar Approach

A straightforward method to avoid volumetric locking effects in finite strain analyses is the $\bar{\mathbf{F}}$-concept. The idea behind this method is simple: use the multiplicative decomposition (11.45) which splits the deformation gradient $\mathbf{F}$ into an isochoric contribution $\mathbf{F}_{\mathrm{iso}}$ and a volumetric contribution $\mathbf{F}_{\text{vol}}$ to define a modified deformation gradient

$$\bar{\mathbf{F}} = \mathbf{F}_{\text{iso}} \cdot (\mathbf{F}_{\text{vol}})_0 \tag{11.146}$$

in each integration point, where $(\mathbf{F}_{\text{vol}})_0$ is computed at the element centroid, i.e. at $\boldsymbol{\xi} = \boldsymbol{\xi}_0$. The Second Piola–Kirchhoff stress tensor is subsequently calculated using the modified deformation gradient: $\boldsymbol{\tau}_{j+1} = \boldsymbol{\tau}(\bar{\mathbf{F}}_{j+1})$ and the internal force vector follows conventionally from:

$$\mathbf{f}_{\text{int},j+1} = \int_{V_0} \mathbf{B}_L^{\mathrm{T}} \boldsymbol{\tau}_{j+1} \mathrm{d}V$$

The $\bar{\mathbf{F}}$-approach bears similarity to the $\bar{\mathbf{B}}$-concept that has been advocated by Hughes (1980) to avoid volumetric locking effects in small-strain elasticity and plasticity, see Chapter 7. Indeed, a deviatoric–volumetric split is made in both cases, where the volumetric contribution is evaluated at the centroid of the element. Nevertheless, the elaboration is rather different, since in the $\bar{\mathbf{B}}$-approach the discrete strain displacement is modified through a redefinition of the $\mathbf{B}$ matrix, cf. Equation (7.215). By contrast, no direct modification to the discrete strain-displacement operators is made for the $\bar{\mathbf{F}}$-approach, since only the Green–Lagrange stress tensor is evaluated in a non-standard manner, namely from the modified deformation gradient $\bar{\mathbf{F}}$, and not from $\mathbf{F}$.

The modification to the computation of the stress tensor shows up in the expression for the tangent stiffness matrix (de Souza Neto *et al.* 1996, 2005, 2008). The derivation is quite lengthy, and results in:

$$\mathbf{K} = \int_{V_0} \mathbf{B}_L^{\mathrm{T}} \mathbf{D} \mathbf{B}_L \mathrm{d}V + \int_{V_0} \mathbf{B}_{NL}^{\mathrm{T}} \boldsymbol{\mathcal{T}} \mathbf{B}_{NL} \mathrm{d}V + \int_{V_0} \mathbf{B}^{\mathrm{T}} \boldsymbol{\mathcal{Q}} (\mathbf{B}_0 - \mathbf{B}) \mathrm{d}V \tag{11.147}$$

The matrix $\mathbf{B}_0$ is as $\mathbf{B}$, cf. Equation (2.19), but evaluated at the element centroid, while $\boldsymbol{\mathcal{Q}}$ is given by:

$$\boldsymbol{\mathcal{Q}} = \begin{bmatrix} \frac{E}{3(1-2\nu)} - \frac{1}{3}\tau_{xx} & \frac{E}{3(1-2\nu)} - \frac{1}{3}\tau_{xx} & \frac{E}{3(1-2\nu)} - \frac{1}{3}\tau_{xx} & 0 & 0 & 0 \\ \frac{E}{3(1-2\nu)} - \frac{1}{3}\tau_{yy} & \frac{E}{3(1-2\nu)} - \frac{1}{3}\tau_{yy} & \frac{E}{3(1-2\nu)} - \frac{1}{3}\tau_{yy} & 0 & 0 & 0 \\ \frac{E}{3(1-2\nu)} - \frac{1}{3}\tau_{zz} & \frac{E}{3(1-2\nu)} - \frac{1}{3}\tau_{zz} & \frac{E}{3(1-2\nu)} - \frac{1}{3}\tau_{zz} & 0 & 0 & 0 \\ -\frac{1}{3}\tau_{xy} & -\frac{1}{3}\tau_{xy} & -\frac{1}{3}\tau_{xy} & 0 & 0 & 0 \\ -\frac{1}{3}\tau_{yz} & -\frac{1}{3}\tau_{yz} & -\frac{1}{3}\tau_{yz} & 0 & 0 & 0 \\ -\frac{1}{3}\tau_{zx} & -\frac{1}{3}\tau_{zx} & -\frac{1}{3}\tau_{zx} & 0 & 0 & 0 \end{bmatrix} \tag{11.148}$$

The first two terms in Equation (11.147) are standard, cf. Chapter 3, but the third term is additional and originates from the modification of the stress computation. It is evident that the tangent stiffness matrix for the $\bar{\mathbf{F}}$ method becomes non-symmetric, which is due to the fact that the $\bar{\mathbf{F}}$-approach does not have a proper variational basis.

11.3.4 Corotational Approach

Corotational approaches have been used primarily for structural elements like truss elements (Chapter 3), beam elements or plate and shell elements (Chapters 9 and 10). Crisfield and

Moita (1996); Moita and Crisfield (1996) have developed a corotational formulation for two- and three-dimensional continuum elements, which allows for the use of simple, low-order elements such as four-noded quadrilaterals, or eight-noded brick elements in large-strain analyses by incorporating them in a corotating framework. One advantage of the corotational formulation is its conceptual simplicity. Another advantage is that concepts that have been derived for small displacement gradients, can be included straightforwardly, without complications. An example is the incompatible modes element for improving the bending behaviour and for mitigating volumetric locking of four-noded two-dimensional, and eight-noded three-dimensional elements (Crisfield and Moita 1996).

Hyperelasticity can be introduced in the corotational formulation as follows. In the centroid of the element we have:

$$\mathbf{F}_0 = \mathbf{R}_0 \cdot \mathbf{U}_0 \tag{11.149}$$

where the local rotating base vectors $\bar{\mathbf{n}}_i$ constitute the columns of the rotation matrix $\mathbf{R}_0$, as can be observed from Figure 3.7. Equation (11.149) can be conceived as the polar decomposition in the centroid of the element, and $\mathbf{U}_0$ therefore serves as the right stretch tensor at this point. Its principal values, the stretches λ_i, minus one can be identified as the principal values $\bar{\epsilon}_i$ of the local engineering strain tensor $\bar{\boldsymbol{\epsilon}}$:

$$\bar{\epsilon}_i = \lambda_i - 1 \tag{11.150}$$

Hence, we have the following relation between the local engineering strain and the right stretch tensor at the element centroid:

$$\bar{\boldsymbol{\epsilon}} = \mathbf{U}_0 - \mathbf{I} \tag{11.151}$$

where the right-hand side can be identified as the Biot strain tensor at the centroid. Accordingly, the Biot strain tensor $\mathbf{U} - \mathbf{I}$ reduces to the local engineering strain $\bar{\boldsymbol{\epsilon}}$ in the centre of the element. It is emphasised that this holds only at the centroid, but for gradually varying strain fields the local engineering strain serves as a good approximation of the Biot strain also elsewhere within the element.

Since the Biot stress tensor $\boldsymbol{T}$ is energetically conjugate to the right stretch tensor $\mathbf{U}$, Equation (11.16), it can be derived from a strain energy function $\mathcal{W}$, as follows:

$$\boldsymbol{T} = \frac{\partial \mathcal{W}}{\partial \mathbf{U}} \tag{11.152}$$

or, in the principal directions,

$$\boldsymbol{T} = \sum_{i=1}^{3} \frac{\partial \mathcal{W}}{\partial \lambda_i} \boldsymbol{E}_i = \sum_{i=1}^{3} \frac{\partial \mathcal{W}}{\partial \lambda_i} \boldsymbol{e}_i \otimes \boldsymbol{e}_i \tag{11.153}$$

with $\boldsymbol{e}_i$ the eigenvectors of the right stretch tensor $\mathbf{U}$, or equivalently, of the Biot strain tensor. As a further approximation the Biot stress tensor can be replaced by the local Cauchy stress tensor:

$$\bar{\boldsymbol{\sigma}} = \sum_{i=1}^{3} \bar{\sigma}_i \, \boldsymbol{a}_i \otimes \boldsymbol{a}_i, \quad \bar{\sigma}_i = \frac{\partial \mathcal{W}}{\partial \lambda_i} \tag{11.154}$$

where $\boldsymbol{a}_i$ are the eigenvectors of the local engineering strain $\bar{\epsilon}$. In a similar manner, the components of the material tangential stiffness tensor can be approximated by:

$$\begin{cases} D_{iijj} = \frac{\partial T_i}{\partial \lambda_j} & , \quad i = j \\ D_{ijij} = D_{ijji} = D_{jiij} = D_{jiji} = \frac{T_i - T_j}{2(\lambda_i - \lambda_j)} & , \quad i \neq j \end{cases} \tag{11.155}$$

References

Bathe KJ 1982 *Finite Element Procedures in Engineering Analysis*. Prentice Hall, Inc.

Bonet J and Wood RD 1997 *Nonlinear Continuum Mechanics for Finite Element Analysis*. Cambridge University Press.

Brink U and Stein E 1996 On some mixed finite element methods for incompressible and nearly incompressible finite elasticity. *Computational Mechanics* **19**, 105–119.

Cescotto S and Fonder G 1979 A finite element approach for large strains of nearly incompressible rubber-like materials. *International Journal of Solids and Structures* **15**, 589–605.

Crisfield MA and Moita GF 1996 A co-rotational formulation for 2-D continua including incompatible modes. *International Journal for Numerical Methods in Engineering* **39**, 2619–2633.

de Borst R, van den Bogert PAJ and Zeilmaker J 1988 Modelling and anlaysis of rubberlike materials. *Heron* **33(1)**, 1–57.

de Souza Neto EA, Perić D and Owen DRJ 2008 *Computational Methods for Plasticity: Theory and Applications*. John Wiley & Sons, Ltd.

de Souza Neto EA, Perić D, Dutko M and Owen DRJ 1996 Design of simple low order finite elements for large strain analysis of nearly incompressible solids. *International Journal of Solids and Structures* **33**, 3277–3296.

de Souza Neto EA, Perić D, Huang GC and Owen DRJ 1995 Remarks on the stability of enhanced strain elements in finite elasticity and elastoplasticity. *Communications in Numerical Methods in Engineering* **11**, 951–961.

de Souza Neto EA, Pires FMA and Owen DRJ 2005 F-bar-based linear triangles and tetrahedra for finite strain analysis of nearly incompressible solids. Part 1: Formulation and benchmarking. *International Journal for Numerical Methods in Engineering* **62**, 353–383.

Dienes JK 1979 On the analysis of rotation and stress rate in deforming bodies. *Acta Mechanica* **32**, 217–232.

Dienes JK 1986 A discussion of material rotation and stress rate. *Acta Mechanica* **65**, 1–11.

Doll S, Schweizerhof K, Hauptmann R and Freischläger C 2000 On volumetric locking of low-order solid and solid-shell elements for finite elastoviscoplastic deformations and selective reduced integration. *Engineering Computations* **17**, 874–902.

Glaser S and Armero F 1997 On the formulation of enhanced strain finite elements in finite deformations. *Engineering Computations* **14**, 759–791.

Glowinski R and Tallec PL 1985 Finite elements in nonlinear incompressible elasticity, in *Finite Elements: Special Problems in Solid Mechanics* (eds Oden JT and Carey GF), pp. 67–93. Prentice Hall, Inc.

Hencky H 1933 The elastic behaviour of vulcanized rubber. *Journal of Applied Mechanics* **1**, 45–53.

Hughes TJR 1980 Generalization of selective integration procedures to anisotropic and nonlinear media. *International Journal for Numerical Methods in Engineering* **15**, 1413–1418.

Hughes TJR 1987 *The Finite Element Methods: Linear Static and Dynamic Finite Element Analysis*. Prentice Hall, Inc.

Jankovich E, Leblanc F, Durand M and Bercovier M 1981 A finite element method for the analysis of rubber parts: experimental and numerical assessment. *Computers and Structures* **14**, 385–391.

Jaumann G 1911 Geschlossenes System physikalischer und chemischer Differentialgesetze. *Sitzbereich Akademischen Wissenschaften Wien (IIa)* **120**, 385–530.

Johnson CG and Bamman DJ 1984 A discussion of stress rates in finite deformation problems. *International Journal of Solids and Structures* **20**, 725–737.

Korelc J and Wriggers P 1996 Consistent gradient formulation for a stable enhanced strain method for large deformations. *Engineering Computations* **13**, 103–123.

Moita GF and Crisfield MA 1996 A finite element formulation for 3-D continua using the co-rotational technique. *International Journal for Numerical Methods in Engineering* **39**, 3775–3792.

Mooney M 1940 A theory of large elastic deformations. *Journal of Applied Mechanics* **11**, 582–592.

Moran B, Ortiz M and Shih CF 1990 Formulation of implicit finite element methods for multiplicative finite deformation plasticity. *International Journal for Numerical Methods in Engineering* **29**, 483–514.

Nagtegaal JC and de Jong JE 1982 Some aspects of nonisotropic work-hardening in finite strain plasticity, in *Plasticity of Metals at Finite Deformation* (eds Lee EH and Mallet RL), p. 65. Stanford University.

Nagtegaal JC, Parks DM and Rice JR 1974 On numerically accurate finite element solutions in the fully plastic range. *Computer Methods in Applied Mechanics and Engineering* **4**, 113–135.

Ogden RW 1972 Large deformation isotropic elasticity: on the correlation of theory and experiment for incompressible rubberlike solids. *Proceedings of the Royal Society of London* **A326**, 565–584.

Ogden RW 1984 *Non-linear Elastic Deformations*. Ellis Horwood.

Peng STJ 1979 The elastic potential function of slightly compressible rubberlike materials. *Journal of Polymer Science* **17**, 345–350.

Peng STJ and Landel RF 1975 Stored energy function and compressibility of compressible rubberlike materials under large strain. *Journal of Applied Physics* **46**, 2599–2604.

Penn RW 1970 Volume changes accompanying the extension of rubber. *Transactions of the Society of Rheology* **14**, 509–517.

Reese S and Wriggers P 2000 A stabilization technique to avoid hourglassing in finite elasticity. *International Journal for Numerical Methods in Engineering* **48**, 79–109.

Rivlin RS 1948 Large elastic deformations of isotropic materials IV. Further developments of the general theory. *Philosophical Transactions of the Royal Society* **A241**, 379–397.

Scharnhorst T and Pian THH 1978 Finite element analysis of rubber-like materials by a mixed model. *International Journal for Numerical Methods in Engineering* **12**, 665–676.

Simo JC and Armero F 1992 Geometrically non-linear enhanced strain mixed methods and the method of incompatible modes. *International Journal for Numerical Methods in Engineering* **33**, 1413–1449.

Simo JC and Pister KS 1984 Remarks on rate constitutive equations for finite deformation problems: computational implications. *Computer Methods in Applied Mechanics and Engineering* **46**, 201–215.

Simo JC and Rifai MS 1990 A class of mixed assumed strain methods and the method of incompatible modes. *International Journal for Numerical Methods in Engineering* **29**, 1595–1638.

Simo JC and Taylor RL 1982 Penalty function formulations for incompressible nonlinear elastostatics. *Computer Methods in Applied Mechanics and Engineering* **35**, 107–118.

Simo JC, Taylor RL and Pister KS 1985 Variational and projection methods for the volume constraint in finite deformation elasto-plasticity. *Computer Methods in Applied Mechanics and Engineering* **51**, 177–208.

Sussmann T and Bathe KJ 1987 A finite element formulation for nonlinear incompressible elastic and inelastic analysis. *Computers & Structures* **29**, 357–409.

Taylor RL, Beresford PJ and Wilson EL 1976 A non-conforming element for stress analysis. *International Journal for Numerical Methods in Engineering* **10**, 1211–1219.

Valanis KC and Landel RF 1967 The strain-energy function of a hyperelastic mateiral in terms of extension ratios. *Journal of Applied Physics* **38**, 2997–3002.

van den Bogert PAJ, de Borst R, Luiten GT and Zeilmaker J 1991 Robust finite elements for 3D-analysis of rubber-like materials. *Engineering Computations* **8**, 3–17.

Varga OH 1966 *Stress–Strain Behaviour of Elastic Materials*. John Wiley & Sons, Ltd.

Wall WA, Bischoff M and Ramm E 2000 A deformation dependent stabilization technique, exemplified by EAS elements at large strains. *Computer Methods in Applied Mechanics and Engineering* **188**, 859–871.

Wriggers P and Reese S 1996 A note on enhanced strain methods for large deformations. *Computer Methods in Applied Mechanics and Engineering* **135**, 201–209.

12

Large-strain Elasto-plasticity

In Chapter 3 large-displacement formulations have been developed for continuum elements. Although there is no limitation from the kinematics, these formulations were limited to small strains because of the constitutive assumption, namely that the rate of the Second Piola–Kirchhoff stress could be related to the rate of the Green–Lagrange strain tensor. In Chapter 11 this limitation was relaxed for hyperelastic materials, and the existence of a strain energy function $\mathcal{W}$ then allowed the Second Piola–Kirchhoff stress to be obtained by direct differentiation of the strain energy function in a physically meaningful manner. The existence of a strain energy function is, however, limited to hyperelastic materials, and excludes path dependence, as in plasticity or damage theories. As we have seen in Chapters 6 and 7, constitutive relations are then phrased as a (linear) relation between the stress rate and the strain rate. In a large-strain context such a constitutive relation – sometimes, but not entirely correctly, called a hypoelastic relation – must be generalised to a relation between an objective stress rate and the rate of deformation, see Chapter 11. Since the real, or 'true', stresses are related to the actual deformation there are advantages of formulating the finite element equations in the current configuration, thus using an Eulerian approach, rather than a Lagrangian approach.

Indeed, the pioneering works of Hibbit *et al.* (1970), McMeeking and Rice (1975), Nagtegaal (1982), and Nagtegaal and de Jong (1981) have exploited such an approach. Since then, there has been a plethora of publications related to large elasto-plastic strains. Although we shall focus on the formulation of proper algorithms and discretisations in this chapter, we note that a large body of literature has, necessarily, focused on the correct representation of the underlying physics (Asaro 1983; Atluri 1984; Dafalias 1984, 1985, 1998; Kratochvil 1973; Mandel 1974; Nemat-Nasser 1982; Rice 1975), including the issue of the so-called 'plastic' spin.

When adopting an Eulerian approach, and a 'hypoelastic' format for the relation between the stress rate and the rate of deformation, the issue arises as to which objective stress rate is best used. In the preceding chapter we have discussed a number of possibilities, including some of the anomalies that may arise, for instance when shearing an elementary cube of material. An oscillatory behaviour can be observed already for elastic constitutive relations – indeed a true hypoelastic relation was used in Equation (11.38) to arrive at the results in Figure 11.1 – which can be aggravated when using a kinematic-hardening elasto-plastic relation. But, irrespective

Non-linear Finite Element Analysis of Solids and Structures, Second Edition.
René de Borst, Mike A. Crisfield, Joris J.C. Remmers and Clemens V. Verhoosel.

of the choice of the objective stress rate, it proved difficult to ensure objectivity under finite rotations, including the avoidance of 'self-straining' under rigid rotations. While this issue may be not so prominent in explicit dynamics codes, see Chapter 5, where time steps have to be taken very small anyway, approaches have been pursued to minimise 'self-straining' under rigid rotations and to ensure objectivity for finite increments (Flanagan and Taylor 1987; Hughes and Winget 1980; Key and Krieg 1982; Pinsky *et al.* 1983).

The difficulties that are associated with the use of 'hypoelastic' type constitutive models for the analysis of large elasto-plastic strains have led to an alternative, more rigorous approach, which is based on a multiplicative decomposition of the deformation gradient into a deformation gradient that describes the elastic deformations, and a deformation gradient that captures the plastic component (Atluri 1983; Lee 1969; Nemat-Nasser 1983). Early work on large-strain finite element implementations that exploit the multiplicative elasto-plastic decomposition includes that by Armero (2004), de Souza Neto *et al.* (2008), Healey and Dodds (1992), Moran *et al.* (1990), Simo (1985, 1988a,b), Simo and Hughes (1998) and Simo and Ortiz (1985). Typically, a hyperelastic constitutive relation governs the elastic deformations, but the main purpose of this approach is to by-pass the integration of the stress rates in the plasticity part, rather than considering large elastic strains. Indeed, most approaches that have been presented assume that the elastic strains remain small, at least compared with the plastic strains.

In this chapter we will first outline the Eulerian approach discussed above, and then directly proceed with the multiplicative decomposition. From this decomposition, the relation to rate formulations will be made, and algorithms will be described that build upon this decomposition.

12.1 Eulerian Formulations

An Eulerian finite element formulation is most conveniently developed starting from the virtual work expression in the current configuration, Equation (2.37):

$$\int_V \delta\boldsymbol{\epsilon} : \boldsymbol{\sigma} \mathrm{d}V = \int_V \rho \delta\mathbf{u} \cdot \mathbf{g} \mathrm{d}V + \int_S \delta\mathbf{u} \cdot \mathbf{t} \mathrm{d}S \tag{12.1}$$

Using Equations (3.74) and (11.15) this identity can also be expressed in terms of the Kirchhoff stress tensor:

$$\int_{V_0} \delta\boldsymbol{\epsilon} : \boldsymbol{\kappa} \mathrm{d}V = \int_{V_0} \rho_0 \delta\mathbf{u} \cdot \mathbf{g} \mathrm{d}V + \int_{S_0} \delta\mathbf{u} \cdot \mathbf{t}_0 \mathrm{d}S \tag{12.2}$$

which has the advantage that the integration can now be carried out for the known volume V_0. As with the Lagrange formulations we use the kinematic relation between the virtual strains and the virtual displacements, Equation (2.36), and the discretisation, Equation (2.12), to arrive at:

$$\delta\boldsymbol{\epsilon} = \mathbf{B}\delta\mathbf{a} \tag{12.3}$$

where for notational simplicity, the element index e has been omitted. For general three-dimensional conditions we have

$$\mathbf{B} = \mathbf{J}^{-1} \begin{bmatrix} \frac{\partial h_1}{\partial \xi} & 0 & 0 & \dots & \dots & \dots & \frac{\partial h_n}{\partial \xi} & 0 & 0 \\ 0 & \frac{\partial h_1}{\partial \eta} & 0 & \dots & \dots & \dots & 0 & \frac{\partial h_n}{\partial \eta} & 0 \\ 0 & 0 & \frac{\partial h_1}{\partial \zeta} & \dots & \dots & \dots & 0 & 0 & \frac{\partial h_n}{\partial \zeta} \\ \frac{\partial h_1}{\partial \eta} & \frac{\partial h_1}{\partial \xi} & 0 & \dots & \dots & \dots & \frac{\partial h_n}{\partial \eta} & \frac{\partial h_n}{\partial \xi} & 0 \\ 0 & \frac{\partial h_1}{\partial \zeta} & \frac{\partial h_1}{\partial \eta} & \dots & \dots & \dots & 0 & \frac{\partial h_n}{\partial \zeta} & \frac{\partial h_n}{\partial \eta} \\ \frac{\partial h_1}{\partial \zeta} & 0 & \frac{\partial h_1}{\partial \xi} & \dots & \dots & \dots & \frac{\partial h_n}{\partial \zeta} & 0 & \frac{\partial h_n}{\partial \xi} \end{bmatrix} \tag{12.4}$$

with $\mathbf{J}^{-1}$ the 6×6 inverse of the Jacobian matrix, see also Box 2.2. Substitution of the discrete kinematic relation, and requiring that the resulting discrete equations hold for any virtual nodal displacement $\delta\mathbf{a}$, yields the discrete equilibrium equation:

$$\mathbf{f}_{\text{int}} = \mathbf{f}_{\text{ext}} \tag{12.5}$$

and, for the Cauchy stress tensor, the internal force vector reads:

$$\mathbf{f}_{\text{int}} = \int_V \mathbf{B}^{\mathrm{T}} \boldsymbol{\sigma} \mathrm{d}V \tag{12.6}$$

whereas for the Kirchhoff stress tensor we have:

$$\mathbf{f}_{\text{int}} = \int_{V_0} \mathbf{B}^{\mathrm{T}} \boldsymbol{\kappa} \mathrm{d}V \tag{12.7}$$

The tangential stiffness matrix can be obtained in a standard manner, namely by differentiating the internal virtual work. For the Kirchhoff stress this results in:

$$\int_{V_0} \dot{\overline{\delta\boldsymbol{\epsilon} : \boldsymbol{\kappa}}} \mathrm{d}V = \int_{V_0} \delta\dot{\boldsymbol{\epsilon}} : \boldsymbol{\kappa} \mathrm{d}V + \int_{V_0} \delta\boldsymbol{\epsilon} : \dot{\boldsymbol{\kappa}} \mathrm{d}V = \int_{V_0} \delta\boldsymbol{\ell} : \boldsymbol{\kappa} \mathrm{d}V + \int_{V_0} \delta\boldsymbol{\epsilon} : \dot{\boldsymbol{\kappa}} \mathrm{d}V \tag{12.8}$$

where the latter equality sign holds because of the symmetry of the Kirchhoff stress tensor. We now elaborate $\delta\boldsymbol{\ell}$ and first note that because $\boldsymbol{\xi}$ is fixed,

$$\delta\left(\frac{\partial \dot{\mathbf{x}}}{\partial \boldsymbol{\xi}}\right) = \delta\left(\frac{\partial \dot{\mathbf{x}}}{\partial \mathbf{x}}\right) \cdot \frac{\partial \mathbf{x}}{\partial \boldsymbol{\xi}} + \frac{\partial \dot{\mathbf{x}}}{\partial \mathbf{x}} \cdot \delta\left(\frac{\partial \mathbf{x}}{\partial \boldsymbol{\xi}}\right) = \mathbf{0}$$

so that, using Equation (11.26):

$$\delta\boldsymbol{\ell} = -\boldsymbol{\ell} \cdot \delta\left(\frac{\partial \mathbf{x}}{\partial \boldsymbol{\xi}}\right) \cdot \left(\frac{\partial \mathbf{x}}{\partial \boldsymbol{\xi}}\right)^{-1} = -\boldsymbol{\ell} \cdot \frac{\partial \delta\mathbf{x}}{\partial \boldsymbol{\xi}} \cdot \left(\frac{\partial \mathbf{x}}{\partial \boldsymbol{\xi}}\right)^{-1} = -\boldsymbol{\ell} \cdot \frac{\partial \delta\mathbf{u}}{\partial \mathbf{x}} \tag{12.9}$$

Next, we choose the Truesdell rate of the Kirchhoff stress as the objective stress rate, Equation (11.37), and relate this stress rate to the rate of deformation tensor via the material tangential stiffness tensor $\mathbf{D}^{\mathrm{TK}}$, where the superscript 'TK' stands for the Truesdell rate of the Kirchhoff stress tensor, so that:

$$\dot{\boldsymbol{\kappa}} = \mathbf{D}^{\mathrm{TK}} : \dot{\boldsymbol{\epsilon}} + \boldsymbol{\ell} \cdot \boldsymbol{\kappa} + \boldsymbol{\kappa} \cdot \boldsymbol{\ell}^{\mathrm{T}} \tag{12.10}$$

Note that the superscript 'T' continues to denote the transpose of a quantity. We next substitute Equation (12.9) and this identity into Equation (12.8) to obtain:

$$\int_{V_0} \overline{\delta\epsilon : \kappa}^{\cdot} \mathrm{d}V = -\int_{V_0} \left(\boldsymbol{\ell} \cdot \frac{\partial \delta \mathbf{u}}{\partial \mathbf{x}} \right) : \kappa \mathrm{d}V + \int_{V_0} \delta\epsilon : \left(\mathbf{D}^{\mathrm{TK}} : \dot{\epsilon} + \boldsymbol{\ell} \cdot \kappa + \kappa \cdot \boldsymbol{\ell}^{\mathrm{T}} \right) \mathrm{d}V \quad (12.11)$$

Exploiting the symmetry of the Kirchhoff stress tensor this identity can be reworked to give:

$$\int_{V_0} \overline{\delta\epsilon : \kappa}^{\cdot} \mathrm{d}V = \int_{V_0} \delta\epsilon : \mathbf{D}^{\mathrm{TK}} : \dot{\epsilon} \mathrm{d}V + \int_{V_0} \left(\frac{\partial \delta \mathbf{u}}{\partial \mathbf{x}} \right)^{\mathrm{T}} \cdot \kappa \cdot \boldsymbol{\ell} \, \mathrm{d}V \quad (12.12)$$

This equation can be discretised in a straightforward manner, yielding:

$$\int_{V_0} \delta\epsilon : \mathbf{D}^{\mathrm{TK}} : \dot{\epsilon} \mathrm{d}V + \int_{V_0} \left(\frac{\partial \delta \mathbf{u}}{\partial \mathbf{x}} \right)^{\mathrm{T}} \cdot \kappa \cdot \boldsymbol{\ell} \, \mathrm{d}V = \delta \mathbf{a}^{\mathrm{T}} \mathbf{K} \dot{\mathbf{a}}$$

with

$$\mathbf{K} = \int_{V_0} \mathbf{B}^{\mathrm{T}} \mathbf{D}^{\mathrm{TK}} \mathbf{B} \mathrm{d}V + \int_{V_0} \mathbf{G}^{\mathrm{T}} \mathcal{K} \mathbf{G} \mathrm{d}V \quad (12.13)$$

the tangential stiffness matrix, and **B** given in Equation (12.4), while $\mathcal{K}$ and **G** are given by

$$\mathcal{K} = \begin{bmatrix} \kappa_{xx} & \kappa_{xy} & \kappa_{zx} & 0 & 0 & 0 & 0 & 0 & 0 \\ \kappa_{xy} & \kappa_{yy} & \kappa_{yz} & 0 & 0 & 0 & 0 & 0 & 0 \\ \kappa_{zx} & \kappa_{yz} & \kappa_{zz} & 0 & 0 & 0 & 0 & 0 & 0 \\ 0 & 0 & 0 & \kappa_{xx} & \kappa_{xy} & \kappa_{zx} & 0 & 0 & 0 \\ 0 & 0 & 0 & \kappa_{xy} & \kappa_{yy} & \kappa_{yz} & 0 & 0 & 0 \\ 0 & 0 & 0 & \kappa_{zx} & \kappa_{yz} & \kappa_{zz} & 0 & 0 & 0 \\ 0 & 0 & 0 & 0 & 0 & 0 & \kappa_{xx} & \kappa_{xy} & \kappa_{zx} \\ 0 & 0 & 0 & 0 & 0 & 0 & \kappa_{xy} & \kappa_{yy} & \kappa_{yz} \\ 0 & 0 & 0 & 0 & 0 & 0 & \kappa_{zx} & \kappa_{yz} & \kappa_{zz} \end{bmatrix} \quad (12.14)$$

and

$$\mathbf{G} = \begin{bmatrix} \frac{\partial h_1}{\partial x_1} & 0 & 0 & \frac{\partial h_2}{\partial x_1} & 0 & 0 & \dots & \dots & \dots \\ \frac{\partial h_1}{\partial x_2} & 0 & 0 & \frac{\partial h_2}{\partial x_2} & 0 & 0 & \dots & \dots & \dots \\ \frac{\partial h_1}{\partial x_3} & 0 & 0 & \frac{\partial h_2}{\partial x_3} & 0 & 0 & \dots & \dots & \dots \\ 0 & \frac{\partial h_1}{\partial x_1} & 0 & 0 & \frac{\partial h_2}{\partial x_1} & 0 & \dots & \dots & \dots \\ 0 & \frac{\partial h_1}{\partial x_2} & 0 & 0 & \frac{\partial h_2}{\partial x_2} & 0 & \dots & \dots & \dots \\ 0 & \frac{\partial h_1}{\partial x_3} & 0 & 0 & \frac{\partial h_2}{\partial x_3} & 0 & \dots & \dots & \dots \\ 0 & 0 & \frac{\partial h_1}{\partial x_1} & 0 & 0 & \frac{\partial h_2}{\partial x_1} & \dots & \dots & \dots \\ 0 & 0 & \frac{\partial h_1}{\partial x_2} & 0 & 0 & \frac{\partial h_2}{\partial x_2} & \dots & \dots & \dots \\ 0 & 0 & \frac{\partial h_1}{\partial x_3} & 0 & 0 & \frac{\partial h_2}{\partial x_3} & \dots & \dots & \dots \end{bmatrix} \quad (12.15)$$

Please note the similarity with the matrices $\mathcal{T}$ and $\mathbf{B}_{NL}$ of the Lagrange approach, Equations (3.105) and (3.106), respectively.

Evidently, the choice of the Truesdell rate of the Kirchhoff stress yields a particularly simple expression for the tangential stiffness matrix. This is not so for other choices of the objective stress rate and the stress tensor. For instance, if we take the Jaumann rate in conjunction with the Kirchhoff stress tensor, the resulting expression for the tangential stiffness matrix is:

$$\mathbf{K} = \int_{V_0} \mathbf{B}^{\mathrm{T}} \left(\mathbf{D}^{\mathrm{JK}} - \mathcal{K}^{\mathrm{JK}} \right) \mathbf{B} \mathrm{d}V + \int_{V_0} \mathbf{G}^{\mathrm{T}} \mathcal{K} \mathbf{G} \mathrm{d}V \tag{12.16}$$

where the superscript 'JK' at the constitutive matrix $\mathbf{D}^{\mathrm{JK}}$ denotes the Jaumann derivative of the Kirchhoff stress. Considering the definitions of the Truesdell rate and the Jaumann rate of the Kirchhoff stress, Equations (11.35) and (11.37), and using the fact that $\boldsymbol{w}$ is anti-symmetric, we have

$$\overset{\star}{\boldsymbol{\kappa}} + \boldsymbol{\ell} \cdot \boldsymbol{\kappa} + \boldsymbol{\kappa} \cdot \boldsymbol{\ell}^{\mathrm{T}} = \overset{\circ}{\boldsymbol{\kappa}} + \boldsymbol{w} \cdot \boldsymbol{\kappa} + \boldsymbol{w} \cdot \boldsymbol{\kappa}^{\mathrm{T}}$$

see Section 11.1.2 for the definitions of the symbols $\star$ and $\circ$. Using the decomposition of the velocity gradient, Equation (11.29), and exploiting the symmetry of the rate of deformation, we can explicitly express the Kirchhoff stress rate in terms of the Jaumann stress rate:

$$\overset{\star}{\boldsymbol{\kappa}} = \overset{\circ}{\boldsymbol{\kappa}} - \dot{\boldsymbol{\epsilon}} \cdot \boldsymbol{\kappa} - \boldsymbol{\kappa} \cdot \dot{\boldsymbol{\epsilon}} \tag{12.17}$$

Substitution of the constitutive matrices then gives,

$$\mathbf{D}^{\mathrm{TK}} : \dot{\boldsymbol{\epsilon}} = \mathbf{D}^{\mathrm{JK}} : \dot{\boldsymbol{\epsilon}} - \dot{\boldsymbol{\epsilon}} \cdot \boldsymbol{\kappa} - \boldsymbol{\kappa} \cdot \dot{\boldsymbol{\epsilon}} \tag{12.18}$$

so that, in index notation, both constitutive matrices are related through:

$$D_{ijkl}^{\mathrm{TK}} = D_{ijkl}^{\mathrm{JK}} - \frac{1}{2} \left(\kappa_{il} \delta_{jk} + \kappa_{jl} \delta_{ik} + \kappa_{ik} \delta_{jl} + \kappa_{jk} \delta_{il} \right) \tag{12.19}$$

where it is noted that expressing one objective stress rate into another through a modification of the constitutive matrix is always possible. Rewriting in matrix-vector format yields:

$$\mathbf{D}^{\mathrm{TK}} = \mathbf{D}^{\mathrm{JK}} - \mathcal{K}^{\mathrm{JK}} \tag{12.20}$$

which explains Equation (12.16), with

$$\mathcal{K}^{\mathrm{JK}} = \begin{bmatrix} 2\kappa_{xx} & 0 & 0 & 0 & \kappa_{zx} & \kappa_{xy} \\ 0 & 2\kappa_{yy} & 0 & \kappa_{yz} & 0 & \kappa_{xy} \\ 0 & 0 & 2\kappa_{yy} & \kappa_{yz} & \kappa_{zx} & 0 \\ 0 & \kappa_{yz} & \kappa_{yz} & \frac{1}{2}(\kappa_{yy} + \kappa_{zz}) & \kappa_{xy} & \kappa_{zx} \\ \kappa_{zx} & 0 & \kappa_{zx} & \kappa_{xy} & \frac{1}{2}(\kappa_{zz} + \kappa_{xx}) & \kappa_{yz} \\ \kappa_{xy} & \kappa_{xy} & 0 & \kappa_{zx} & \kappa_{yz} & \frac{1}{2}(\kappa_{zz} + \kappa_{xx}) \end{bmatrix} \tag{12.21}$$

Using a similar procedure the tangential stiffness matrices for other objective rates in conjunction with different stress measures can be derived. For instance, when using the Jaumann rate for the Cauchy stress tensor, one arrives at (McMeeking and Rice 1975):

$$\mathbf{K} = \int_V \mathbf{B}^{\mathrm{T}} \left(\mathbf{D}^{\mathrm{JC}} - \mathcal{S}^{\mathrm{JK}} + \mathcal{S}^{\mathrm{JC}} \right) \mathbf{B} \mathrm{d}V + \int_V \mathbf{G}^{\mathrm{T}} \mathcal{S} \mathbf{G} \mathrm{d}V \tag{12.22}$$

with

$$\mathcal{S} = \begin{bmatrix} \sigma_{xx} & \sigma_{xy} & \sigma_{zx} & 0 & 0 & 0 & 0 & 0 & 0 \\ \sigma_{xy} & \sigma_{yy} & \sigma_{yz} & 0 & 0 & 0 & 0 & 0 & 0 \\ \sigma_{zx} & \sigma_{yz} & \sigma_{zz} & 0 & 0 & 0 & 0 & 0 & 0 \\ 0 & 0 & 0 & \sigma_{xx} & \sigma_{xy} & \sigma_{zx} & 0 & 0 & 0 \\ 0 & 0 & 0 & \sigma_{xy} & \sigma_{yy} & \sigma_{yz} & 0 & 0 & 0 \\ 0 & 0 & 0 & \sigma_{zx} & \sigma_{yz} & \sigma_{zz} & 0 & 0 & 0 \\ 0 & 0 & 0 & 0 & 0 & 0 & \sigma_{xx} & \sigma_{xy} & \sigma_{zx} \\ 0 & 0 & 0 & 0 & 0 & 0 & \sigma_{xy} & \sigma_{yy} & \sigma_{yz} \\ 0 & 0 & 0 & 0 & 0 & 0 & \sigma_{zx} & \sigma_{yz} & \sigma_{zz} \end{bmatrix} \tag{12.23}$$

and $\mathcal{S}^{\mathrm{JK}}$ as $\mathcal{K}^{\mathrm{JK}}$, Equation (12.21), but with Cauchy stresses in lieu of Kirchhoff stresses. The matrix $\mathcal{S}^{\mathrm{JC}}$ is derived from the relation between the Jaumann rates of the Cauchy and the Kirchhoff stress tensors, Equations (11.34) and (11.35), resulting in:

$$\mathbf{D}^{\mathrm{JK}} : \dot{\boldsymbol{\epsilon}} = \det \mathbf{F} \left(\mathbf{D}^{\mathrm{JC}} : \dot{\boldsymbol{\epsilon}} + \boldsymbol{\sigma} \, \mathrm{tr}(\dot{\boldsymbol{\epsilon}}) \right) \tag{12.24}$$

In index notation, this equation can be reworked to give:

$$D^{\mathrm{JK}}_{ijkl} = \det \mathbf{F} \left(D^{\mathrm{JC}}_{ijkl} + \sigma_{ij} \delta_{kl} \right) \tag{12.25}$$

so that we finally obtain the matrix:

$$\mathcal{S}^{\mathrm{JC}} = \begin{bmatrix} \sigma_{xx} & \sigma_{xx} & \sigma_{xx} & 0 & 0 & 0 \\ \sigma_{yy} & \sigma_{yy} & \sigma_{yy} & 0 & 0 & 0 \\ \sigma_{zz} & \sigma_{zz} & \sigma_{zz} & 0 & 0 & 0 \\ \sigma_{xy} & \sigma_{xy} & \sigma_{xy} & 0 & 0 & 0 \\ \sigma_{yz} & \sigma_{yz} & \sigma_{yz} & 0 & 0 & 0 \\ \sigma_{zx} & \sigma_{zx} & \sigma_{zx} & 0 & 0 & 0 \end{bmatrix} \tag{12.26}$$

where one can note a resemblance with the matrix $\mathcal{Q}$ employed in the $\bar{\mathbf{F}}$-approach, Equation (11.148), but, by contrast, in the present case, the matrix $\mathcal{S}^{\mathrm{JC}}$ does not necessarily cause a non-symmetry of the system. Indeed, the non-symmetry in $\mathcal{S}^{\mathrm{JC}}$ can be balanced by a non-symmetry in $\mathbf{D}^{\mathrm{JC}}$, which, for instance, is the case when a hyperelastic constitutive relation is adopted.

By contrast, the use of the Truesdell rate of the Cauchy stress does not result in a non-symmetric tangential stiffness matrix. Indeed, from Equations (11.36) and (11.37) we obtain:

$$\mathbf{D}^{\mathrm{TK}} : \dot{\boldsymbol{\epsilon}} = \det \mathbf{F} \, \mathbf{D}^{\mathrm{TC}} : \dot{\boldsymbol{\epsilon}} \tag{12.27}$$

Inserting this identity into Equation (12.13) and using the relation between the Cauchy and the Kirchhoff stress tensors, Equation (11.15), yields the tangential stiffness matrix:

$$\mathbf{K} = \int_V \mathbf{B}^{\mathrm{T}}\mathbf{D}^{\mathrm{TC}}\mathbf{B}\mathrm{d}V + \int_V \mathbf{G}^{\mathrm{T}}\boldsymbol{\mathcal{S}}\mathbf{G}\mathrm{d}V \tag{12.28}$$

12.2 Multiplicative Elasto-plasticity

The multiplicative elasto-plastic decomposition originates from Lee (1969), and assumes the existence of three configurations: the initial, undeformed configuration, with a line segment $\mathrm{d}\boldsymbol{\xi}$, which is first moved, by a purely plastic deformation, into an intermediate configuration $\mathrm{d}\hat{\mathbf{x}}$, and subsequently, into the final configuration $\mathrm{d}\mathbf{x}$ through a pure elastic deformation (Figure 12.1). In keeping with the notation introduced in Chapter 3 we then have

$$\mathrm{d}\hat{\mathbf{x}} = \frac{\partial \hat{\mathbf{x}}}{\partial \boldsymbol{\xi}} \cdot \mathrm{d}\boldsymbol{\xi} \;\rightarrow\; \mathbf{F}^{\mathrm{p}} = \frac{\partial \hat{\mathbf{x}}}{\partial \boldsymbol{\xi}} \tag{12.29}$$

for the mapping from the initial state to the intermediate configuration, with $\mathbf{F}^{\mathrm{p}}$ the plastic part of the deformation gradient, and

$$\mathrm{d}\mathbf{x} = \frac{\partial \mathbf{x}}{\partial \hat{\mathbf{x}}} \cdot \mathrm{d}\hat{\mathbf{x}} \;\rightarrow\; \mathbf{F}^{\mathrm{e}} = \frac{\partial \mathbf{x}}{\partial \hat{\mathbf{x}}} \tag{12.30}$$

the mapping from the intermediate state to the final configuration, with the elastic part of the deformation gradient, which, in standard manner, can be decomposed into a rotational part,

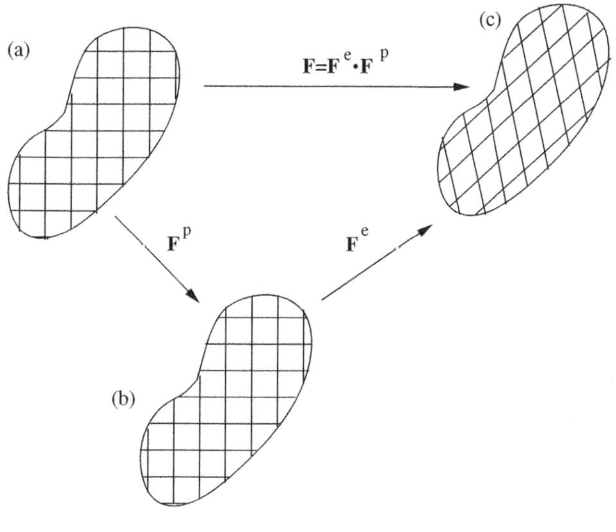

Figure 12.1 The multiplicative elasto-plastic decomposition: (a) initial state; (b) intermediate state; and (c) final state

$\mathbf{R}^e$ and a contribution that stems from a pure deformation, $\mathbf{U}^e$, as follows:

$$\mathbf{F}^e = \mathbf{R}^e \cdot \mathbf{U}^e \tag{12.31}$$

Subsequently, definition (12.30) can be used to define the right Cauchy–Green deformation tensor referred to the intermediate, elastic reference state $\hat{\mathbf{x}}$,

$$\mathbf{C}^e = (\mathbf{F}^e)^T \cdot \mathbf{F}^e \tag{12.32}$$

the 'elastic' Green–Lagrange strain tensor based upon $\mathbf{C}^e$,

$$\boldsymbol{\gamma}^e = \frac{1}{2}\left(\mathbf{C}^e - \mathbf{I}\right) \tag{12.33}$$

and the left Cauchy–Green deformation tensor referred to the intermediate, elastic reference state:

$$\mathbf{B}^e = \mathbf{F}^e \cdot (\mathbf{F}^e)^T \tag{12.34}$$

Considering the definition of the deformation gradient, Equation (3.54), and combining Equations (12.29) and (12.30) gives the multiplicative decomposition of the deformation gradient for elasto-plastic deformations:

$$\mathbf{F} = \frac{\partial \mathbf{x}}{\partial \boldsymbol{\xi}} = \frac{\partial \mathbf{x}}{\partial \hat{\mathbf{x}}} \cdot \frac{\partial \hat{\mathbf{x}}}{\partial \boldsymbol{\xi}} = \mathbf{F}^e \cdot \mathbf{F}^p \tag{12.35}$$

For the one-dimensional case, Equation (12.35) particularises as:

$$\lambda = \frac{\ell}{\ell_0} = \frac{\ell}{\ell^p}\frac{\ell^p}{\ell_0} = \lambda^e \lambda^p \tag{12.36}$$

with λ the stretch ratio, which is multiplicatively decomposed into an elastic and a plastic stretch ratio, λ^e and λ^p, respectively.

The multiplicative decomposition for elasto-plasticity is not unique. For instance, it would be equally possible to rotate the intermediate configuration by $\mathbf{R}$, such that:

$$\mathrm{d}\bar{\mathbf{x}} = \frac{\partial \bar{\mathbf{x}}}{\partial \hat{\mathbf{x}}} \cdot \mathrm{d}\hat{\mathbf{x}} \;\rightarrow\; \mathbf{R} = \frac{\partial \bar{\mathbf{x}}}{\partial \hat{\mathbf{x}}} \tag{12.37}$$

whence, using Equation (12.29),

$$\mathrm{d}\bar{\mathbf{x}} = \frac{\partial \bar{\mathbf{x}}}{\partial \hat{\mathbf{x}}} \cdot \frac{\partial \hat{\mathbf{x}}}{\partial \boldsymbol{\xi}} \cdot \mathrm{d}\boldsymbol{\xi} = \mathbf{R} \cdot \mathbf{F}^p \cdot \mathrm{d}\boldsymbol{\xi} \;\rightarrow\; \bar{\mathbf{F}}^p = \mathbf{R} \cdot \mathbf{F}^p \tag{12.38}$$

We next invoke Equation (12.30) to write:

$$\mathrm{d}\mathbf{x} = \frac{\partial \mathbf{x}}{\partial \hat{\mathbf{x}}} \cdot \frac{\partial \hat{\mathbf{x}}}{\partial \bar{\mathbf{x}}} \cdot \mathrm{d}\bar{\mathbf{x}} = \mathbf{F}^e \cdot \mathbf{R}^T \cdot \mathrm{d}\bar{\mathbf{x}} \;\rightarrow\; \bar{\mathbf{F}}^e = \mathbf{F}^e \cdot \mathbf{R}^T \tag{12.39}$$

and we straightforwardly arrive at the following, alternative elasto-plastic multiplicative decomposition:

$$\mathbf{F} = \frac{\partial \mathbf{x}}{\partial \bar{\mathbf{x}}} \cdot \frac{\partial \bar{\mathbf{x}}}{\partial \boldsymbol{\xi}} = \bar{\mathbf{F}}_e \cdot \bar{\mathbf{F}}_p \tag{12.40}$$

which is equivalent to the decomposition of Equation (12.35), since

$$\bar{\mathbf{F}}_{\mathrm{e}} \cdot \bar{\mathbf{F}}^{\mathrm{p}} = \mathbf{F}^{\mathrm{e}} \cdot \mathbf{R}^{\mathrm{T}} \cdot \mathbf{R} \cdot \mathbf{F}^{\mathrm{p}} = \mathbf{F}^{\mathrm{e}} \cdot \mathbf{F}^{\mathrm{p}}$$

For crystalline materials, it is physically reasonable to consider that the plastic deformation gradient $\mathbf{F}^{\mathrm{p}}$ purely represents the plastic sliding between crystals, while the elastic deformation gradient $\mathbf{F}^{\mathrm{e}}$ includes the distortion of the crystal lattice and its rotation (Asaro 1983; Peirce *et al.* 1982; Rice 1971). This is represented by the decomposition of Equation (12.35) and is shown graphically in Figure 12.1.

Unfortunately, the Second Piola–Kirchhoff stress tensor $\boldsymbol{\tau}$ cannot be related generally to the Green–Lagrange strain tensor $\boldsymbol{\gamma}^{\mathrm{e}}$ that can be constructed on the basis of the elastic deformation gradient (Lubliner 1990):

$$\boldsymbol{\gamma}^{\mathrm{e}} = \frac{1}{2}\left((\mathbf{F}^{\mathrm{e}})^{\mathrm{T}} \cdot \mathbf{F}^{\mathrm{e}} - \mathbf{I}\right)$$

since it is not invariant with respect to a rotation of the intermediate configuration. In particular, using Equation (12.39), one obtains:

$$\bar{\boldsymbol{\gamma}}^{\mathrm{e}} = \mathbf{R} \cdot \boldsymbol{\gamma}^{\mathrm{e}} \cdot \mathbf{R}^{\mathrm{T}}$$

For the special case of isotropy, one also has

$$\bar{\boldsymbol{\tau}} = \mathbf{R} \cdot \boldsymbol{\tau} \cdot \mathbf{R}^{\mathrm{T}}$$

and the stress is not affected by the frame in which the intermediate configuration is represented.

The important consequence of the decomposition of Equation (12.35) is that, although the elastic and plastic deformations are decomposed in a multiplicative sense, this is not so for the strain rates (Atluri 1983; Nemat-Nasser 1983). From Equations (11.27) and (12.35) we infer:

$$\boldsymbol{\ell} = \dot{\mathbf{F}}^{\mathrm{e}} \cdot \mathbf{F}^{\mathrm{p}} \cdot \mathbf{F}^{-1} + \mathbf{F}^{\mathrm{e}} \cdot \dot{\mathbf{F}}^{\mathrm{p}} \cdot \mathbf{F}^{-1} = \underbrace{\dot{\mathbf{F}}^{\mathrm{e}} \cdot (\mathbf{F}^{\mathrm{e}})^{-1}}_{\boldsymbol{\ell}^{\mathrm{e}}} + \underbrace{\mathbf{F}^{\mathrm{e}} \cdot \dot{\mathbf{F}}^{\mathrm{p}} \cdot \mathbf{F}^{-1}}_{\boldsymbol{\ell}^{\mathrm{p}}} \tag{12.41}$$

where, in view of the definition of the velocity gradient, Equation (11.26), the additively decomposed elastic and plastic velocity gradients, $\boldsymbol{\ell}^{\mathrm{e}}$ and $\boldsymbol{\ell}^{\mathrm{p}}$, refer to the current configuration. It is emphasised that this additive decomposition of the velocity gradient depends crucially on the definition for $\boldsymbol{\ell}^{\mathrm{p}}$ as given in Equation (12.41). For instance, when

$$\boldsymbol{L}^{\mathrm{p}} \equiv \dot{\mathbf{F}}^{\mathrm{p}} \cdot (\mathbf{F}^{\mathrm{p}})^{-1} \tag{12.42}$$

is substituted for $\boldsymbol{\ell}^{\mathrm{p}}$, which would then be similar to the definition of $\boldsymbol{\ell}^{\mathrm{e}}$, an additive decomposition is not obtained. Nevertheless, the symmetric part of $\boldsymbol{L}$,

$$\boldsymbol{D}^{\mathrm{p}} = \frac{1}{2}\left(\boldsymbol{L}^{\mathrm{p}} + (\boldsymbol{L}^{\mathrm{p}})^{\mathrm{T}}\right) \tag{12.43}$$

is a measure for the plastic stretching, as the eigenvalues D_i^{p} of the spectral decomposition,

$$\boldsymbol{D}^{\mathrm{p}} = \sum_{i=1}^{3} D_i^{\mathrm{p}} \boldsymbol{e}_i \otimes \boldsymbol{e}_i$$

see Equation (1.69), represent the principal, instantaneous rates of plastic stretching of the intermediate configuration. The anti-symmetric part of $\boldsymbol{L}^{\mathrm{p}}$ is named the plastic spin tensor,

$$\boldsymbol{W}^{\mathrm{p}} = \frac{1}{2}\left(\boldsymbol{L}^{\mathrm{p}} - (\boldsymbol{L}^{\mathrm{p}})^{\mathrm{T}}\right) \tag{12.44}$$

and represents the instantaneous rate of plastic spin of the intermediate configuration. In principle, a constitutive equation must be postulated for the plastic spin tensor (Dafalias 1984, 1985, 1998; Kratochvil 1973), but in this treatment the hypothesis is made that the plastic spin vanishes:

$$\boldsymbol{W}^{\mathrm{p}} = \mathbf{0} \tag{12.45}$$

This hypothesis holds rigorously for plastic isotropy, but not necessarily for plastic anisotropy. Using the hypothesis of Equation (12.45) it directly follows that:

$$\boldsymbol{D}^{\mathrm{p}} = \boldsymbol{L}^{\mathrm{p}} \tag{12.46}$$

Using the definition of Equation (12.42), the plastic strain rate in the current configuration can be written as:

$$\dot{\boldsymbol{\epsilon}}^{\mathrm{p}} = (\boldsymbol{\ell}^{\mathrm{p}})^{\mathrm{sym}} = \frac{1}{2}\left(\mathbf{F}^{\mathrm{e}} \cdot \boldsymbol{L}^{\mathrm{p}} \cdot (\mathbf{F}^{\mathrm{e}})^{-1} + (\mathbf{F}^{\mathrm{e}})^{-\mathrm{T}} \cdot (\boldsymbol{L}^{\mathrm{p}})^{\mathrm{T}} \cdot (\mathbf{F}^{\mathrm{e}})^{\mathrm{T}}\right) \tag{12.47}$$

and, exploiting a transformation similar to that in Equation (3.80), an equivalent expression can be obtained in the intermediate configuration:

$$\dot{\boldsymbol{\gamma}}^{\mathrm{p}} = (\mathbf{F}^{\mathrm{e}})^{\mathrm{T}} \cdot \dot{\boldsymbol{\epsilon}}^{\mathrm{p}} \cdot \mathbf{F}^{\mathrm{e}} = \frac{1}{2}\left((\mathbf{C}^{\mathrm{e}})^{\mathrm{T}} \cdot \boldsymbol{L}^{\mathrm{p}} + (\boldsymbol{L}^{\mathrm{p}})^{\mathrm{T}} \cdot \mathbf{C}^{\mathrm{e}}\right) \tag{12.48}$$

with $\mathbf{C}^{\mathrm{e}}$ as in Equation (12.32). Clearly, there is no unequivocal definition for the plastic strain rate, neither in the current configuration, nor in the intermediate configuration (Lubliner 1990). For instance, under the assumption that the elastic strains remain small, so that $\mathbf{C}^{\mathrm{e}} \approx \mathbf{I}$, Equation (12.48) can be approximated as:

$$\dot{\boldsymbol{\gamma}}^{\mathrm{p}} \approx \frac{1}{2}\left(\boldsymbol{L}^{\mathrm{p}} + (\boldsymbol{L}^{\mathrm{p}})^{\mathrm{T}}\right) \equiv \boldsymbol{D}^{\mathrm{p}} \tag{12.49}$$

which, using Equation (3.80), becomes in the current configuration:

$$\dot{\boldsymbol{\epsilon}}^{\mathrm{p}} = (\mathbf{F}^{\mathrm{e}})^{-\mathrm{T}} \cdot \boldsymbol{D}^{\mathrm{p}} \cdot (\mathbf{F}^{\mathrm{e}})^{-1} \tag{12.50}$$

A similar rate has been proposed by Simo (1992):

$$\dot{\boldsymbol{\epsilon}}^{\mathrm{p}} = \mathbf{F}^{\mathrm{e}} \cdot \boldsymbol{D}^{\mathrm{p}} \cdot (\mathbf{F}^{\mathrm{e}})^{-1} \tag{12.51}$$

which can also be derived as the Lie derivative of the 'elastic' left Cauchy–Green deformation tensor $\mathbf{B}^{\mathrm{e}}$ (Marsden and Hughes 1983). When the elastic strains remain small, $\mathbf{U}^{\mathrm{e}} \approx \mathbf{I}$, and using Equation (12.31), $\mathbf{F}^{\mathrm{e}} \approx \mathbf{R}^{\mathrm{e}}$ and $(\mathbf{F}^{\mathrm{e}})^{-\mathrm{T}} \approx (\mathbf{R}^{\mathrm{e}})^{-\mathrm{T}} = \mathbf{R}^{\mathrm{e}}$. With these approximations, the definitions for $\dot{\boldsymbol{\epsilon}}^{\mathrm{p}}$, Equations (12.47), (12.50) and (12.51) coincide, and reduce to:

$$\dot{\boldsymbol{\epsilon}}^{\mathrm{p}} = \mathbf{R}^{\mathrm{e}} \cdot \boldsymbol{D}^{\mathrm{p}} \cdot (\mathbf{R}^{\mathrm{e}})^{\mathrm{T}} \tag{12.52}$$

which can be interpreted as the plastic stretching $\boldsymbol{D}^{\mathrm{p}}$ which is transformed to the deformed configuration through the elastic rotation.

12.3 Multiplicative Elasto-plasticity *versus* Rate Formulations

As discussed earlier in this chapter, most early large-strain finite element formulations are rooted in a rate formulation, usually based on the Jaumann rate. Such rate formulations can be related to approaches that are based on the multiplicative decomposition (Needleman 1985). The point of departure is that we assume the existence of a strain energy function $\mathcal{W}$, so that the Second Piola–Kirchhoff stress tensor can be derived as:

$$\boldsymbol{\tau} = 2\frac{\partial \mathcal{W}}{\partial \mathbf{C}^{\mathrm{e}}} \tag{12.53}$$

or, equivalently,

$$\boldsymbol{\tau} = \frac{\partial \mathcal{W}}{\partial \boldsymbol{\gamma}^{\mathrm{e}}}$$

Assuming no coupling between the plastic strains and the elastic moduli, the elastic tangential stiffness tensor can be derived as:

$$\mathbf{D}^{\mathrm{e}} = \frac{\partial^2 \mathcal{W}}{\partial \boldsymbol{\gamma}^{\mathrm{e}} \partial \boldsymbol{\gamma}^{\mathrm{e}}} \tag{12.54}$$

In consideration of Equations (3.73) and (11.15) the Kirchhoff stress can be expressed as:

$$\boldsymbol{\kappa} = \mathbf{F}^{\mathrm{e}} \cdot \boldsymbol{\tau} \cdot (\mathbf{F}^{\mathrm{e}})^{\mathrm{T}} \tag{12.55}$$

The Truesdell rate of the Kirchhoff stress tensor with respect to the intermediate configuration can now be expressed as, cf. Equation (11.37):

$$\overset{\star}{\boldsymbol{\kappa}} = \dot{\boldsymbol{\kappa}} - \boldsymbol{\ell}^{\mathrm{e}} \cdot \boldsymbol{\kappa} - \boldsymbol{\kappa} \cdot (\boldsymbol{\ell}^{\mathrm{e}})^{\mathrm{T}} \tag{12.56}$$

or, using the tangential stiffness tensor $\mathbf{D}^{\mathrm{TK}}$ that sets the relation between the Truesdell rate of the Kirchhoff stress and the elastic deformation rate $\dot{\boldsymbol{\epsilon}}^{\mathrm{e}}$ and rearranging,

$$\dot{\boldsymbol{\kappa}} = \mathbf{D}^{\mathrm{TK}} : \dot{\boldsymbol{\epsilon}}^{\mathrm{e}} - \boldsymbol{\ell}^{\mathrm{e}} \cdot \boldsymbol{\kappa} - \boldsymbol{\kappa} \cdot (\boldsymbol{\ell}^{\mathrm{e}})^{\mathrm{T}} \tag{12.57}$$

From the first equality sign in Equation (11.37) and from the relation between the rate of deformation and the rate of the Green–Lagrange strain tensor, Equation (3.80), the following relation can be inferred between $\mathbf{D}^{\mathrm{TK}}$ and the tangential stiffness tensor defined in Equation (12.54):

$$D^{\mathrm{TK}}_{ijkl} = F^{\mathrm{e}}_{im} F^{\mathrm{e}}_{jn} F^{\mathrm{e}}_{ko} F^{\mathrm{e}}_{lp} D^{\mathrm{e}}_{mnop} \tag{12.58}$$

Next, use of Equation (12.19),

$$D^{\mathrm{JK}}_{ijkl} = D^{\mathrm{TK}}_{ijkl} + \frac{1}{2}\left(\kappa_{il}\delta_{jk} + \kappa_{jl}\delta_{ik} + \kappa_{ik}\delta_{jl} + \kappa_{jk}\delta_{il}\right)$$

which sets the relation between the constitutive matrices $\mathbf{D}^{\mathrm{TK}}$ and $\mathbf{D}^{\mathrm{JK}}$ that are used in the Kirchhoff rate and the Jaumann rate of the Kirchhoff stress tensor, respectively, gives a relation between $\mathbf{D}^{\mathrm{e}}$ and $\mathbf{D}^{\mathrm{JK}}$. For most materials, the stresses are small compared with the elastic moduli, and Equation (12.19) can be approximated as

$$\mathbf{D}^{\mathrm{JK}} \approx \mathbf{D}^{\mathrm{TK}} \tag{12.59}$$

If, in addition, it is assumed that the elastic strains remain small, we have $\mathbf{U}^{\mathrm{e}} \approx \mathbf{I}$, so that $\mathbf{F}^{\mathrm{e}} \approx \mathbf{R}^{\mathrm{e}}$. Assuming, furthermore, elastic isotropy, the latter has no effect, so that

$$\mathbf{D}^{\mathrm{TK}} \approx \mathbf{D}^{\mathrm{e}} \tag{12.60}$$

and we have the approximate relation:

$$\dot{\boldsymbol{\kappa}} = \mathbf{D}^{\mathrm{e}} : \dot{\boldsymbol{\epsilon}}^{\mathrm{e}} + \boldsymbol{w}^{\mathrm{e}} \cdot \boldsymbol{\kappa} - \boldsymbol{\kappa} \cdot \boldsymbol{w}^{\mathrm{e}} \tag{12.61}$$

Finally, it is assumed that $\boldsymbol{w} = \boldsymbol{w}^{\mathrm{e}}$, which implies that the plastic spin $\boldsymbol{W}^{\mathrm{p}} = \mathbf{0}$, and we arrive at:

$$\dot{\boldsymbol{\kappa}} = \mathbf{D}^{\mathrm{e}} : \dot{\boldsymbol{\epsilon}}^{\mathrm{e}} + \boldsymbol{w} \cdot \boldsymbol{\kappa} - \boldsymbol{\kappa} \cdot \boldsymbol{w} \tag{12.62}$$

which is at the basis of the Eulerian finite element formulation derived at the beginning of this chapter.

We proceed with the plasticity part of the model. As the Cauchy stresses are the 'true' stresses, they enter the yield criterion, flow rule and hardening model. However, as shown before, Equations (12.22) and (12.26), the use of the Jaumann rate of the Cauchy stress tensor usually leads to a non-symmetric tangential stiffness matrix. By contrast, the objective stress rates, like those of Jaumann and Truesdell, of the Kirchhoff stress tensor result in symmetric tangential stiffness matrices. Now, for the von Mises and Tresca yield functions with an associated flow rule, the plastic deformations are isochoric – see the text after Equation (7.55) – so that $\det \mathbf{F}^{\mathrm{p}} = 1$, and,

$$\det \mathbf{F} = \det \mathbf{F}^{\mathrm{e}} \cdot \det \mathbf{F}^{\mathrm{p}} = \det \mathbf{F}^{\mathrm{e}}$$

Hence, again relying on the assumption that the elastic strains remain small, we have that $\det \mathbf{F} \approx 1$. Considering the relation between the Cauchy and the Kirchhoff stress tensors, Equation (11.15), we observe that we can, under the assumptions made above, replace the Cauchy stress by the Kirchhoff stress, with all the computational conveniences that come with it (Nagtegaal and de Jong 1981).

Accordingly, we can express the yield function f in terms of the Kirchhoff stress and the hardening variables $\boldsymbol{\alpha}$ – note that in order to avoid confusion, the symbol $\boldsymbol{\alpha}$ is now used, instead of $\boldsymbol{\kappa}$, which was used consistently in Chapter 7:

$$f = f(\boldsymbol{\kappa}, \boldsymbol{\alpha})$$

This format of the yield function is limited to isotropic hardening plasticity. The generalisation to kinematic hardening, however, is straightforward, and runs along the lines indicated in Chapter 7. Following Equation (12.41) we adopt an additive decomposition of the rate of deformation:

$$\dot{\boldsymbol{\epsilon}} = \dot{\boldsymbol{\epsilon}}^{\mathrm{e}} + \dot{\boldsymbol{\epsilon}}^{\mathrm{p}} \tag{12.63}$$

and a flow rule as in Equation (7.43), but now phrased in terms of Kirchhoff stresses:

$$\dot{\boldsymbol{\epsilon}}^{\mathrm{p}} = \dot{\lambda}\mathbf{m} , \qquad \mathbf{m} = \frac{\partial g}{\partial \boldsymbol{\kappa}} \tag{12.64}$$

with g the plastic potential function, see Chapter 7. Substitution of the additive decomposition of the strain rates and the flow rule into Equation (12.62) yields:

$$\dot{\boldsymbol{\kappa}} = \mathbf{D}^{\mathrm{e}} : (\dot{\boldsymbol{\epsilon}} - \dot{\lambda}\mathbf{m}) + \boldsymbol{w} \cdot \boldsymbol{\kappa} - \boldsymbol{\kappa} \cdot \boldsymbol{w} \tag{12.65}$$

With $\mathbf{n}$ the gradient to the yield surface, cf. Equation (7.45), but also expressed in Kirchhoff stresses, Prager's consistency condition $\dot{f} = 0$ becomes:

$$\mathbf{n} : \dot{\boldsymbol{\kappa}} - h\dot{\lambda} = 0 \tag{12.66}$$

with h the hardening modulus, Equation (7.74). A double contraction of $\dot{\boldsymbol{\kappa}}$ as expressed in Equation (12.65) by $\mathbf{n}$, and exploiting the consistency condition gives:

$$h\dot{\lambda} = \mathbf{n} : \mathbf{D}^{\mathrm{e}} : (\dot{\boldsymbol{\epsilon}} - \dot{\lambda}\mathbf{m}) + \mathbf{n} : (\boldsymbol{w} \cdot \boldsymbol{\kappa} - \boldsymbol{\kappa} \cdot \boldsymbol{w}) \tag{12.67}$$

This equation can be brought into a small-strain format, i.e.

$$h\dot{\lambda} = \mathbf{n} : \mathbf{D}^{\mathrm{e}} : (\dot{\boldsymbol{\epsilon}} - \dot{\lambda}\mathbf{m}) \tag{12.68}$$

under the condition that

$$\mathbf{n} : (\boldsymbol{w} \cdot \boldsymbol{\kappa} - \boldsymbol{\kappa} \cdot \boldsymbol{w}) = 0 \tag{12.69}$$

Since the second-order tensors $\mathbf{n}$ and $\boldsymbol{\kappa}$ are symmetric, and since $\boldsymbol{w}$ is anti-symmetric, this relation indeed holds true. In fact, this requirement is satisfied for any objective stress rate of this format where $\boldsymbol{w}$ is replaced by another anti-symmetric second-order tensor. For instance, the Green–Naghdi rate, Equation (11.25), would also be suitable but the Truesdell rate would not. Noting that a comparison of Equations (11.35) and (12.62) results in the following relation between the Jaumann rate of the Kirchhoff stress, and the elastic part of the rate of deformation,

$$\overset{\circ}{\boldsymbol{\kappa}} = \mathbf{D}^{\mathrm{e}} : \dot{\boldsymbol{\epsilon}}^{\mathrm{e}}$$

we can derive the tangential stiffness expression:

$$\overset{\circ}{\boldsymbol{\kappa}} = \mathbf{D} : \dot{\boldsymbol{\epsilon}} \tag{12.70}$$

where the tangential stiffness matrix

$$\mathbf{D} = \mathbf{D}^{\mathrm{e}} - \frac{(\mathbf{D}^{\mathrm{e}} : \mathbf{m}) \otimes (\mathbf{D}^{\mathrm{e}} : \mathbf{n})}{h + \mathbf{n} : \mathbf{D}^{\mathrm{e}} : \mathbf{m}} \tag{12.71}$$

has the same format as in small-strain plasticity, cf. Equation (7.76).

In the preceding it has been shown that results that are based on the multiplicative decomposition should be close to results of formulations that are based on the Jaumann rate, provided that the elastic strains remain small. This will now be illustrated for the case of a square element that is subjected to simple shear, as in Figure 11.1. Closed-form expressions for various rates – including the Jaumann rate, the Truesdell rate and the Green–Naghdi rate – have been

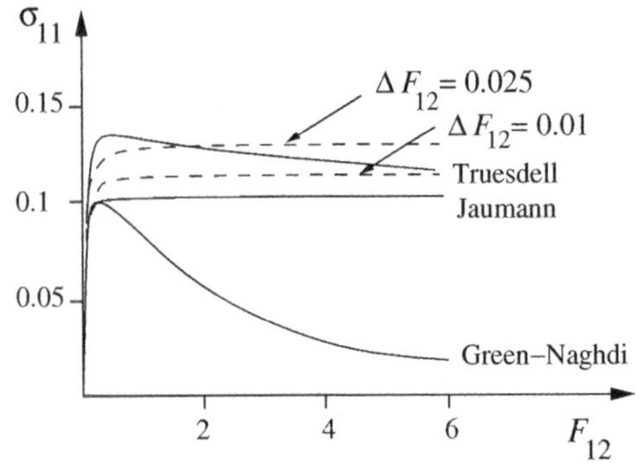

Figure 12.2 Normalised normal stress as a function of the shear strain. The dashed lines give the results for the multiplicative decomposition

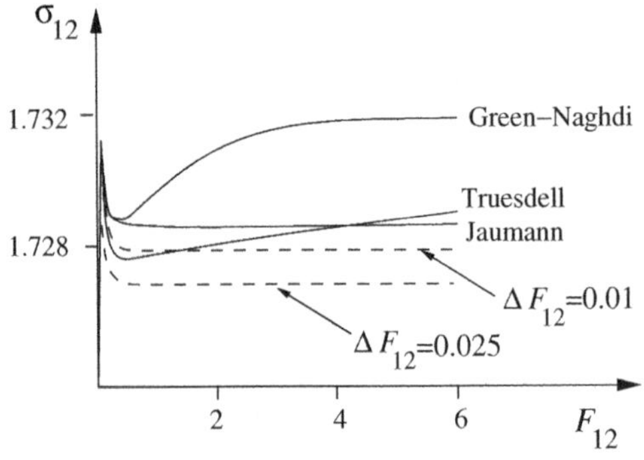

Figure 12.3 Normalised shear stress as a function of the shear strain. The dashed lines give the results for the multiplicative decomposition

derived by Moss (1984). For simplicity, and without loss of generality, the shear modulus μ has been set to unity, the yield strength $\bar{\sigma} = 0.1\ \text{N/mm}^2$, and hardening is omitted. Since the deformation is isochoric, the value of the bulk modulus is irrelevant, and also, there is no difference between the Cauchy stress and the Kirchhoff stress. While the solutions for the rates are exact, the solutions that are based on the multiplicative decomposition referenced to the current configuration have been integrated numerically. Figures 12.2 and 12.3 show that, as expected, they converge towards the solution based on the Jaumann rate, upon refinement of the step size, here ΔF_{12}, which is taken as measure of the shear strain increment.

12.4 Integration of the Rate Equations

In the previous section, we have shown that for small elastic strains, a Jaumann rate formulation is effectively equivalent to a multiplicative decomposition approach (Needleman 1985). However, unless the increments are very small, the integration of the resulting rate equations can entail large errors. This is exactly the case for explicit dynamics computations, where stability requirements impose a strict upper bound to the allowable time step, see Chapter 5. For such codes, an explicit scheme as shown in Box 5.2 is usually applied, so that the nodal velocities are known at the mid-interval, and the stresses and displacements are known at the interval borders; see Box 12.1 for an algorithm that is based on the Jaumann rate of the Cauchy stress.

Box 12.1 Integration of the rate equations for explicit dynamics

1. For given $\dot{\mathbf{a}}^{t+\frac{1}{2}\Delta t}$ compute $\boldsymbol{\ell}^{t+\frac{1}{2}\Delta t} = \sum_{k=1}^{n} \left(\frac{\partial h_k}{\partial \mathbf{x}}\right)^t \dot{\mathbf{a}}_k^{t+\frac{1}{2}\Delta t}$
2. Using Equations (11.30) and (11.31), compute:
$$\dot{\boldsymbol{\epsilon}}^{t+\frac{1}{2}\Delta t} = \frac{1}{2}(\boldsymbol{\ell}^{t+\frac{1}{2}\Delta t} + (\boldsymbol{\ell}^{t+\frac{1}{2}\Delta t})^{\mathrm{T}})$$
$$\boldsymbol{w}^{t+\frac{1}{2}\Delta t} = \frac{1}{2}(\boldsymbol{\ell}^{t+\frac{1}{2}\Delta t} - (\boldsymbol{\ell}^{t+\frac{1}{2}\Delta t})^{\mathrm{T}})$$
3. Using $\mathbf{D}$ from Equation (12.71), compute the stress rate at the midpoint:
$$\dot{\boldsymbol{\sigma}}^{t+\frac{1}{2}\Delta t} = \mathbf{D}(\boldsymbol{\sigma}^t, \kappa^t)\, \dot{\boldsymbol{\epsilon}}^{t+\frac{1}{2}\Delta t}$$
4. Compute the stress increment:
$$\Delta\boldsymbol{\sigma} = \Delta t\, \dot{\boldsymbol{\sigma}}^{t+\frac{1}{2}\Delta t}$$
5. Update the stress at the end of the time step:
$$\boldsymbol{\sigma}^{t+\Delta t} = \boldsymbol{\sigma}^t + \Delta\boldsymbol{\sigma} + \Delta t \left(\boldsymbol{w}^{t+\frac{1}{2}\Delta t}\boldsymbol{\sigma}^t - \boldsymbol{\sigma}^t \boldsymbol{w}^{t+\frac{1}{2}\Delta t}\right)$$
6. Compute the internal force vector:
$$\mathbf{f}_{\mathrm{int}}^{t+\Delta t} = \int_{V^{t+\Delta t}} \mathbf{B}^{\mathrm{T}}(\mathbf{x}^{t+\Delta t})\boldsymbol{\sigma}^{t+\Delta t} \mathrm{d}V$$

As an alternative to steps 3 and 4 a small-strain return-mapping algorithm can be adopted, see Chapter 6.

For cases other than in explicit dynamics the integration of the rate equations must be approached in a more rigorous way. In addition to the accurate integration of the plastic flow, which is already important in small-strain plasticity, see Chapter 7, the accurate treatment of the rotation becomes an issue for large strains. The purpose of objective stress rates such as the Jaumann rate is to obviate the emergence of straining under pure rotations. While the use of an objective stress rate guarantees this for infinitesimal rotations, this is not necessarily so for finite rotation increments. The development of algorithms that set out to achieve incremental objectivity has been the subject of a considerable body of work (Flanagan and Taylor 1987; Hughes and Winget 1980; Key and Krieg 1982; Pinsky *et al.* 1983).

Herein, we will outline the algorithm proposed by Hughes and Winget (1980), which starts from the assumption that the update process can be additively decomposed in a rotation and in a stress update in a fixed coordinate frame. In keeping with Equation (1.55) for the transformation of second-order tensors, we have for the Cauchy stress tensor:

$$\bar{\boldsymbol{\sigma}} = \mathbf{R} \cdot \boldsymbol{\sigma} \cdot \mathbf{R}^{\mathrm{T}}$$

while the stress update follows from:

$$\Delta \boldsymbol{\sigma} = \Delta t \mathbf{D} : \dot{\boldsymbol{\epsilon}} \tag{12.72}$$

with $\mathbf{D}$ the tangential stiffness tensor, Equation (12.71). Accordingly, the stress at $t + \Delta t$ is found as the sum of both operations:

$$\boldsymbol{\sigma}^{t+\Delta t} = \mathbf{R} \cdot \boldsymbol{\sigma}^{t} \cdot \mathbf{R}^{\mathrm{T}} + \Delta t \mathbf{D} : \dot{\boldsymbol{\epsilon}} \tag{12.73}$$

In the algorithm proposed by Hughes and Winget (1980), Equation (12.73) is replaced by:

$$\boldsymbol{\sigma}^{t+\Delta t} = \widehat{\mathbf{R}}(\theta_{\mathrm{m}}) \cdot \boldsymbol{\sigma}^{t} \cdot \widehat{\mathbf{R}}^{\mathrm{T}}(\theta_{\mathrm{m}}) + \mathbf{D} : \Delta \boldsymbol{\epsilon}_{\mathrm{m}} \tag{12.74}$$

where $\widehat{\mathbf{R}}(\theta_{\mathrm{m}})$ is an approximation to $\mathbf{R}$, and is a function of the angle θ at the midpoint of the time interval, i.e. at $t + \frac{1}{2}\Delta t$. When the incremental motion involves a pure rotation, we must have that $\widehat{\mathbf{R}}(\theta_{\mathrm{m}}) \rightarrow \mathbf{R}$. Defining:

$$\Delta \boldsymbol{\ell}_{\mathrm{m}} = \left(\frac{\partial \Delta \mathbf{u}}{\partial \mathbf{x}} \right)_{\mathrm{m}} \tag{12.75}$$

or

$$\Delta \boldsymbol{\ell}_{\mathrm{m}} = \sum_{k=1}^{n} \left(\frac{\partial h_k}{\partial \mathbf{x}} \right)_{\mathrm{m}} \Delta \mathbf{a}_k \tag{12.76}$$

the incremental quantities $\Delta \boldsymbol{\epsilon}_{\mathrm{m}}$ and $\Delta \boldsymbol{w}_{\mathrm{m}}$ can be computed, cf. Equations (11.30) and (11.31), as:

$$\Delta \boldsymbol{\epsilon}_{\mathrm{m}} = \frac{1}{2}(\Delta \boldsymbol{\ell}_{\mathrm{m}} + \Delta \boldsymbol{\ell}_{\mathrm{m}}^{\mathrm{T}}) \tag{12.77}$$

and

$$\Delta \boldsymbol{w}_{\mathrm{m}} = \frac{1}{2}(\Delta \boldsymbol{\ell}_{\mathrm{m}} - \Delta \boldsymbol{\ell}_{\mathrm{m}}^{\mathrm{T}}) \tag{12.78}$$

For incremental objectivity we must ensure that under a rigid body motion, no straining occurs, and that $\widehat{\mathbf{R}}(\theta_{\mathrm{m}}) \rightarrow \mathbf{R}$. For such a rigid rotation we have:

$$\mathbf{x}^{t+\Delta t} = \mathbf{R} \cdot \mathbf{x}^{t} \tag{12.79}$$

With

$$\mathbf{x}_{\mathrm{m}} = \frac{1}{2}\left(\mathbf{x}^{t} + \mathbf{x}^{t+\Delta t}\right) = \frac{1}{2}(\mathbf{R} + \mathbf{I}) \cdot \mathbf{x}^{t} \tag{12.80}$$

we obtain

$$\Delta \mathbf{u} = (\mathbf{R} - \mathbf{I}) \cdot \mathbf{x}^{t} = 2(\mathbf{R} - \mathbf{I}) \cdot (\mathbf{R} + \mathbf{I})^{-1} \cdot \mathbf{x}_{\mathrm{m}} \tag{12.81}$$

so that:

$$\Delta \boldsymbol{\ell}_{\mathrm{m}} = 2\,(\mathbf{R} - \mathbf{I}) \cdot (\mathbf{R} + \mathbf{I})^{-1} \tag{12.82}$$

Substitution into Equation (12.77) then gives after elaboration:

$$\Delta \boldsymbol{\epsilon}_{\mathrm{m}} = \mathbf{R} \cdot (\mathbf{R} + \mathbf{I})^{-1} - (\mathbf{R} + \mathbf{I})^{-\mathrm{T}} + (\mathbf{R} + \mathbf{I})^{-\mathrm{T}} \cdot \mathbf{R}^{\mathrm{T}} - (\mathbf{R} + \mathbf{I})^{-1} \tag{12.83}$$

By virtue of the fact that $\mathbf{R}$ is anti-symmetric, we have

$$(\mathbf{R} + \mathbf{I})^{-\mathrm{T}} = \left(\mathbf{R}^{\mathrm{T}} + \mathbf{I}\right)^{-1} = \left(\mathbf{R}^{\mathrm{T}} + \mathbf{R} \cdot \mathbf{R}^{\mathrm{T}}\right)^{-1} = \left((\mathbf{I} + \mathbf{R}) \cdot \mathbf{R}^{\mathrm{T}}\right)^{-1} = \mathbf{R} \cdot (\mathbf{R} + \mathbf{I})^{-1}$$

and, hence, in Equation (12.83) the first two terms cancel each other, as well as the last two terms. The requirement that a pure rigid rotation does not produce straining is thus satisfied.

With $\Delta \boldsymbol{\epsilon}_{\mathrm{m}} = \mathbf{0}$, we can elaborate Equation (12.78) as:

$$\Delta \boldsymbol{w}_{\mathrm{m}} = \Delta \boldsymbol{\ell}_{\mathrm{m}} = 2(\widehat{\mathbf{R}} - \mathbf{I}) \cdot (\widehat{\mathbf{R}} + \mathbf{I})^{-1} \tag{12.84}$$

cf. Equation (12.82). This equation can be solved for $\mathbf{R}$ to give:

$$\widehat{\mathbf{R}} = \left(\mathbf{I} - \frac{1}{2}\Delta \boldsymbol{w}_{\mathrm{m}}\right)^{-1} \cdot \left(\mathbf{I} + \frac{1}{2}\Delta \boldsymbol{w}_{\mathrm{m}}\right) \tag{12.85}$$

Using $\Delta \boldsymbol{\theta}_{\mathrm{m}}$, the axial vector of $\Delta \boldsymbol{w}_{\mathrm{m}}$, cf. Equations (9.133) and (9.134), an alternative expression can be constructed that obviates the matrix inversions of the preceding equation:

$$\widehat{\mathbf{R}} = \mathbf{I} + \frac{\Delta \boldsymbol{w}_{\mathrm{m}}}{1 + \frac{1}{4}\Delta \boldsymbol{\theta}_{\mathrm{m}} \cdot \Delta \boldsymbol{\theta}_{\mathrm{m}}} \cdot \left(\mathbf{I} + \frac{1}{2}\Delta \boldsymbol{w}_{\mathrm{m}}\right) \tag{12.86}$$

While the Hughes–Winget algorithm gives the correct solution for a pure rotation, it can introduce inaccuracies when the incremental motion involves stretching as well as rotation. More sophisticated midpoint procedures have therefore been advocated by Hughes and Winget (1980) and Key and Krieg (1982); see de Souza Neto *et al.* (2008) for an overview. In essence, these techniques involve first rotating the stresses to the midpoint configuration, then applying the stress update, for instance using a standard small-strain return-mapping algorithm, and finally rotating on to the final configuration. Such a technique therefore requires the midpoint rotation matrix. For the rotation matrix $\mathbf{R}$ this implies that the matrix $\mathbf{R}^{\frac{1}{2}}$ is required, such that:

$$\mathbf{R}^{\frac{1}{2}}\mathbf{R}^{\frac{1}{2}} = \mathbf{R} \tag{12.87}$$

For a two-dimensional configuration the rotation matrix can be expressed as

$$\mathbf{R} = \begin{bmatrix} \cos\phi & -\sin\phi \\ \sin\phi & \cos\phi \end{bmatrix}$$

and $\mathbf{R}^{\frac{1}{2}}$ then follows from:

$$\mathbf{R}^{\frac{1}{2}} = \begin{bmatrix} \cos(\phi/2) & -\sin(\phi/2) \\ \sin(\phi/2) & \cos(\phi/2) \end{bmatrix}$$

In three dimensions, the most straightforward way to obtain $\mathbf{R}^{\frac{1}{2}}$ is through the use of quaternions.

12.5 Exponential Return-mapping Algorithms

We return to the multiplicative elasto-plastic decomposition of the deformation gradient, and pick up the discussion at the elastically rotated plastic stretching. We will henceforth limit the discussion to isotropy, so that the plastic spin tensor vanishes rigorously: $\boldsymbol{W}^{\mathrm{p}} = \mathbf{0}$, cf. Equation (12.45). Consequently, Equation (12.52) can be rewritten as:

$$\dot{\boldsymbol{\epsilon}}^{\mathrm{p}} = \mathbf{R}^{\mathrm{e}} \cdot \boldsymbol{L}^{\mathrm{p}} \cdot (\mathbf{R}^{\mathrm{e}})^{\mathrm{T}} \tag{12.88}$$

Using Equation (12.42) and rearranging gives:

$$\dot{\mathbf{F}}^{\mathrm{p}} \cdot (\mathbf{F}^{\mathrm{p}})^{-1} = (\mathbf{R}^{\mathrm{e}})^{\mathrm{T}} \cdot \dot{\boldsymbol{\epsilon}}^{\mathrm{p}} \cdot \mathbf{R}^{\mathrm{e}} \tag{12.89}$$

or, upon substitution of the flow rule, Equation (12.64), but rewritten in tensor notation instead of in Voigt notation,

$$\dot{\boldsymbol{\epsilon}}^{\mathrm{p}} = \dot{\lambda}\mathbf{M}\,, \quad \mathbf{M} = \frac{\partial g}{\partial \boldsymbol{\kappa}} \tag{12.90}$$

we obtain:

$$\dot{\mathbf{F}}^{\mathrm{p}} \cdot (\mathbf{F}^{\mathrm{p}})^{-1} = (\mathbf{R}^{\mathrm{e}})^{\mathrm{T}} \cdot (\dot{\lambda}\mathbf{M}) \cdot \mathbf{R}^{\mathrm{e}} \tag{12.91}$$

This evolution equation for the plastic flow can be integrated accurately using exponential map integrators (Cuitino and Ortiz 1992; Eterovic and Bathe 1990; Perić *et al.* 1992; Simo 1992; Weber and Anand 1990), and, at iteration $j+1$ of the time step $t \rightarrow t + \Delta t$, for a backward exponential integrator, results in:

$$\mathbf{F}^{\mathrm{p}}_{j+1} = \exp\left((\mathbf{R}^{\mathrm{e}}_{j+1})^{\mathrm{T}} \cdot (\Delta\lambda \mathbf{M}_{j+1}) \cdot \mathbf{R}^{\mathrm{e}}_{j+1}\right) \cdot \mathbf{F}^{\mathrm{p}}_{0} \tag{12.92}$$

where, as usual, the subscript ‘0’ denotes the value of $\mathbf{F}^{\mathrm{p}}$ at time t. Assuming isotropy, this equation can be simplified to:

$$\mathbf{F}^{\mathrm{p}}_{j+1} = \left(\mathbf{R}^{\mathrm{e}}_{j+1}\right)^{\mathrm{T}} \cdot \exp\left(\Delta\lambda \mathbf{M}_{j+1}\right) \cdot \mathbf{R}^{\mathrm{e}}_{j+1} \cdot \mathbf{F}^{\mathrm{p}}_{0} \tag{12.93}$$

For metals, the plastic flow is isochoric, as in von Mises and Tresca plasticity, see Chapter 7. This property is exactly preserved by the integrator of Equation (12.93), since then $\det[\exp(\Delta\lambda\mathbf{M})] = 1$, so that

$$\det\left[\mathbf{F}^{\mathrm{p}}_{j+1}\right] = \det\left[\left(\mathbf{R}^{\mathrm{e}}_{j+1}\right)^{\mathrm{T}}\right] \cdot \det\left[\exp\left(\Delta\lambda\mathbf{M}_{j+1}\right)\right] \cdot \det\left[\mathbf{R}^{\mathrm{e}}_{j+1}\right] \cdot \det\left[\mathbf{F}^{\mathrm{p}}_{0}\right] = \det\left[\mathbf{F}^{\mathrm{p}}_{0}\right]$$

where, for isochoric plastic flow, also $\det[\mathbf{F}^{\mathrm{p}}_{0}] = 1$. This favourable property is not obtained for a standard Euler backward method, where

$$\mathbf{F}^{\mathrm{p}}_{j+1} = \left(\mathbf{I} - \Delta\lambda(\mathbf{R}^{\mathrm{e}}_{j+1})^{\mathrm{T}} \cdot \mathbf{M}_{j+1} \cdot \mathbf{R}^{\mathrm{e}}_{j+1}\right)^{1} \cdot \mathbf{F}^{\mathrm{p}}_{0} \tag{12.94}$$

It is noted that, while most work has been directed towards J_2-plasticity, applications to crystal plasticity (Miehe 1996; Miehe and Schotte 2004), and to cohesive-frictional materials, which are plastically dilatant or contractant, rather than plastically volume-preserving, have been reported as well (Meschke and Liu 1999; Rouainia and Wood 2000).

We next substitute the exponential update formula, Equation (12.93), into the multiplicative elasto-plastic decomposition, Equation (12.35). After inversion, using the property that

$$(\exp(\mathbf{A}))^n = \exp(n\mathbf{A}) \tag{12.95}$$

with n an integer, and introducing

$$\Delta\mathbf{F} = \frac{\partial \mathbf{x}^{t+\Delta t}}{\partial \mathbf{x}^t} = \mathbf{I} + \frac{\partial \Delta\mathbf{u}}{\partial \mathbf{x}^t} \tag{12.96}$$

as the deformation gradient that maps the configuration at t onto that at $t+\Delta t$, we obtain for the update of the elastic deformation gradient (for $n = -1$):

$$\mathbf{F}^{\mathrm{e}}_{j+1} = \Delta\mathbf{F} \cdot \mathbf{F}^{\mathrm{e}}_0 \cdot \left(\mathbf{R}^{\mathrm{e}}_{j+1}\right)^{\mathrm{T}} \cdot \exp\left(-\Delta\lambda \mathbf{M}_{j+1}\right) \cdot \mathbf{R}^{\mathrm{e}}_{j+1} \tag{12.97}$$

Equation (12.97), together with the evolution law for the internal variables $\boldsymbol{\alpha}$, the elastic relation between the Kirchhoff stress tensor and the logarithmic strain tensor $\boldsymbol{\varepsilon}$, cf. Equation (11.112),

$$\boldsymbol{\kappa}_{j+1} = \mathbf{D}^{\mathrm{e}} : \boldsymbol{\varepsilon}^{\mathrm{e}}_{j+1} \tag{12.98}$$

and the kinematic relation,

$$\boldsymbol{\varepsilon}^{\mathrm{e}}_{j+1} = \frac{1}{2} \ln \mathbf{B}^{\mathrm{e}}_{j+1} = \frac{1}{2} \ln \left(\mathbf{F}^{\mathrm{e}}_{j+1} \cdot \left(\mathbf{F}^{\mathrm{e}}_{j+1}\right)^{\mathrm{T}}\right) \tag{12.99}$$

cf. Equation (11.109), completely define the incremental update problem, where it is noted that the elastic rotation can be obtained directly from the elastic deformation gradient $\mathbf{F}^{\mathrm{e}}_{j+1}$ via

$$\mathbf{R}^{\mathrm{e}}_{j+1} = \left(\mathbf{B}^{\mathrm{e}}_{j+1}\right)^{-\frac{1}{2}} \cdot \mathbf{F}^{\mathrm{e}}_{j+1} = \left(\mathbf{F}^{\mathrm{e}}_{j+1} \cdot \left(\mathbf{F}^{\mathrm{e}}_{j+1}\right)^{\mathrm{T}}\right)^{-\frac{1}{2}} \cdot \mathbf{F}^{\mathrm{e}}_{j+1} \tag{12.100}$$

We next note that the elastic deformation gradient

$$\mathbf{F}^{\mathrm{e}}_e = \Delta\mathbf{F} \cdot \mathbf{F}^{\mathrm{e}}_0 \tag{12.101}$$

can be conceived as the elastic deformation gradient in the trial state, since the Kirchhoff stress that is computed on basis of it,

$$\boldsymbol{\kappa}_e = \mathbf{D}^{\mathrm{e}} : \boldsymbol{\varepsilon}^{\mathrm{e}}_e = \frac{1}{2}\mathbf{D}^{\mathrm{e}} : \ln\left(\mathbf{F}^{\mathrm{e}}_e \cdot (\mathbf{F}^{\mathrm{e}}_e)^{\mathrm{T}}\right) \tag{12.102}$$

plays the role of the trial stress that enters the yield function $f = f(\boldsymbol{\kappa}_e, \boldsymbol{\alpha}_0)$ to check whether plasticity occurs in an integration point. If $f(\boldsymbol{\kappa}_e, \boldsymbol{\alpha}_0) \geq 0$ plastic flow can occur, and an iterative procedure is invoked to solve the local set of non-linear equations:

$$\begin{pmatrix} \mathbf{r}_{\mathbf{F}^{\mathrm{e}}} \\ \mathbf{r}_{\alpha} \\ r_f \end{pmatrix} = \mathbf{0} \tag{12.103}$$

with the local residuals defined as:

$$\begin{cases} \mathbf{r}_{\mathrm{F}^{\mathrm{e}}} = \mathbf{F}^{\mathrm{e}}_{j+1} - \mathbf{F}^{\mathrm{e}}_{e} \cdot (\mathbf{R}^{\mathrm{e}}_{j+1})^{\mathrm{T}} \cdot \exp\left(-\Delta\lambda \mathbf{M}_{j+1}\right) \cdot \mathbf{R}^{\mathrm{e}}_{j+1} \\ \mathbf{r}_{\alpha} = \boldsymbol{\alpha}_{j+1} - \boldsymbol{\alpha}_0 - \Delta\lambda \mathbf{p}(\boldsymbol{\kappa}_{j+1}, \boldsymbol{\alpha}_{j+1}) \\ r_f = f(\boldsymbol{\kappa}_{j+1}, \boldsymbol{\alpha}_{j+1}) \end{cases} \tag{12.104}$$

When $f(\boldsymbol{\kappa}_e, \boldsymbol{\alpha}_0) < 0$ we have purely elastic behaviour, and the update follows simply as $\boldsymbol{\kappa}_{j+1} = \boldsymbol{\kappa}_e$ and $\boldsymbol{\alpha}_{j+1} = \boldsymbol{\alpha}_0$.

A further simplification can be obtained by solving for the elastic logarithmic strain $\boldsymbol{\varepsilon}^{\mathrm{e}}_{j+1}$, rather than solving for the elastic deformation gradient $\mathbf{F}^{\mathrm{e}}_{j+1}$ (de Souza Neto *et al.* 2008). To this end we post-multiply Equation (12.97) by $\mathbf{R}^{\mathrm{e}}_{j+1}$ and, subsequently, by $\exp\left(-\Delta\lambda \mathbf{M}_{j+1}\right)$. Again making use of the property in Equation (12.95) we obtain:

$$\mathbf{V}^{\mathrm{e}}_{j+1} \cdot \exp\left(\Delta\lambda \mathbf{M}_{j+1}\right) = \mathbf{F}^{\mathrm{e}}_{e} \cdot (\mathbf{R}^{\mathrm{e}}_{j+1})^{\mathrm{T}} \tag{12.105}$$

A post-multiplication of each side of this equation by its transpose, and exploiting Equation (12.95) for $n = 2$ yields:

$$\mathbf{V}^{\mathrm{e}}_{j+1} \cdot \exp\left(2\Delta\lambda \mathbf{M}_{j+1}\right) \cdot \mathbf{V}^{\mathrm{e}}_{j+1} = \mathbf{F}^{\mathrm{e}}_{e} \cdot (\mathbf{F}^{\mathrm{e}}_{e})^{\mathrm{T}} \tag{12.106}$$

The tensors $\mathbf{V}^{\mathrm{e}}$ and $\mathbf{M}$ are coaxial because of the assumed elastic and plastic isotropy, and therefore commute. Making use of this property, taking the square root, and solving for $\mathbf{V}^{\mathrm{e}}_{j+1}$ gives:

$$\mathbf{V}^{\mathrm{e}}_{j+1} = \mathbf{V}^{\mathrm{e}}_{e} \cdot \exp\left(-\Delta\lambda \mathbf{M}_{j+1}\right) \tag{12.107}$$

The logarithm of this expression

$$\boldsymbol{\varepsilon}^{\mathrm{e}}_{j+1} = \boldsymbol{\varepsilon}^{\mathrm{e}}_{e} - \Delta\lambda \mathbf{M}_{j+1} \tag{12.108}$$

has a format that is identical to that in small-strain plasticity. The similarity can be brought out even more clearly by transforming Equation (12.108) into the (Kirchhoff) stress space by pre-multiplying by $\mathbf{D}^{\mathrm{e}}$, so that, using Equation (11.112),

$$\boldsymbol{\tau}^{\mathrm{e}}_{j+1} = \boldsymbol{\tau}^{\mathrm{e}}_{e} - \Delta\lambda \mathbf{D}^{\mathrm{e}} : \mathbf{M}_{j+1} \tag{12.109}$$

It is emphasised that the simplicity of the approach crucially depends on the assumed elastic and plastic isotropy. If these conditions are not fulfilled a more complicated algorithm evolves. An algorithm based on this derivation is shown in Box 12.2.

The 'material' part of the tangential stiffness matrix that is consistent with the update algorithm based on the preceding derivation, and summarised in Box 12.2, can be obtained by straightforward differentiation. The Kirchhoff stress is obtained from the elastic logarithmic trial strain and from the internal variables, hence:

$$\boldsymbol{\kappa}_{j+1} = \boldsymbol{\kappa}(\boldsymbol{\varepsilon}^{\mathrm{e}}_{e}, \boldsymbol{\alpha}_0) \tag{12.110}$$

From Equation (11.109) the elastic logarithmic trial strain derives from the elastic trial left Cauchy–Green deformation tensor $\mathbf{B}^{\mathrm{e}}_{e}$, and therefore, from the deformation gradient $\mathbf{F}_{j+1}$, since the plastic part, $\mathbf{F}^{\mathrm{F}}_{0}$, is 'frozen', and does not change during the trial step. Accordingly,

Box 12.2 Stress update for multiplicative finite strain plasticity

1. Initialise:
   - Deformation gradient $\mathbf{F}_0 = \mathbf{F}^t$
   - Elastic deformation gradient $\mathbf{F}_0^{\mathrm{e}} = (\mathbf{F}^{\mathrm{e}})^t$
   - Hardening variables: $\boldsymbol{\alpha}_0 = \boldsymbol{\alpha}^t$
2. Compute $\Delta\mathbf{F} = \frac{\partial \mathbf{x}^{t+\Delta t}}{\partial \mathbf{x}^t} = \mathbf{I} + \frac{\partial \Delta\mathbf{u}}{\partial \mathbf{x}^t}$, $\mathbf{F}^{t+\Delta t} = \Delta\mathbf{F} \cdot \mathbf{F}_0$
3. Compute the trial elastic deformation gradient: $\mathbf{F}_e^{\mathrm{e}} = \Delta\mathbf{F} \cdot \mathbf{F}_0^{\mathrm{e}}$
4. Compute the trial stress, Equation (12.102): $\boldsymbol{\kappa}_e = \mathbf{D}^{\mathrm{e}} : \boldsymbol{\varepsilon}_e^{\mathrm{e}} = \frac{1}{2}\mathbf{D}^{\mathrm{e}} : \ln\left(\mathbf{F}_e^{\mathrm{e}} \cdot (\mathbf{F}_e^{\mathrm{e}})^{\mathrm{T}}\right)$
5. Compute the Kirchhoff stress $\boldsymbol{\kappa}_{j+1}$, the hardening variables $\boldsymbol{\alpha}_{j+1}$ and the plastic flow increment $\Delta\lambda$ according to a (small-strain) return-mapping algorithm:

   If $f(\boldsymbol{\kappa}_e, \boldsymbol{\alpha}_0) \geq 0$:

   – Solve the system of local residuals: $\begin{pmatrix} \mathbf{r}_\kappa \\ \mathbf{r}_\alpha \\ r_f \end{pmatrix} = \mathbf{0}$

   using a local iterative procedure with $j = 0, \ldots, n$, cf. Equation (7.96):

$$\begin{cases} \mathbf{r}_\kappa = \boldsymbol{\kappa}_{j+1} - \boldsymbol{\kappa}_e + \Delta\lambda \mathbf{D}^{\mathrm{e}} \mathbf{M}(\boldsymbol{\kappa}_{j+1}, \boldsymbol{\alpha}_{j+1}) \\ \mathbf{r}_\alpha = \boldsymbol{\alpha}_{j+1} - \boldsymbol{\alpha}_0 - \Delta\lambda \mathbf{p}(\boldsymbol{\kappa}_{j+1}, \boldsymbol{\alpha}_{j+1}) \\ r_f = f(\boldsymbol{\kappa}_{j+1}, \boldsymbol{\alpha}_{j+1}) \end{cases}$$

   Else $\boldsymbol{\kappa}_{n+1} = \boldsymbol{\kappa}_e$, $\boldsymbol{\alpha}_{n+1} = \boldsymbol{\alpha}_0$
6. Compute the elastic deformation gradient: $(\mathbf{F}^{\mathrm{e}})^{t+\Delta t} = \exp\left((\mathbf{D}^{\mathrm{e}})^{-1} : \boldsymbol{\kappa}_{n+1}\right)$
7. Compute the Cauchy stress tensor: $\boldsymbol{\sigma}^{t+\Delta t} = (\det \mathbf{F}^{t+\Delta t})^{-1} \boldsymbol{\kappa}_{n+1}$

we have for the material part of the tangential stiffness tensor:

$$\frac{\partial \boldsymbol{\kappa}}{\partial \mathbf{F}_{j+1}} = \mathbf{D} : \frac{\partial \boldsymbol{\varepsilon}_e^{\mathrm{e}}}{\partial \mathbf{B}_e^{\mathrm{e}}} : \frac{\partial \mathbf{B}_e^{\mathrm{e}}}{\partial \mathbf{F}_{j+1}} \tag{12.111}$$

where

$$\mathbf{D} = \frac{\partial \boldsymbol{\kappa}}{\partial \boldsymbol{\varepsilon}_e^{\mathrm{e}}} \tag{12.112}$$

is the small-strain elasto-plastic consistent tangent operator, cf. Equation (12.71),

$$\frac{\partial \boldsymbol{\varepsilon}_e^{\mathrm{e}}}{\partial \mathbf{B}_e^{\mathrm{e}}} = \frac{1}{2} \frac{\partial \ln[\mathbf{B}_e^{\mathrm{e}}]}{\partial \mathbf{B}_e^{\mathrm{e}}} \tag{12.113}$$

and the components of $\frac{\partial \mathbf{B}^{\mathrm{e}}_{e}}{\partial \mathbf{F}_{j+1}}$ are given by:

$$\left[\frac{\partial \mathbf{B}^{\mathrm{e}}_{e}}{\partial \mathbf{F}_{j+1}}\right]_{ijkl} = \delta_{ik}(B^{\mathrm{e}}_{e})_{jl} + \delta_{jk}(B^{\mathrm{e}}_{e})_{il} \tag{12.114}$$

Clearly, only $\mathbf{D}$ is dependent on the constitutive model, since the other factors derive from the large-strain effects.

Alternatively, we can define the elastic left Cauchy–Green deformation tensor in the principal directions. To this end, we consider its spectral decomposition in the trial state:

$$\mathbf{B}^{\mathrm{e}}_{e} = \sum_{i=1}^{3} (\lambda^{\mathrm{e}}_{e})^2_i \, (\boldsymbol{e}_e)_i \otimes (\boldsymbol{e}_e)_i \tag{12.115}$$

and in the state after the return map:

$$\mathbf{B}^{\mathrm{e}}_{j+1} = \sum_{i=1}^{3} (\lambda^{\mathrm{e}}_{j+1})^2_i \boldsymbol{e}_i \otimes \boldsymbol{e}_i \tag{12.116}$$

with $\left(\lambda^{\mathrm{e}}_{e}\right)_i$, $\left(\lambda^{\mathrm{e}}_{j+1}\right)_i$ the principal elastic stretches in the trial and the final state, respectively. As in the small-strain case, cf. Chapter 7, the (Kirchhoff) stress in the trial state, $\boldsymbol{\kappa}_e$, is coaxial with the trial elastic left Cauchy–Green strain tensor, $\mathbf{B}^{\mathrm{e}}_{e}$, by virtue of isotropy, and since coaxiality also holds between the stress tensor and the plastic flow tensor $\mathbf{M}$ for coaxial flow rules, by the same reasoning as in Chapter 7 we arrive at the conclusion that $\boldsymbol{\kappa}_{j+1}$ and $\mathbf{B}^{\mathrm{e}}_{e}$ are coaxial, and hence: $(\boldsymbol{e}_e)_i = \boldsymbol{e}_i$. The return map can therefore be carried out fully in the principal space, and using the spectral decomposition, the Kirchhoff stress tensor can be recovered, and subsequently by a standard transformation, the Cauchy stress tensor. The algorithmic development in the principal space can bring complications, for instance when two or more principal stretches coincide, where it is recalled that similar problems arise in (small-strain) plasticity with singular yield surfaces, e.g. Box 7.8, or in hyperelasticity when the strain energy function is phrased in terms of principal stretches, such as for the Ogden model (Box 11.3). For details regarding the implementation of large-strain elasto-plasticity in the principal space, including the construction of consistently linearised tangential stifness matrices, reference is made to Miehe (1998), Rosati and Valoroso (2004) and Simo (1998); see Simo and Hughes (1998) for an overview.

References

Armero F 2004 Elastoplastic and viscoplastic deformations in solids and structures, in *The Encyclopedia of Computational Mechanics* (eds Stein E, de Borst R and Hughes TJR), vol. II, pp. 227–266. John Wiley & Sons, Ltd.

Asaro RJ 1983 Crystal plasticity. *Journal of Applied Mechanics* **50**, 921–934.

Atluri SN 1983 Alternate stress and conjugate strain measures and mixed variational formulations involving rigid rotations, for computational analysis of finitely deformed solids, with applications to planes and shells – I, Theory. *Computers & Structures* **18**, 93–116.

Atluri SN 1984 On constitutive relations at finite strain. *Computer Methods in Applied Mechanics and Engineering* **43**, 137–171.

Cuitino AM and Ortiz M 1992 A material-independent method for extending stress update algorithms from small-strain plasticity to finite plasticity with multiplicative kinematics. *Engineering Computations* **9**, 437–451.

Dafalias YF 1984 The plastic spin concept and a simple illustration of its role in finite plastic transformations. *Mechanics of Materials* **3**, 223–233.

Dafalias YF 1985 The plastic spin. *Journal of Applied Mechanics* **52**, 865–871.

Dafalias YF 1998 The plastic spin: necessity or redundancy? *International Journal of Plasticity* **14**, 909–931.

de Souza Neto EA, Perić D and Owen DRJ 2008 *Computational Methods for Plasticity: Theory and Applications*. John Wiley & Sons, Ltd.

Eterovic AL and Bathe KJ 1990 A hyperelastic based large strain elasto-plastic constitutive formulation with combined isotropic-kinematic hardening using logarithmic stress and strain measures. *International Journal for Numerical Methods in Engineering* **30**, 1099–1114.

Flanagan DP and Taylor LM 1987 An accurate numerical algorithm for stress integration with finite rotations. *Computer Methods in Applied Mechanics and Engineering* **62**, 305–320.

Healey BE and Dodds RH 1992 A large-strain plasticity model for implicit finite element analyses. *Computational Mechanics* **9**, 95–112.

Hibbit HD, Marcal PV and Rice JR 1970 A finite element formulation for problems for large strain and large displacement. *International Journal of Solids and Structures* **6**, 1069–1086.

Hughes TJR and Winget J 1980 Finite rotation effects in numerical integration of rate constitutive equations arising in large-deformation analysis. *International Journal for Numerical Methods in Engineering* **15**, 1862–1867.

Key SW and Krieg RD 1982 On the numerical implementation of inelastic time dependent and time independent, finite strain constitutive equations in structural mechanics. *Computer Methods in Applied Mechanics and Engineering* **33**, 439–452.

Kratochvil J 1973 On a fintie strain theory of elastic-inelastic materials. *Acta Mechanica* **16**, 127–142.

Lee EH 1969 Elastic-plastic deformation at finite strain. *Journal of Applied Mechanics* **36**, 1–6.

Lubliner J 1990 *Plasticity Theory*. Macmillan.

Mandel J 1974 Thermodynamics and plasticity, in *Foundations of Continuum Thermodynamics* (ed. Domingues JJD), pp. 283–304. Macmillan.

Marsden JE and Hughes TJR 1983 *Mathematical Foundations of Elasticity*. Prentice Hall, Inc.

McMeeking RM and Rice JR 1975 Finite-element formulations for problems of large elastic-plastic deformation. *International Journal of Solids and Structures* **11**, 601–616.

Meschke G and Liu WN 1999 A reformulation of the exponential algorithm for finite strain plasticity in terms of Cauchy stresses. *Computer Methods in Applied Mechanics and Engineering* **173**, 167–187.

Miehe C 1996 Exponential map algorithm for stress updates in anisotropic multiplicative elastoplasticity for single crystals. *International Journal for Numerical Methods in Engineering* **39**, 3367–3390.

Miehe C 1998 A formulation of finite elastoplasticity based on dual co- and contra-variant eigenvector triads normalized with respect to a plastic metric. *Computer Methods in Applied Mechanics and Engineering* **159**, 223–260.

Miehe C and Schotte J 2004 Crystal plasticity and the evolution of polycrystalline microstructure, in *The Encyclopedia of Computational Mechanics* (eds Stein E, de Borst R and Hughes TJR), vol. II, pp. 267–289. John Wiley & Sons, Ltd.

Moran B, Ortiz M and Shih CF 1990 Formulation of implicit finite element methods for multiplicative finite deformation plasticity. *International Journal for Numerical Methods in Engineering* **29**, 483–514.

Moss WC 1984 On instabilities in large deformation simple shear loading. *Computer Methods in Applied Mechanics and Engineering* **46**, 329–338.

Nagtegaal JC 1982 On the implementation of inelastic constitutive equations with special reference to large deformation problems. *Computer Methods in Applied Mechanics and Engineering* **33**, 469–484.

Nagtegaal JC and de Jong JE 1981 Some computational aspects of elastic-plastic large strain analysis. *International Journal for Numerical Methods in Engineering* **17**, 15–41.

Needleman A 1985 On finite element formulations for large elastic-plastic deformations. *Computers & Structures* **20**, 247–257.

Nemat-Nasser S 1982 On finite deformation elasto-plasticity. *International Journal of Solids and Structures* **18**, 857–872.

Nemat-Nasser S 1983 On finite plastic flow of crystalline solids and geomaterials. *Journal of Applied Mechanics* **35**, 151–183.

Peirce D, Asaro RJ and Needleman A 1982 An analysis of nonuniform and localized deformation in crystalline solids. *Acta Metallurgica* **30**, 1087–1119.

Perić D, Owen DRJ and Honnor ME 1992 A model for finite strain elasto-plasticity based on logarithmic strains: computational issues. *Computer Methods in Applied Mechanics and Engineering* **94**, 35–61.

Pinsky PM, Ortiz M and Pister KS 1983 Numerical integration of rate constitutive equations in finite deformation analysis. *Computer Methods in Applied Mechanics and Engineering* **40**, 137–158.

Rice JR 1971 Inelastic constitutive relations for solids: an internal variable theory and its application to metal plasticity. *Journal of the Mechanics and Physics of Solids* **19**, 433–455.

Rice JR 1975 Continuum mechanics and thermodynamics of plasticity in relation to microscale deformation mechanisms, in *Constitutive Equations in Plasticity* (ed. Argon AS), pp. 23–75. MIT Press.

Rosati L and Valoroso N 2004 A return map algorithm for general isotropic elasto/visco-plastic materials in principal space. *International Journal for Numerical Methods in Engineering* **60**, 461–498.

Rouainia M and Wood DM 2000 An implicit constitutive algorithm for finite strain cam-clay elasto-plastic model. *Mechanics of Cohesive-frictional Material* **5**, 469–489.

Simo JC 1985 A finite strain beam formulation. The three-dimensional dynamic problem. Part I. *Computer Methods in Applied Mechanics and Engineering* **49**, 55–70.

Simo JC 1988a A framework for finite strain elastoplasticity based on maximum plastic dissipation and the multiplicative decomposition. Part I. Continuum formulation. *Computer Methods in Applied Mechanics and Engineering* **66**, 199–219.

Simo JC 1988b A framework for finite strain elastoplasticity based on maximum plastic dissipation and the multiplicative decomposition. Part II. Computational aspects. *Computer Methods in Applied Mechanics and Engineering* **68**, 1–31.

Simo JC 1992 Algorithms for static and dynamic multiplicative plasticity that preserve the classical return mapping schemes of the infinitesimal theory. *Computer Methods in Applied Mechanics and Engineering* **99**, 61–112.

Simo JC 1998 Numerical analysis of classical plasticity, in *Handbook for Numerical Analysis* (eds Ciarlet PG and Lions JJ), vol. IV, pp. 183–499. Elsevier.

Simo JC and Hughes TJR 1998 *Computational Inelasticity: Theory and Applications*. Springer.

Simo JC and Ortiz M 1985 A unified approach to finite deformation elastoplastic analysis based on the use of hyperelastic constitutive equations. *Computer Methods in Applied Mechanics and Engineering* **49**, 221–245.

Weber G and Anand L 1990 Finite deformation constitutive equations and a time integration procedure for isotropic hyperelastic viscoplastic solids. *Computer Methods in Applied Mechanics and Engineering* **79**, 173–202.

Part V

Advanced Discretisation Concepts

13

Interfaces and Discontinuities

Interfaces occur in a wide range of structures. In civil engineering they are for instance encountered in reinforced soils, as the intermediate layer between rock and concrete, or in the analysis of rock joints (Goodman *et al.* 1968). Applications in reinforced and prestressed concrete include the modelling of discrete cracking (Ingraffea and Saouma 1985; Rots 1991), and aggregate interlock and bond between concrete and reinforcement (Feenstra *et al.* 1991). In rubber parts interfaces can be of importance when disintegration of rubber and texture is concerned such as in conveyor belts. Furthermore, interfaces occur in delamination and fibre pull-out of composite structures (Allix and Ladevèze 1992; Schellekens and de Borst 1993a, 1994), frictional contact in forming processes (Qiu *et al.* 1991; Wriggers 2006), and in coatings (van den Bosch *et al.* 2007, 2008).

Often, there is a discontinuity in the material properties at an interface, such as the elastic stiffness or other physical properties such as conductivity. Such discontinuities tend to be among the most critical parts in a structure, and can act as a precursor to failure. Two types of fracture analyses can be distinguished. The first class of methods considers the computation of fracture properties for a given, stationary crack. Typically, this relates to properties like stress intensity factors or to the J-integral. Often, linear elastic fracture mechanics is used as the underlying theory. With the proper knowledge of these quantities, fracture mechanics makes it possible to determine if a crack will propagate, in which direction – although a number of different hypotheses exist – and, for dynamic problems, at what speed the crack will propagate. Since the stress field is singular at the crack tip in linear elastic fracture mechanics, tailored numerical schemes have been developed for capturing this singularity, especially for coarse discretisations (Barsoum 1976; Henshell and Shaw 1976).

More difficult is the simulation of crack propagation. In the first approaches, a stress intensity factor was computed, and on the basis of this information it was decided whether, and if yes, how much, the crack would propagate. After propagation of the crack, a new mesh was generated for the new geometry and the process was repeated (Ingraffea and Saouma 1985). This approach consists of a series of computations for a stationary crack using linear-elastic fracture mechanics, and can be useful if the direction of crack propagation is known, and if the fracture process zone is sufficiently small compared with the structural dimensions. Evidently,

Non-linear Finite Element Analysis of Solids and Structures, Second Edition.
René de Borst, Mike A. Crisfield, Joris J.C. Remmers and Clemens V. Verhoosel.

the remeshing procedures that are inherent in this approach can be difficult and cumbersome, especially for three-dimensional analyses.

13.1 Interface Elements

When the fracture process zone is not small compared with the structural dimensions, cohesive-zone models have to be used to properly simulate crack propagation. To accommodate cohesive-zone models in finite element analysis, interface elements have found widespread use. Interface elements have to be inserted a priori in the finite element mesh. When the direction of crack propagation is known interface elements can be used in a straightforward manner (Rots 1991). An example where interface elements have been used successfully, fully exploiting the potential of cohesive-zone models, is the analysis of delamination in layered composite materials (Allix and Corigliano 1999; Allix and Ladevèze 1992; Schellekens and de Borst 1993a, 1994). Since the propagation of delaminations is then restricted to the interfaces between the plies, inserting interface elements at these locations can capture the kinematics of the failure mode in an exact manner.

To allow for a more arbitrary direction of crack propagation Xu and Needleman (1994, 1996) have inserted interface elements equipped with a cohesive zone model between *all* continuum elements. A related method, using remeshing, was proposed by Camacho and Ortiz (1996). Although such analyses provide much insight, they suffer from a certain mesh bias, since the direction of crack propagation is not entirely free, but is restricted to interelement boundaries. This has been demonstrated in Tijssens *et al.* (2000), where the single-edge notched beam of Figure 6.13 has also been analysed, but now with a finite element model in which interface elements equipped with a quasi-brittle decohesion relation were inserted between all continuum elements (Figure 13.1).

The governing kinematic quantities in continuous interface elements are a set of mutually orthogonal, relative displacements: v_n, v_s, v_t for the normal and the two sliding modes, respectively. When collecting the relative displacements in a relative displacement vector $\mathbf{v}$, they can be related to the displacements at the upper (+) and the lower sides (−) of the interface, $u_n^-, u_n^+, u_s^-, u_s^+, u_t^-, u_t^+$, via (Figure 13.2)

$$\mathbf{v} = \mathbf{L}\mathbf{u} \tag{13.1}$$

with

$$\mathbf{u}^{\mathrm{T}} = (u_n^-, u_n^+, u_s^-, u_s^+, u_t^-, u_t^+) \tag{13.2}$$

and

$$\mathbf{L} = \begin{bmatrix} -1 & +1 & 0 & 0 & 0 & 0 \\ 0 & 0 & -1 & +1 & 0 & 0 \\ 0 & 0 & 0 & 0 & -1 & +1 \end{bmatrix} \tag{13.3}$$

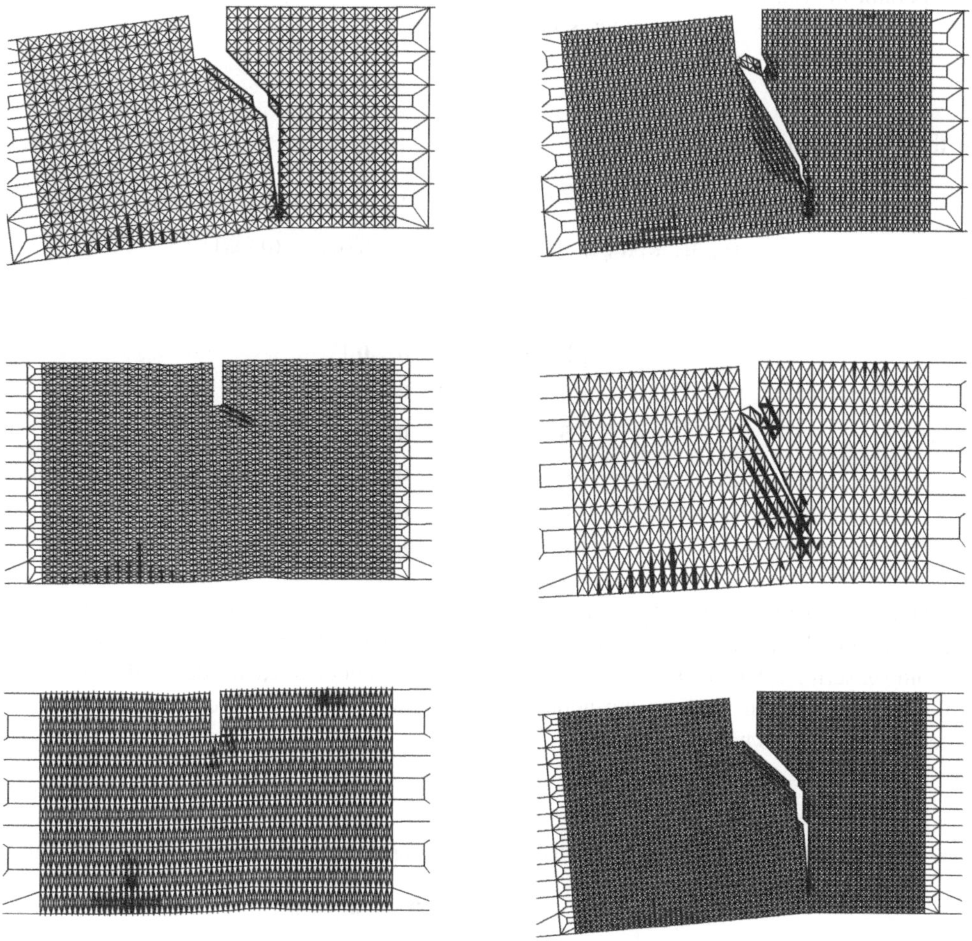

Figure 13.1 Crack patterns for different discretisations using interface elements between all solid elements. Only the part of the single-edge notched beam near the notch is shown (Tijssens *et al.* 2000)

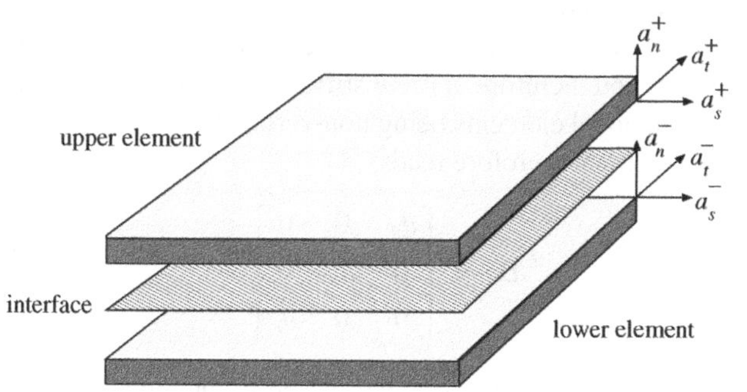

Figure 13.2 Planar interface element between two three-dimensional finite elements

an operator matrix. For each side of the interface element, the displacements contained in **u** are interpolated in a standard manner, see Chapter 2, as

$$\mathbf{u} = \mathbf{H}\mathbf{a}$$

with **a** the nodal displacement array,

$$\mathbf{a} = \left((a_n^-)_1, \ldots, (a_n^-)_n, (a_s^-)_1, \ldots, (a_s^-)_n, (a_t^-)_1, \ldots, (a_t^-)_n, \right.$$
$$\left. (a_n^+)_1, \ldots, (a_n^+)_n, (a_s^+)_1, \ldots, (a_s^+)_n, (a_t^+)_1, \ldots, (a_t^+)_n\right)^{\mathrm{T}}$$

and

$$\mathbf{H} = \begin{bmatrix} \mathbf{h} & \mathbf{0} & \mathbf{0} & \mathbf{0} & \mathbf{0} & \mathbf{0} \\ \mathbf{0} & \mathbf{h} & \mathbf{0} & \mathbf{0} & \mathbf{0} & \mathbf{0} \\ \mathbf{0} & \mathbf{0} & \mathbf{h} & \mathbf{0} & \mathbf{0} & \mathbf{0} \\ \mathbf{0} & \mathbf{0} & \mathbf{0} & \mathbf{h} & \mathbf{0} & \mathbf{0} \\ \mathbf{0} & \mathbf{0} & \mathbf{0} & \mathbf{0} & \mathbf{h} & \mathbf{0} \\ \mathbf{0} & \mathbf{0} & \mathbf{0} & \mathbf{0} & \mathbf{0} & \mathbf{h} \end{bmatrix} \tag{13.4}$$

with **h** a $1 \times n$ matrix containing the shape functions $h_1, \ldots, h_n$. It is noted that some confusion can arise from this notation, since the first subscript n relates to the normal direction, while the second subscript n denotes the number of node pairs (one node at each side) of the interface element. The relation between the nodal displacements and the relative displacements for interface elements is now derived as:

$$\mathbf{v} = \mathbf{L}\mathbf{H}\mathbf{a} = \mathbf{B}_d\mathbf{a} \tag{13.5}$$

where the relative displacement–nodal displacement matrix $\mathbf{B}_d$ for the interface element reads:

$$\mathbf{B}_d = \begin{bmatrix} -\mathbf{h} & +\mathbf{h} & \mathbf{0} & \mathbf{0} & \mathbf{0} & \mathbf{0} \\ \mathbf{0} & \mathbf{0} & -\mathbf{h} & +\mathbf{h} & \mathbf{0} & \mathbf{0} \\ \mathbf{0} & \mathbf{0} & \mathbf{0} & \mathbf{0} & -\mathbf{h} & +\mathbf{h} \end{bmatrix} \tag{13.6}$$

For an arbitrarily oriented interface element the matrix $\mathbf{B}_d$ subsequently has to be transformed to the local coordinate system of the parent interface element.

Conventional interface elements have to be inserted in the finite element mesh at the beginning of the computation, and therefore, a finite stiffness must be assigned in the pre-cracking phase with at least the diagonal elements being non-zero. Prior to crack initiation, the stiffness matrix in the interface element therefore reads:

$$\mathbf{D}_d = \begin{bmatrix} d_n & 0 & 0 \\ 0 & d_s & 0 \\ 0 & 0 & d_t \end{bmatrix} \tag{13.7}$$

with d_n the stiffness normal to the interface and d_s and d_t the tangential stiffnesses. With the material tangential stiffness matrix $\mathbf{D}_d$ the element tangential stiffness matrix can be derived in a straightforward fashion, leading to:

$$\mathbf{K} = \int_{S_d} \mathbf{B}_d^{\mathrm{T}} \mathbf{D}_d \mathbf{B}_d \mathrm{d}S \tag{13.8}$$

where the integration domain extends over the surface of the interface S_d. With the aim of numerical integration the integral of Equation (13.8) is replaced by an integral over the isoparametric coordinates ξ, η, cf. Equation (2.21) for the internal force vector of a continuum element. Then, the element stiffness matrix becomes:

$$\mathbf{K} = \int_{-1}^{+1} \int_{-1}^{+1} (\det \mathbf{J}) \mathbf{B}_d^{\mathrm{T}} \mathbf{D}_d \mathbf{B}_d \mathrm{d}\xi \mathrm{d}\eta \tag{13.9}$$

with $\mathbf{J}$ the Jacobian matrix. Numerical integration subsequently results in:

$$\mathbf{K} = \sum_{i=1}^{n_i} w_i \det \mathbf{J}_i (\mathbf{B}_d^{\mathrm{T}})_i (\mathbf{D}_d)_i (\mathbf{B}_d)_i \tag{13.10}$$

with w_i the weight factor of integration point i, and n_i the number of integration points in the interface element.

For comparison with other approaches and to elucidate the phenomenon of traction oscillations in numerically integrated interface elements we expand the stiffness matrix in the pre-cracking phase as (Schellekens and de Borst 1993b):

$$\mathbf{K} = \begin{bmatrix} \mathbf{K}_n & \mathbf{0} & \mathbf{0} \\ \mathbf{0} & \mathbf{K}_s & \mathbf{0} \\ \mathbf{0} & \mathbf{0} & \mathbf{K}_t \end{bmatrix} \tag{13.11}$$

with the submatrices $\mathbf{K}_\pi$, $\pi = n, s, t$ defined as:

$$\mathbf{K}_\pi = d_\pi \begin{bmatrix} \mathbf{h}^{\mathrm{T}}\mathbf{h} & -\mathbf{h}^{\mathrm{T}}\mathbf{h} \\ -\mathbf{h}^{\mathrm{T}}\mathbf{h} & \mathbf{h}^{\mathrm{T}}\mathbf{h} \end{bmatrix} \tag{13.12}$$

with d_π the stiffnesses in the interface prior to crack initiation. For example, when using a 2×2 Gauss integration scheme for a linear, plane interface element with four node pairs and a surface A, one obtains:

$$\mathbf{h}^{\mathrm{T}}\mathbf{h} = \frac{A}{36} \begin{bmatrix} 4 & 2 & 1 & 2 \\ 2 & 4 & 2 & 1 \\ 1 & 2 & 4 & 2 \\ 2 & 1 & 2 & 4 \end{bmatrix} \tag{13.13}$$

A basic requirement of interface elements is that during the elastic stage of the loading process no significant additional deformations occur due to the presence of these elements in the finite element model. Therefore, sufficiently high initial stiffnesses d_π have to be supplied for the interface elements. Depending on the applied numerical integration scheme, however, this high dummy stiffness can result in undesired spurious oscillations of the stress field. The off-diagonal coupling terms of the submatrix $\mathbf{h}^{\mathrm{T}}\mathbf{h}$ that features in the stiffness matrix of the

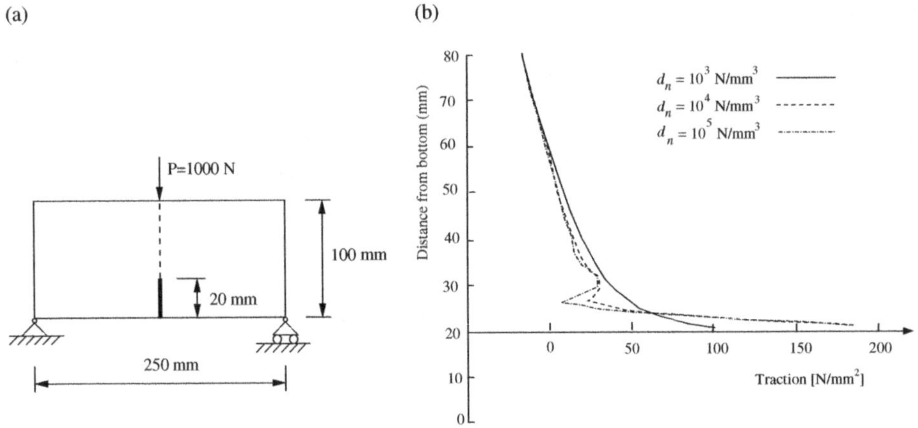

Figure 13.3 (a) Geometry of a symmetric, notched, three-point bending beam. (b) Traction profiles ahead of the notch using linear interface elements with Gauss integration. Results are shown for different values of a 'dummy' normal stiffness d_n in the pre-cracking phase (Schellekens and de Borst 1993b)

interface elements, cf. Equation (13.12), can lead to spurious traction oscillations in the pre-cracking phase for high stiffness values (Schellekens and de Borst 1993b). An example of an oscillatory traction pattern ahead of a notch is given in Figure 13.3. When analysing dynamic fracture, spurious wave reflections can occur as a result of the introduction of artificially high stiffness values prior to the onset of delamination.

The traction oscillations shown in Figure 13.3 are typical when Gauss integration is used for the numerical integration of the stiffness matrix and the internal force vector of interface elements. More precisely, it is the combination of high traction gradients and a Gauss integration scheme in interface elements which causes the oscillations. Experience shows that oscillations disappear when Newton–Cotes integration or Lobatto integration – which is identical for low-order schemes – is used instead. This can be shown rather easily for the linear, plane interface element with four node pairs. Eigenvalue analyses show that the coupling between the degrees of freedom present for Gauss integration then disappears. For interface elements with a quadratic interpolation an improved behaviour is also observed, although this cannot be explained readily from eigenvalue analyses (Schellekens and de Borst 1993b). To further investigate the relatively good performance of Newton–Cotes integration in this case, the submatrix $\mathbf{h}^{\mathrm{T}}\mathbf{h}$ of a quadratic, plane interface element is elaborated as:

$$\mathbf{h}^{\mathrm{T}}\mathbf{h} = \frac{A}{180}\begin{bmatrix} 10 & -10 & 5 & -10 & 5 & -10 & 5 & -10 \\ -10 & 40 & -10 & 20 & -10 & 20 & -10 & 20 \\ 5 & -10 & 10 & -10 & 5 & -10 & 5 & -10 \\ -10 & 20 & -10 & 40 & -10 & 20 & -10 & 20 \\ 5 & -10 & 5 & -10 & 10 & -10 & 5 & -10 \\ -10 & 20 & -10 & 20 & -10 & 40 & -10 & 20 \\ 5 & -10 & 5 & -10 & 5 & -10 & 10 & -10 \\ -10 & 20 & -10 & 20 & -10 & 20 & -10 & 40 \end{bmatrix} \tag{13.14}$$

This submatrix is subsequently decomposed in a matrix which contains the contributions of the integration points that coincide with the element nodes:

$$(\mathbf{h}^{\mathrm{T}}\mathbf{h})_1 = \frac{A}{180}\begin{bmatrix} 5 & 0 & 0 & 0 & 0 & 0 & 0 & 0 \\ 0 & 20 & 0 & 0 & 0 & 0 & 0 & 0 \\ 0 & 0 & 5 & 0 & 0 & 0 & 0 & 0 \\ 0 & 0 & 0 & 20 & 0 & 0 & 0 & 0 \\ 0 & 0 & 0 & 0 & 5 & 0 & 0 & 0 \\ 0 & 0 & 0 & 0 & 0 & 20 & 0 & 0 \\ 0 & 0 & 0 & 0 & 0 & 0 & 5 & 0 \\ 0 & 0 & 0 & 0 & 0 & 0 & 0 & 20 \end{bmatrix} \tag{13.15}$$

and a matrix that contains the contribution of the integration point located in the centre of the element:

$$(\mathbf{h}^{\mathrm{T}}\mathbf{h})_2 = \frac{A}{180}\begin{bmatrix} 5 & -10 & 5 & -10 & 5 & -10 & 5 & -10 \\ -10 & 20 & -10 & 20 & -10 & 20 & -10 & 20 \\ 5 & -10 & 5 & -10 & 5 & -10 & 5 & -10 \\ -10 & 20 & -10 & 20 & -10 & 20 & -10 & 20 \\ 5 & -10 & 5 & -10 & 5 & -10 & 5 & -10 \\ -10 & 20 & -10 & 20 & -10 & 20 & -10 & 20 \\ 5 & -10 & 5 & -10 & 5 & -10 & 5 & -10 \\ -10 & 20 & -10 & 20 & -10 & 20 & -10 & 20 \end{bmatrix} \tag{13.16}$$

which confirms that the coupling is entirely due to the contribution to the stiffness matrix of the centre integration point. Evidently, the fact that only a single integration point causes a coupling effect, significantly reduces the oscillations when compared with Gauss integration, where all integration points contribute to this undesired phenomenon. The good performance of the Newton–Cotes scheme in the case of two-dimensional, quadratic interface elements, which is often observed, is also caused by the following property. The part of the **B** matrix that stems from the centre integration point, and which relates nodal displacements normal to the element to the normal relative displacements in the integration point is proportional to:

$$\mathbf{B}^* \sim \begin{bmatrix} -\frac{1}{4} & \frac{1}{2} & -\frac{1}{4} & \frac{1}{2} & -\frac{1}{4} & \frac{1}{2} & -\frac{1}{4} & \frac{1}{2} \\ 0 & 0 & 0 & 0 & 0 & 0 & 0 & 0 \\ 0 & 0 & 0 & 0 & 0 & 0 & 0 & 0 \end{bmatrix} \tag{13.17}$$

The relation between the non-zero components in the **B** matrix is therefore such that when a displacement field over an element has a gradient in either the ξ- or in the η-direction, the nodes in the direction in which the displacement field varies do not have a resulting contribution to the tractions and relative displacements in the central integration point. The tractions and relative displacements in the central point of the element are only dependent on the nodal values of the nodes which are located in the direction in which no gradient exists. In this special case the element acts as if no coupling between the node pairs exists and no oscillations will occur.

As an alternative interface elements with lumped integration have been used frequently. In this class of interface elements relative displacements at the n node pairs are used instead of an interpolated relative displacement field in integration points. The element stiffness matrix is now expanded as:

$$\mathbf{K} = \sum_{i=1}^{n_i} A_i(\mathbf{B}_d^{\mathrm{T}})_i(\mathbf{D}_d)_i(\mathbf{B}_d)_i \tag{13.18}$$

where the relative displacement–nodal displacement matrix $(\mathbf{B}_d)_i$ for the interface element at node set i reads:

$$(\mathbf{B}_d)_i = \begin{bmatrix} -1 & +1 & 0 & 0 & 0 & 0 \\ 0 & 0 & -1 & +1 & 0 & 0 \\ 0 & 0 & 0 & 0 & -1 & +1 \end{bmatrix} \tag{13.19}$$

and A_i, is the surface that can be attributed to the node set i. Upon elaboration the submatrix $\mathbf{K}_\pi$ for a linear, plane interface element with four node pairs and a surface A now becomes:

$$\mathbf{K}_\pi = \frac{Ad_\pi}{4}\begin{bmatrix} 1 & 0 & 0 & 0 & -1 & 0 & 0 & 0 \\ 0 & 1 & 0 & 0 & 0 & -1 & 0 & 0 \\ 0 & 0 & 1 & 0 & 0 & 0 & -1 & 0 \\ 0 & 0 & 0 & 1 & 0 & 0 & 0 & -1 \\ -1 & 0 & 0 & 0 & 1 & 0 & 0 & 0 \\ 0 & -1 & 0 & 0 & 0 & 1 & 0 & 0 \\ 0 & 0 & -1 & 0 & 0 & 0 & 1 & 0 \\ 0 & 0 & 0 & -1 & 0 & 0 & 0 & 1 \end{bmatrix} \tag{13.20}$$

Evidently, no coupling exists between the degrees of freedom of the different node pairs. It is noted that interface elements with lumped integration are close to so-called nodal or point interface elements (Ngo and Scordelis 1967), which can, in a mechanical sense, be interpreted as interfaces that are connected by (non-linear) springs at the nodes.

As a further example of the use of interface elements to simulate fracture we consider the cantilever beam of Figure 13.4. The beam is 7.5 mm long and 1 mm thick, and is perforated across its entire length by holes with a diameter of 0.2 mm and a spacing of 0.375 mm. The beam is loaded by two forces $\lambda\hat{f}$ as shown in the figure. The problem is discretised using 9688 six-noded triangles with a seven-point Gauss integration scheme. Plane-strain conditions have been assumed, with linear elasticity prior to fracture with a Young's modulus $E = 100\,\mathrm{N/mm^2}$ and a Poisson's ratio $\nu = 0.3$. Because of symmetry with respect to the x-axis, it is assumed that mode-I fracture takes place along this axis. This predefined interface is discretised using 161 six-noded interface elements with a three-point Newton–Cotes integration scheme. Prior to cracking, a dummy stiffness $d_n = 1.0 \times 10^4\,\mathrm{N/mm^3}$ has been adopted. The ultimate traction is $t_{\mathrm{ult}} = 1\,\mathrm{N/mm^2}$, and a bilinear decohesion relation has been adopted with a fracture toughness $\mathcal{G}_c = 2.5 \times 10^{-3}\,\mathrm{N/mm}$.

The path-following method based on the energy release rate, Equation (4.32), has been used at turning points in the load–displacement curve. However, since fracture propagation

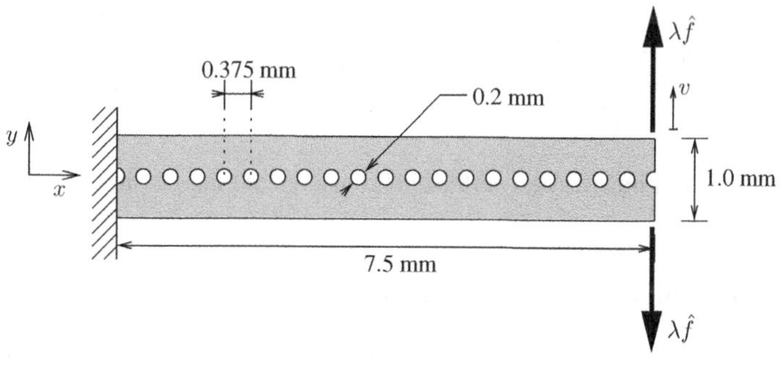

Figure 13.4 Geometry and loading conditions of a perforated cantilever beam

between two holes is followed by a traject without energy dissipation, this constraint cannot be used throughout the simulation. A force control has been used initially. When the dissipation increment exceeds a threshold value, the solver switches to the energy release constraint. When the energy release constraint is active, the step size is adjusted by aiming for five Newton–Raphson iterations per step (Verhoosel *et al.* 2009). The algorithm switches back to force control when the dissipation increment falls below another threshold.

The part of the force–displacement curve that describes the fracture of the first four segments is shown in Figure 13.5. The shaded area corresponds to half the amount of energy that is dissipated during the complete fracture of a segment between two consecutive holes, which equals the length of a segment times the fracture toughness: $0.175 \cdot \mathcal{G}_c = 4.375 \times 10^{-4}$ N/mm. The dashed lines A, B and C represent the elastic load–displacement curves for the cases where the crack has propagated through one, two and three segments, respectively; see Figure 13.6 for a partially fractured beam.

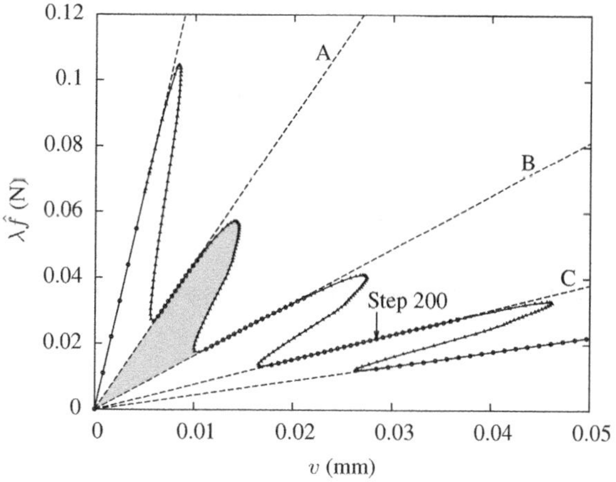

Figure 13.5 Load–displacement curve of the perforated cantilever beam

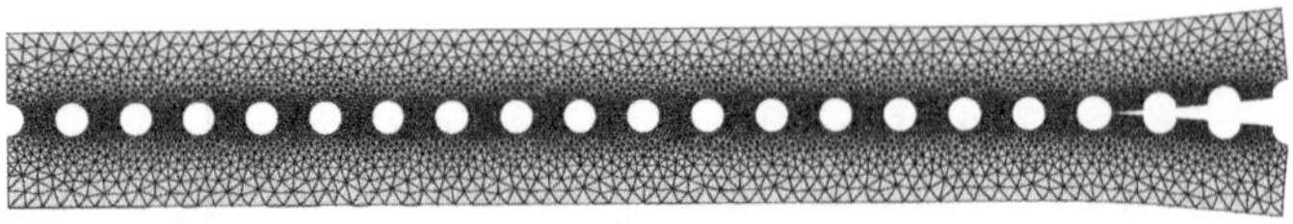

Figure 13.6 Deformation of a perforated beam after 200 steps (scaled by a factor 5)

13.2 Discontinuous Galerkin Methods

Discontinuous Galerkin methods have classically been employed for the computation of fluid flow (Cockburn 2004). More recently, attention has been given to their potential use in solid mechanics, and especially for problems involving cracks (Mergheim *et al.* 2004), or for constitutive models that incorporate spatial gradients (Wells *et al.* 2004). Finally, the use of a discontinuous Galerkin formalism can be a way to avoid traction oscillations in the pre-cracking phase.

For a discussion on the application of spatially discontinuous Galerkin to fracture it suffices to divide the domain into two subdomains, V^- and V^+, separated by an interface S_d. In a standard manner, the balance of linear momentum (2.4) is multiplied by test functions $\delta\mathbf{u}$, and after application of the divergence theorem, we obtain:

$$\begin{aligned}\int_{V/S_d} \nabla\delta\mathbf{u} : \boldsymbol{\sigma}\mathrm{d}V - \int_{S_d} \delta\mathbf{u}^+ \cdot \mathbf{t}_d^+ \mathrm{d}S - \int_{S_d} \delta\mathbf{u}^- \cdot \mathbf{t}_d^- \mathrm{d}S = \\ \int_V \delta\mathbf{u} \cdot \rho\mathbf{g}\mathrm{d}V + \int_S \delta\mathbf{u} \cdot \mathbf{t}\mathrm{d}S\end{aligned} \tag{13.21}$$

where the surface (line) integral on the external boundary S has been explicitly separated from that on the interface S_d. Note that restriction has been made to quasi-static loading conditions, but this is not essential. Prior to crack initiation, continuity of displacements and tractions must be enforced along S_d, at least in an approximate sense:

$$\begin{aligned}\mathbf{u}^+ - \mathbf{u}^- &= \mathbf{0} \\ \mathbf{t}_d^+ + \mathbf{t}_d^- &= \mathbf{0}\end{aligned} \tag{13.22}$$

with $\mathbf{t}_d^+ = \mathbf{n}_{S_d}^+ \cdot \boldsymbol{\sigma}^+$ and $\mathbf{t}_d^- = \mathbf{n}_{S_d}^- \cdot \boldsymbol{\sigma}^-$. Assuming small displacement gradients, we can set $\mathbf{n}_{S_d} = \mathbf{n}_{S_d}^+ = -\mathbf{n}_{S_d}^-$, so that the expressions for the interface tractions reduce to $\mathbf{t}_d^+ = \mathbf{n}_{S_d} \cdot \boldsymbol{\sigma}^+$ and $\mathbf{t}_d^- = -\mathbf{n}_{S_d} \cdot \boldsymbol{\sigma}^-$.

A classical procedure to enforce conditions (13.22) is to use Lagrange multipliers. Then,

$$\boldsymbol{\lambda} = \mathbf{t}_d^+ = -\mathbf{t}_d^- \tag{13.23}$$

along S_d, and Equation (13.21) transforms into:

$$\int_{V/S_d} \nabla\delta\mathbf{u} : \boldsymbol{\sigma}\mathrm{d}V - \int_{S_d} (\delta\mathbf{u}^+ - \delta\mathbf{u}^-) \cdot \boldsymbol{\lambda}\mathrm{d}S = \int_V \delta\mathbf{u} \cdot \rho\mathbf{g}\mathrm{d}V + \int_S \delta\mathbf{u} \cdot \mathbf{t}\mathrm{d}S \tag{13.24}$$

augmented with:

$$\int_{S_d} \delta\boldsymbol{\lambda} \cdot (\mathbf{u}^+ - \mathbf{u}^-)\mathrm{d}S = 0 \tag{13.25}$$

$\delta\boldsymbol{\lambda}$ being the test function for the Lagrange multiplier field $\boldsymbol{\lambda}$. After discretisation, Equations (13.24) and (13.25) result in a set of algebraic equations that are of a standard mixed format and therefore can give rise to difficulties. For this reason, alternative expressions are often sought, in which $\boldsymbol{\lambda}$ is directly expressed in terms of the interface tractions $\mathbf{t}_d^-$ and $\mathbf{t}_d^+$. One such possibility is to enforce Equation (13.22) pointwise, so that Equation (13.23) is replaced by:

$$\boldsymbol{\lambda} = -\mathbf{t}_d \tag{13.26}$$

and one obtains:

$$\int_{V/S_d} \nabla\delta\mathbf{u} : \boldsymbol{\sigma}\mathrm{d}V - \int_{S_d} (\delta\mathbf{u}^+ - \delta\mathbf{u}^-) \cdot \mathbf{t}_d\mathrm{d}S = \int_V \delta\mathbf{u} \cdot \rho\mathbf{g}\mathrm{d}V + \int_S \delta\mathbf{u} \cdot \mathbf{t}\mathrm{d}S \tag{13.27}$$

With the aid of relation (13.5) between the relative displacements $\mathbf{v} = \mathbf{u}^+ - \mathbf{u}^-$ and the nodal displacements at both sides of the interface S_d, and the linearised interface traction–relative displacement relation (6.74), the second term on the left-hand side can be elaborated in a discrete format as:

$$\int_{S_d} (\delta\mathbf{u}^+ - \delta\mathbf{u}^-) \cdot \mathbf{t}_d\mathrm{d}S = \delta\mathbf{a}^{\mathrm{T}} \left(\int_{S_d} \mathbf{B}_d^{\mathrm{T}}\mathbf{D}_d\mathbf{B}_d\mathrm{d}S \right) \mathbf{a} \tag{13.28}$$

which, not surprisingly, has exactly the same format as obtained for a conventional interface element. An example of a calculation using cohesive zones within the framework of a discontinuous Galerkin method is given in Figure 13.7.

Another possibility for the replacement of $\boldsymbol{\lambda}$ by an explicit function of the tractions is to take the average of the stresses at both sides of the interface:

$$\boldsymbol{\lambda} = \frac{1}{2}\mathbf{n}_{S_d} \cdot (\boldsymbol{\sigma}^+ + \boldsymbol{\sigma}^-) \tag{13.29}$$

The surface integrals for the interface in Equation (13.24) can now be reworked as:

$$\int_{S_d} (\delta\mathbf{u}^+ - \delta\mathbf{u}^-) \cdot \boldsymbol{\lambda}\mathrm{d}S = \int_{S_d} \frac{1}{2}(\delta\mathbf{u}^+ - \delta\mathbf{u}^-) \cdot \mathbf{n}_{S_d} \cdot (\boldsymbol{\sigma}^+ + \boldsymbol{\sigma}^-)\mathrm{d}S \tag{13.30}$$

To ensure a proper conditioning of the discretised equations, one has to add Equation (13.25), so that the modified form of Equation (13.21) finally becomes:

$$\begin{aligned}
&\int_{V/S_d} \nabla\delta\mathbf{u} : \boldsymbol{\sigma}\mathrm{d}V - \int_{S_d} \frac{1}{2}(\delta\mathbf{u}^+ - \delta\mathbf{u}^-) \cdot \mathbf{n}_{S_d} \cdot (\boldsymbol{\sigma}^+ + \boldsymbol{\sigma}^-)\mathrm{d}S - \\
&\alpha\int_{S_d} \frac{1}{2}(\nabla^{\mathrm{sym}}\delta\mathbf{u}^+ + \nabla^{\mathrm{sym}}\delta\mathbf{u}^-) : \mathbf{D}^{\mathrm{e}} \cdot \mathbf{n}_{S_d} \cdot (\mathbf{u}^+ - \mathbf{u}^-)\mathrm{d}S = \\
&\int_{V/S_d} \delta\mathbf{u} \cdot \rho\mathbf{g}\mathrm{d}S + \int_S \delta\mathbf{u} \cdot \mathbf{t}\mathrm{d}S
\end{aligned} \tag{13.31}$$

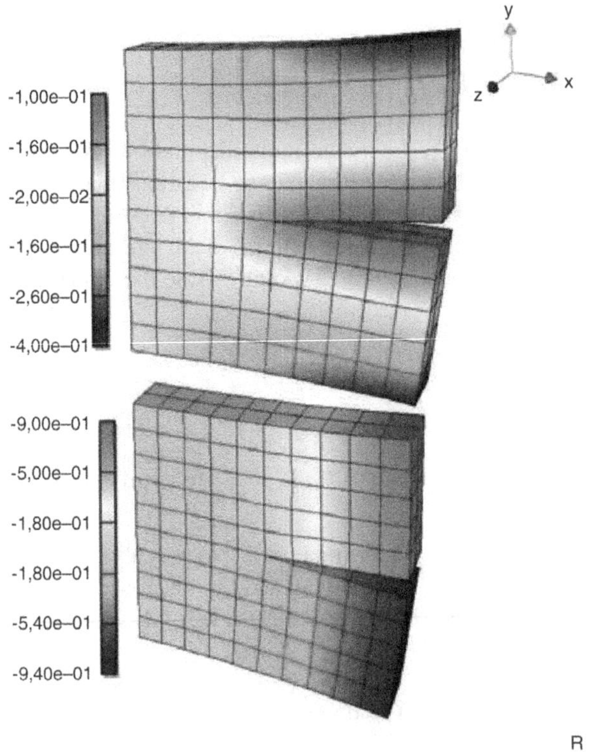

Figure 13.7 Three-dimensional simulations of crack propagation using a cohesive-zone model and a discontinuous Galerkin method

To ensure symmetry, $\alpha = 1$, but then a diffusionlike term, $\int_{S_d} \tau(\delta\mathbf{u}^+ - \delta\mathbf{u}^-) \cdot (\mathbf{u}^+ - \mathbf{u}^-)\mathrm{d}S$ has to be added to ensure numerical stability (Nitsche 1970). The numerical parameter $\tau = \mathcal{O}(|k|/w)$, with $|k|$ a suitable norm of the diffusionlike matrix that results from elaborating this term and w a measure of the grid density. For the unsymmetric choice $\alpha = -1$, addition of a diffusionlike term may not be necessary (Baumann and Oden 1999).

With a standard interpolation on both V^- and V^+ and requiring that the resulting equations hold for any admissible $\delta\mathbf{a}$, we obtain the discrete format:

$$
\begin{aligned}
&\int_{V^-} \mathbf{B}^{\mathrm{T}}\boldsymbol{\sigma}\mathrm{d}V + \int_{S_d} \frac{1}{2}\mathbf{H}^{\mathrm{T}}\mathbf{n}_{S_d}^{\mathrm{T}}(\boldsymbol{\sigma}^+ + \boldsymbol{\sigma}^-)\mathrm{d}S - \alpha\int_{S_d} \frac{1}{2}\mathbf{B}^{\mathrm{T}}\mathbf{D}^{\mathrm{e}}\mathbf{n}_{S_d}^{\mathrm{T}}(\mathbf{u}^+ - \mathbf{u}^-)\mathrm{d}S = \\
&\qquad\qquad \int_{V^-} \mathbf{B}^{\mathrm{T}}\rho\mathbf{g}\mathrm{d}V + \int_S \mathbf{H}^{\mathrm{T}}\mathbf{t}\mathrm{d}S \\
&\int_{V^+} \mathbf{B}^{\mathrm{T}}\boldsymbol{\sigma}\mathrm{d}V - \int_{S_d} \frac{1}{2}\mathbf{H}^{\mathrm{T}}\mathbf{n}_{S_d}^{\mathrm{T}}(\boldsymbol{\sigma}^+ + \boldsymbol{\sigma}^-)\mathrm{d}S - \alpha\int_{S_d} \frac{1}{2}\mathbf{B}^{\mathrm{T}}\mathbf{D}^{\mathrm{e}}\mathbf{n}_{S_d}^{\mathrm{T}}(\mathbf{u}^+ - \mathbf{u}^-)\mathrm{d}S = \\
&\qquad\qquad \int_{V^+} \mathbf{B}^{\mathrm{T}}\rho\mathbf{g}\mathrm{d}S + \int_S \mathbf{H}^{\mathrm{T}}\mathbf{t}\mathrm{d}S
\end{aligned}
\tag{13.32}
$$

with $\mathbf{n}_{S_d}$ written in a matrix form:

$$\mathbf{n}_{S_d}^{\mathrm{T}} = \begin{bmatrix} n_x & 0 & 0 & n_y & 0 & n_z \\ 0 & n_y & 0 & n_x & n_y & 0 \\ 0 & 0 & n_z & 0 & n_z & n_x \end{bmatrix} \tag{13.33}$$

where n_x, n_y, n_z are the components of the vector $\mathbf{n}_{S_d}$. After linearisation, one obtains:

$$\begin{bmatrix} \mathbf{K}^{--} & \mathbf{K}^{-+} \\ \mathbf{K}^{+-} & \mathbf{K}^{++} \end{bmatrix} \begin{pmatrix} \mathrm{d}\mathbf{a}^{-} \\ \mathrm{d}\mathbf{a}^{+} \end{pmatrix} = \begin{pmatrix} \mathbf{f}_{\mathrm{ext}}^{\mathrm{a}^-} - \mathbf{f}_{\mathrm{int}}^{\mathrm{a}^-} \\ \mathbf{f}_{\mathrm{ext}}^{\mathrm{a}^+} - \mathbf{f}_{\mathrm{int}}^{\mathrm{a}^+} \end{pmatrix} \tag{13.34}$$

with $\mathbf{a}^+$, $\mathbf{a}^-$ arrays that contain the nodal values of the displacements at the minus and the plus side of the interface, respectively, with $\mathbf{f}_{\mathrm{ext}}^{\mathrm{a}^-}$, $\mathbf{f}_{\mathrm{ext}}^{\mathrm{a}^+}$ and $\mathbf{f}_{\mathrm{int}}^{\mathrm{a}^-}$, $\mathbf{f}_{\mathrm{int}}^{\mathrm{a}^+}$ the right-hand and left-hand sides of Equations (13.32), respectively, and the submatrices defined by:

$$\begin{aligned}
\mathbf{K}^{--} &= \int_{V^-} \mathbf{B}^{\mathrm{T}}\mathbf{D}^{\mathrm{e}}\mathbf{B}\mathrm{d}V + \frac{1}{2}\int_{S_d} \mathbf{H}^{\mathrm{T}}\mathbf{n}_{S_d}^{\mathrm{T}}\mathbf{D}^{\mathrm{e}}\mathbf{B}\mathrm{d}S + \frac{1}{2}\alpha\int_{S_d} \mathbf{B}^{\mathrm{T}}\mathbf{D}^{\mathrm{e}}\mathbf{n}_{S_d}\mathbf{H}\mathrm{d}S \\
\mathbf{K}^{-+} &= \frac{1}{2}\int_{S_d} \mathbf{H}^{\mathrm{T}}\mathbf{n}_{S_d}^{\mathrm{T}}\mathbf{D}^{\mathrm{e}}\mathbf{B}\mathrm{d}S - \frac{1}{2}\alpha\int_{S_d} \mathbf{B}^{\mathrm{T}}\mathbf{D}^{\mathrm{e}}\mathbf{n}_{S_d}\mathbf{H}\mathrm{d}S \\
\mathbf{K}^{+-} &= \frac{1}{2}\alpha\int_{S_d} \mathbf{B}^{\mathrm{T}}\mathbf{D}^{\mathrm{e}}\mathbf{n}_{S_d}\mathbf{H}\mathrm{d}S - \frac{1}{2}\int_{S_d} \mathbf{H}^{\mathrm{T}}\mathbf{n}_{S_d}^{\mathrm{T}}\mathbf{D}^{\mathrm{e}}\mathbf{B}\mathrm{d}S \\
\mathbf{K}^{++} &= \int_{V^+} \mathbf{B}^{\mathrm{T}}\mathbf{D}^{\mathrm{e}}\mathbf{B}\mathrm{d}V - \frac{1}{2}\int_{S_d} \mathbf{H}^{\mathrm{T}}\mathbf{n}_{S_d}^{\mathrm{T}}\mathbf{D}^{\mathrm{e}}\mathbf{B}\mathrm{d}S - \frac{1}{2}\alpha\int_{S_d} \mathbf{B}^{\mathrm{T}}\mathbf{D}^{\mathrm{e}}\mathbf{n}_{S_d}\mathbf{H}\mathrm{d}S
\end{aligned} \tag{13.35}$$

References

Allix O and Corigliano A 1999 Geometrical and interfacial non-lineararities in the analysis of delamination in composites. *International Journal of Solids and Structures* **36**, 2189–2216.

Allix O and Ladevèze P 1992 Interlaminar interface modelling for the prediction of delamination. *Composite Structures* **22**, 235–242.

Barsoum RS 1976 On the use of isoparametric finite elements in linear fracture mechanics. *International Journal for Numerical Methods in Engineering* **10**, 225–237.

Baumann CE and Oden JT 1999 A discontinuous *hp* finite element method for the euler and Navier-Stokes problems. *International Journal for Numerical Methods in Fluids* **31**, 79–95.

Camacho GT and Ortiz M 1996 Computational modelling of impact damage in brittle materials. *International Journal of Solids and Structures* **33**, 2899–2938.

Cockburn B 2004 Discontinuous Galerkin methods for computational fluid dynamics, in *The Encyclopedia of Computational Mechanics* (eds Stein E, de Borst R and Hughes TJR), vol. III, pp. 417–471. John Wiley & Sons, Ltd.

Feenstra PH, de Borst R and Rots JG 1991 Numerical study on crack dilatancy; part 1: Models and stability analysis; part 2: Applications. *Journal of Engineering Mechanics* **117**, 733–769.

Goodman RE, Taylor RL and Brekke TL 1968 A model for the mechanics of jointed rock. *ASCE Journal of the Soil Mechanics and Foundation Engineering Division* **94**, 637–659.

Henshell RD and Shaw KG 1976 Crack tip finite elements are unnecessary. *International Journal for Numerical Methods in Engineering* **9**, 495–507.

Ingraffea AR and Saouma V 1985 Numerical modelling of discrete crack propagation in reinforced and plain concrete, in *Fracture Mechanics of Concrete* (eds Sih GC and DiTommaso), Martinus Nijhoff Publishers Dordrecht pp. 171–225. Martinus Nijhoff.

Mergheim J, Kuhl E and Steinmann P 2004 A hybrid discontinuous Galerkin/interface method for the computational modelling of failure. *Communications in Numerical Methods in Engineering* **20**, 511–519.

Ngo D and Scordelis AC 1967 Finite element analysis of reinforced concrete beams. *Journal of the American Concrete Institute* **64**, 152–163.

Nitsche JA 1970 Über ein variationsprinzip zur lösung dirichlet–problemen bei verwendung von teilraümen, die keinen randbedingungen unterworfen sind. *Abhandlungen des Mathematischen Seminars Universität Hamburg* **36**, 9–15.

Qiu X, Plesha ME and Meyer DW 1991 Stiffness matrix integration rules for contact-friction finite elements. *Computer Methods in Applied Mechanics and Engineering* **93**, 385–399.

Rots JG 1991 Smeared and discrete representations of localized fracture.. *International Journal of Fracture* **51**, 45–59.

Schellekens JCJ and de Borst R 1993a A nonlinear finite-element approach for the analysis of mode 1 free edge delamination in composites. *International Journal of Solids and Structures* **30**, 1239–1253.

Schellekens JCJ and de Borst R 1993b On the numerical integration of interface elements. *International Journal for Numerical Methods in Engineering* **36**, 43–66.

Schellekens JCJ and de Borst R 1994 Free edge delamination in carbon-epoxy laminates: a novel numerical/experimental approach. *Composite Structures* **28**, 357–373.

Tijssens MGA, Sluys LJ and van der Giessen E 2000 Numerical simulation of quasi-brittle fracture using damaging cohesive surfaces. *European Journal of Mechanics: A/Solids* **19**, 761–779.

van den Bosch MJ, Schreurs PJG and Geers MGD 2007 A cohesive zone model with a large displacement formulation accounting for interfacial fibrilation. *European Journal of Mechanics: A/Solids* **26**, 1–19.

van den Bosch MJ, Schreurs PJG, Geers MGD and van Maris MPHFL 2008 Interfacial characterization of pre-strained polymer coated steel by a numerical-experimental approach. *Mechanics of Materials* **40**, 302–317.

Verhoosel CV, Remmers JJC and Gutiérrez MA 2009 A dissipation-based arc-length method for robust simulation of brittle and ductile failure. *International Journal for Numerical Methods in Engineering* **77**, 1290–1321.

Wells GN, Garikipati K and Molari L 2004 A discontinuous Galerkin formulation for a strain gradient-dependent damage model. *Computer Methods in Applied Mechanics and Engineering* **193**, 3633–3645.

Wriggers P 2006 *Computational Contact Mechanics, 2nd edn*. Springer.

Xu XP and Needleman A 1994 Numerical simulations of fast crack growth in brittle solids. *Journal of the Mechanics and Physics of Solids* **42**, 1397–1434.

Xu XP and Needleman A 1996 Numerical simulations of dynamic crack growth along an interface. *International Journal of Fracture* **74**, 289–324.

14

Meshless and Partition-of-unity Methods

The basic idea of the finite element method is to divide a body into a number of elements. Within each element the primal variable, e.g. the displacement $u(\mathbf{x})$, is interpolated according to Equation (2.9). Outside the element the interpolant is defined to be strictly zero. Typically, and this is advantageous for imposing essential boundary conditions, the shape functions are defined such that they attain a unit value at the node to which they are attached, while they are zero at all other nodes. The part of the body on which the interpolants are non-zero is called the domain of influence or the support of a node. In the finite element method the support of a node is compact, which leads to stiffness matrices with a narrow bandwidth. This is advantageous for the Gauss elimination process, which can then be carried out efficiently.

While the finite element method is generally considered as the most versatile approximation method for stress analysis, it suffers from certain drawbacks. First, there is the low accuracy of finite elements when steep stress gradients have to be approximated. This is a drawback, especially in linear elastic fracture mechanics, since the elasticity solution exhibits a singularity at edges in the domain and at the tip of a crack. Evidently, the low-order polynomials that are used in the finite element method cannot describe such a singularity accurately. Accordingly, crack initiation is usually not well predicted when using standard finite elements. To ameliorate this poor performance, modifications have been proposed, such as the use of quarter-point elements (Barsoum 1976; Henshell and Shaw 1976). In these elements, the singularity enters the interpolant as a consequence of a shift of the mid-side nodes to a quarter of the side.

For crack propagation there is the additional problem that, when a crack has been advanced in the simulation, a new discretisation has to be set up because, essentially, the geometry of the body has changed. This necessitates meshing of the new domain, where care should be taken that a sufficiently fine mesh is applied around the crack tip. Sophisticated meshing and remeshing algorithms have been developed, but problems persist, in particular for three-dimensional problems, or when inelastic effects have to be taken into account. The transport of state variables which is then necessary, tends to diffuse the solution and to cause a temporary loss of satisfaction of the equilibrium equations.

Non-linear Finite Element Analysis of Solids and Structures, Second Edition.
René de Borst, Mike A. Crisfield, Joris J.C. Remmers and Clemens V. Verhoosel.

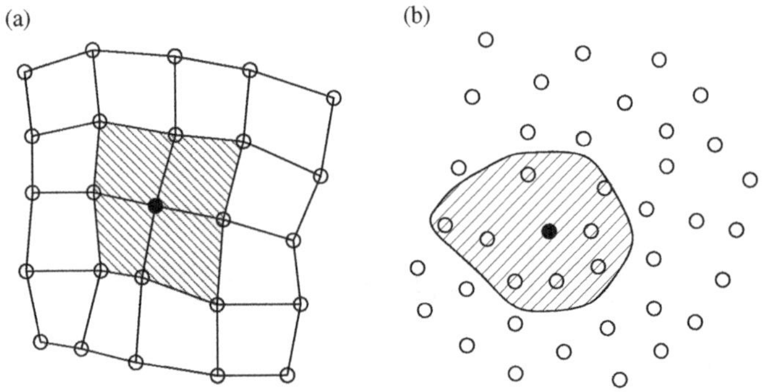

Figure 14.1 Typical domains of influence in a numerical method with nodal connectivity (a) and a meshless method (b). The domains of influence of the solid nodes are shaded

In view of the above limitations, discretisation methods have been sought that facilitate an improved resolution in the presence of stress singularities for crack initiation and that obviate the need for remeshing after crack propagation.

14.1 Meshless Methods

Meshless or meshfree methods do not require an explicitly defined connectivity between nodes for the definition of the shape functions. Instead, each node has a domain of influence which does not depend on the arrangement of the nodes. The domain of influence of a node is the part of the domain over which the shape function of that specific node is non-zero. In finite element methods the domain of influence is set by node connections, whereas in a meshless method the domain of influence can have more arbitrary shapes. Figure 14.1 shows the domains of influence of nodes for a (finite element) method that requires a nodal connectivity [Figure 14.1(a)] and for a meshless method [Figure 14.1(b)]. A host of meshless methods have been proposed, e.g. the element-free Galerkin method (Belytschko *et al.* 1994, 1996; Nayroles *et al.* 1992), the material point method (Sulsky *et al.* 1994), the reproducing kernel particle method (Liu *et al.* 1995), hp-clouds (Duarte and Oden 1996), finite spheres (De and Bathe 2000), the finite point method (Onate *et al.* 1996), and the natural neighbour Galerkin method (Sukumar *et al.* 2001; Yvonnet *et al.* 2004; Chinesta *et al.* 2010). Atluri and Zhu (1998, 2000) have developed a meshless method departing from a Petrov–Galerkin approach, in which the shape functions and the test functions are chosen from different spaces. The partition-of-unity method (Babuška and Melenk 1997; Melenk and Babuška 1996) was originally considered as a meshless method, but has opened ways to reconsider finite element methods, see Section 14.2.

14.1.1 The Element-free Galerkin Method

In the element-free Galerkin method shape functions are formulated by applying a moving least squares approximation (Lancaster and Salkauskas 1981). The approximation function for a node, $u(\mathbf{x})$, is restricted to its domain of influence and is expressed as the inner product of a vector $\mathbf{p}(\mathbf{x})$ and a vector $\mathbf{a}(\mathbf{x})$,

$$u(\mathbf{x}) = \mathbf{p}^{\mathrm{T}}(\mathbf{x})\mathbf{a}(\mathbf{x}) \tag{14.1}$$

in which $\mathbf{p}(\mathbf{x})$ contains basis terms that are functions of the coordinates $\mathbf{x}$. Normally, monomials such as $1, x, y, z, x^2, xy, \ldots$ are chosen, although also more sophisticated functions can be taken. The array $\mathbf{a}(\mathbf{x})$ contains the coefficients of the basis terms. In a moving least squares interpolation each node k is assigned a weight function w_k which renders the coefficients non-uniform. These weight functions appear in the sum J_{mls} as:

$$J_{\mathrm{mls}} = \sum_{k=1}^{n} w_k(\mathbf{x}) \left(\mathbf{p}^{\mathrm{T}}(\mathbf{x}_k)\mathbf{a}(\mathbf{x}) - u_k\right)^2 \tag{14.2}$$

with u_k the value of $u(\mathbf{x})$ at node k, and J_{mls} has to be minimised with respect to $\mathbf{a}(\mathbf{x})$. Typical choices for the weight functions are Gauss distributions or splines, whereby the domain of influence may take the shape of a disc (sphere) or rectangle (brick) in two (three) dimensions. Elaboration of the stationarity requirement of J_{mls} with respect to $\mathbf{a}(\mathbf{x})$ gives:

$$\frac{\partial J_{\mathrm{mls}}}{\partial \mathbf{a}(\mathbf{x})} = \sum_{k=1}^{n} w_k(\mathbf{x}) \left[2\mathbf{p}(\mathbf{x}_k)\mathbf{p}^{\mathrm{T}}(\mathbf{x}_k)\mathbf{a}(\mathbf{x}) - 2\mathbf{p}(\mathbf{x}_k)u_k\right] = \mathbf{0} \tag{14.3}$$

Thus, $\mathbf{a}(\mathbf{x})$ can be obtained as

$$\mathbf{a}(\mathbf{x}) = \mathbf{A}^{-1}(\mathbf{x})\mathbf{C}(\mathbf{x})\mathbf{u} \tag{14.4}$$

where $\mathbf{u}$ contains all u_k, and

$$\mathbf{A}(\mathbf{x}) = \sum_{k=1}^{n} w_k(\mathbf{x})\mathbf{p}(\mathbf{x}_k)\mathbf{p}^{\mathrm{T}}(\mathbf{x}_k) \tag{14.5a}$$

$$\mathbf{C}(\mathbf{x}) = \left[w_1(\mathbf{x})\mathbf{p}(\mathbf{x}_1), w_2(\mathbf{x})\mathbf{p}(\mathbf{x}_2), \ldots, w_n(\mathbf{x})\mathbf{p}(\mathbf{x}_n)\right] \tag{14.5b}$$

Equation (14.4) is substituted into Equation (14.1), which leads to:

$$u(\mathbf{x}) = \underbrace{\mathbf{p}^{\mathrm{T}}(\mathbf{x})\mathbf{A}^{-1}(\mathbf{x})\mathbf{C}(\mathbf{x})}_{\mathbf{H}(\mathbf{x})}\mathbf{u} \tag{14.6}$$

and the matrix $\mathbf{H}(\mathbf{x})$ that contains the shape functions can be identified as:

$$\mathbf{H}(\mathbf{x}) = \mathbf{p}^{\mathrm{T}}(\mathbf{x})\mathbf{A}^{-1}(\mathbf{x})\mathbf{C}(\mathbf{x}) \tag{14.7}$$

Shape functions which are generated in this manner, are usually not of a polynomial form, even though $\mathbf{p}(\mathbf{x})$ contains only polynomial terms. When moving least squares shape functions are used, the weight functions that are attached to each node determine the degree of continuity of the interpolants and the extent of the support of the node. A high degree of continuity can thus be achieved, and steep stress gradients can be captured accurately, which is beneficial for the proper prediction of crack initiation. The fact that the extent of the support is determined by the weight function w_k stands in contrast to the finite element method. Consequently, there are

Box 14.1 Algorithm to compute shape functions in the element-free Galerkin method

Loop on integration points, counter i:
  Set $\mathbf{A}(\mathbf{x}_i)$ and $\mathbf{B}(\mathbf{x}_i)$ equal to zero
  Loop on nodes, counter k:
    Extract size of support d_k for node k
    Compute $s = ||\mathbf{x}_k - \mathbf{x}_i||$:
    If $s \leq d_k$:
      Compute $w_k(s)$
      Evaluate $\mathbf{p}(\mathbf{x}_k)$
      Compute contributions to $\mathbf{A}(\mathbf{x}_i)$, $\mathbf{B}(\mathbf{x}_i)$
    End if
    Evaluate $\mathbf{p}(\mathbf{x}_i)$
    Compute $\mathbf{H}(\mathbf{x}_i) = \mathbf{p}^{\mathrm{T}}(\mathbf{x}_i)\mathbf{A}^{-1}(\mathbf{x}_i)\mathbf{B}(\mathbf{x}_i)$
  End loop
End loop

no elements needed to define the support of a node. A mesh is not necessary and approximation methods based on moving least squares functions therefore belong to the class of meshless or meshfree methods. The support of one node normally includes several other nodes and is therefore less compact than with finite element methods, and leads to a larger bandwidth of the system of equations.

Specific routines must be programmed for the construction of the shape functions in an element-free Galerkin method. As with the calculation of finite element shape functions, distinction must be made between the reference point of the shape function, i.e. the node, and the point at which the shape function is evaluated, usually an integration point. Herein, the index k will be used to denote a node, and the index i will denote an integration point. A generic algorithm to compute the shape functions within an element-free Galerkin method is given in Box 14.1, see also Krysl and Belytschko (2001). Similar to routines for the construction of finite element shape functions, the algorithm is 'integration point' based. Within a finite element method it is known a priori which nodes have non-zero shape functions at an integration point. This facilitates storage, and efficient programming can be done on an element-by-element basis. In a meshless method the situation is different, and an efficient manner to store the shape functions is less straightforward.

The shape functions that arise in the element-free Galerkin approach are not interpolating, and the nodal parameters u_k contained in the array $\mathbf{u}$ in Equation (14.4) are not the nodal values of the approximant function $u(\mathbf{x})$. Therefore, the imposition of essential boundary conditions, constraint equations and point loads is not trivial. A range of methods have been proposed in the literature, such as modified collocation methods (Atluri and Zhu 1999; Chen and Wang 2000; Wagner and Liu 2000; Wu and Plesha 2002), where the nodal parameter u_k is expressed in terms of the value of the approximant function at the node $u(\mathbf{x}_k)$, the weak imposition of essential boundary conditions via additional integrals in the Galerkin formulation, Lagrange multipliers (Belytschko *et al.* 1994; Wu and Plesha 2002), the use

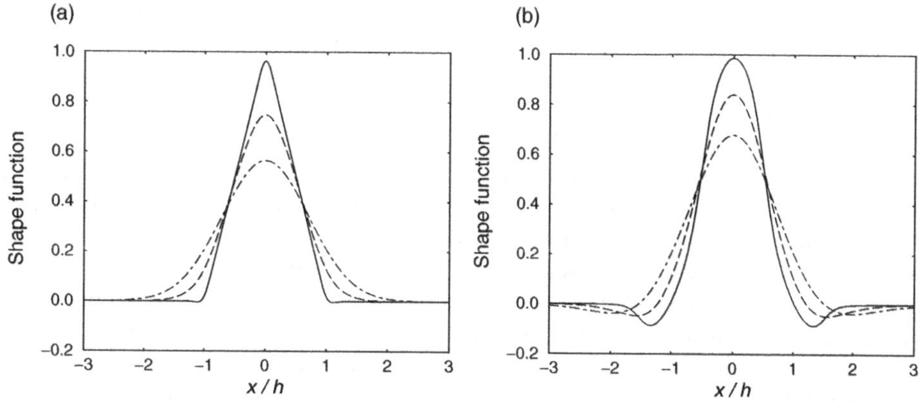

Figure 14.2 Shape functions that arise in the element-free Galerkin method. (a) A monomial base vector with $d/h = 2$ (solid line), $d/h = 3$ (dashed line), and $d/h = 4$ (dash-dotted line). (b) A quadratic monomial base vector with $d/h = 3$ (solid line), $d/h = 4$ (dashed line), and $d/h = 5$ (dash-dotted line)

of a modified variational principle in which the Lagrange multipliers are replaced by the tractions on the boundary (Lu *et al.* 1994), penalty formulations (Gavete *et al.* 2000), and the augmented Lagrangian approach (Ventura 2002), which eliminates the shortcomings of Lagrange multipliers and penalty functions, while preserving their respective advantages.

A common expression for circular domains of influence is the exponential weight function (Belytschko *et al.* 1994; Krysl and Belytschko 1999)

$$w_k(s) = \frac{\exp\left(-\frac{s^2}{\alpha^2}\right) - \exp\left(-\frac{1}{\alpha^2}\right)}{1 - \exp\left(-\frac{1}{\alpha^2}\right)} \tag{14.8}$$

with $s = ||\mathbf{x} - \mathbf{x}_k||/d \leq 1$, and with α the relative weights inside the domain of influence. In Figure 14.2 the effects of the size of the domain of influence d relative to the nodal spacing h are given for base vectors $\mathbf{p}(\mathbf{x})$ that contain up to linear and up to quadratic monomials, respectively. Figure 14.2(a) gives the shape functions for a base vector $\mathbf{p}(\mathbf{x}) = (1, x)^{\mathrm{T}}$, for which $d_{\min}/h = 1$, and Figure 14.2(b) is for $\mathbf{p}(\mathbf{x}) = (1, x, x^2)^{\mathrm{T}}$, where $d_{\min}/h = 2$. Small values of d/h result in shape functions that are very similar to finite element shape functions, and are almost interpolating. Evidently, the differences between the shape functions that stem from a linear and from a quadratic base vector become smaller for larger values of d/h.

Clearly, the size of the support of a node relative to the nodal spacing determines the properties of a meshfree method, and largely influences the quality and the efficiency of the resolution. When the support is made equal to the nodal spacing, shape functions obtained in meshfree methods can become identical to finite element shape functions, thus showing that meshfree methods *encompass* finite element methods. On the other hand, a larger support leads to shape functions in meshfree methods that can be similar to higher-order polynomials, even if the base vector $\mathbf{p}(\mathbf{x})$ contains only constant and linear terms. When the weight function extends to infinity, the moving least squares approximation degenerates to the classical least squares approximation. The optimal choice is an intermediate size of the domain of influence, so that the shape functions are richer than finite element shape functions, but more spatial variation is permitted compared with classical least squares approximations. When such a more compact support is used within a Galerkin formulation, a smaller bandwidth of the stiffness matrix is obtained. A classical least squares approach would lead to a full stiffness matrix.

Since the shape functions in meshless methods can be rational functions, they normally cannot be integrated analytically. In the first studies, a background cell structure was used, whereby in each cell Gauss integration was applied (Belytschko *et al.* 1994), which is the most widely used integration method in applications of the element-free Galerkin method. Two key considerations are the size of the integration cells, and the number of integration points per cell. Obviously, these issues are related to the number of nodes used. At least conceptually, the integration cell structure violates the original idea of a meshless method. Therefore, alternative integration methods have been studied, such as nodal integration methods (Beissel and Belytschko 1996; Chen *et al.* 2001) and Gauss integration methods that avoid the need for an underlying cell structure by employing the support of the node as the integration domain (Atluri and Zhu 1998, 2000). The latter class of methods has been denoted as truly meshless, since no mesh is needed for the formulation of the shape functions or for the evalution of the integrals. However, the constraints that are imposed on the background mesh of cells with integration points are much less strict than those on a finite element mesh. For instance, compatibility is not needed. Therefore, the differences between a truly meshless method and a meshless method that uses a grid of integration cells are perhaps smaller than they may seem at first sight.

The numerical integration scheme that is to be preferred largely depends on the application. When the element-free Galerkin method is chosen to benefit from the high continuity of its shape functions, then Gauss integration seems to be the most robust and flexible option. When the meshing of complicated, three-dimensional structures is to be avoided, the shape functions of a meshless method can still be integrated by means of Gauss quadrature, since the only constraint put on the integration cells is that they cover the whole domain. By contrast, when propagating discontinuities are to be modelled, the subdivision of integration cells can become a cumbersome task, and nodal integration can offer more flexibility.

14.1.2 Application to Fracture

Discontinuous shape functions for use in fracture mechanics applications can be obtained in a straightforward manner by truncating the appropriate weight functions. Implicitly, the same procedure is applied as for nodes close to the boundary of the domain: the part of the domain of influence that falls outside the computational domain is simply not taken into account in the integration.

A different situation arises when the crack does not pass completely through a domain of influence, so that the crack tip lies inside the support. Figure 14.3 illustrates three different procedures on how to truncate the domain of influence in the case of intersection by a crack, see also Fleming *et al.* (1997). In the visibility criterion the connectivity between an integration point and a node is taken into account if and only if a line can be drawn that is not intersected by a non-convex boundary. The resulting shape functions are not only discontinuous over the crack path, but also over the line that connects the node and crack tip. Although convergent results can be obtained, the presence of discontinuities in the shape functions beyond the crack path is less desirable. As an alternative, it has been suggested to redefine the weight function.

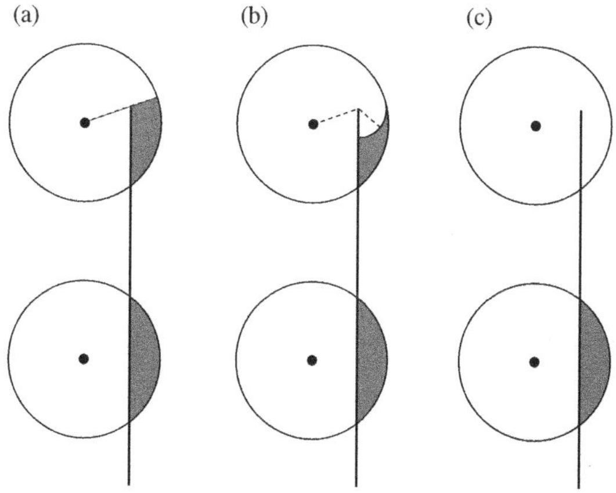

Figure 14.3 Domains of influence intersected by a crack or the crack tip: truncation of the weight function according to the visibility criterion (a), the diffraction criterion (b) and the see-through criterion (c) – the shaded areas denote the neglected part of the domain of influence

For instance, the line that connects the node and integration point can be wrapped around the crack tip, in a similar way to light diffracting around sharp edges – hence the name diffraction criterion – or, the visibility criterion can be adapted such that some transparency is assigned to the part of the crack close to the crack tip. In either way, shape functions are obtained that are smooth and continuous for the part of the domain not intersected by the crack. In the see-through or continuous path criterion, truncation of the weight function only occurs when the domain of influence is completely intersected by the crack path. In this manner, the effect of the crack propagation is delayed, and inaccuracies have been reported (Fleming *et al.* 1997).

Another issue is the spatial resolution around the crack path and the crack tip. For linear elastic fracture mechanics applications, the shape functions should properly capture the $r^{-1/2}$–stress singularity near the crack tip in order to accurately compute the stress intensity factors. Apart from a nodal densification around the crack tip, this can be achieved by locally enriching the base vector **p** through the addition of the set

$$\boldsymbol{\psi} = \left(\sqrt{r}\cos(\theta/2)\,,\ \sqrt{r}\sin(\theta/2)\,,\ \sqrt{r}\sin(\theta/2)\sin(\theta)\,,\ \sqrt{r}\cos(\theta/2)\sin(\theta)\right)^{\mathrm{T}} \tag{14.9}$$

with r the distance from the crack tip and θ is measured from the current direction of crack propagation (Fleming *et al.* 1997). Alternatively, these functions can be added to the sum of Equation (14.2). This is possible by virtue of the fact that, similar to conventional finite element shape functions, shape functions obtained from a moving least squares approximation satisfy the partition-of-unity property, an issue which we will return to in the next section.

As an example of a meshfree simulation of crack propagation using linear elastic fracture mechanics, dynamic crack extension in a three-dimensional cube is considered (Krysl and Belytschko 1999). A penny-shaped crack is initially present, which extends internally in the cube. When the crack reaches the free surfaces of the cube, a full separation of the cube takes place. In Figure 14.4 the development of the crack is plotted for eight successive stages.

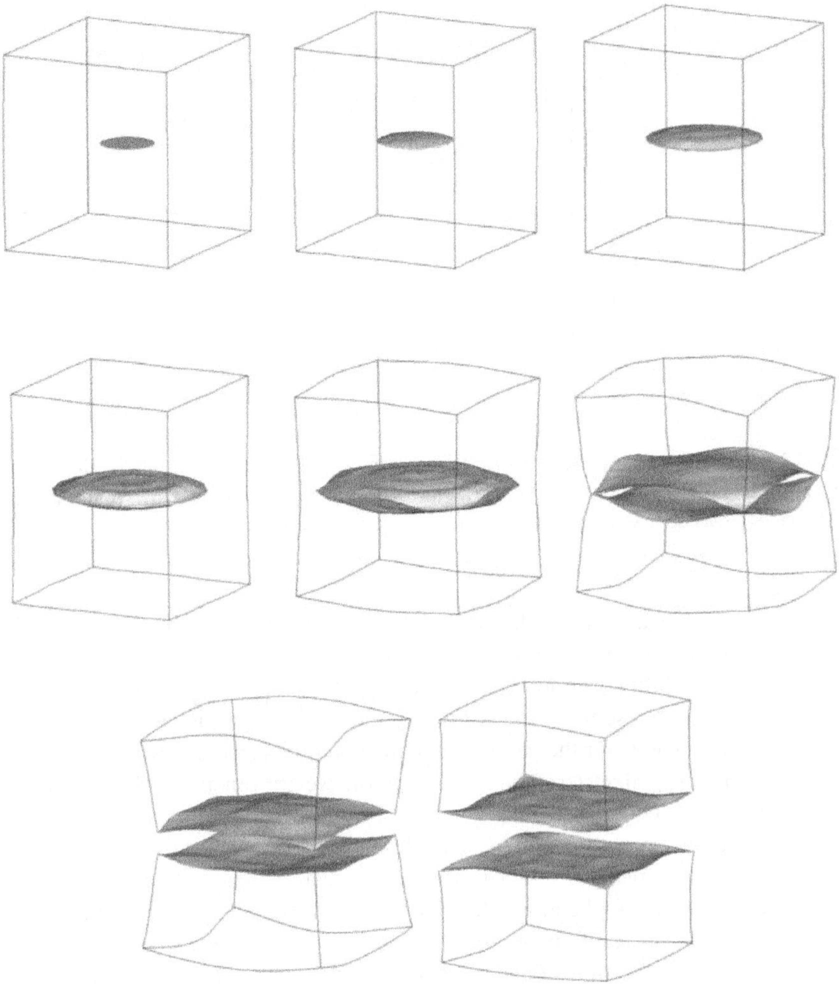

Figure 14.4 Cube with an initial penny-shape crack: propagation of the crack towards the free surfaces of the specimen (Krysl and Belytschko 1999)

It shows the ability of meshfree methods to describe not only cracks as line segments, but also as faces in three-dimensional analyses.

14.1.3 Higher-order Damage Mechanics

The high degree of continuity that is incorporated in meshfree methods makes them ideally suited for localisation and failure analyses that adopt higher-order continuum models. Also, the flexibility is increased compared with conventional finite element methods, since there is no direct connectivity, which makes placing additional nodes in regions with high strain

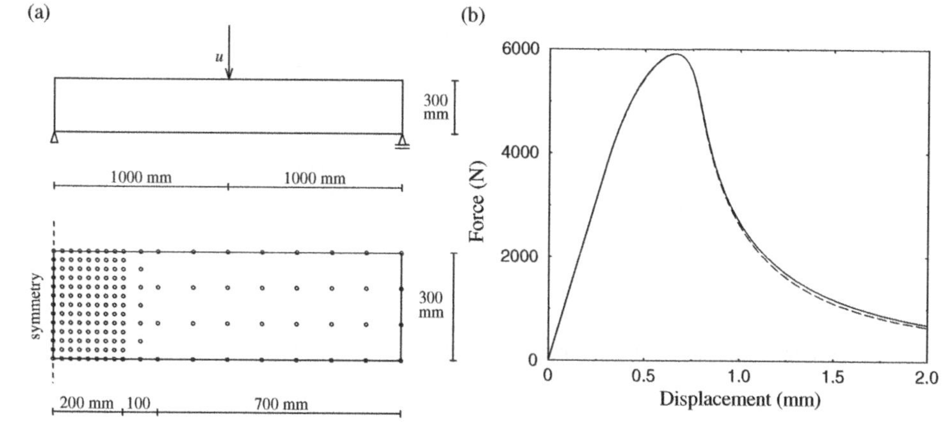

Figure 14.5 (a) Three-point bending beam and node distribution (for half of the symmetric beam). (b) Load–displacement curves for three-point bending beam. Comparison between the second-order implicit gradient damage model (dashed line) and the fourth-order implicit gradient damage model (solid line) (Askes *et al.* 2000)

gradients particularly simple. An example is offered in Figure 14.5 for a fourth-order gradient damage model, in which the non-local equivalent strain $\bar{\epsilon}$ follows from the solution of the partial differential equation:

$$\bar{\epsilon} - c_1 \nabla^2 \bar{\epsilon} - c_2 \nabla^4 \bar{\epsilon} = \tilde{\epsilon} \tag{14.10}$$

where ∇^4 is a short-hand notation for $\frac{\partial^4}{\partial x^4} + \frac{\partial^4}{\partial x^2 \partial y^2} + \frac{\partial^4}{\partial y^4}$, and c_1 and c_2 are material parameters. Equation (14.10) is assumed to hold on the entire domain. Evidently, even after order reduction by partial integration, $\mathcal{C}^1$-continuous shape functions are necesary for the interpolation of the non-local strain $\bar{\epsilon}$, with all the computational inconveniences that come with it when finite elements are employed. Here, meshfree methods offer a distinct advantage, since they can be easily constructed such that they incorporate $\mathcal{C}^\infty$-continuous shape functions. In Figure 14.5 the element-free Galerkin method has been used to solve the damage evolution that is described by the fourth order gradient scalar damage model of Equations (6.12), (6.16), (6.147) and (14.10) to predict the damage evolution in a three-point bending beam (Askes *et al.* 2000). In both cases, a quadratic convergence behaviour of the Newton–Raphson iterative method was obtained when using a properly linearised tangent stiffness matrix. The differences between the fourth-order and the second-order ($c_2 = 0$) gradient damage models appear to be minor, an issue which we will come back to in the next chapter.

We finally remark that the larger connectivity that is inherent in meshless methods tends to diffuse and broaden bands of localised deformations. However, as explained in Chapter 6 it would be incorrect to conclude that meshless methods have a regularising influence. Indeed, Pamin *et al.* (2003) have shown that meshless methods exhibit the same mesh dependence as finite element methods when strain softening is introduced in constitutive models without regularisation (Figure 6.9).

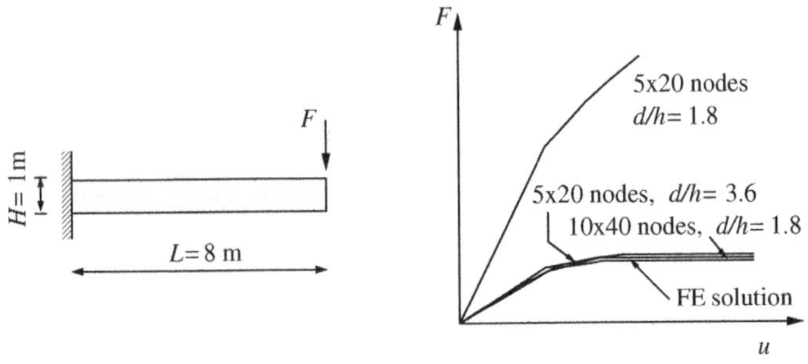

Figure 14.6 Load–displacement curves for a plane-strain cantilever beam

14.1.4 Volumetric Locking

It has been shown in Chapters 7 and 11 that low-order displacement based finite elements exhibit volumetric locking, i.e. an overstiff response is obtained for incompressible or, in the case of plasticity, also for dilatant/contractant material behaviour. An overly stiff behaviour also occurs in linear elastic planar problems where bending is dominant (shear locking), and in shells, where the difference in the order of approximation of the transverse and the membrane strains can lead to membrane locking, see Chapters 9 and 10. In Belytschko *et al.* (1994) and Krysl and Belytschko (1996) it has been shown that the higher-order shape functions of the element-free Galerkin method make it possible for shear locking and membrane locking to be avoided in elasticity, while Askes *et al.* (1999) and Dolbow and Belytschko (1999) have shown that this favourable property also holds for volumetric locking in incompressible elasticity and elasto-plasticity. Later, this property was investigated in greater detail by, among others, González *et al.* (2004) and Huerta and Fernández-Méndez (2001).

As a first example we take the plane-strain cantilever beam of Figure 14.6, which has been used by Belytschko *et al.* (1994) to demonstrate that the element-free Galerkin method can avoid shear locking. In Askes *et al.* (1999) the same beam has been used in an elasto-plastic analysis to show that volumetric locking can be avoided as well. Incompressibility was enforced both in the elastic regime (with Poisson's ratio $\nu = 0.4999$) and in the elasto-plastic regime, where a von Mises associated plasticity model was used. Two different discretisations were used, a coarse one with 5 nodes over the height and 20 nodes along the beam, and a finer discretisation with 10×40 nodes. Background meshes of 10×40 and 20×40 integration cells have been used for the coarser and for the finer discretisation, respectively. For the coarser mesh the effect of the size of the support has been studied, and analyses have been carried out for $d/h \approx 1.8$ and for $d/h \approx 3.6$. A clear limit load is obtained for the larger support ($d/h \approx 3.6$). Its correctness has been verified by the analysis with the finer discretisation and by a finite element analysis with a mesh of 5×40 quadratic triangles in a crossed configuration (resulting in 800 quadratic triangles). The analysis with the coarse discretisation and the smaller support also shows an overstiff behaviour in the elastic regime, which is mainly due to shear locking.

A second example concerns the elasto-plastic analysis of a infinite strip that is pushed into a half-space (Figure 14.7) (Askes *et al.* 1999). With a Poisson's ratio of $\nu = 0.49$ and a von Mises plasticity model with an associated flow rule (nearly) incompressible conditions are present in the elastic and in the elasto-plastic regime. The half-space has been modelled using 17×9

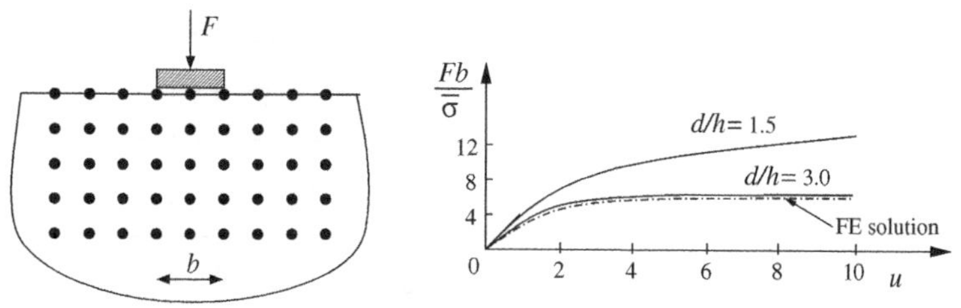

Figure 14.7 Load–displacement curves for a strip foundation on an elasto-plastic, cohesive soil

nodes and 64×32 integration cells. For a smaller support ($d/h = 1.5$) volumetric locking is observed again, which disappeared for a larger domain of influence ($d/h = 3$), where a clear limit load was found that is within the bounds of the analytical solution. A computation with a mesh of 2048 quadratic triangular elements in a crossed configuration confirmed the correctness of the limit load.

14.2 Partition-of-unity Approaches

A unifying approach to discretisation methods that accommodates crack initiation and crack propagation is enabled by the partition-of-unity concept (Babuška and Melenk 1997; Melenk and Babuška 1996). A collection of functions h_k, associated with node k, form a partition of unity if

$$\sum_{k=1}^{n} h_k(\mathbf{x}) = 1 \tag{14.11}$$

with n the number of discrete nodal points. For a set of shape functions h_k that satisfy the partition-of-unity property, a field u can be interpolated as follows:

$$u(\mathbf{x}) = \sum_{k=1}^{n} h_k(\mathbf{x}) \left(\bar{a}_k + \sum_{l=1}^{m} \psi_l(\mathbf{x}) \hat{a}_{kl} \right) \tag{14.12}$$

with $\bar{a}_k$ the 'regular' nodal degrees of freedom, $\psi_l(\mathbf{x})$ the enhanced basis terms, and $\hat{a}_{kl}$ the additional degrees of freedom at node k, which represent the amplitudes of the l-th enhanced basis term $\psi_l(\mathbf{x})$. A basic requirement of the enhanced basis terms ψ_l is that they are linearly independent, mutually, but also with respect to the set of functions h_k. For completeness, we note that this requirement also holds for the enrichment functions that were used in meshless methods to account for the singularities in linear-elastic fracture mechanics, Equation (14.9). This is because the basis functions used in meshless methods also form a partition of unity.

In conventional finite element notation we thus interpolate a displacement field as:

$$\mathbf{u} = \mathbf{H}(\bar{\mathbf{a}} + \boldsymbol{\Psi}\hat{\mathbf{a}}) \tag{14.13}$$

where $\mathbf{H}$ contains the standard shape functions, and $\boldsymbol{\Psi}$ the enhanced basis terms. The arrays $\bar{\mathbf{a}}$ and $\hat{\mathbf{a}}$ collect the standard and the additional nodal degrees of freedom, respectively. A displacement field that contains a single discontinuity can be represented by choosing (Belytschko

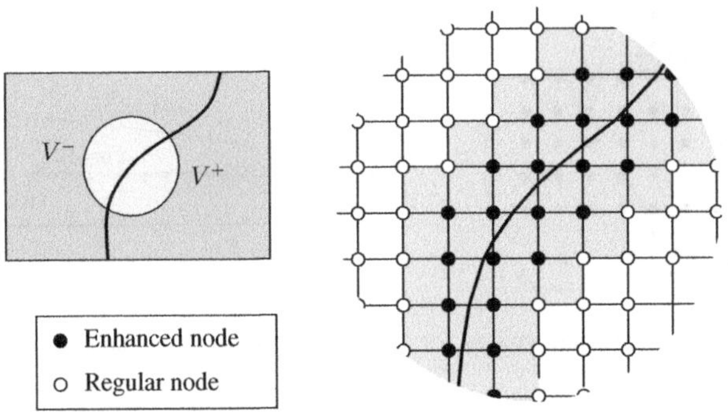

Figure 14.8 Two-dimensional finite element mesh with a discontinuity denoted by the bold line. The grey elements contain additional terms in the stiffness matrix and the internal force vector.

and Black 1999; Hansbo and Hansbo 2004; Moës *et al.* 1999; Wells and Sluys 2001):

$$\mathbf{\Psi} = \mathcal{H}_{S_d}\mathbf{I} \tag{14.14}$$

with $\mathcal{H}_{S_d}$ the Heaviside function, which separates the V^--domain from the V^+-domain (Figure 14.8). Substitution into Equation (14.13) gives:

$$\mathbf{u} = \underbrace{\mathbf{H}\bar{\mathbf{a}}}_{\bar{\mathbf{u}}} + \mathcal{H}_{S_d} \underbrace{\mathbf{H}\hat{\mathbf{a}}}_{\hat{\mathbf{u}}} \tag{14.15}$$

Identifying the continuous fields $\bar{\mathbf{u}} = \mathbf{H}\bar{\mathbf{a}}$ and $\hat{\mathbf{u}} = \mathbf{H}\hat{\mathbf{a}}$ we observe that Equation (14.15) exactly describes a displacement field that is crossed by a single discontinuity, but is otherwise continuous. Accordingly, the partition-of-unity property of finite element shape functions can be used in a straightforward fashion to incorporate discontinuities in a continuum such that their discontinuous character is preserved.

We take the balance of momentum, Equation (2.4), as point of departure, so that, neglecting inertia and body forces, we have:

$$\nabla \cdot \boldsymbol{\sigma} = \mathbf{0}$$

We multiply this identity by test functions $\delta\mathbf{u}$, and take them from the same space as the trial functions for $\mathbf{u}$:

$$\delta\mathbf{u} = \delta\bar{\mathbf{u}} + \mathcal{H}_{S_d}\delta\hat{\mathbf{u}} \tag{14.16}$$

Applying the divergence theorem and requiring that this identity holds for arbitrary $\delta\bar{\mathbf{u}}$ and $\delta\hat{\mathbf{u}}$ yields the following set of coupled equations:

$$\int_V \nabla(\delta\bar{\mathbf{u}}) : \boldsymbol{\sigma} \mathrm{d}V = \int_S \delta\bar{\mathbf{u}} \cdot \mathbf{t} \mathrm{d}S \tag{14.17a}$$

$$\int_{V^+} \nabla(\delta\hat{\mathbf{u}}) : \boldsymbol{\sigma} \mathrm{d}V + \int_{S_d} \delta\hat{\mathbf{u}} \cdot \mathbf{t}_d \mathrm{d}S = \int_S \mathcal{H}_{S_d}\delta\hat{\mathbf{u}} \cdot \mathbf{t} \mathrm{d}S \tag{14.17b}$$

where in the volume integrals the Heaviside function has been eliminated by a change of the integration domain from V to V^+. Interpolating the trial and the test functions in the same space,

$$\begin{cases} \bar{\mathbf{u}} = \mathbf{H}\bar{\mathbf{a}} & , \quad \hat{\mathbf{u}} = \mathbf{H}\hat{\mathbf{a}} \\ \delta\bar{\mathbf{u}} = \mathbf{H}\delta\bar{\mathbf{a}} & , \quad \delta\hat{\mathbf{u}} = \mathbf{H}\delta\hat{\mathbf{a}} \end{cases} \tag{14.18}$$

and requiring that the resulting equations must hold for any admissible $\delta\bar{\mathbf{a}}$ and $\delta\hat{\mathbf{a}}$, we obtain the discrete format:

$$\int_V \mathbf{B}^{\mathrm{T}}\boldsymbol{\sigma}\mathrm{d}V = \int_S \mathbf{H}^{\mathrm{T}}\mathbf{t}\mathrm{d}S \tag{14.19a}$$

$$\int_{V^+} \mathbf{B}^{\mathrm{T}}\boldsymbol{\sigma}\mathrm{d}V + \int_{S_d} \mathbf{H}^{\mathrm{T}}\mathbf{t}_d\mathrm{d}S = \int_S \mathcal{H}_{S_d}\mathbf{H}^{\mathrm{T}}\mathbf{t}\mathrm{d}S \tag{14.19b}$$

After linearisation, the following matrix-vector equation is obtained:

$$\begin{bmatrix} \mathbf{K}_{\bar{a}\bar{a}} & \mathbf{K}_{\bar{a}\hat{a}} \\ \mathbf{K}_{\bar{a}\hat{a}}^{\mathrm{T}} & \mathbf{K}_{\hat{a}\hat{a}} \end{bmatrix} \begin{pmatrix} \mathrm{d}\bar{\mathbf{a}} \\ \mathrm{d}\hat{\mathbf{a}} \end{pmatrix} = \begin{pmatrix} \mathbf{f}_{\mathrm{ext}}^{\bar{a}} - \mathbf{f}_{\mathrm{int}}^{\bar{a}} \\ \mathbf{f}_{\mathrm{ext}}^{\hat{a}} - \mathbf{f}_{\mathrm{int}}^{\hat{a}} \end{pmatrix} \tag{14.20}$$

with $\mathbf{f}_{\mathrm{int}}^{\bar{a}}, \mathbf{f}_{\mathrm{int}}^{\hat{a}}$ given by the left-hand sides of Equations (14.19), with $\mathbf{f}_{\mathrm{ext}}^{\bar{a}}, \mathbf{f}_{\mathrm{ext}}^{\hat{a}}$ given by the right-hand sides of Equations (14.19), and

$$\begin{aligned} \mathbf{K}_{\bar{a}\bar{a}} &= \int_V \mathbf{B}^{\mathrm{T}}\mathbf{D}\mathbf{B}\mathrm{d}V \\ \mathbf{K}_{\bar{a}\hat{a}} &= \int_{V^+} \mathbf{B}^{\mathrm{T}}\mathbf{D}\mathbf{B}\mathrm{d}V \\ \mathbf{K}_{\hat{a}\hat{a}} &= \int_{V^+} \mathbf{B}^{\mathrm{T}}\mathbf{D}\mathbf{B}\mathrm{d}V + \int_{S_d} \mathbf{H}^{\mathrm{T}}\mathbf{D}_d\mathbf{H}\mathrm{d}S \end{aligned} \tag{14.21}$$

If the material tangential stiffness matrices of the bulk and the interface, $\mathbf{D}$ and $\mathbf{D}_d$, respectively, are symmetric, the total tangential stiffness matrix remains symmetric. It is emphasised that in this concept, the additional degrees of freedom cannot be condensed at element level, if one wishes to represent a discontinuity that it is continuous at interelement boundaries.

When the discontinuity coincides with a side of the element, the formulation of Chapter 13 for interface elements is retrieved. For this, we expand the term in $\mathbf{K}_{\hat{a}\hat{a}}$ which relates to the discontinuity as:

$$\int_{S_d} \mathbf{H}^{\mathrm{T}}\mathbf{D}_d\mathbf{H}\mathrm{d}S_d = \begin{bmatrix} \mathbf{K}_n & \mathbf{0} & \mathbf{0} \\ \mathbf{0} & \mathbf{K}_s & \mathbf{0} \\ \mathbf{0} & \mathbf{0} & \mathbf{K}_t \end{bmatrix} \tag{14.22}$$

with $\mathbf{K}_\pi = d_\pi \mathbf{h}^{\mathrm{T}}\mathbf{h}$ (Simone 2004), which closely resembles Equations (13.11) and (13.12). Defining the sum of the nodal displacements $\bar{\mathbf{a}}$ and $\hat{\mathbf{a}}$ as primary variable $\mathbf{a}$ on the + side of the interface and setting $\mathbf{a} = \bar{\mathbf{a}}$ on the – side and rearranging then leads to the standard interface formulation. However, even though formally the matrices can coincide for the partition-of-unity based method and the conventional interface formulation, the former does *not* share the disadvantages of traction oscillations and spurious wave reflections prior to the onset of decohesion, simply because the partition-of-unity concept permits the placement of cohesive surfaces in the mesh only at onset of decohesion, thereby by-passing the whole problem of having to assign a high (dummy) stiffness to the interface prior to crack initiation.

We next consider a structured mesh, composed of four-noded elements, shown in Figure 14.8. In the finite element model, the nodes with a support that is crossed by a discontinuity are enhanced. These are marked by filled circles. The other nodes, denoted by open circles, remain unchanged. Since only the nodes of elements that are crossed by the discontinuity have additional degrees of freedom $\hat{\mathbf{a}}$, the total number of degrees of freedom of the system is just marginally higher than without a discontinuity. When an element is supported by one or more enhanced nodes, the additional terms will emerge in the stiffness matrix and the force vector, Equation (14.19). The elements that contain a discontinuity will also be augmented with the surface integrals for the cohesive behaviour.

The integrals in the equilibrium equations, Equations (14.19) are integrated numerically, e.g. using Newton–Cotes or Gauss integration. A requirement for the use of a Gauss scheme is that the field is continuous and smooth. In the present case, the stress field in the elements that are crossed by a discontinuity is only piecewise continuous. Although the accuracy will increase when more sampling points are used, the result of the numerical procedure will not be exact, and many integration points are needed to obtain good accuracy. Since the stress field is continuous and smooth on either side of a discontinuity within an element, and since the position of the discontinuity is known, the terms in the equilibrium equation can be integrated in parts. The element is divided into a number of triangular or quadrilateral subelements, which are integrated in a standard manner, e.g. using a Gauss integration scheme (Figure 14.9). The integration of the contributions due to the discontinuity at S_d in Equations (14.19) is straightforward. For instance, for a two-dimensional situation, the discontinuity is represented by a straight line and is integrated using a one or two-point Gauss integration scheme.

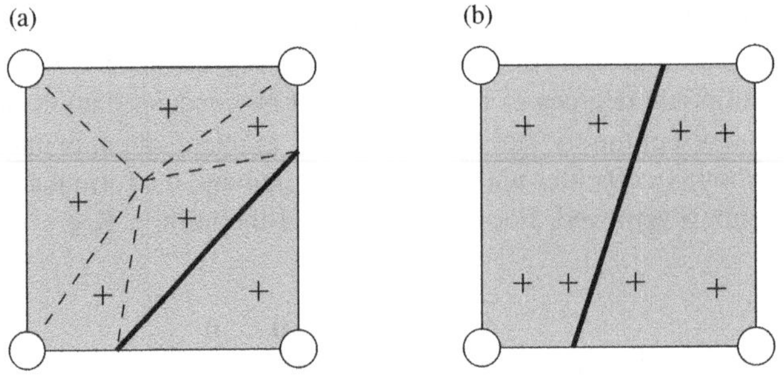

Figure 14.9 Numerical integration of quadrilateral elements crossed by a discontinuity (bold line). (a) The sample points are denoted by a +. (a) The element is split into a sub element with five vertices and one with three vertices. The first part is triangulated into five areas, denoted by the dashed lines. Each of these areas is integrated using a standard one-point integration scheme. (b) The element is split into two quadrilateral sub elements. Each of these parts can be integrated with a standard 2 × 2 Gauss integration

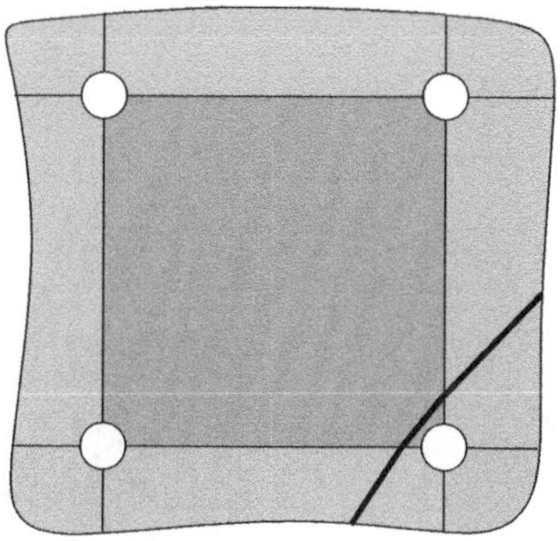

Figure 14.10 An element is crossed by a discontinuity near one of the nodes

The introduction of the enhanced basis terms deteriorates the condition of the stiffness matrix. In particular when the discontinuity crosses an element in the vicinity of a node (Figure 14.10), the contributions of the various terms in the stiffness matrix will have different magnitudes, which can lead to a stiffness matrix that is less well-conditioned. This problem can be ameliorated by only enhancing a node when it has a significant contribution to the stiffness matrix. Therefore, when the discontinuity splits an element such that a part of the element is much smaller than the other part, the node that supports the smallest part is not enhanced if (Wells and Sluys 2001):

$$\frac{\min(V^+, V^-)}{V} < \varepsilon \tag{14.23}$$

with ε a tolerance. Evidently, this will affect the computational results, but for reasonable values of ε, e.g. $\varepsilon \approx 0.05$, numerical experience shows that these effects are small.

14.2.1 Application to Fracture

An important advantage of the partition-of-unity approach is the possibility to extend a crack during a calculation, unbiased by the original discretisation (Figure 14.8). This provides an increased flexibility compared with the use of classical (predefined) interface elements. Moreover, as noted before, it eliminates the necessity to use dummy stiffnesses prior to crack initiation, since the enhanced degrees of freedom are generated only upon crack propagation, i.e. in the course of the computation.

Two approaches to crack initiation and crack propagation can be followed. In linear elastic fracture mechanics singular terms arise. As in the element-free Galerkin approach, the partition-of-unity property of the shape functions allows for the addition of functions at the crack tip that capture the stress singularity, Equation (14.9). For the description of this singular field a second set of enhanced nodes is necessary in addition to the enhancement of the nodes in the wake of the crack tip, that support a traction-free discontinuity (Belytschko and Black 1999; Moës *et al.* 1999). In this model, the exact moment of crack extension is important, and therefore, a stress criterion must be used that is based on the full singular stress field around the crack tip. Along this line of reasoning the precise location of the crack tip is also of importance.

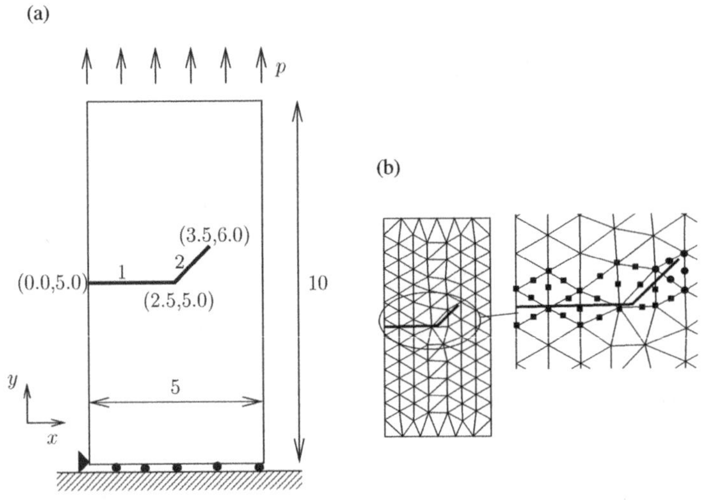

Figure 14.11 (a) Geometry of a block that contains a stationary, kinked crack. (b) Nodes enhanced with jump functions (squares), or with crack-tip functions (circles)

To enable the use for relatively coarse discretisations it is therefore necessary to be able to advance the crack tip to any position within an element. This can be achieved using a level set function to track the exact location of the crack tip (Gravouil *et al.* 2002), which, however, is at the expense of the solution of an additional problem.

To illustrate the potential of the method for linear elastic fracture mechanics problems, we consider a block that contains a stationary, kinked crack (Figure 14.11) (Askes *et al.* 2003). When utilising linear elastic fracture mechanics, the nodes for which the support is crossed by a crack are enhanced with a Heaviside function and those for which the support contains a crack tip are enhanced by near-tip terms as in Equation (14.9), similar to the procedure used in meshfree methods, see also Figure 14.11. The problem has been analysed with a relatively coarse mesh with a uniform element size, and with a finer, reference mesh, with the crack explicitly built in the mesh and refined around the crack tip. The bottom edge of the block is restrained and a uniformly distributed load p is applied to the top edge. The deformed meshes that result from the computations are shown in Figure 14.12, which also gives the contours of the normal stress in the y-direction. The general form of the contour plot is the same for both computations. The resolution of the contours for the enhanced mesh is smaller, since stresses have been post-processed at nodal points only. This has a smoothing effect, with the stress singularity obvious only when the crack tip lies close to an element node.

The use of linear elastic fracture mechanics suffers from a certain overhead in the computational costs, because it requires the temporary enhancement of nodes to accommodate the singular terms in the stress field in addition to the enhancement needed to describe the discontinuity, and is less general since it cannot accommodate non-linear behaviour of the bulk material. A more general approach is to use cohesive zone models (Chapter 6). When applying the partition-of-unity approach to cohesive crack propagation the discontinuity is normally extended across an element, such that the tip touches the boundary of the next element (Wells and Sluys 2001). In order to enforce a zero displacement jump at the tip, the nodes which

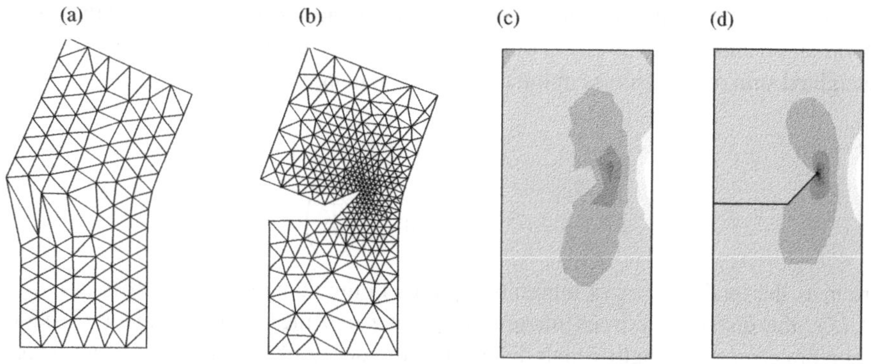

Figure 14.12 Deformed configurations for enhanced mesh (a) and reference mesh with explicit discontinuity (b), and contour plots of σ_{yy} for enhanced (c) and reference mesh with explicit discontinuity (d)

support this boundary are not enhanced (Figure 14.13). Since, in the cohesive approach, the stress field is non-singular at the crack tip, there is no need to enhance the nodes around the crack tip with terms that accommodate the $r^{-1/2}$–stress singularity, and standard polynomial shape functions suffice. Nevertheless, Karihaloo and Xiao (2010), Moës and Belytschko (2002) and Xiao and Karihaloo (2006) have augmented the displacement field near cohesive cracks in order to improve the stress prediction. This can be beneficial for properly determining the direction of crack propagation. Indeed, in the cohesive approach the onset of crack initiation is less critical, as crack propagation is governed by the fracture energy $\mathcal{G}_c$. Crack initiation that is slightly too early, or slightly too late, will be corrected during propagation. A wrong estimate of the direction of crack propagation, however, is not corrected in later stages of the fracture process. In Wells and Sluys (2001) it has been assumed that the crack is extended when the major principal stress exceeds the tensile strength, with a crack propagation direction that is

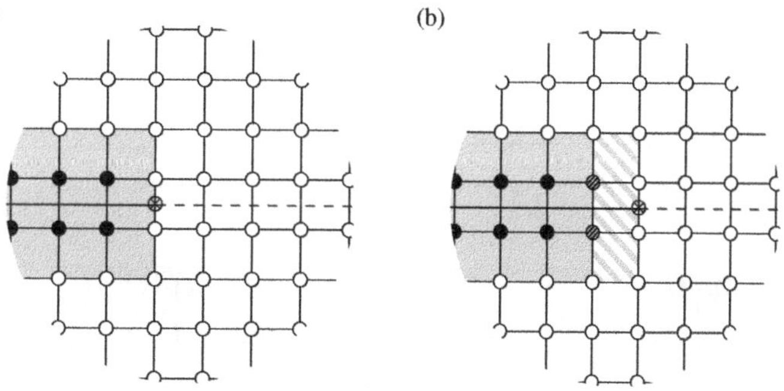

Figure 14.13 Extension of a discontinuity. (a) When the stress in the sample point that is denoted by ⊗ exceeds the threshold value, the crack is extended into the next element. (b) The hashed nodes and elements have been enhanced. The nodes that support the current crack tip are not enhanced in order to enforce a zero opening at the crack tip

normal to the corresponding principal axis. To improve the stress that is used in the criteria for crack initiation and for the direction of crack propagation, the stress at the tip is computed as the weighted sum of the stresses in the neighbouring integration points:

$$\boldsymbol{\sigma}_{\mathrm{tip}} = \frac{\sum_{i=1}^{n_i} \boldsymbol{\sigma}_i \exp\left(-\frac{r_i^2}{2\ell_e^2}\right)}{\sum_{i=1}^{n_i} \exp\left(-\frac{r_i^2}{2\ell_e^2}\right)} \tag{14.24}$$

where n_i is the total number of integration points in the domain, $\boldsymbol{\sigma}_i$ the stress in integration point i, r_i the distance between integration point i and the crack tip, and ℓ_e a parameter with the dimension of length which determines the decay of the influence of an integration point. In Wells and Sluys (2001) ℓ_e has been taken to be approximately equal to three times a characteristic element size h. With this modification the stress that is used in the criterion for crack initation is usually smaller than the actual tip stress. As a result, the crack can be extended slightly too late.

Using the interpolation of Equation (14.15) the relative displacement at the discontinuity S_d is obtained as:

$$\mathbf{v} = \hat{\mathbf{u}} \mid_{\mathbf{x} \in S_d} \tag{14.25}$$

When using a cohesive zone model, the tractions $\mathbf{t}_d$ at the discontinuity S_d can directly be derived from Equation (6.46).

The objectivity of computations with respect to mesh refinement is now demonstrated for a three-point bending beam of unit thickness. The beam is loaded quasi-statically by means of an imposed displacement at the centre of the beam on the top edge. The geometric and material data can be found in Wells and Sluys (2001). Figure 14.14 shows the crack after propagation

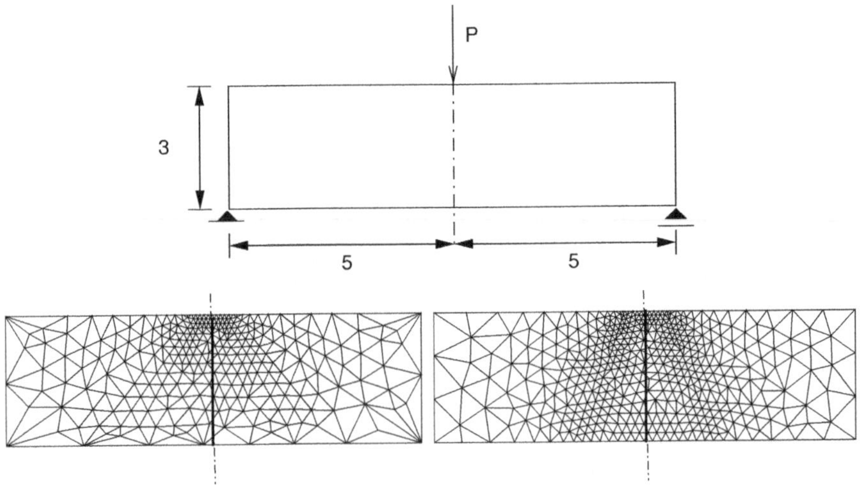

Figure 14.14 Crack path at the final stage of loading for the coarse mesh (523 elements) and the fine mesh (850 elements) (Wells and Sluys 2001)

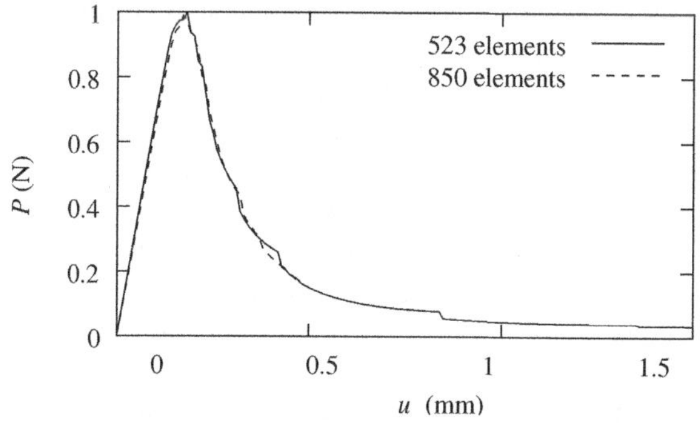

Figure 14.15 Load–displacement diagrams for the analysis of the symmetrically loaded beam using two meshes (Wells and Sluys 2001)

throughout the beam. Two meshes are shown, one with 523 elements and the other with 850 elements. Clearly, in both cases the crack propagates from the centre at the bottom of the beam in a straight line towards the loading point, and is not influenced by the mesh structure. The load–displacement responses of Figure 14.15 confirm objectivity with respect to mesh refinement. From the curve for the coarser mesh the energy dissipation is calculated as 0.308 J, which only slightly exceeds the fracture energy multiplied by the depth and the thickness of the beam (0.3 J). Some small irregularities are observed in the load–displacement curve, especially for the coarser mesh. These are caused by the fact that in this implementation a cohesive zone is inserted entirely in an element when the tensile strength has been exceeded.

The requirement that the crack path is not biased by the direction of the mesh lines is normally even more demanding than the requirement of objectivity with respect to mesh refinement. Figure 14.16 shows that the approach also fully satisfies this requirement, since the numerically predicted crack path of the single-edge notched beam of Figure 6.13 is in excellent agreement with experimental observations.

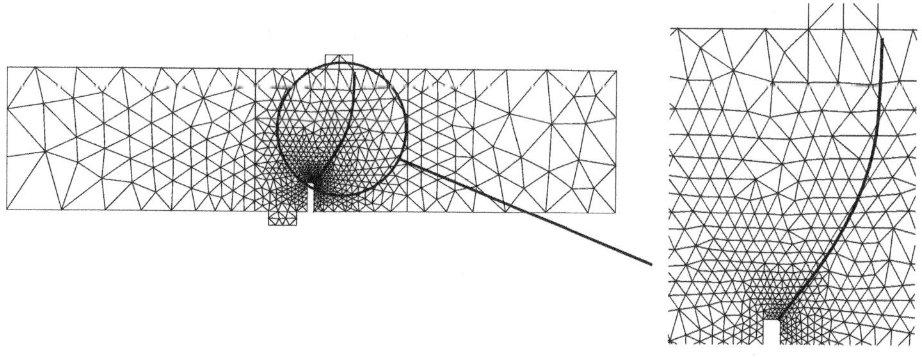

Figure 14.16 Crack path that results from the analysis of the single-edge notched beam using the partition-of-unity method (Wells and Sluys 2001)

In the above examples, growth of a single, continuous cohesive crack was simulated. Crack propagation in heterogeneous materials, but also fast crack growth in more homogeneous materials are often characterised by the nucleation of microcracks at several locations, which can grow, branch and eventually link up to form macroscopically observable cracks. To accommodate this observation, the concept of cohesive segments has been proposed in Remmers *et al.* (2003a, 2008). Exploiting the partition-of-unity property of finite element shape functions, crack segments equipped with a cohesive law are placed over a patch of elements when a loading criterion is met at an integration point. Since the cohesive segments can grow and eventually coalesce, they can also simulate a single, dominant crack.

The partition-of-unity approach to cohesive fracture has a number of advantages over interface elements. The cohesive surface can be placed as a discontinuity anywhere in the model, irrespective of the structure of the underlying finite element mesh. Moreover, it is possible to extend a cohesive surface during the simulation by adding additional degrees of freedom. This avoids the use of high dummy stiffnesses to model a perfect bond prior to cracking and prevents numerical problems such as stress oscillations (Chapter 13), or spurious stress wave reflections in structural dynamics. Since degrees of freedom are only added when a cohesive surface is extended, the additional number of degrees of freedom remains limited.

14.2.2 Extension to Large Deformations

The partition-of-unity approach to model crack propagation can be extended to large deformations in an elegant manner that naturally fits within standard continuum mechanics concepts. As point of departure we take the balance of momentum in the current configuration, transform it to a weak format, substitute the trial and test functions in the form of (14.15) and (14.16), and require that the resulting identity holds for all test functions $\delta\bar{\mathbf{u}}$ and $\delta\hat{\mathbf{u}}$. This results in Equations (14.17).

In this derivation Equation (1.75) has been used, and the product of the normal vector $\mathbf{n}$ and the Cauchy stress tensor $\boldsymbol{\sigma}$ has been replaced by the stress vector $\mathbf{t}$. At the (external) boundary S of the body $\mathcal{B}$ the normal is defined unambiguously, also for large displacements. This is not so for the internal boundary, where the original surface $S_{d,0}$ is split into two distinct surfaces, S_d^- and S_d^+, with different normals, $\mathbf{n}_d^-$ and $\mathbf{n}_d^+$, respectively (Figure 14.17). Noting that for the position vector we can write:

$$\mathbf{x} = \boldsymbol{\xi} + \bar{\mathbf{u}} + \mathcal{H}_{S_{d,0}}\hat{\mathbf{u}} \tag{14.26}$$

where $\mathcal{H}_{S_{d,0}}$ is the Heaviside function centred at the discontinuity $S_{d,0}$, the deformation gradient can be derived as:

$$\mathbf{F} = \mathbf{I} + \frac{\partial\bar{\mathbf{u}}}{\partial\boldsymbol{\xi}} + \mathcal{H}_{S_{d,0}}\frac{\partial\hat{\mathbf{u}}}{\partial\boldsymbol{\xi}} + \delta_{S_{d,0}}\hat{\mathbf{u}} \otimes \mathbf{n}_{d,0} \tag{14.27}$$

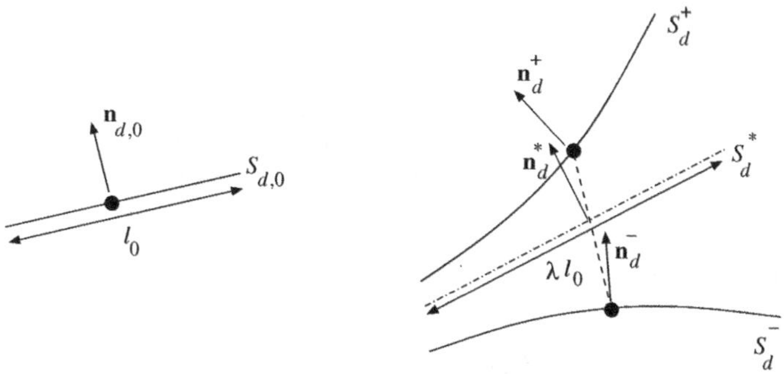

Figure 14.17 An interface length associated with an integration point in the reference and the current configurations. The solid dot represents the integration point

with $\delta_{S_{d,0}}$ the Dirac function centred at $S_{d,0}$. The deformation gradient on the $-$ side denoted by $\mathbf{F}^-$ and that on the $+$ side denoted by $\mathbf{F}^+$, therefore read:

$$\begin{cases} \mathbf{F}^- = \mathbf{I} + \frac{\partial \bar{\mathbf{u}}}{\partial \xi} \\ \mathbf{F}^+ = \mathbf{I} + \frac{\partial \bar{\mathbf{u}}}{\partial \xi} + \frac{\partial \hat{\mathbf{u}}}{\partial \xi} \end{cases} \tag{14.28}$$

Using Nanson's formula, Equation(11.5), the normal on the $-$ side and that on the $+$ side, $\mathbf{n}_d^-$ $\mathbf{n}_d^+$, respectively, can now be related to that in the original configuration $\mathbf{n}_{d,0}$, as follows:

$$\begin{cases} \mathbf{n}_d^- = \det(\mathbf{F}^-)\mathbf{n}_{d,0} \cdot (\mathbf{F}^-)^{-1} \frac{\mathrm{d}S_{d,0}}{\mathrm{d}S_d^-} \\ \mathbf{n}_d^+ = \det(\mathbf{F}^+)\mathbf{n}_{d,0} \cdot (\mathbf{F}^+)^{-1} \frac{\mathrm{d}S_{d,0}}{\mathrm{d}S_d^+} \end{cases} \tag{14.29}$$

In a cohesive zone model it is not trivial to relate cohesive forces on surfaces with different normals. For this reason Wells *et al.* (2002) have assumed that they work on an intermediate surface S_d^* (Figure 14.17), with a normal $\mathbf{n}_d^*$ defined as an average of $\mathbf{n}_d^-$ and $\mathbf{n}_d^+$:

$$\mathbf{n}_d^* = \det(\mathbf{F}^*)\mathbf{n}_{d,0} \cdot (\mathbf{F}^*)^{-1} \frac{\mathrm{d}S_{d,0}}{\mathrm{d}S_d^*} \tag{14.30}$$

where

$$\begin{cases} \mathrm{d}S_d^* = \frac{1}{2}(\mathrm{d}S_d^- + \mathrm{d}S_d^+) \\ \mathbf{F}^* = \frac{1}{2}\left(\mathbf{F}^- + \mathbf{F}^+\right) \end{cases} \tag{14.31}$$

We now take the Eulerian finite element formulation outlined in Chapter 12 as point of departure, and we employ the version in which the Truesdell rate of the Cauchy stress tensor is used. The rationale for preferring the Cauchy stress is that for a proper prediction of delamination and crack growth this stress enters the crack initiation criterion. Also, different from isochoric plasticity, it cannot be replaced by the Kirchhoff stress, since we now cannot exploit the incompressibility property which makes $\det \mathbf{F} \approx 1$. Partitioning as in Equation (14.19) we obtain the following set of linearised equations:

$$\begin{bmatrix} \mathbf{K}_{\bar{a}\bar{a}} & \mathbf{K}_{\bar{a}\hat{a}} \\ \mathbf{K}_{\hat{a}\bar{a}} & \mathbf{K}_{\hat{a}\hat{a}} \end{bmatrix} \begin{pmatrix} \mathrm{d}\bar{\mathbf{a}} \\ \mathrm{d}\hat{\mathbf{a}} \end{pmatrix} = \begin{pmatrix} \mathbf{f}_{\mathrm{ext}}^{\bar{a}} - \mathbf{f}_{\mathrm{int}}^{\bar{a}} \\ \mathbf{f}_{\mathrm{ext}}^{\hat{a}} - \mathbf{f}_{\mathrm{int}}^{\hat{a}} \end{pmatrix} \tag{14.32}$$

with external force vectors

$$\begin{cases} \mathbf{f}_{\mathrm{ext}}^{\bar{a}} = \int_S \mathbf{H}^{\mathrm{T}} \mathbf{t} \mathrm{d}S \\ \mathbf{f}_{\mathrm{ext}}^{\hat{a}} = \int_S \mathcal{H}_{S_d} \mathbf{H}^{\mathrm{T}} \mathbf{t} \mathrm{d}S \end{cases} \tag{14.33}$$

the internal force vectors

$$\begin{cases} \mathbf{f}_{\mathrm{int}}^{\bar{a}} = \int_V \mathbf{B}^{\mathrm{T}} \boldsymbol{\sigma} \mathrm{d}V \\ \mathbf{f}_{\mathrm{int}}^{\hat{a}} = \int_{V^+} \mathbf{B}^{\mathrm{T}} \boldsymbol{\sigma} \mathrm{d}V + \int_{S_d} \mathbf{H}^{\mathrm{T}} \mathbf{t}_d \mathrm{d}S \end{cases} \tag{14.34}$$

and the tangential stiffness submatrices, cf. Equation (12.28):

$$\begin{aligned}
\mathbf{K}_{\bar{a}\bar{a}} &= \int_V \mathbf{B}^{\mathrm{T}} \mathbf{D}^{\mathrm{TC}} \mathbf{B} \mathrm{d}V + \int_V \mathbf{G}^{\mathrm{T}} \boldsymbol{\mathcal{S}} \mathbf{G} \mathrm{d}V \\
\mathbf{K}_{\bar{a}\hat{a}} &= \int_{V^+} \mathbf{B}^{\mathrm{T}} \mathbf{D}^{\mathrm{TC}} \mathbf{B} \mathrm{d}V + \int_{V^+} \mathbf{G}^{\mathrm{T}} \boldsymbol{\mathcal{S}} \mathbf{G} \mathrm{d}V \\
\mathbf{K}_{\hat{a}\bar{a}} &= \int_{V^+} \mathbf{B}^{\mathrm{T}} \mathbf{D}^{\mathrm{TC}} \mathbf{B} \mathrm{d}V + \int_{V^+} \mathbf{G}^{\mathrm{T}} \boldsymbol{\mathcal{S}} \mathbf{G} \mathrm{d}V + \int_{S_d} \boldsymbol{H} \boldsymbol{T} \mathbf{G} \mathrm{d}S \\
\mathbf{K}_{\hat{a}\hat{a}} &= \int_{V^+} \mathbf{B}^{\mathrm{T}} \mathbf{D}^{\mathrm{TC}} \mathbf{B} \mathrm{d}V + \int_{V^+} \mathbf{G}^{\mathrm{T}} \boldsymbol{\mathcal{S}} \mathbf{G} \mathrm{d}V + \frac{1}{2} \int_{S_d} \boldsymbol{H} \boldsymbol{T} \mathbf{G} \mathrm{d}S
\end{aligned} \tag{14.35}$$

with the matrices **B**, **G** and $\boldsymbol{\mathcal{S}}$ given by Equations (12.4), (12.15) and (12.23), respectively,

$$\boldsymbol{H} = \begin{bmatrix} h_1 & 0 & 0 & \dots & 0 & 0 & 0 \\ 0 & h_1 & 0 & \dots & 0 & 0 & 0 \\ 0 & 0 & h_1 & \dots & 0 & 0 & 0 \\ \vdots & \vdots & \vdots & \ddots & \vdots & \vdots & \vdots \\ 0 & 0 & 0 & \dots & h_n & 0 & 0 \\ 0 & 0 & 0 & \dots & 0 & h_n & 0 \\ 0 & 0 & 0 & \dots & 0 & 0 & h_n \end{bmatrix} \tag{14.36}$$

and

$$
\boldsymbol{T} = \begin{bmatrix} t_x & t_y & t_z & 0 & 0 & 0 & 0 & 0 & 0 \\ 0 & 0 & 0 & 0 & 0 & 0 & 0 & 0 & 0 \\ 0 & 0 & 0 & 0 & 0 & 0 & 0 & 0 & 0 \\ 0 & 0 & 0 & t_x & t_y & t_z & 0 & 0 & 0 \\ 0 & 0 & 0 & 0 & 0 & 0 & 0 & 0 & 0 \\ 0 & 0 & 0 & 0 & 0 & 0 & 0 & 0 & 0 \\ 0 & 0 & 0 & 0 & 0 & 0 & t_x & t_y & t_z \\ 0 & 0 & 0 & 0 & 0 & 0 & 0 & 0 & 0 \\ 0 & 0 & 0 & 0 & 0 & 0 & 0 & 0 & 0 \end{bmatrix} \tag{14.37}
$$

containing the shape functions h_k and the nominal stresses t_x, t_y and t_z, respectively. The third contribution to the tangential submatrices $\mathbf{K}_{\hat{a}\bar{a}}$ and $\mathbf{K}_{\hat{a}\hat{a}}$ stems from the consistent linearisation of the traction at the discontinuity S_d (Wells *et al.* 2002). Attention is drawn to the factor $\frac{1}{2}$ which arises because of the definition of $\mathbf{n}_d^*$, Equation (14.30). The third term in $\mathbf{K}_{\hat{a}\bar{a}}$ causes the tangential stiffness matrix to be non-symmetric. However, numerical experience indicates that the effect of the non-symmetric terms on the convergence speed of the Newton–Raphson method is usually not significant (Wells *et al.* 2002).

To test the geometrically non-linear model a double-cantilever beam, shown in Figure 14.18, is analysed. The beam consists of two layers of the same material. The dashed line in Figure 14.18 shows the interface between the two layers. The parameter a is the initial delamination length ($a = 1$ mm), where the interface is assumed to be traction-free. The following material properties have been adopted: Young's modulus $E = 100$ MPa and Poisson's ratio $\nu = 0.3$ for the continuum and at the interface a tensile strength $f_t = 1$ MPa and fracture energy $\mathcal{G}_c = 0.05$ N/mm have been adopted.

To test the objectivity of the model with respect to spatial discretisation, the peel test is analysed using two different, unstructured meshes. The first mesh consists of 781 elements, and the second mesh is composed of 2896 elements. The deformed configurations for both discretisations are shown for a displacement $u = 6$ mm in Figure 14.19. Please note that the actual strains are small, since the majority of the deformation concerns the crack opening at the discontinuity. The load–displacement responses for the two meshes are shown in Figure 14.20.

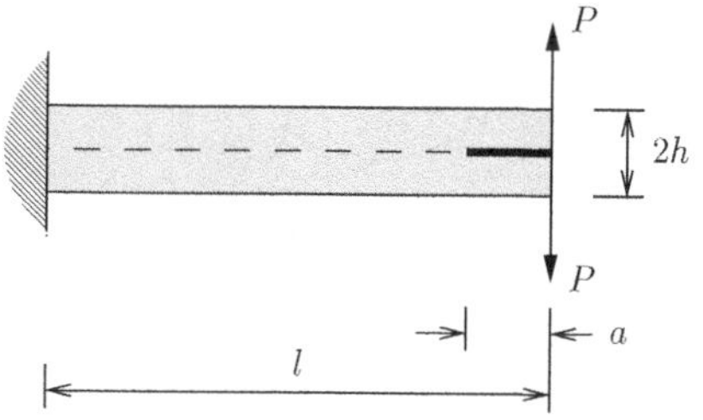

Figure 14.18 Peel test geometry for a two-layer laminate. The dashed line is the interlaminar boundary and the initial delamination length is denoted by a

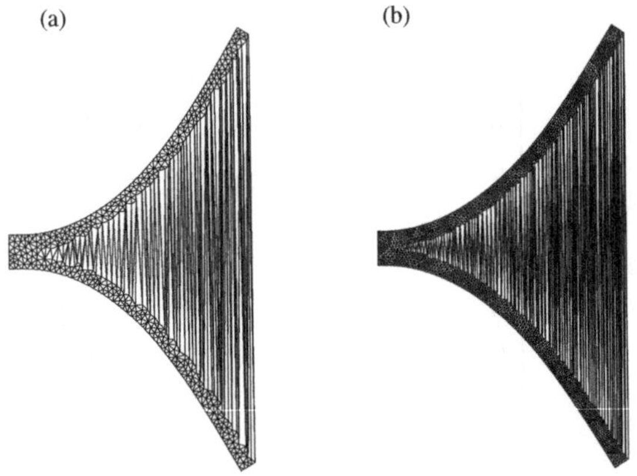

Figure 14.19 Peel test in the deformed configuration at different displacements with (a) 781 elements and (b) 2896 elements. The deformations are not magnified (Wells *et al.* 2002)

The roughness of the response for the coarser mesh is due to the extension of a discontinuity through an entire element and the 'jumping' of inelastic deformation from integration point to integration point. Nevertheless, a properly converged solution was obtained for all load increments. The stress ahead of the delamination tip is complex, and has a high gradient. As the mesh is refined, the response becomes smoother. Although the response is rough for the coarse mesh, the response generally follows the response for the finer mesh, indicating that the computed result is not dependent on the spatial discretisation.

Next, a combination of delamination growth and structural instability is considered (Allix and Corigliano 1999; Remmers *et al.* 2003b; Wells *et al.* 2002). The double cantilever beam in Figure 14.21 has an initial delamination length of $a_0 = 10$ mm and is subjected to an axial compressive load $2P$. Two small forces, denoted by P_0, are applied to trigger the buckling mode. Both layers are made of the same material with a Young's modulus $E = 135$ GPa and a Poisson's ratio $\nu = 0.18$. Due to symmetry in the geometry of the model and in the applied

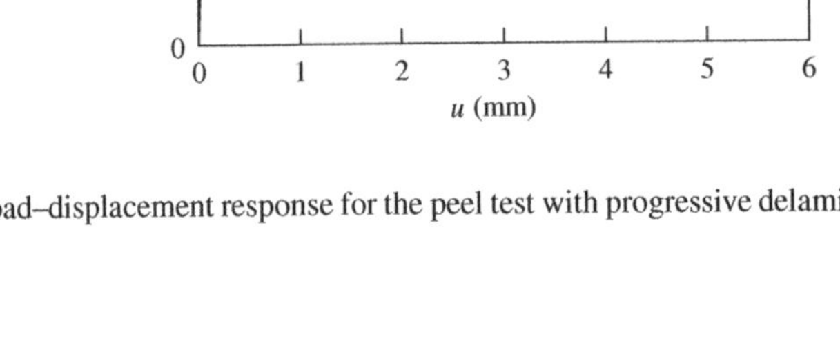

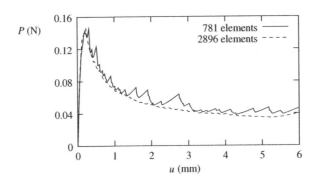

Figure 14.20 Load–displacement response for the peel test with progressive delamination (Wells *et al.* 2002)

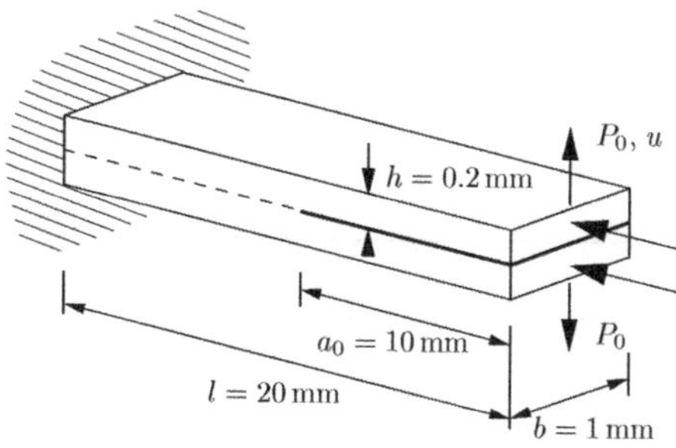

Figure 14.21 Geometry of a double cantilever beam with initial delamination a_0 under compression

loading, delamination propagation can be modelled with an exponential mode-I decohesion law. The tensile strength f_t is equal to $50\,\mathrm{N/mm^2}$, and the fracture energy is $\mathcal{G}_c = 0.8\,\mathrm{N/mm}$. The finite element mesh is composed of eight-noded enhanced solid-like shell elements (Figure 14.22) (Remmers *et al.* 2003b). It consists of just one element in the thickness direction, but is locally refined to capture delamination growth correctly. Figure 14.22 shows the lateral displacement u of the beam as a function of the external force P. The load–displacement response for a specimen with perfect bond (no delamination growth) is given as a reference.

14.2.3 Dynamic Fracture

In Chapter 5 the basic solution algorithms for non-linear dynamics have been laid out. We will now apply the partition-of-unity approach to dynamic fracture. The acceleration can then be found by differentiating the displacement field, Equation (14.15), twice with respect to time:

$$\ddot{\mathbf{u}} = \ddot{\bar{\mathbf{u}}} + \mathcal{H}_{S_d}\ddot{\hat{\mathbf{u}}} \tag{14.38}$$

Inserting this expression into the equation of motion, Equation (2.4), multiplying by the corresponding test functions, Equation (14.16), integrating over the domain V, applying the

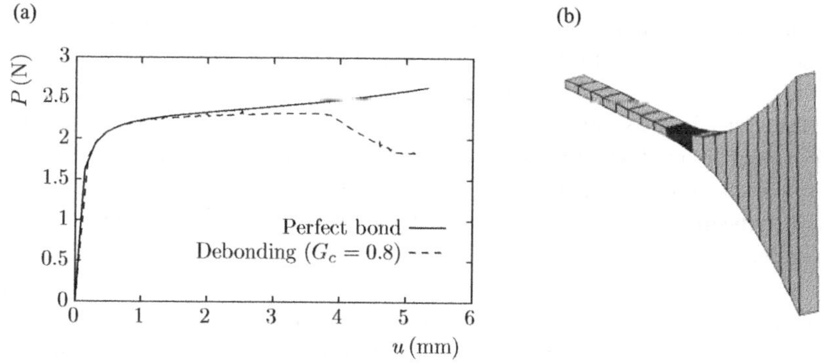

Figure 14.22 Delamination buckling test. (a) Tip displacement as a function of the applied axial load P. (b) Deformation in final state (Remmers *et al.* 2003b)

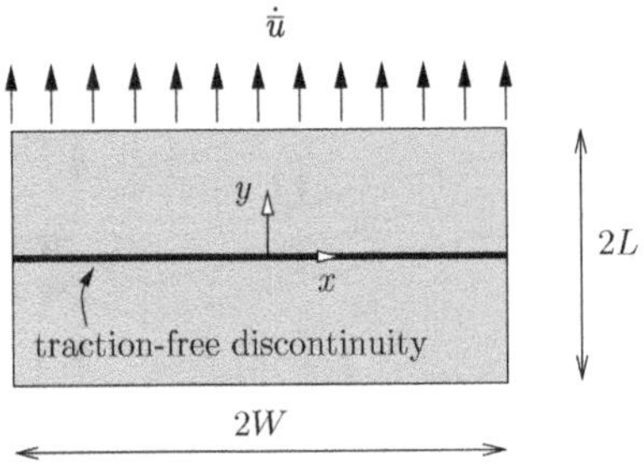

Figure 14.23 Geometry and loading condition of the block with a traction-free discontinuity

divergence theorem, discretising according to Equation (14.18), and requiring that the result holds for all admissible $\delta\bar{\mathbf{a}}$ and $\delta\hat{\mathbf{a}}$ results in the semi-discrete set of balance of momentum equations:

$$\begin{bmatrix} \mathbf{M}_{\bar{a}\bar{a}} & \mathbf{M}_{\bar{a}\hat{a}} \\ \mathbf{M}_{\hat{a}\bar{a}} & \mathbf{M}_{\hat{a}\hat{a}} \end{bmatrix} \begin{pmatrix} \mathrm{d}\ddot{\bar{\mathbf{a}}} \\ \mathrm{d}\ddot{\hat{\mathbf{a}}} \end{pmatrix} = \begin{pmatrix} \mathbf{f}^{\bar{a}}_{\text{ext}} - \mathbf{f}^{\bar{a}}_{\text{int}} \\ \mathbf{f}^{\hat{a}}_{\text{ext}} - \mathbf{f}^{\hat{a}}_{\text{int}} \end{pmatrix} \tag{14.39}$$

where the internal force vectors, $\mathbf{f}^{\bar{a}}_{\text{int}}$ and $\mathbf{f}^{\hat{a}}_{\text{int}}$ and the external force vectors, $\mathbf{f}^{\bar{a}}_{\text{ext}}$ and $\mathbf{f}^{\hat{a}}_{\text{ext}}$, follow from Equation (14.19), as for the quasi-static simulation, and the mass matrices are given by:

$$\mathbf{M}_{\bar{a}\bar{a}} = \int_V \rho \mathbf{H}^{\mathrm{T}} \mathbf{H} \mathrm{d}V$$

$$\mathbf{M}_{\bar{a}\hat{a}} = \mathbf{M}_{\hat{a}\hat{a}} = \int_{V^+} \rho \mathbf{H}^{\mathrm{T}} \mathbf{H} \mathrm{d}V \tag{14.40}$$

For fast crack propagation, explicit time integration schemes are often preferred. To fully exploit their efficiency, they are normally used in conjuction with a lumped mass matrix, see Chapter 5. The off-diagonal submatrix $\mathbf{M}_{\bar{a}\hat{a}}$ contains terms that couple the regular and the additional degrees of freedom. By lumping the mass matrix, this information is lost. The effects of this loss of information have been assessed for the block of Figure 14.23 (Remmers *et al.* 2008), which has been used in Chapter 5 to demonstrate the explicit solver.

For the present purpose the block is divided into two parts by means of a horizontal crack at $y = 0$, which crosses the entire width of the specimen (Figure 14.23). In the finite element analysis, the crack is represented as traction-free by exploiting the partition-of-unity property of the shape functions. In the simulations, the stress wave that carries the tensile stress of 25 MPa propagates from the top of the specimen, and reaches the discontinuity after $t \approx 2\ \mu$s. Since the crack acts as a traction-free (internal) boundary, the wave reflects at $y = 0$ and subsequently travels back to the top. Obviously, the stresses in the lower part of the specimen should remain zero throughout the simulation.

Figure 14.24 shows the values of σ_{yy} along the vertical centre line of the specimen for a consistent mass matrix and for a lumped mass matrix when the stress wave has been reflected by the traction-free crack. The simulations are compared with a benchmark calculation in which the slit is modelled by disconnecting adjacent elements. The simulation with the consistent mass matrix shows an excellent agreement with the benchmark simulation [Figure 14.24(a)]. The stress wave is properly reflected and, more importantly, the lower part of the specimen

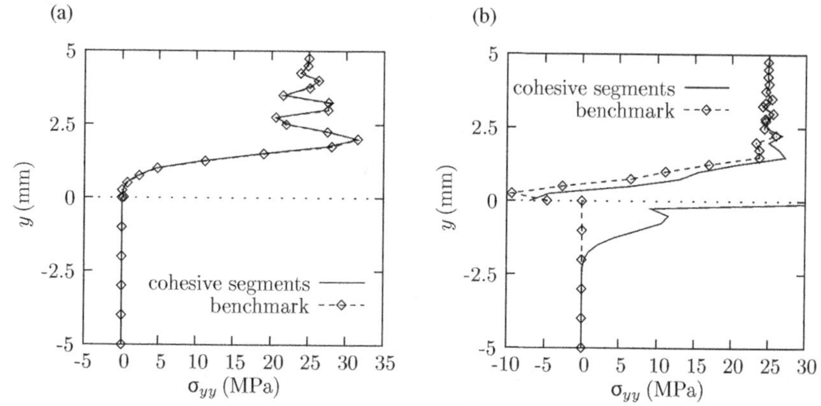

Figure 14.24 σ_{yy} as a function of y along the centre line of the specimen at $t = 3\ \mu$s. (a) Consistent mass matrix. (b) Lumped mass matrix. The traction-free crack (dotted line) is modelled using a Heaviside function

remains stress free. This is not so for a lumped mass matrix [Figure 14.24(b)]. A considerable amount of the energy is transferred across the crack, which results in a stress wave in the lower part of the specimen, and a very high stress of over 42 MPa in the integration points just below the discontinuity.

The simulations have been repeated with a different jump function. Instead of the Heaviside function a symmetric jump function is used, $\mathcal{H}^- = -1$ and $\mathcal{H}^+ = +1$. The corresponding traction profiles are shown in Figure 14.25. The spurious stress wave reflections in the simulations with the lumped mass matrix representation have disappeared [Figure 14.25(b)] and the stresses in the lower part of the specimen remain zero.

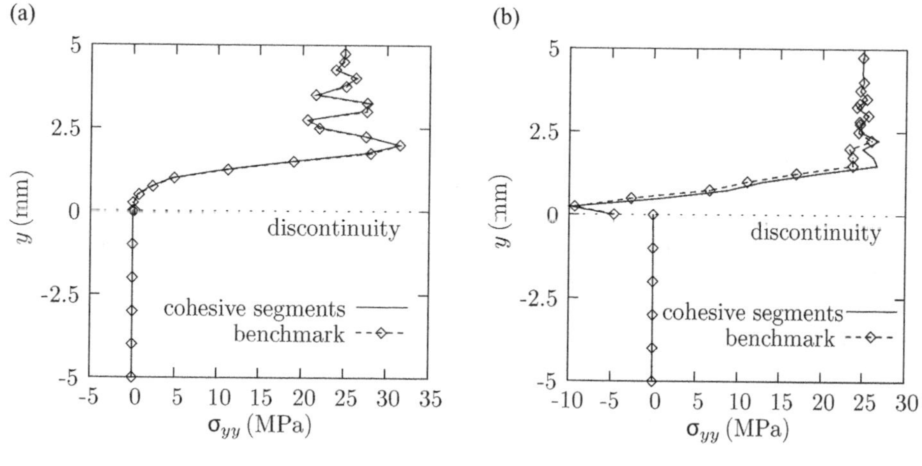

Figure 14.25 σ_{yy} as a function of y along the centre line of the specimen at $t = 3\ \mu$s. (a) Consistent mass matrix. (b) Lumped mass matrix. The traction-free crack (dotted line) is modelled using a symmetric jump function

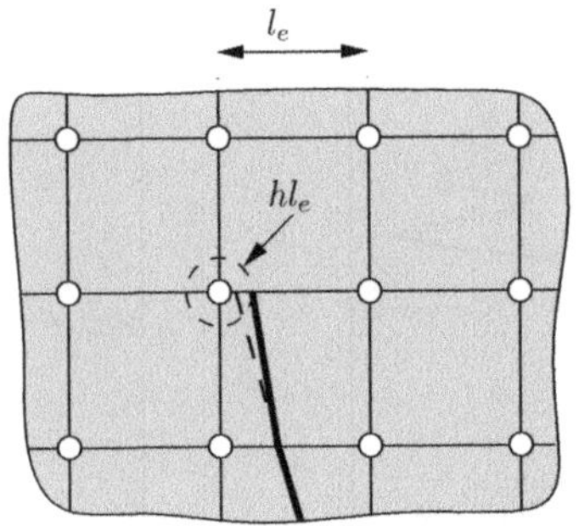

Figure 14.26 Artificial deflection of a crack which would otherwise cross a supporting node within a distance $h\ell_e$. The dashed line denotes the original position of the discontinuity

The use of an explicit time integration scheme has another consequence. When an element is crossed by a discontinuity, the two parts can be considered as individual elements, each with a smaller effective length ℓ_e than the original element. An element can be crossed by a discontinuity in such a way that one of the two resulting parts of the element becomes so small that the critical time step for a stable solution procedure will almost become zero, cf. Chapter 5. To avoid this situation, a discontinuity is not allowed to cross an element boundary when the distance to a node is less than $h\ell_e$, with h an offset factor (Figure 14.26). Although the position, and therefore the further crack extension, is slightly modified, the deflection turns out to be minimal for small values of the offset factor, e.g. $h \approx 0.1$, typically in the order of a few degrees only. Because of the nearly linear relation between the critical time step and the distance between the discontinuity, the stability requirement can now be expanded as:

$$\Delta t = \alpha \frac{h\ell_e}{c_d} \tag{14.41}$$

where $\alpha < 1$. In the above calculations $\alpha = 0.1$.

14.2.4 Weak Discontinuities

While discontinuities like cracks involve a jump in the displacement, other physical phenomena exist for which the displacements at the discontinuity remain continuous, but for which the displacement gradient experiences a finite jump. Typical examples are solid–solid phase boundaries, e.g. between martensite and austenite, or between blades in twinned martensite [Figure 14.27(a)] (Bhattacharya 2003). In such cases, the interface conditions are characterised by:

$$[\![\nabla \mathbf{u}]\!] = \mathbf{c} \otimes \mathbf{n}_d \tag{14.42}$$

with $\mathbf{c}$ a non-zero vector. Instead of taking the Heaviside function $\mathcal{H}_{S_d}$ at the discontinuity S_d as the enhanced basis function ψ_l, the distance function $\mathcal{D}_{S_d}$ is substituted for ψ_l (Belytschko *et al.* 2001). This function is continuous at the discontinuity S_d, but its normal derivative is

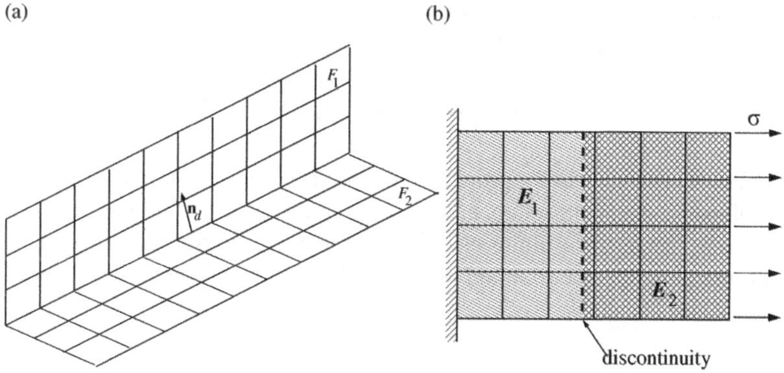

Figure 14.27 (a) Twinning in martensite as an example of a solid–solid phase boundary. (b) Plate under uniaxial tension with different stiffness moduli

discontinuous and equal to $\mathcal{H}_{S_d}$:

$$\mathbf{n}_{S_d} \cdot \nabla \mathcal{D}_{S_d} = \mathcal{H}_{S_d} \tag{14.43}$$

thus meeting condition (14.42). The enhanced displacement field is subsequently obtained by substituting $\psi_l = \mathcal{D}_{S_d}$ into Equation (14.13), so that:

$$\mathbf{u} = \underbrace{\mathbf{H}\bar{\mathbf{a}}}_{\bar{\mathbf{u}}} + \mathcal{D}_{S_d} \underbrace{\mathbf{H}\hat{\mathbf{a}}}_{\hat{\mathbf{u}}} \tag{14.44}$$

A simple example of a weak discontinuity is given in Figure 14.27(b), which shows a plate subjected to a uniaxial stress. Because of the different values of Young's modulus on the left and the right parts of the plate, there will be a discontinuity in the displacement gradient. In standard finite element analysis this discontinuity is captured by letting it coincide with the boundaries of $\mathcal{C}^0$-continuous finite elements. However, the enhanced displacement interpolation of Equation (14.44) can capture this weak discontinuity in an exact manner without the need to align the boundaries of finite elements with the discontinuity. Indeed, this concept can be taken further to the limiting case that one of the Young's moduli vanishes. Then, the boundary of a structure can be modelled without the need to align finite element boundaries with the structure boundaries. This advantage can be exploited to model complex geometries and microstructures while avoiding the need to carry out a complicated meshing operation, e.g. Moës *et al.* (2003), where the complex geometry of a woven composite has been modelled in three dimensions using the partition-of-unity approach.

While the example of Figure 14.27 is simple and can be solved using a standard interpolation, this is not so for evolving discontinuities as solid–solid phase boundaries. Level sets (Gravouil *et al.* 2002; Hou *et al.* 1999; Osher and Paragios 2003) are the most common approach to track the propagating discontinuities. The idea is that the position of the discontinuity S_d coincides with the zero level set of a smooth, scalar-valued function f,

$$S_d = \{\mathbf{x} \in V : f(\mathbf{x}, t) = 0\} \tag{14.45}$$

The distance function of Equation (14.43) can be chosen as a level set function: $f = \mathcal{D}_{S_d}$, and its evolution is then governed by the Hamilton–Jacobi equation:

$$\dot{\mathcal{D}}_{S_d}(\mathbf{x}, t) + v_n \|\nabla \mathcal{D}_{S_d}(\mathbf{x}, t)\| = 0 \tag{14.46}$$

with v_n the normal component of the propagation velocity of the discontinuity S_d. This equation needs to be solved for every time step, and its solution gives the position of the discontinuity, see Valance *et al.* (2008) for a finite element implementation. Subsequently, the enhanced displacement interpolation of Equation (14.44) can be used to carry out the stress analysis on a fixed grid.

References

Allix O and Corigliano A 1999 Geometrical and interfacial non-lineararities in the analysis of delamination in composites. *International Journal of Solids and Structures* **36**, 2189–2216.

Askes H, de Borst R and Heeres OM 1999 Conditions for locking-free elasto-plastic analyses in the element-free Galerkin method. *Computer Methods in Applied Mechanics and Engineering* **173**, 99–109.

Askes H, Pamin J and de Borst R 2000 Dispersion analysis and element-free Galerkin solutions of second- and fourth-order gradient-enhanced damage models. *International Journal for Numerical Methods in Engineering* **49**, 811–832.

Askes H, Wells GN and de Borst R 2003 Novel discretization concepts, in *Comprehensive Structural Integrity* (eds Milne I, Ritchi RO and Karihaloo B), vol. III, pp. 377–425. Elsevier.

Atluri SN and Zhu T 1998 A new Meshless Local Petrov-Galerkin (MLPG) approach in computational mechanics. *Computational Mechanics* **22**, 117–127.

Atluri SN and Zhu T 1999 A critical assessment of the truly new Meshless Local Petrov-Galerkin (MLPG), and Local Boundary Integral Equation methods. *Computational Mechanics* **24**, 348–372.

Atluri SN and Zhu T 2000 New concepts in meshless methods. *International Journal for Numerical Methods in Engineering* **47**, 537–556.

Babuška I and Melenk JM 1997 The partition of unity method. *International Journal for Numerical Methods in Engineering* **40**, 727–758.

Barsoum RS 1976 On the use of isoparametric finite elements in linear fracture mechanics. *International Journal for Numerical Methods in Engineering* **10**, 225–237.

Beissel S and Belytschko T 1996 Nodal integration in the element-free Galerkin method. *Computer Methods in Applied Mechanics and Engineering* **139**, 49–74.

Belytschko T and Black T 1999 Elastic crack growth in finite elements with minimal remeshing. *International Journal for Numerical Methods in Engineering* **45**, 601–620.

Belytschko T, Krongauz Y, Organ D, Fleming M and Krysl P 1996 Meshless methods: an overview and recent developments. *Computer Methods in Applied Mechanics and Engineering* **193**, 3–47.

Belytschko T, Lu YY and Gu L 1994 Element-free Galerkin methods. *International Journal for Numerical Methods in Engineering* **37**, 229–256.

Belytschko T, Moës N, Usui S and Parimi C 2001 Arbitrary discontinuities in finite elements. *International Journal for Numerical Methods in Engineering* **50**, 993–1013.

Bhattacharya K 2003 *Microstructure of Martensite: Why it Forms and How it Gives Rise to the Shape-Memory Effect*. Oxford University Press.

Chen JS and Wang HP 2000 New boundary condition treatments in meshfree computation of contact problems. *Computer Methods in Applied Mechanics and Engineering* **187**, 441–468.

Chen JS, Wu CT, Yoon S and You Y 2001 A stabilized conforming nodal integration for Galerkin mesh-free methods. *International Journal for Numerical Methods in Engineering* **50**, 435–466.

Chinesta F, Cescotto S, Cueto E and Lorong P 2010 *Natural Element Methods for the Simulation of Structures and Processes*. ISTE and John Wiley & Sons, Ltd.

De S and Bathe KJ 2000 The method of finite spheres. *Computational Mechanics* **25**, 329–345.

Dolbow J and Belytschko T 1999 Volumetric locking in the element free Galerkin method. *International Journal for Numerical Methods in Engineering* **46**, 925–942.

Duarte CA and Oden J 1996 H-p clouds – an h-p meshless method. *Numerical Methods for Partial Differential Equations* **12**, 673–705.

Fleming M, Chu YA, Moran B and Belytschko T 1997 Enriched element-free Galerkin methods for crack tip fields. *International Journal for Numerical Methods in Engineering* **40**, 1483–1504.

Gavete L, Benito JJ, Falcón S and Ruiz A 2000 Implementation of essential boundary conditions in a meshless method. *Communications in Numerical Methods in Engineering* **16**, 409–421.

González D, Cueto E and Doblaré M 2004 Volumetric locking in natural neighbour Galerkin methods. *International Journal for Numerical Methods in Engineering* **61**, 611–632.

Gravouil A, Moës N and Belytschko T 2002 Non-planar 3D crack growth by the extended finite element and level sets – Part I: Mechanical model and Part II: Level set update. *International Journal for Numerical Methods in Engineering* **53**, 2549–2586.

Hansbo A and Hansbo P 2004 A finite element method for the simulation of strong and weak discontinuities in solid mechanics. *Computer Methods in Applied Mechanics and Engineering* **193**, 3523–3540.

Henshell RD and Shaw KG 1976 Crack tip finite elements are unnecessary. *International Journal for Numerical Methods in Engineering* **9**, 495–507.

Hou T, Rosakis P and Lefloch P 1999 A level-set approach to the computation of twinning and phase-transition dynamics. *Journal of Computational Physics* **150**, 302–331.

Huerta A and Fernández-Méndez S 2001 Locking in the incompressible limit for the element-free Galerkin method. *International Journal for Numerical Methods in Engineering* **51**, 1361–1383.

Karihaloo BL and Xiao QZ 2010 Asymptotic fields ahead of mixed mode frictional cohesive cracks. *Zeitschrift für Angewandte Mathematik und Mechanik* **90**, 121–129.

Krysl P and Belytschko T 1996 Analysis of thin shells by the element-free Galerkin method. *International Journal of Solids and Structures* **33**, 3057–3080.

Krysl P and Belytschko T 1999 The element-free Galerkin method for dynamic propagation of arbitrary 3-D cracks. *International Journal for Numerical Methods in Engineering* **44**, 767–800.

Krysl P and Belytschko T 2001 ESFLIB: a library to compute the element-free Galerkin shape functions. *Computer Methods in Applied Mechanics and Engineering* **190**, 2181–2205.

Lancaster P and Salkauskas K 1981 Surfaces generated by moving least squares. *Mathematics of Computation* **37**, 141–158.

Liu WK, Jun S and Zhang YF 1995 Reproducing kernel particle methods. *International Journal for Numerical Methods in Fluids* **20**, 1081–1106.

Lu YY, Belytschko T and Gu L 1994 A new implementation of the element free Galerkin method. *Computer Methods in Applied Mechanics and Engineering* **113**, 229–256.

Melenk JM and Babuška I 1996 The partition of unity finite element method: basic theory and applications. *Computer Methods in Applied Mechanics and Engineering* **139**, 289–314.

Moës N and Belytschko T 2002 Extended finite element method for cohesive crack growth. *Engineering Fracture Mechanics* **69**, 813–833.

Moës N, Cloirec M, Cartraud P and Remacle J 2003 A computational approach to handle complex microstructure geometries. *Computer Methods in Applied Mechanics and Engineering* **192**, 3163–3177.

Moes N, Dolbow J and Belytschko T 1999 A finite element method for crack growth without remeshing. *International Journal for Numerical Methods in Engineering* **46**, 131–150.

Nayroles B, Touzot G and Villon P 1992 Generalizing the finite element method: diffuse approximations and diffuse elements. *Computational Mechanics* **10**, 307–318.

Onate E, Idelsohn S, Zienkiewicz OC and Taylor RL 1996 A finite point method in computational mechanics. applications to convective transport and fluid flow. *International Journal for Numerical Methods in Engineering* **39**, 3839–3866.

Osher S and Paragios N 2003 *Geometric Level Set Methods in Imaging, Vision, and Graphics*. Springer-Verlag.

Pamin J, Askes H and de Borst R 2003 Two gradient plasticity theories discretized with the element-free Galerkin method. *Computer Methods in Applied Mechanics and Engineering* **192**, 2377–2407.

Remmers JJC, de Borst R and Needleman A 2003a A cohesive segments method for the simulation of crack growth. *Computational Mechanics* **31**, 69–77.

Remmers JJC, de Borst R and Needleman A 2008 The simulation of dynamic crack propagation using the cohesive segments method. *Journal of the Mechanics and Physics of Solids* **56**, 70–92.

Remmers JJC, Wells GN and de Borst R 2003b A discontinuous solid-like shell element for arbitrary delaminations. *International Journal for Numerical Methods in Engineering* **58**, 2013–2040.

Simone A 2004 Partition of unity-based discontinuous elements for interface phenomena: computational issues. *Communications in Numerical Methods in Engineering* **20**, 465–478.

Sukumar N, Moran B, Semenov AY and Belikov VV 2001 Natural neighbour Galerkin methods. *International Journal for Numerical Methods in Engineering* **50**, 1–27.

Sulsky D, Chen Z and Schreyer HL 1994 A particle method for history-dependent materials. *Computer Methods in Applied Mechanics and Engineering* **118**, 179–196.

Valance S, de Borst R, Réthoré J and Coret M 2008 A partition-of-unity based finite element method for level sets. *International Journal for Numerical Methods in Engineering* **76**, 1513–1527.

Ventura G 2002 An augmented Lagrangian approach to essential boundary conditions in meshless methods. *International Journal for Numerical Methods in Engineering* **53**, 825–842.

Wagner GJ and Liu WK 2000 Application of essential boundary conditions in mesh-free methods: a corrected collocation method. *International Journal for Numerical Methods in Engineering* **47**, 1367–1370.

Wells GN and Sluys LJ 2001 A new method for modelling cohesive cracks using finite elements. *International Journal for Numerical Methods in Engineering* **50**, 2667–2682.

Wells GN, de Borst R and Sluys LJ 2002 A consistent geometrically non-linear approach for delamination. *International Journal for Numerical Methods in Engineering* **54**, 1333–1355.

Wu CK and Plesha ME 2002 Essential boundary condition enforcement in meshless methods: boundary flux collocation method. *International Journal for Numerical Methods in Engineering* **53**, 499–514.

Xiao QZ and Karihaloo BL 2006 Improving the accuracy of XFEM crack field using higher order quadrature and statically admissible stress recovery. *International Journal for Numerical Methods in Engineering* **66**, 1378–1410.

Yvonnet J, Ryckelynck D, Lorong P and Chinesta F 2004 A new extension of the natural element method for non-convex and discontinuous problems: the constrained natural element method (C-NEM). *International Journal for Numerical Methods in Engineering* **60**, 1451–1474.

15

Isogeometric Finite Element Analysis

Finite element methods are nowadays used for the analysis of (engineering) structures of tremendous complexity. The difficulties associated with transforming complex design models into analysis models has led to the development of an innovative design-through-analysis concept (Cottrell *et al.* 2009). *Isogeometric analysis* was proposed by Hughes *et al.* (2005) as a novel analysis strategy that integrates computer aided geometric design and finite element analyses, and thereby rigorously eliminates the difficulties in creating suitable analysis models for complex designs.

The fundamental idea of isogeometric analysis is to directly use the design model for analysis purposes, thereby by-passing the need for geometry clean-up or meshing operations. The current industry standard in design is based on non-uniform rational B-splines (NURBS), a technology that has superseded the earlier developments in B-splines. In recent years, T-splines have been introduced to overcome some deficiencies in the analysis of NURBS. Whereas B-splines, NURBS and T-splines form a hierarchy of design-oriented spline technologies, alternative techniques such as subdivision surfaces have their own merits and can be preferred for specific applications. However, for engineering design, B-splines, NURBS and T-splines are the dominant technology. For this reason we restrict ourselves to these spline technologies in the remainder of this chapter.

15.1 Basis Functions in Computer Aided Geometric Design

As we have seen in Chapter 2 the parametrisation of the geometry is a fundamental aspect of finite element methods. In finite elements it is customary to employ C^0-continuous basis functions to parametrise the geometry, see Equation (2.24), and isoparametric elements, where the same basis functions are used for the geometry parametrisation and for the trial and test spaces, are predominantly present in analyses.

Historically, the research in geometry parametrisation is attributed to the field of computer aided geometric design. Although the requirements that derive from design differ from those

Non-linear Finite Element Analysis of Solids and Structures, Second Edition.
René de Borst, Mike A. Crisfield, Joris J.C. Remmers and Clemens V. Verhoosel.

for analysis, the conceptual idea of many geometry parametrisation methods employed in computer aided geometric design resembles that of finite element analysis. In both fields, the geometry is parametrised by a linear combination of N global basis functions:

$$\mathbf{x}(\boldsymbol{\xi}) = \sum_{k=1}^{N} h_k(\boldsymbol{\xi})\mathbf{p}_k \tag{15.1}$$

with the basis function $h_k : \widehat{V} \rightarrow V$ mapping a coordinate $\boldsymbol{\xi}$ in the parameter domain $\widehat{V}$, which has the dimension d_p, onto a coordinate in the physical domain V, with dimension d_s. The coefficients assembled in $\mathbf{p}_k$ are referred to as control points. For notational convenience we re-express Equation (15.1) using matrix-vector notation as:

$$\mathbf{x}(\boldsymbol{\xi}) = \mathbf{P}^{\mathrm{T}}\mathbf{h}(\boldsymbol{\xi}) \tag{15.2}$$

with

$$\mathbf{P} = \begin{bmatrix} \mathbf{p}_1^{\mathrm{T}} \\ \vdots \\ \mathbf{p}_N^{\mathrm{T}} \end{bmatrix} \qquad \mathbf{h} = \begin{pmatrix} h_1 \\ \vdots \\ h_N \end{pmatrix} \tag{15.3}$$

with $\mathbf{P}$ a $N \times d_s$ matrix. The matrix-vector product of Equation (15.2) implies that the physical coordinate $\mathbf{x}$ and the vectors that contain the control points $\mathbf{p}_k$ are defined as column vectors. The control points play a role that is similar to the nodes in finite element simulations, see Equation (2.24). The structured collection of control points is referred to as the control net, which can be interpreted as the equivalent of the mesh in finite elements.

It is noted that a distinction is made between the dimension d_s of the physical domain, and the dimension d_p of the parameter domain. For many problems this distinction is not necessary, since the dimensions of the parameter domain and physical domain coincide. However, there are problems for which both dimensions differ, the most important class for which this holds being thin-walled structures such as beams, shells and plates, see Chapters 9 and 10. In order to avoid confusion, we exclusively use the term 'dimension' in relation to the physical domain. For the parameter domain, we will use the terminology univariate for one 'dimension', bivariate for two 'dimensions' and trivariate for three 'dimensions'.

15.1.1 *Univariate B-splines*

The fundamental building block of isogeometric analysis is the univariate B-spline (Cottrell *et al.* 2009; Rogers 2001). A univariate B-spline is a parametrised curve, according to Equation (15.1), with piecewise polynomial basis functions $\{h_{k,p}(\xi)\}_{k=1}^{N}$, with N and p denoting the number and order of global basis functions, respectively. The basis functions are defined over a knot vector $\Xi = \{\xi_1, \xi_2, \ldots, \xi_{N+p+1}\}$. The knot values ξ_k are non-decreasing with increasing knot index k, i.e. $\xi_1 \leq \xi_2 \leq \ldots \leq \xi_{N+p+1}$. Consequently, the knots divide the parameter domain $\widehat{V} = [\xi_1, \xi_{N+p+1}]$ in knot intervals of non-negative length. In the remainder of this chapter we will restrict the discussion to open B-splines, the class of B-splines which is created using knot vectors in which the first and the last knot values are repeated $p+1$ times,

Box 15.1 Univariate B-spline

The third-order B-spline basis functions constructed over the knot vector $\Xi = \{0, 0, 0, 0, 1, 2, 2, 3, 3, 3, 4, 4, 4, 4\}$ are shown in the left-hand figure below. Since the first and last knot values are repeated $p + 1$ times, this is an open B-spline. Although the internal knot values are equally spaced, their repetition makes this B-spline non-uniform. Using this basis, a B-spline (the smooth curve in the right-hand figure) can be constructed using the (non-smooth) control net.

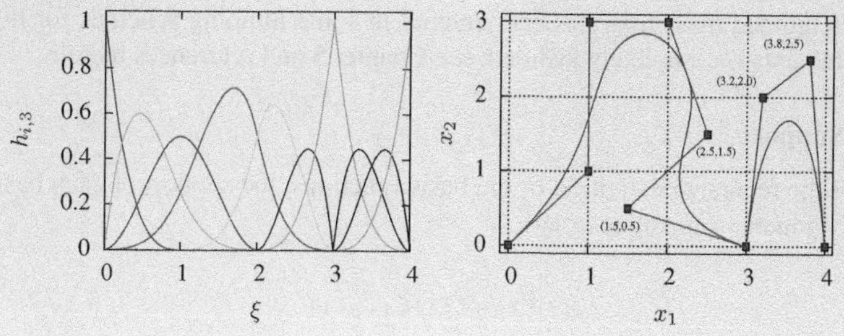

see also Box 15.1. Open B-splines are called *uniform* when based on a knot vector with equally distributed knots, so that $\Delta_e = \xi_{k+1} - \xi_k$ is a constant for $k = p + 1, \ldots, N$. Otherwise, we speak of a non-uniform B-spline. For a univariate B-spline, the *elements* are defined as the knot intervals of positive length. See the next section for a more detailed discussion on the element definition for splines. The number of elements is n_e.

The B-spline basis $\{h_{k,p}(\xi)\}_{k=1}^{N}$ is defined recursively, starting with piecewise constant ($p = 0$) functions:

$$h_{k,0}(\xi) = \begin{cases} 1 & \xi_k \leq \xi < \xi_{k+1} \\ 0 & \text{otherwise} \end{cases} \tag{15.4}$$

from which the higher-order ($p = 1, 2, \ldots$) basis functions follow from the Cox–de Boor recursion formula (Cox 1972; de Boor 1972):

$$h_{k,p}(\xi) = \frac{\xi \quad \xi_k}{\xi_{k+p} - \xi_k} h_{k,p-1}(\xi) + \frac{\xi_{k+p+1} - \xi}{\xi_{k+p+1} - \xi_{k+1}} h_{k+1,p-1}(\xi) \tag{15.5}$$

Efficient and robust algorithms have been developed for the evaluation of these basis functions and their derivatives (Piegl and Tiller 1997). The definition of the linear B-spline basis functions ($p = 1$) results in the first-order Lagrange elements. As a result, first-order B-splines are identical to standard linear finite elements. C^0-continuous B-splines, that is splines for which all the internal knots have a multiplicity p, span the same space as p-order Lagrange polynomials. Accordingly, they share many characteristics with standard finite elements. However, B-spline basis functions possess some properties which differ fundamentally from those of Lagrange elements commonly used in finite element formulations. They are listed in the following, where it is emphasised that we restrict our discussion to open B-splines.

15.1.1.1 Non-negativity

All spline basis functions are non-negative over the whole parameter domain $\widehat{V}$:

$$h_{k,p}(\xi) \geq 0 \tag{15.6}$$

with $p \geq 0$. This is an advantage when constructing lumped mass matrices in (explicit) dynamic analysis, as the resulting entries in these matrices are positive by definition, and hence the matrices are positive definite (Cottrell *et al.* 2009). Accordingly, the possible occurrence of non-positive lumped mass matrices encountered in some lumping schemes for higher-order Lagrange elements is completely avoided, see Chapter 5 and references therein.

15.1.1.2 Support

By virtue of the recursive definition of the basis functions, the support of each basis function increases with increasing spline order:

$$\widehat{V}_A = (\xi_A, \xi_{A+p+1}) \tag{15.7}$$

with $A \in \{1, \ldots, N\}$. We also refer to this domain as the local basis function domain. The number of basis functions which have a support over each knot interval, and thus over each element, equals $p + 1$. In this sense, univariate B-splines do not differ from Lagrange elements. Since the dimensions of the element matrices, and thus, the bandwidth of the global matrix, depend on the number of basis functions supported per element, this is an important similarity.

15.1.1.3 Continuity

By virtue of the construction of the basis functions, the continuity of the piecewise polynomial basis functions is reduced at the element intersections. Obviously, the zero-order functions are discontinuous. In the case of an increasing internal knot vector – $\xi_{p+1} < \xi_{p+2} < \ldots < \xi_{N+1}$ – the shape functions are $(p - 1)$-times continuously differentiable: $h_k(\xi) \in \mathcal{C}^{p-1}$. The continuity of splines is controlled by the knot multiplicities. If a knot value is repeated m times, the continuity of the basis functions at that coordinate is $\mathcal{C}^{p-m}$. Open B-splines are therefore $\mathcal{C}^{-1}$, or discontinuous, at the domain boundaries. From an analysis perspective this is a prerequisite as the interpolatory basis functions at the boundaries are suitable for the application of essential boundary conditions.

15.1.1.4 Partition-of-unity Property

Univariate B-spline basis functions satisfy the partition-of-unity property:

$$\sum_{k=1}^{N} h_{k,p}(\xi) = 1 \tag{15.8}$$

for open B-splines. This property is a consequence of the repeated boundary knot values, and, from the vantage of analysis, is required to permit affine transformations of the B-spline.

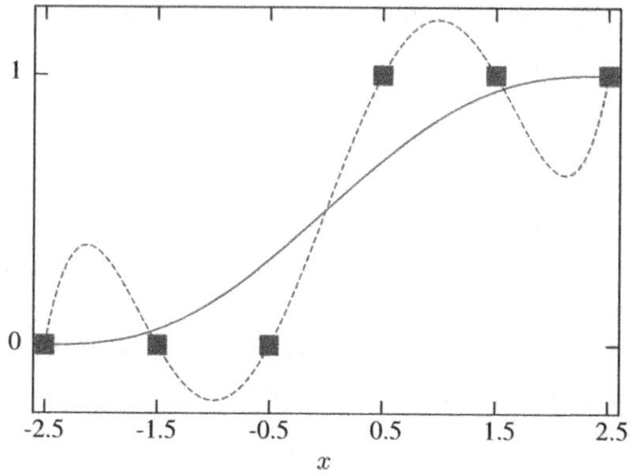

Figure 15.1 Interpolation of discontinuous data (squares) by a fifth-order B-spline (solid line) and a fifth-order Lagrange polynomial (dashed line). The oscillatory behaviour of the Lagrange interpolation is absent in the B-spline interpolation

15.1.1.5 Variation Diminishing Property

B-splines are variation diminishing in the neighbourhood of discontinuous data (Farin 1993). Most importantly, the Gibbs effect observed with Lagrange elements is not present. This property is illustrated in Figure 15.1, and generally results in more stable discretisations, particularly in the presence of sharp gradients.

15.1.1.6 Refinement

B-splines can be refined in various ways. First, the number of basis functions can be increased by inserting additional knots in the knot vector. Adding a single knot increases the number of basis functions by one. As this effectively subdivides an element, at least in the case that a unique knot is inserted, the process of knot insertion is closely related to h-refinement in finite elements. A second refinement strategy is provided by first increasing the multiplicity of all existing knots and subsequently increasing the polynomial order by one. In this way, a basis function is added for each element. This refinement strategy, referred to as order elevation, is closely related to p-refinement.

Knot insertion and order elevation are hierarchical refinement schemes. That is, each original basis function can be represented by a linear combination of refined basis functions. In addition, isogeometric analysis provides a third 'refinement' strategy, referred to as k-refinement, which does not create a nested sequence of spaces. In this strategy, an increase of the spline order is followed by the insertion of knots. This scheme adds fewer basis functions than order elevation. In contrast to p-refinement the continuity, or smoothness, of the basis functions is affected by this strategy. In fact, the continuity (using pure k-refinement) is increased by an order from C^{p-1} to C^p, with p the order of the original spline.

Efficient and robust algorithms exist for performing the various refinement operations. These algorithms compute the control points corresponding to the refined basis. In the case of knot insertion and order elevation, the newly computed control point positions yield exactly the same parametrisation of the geometry as the unrefined B-spline. In the case of k-refinement, this is generally not possible, and some approximation needs to be made.

15.1.2 Univariate NURBS

A drawback of B-splines is their inability to exactly represent a number of objects that are of engineering interest, for instance, conic sections (Box 15.2). For this reason, NURBS, which are a rational generalisation of B-splines, have superseded B-splines in computer aided geometric design, and have become the industry standard. NURBS parametrise geometric objects according to Equation (15.1), with basis functions:

$$r_k(\xi) = \frac{h_k(\xi)W_k}{w(\xi)} \tag{15.9}$$

where $w(\xi) = \sum_{k=1}^{N} h_k(\xi)W_k$ is the weighting function. Defining a NURBS requires the control net $\{\mathbf{p}_k\}_{k=1}^{N}$ to be supplemented with a set of scalar control point weights, $\{W_k\}_{k=1}^{N}$. Singularities in the rational basis functions are avoided by requiring all control point weight to be positive, which we will adhere to in the remainder of the chapter. In the special case that $W_k = c$, where c is an arbitrary positive real number, the NURBS basis reduces to the B-spline basis. NURBS share the properties of B-splines discussed in the previous section. Note that for the sake of notational convenience we will omit the order of the basis function. The basis function $h_{k,p}(\xi)$ will therefore simply be denoted as $h_k(\xi)$, and the order will be clear from the context.

15.1.3 Multivariate B-splines and NURBS Patches

Multivariate B-splines are created by means of a tensor product structure. Surfaces and volumes constructed in this way using the parametric map, Equation (15.1), are referred to as bivariate and trivariate patches, respectively. The required bivariate basis functions defined over the parameter domain $\widehat{V} \subset \mathbb{R}^2$ with parametric coordinate $\boldsymbol{\xi} = (\xi, \eta)$ are given by:

$$h_a(\boldsymbol{\xi}) = h_k(\xi)h_l(\eta) \tag{15.10}$$

with $a = (l-1)N_2 + k$ and univariate B-spline basis functions $h_k(\xi)$ and $h_l(\eta)$ defined over the knot vectors Ξ_ξ and Ξ_η, respectively. Note that we distinguish the bivariate basis functions, $\{h_a(\boldsymbol{\xi})\}_{a=1}^{N}$, from the univariate functions $\{h_k(\xi)\}_{k=1}^{N_1}$ and $\{h_l(\eta)\}_{l=1}^{N_2}$, by means of its argument, which is a scalar in the latter case, and a vector in the former case. By extension, the trivariate basis functions defined over the parameter domain $\widehat{V} \subset \mathbb{R}^3$ with parametric coordinate $\boldsymbol{\xi} = (\xi, \eta, \zeta)$ are

$$h_a(\boldsymbol{\xi}) = h_k(\xi)h_l(\eta)h_m(\zeta) \tag{15.11}$$

with $a = (m-1)N_1N_2 + (l-1)N_2 + k$, and univariate B-spline basis function $h_m(\zeta)$ defined over the knot vector Ξ_ζ.

Supplemented with a control net, $\{\mathbf{p}_k\}_{k=1}^{N}$, the mapping of Equation (15.1) can be used to parametrise a wide range of surfaces and volumes by bivariate and trivariate splines,

Box 15.2 Quarter circle and quarter hemisphere

A quarter of a circle with radius R can be represented exactly by defining a second-order NURBS over the knot vector $\Xi = \{0, 0, 0, 1, 1, 1\}$ with control points $\mathbf{P} = \{(R, 0), (R, R), (0, R)\}$ and corresponding weights $\mathbf{w} = \{1, \frac{1}{2}\sqrt{2}, 1\}$. The exact NURBS representation is shown in the left-hand figure below by the solid curve. The approximate B-spline representation is shown for comparison, and is represented by the dashed curve.

The quarter hemisphere shown in the right-hand figure below is exactly parametrised by a bivariate second-order NURBS patch with knot vectors $\Xi_\xi = \{0, 0, 0, 1, 1, 1\}$ and $\Xi_\eta = \{0, 0, 0, 1, 1, 1\}$, control net $\{(R, 0, 0), (R, R, 0), (0, R, 0), (R, 0, R), (R, R, R), (0, R, R), (0, 0, R), (0, 0, R), (0, 0, R)\}$, and control point weights $\{1, \frac{1}{2}\sqrt{2}, 1, \frac{1}{2}\sqrt{2}, \frac{1}{2}, \frac{1}{2}\sqrt{2}, 1, \frac{1}{2}\sqrt{2}, 1\}$. Note that at $\mathbf{x} = (0, 0, R)$ three control points coincide.

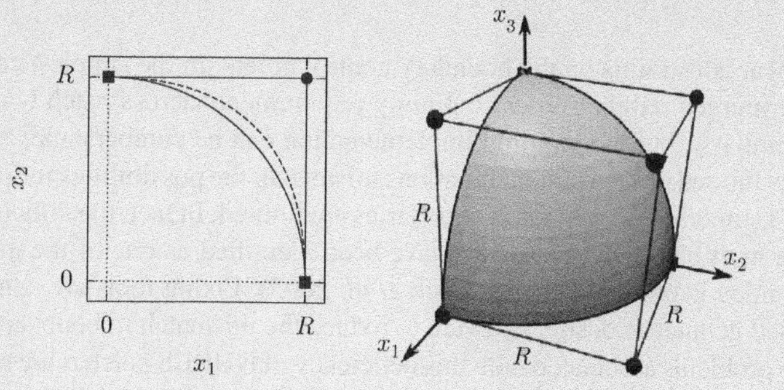

respectively. Together with control point weights, the NURBS functions can then be constructed as:

$$r_k(\boldsymbol{\xi}) = \frac{h_k(\boldsymbol{\xi}) W_k}{w(\boldsymbol{\xi})} \tag{15.12}$$

which, except for its vector argument, equals the univariate definition in Equation (15.9).

By virtue of the tensor product structure of NURBS patches, the parameter domain is a square, or a cube in three dimensions. Although the physical domain can attain significant geometric complexity by the parametric map of Equation (15.1), the objects that can be described by a single patch need to be topologically equivalent to a square or a cube. Geometric objects topologically different from squares or cubes, including many engineering designs, can be represented by an assembly of multiple NURBS patches. Morever, the use of multiple NURBS patches has other advantages, the most important the ability to perform local refinements. Local refinement is not possible in a single patch due to the tensor product structure, but can be achieved by combining multiple patches, as illustrated in Figure 15.2.

There is, however, a disadvantage to combining multiple NURBS patches. If two NURBS patches – and, by induction, multiple patches – are conforming, so that the parametrisation of the connecting boundaries matches, a C^0-continuous connection can easily be established

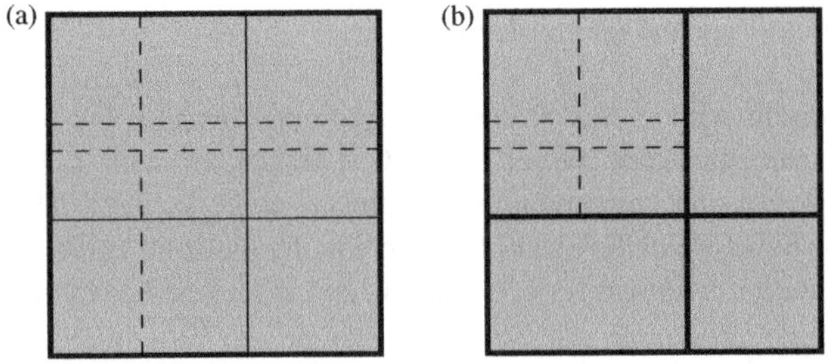

Figure 15.2 Single-patch (a) *vs* multi-patch (b) refinement. The thick lines indicate patch boundaries. In the single-patch case, the refinements applied to the upper-left quadrant propagate into the adjacent quadrants, whereas in the multi-patch case the refinement remains local

by putting linear constraints on the boundary control points. In the simplest case, two control points can be merged. Higher-order continuity requirements across patch boundaries can be established similarly, but the required implementation can be cumbersome, and the range of possibilities is limited. When parametrisations mismatch, the possibilities for obtaining the required level of smoothness over patch boundaries are limited. In fact, the difficulties associated with merging multiple NURBS patches have been identified as one of the major challenges in computer aided geometric design (Kasik *et al.* 2005). Fixing tools are commonly used in computer aided geometric design software to reduce the mismatch in boundary compatibility. However, the problems associated with this deficiency of NURBS patches are more fundamental for analysis, and cannot be fixed heuristically. For example, a gap between two NURBS patches can be reduced by refinement of the patches, such that it is no longer visible to the naked eye. Such a fix can be sufficient in design, but is not good enough for analysis as it fails to resolve the fundamental compatibility problem, and consequently, such a NURBS mesh is unsuitable for analysis purposes.

15.1.4 T-splines

The difficulties associated with the tensor product structure of NURBS patches have led to the development of T-splines, a spline technology that rigorously resolves these problems (Sederberg *et al.* 2003). Abandoning the global tensor product structure of NURBS, T-splines can represent geometric objects of arbitrary topological complexity without the need for multiple patches. Furthermore, T-splines can be refined locally. These advantages make T-splines ideal for isogeometric finite element analysis. It is important to note that T-splines are a generalisation of NURBS, or, put differently, NURBS are a special form of T-splines.

We introduce the fundamentals of T-splines based on Scott *et al.* (2011a). This is because T-splines as introduced in Scott *et al.* (2011a) more naturally connect to standard finite element technology. As T-splines are rooted in computer aided geometric design, they have so far only been used as a technology to represent surfaces, i.e. $d_p = 2$ and $d_s = 3$. Although there are no roadblocks to extending the technology to the volumetric case, $d_p = 3$, we restrict the discussion to bivariate T-splines.

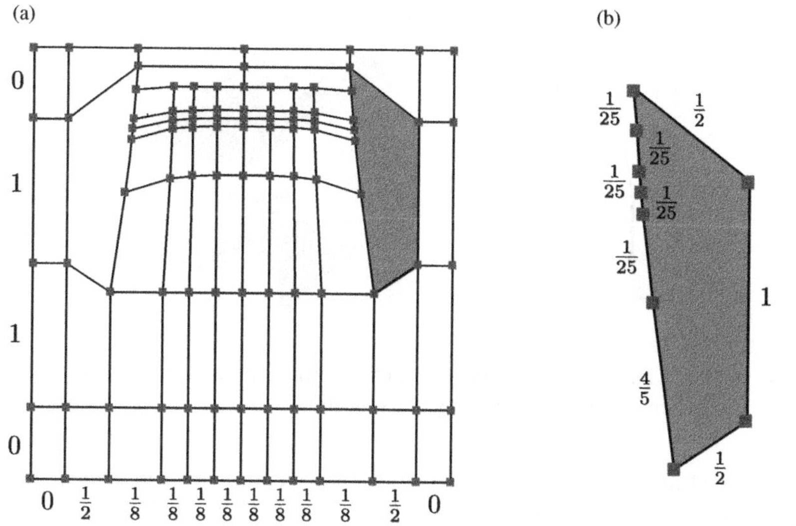

Figure 15.3 (a) A bivariate T-mesh with multiple T-junctions. Knot intervals are assigned to all edges in the T-mesh. (b) T-mesh face zoom. The knot intervals sum up to the same value on opposing sides of the the T-mesh faces

A T-spline is constructed from a T-mesh or T-spline control mesh, a mesh of quadrilateral faces with four corner vertices and an arbitrary number of vertices at each side of the quadrilateral (Figure 15.3). The T-mesh contains the topology information of the T-spline. As for NURBS patches, a control point coordinate $\mathbf{p}_k$ and control point weight W_k are assigned to each vertex in the T-spline control mesh.

The name T-splines is derived from the T-junctions that appear at the vertices on the sides of the quadrilaterals. We restrict the discussion here to T-meshes with valence four corner vertices, i.e. four quadrilaterals come together at every corner vertex in the interior of the mesh, and T-junction vertices.

From the viewpoint of isogeometric analysis it is sufficient to consider a subset of T-splines, referred to as analysis-suitable T-splines (Scott *et al.* 2011b). Analysis-suitable T-splines are constructed over T-meshes with some minor restrictions on their topology, referred to as analysis-suitable T-meshes. Analysis-suitable T-splines guarantee the most important properties of B-splines discussed in Section 15.1.1 and, most importantly, guarantee linear independence of the T-spline basis. The minor restrictions imposed on the topology of analysis-suitable T-splines are in practice irrelevant.

As for any of the spline technologies discussed in this chapter, T-splines are based on the parametric map of Equation (15.1). With the introduction of the control points, $\{\mathbf{p}_k\}_{k=1}^{N}$, only the definition of the T-spline basis, $\{h_k(\boldsymbol{\xi})\}_{k=1}^{N}$, remains. To this end we supplement the T-mesh with knot intervals, non-negative real numbers assigned to each segment between two vertices (Figure 15.3). In order to obtain the basis functions, knot intervals need to sum up to the same value on opposing sides of the quadrilaterals.

Using the T-mesh supplemented with knot intervals, so-called local knot interval vectors $\{\boldsymbol{\Delta\Xi}_k\}_{k=1}^{N}$ can be assigned to all vertices. For T-spline bases of an odd polynomial order, these

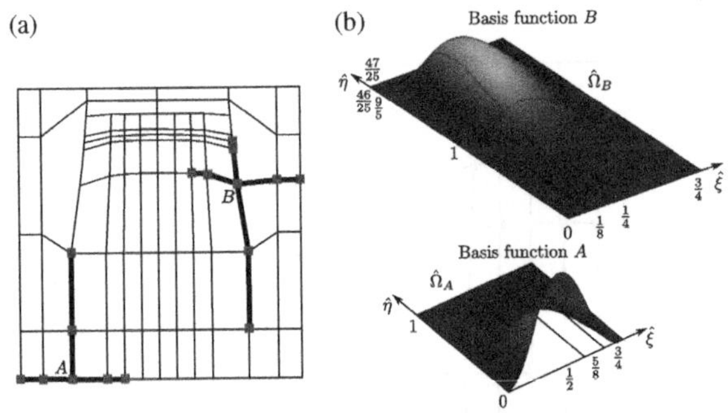

Figure 15.4 (a) Local knot interval construction for two vertices for the T-mesh in Figure 15.3. (b) The local basis function domains and corresponding basis functions

local knot interval vectors are constructed by traversing the mesh in each of the four directions of the mesh, until $\frac{1}{2}(p+1)$ vertices or sides are intersected. Figure 15.4 shows this process for the T-mesh introduced in Figure 15.3. The local knot interval vectors corresponding to the vertices A and B are found as

$$\begin{aligned}\boldsymbol{\Delta\Xi}_A &= \left\{\left(0,\frac{1}{2},\frac{1}{8},\frac{1}{8}\right),\left(0,0,0,1\right)\right\}\\ \boldsymbol{\Delta\Xi}_B &= \left\{\left(\frac{1}{8},\frac{1}{8},\frac{1}{2},0\right),\left(1,\frac{4}{5},\frac{1}{25},\frac{1}{25}\right)\right\}\end{aligned} \tag{15.13}$$

We restrict our discussion to T-splines of an odd polynomial order, where the basis functions can be regarded as vertex-centred, since $\frac{1}{2}(p+1)$ knot intervals are found in each direction starting from the central vertex. By contrast, basis functions for even-order T-splines can be regarded as face-centred. A study of the use of even-order T-splines in isogeometric analysis can be found in Bazilevs *et al.* (2010).

Setting the origin to $(0,0)$, the knot interval vectors can be transformed into local knot vectors, $\{\boldsymbol{\Xi}_k\}_{k=1}^N$, which define the local basis function domains $\{\widehat{V}_k\}_{k=1}^N$. For the two cases in (15.13) we then obtain the knot vectors

$$\begin{aligned}\boldsymbol{\Xi}_A &= \left\{\left(0,0,\frac{1}{2},\frac{5}{8},\frac{3}{4}\right),\left(0,0,0,0,1\right)\right\}\\ \boldsymbol{\Xi}_B &= \left\{\left(0,\frac{1}{8},\frac{1}{4},\frac{3}{4},\frac{3}{4}\right),\left(0,1,\frac{9}{5},\frac{46}{25},\frac{47}{25}\right)\right\}\end{aligned} \tag{15.14}$$

Using these knot vectors of length $p+2$, we can use the Cox–de Boor recursion formula, Equation (15.5), to obtain a single basis function associated with the central vertex with support over the local basis function domains, $\widehat{V}_A = [0, \frac{3}{4}] \otimes [0, 1]$ and $\widehat{V}_B = [0, \frac{3}{4}] \otimes [0, \frac{47}{25}]$. The basis functions constructed for the vertices A and B are shown in Figure 15.4. Since the T-spline basis functions are constructed in the same way as B-spline basis functions, T-splines inherit the properties of B-splines as discussed in Section 15.1.1.

Using the control point weights assigned to all vertices, $\{W_k\}_{k=1}^N$, the T-spline basis functions can be made rational using Equation (15.9) in exactly the same way as was done for B-spline

Figure 15.5 Third-order T-spline representation of a stiffener based on the T-mesh shown in Figure 15.3

basis functions. In combination with the parametric map of Equation (15.1) the control points, $\{\mathbf{p}_k\}_{k=1}^{N}$, then define the T-spline. The T-spline resulting from the T-mesh in Figure 15.3 is shown in Figure 15.5.

15.2 Isogeometric Finite Elements

So far, spline basis functions have been considered from a global perspective. That is, to construct a spline basis function we relied on the definition of global knot vectors or T-spline meshes. This is in contrast to the way in which we normally construct finite element basis functions, which we induce from a canonical set of shape functions defined over a parent element. The availability of an element (data) structure for splines is of pivotal importance for the success of isogeometric analysis, as it provides a unified approach to spline technologies that is compatible with standard finite element technology.

15.2.1 Bézier Element Representation

To develop an element structure for splines, we need a precise definition of an 'element'. Henceforth, we shall refer to an element as a region in the physical space with a volume, V_e, which is mapped from a parametric element domain, $\widehat{V}_e$, that is bounded by lines of reduced continuity. This means that the basis functions on an element are $\mathcal{C}^\infty$-continuous functions. The union of all parametric elements constitutes the parameter domain, and, as a consequence, the union of all elements yields the physical domain.

This definition encapsulates the elements used in traditional $\mathcal{C}^0$-continuous finite elements, where $\mathcal{C}^\infty$-continuous polynomials are used as the basis functions which are restricted to an element. These $\mathcal{C}^\infty$-continuous regions are bounded by mesh lines with $\mathcal{C}^0$-continuity. By virtue of the construction of univariate B-spline basis functions discussed in Section 15.1.1, lines of reduced continuity can be inserted at the parametric coordinates that correspond to the knot values. As a result, a univariate B-spline element is defined as a curve segment of positive length in physical space mapped onto from a knot interval, which is also of positive length. As in standard finite elements, the basis functions restricted to the elements are polynomials. NURBS elements are essentially not different from B-spline elements, and the tensor product structure of multivariate splines provides a natural definition of elements in NURBS patches.

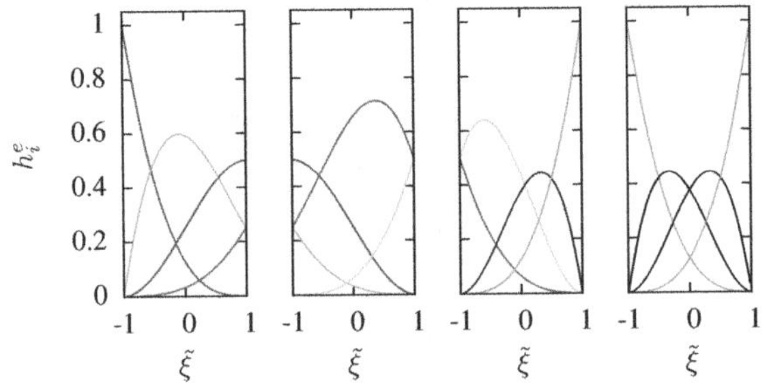

Figure 15.6 Restrictions of the B-spline basis functions in Box 15.1 to the elements. The restricted basis functions are plotted over the parent element domain $\widehat{V}_e = [-1, 1]$. In contrast to traditional finite elements the restricted basis functions generally differ per element

As a consequence of their construction, however, the definition of T-spline elements contains some subtleties (Scott *et al.* 2011a).

B-spline basis functions, either univariate or multivariate, are piecewise polynomials defined over the parameter domain. In contrast to standard finite elements, the restrictions of these functions to the elements are not the same for all elements. This lack of a canonical set of parent element basis functions is illustrated in Figure 15.6 for the B-spline basis shown in Box 15.1. It is, however, possible to restrict the basis functions with a support over an element to that element, and to express them as a linear combination of a canonical set of basis functions

$$\mathbf{h}_e = \mathcal{C}_e \mathbf{B} \tag{15.15}$$

For reasons that will be discussed in the next section, we will use Bernstein polynomials of order p, collected in $\mathbf{B}$, as the canonical set of element basis functions. Any other set that spans the same space, such as Lagrange polynomials of order p, is, in principle, equally suited. We refer to the matrix $\mathcal{C}_e$ as the element extraction operator. The element extraction operator lumps all global information onto the element. A schematic representation of the action of the element extraction operators is shown in Figure 15.7 for the univariate B-spline basis functions introduced in Box 15.1. In standard finite element technology, all extraction operators equal the identity matrix, and normally do not appear in the formulation. We note that the element extraction operators can be interpreted as matrices that establish linear constraints between

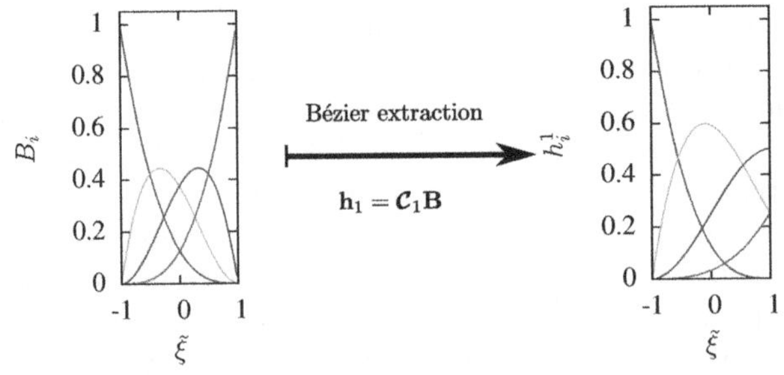

Figure 15.7 Schematic representation of the Bézier extraction operator

the shape functions in adjacent elements. By supplying the correct constraints, interelement smoothness can be obtained.

When the element extraction operators are available, the basis functions can be constructed from the canonical set of elements. This makes isogeometric analysis a real element technology, suitable to integrate in existing finite element codes. All information required for carrying out an analysis is assembled in a Bézier mesh, which contains:

- The (global) control net, $\{\mathbf{p}_k\}_{k=1}^{N}$, supplemented with control point weights, $\{W_k\}_{k=1}^{N}$, in the case of NURBS.
- A set of Bézier elements, each supplemented with a list containing the global indices of the basis functions with support over the element (commonly referred to as the connectivity array), and an element extraction operator, $\boldsymbol{C}_e$.

The Bézier mesh can be regarded as an extension of the mesh used in standard finite element analysis, in which the control point weights and Bézier extraction operators are generally omitted.

In the next section we will introduce Bézier extraction as a tool for obtaining the extraction operators. It is emphasised that the Bézier representation decouples the geometry problem from the analysis problem. In fact, Bézier meshes can already be obtained from computer aided geometric design software. An additional advantage of the Bézier representation is that it provides a unified interface to a variety of spline technologies, which prevents ad-hoc implementations on the analysis side.

15.2.2 Bézier Extraction

We refer to the process of obtaining the Bézier mesh from a spline object as Bézier extraction. The extraction process is best illustrated by a univariate B-spline. We again consider the third-order B-spline basis of Box 15.1, which was constructed using the knot vector $\Xi = \{0, 0, 0, 0, 1, 2, 2, 3, 3, 3, 4, 4, 4, 4\}$. In combination with the control points $\{\mathbf{p}_k\}_{k=1}^{N}$ this basis defines the B-spline curve shown in Box 15.1.

The corner stone of Bézier extraction is the process of knot insertion. With the insertion of a knot value $\bar{\xi}$ in the knot vector Ξ, the control points change according to

$$\bar{\mathbf{P}} = [\boldsymbol{C}^1]^{\mathrm{T}}\mathbf{P} \tag{15.16}$$

in order to preserve the parametrisation; see Piegl and Tiller (1997) for general knot insertion algorithms to determine the operator $\boldsymbol{C}^1$, and Borden *et al.* (2011) for knot algorithms that are designed for knot multiplication, where an existing knot value is repeated. We now use knot insertion to express the control point positions, $\{\bar{\mathbf{p}}_k\}_{k=1}^{\bar{N}}$, with $\bar{N} = n_e p + 1$, for the knot vector $\bar{\Xi} = \{0, 0, 0, 0, 1, 1, 1, 2, 2, 2, 3, 3, 3, 4, 4, 4\}$ with all internal knot values repeated p times as

$$\bar{\mathbf{P}} = [\boldsymbol{C}^{\bar{N}-N}]^{\mathrm{T}}[\boldsymbol{C}^{\bar{N}-N-1}]^{\mathrm{T}} \ldots [\boldsymbol{C}^2]^{\mathrm{T}}[\boldsymbol{C}^1]^{\mathrm{T}}\mathbf{P} = \boldsymbol{C}^{\mathrm{T}}\mathbf{P} \tag{15.17}$$

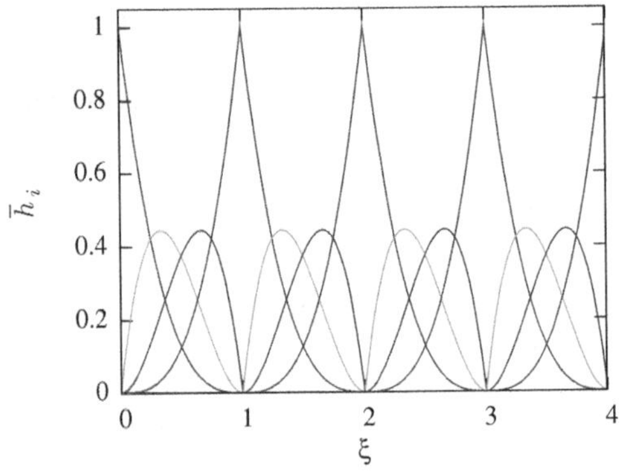

Figure 15.8 B-spline basis after the knot insertion process

The parametrisation of the B-spline remains unchanged upon the insertion of additional knots. In accordance with Equation (15.2) this is expressed by the equality:

$$\mathbf{P}^{\mathrm{T}}\mathbf{h} = \bar{\mathbf{P}}^{\mathrm{T}}\bar{\mathbf{h}} \tag{15.18}$$

with $\{\bar{h}_k\}_{k=1}^{\bar{N}}$ the B-spline basis in accordance with the knot vector $\bar{\boldsymbol{\Xi}}$. Substitution of $\bar{\mathbf{P}}$ as in Equation (15.17) then yields:

$$\mathbf{P}^{\mathrm{T}}\mathbf{h} = \mathbf{P}^{\mathrm{T}}\mathcal{C}\bar{\mathbf{h}} \tag{15.19}$$

Since this equality holds for arbitrary control points, the B-spline basis $\{\bar{h}_k\}_{k=1}^{\bar{N}}$ is related to the original basis $\{h_k\}_{k=1}^{N}$ through

$$\mathbf{h} = \mathcal{C}\bar{\mathbf{h}} \tag{15.20}$$

Hence, every original basis function can be expressed as a linear combination of refined basis functions. Figure 15.8 shows the basis $\{\bar{h}_k\}_{k=1}^{\bar{N}}$ according to the knot vector $\bar{\boldsymbol{\Xi}}$. In contrast to the original B-spline basis, $\{h_k\}_{k=1}^{N}$, this refined basis is comprised of a canonical set of element basis functions, the Bernstein polynomials. We now define the operators $\mathbf{Z}_e$ and $\bar{\mathbf{Z}}_e$ to select the basis function $\mathbf{h}_e$ and $\bar{\mathbf{h}}_e$ with support over $\widehat{V}_e$, to obtain:

$$\mathbf{h}_e = \mathbf{Z}_e\mathbf{h} = \mathbf{Z}_e\mathcal{C}\bar{\mathbf{Z}}_e^{\mathrm{T}}\bar{\mathbf{h}}_e \tag{15.21}$$

Since $\bar{\mathbf{h}}_e$ contains the same set of Bernstein polynomials for every element, we can elaborate the element extraction operator introduced in Equation (15.15) as:

$$\mathcal{C}_e = \mathbf{Z}_e\mathcal{C}\bar{\mathbf{Z}}_e^{\mathrm{T}} \tag{15.22}$$

By virtue of their tensor product structure, multivariate B-spline extraction operators can be inferred directly from the univariate operators in the independent directions. For instance, we consider the bivariate basis functions in Equation (15.10), and let $[\mathcal{C}_e]_1$ and $[\mathcal{C}_e]_2$ be the extraction operators in the ξ- and η-directions, respectively. Note that we consider the basis functions over the parent element domain $\widehat{V}_e = [-1, 1] \otimes [-1, 1]$ (with coordinate $\boldsymbol{\xi} = (\xi, \eta)$).

Considering the local element basis functions

$$h_{e,a}(\tilde{\boldsymbol{\xi}}) = h_{e,k}(\tilde{\xi})h_{e,l}(\tilde{\eta}) \tag{15.23}$$

with $a = (l-1)(p+1)+k$, we obtain

$$h_{e,a}(\tilde{\boldsymbol{\xi}}) = [\boldsymbol{C}_e]_{ab}B_b(\tilde{\boldsymbol{\xi}}) \tag{15.24}$$

with $[\boldsymbol{C}_e]_{ab}$ the bivariate extraction operator:

$$[\boldsymbol{C}_e]_{ab} = [\boldsymbol{C}_e]_{1,km}[\boldsymbol{C}_e]_{2,ln} \tag{15.25}$$

and $b = (n-1)(p+1)+m$. Since the Bézier extraction provides an element-based construction of the global B-spline basis functions, the global NURBS basis functions can be constructed subsequently. That is, the Bézier representation is used to construct the global B-spline basis functions restricted to the elements, $\mathbf{h}_e$, from which the NURBS basis functions restricted to the elements, $\mathbf{r}_e$, are then obtained using Equation (15.9).

The extraction process for T-splines does not differ fundamentally from the process for B-splines, but requires consideration of some details that are specific for T-splines. A particularly interesting aspect is that Bézier elements in T-splines can support more global basis functions than its number of Bernstein polynomials. This property, which is a consequence of the appearance of T-junctions, results in non-square element extraction operators (Scott *et al.* 2011a).

15.3 PyFEM: Shape Functions for Isogeometric Analysis

The Bézier representation of the B-spline shape functions is implemented in the file `BezierShapeFunctions.py` which can be found in the directory `pyfem/util`. The main structure of the file is as follows:

> ⟨*Bézier shape functions* ⟩ ≡
>
> ⟨*Bézier shape function algorithms* 488⟩
> ⟨*Bézier shape function main routine* 488⟩

The fragment ⟨*Bézier shape function algorithms*⟩ contains the implementation of the shape function routines for various integration orders, both for univariate and bivariate B-splines. The fragment ⟨*Bézier shape function main routine*⟩ contains a single function `getElemBezierData`, which calculates the shape functions and their derivatives for a single Bézier element. This function is similar to the function `getElemShapeData` on page 39 apart from the fact that in this case, the element extraction operator $\boldsymbol{C}_e$ is needed as input:

⟨*Bézier shape function main routine* ⟩≡ 487

```
def getElemBezierData( elemCoords , C , order = 0 , \
                       method = 'Gauss' , elemType = 'default' ):

  elemData = elemShapeData()                                        38
  (Calculation of Bézier shape functions in all integration points)
  elemData.sData.append( sData )                                    38

  return elemData
```

The extraction operator is passed as a two-dimensional array `C`. The other arguments of the function have the same purpose as in the function `getElemShapeData`. The shape functions and their derivatives in a single integration point are stored in the data container `sData`. The integration point data are collected in the instance `elemData`. The classes `shapeData` and `elemShapeData` are implemented in the file `shapeFunctions.py`.

The shape functions and their derivatives are calculated in a collection of routines, which are implemented in the fragment ⟨*Bézier shape function utility routines*⟩. The shape functions for a third-order, univariate B-spline are calculated in the function `getBezierLine4`, where the number 4 refers to the number of supported shape functions by a single element:

⟨*Bézier shape function algorithms* ⟩≡ 487

```
def getBezierLine4( xi , C ):

  sData     = shapeData()                                           38
  sData.xi  = xi

  B     = empty( 4 )
  dBdxi = empty( shape = (4,1) )
```

In addition to the parametric coordinate `xi`, the element extraction matrix `C` is also an argument. The Bézier shape functions and derivatives are stored in temporary arrays `B` and `dBdxi` which are initialised as arrays of length 4. The actual values of the shape functions are calculated next:

⟨*Bézier shape function algorithms* ⟩+≡ 488

```
  B[0] = -0.125*(xi-1.)**3
  B[1] =  0.375*(xi-1.)**2*(xi+1.)
  B[2] = -0.375*(xi-1.)*(xi+1.)**2
  B[3] =  0.125*(xi+1.)**3

  dBdxi[0,0] = -0.375*(xi-1.)**2
  dBdxi[1,0] =  0.75 *(xi-1.) *  (xi+1.) + 0.375*(xi-1.)**2
  dBdxi[2,0] = -0.375*(1.+xi)**2 - 0.75*(1.+xi)*(xi-1.)
  dBdxi[3,0] =  0.375*(xi+1.)**2
```

The final step is the calculation of the value of the B-spline shape functions over the element by means of Equation (15.15):

⟨*Bézier shape function algorithms* ⟩+≡ 488

```
sData.h      = dot( C , B )
sData.dhdxi  = dot( C , dBdxi )

return sData
```

The shape functions `h` and their parametric derivatives `dhdxi` are stored in the container `sData` and are returned to the main function. Here, additional data are processed, such as the derivative with respect to the physical coordinates, and the physical integration weight. This is identical to the process described in fragment ⟨*Shape function main routine*⟩ on page 39.

The Bézier shape function utility is demonstrated in the program `beziertest.py` in the directory `examples/ch15`. In this example, the curve shown in Box 15.1 is constructed using a Bézier mesh. The curve in this box is defined by a control net that consists of 10 control points, and is represented by 4 Bézier elements. Each element, e, supports 4 shape functions. The corresponding extraction matrices $\mathcal{C}_e$ are:

$$\mathcal{C}_0 = \begin{bmatrix} 1 & 0 & 0 & 0 \\ 0 & 1 & \frac{1}{2} & \frac{1}{4} \\ 0 & 0 & \frac{1}{2} & \frac{1}{2} \\ 0 & 0 & 0 & \frac{1}{4} \end{bmatrix} \quad \mathcal{C}_1 = \begin{bmatrix} \frac{1}{4} & 0 & 0 & 0 \\ \frac{1}{2} & \frac{1}{2} & 0 & 0 \\ \frac{1}{4} & \frac{1}{2} & 1 & \frac{1}{2} \\ 0 & 0 & 0 & \frac{1}{2} \end{bmatrix} \quad \mathcal{C}_2 = \begin{bmatrix} \frac{1}{2} & 0 & 0 & 0 \\ \frac{1}{2} & 1 & 0 & 0 \\ 0 & 0 & 1 & 0 \\ 0 & 0 & 0 & 1 \end{bmatrix} \quad \mathcal{C}_3 = \begin{bmatrix} 1 & 0 & 0 & 0 \\ 0 & 1 & 0 & 0 \\ 0 & 0 & 1 & 0 \\ 0 & 0 & 0 & 1 \end{bmatrix}$$

The structure of the file `beziertest.py` is given by:

⟨*Bézier curve example* ⟩≡

⟨*Bézier extraction example initialisation* 489⟩
⟨*Bézier extraction example main calculation* 490⟩
⟨*Bézier extraction example print curve*⟩

The geometry of the curve is specified in the fragment⟨*Bézier extraction example initialisation*⟩.

⟨*Bézier curve example initialisation* ⟩≡ 489

⟨*Initialisation of control net coordinates*⟩
⟨*Initialisation of element connectivity*⟩
⟨*Initialisation of extraction matrices*⟩

The control net coordinates, the element connectivity and the extraction operators are stored as arrays `coords`, `elems` and `C`, respectively. The positions of the points $\mathbf{x}$ as a function of the parametric coordinate ξ are obtained by a nested loop over the elements and the integration points:

⟨*Bézier curve example main calculation* ⟩≡ 489

```
for elemNodes,Celem in zip(elems,C):

  sdata = getElemBezierData ( coords[elemNodes,:] , Celem , \     488
                              order = 100 , elemType = "line4" )

  for idata in sdata:
    x = dot( idata.h , coords[ elemNodes , : ] )
    output.append( x )
    length += idata.weight
```

For each element, the element coordinates and the element extraction operator `Celem` are used to determine the element shape function data. In this case, we indicate that each element contains 100 integration points. Since the dimensions of the physical and parameter domain are different, we have to explicitly indicate that a `'Line4'` integration scheme is required. The position of a point `x` on the curve is determined by multiplying the shape function and the element coordinates according to Equation (15.1). This position is appended to the list `output`. The weight `idata.weight` is a measure for the length of the section of the curve corresponding to an integration point. As a result, the sum of all integration weights is equal to the total length of the curve. The length and a graph of the curve are printed to the screen in the fragment ⟨*Bézier extraction example print curve*⟩.

15.4 Isogeometric Analysis in Non-linear Solid Mechanics

The fundamental idea of isogeometric analysis is to use spline basis functions for both the geometry parametrisation and for the discretisation of the approximate solutions. Hence, the elements provided by a Bézier mesh are isoparametric. Indeed, isogeometric analysis merely provides a basis for Galerkin discretisations, and, as such, does not differ from standard finite elements regarding the way they are used for the discretisation of weak forms. The availability of a proper element definition in conjunction with the Bézier element data structure provides an interface to isogeometric analysis that is compatible with standard finite element technology.

The advantages of isogeometric analysis compared with standard finite element technology are twofold. First, the possibility to directly use computer aided geometric design models for the analysis streamlines the design-through-analysis concept. This is particularly useful for complex geometries. In Section 15.4.1 we will illustrate this advantage for the design and the analysis of shell structures. The other advantage of isogeometric analysis results from the fundamentally different nature of the spline basis functions compared with Lagrange basis functions. The control over inter-element continuity allows for the direct discretisation of higher-order differential equations, such as the Kirchhoff–Love shells considered in Section 15.4.1 or the higher-order gradient damage models in Section 15.4.2, see also Chapters 6 and 14 for finite element and meshless solutions to the latter problem. At the same time, splines allow for the flexible insertion of discontinuities, making it a viable discretisation tool for fracture models, including cohesive zone approaches, which will be discussed in Section 15.4.3.

With the examples of isogeometric analysis for non-linear solid mechanics problems discussed in the remainder of this chapter we give a flavour of its possibilities. By no means is this overview meant to be complete. Solid mechanics problems for which isogeometric analysis has been shown to have advantages also include contact and friction problems (Temizer *et al.* 2011) and structural optimization problems (Wall *et al.* 2008).

15.4.1 Design-through-analysis of Shell Structures

Shell formulations generally rely on the presence of a parametric mid-surface, see Chapter 10. The geometry of the mid-surface is governed by a mapping from a parameter domain onto the physical domain. This mapping enters the formulation with the definition of the basis vectors, which are the gradients of the physical coordinate with respect to the parametric coordinate, and governs all subsequently derived quantities, including strains and curvatures. In standard finite element technology this surface is defined through the elements, and hence, an element-wise parametric description of the surface is obtained. In such a case the parent element domain serves as an element-wise parameter domain.

In computer aided geometric design the mid-surface parametrisation is provided by Equation (15.1), with a bivariate parameter domain ($d_p = 2$) mapped onto a subset of the three-dimensional space ($d_s = 3$). In contrast to standard finite element analyses, isogeometric analysis directly uses the spline basis functions to discretise the fields that describe the deformation. Considering the process of converting a geometric design into an analysis-suitable object, isogeometric analysis offers two major advantages. First, no meshing or geometry clean-up procedures are required to construct the analysis model. A significant reduction of the time required to perform a design-through-analysis cycle can thus be obtained, particularly when complex geometries are considered (Cottrell *et al.* 2009). Evidently, isogeometric analysis requires the geometric object to be analysis-suitable. Consequently, problems in the parametric design are now solved where they should be solved, namely in the geometric object. As a consequence, improper geometry definitions that result in meshing problems and the requirement of geometry clean-up tools are no longer solved in an ad-hoc fashion, but are eliminated rigorously. The second major advantage of an isogeometric analysis of shells is that splines are very efficient in parametrising curved surfaces. In standard finite element analyses of shells, the number of elements can be dictated by the geometry, which can render the approach inefficient.

The above considerations highlight the potential advantages of using isogeometric finite elements for shell formulations of any kind. The benefits of isogeometric analysis for Reissner–Mindlin shells have been stipulated by Benson *et al.* (2010). Since the Reissner–Mindlin formulations only require C^0-continuity, they have been used extensively in finite element analysis. The advantages of isogeometric analysis are then as discussed above, although the use of splines has been found to improve robustness in computations.

This is different for Kirchhoff–Love shells. From Chapter 10 we recall that, in general, the strain in a plate/shell is given, cf. Equation (10.1), by:

$$\boldsymbol{\epsilon} = \boldsymbol{\epsilon}_\ell + z_\ell \boldsymbol{\chi}$$

where the array $\boldsymbol{\epsilon}_\ell$ contains the in-plane strains at the reference plane, and is given by Equation (10.2), and z_ℓ is the vertical coordinate with respect to the shell reference plane (Figure 9.1). The fundamental assumption in the Kirchhoff–Love formulation is that the director is taken to be equal to the unit normal vector to the shell reference plane:

$$\mathbf{d} = \frac{\frac{\partial \mathbf{x}}{\partial \xi} \times \frac{\partial \mathbf{x}}{\partial \eta}}{\left\| \frac{\partial \mathbf{x}}{\partial \xi} \times \frac{\partial \mathbf{x}}{\partial \eta} \right\|} \tag{15.26}$$

with ξ, η the local coordinates of the mid-surface (Figure 10.3). As a consequence the rotations θ_x and θ_y, which are independent variables in the Reissner–Mindlin shell formulation, are constrained, cf. Equation (9.2), by:

$$\theta_x = -\frac{\partial w}{\partial x} \quad \text{and} \quad \theta_y = -\frac{\partial w}{\partial y} \tag{15.27}$$

with w the out-of-plane displacement of the shell. Substitution of these constraints into Equation (10.3) results in the curvatures for a Kirchhoff–Love shell theory:

$$\boldsymbol{\chi} = \begin{pmatrix} -\frac{\partial^2 w}{\partial x^2} \\ -\frac{\partial^2 w}{\partial y^2} \\ -2\frac{\partial^2 w}{\partial x \partial y} \end{pmatrix} \tag{15.28}$$

The second-order derivatives which now evolve require a $\mathcal{C}^1$-continuity, similar to the Euler–Bernoulli beams (Chapter 9). For beam elements, this requirement can be met relatively easily, e.g. using Hermite interpolations, Equation (9.14b), but this is less so for shell elements of arbitrary shapes.

Since $\mathcal{C}^1$-continuity is naturally obtained using splines of order two or higher, the Kirchhoff–Love shell formulation can be discretised straightforwardly in isogeometric analysis. We illustrate the potential of isogeometric finite element analysis for the Kirchhoff–Love shell formulation with two examples. First, we shall show results for the Scordelis-Lo roof, which has evolved as a standard benchmark problem for linear shell analysis (MacNeal and Harder 1985). Next, the potential for geometrically non-linear analysis will be demonstrated for a channel-section beam problem (Chróscielewski *et al.* 1992).

The Scordelis-Lo roof has been used in Chapter 4 to assess the performance of line searches, and was modelled using eight-noded shell elements. Now, the exact geometry of the roof is described by a single-element third-order NURBS (Figure 15.9), with control points and weights given in Table 15.1. Various refinements of this single-element NURBS have been considered. The used meshes are shown in Figure 15.10. The mesh shown in Figure 15.10(a) has been obtained by a single subdivision in both directions of the original object. The mesh shown in Figure 15.10(b) has been obtained by applying further subdivisions. Evidently, the mesh shown in Figure 15.10(b) is a T-spline, since it contains T-junctions. The T-spline discretisation consists of 224 elements and 257 basis functions, and contains $3 \times 257 = 771$ degrees of freedom.

The deflection of the roof due to the gravity load is determined using the Kirchhoff–Love formulation. In this linear computation, both the external force vector and stiffness matrix have been computed in the undeformed state. A fourth-order Gauss integration appeared to yield sufficient accuracy, see Hughes *et al.* (2010) for a detailed study of the numerical integration of rational basis functions. For the T-spline discretisation in Figure 15.10, the downward

Table 15.1 Control point positions and weights for the Scordelis-Lo roof, with $i = 1, \ldots, 4$. A linear parametrisation in the x_3-direction is obtained by selecting the control points according to $\Xi^g = \{0, \frac{1}{3}, \frac{2}{3}, 1\}$ which corresponds to the knot vector $\Xi_\zeta = \{0, 0, 0, 0, 1, 1, 1, 1\}$

i	$p_{i,1}/R$	$p_{i,2}/R$	$p_{i,3}/L$	W_i
$4(i-1)+1$	$-\sin\theta$	$\cos\theta$	Ξ_1^g	1
$4(i-1)+2$	$-\dfrac{\sin\theta}{1+2\cos\theta}$	$\dfrac{2+\cos\theta}{1+2\cos\theta}$	Ξ_2^g	$\dfrac{1+2\cos\theta}{3}$
$4(i-1)+3$	$\dfrac{\sin\theta}{1+2\cos\theta}$	$\dfrac{2+\cos\theta}{1+2\cos\theta}$	Ξ_3^g	$\dfrac{1+2\cos\theta}{3}$
$4(i-1)+4$	$\sin\theta$	$\cos\theta$	Ξ_4^g	1

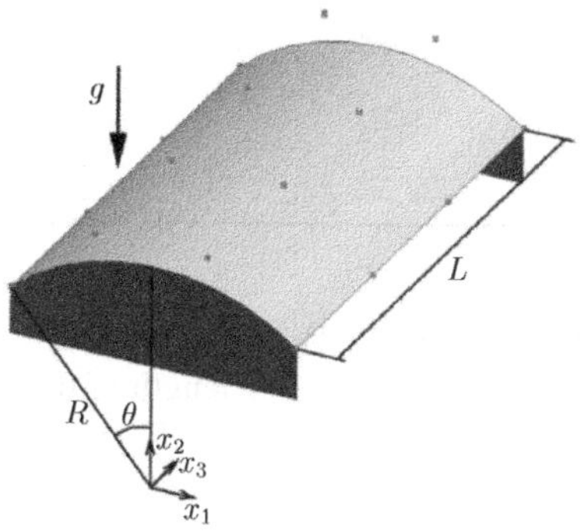

Figure 15.9 A single-element NURBS representation of the Scordelis–Lo roof. The control points are indicated by squares. At the supports displacements are constrained in x_1- and x_2-direction only

deflection of the mid-point on either of the free edges is computed as 0.301, which is in excellent agreement with results reported in the literature (MacNeal and Harder 1985). Mesh convergence studies have been presented by Kiendl *et al.* (2009).

The ability of the isogeometric Kirchhoff–Love shell formulation to accurately capture geometrically non-linear behaviour is now shown for the channel-section beam problem of

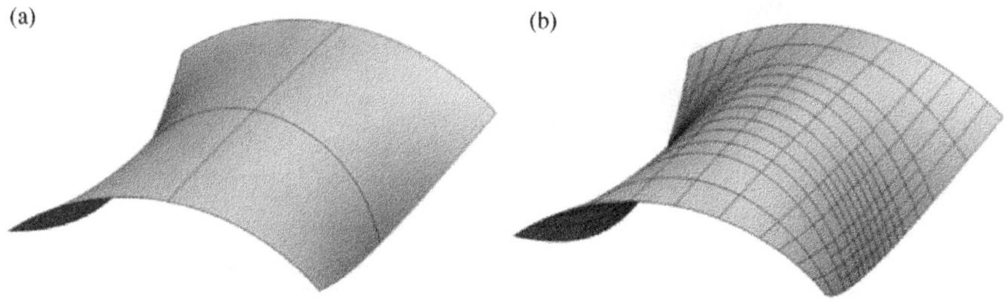

Figure 15.10 (a) Third-order NURBS discretisation of the Scordelis–Lo roof with 4 Bézier elements and 25 basis functions. (b) Third-order T-spline with 224 elements and 257 basis functions

Table 15.2 Control point positions for the channel section beam, with $i = 1, \ldots, 5$. On account of the beam's symmetry with respect to the x_1–x_3-plane, only the control points with non-negative x_2-coordinate are shown in the table. A linear parametrisation in the x_1-direction is established by selecting the control points according to: $\Xi^g = \{0, \frac{1}{6}, \frac{1}{2}, \frac{5}{6}, 1\}$ corresponding to the knot vector $\Xi = \{0, 0, 0, 0, , \frac{1}{2}, 1, 1, 1, 1\}$

i	$p_{i,1}/L$	$p_{i,2}/(2b)$	$p_{i,3}/a$
$11(i-1)+1$	Ξ_1^g	1	1
$11(i-1)+2$	Ξ_2^g	1	$\frac{11}{20}$
$11(i-1)+3$	Ξ_3^g	1	$\frac{1}{10}$
$11(i-1)+4$	Ξ_4^g	1	0
$11(i-1)+5$	Ξ_5^g	$\frac{14}{15}$	0
$11(i-1)+6$	Ξ_6^g	0	0

Figure 15.11 (Chróscielewski *et al.* 1992). The height of the web, in dimensionless form, is $b = 6$, and the width of the flanges is $a = 2$. The length of the beam is equal to $L = 36$ and the thickness of the web and the flanges equals $t = 0.05$. The Young's modulus is equal to 10^7 and the Poisson's ratio is $\nu = 0.333$. The beam is loaded by a downward point force applied to the tip of the beam.

The geometry of the beam is described by a bivariate uniform B-spline of order three, with the control points shown in Table 15.2. In the x_1-direction a linear parametrisation is enforced, but note that the control points themselves are not distributed linearly in the x_1-direction. Attention is drawn to the fact that the corners of the channel have been rounded (although not circular). Indeed, the B-spline surface can efficiently represent rounded corners. In order to accurately capture the deformation pattern of the beam for geometrically non-linearities, a refinement to the original geometry is made with 2048 Bézier elements and 2345 basis functions, which results in $3 \times 2345 = 7035$ degrees of freedom.

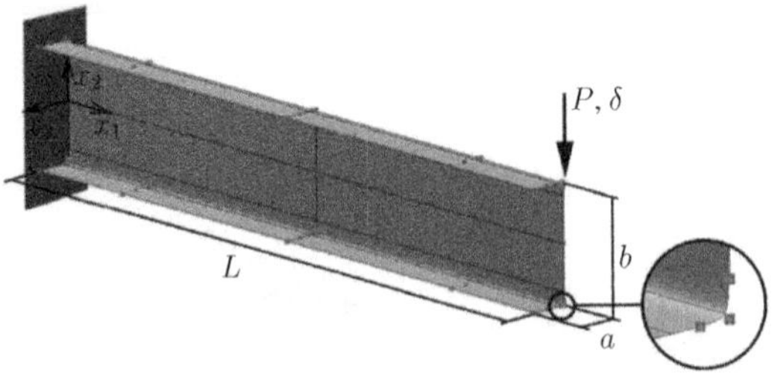

Figure 15.11 Uniform B-spline representation of a channel-section beam with rounded corners. The control points are indicated by squares. At the left support the displacements and the rotations are fully constrained

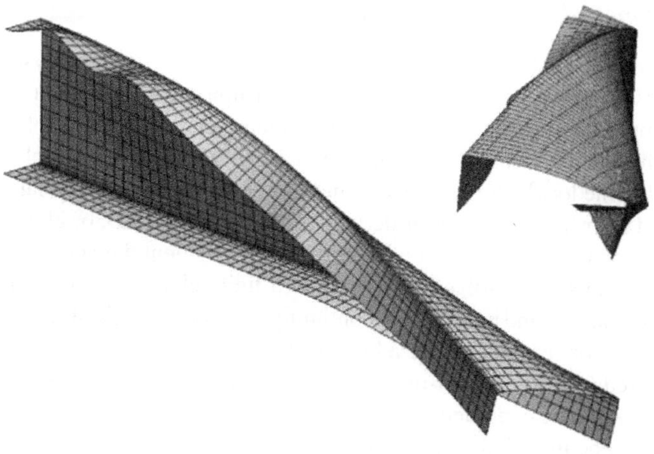

Figure 15.12 Deformed channel-section beam at $\delta = 4$ computed using 2048 Bézier elements

The load–deformation behaviour of the beam has been traced using displacement control with the downward deflection δ of the tip of the beam being increased with steps of 0.1 until $\delta = 4$. The point load is applied to the element vertex closest to (36, 3, 0). Note that, in contrast to standard finite elements, multiple global basis functions are supported at this element vertex, and as a consequence, displacement control is implemented as a linear constraint between the supported basis functions, rather than as a constraint to a single degree of freedom. The final deformation of the beam is shown in Figure 15.12 and is in excellent agreement with other results reported in the literature (Betsch *et al.* 1996). The most striking non-linear effect is the buckling of the upper flange. This buckling phenomenon can also be inferred from the force–displacement curve in Figure 15.13. At the peak load the upper flange buckles, followed by a post-buckling behaviour.

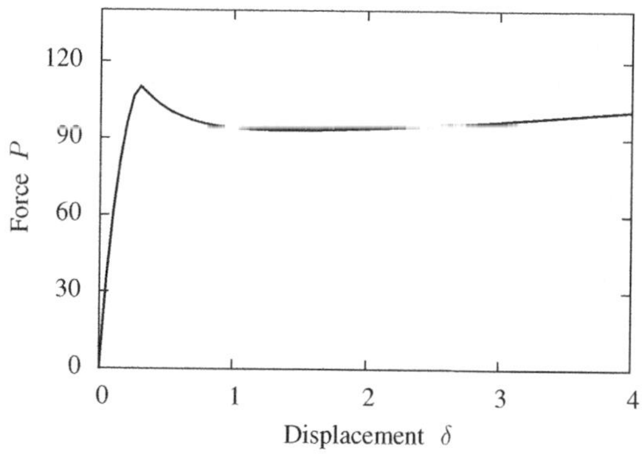

Figure 15.13 Load–displacement curve for the channel-section beam problem obtained using 2048 Bézier elements

15.4.2 Higher-order Damage Models

In Chapter 6 it has been shown that the spurious mesh sensitivity in analyses using damage theories can be overcome effectively by adopting a non-local equivalent strain definition. The gradient approximations to this non-local damage formulation were found to be more efficient from the vantage point of numerical modelling. Then, the non-local equivalent strain $\bar{\epsilon}$ can be related to the local equivalent strain measure $\tilde{\epsilon}$ through a differential equation, which can be truncated after the second-order derivative, Equation (6.152), or after the fourth-order derivative, Equation (14.10). The second-order implicit gradient damage model described by Equation (6.152) can still be solved using standard finite elements, since partial integration allows an order reduction, and merely a $\mathcal{C}^0$-continuity is required for the interpolation functions. However, this is no longer the case for a higher-order gradient damage model. In Chapter 14 this has been solved using the higher-order continuity property of meshless methods. Now, this property of splines will be exploited (Verhoosel *et al.* 2011b). Moreover, a sixth-order gradient damage model will be included in the analysis:

$$\bar{\epsilon} - c_1\nabla^2\bar{\epsilon} - c_2\nabla^4\bar{\epsilon} - c_3\nabla^6\bar{\epsilon} = \tilde{\epsilon} \tag{15.29}$$

where ∇^6 is a short-hand notation similar to that for ∇^4, which was used in Equation (14.10). The fourth-order gradient damage formulation ($c_3 = 0$) requires $\mathcal{C}^1$-continuity, and hence can be discretised using quadratic splines. The $\mathcal{C}^2$-continuity required for the sixth-order formulation is provided by cubic splines.

In this section we consider two numerical simulations that illustrate the capability of isogeometric analysis to discretise the gradient damage formulations of order two, four and six. In the first simulation we revisit the uniaxial rod problem discussed in Chapter 6 in the context of gradient damage formulations, and in Chapter 14 in the context of meshless methods. The second simulation considers a two-dimensional problem in which a diagonal failure band develops, which is captured by local T-spline refinements.

We again consider the one-dimensional bar problem first discussed in Chapter 6 and shown schematically in Figure 6.18. Force–displacement curves have been determined for the non-local damage formulation, and for the second-, fourth- and sixth-order implicit gradient models. Figure 15.14 shows the force–displacement curves for the second-order gradient formulation and for the non-local formulation, and was obtained using 80 Bézier elements of order one and of order three. Meaningful results for the higher-order formulations cannot be obtained using linear elements. A comparison of the results for linear B-splines with the results obtained using cubic B-splines shows the superior convergence behaviour of cubic basis functions. In Figure 15.15 we show the axial stress in the bar as a function of the location. For linear basis functions, i.e. standard finite elements, an oscillatory behaviour is observed (Simone *et al.* 2003). When using higher-order basis functions for the second-order formulation, these oscillations reduce drastically, which can be attributed to the variation diminishing property of splines.

Figure 15.16 further shows a comparison between the gradient formulations and a non-local formulation of the integral type. All results are obtained on a cubic Bézier mesh with 1280 elements. The results are in excellent agreement with results obtained using finite elements or using meshless methods (Askes *et al.* 2000; Peerlings *et al.* 1996). As in Askes *et al.* (2000), see also Chapter 14, the incorporation of fourth-order derivatives in the implicit scheme improves the results, in the sense that the computed force–displacement curve is closer to that of the non-

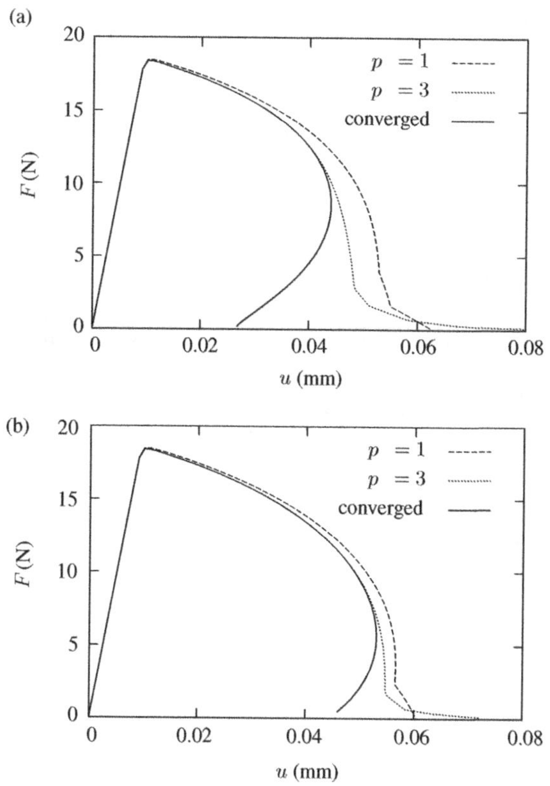

Figure 15.14 Comparison of results obtained with 80 Bézier elements of order one ($p = 1$) and order three ($p = 3$) for the second-order gradient (a) and non-local damage formulation (b) (Verhoosel *et al.* 2011b)

local formulation. In line with this observation the sixth-order formulation gives an even closer approximation of the non-local result. Indeed, the sixth-order gradient damage formulation is very efficient, since the results are close to the non-local formulation, while the computational effort is very much reduced compared with the non-local formulation.

As a second example we consider the L shaped specimen of Figure 15.17. The free rotation of the rigid end-plates is incorporated by means of linear constraints on the boundary control points, which is possible due to the fact that the basis functions on the corresponding boundaries can exactly represent all affine motions, and, in particular, the rigid rotations and translations. The diagonal failure zone requires mesh refinements, which can be achieved using T-splines.

An isotropic elastic-damaging material model is used with modulus of elasticity $E = 10\,\text{GPa}$ and a Poisson's ratio $\nu = 0.2$. Plane-stress conditions have been adopted. The modified von Mises local equivalent strain, Equation (6.21), has been used in combination with an exponential damage relation (Geers *et al.* 1998). The force–displacement curves have been obtained using the cubic Bézier T-spline mesh of Figure 15.18 (Verhoosel *et al.* 2011b). The mesh consists of 1686 Bézier elements and has 1543 basis functions. It is noted that for the third-order T-splines the number of basis functions is in the same range as the number of elements, which

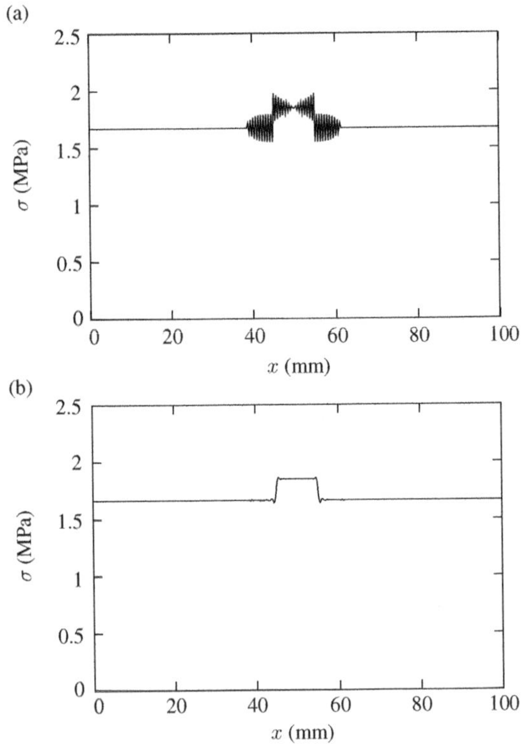

Figure 15.15 Axial stress in the one-dimensional bar at $u = 0.023$ mm for the second-order gradient damage formulation with linear basis functions (a), and for cubic basis functions (b). The stress is plotted in the integration points (Verhoosel *et al.* 2011b)

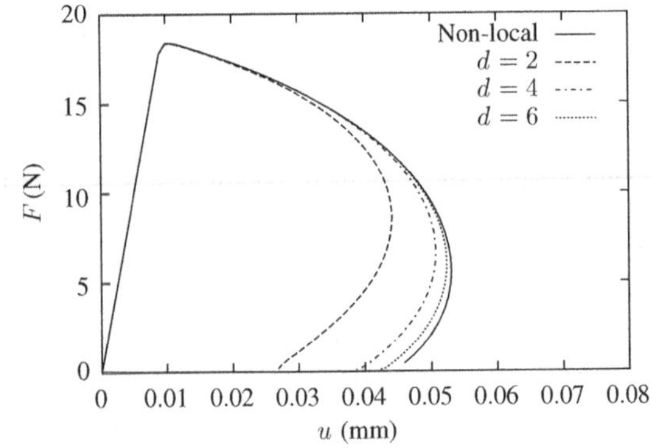

Figure 15.16 Force–displacement diagrams for the bar loaded in tension using the non-local formulation and d-th order gradient formulations. All results are obtained using cubic Bézier meshes with 1280 elements (Verhoosel *et al.* 2011b)

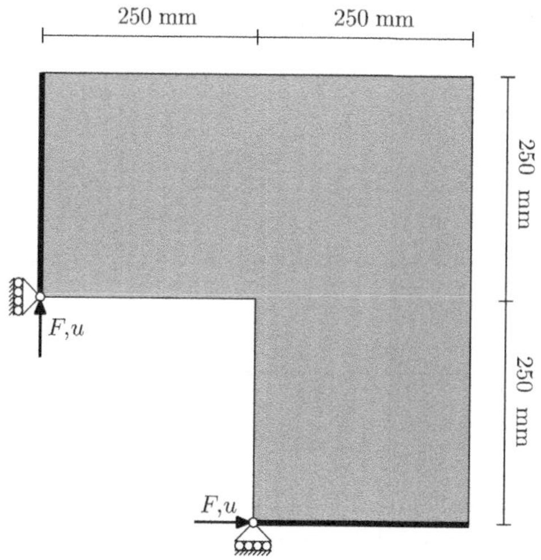

Figure 15.17 L-shaped specimen. The thickness of the specimen is 200 mm (Verhoosel *et al.* 2011b)

contrasts standard finite elements with a cubic interpolation. A $\mathcal{C}^2$-continuous base mesh is created using a non-tensor product T-spline. The $\mathcal{C}^2$ basis function centred around the re-entrant corner is shown in Figure 15.18.

Figure 15.19 compares the results of the various formulations. Upon an increase of the order of the formulation the approximation of the non-local result improves. An increase of the order of the formulation also increases the total amount of dissipated energy. This is caused by the smoothing effect of the higher-order derivatives. In Figure 15.20 the maximum principal stress contours are shown for the second-order and for the sixth-order formulations. No substantial stress oscillations are observed, which is consistent with the observations on the one-dimensional model problem discussed before. Again, the sixth-order formulation is highly efficient, since it closely approximates the result for the non-local model, with a fraction of the computational effort.

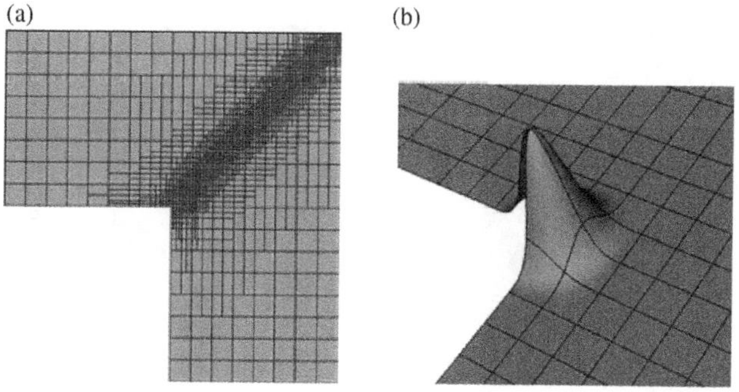

Figure 15.18 (a) Bézier mesh for the L-shaped specimen. (b) Smooth $\mathcal{C}^2$ basis function centred around the re-entrant corner of the L-shaped domain (Verhoosel *et al.* 2011b)

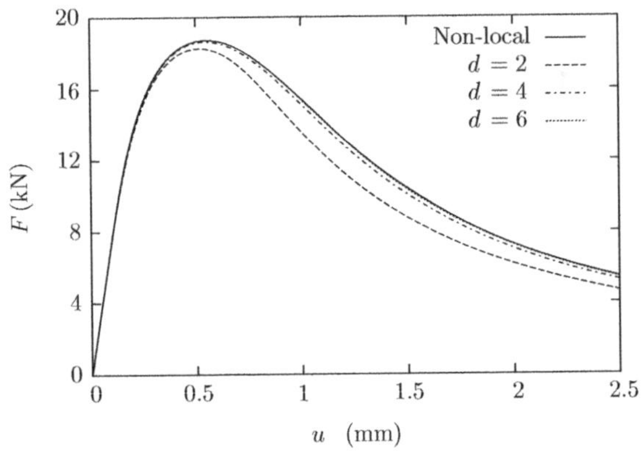

Figure 15.19 Force–displacement results for the L-shaped specimen using the non-local formulation and d-th order gradient formulations (Verhoosel *et al.* 2011b)

15.4.3 Cohesive Zone Models

The control over basis function continuity makes isogeometric analysis a viable candidate for the modelling of evolving and propagating discontinuities, including crack models based on the cohesive zone concept. On the one hand, the higher-order continuity of spline basis functions results in a more accurate representation of the stresses, which results in a better prediction of the direction in which a discontinuity propagates, see also the discussion around Equation (14.24). And, on the other hand, discontinuities can be inserted arbitrarily by means of knot insertion, which lowers the continuity and permits jumps in the displacement field (Verhoosel *et al.* 2011a).

In this section, we first illustrate the concept of inserting a discontinuity in isogeometric analysis by means of knot insertion in a univariate B-spline setting. With the insight that we obtain through this simple example, we can proceed to two typical fracture simulations. First, we use a NURBS-based discretisation for modelling debonding between a circular fibre and

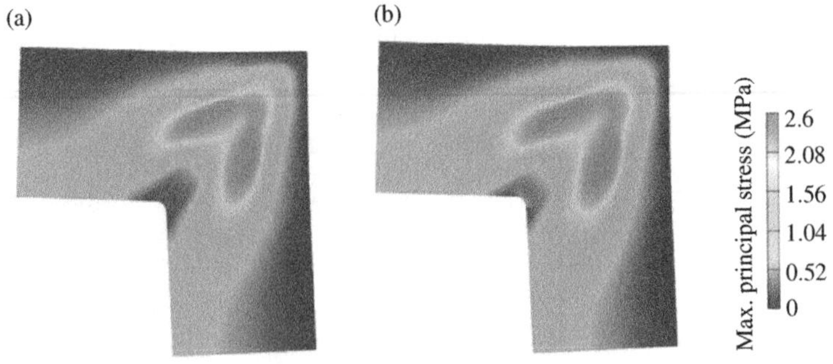

Figure 15.20 Maximum principal stress contours at $u = 1.25$ mm as computed by the second-order formulation (a) and the sixth-order formulation (b). Displacements are amplified by a factor of 15 (Verhoosel *et al.* 2011b)

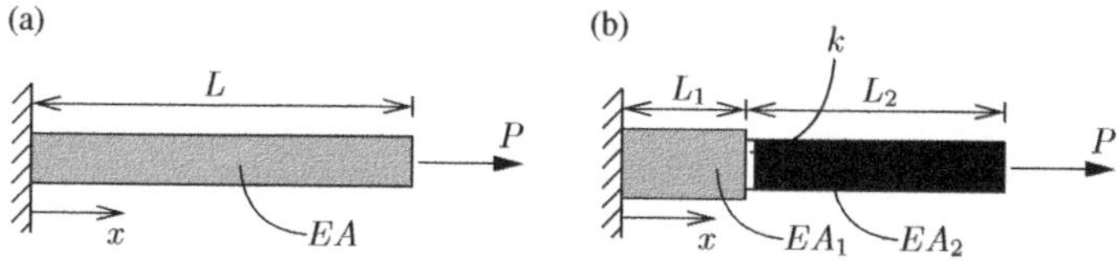

Figure 15.21 Schematic representation of two rods loaded in tension. The two segments of the composite rod are connected by a zero-thickness adhesive layer with stiffness k

the epoxy matrix in which it is embedded. Secondly, we consider propagating cohesive cracks in a single-edge notched beam, which requires the use of T-splines.

We consider the one-dimensional bar of Figure 15.21, which is loaded in tension. The bar has a length L, a stiffness EA, and is loaded by a force P. The bar is parametrised by a quadratic B-spline with a knot vector $\Xi = \{0, 0, 0, 1, 1, 1\}$, and the control points are given by $p_1 = 0$, $p_2 = \frac{1}{2}L$ and $p_3 = L$ with uniform weights $W_1 = W_2 = W_3 = 1$. The corresponding basis functions are shown in Figure 15.22. This choice of control points results in a linear parametrisation of the rod: $x = L\xi$. Using the shape functions, the displacement field can be approximated as:

$$u = \sum_{k=1}^{N} h_k(\xi)u_k \tag{15.30}$$

Any solution method will then give the coefficients $u_1 = 0$, $u_2 = \frac{1}{2}\frac{PL}{EA}$ and $u_3 = \frac{PL}{EA}$, so that $u(\xi) = \frac{PL\xi}{EA}$, which can obviously be rewritten as the exact solution $u(x) = \frac{Px}{EA}$.

Now consider the composite bar shown in Figure 15.21(b). The two segments of the bar, with stiffnesses EA_1 and EA_2, respectively, and lengths L_1 and L_2, such that $L_1 + L_2 = L$, are connected by an adhesive layer at $x = L_1$. The infinitely thin adhesive layer is assumed to have a stiffness k, such that the displacement jump over the layer equals $[\![u]\!] = P/k$. Since the basis functions corresponding to $\Xi = \{0, 0, 0, 1, 1, 1\}$ are $\mathcal{C}^1$-continuous on $(0, L)$, the discontinuous deformation of the composite bar cannot be represented exactly by these basis functions. In order to obtain the exact solution, we enhance the solution space such that we allow for a discontinuity in the displacement field at $x_d = L_1$. From the parametrisation of the bar this physical position is known to coincide with the point $\xi_d = \frac{L_1}{L}$ in the parametric domain. We now create a discontinuity at $x_d = L_1$ by inserting a knot with multiplicity $p + 1 = 3$ at $\xi = \frac{L_1}{L}$, which changes the knot vector to $\Xi = \left\{0, 0, 0, \frac{L_1}{L}, \frac{L_1}{L}, \frac{L_1}{L}, 1, 1, 1\right\}$. The corresponding basis functions for the case that $L_1 = \frac{1}{3}L$ are shown in Figure 15.22(b). When the corresponding control points are taken as $p_1 = 0$, $p_2 = \frac{1}{2}L_1$, $p_3 = L_1$, $p_4 = L_1$, $p_5 = L_1 + \frac{1}{2}L_2$ and $p_6 = L$, the original parametrisation is preserved. When determining the deformation of the bar using the new basis functions, the coefficients $u_1 = 0$, $u_2 = \frac{1}{2}\frac{PL_1}{EA_1}$, $u_3 = \frac{PL_1}{EA_1}$, $u_4 = \frac{PL_1}{EA_1} + \frac{P}{k}$, $u_5 = \frac{PL_1}{EA_1} + \frac{P}{k} + \frac{1}{2}\frac{PL_2}{EA_2}$ and $u_6 = \frac{PL_1}{EA_1} + \frac{P}{k} + \frac{PL_2}{EA_2}$ are obtained, which yields the exact solution.

We now illustrate the use of NURBS to simulate adhesive fracture by considering a fibre with a radius of 5 μm embedded in square block with dimensions $30 \times 30\,\mu m$, made out of epoxy (Figure 15.23). A plane-strain assumption is adopted. The specimen is loaded in the horizontal direction by gradually increasing the horizontal displacement $\bar{u}$ on the left and right edges.

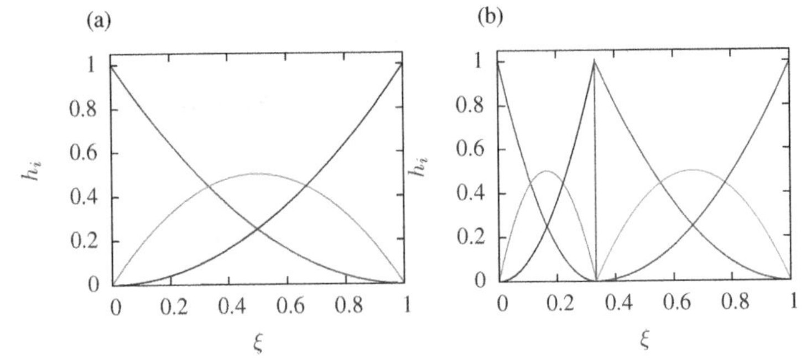

Figure 15.22 Quadratic B-spline basis functions used for the one-dimensional rod example without (a) and with (b) a discontinuity at $x_d = L_1 = \frac{1}{3}L$

Contraction of the epoxy specimen in the vertical direction is prevented by roller supports on the upper and lower edges of the block. By virtue of the two-fold symmetry of the specimen, only a quarter of the specimen needs to be discretised.

A linear elastic isotropic material description is used for both the fibre and the epoxy. For the epoxy a Young's modulus $E = 4.3$ GPa and Poisson's ratio $\nu = 0.34$ have been adopted. The fibre is much stiffer with a modulus of elasticity $E = 225.0$ GPa and a Poisson's ratio $\nu = 0.2$. The traction on the fibre–epoxy interface is related to its opening by means of the Xu–Needleman decohesion relation (Xu and Needleman 1993). The tensile strength and the fracture toughness are taken equal to $f_t = 50$ MPa and to $\mathcal{G}_c = 4 \times 10^{-3}$ N/mm, respectively. Equation (6.49) has been used to define the traction, with $\alpha = 2.3$ the mode-mixity parameter. To prevent the occurence of a negative crack opening a penalty parameter $k_p = 10^5$ MPa/mm is used.

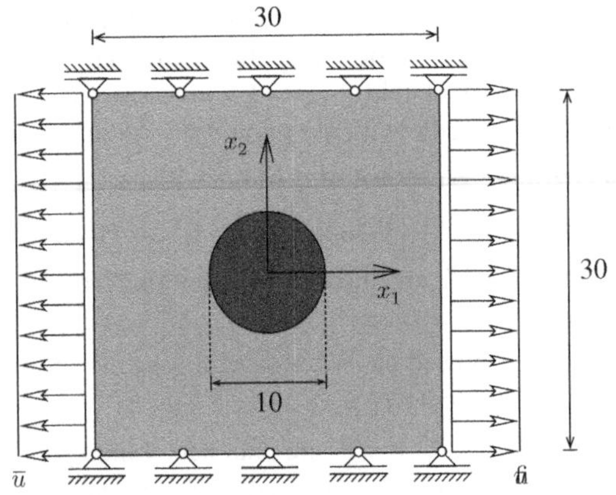

Figure 15.23 Schematic representation of a fibre with a circular cross section embedded in a square block of epoxy. All dimensions are in micrometres (Verhoosel *et al.* 2011a)

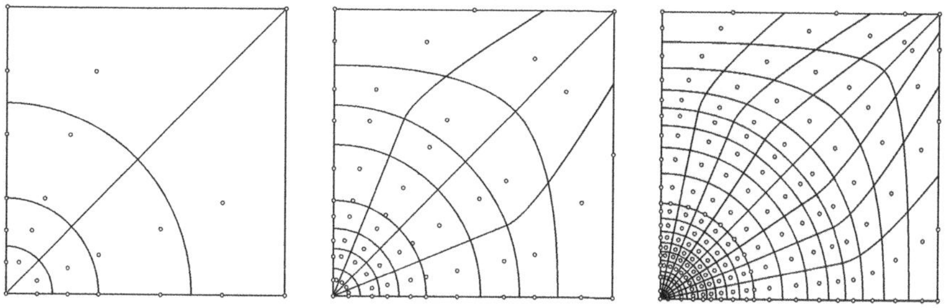

Figure 15.24 NURBS meshes used for the fibre–epoxy simulations. Note that normally the control nodes do not coincide with the element vertices (Verhoosel *et al.* 2011a)

Four different quadratic NURBS meshes have been used. The coarsest mesh, consisting of only 8 elements (64 degrees of freedom), is shown in Figure 15.24. An attractive feature of this discretisation is that the geometry is represented exactly with only 8 elements. To create a discontinuity in the radial direction, the knot that coincides with the interface is assigned a multiplicity of $p+1=3$. Figure 15.24 also shows two uniformly refined meshes, with 32 elements (144 degrees of freedom) and 128 elements (400 degrees of freedom), respectively. Moreover, the response of the system was determined using a mesh consisting of 2048 elements (4644 degrees of freedom), which we will henceforth refer to as the reference solution.

The response of the system is characterised in terms of σ_{xx} at $\mathbf{x}=(15,0)\,\mu$m *vs* the prescribed displacement at the left and the right edges. The response curves for the different meshes are shown in Figure 15.25. The result for the 128-element mesh coincides with that of the reference solution and is therefore not visible in Figure 15.25, and also the curve obtained using 32 elements is already close to the reference solution. The present solutions compare favourably with finite element solutions in the literature which are either obtained

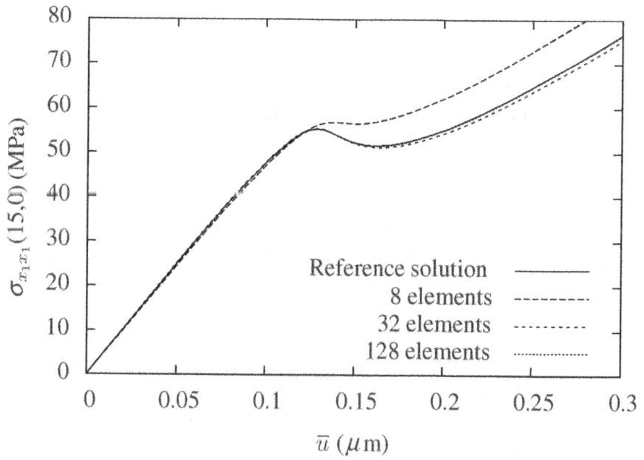

Figure 15.25 Response curves for the fibre–epoxy system determined using various meshes. σ_{xx} at $\mathbf{x}=(15,0)$ μm has been plotted *vs* the horizontal displacement $\bar{u}$. Note that the response using 128 elements coincides with that of the reference solution (Verhoosel *et al.* 2011a)

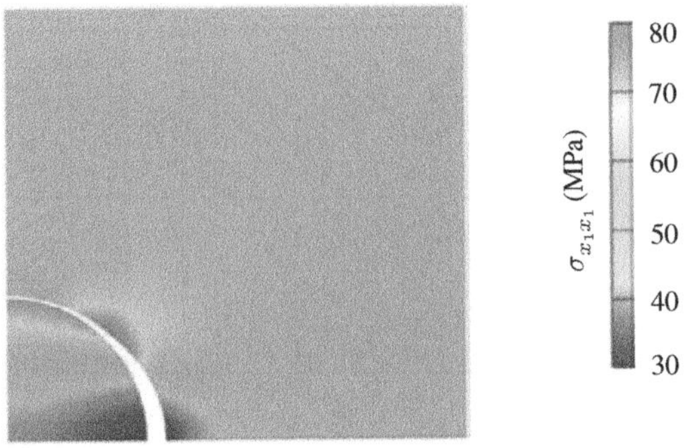

Figure 15.26 Contour plot showing σ_{xx} in the fibre–epoxy system at $\bar{u} = 0.165\,\mu$m using the 128 elements discretisation. The displacements are amplified by a factor of 10 (Verhoosel *et al.* 2011a)

using interface elements, or using a partition-of-unity approach. Partly, this can be attributed to the ability of NURBS to accurately parametrise the circular geometry of the fibre. On the other hand, the continuous stress field (Figure 15.26), that results from the C^1-continuous displacement field obtained using second-order NURBS, also has a favourable effect. When the direction of crack propagation is not prescribed along an interface, as in the present example, but has to be computed on the basis of the computed local stresses, the effect in terms of accuracy, in particular the location of the computed crack path, is even more pronounced. Such an example will be treated next.

For propagating discontinuities where the crack path is not predefined, NURBS no longer satisfy, and the ability of T-splines to flexibly modify the mesh topology is key to a successful application of isogeometric analysis. As an example, we consider the Single-Edge Notched Beam of Figure 6.13, that has also been used in the preceding chapters to test the performance of embedded discontinuities (Figure 6.14), of standard finite elements for the implicit second-order gradient damage model (Figure 6.20), of interface elements and of partition-of-unity finite element models to simulate cohesive fracture (Figures 13.1 and 14.16). Now, the single-edge notched beam geometry and the deformation are described using cubic T-splines. Figure 15.27 shows that the T-spline mesh allows for local control point insertion to represent the loading plates and the initial notch. The coarsest mesh which has been considered consists of 130 elements (402 degrees of freedom). Two uniform mesh refinements have been applied, with 334 elements (868 degrees of freedom) and 1204 elements (2734 degrees of freedom), respectively.

The response of the beam is shown in Figure 15.28 for the three meshes. The response is characterised in terms of the force P *vs* the crack mouth sliding displacement. From Figure 15.28 we observe that the result using the intermediate mesh practically coincides with that for the fine mesh. Accordingly, an accurate solution is already obtained with just 334 elements. We also observe that the coarsest mesh experiences significant bumps in the response curve (Figure 15.28). As in Figure 14.15, which was obtained using a partition-of-unity finite element method applied to cohesive fracture, these bumps are attributable to the fact that the crack is abruptly extended when the propagation criterion is violated. A more gradual extension of the crack, which takes place for the finer meshes, reduces this effect significantly.

Figure 15.29 shows a contour plot of the cracked beam. The dominant crack nucleates at the bottom right corner of the initial notch at an angle of 37° with the vertical axis. Upon

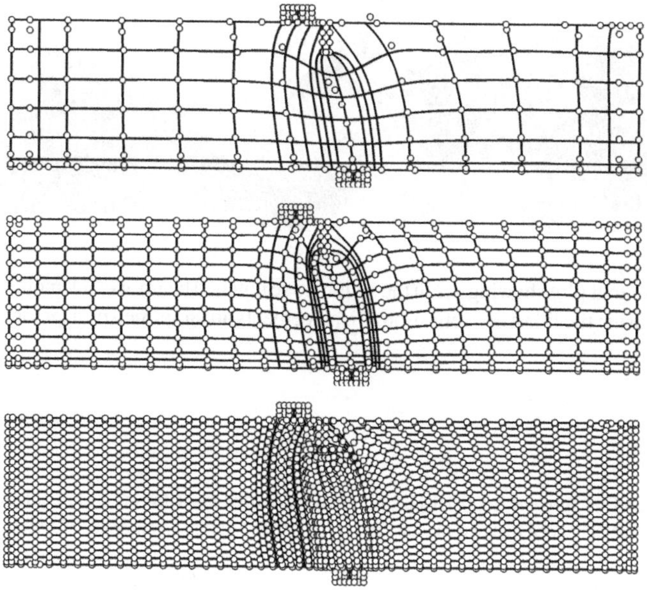

Figure 15.27 Meshes used for the single-edge notched beam simulations. Note that the control points do not need to coincide with the element vertices (Verhoosel *et al.* 2011a)

extension, the crack gradually deflects to eventually propagate parallel to the vertical axis. It is also observed that a secondary crack nucleates at the bottom edge of the specimen. Note from the contour plot that both crack paths are smooth since the directions of the normal vectors from one segment to another have been matched.

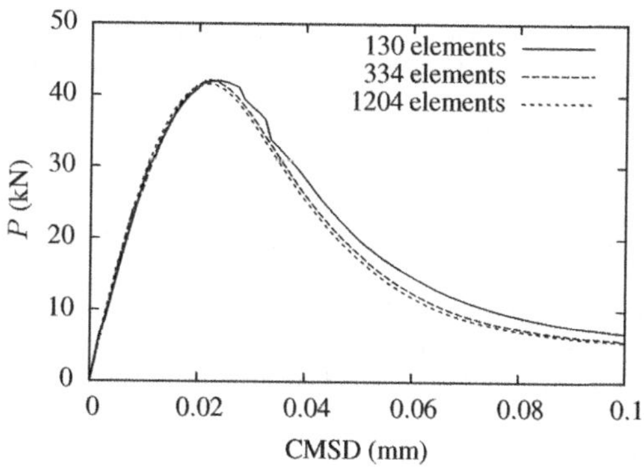

Figure 15.28 Response curves for the single-edge notched beam simulations. The response is measured in terms of the applied force P versus the crack mouth sliding displacement (CMSD), which equals the vertical displacement difference between the left and the right notches (Verhoosel *et al.* 2011a)

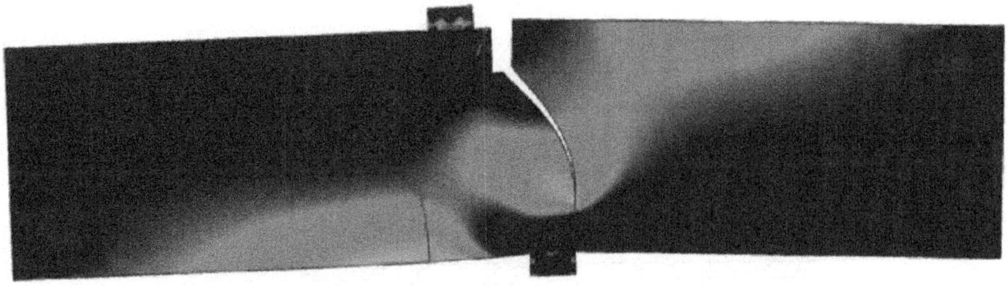

Figure 15.29 Contour plot showing σ_{xx} in the single-edge notched beam at crack mouth sliding displacement = 0.033 mm using the finest discretisation. Displacements are amplified by a factor of 100 (Verhoosel *et al.* 2011a)

References

Askes H, Pamin J and de Borst R 2000 Dispersion analysis and element-free Galerkin solutions of second-and fourth-order gradient-enhanced damage models. *International Journal for Numerical Methods in Engineering* **49**, 811–832.

Bazilevs Y, Calo VM, Cottrell JA, Evans JA, Hughes TJR, Lipton S, Scott MA and Sederberg TW 2010 Isogeometric analysis using T-splines. *Computer Methods in Applied Mechanics and Engineering* **199**, 229–263.

Benson DJ, Bazilevs Y, Hsu MC and Hughes TJR 2010 Isogeometric shell analysis: the Reissner-Mindlin shell. *Computer Methods in Applied Mechanics and Engineering* **199**, 276–289.

Betsch P, Gruttmann F and Stein E 1996 A 4-node finite shell element for the implementation of general hyperelastic 3d-elasticity at finite strains. *Computer Methods in Applied Mechanics and Engineering* **130**, 57–79.

Borden MJ, Scott MA, Evans JA and Hughes TJR 2011 Isogeometric finite element data structures based on Bézier extraction of NURBS. *International Journal for Numerical Methods in Engineering* **87**, 15–47.

Chróscielewski J, Makowski J and Stumpf H 1992 Genuinely resultant shell finite elements accounting for geometric and material non-linearity. *International Journal for Numerical Methods in Engineering* **35**, 63–94.

Cottrell JA, Hughes TJR and Bazilevs Y 2009 *Isogeometric Analysis: Toward Integration of CAD and FEA*. John Wiley & Sons, Ltd.

Cox MG 1972 The numerical evaluation of B-splines. *IMA Journal of Applied Mathematics* **10**, 134–149.

de Boor C 1972 On calculating with B-splines. *Journal of Approximation Theory* **6**, 50–62.

Farin G 1993 *Curves and Surfaces for CAGD*. Academic Press, Inc.

Geers MGD, de Borst R, Brekelmans WAM and Peerlings RHJ 1998 Strain-based transient-gradient damage model for failure analyses. *Computer Methods in Applied Mechanics and Engineering* **160**, 133–153.

Hughes T, Cottrell J and Bazilevs Y 2005 Isogeometric analysis: CAD, finite elements, NURBS, exact geometry and mesh refinement. *Computer Methods in Applied Mechanics and Engineering* **194**, 4135–4195.

Hughes TJR, Reali A and Sangalli G 2010 Efficient quadrature for NURBS-based isogeometric analysis. *Computer Methods in Applied Mechanics and Engineering* **199**, 301–313.

Kasik DJ, Buxton W and Ferguson DR 2005 Ten CAD challenges. *Computer Graphics and Applications, IEEE* **25**, 81–92.

Kiendl J, Bletzinger KU, Linhard J and Wüchner R 2009 Isogeometric shell analysis with Kirchhoff-Love elements. *Computer Methods in Applied Mechanics and Engineering* **198**, 3902–3914.

MacNeal R and Harder R 1985 A proposed standard set of problems to test finite element accuracy. *Finite Elements in Analysis and Design* **1**, 3–20.

Peerlings RHJ, de Borst R, Brekelmans WAM and de Vree HPJ 1996 Gradient-enhanced damage for quasi-brittle materials. *International Journal for Numerical Methods in Engineering* **39**, 3391–3403.

Piegl L and Tiller W 1997 *The NURBS Book*, 2nd edn. Springer-Verlag.

Rogers DF 2001 *An Introduction to NURBS*. Academic Press.

Scott MA, Borden MJ, Verhoosel CV, Sederberg TW and Hughes TJR 2011a Isogeometric finite element data structures based on Bézier extraction of T-splines. *International Journal for Numerical Methods in Engineering* **88**, 126–156.

Scott MA, Li X, Sederberg TW and Hughes TJR 2011b Local refinement of analysis-suitable T-splines. ICES report 11-06, University of Texas at Austin.

Sederberg TW, Zheng J, Bakenov A and Nasri A 2003 T-splines and T-NURCCs. *ACM Transactions on Graphics* **22**, 477–484.

Simone A, Askes H, Peerlings RHJ and Sluys LJ 2003 Interpolation requirements for implicit gradient-enhanced continuum damage models. *Communications in Numerical Methods in Engineering* **19**, 563–572.

Temizer I, Wriggers P and Hughes TJR 2011 Contact treatment in isogeometric analysis with NURBS. *Computer Methods in Applied Mechanics and Engineering* **200**, 1100–1112.

Verhoosel CV, Scott MA, de Borst R and Hughes TJR 2011a An isogeometric approach to cohesive zone modeling. *International Journal for Numerical Methods in Engineering* **87**, 336–360.

Verhoosel CV, Scott MA, Hughes TJR and de Borst R 2011b An isogeometric analysis approach to gradient damage models. *International Journal for Numerical Methods in Engineering* **86**, 115–134.

Wall WA, Frenzel MA and Cyron C 2008 Isogeometric structural shape optimization. *Computer Methods in Applied Mechanics and Engineering* **197**, 2976–2988.

Xu XP and Needleman A 1993 Void nucleation by inclusion debonding in a crystal matrix. *Modelling and Simulation in Materials Science and Engineering* **1**, 111–132.

Index

Non-linear Finite Element Analysis of Solids and Structures, Second Edition.
René de Borst, Mike A. Crisfield, Joris J.C. Remmers and Clemens V. Verhoosel.

Printed and bound by CPI Group (UK) Ltd, Croydon, CR0 4YY
12/01/2025
14624503-0003